Vogels's Quantitative Chemical Analysis

Vogel's

Textbook of Quantitative Chemical Analysis

SIXTH EDITION

Revised by the following members of
The School of Chemical and Life Sciences
University of Greenwich, London

J MENDHAM, MA, MSc, CChem, MRSC
Former Principal Lecturer in Analytical Chemistry

R C DENNEY, BSc, PhD, CChem, FRSC, FRSM
Former Reader in Analytical Chemistry

J D BARNES, BSc, PhD
Academic Development Officer

M THOMAS, BSc, PhD, CChem, MRSC
Senior Lecturer in Analytical Chemistry

PRENTICE HALL

An imprint of **PEARSON EDUCATION**

Harlow, England · London · New York · Reading, Massachusetts · San Francisco · Toronto · Don Mills, Ontario · Sydney
Tokyo · Singapore · Hong Kong · Seoul · Taipei · Cape Town · Madrid · Mexico City · Amsterdam · Munich · Paris · Milan

Pearson Education Limited
Edinburgh Gate
Harlow
Essex CM20 2JE
England
and Associated Companies throughout the world

Visit us on the world wide web at:
http://www.awl-he.com

First published 1939
New impressions 1941, 1942, 1943, 1944, 1945, 1946, 1947, 1948
Second edition 1951
New impressions 1953, 1955, 1957, 1958, 1959, 1960
Third edition 1961 (published under the title *A Textbook of Quantitative Inorganic Analysis Including Elementary Instrumental Analysis*)
New impressions 1962, 1964, 1968, 1969, 1971, 1974, 1975
Fourth edition 1978
New impressions 1979, 1981, 1983, 1985, 1986, 1987
Fifth edition 1989
New impressions 1991, 1998
This edition 2000

ISBN 0 582 22628 7

British Library Cataloguing-in-Publication Data
A catalogue record for this book is available from the British Library

Library of Congress Cataloging-in-Publication Data
A catalog record for this book is available from the Library of Congress

Typeset by 35 in $9\frac{1}{2}$/11pt Times New Roman
Produced by Addison Wesley Longman Singapore (Pte) Ltd.
Printed in Singapore

Contents

Detailed contents

Detailed contents

Detailed contents

Detailed contents

Detailed contents

Preface to the sixth edition

Very few chemistry books survive to a sixth edition having had a lifetime of sixty years. It says much for the original quality of work and foresight of Dr Arthur I. Vogel that what he started so long ago has become accepted and acknowledged as providing a basis for the teaching, learning and application of analytical chemistry throughout the world in many different languages. We consider it a privilege to have been given the opportunity to add to this work and to extend it into the new and rapidly developing areas which in many instances Dr Vogel saw beginning before he died. As long ago as 1961 he had the foresight to anticipate the value of infrared spectroscopy for its analytical potential when he included it in the third edition. We have now built further on his work and whilst seeking to modernise the approach, the text and the analytical concepts, have tried to retain the approach and ideas which he built into the original publication. We are well aware of the quality of advanced electronic, computerised equipment available in many analytical laboratories, but are only too conscious of the fact that such equipment is often abused or underperforms due to the lack of basic analytical and chemical knowledge of the user in many instances. We believe there is still a need for the analyst to understand the concepts behind the analytical methods and to provide this information for less well-equipped laboratories.

As there has been an increasing emphasis on the importance of sampling and the evaluation of data, these sections have been extensively revised and sections on calibration methods and the analysis of variance expanded. Experimental design, optimisation techniques and examples of multivariate analysis have been included in an introduction to chemometrics, and the chapter on sampling has been enlarged to include more details on sample methodology and the physical state of samples. The increasing roles of legislation, safety and analytical criteria have been emphasised.

In earlier editions 'wet' chemical analysis formed the cornerstones of quantitative analysis, but there has been a considerable reduction in the use of gravimetric methods in the analytical laboratory. This section has, therefore, been substantially reduced, but sufficient information has been retained to enable the valuable techniques and processes to be used for instructing in laboratory skills. Under titrimetry we have brought together both classical and instrumental procedures in order to avoid duplication of theoretical principles and to assist the reader in making ready comparisons between methods along with an evaluation of the instrumental techniques used together with the way in which these are applied in automatic systems.

The large section on titrimetry is well justified on the grounds that it is a valuable source of classical reactions that many analysts have informed us are frequently ignored in other books. Such reactions are readily adapted for use in automatic analysers and for training new analysts.

In response to suggestions made by reviewers of the fifth edition we have reintroduced a separate chapter on thermal methods. Instrumentation and applications in this analytical area have developed extensively in recent years, especially for studies on materials and pharmaceutical products. Experiments incorporating these disciplines have been included.

The chapters on separative techniques have been expanded to deal not only with the theory of the subject but also to show the breadth of the variations of techniques and procedures and the extent of the applications. Improvements in detector and column design have been included and the importance of coupled systems between chromatography and mass spectrometry explained. Capillary electrophoresis has been included for the first time, indicating the value of this rapidly developing area.

Electroanalytical methods are of increasing value in many applications, especially in electrode systems. These methods have been collected together in a single chapter and the theory, instrumentation and applications for each system can be readily compared.

The use of spectrometric methods for both qualitative and quantitative analysis is now vast and we have once again devoted a large portion of the book to these, in all cases dealing with the theory in detail and providing many examples of applications in both inorganic and organic areas. Sampling and instrumentation have been up-dated in the chapter on Atomic Spectroscopy. As a result of reviewers' suggestions we have also included a detailed discussion on limits of detection and sensitivity as appropriate for this part of the book. For the first time we have introduced chapters on nuclear magnetic resonance spectroscopy (NMR) and on mass spectrometry. Improved instrumentation has led to increasing uses of NMR for quantitative analysis, especially for pharmaceutical products and illegal drugs. Similarly, the applications of mass spectrometry are increasing daily with the continual developments and improvements in coupled systems and these have been reflected in the structure of this chapter.

We again owe a great debt to the many companies and individuals who have been prepared to give their time to discuss our ideas with us, lend us equipment and allow us to use diagrams, illustrations and tables for no charge. Throughout our work we have met only kindness, encouragement and cooperation from the people we have approached. If we have failed to give due recognition to them in the text we apologise for any oversight but wish to record our collective thanks to everyone who may have assisted us in any way. We are also grateful to those people who wrote to us after the publication of the fifth edition to draw our attention to errors which had been introduced in the text. As far as possible we have sought to ensure that those mistakes have not been caried over into this new edition, but will be equally pleased to learn of any errors or omissions in order that they can be rectified in any later re-print. We hope that we have managed to maintain the high standards that Dr Arthur I. Vogel helped to establish sixty years ago.

Throughout our work we have continued to receive encouragement and support from the University of Greenwich and from our many friends and colleagues there who have frequently provided ideas and data to assist us. As usual we are indebted to our ever tolerant wives who have accepted the stress, moods and words of writers seeking to bring concepts to completion and to meet deadlines. Finally we acknowledge our debt to Dr G. H. Jeffery and to J. Bassett who played a major role in assisting Dr Vogel in the earlier editions of this book and in the later preparation of the fourth and fifth editions. We hope that we have managed to maintain the standard of the workmanship in this edition and that it will continue to serve as a valuable source for teaching and the practical application of analytical chemistry for many years to come.

J. Mendham, R. C. Denney, J. D. Barnes, M. Thomas
September 1998

Preface to first edition

In writing this book, the author had as his primary object the provision of a complete up-to-date text-book of quantitative inorganic analysis, both theory and practice, at a moderate price to meet the requirements of University and College students of all grades. It is believed that the material contained therein is sufficiently comprehensive to cover the syllabuses of all examinations in which quantitative inorganic analysis plays a part. The elementary student has been provided for, and those sections devoted to his needs have been treated in considerable detail. The volume should therefore be of value to the student throughout the whole of his career. The book will be suitable *inter alia* for students preparing for the various Intermediate B.Sc. and Higher School Certificate Examinations, the Ordinary and Higher National Certificates in Chemistry, the Honours and Special B.Sc. of the Universities, the Associateship of the Institute of Chemistry, and other examinations of equivalent standard. It is hoped, also, that the wide range of subjects discussed within its covers will result in the volume having a special appeal to practising analytical chemists and to all those workers in industry and research who have occasion to utilise methods of inorganic quantitative analysis.

The kind reception accorded to the author's *Text Book of Qualitative Chemical Analysis* by teachers and reviewers seems to indicate that the general arrangement of that book has met with approval. The companion volume on *Quantitative Inorganic Analysis* follows essentially similar lines. Chapter I is devoted to the theoretical basis of quantitative inorganic analysis, Chapter II to the experimental technique of quantitative analysis, Chapter III to volumetric analysis, Chapter IV to gravimetric analysis (including electro-analysis), Chapter V to colorimetric analysis, and Chapter VI to gas analysis; a comprehensive Appendix has been added, which contains much useful matter for the practising analytical chemist. The experimental side is based essentially upon the writer's experience with large classes of students of various grades. Most of the determinations have been tested out in the laboratory in collaboration with the author's colleagues and senior students, and in some cases this has resulted in slight modifications of the details given by the original authors. Particular emphasis has been laid upon recent developments in experimental technique. Frequently the source of certain apparatus or chemicals has been given in the text; this is not intended to convey the impression that these materials cannot be obtained from other sources, but merely to indicate that the author's own experience is confined to the particular products mentioned.

The ground covered by the book can best be judged by perusal of the Table of Contents. An attempt has been made to strike a balance between the classical and modern procedures, and to present the subject of analytical chemistry as it is today. The theoretical aspect has been stressed throughout, and numerous cross-references are given to Chapter I (the theoretical basis of quantitative inorganic analysis).

No references to the original literature are given in the text. This is because the introduction of such references would have considerably increased the size and therefore the price of the book. However, a discussion on the literature of analytical chemistry is given in the Appendix. With the aid of the various volumes mentioned therein – which should be available in all libraries of analytical chemistry – and the Collective Indexes of *Chemical Abstracts* or of *British Chemical Abstracts*, little difficulty will, in general, be experienced in finding the original sources of most of the determinations described in the book.

In the preparation of this volume, the author has utilised pertinent material wherever it was to be found. While it is impossible to acknowledge every source individually, mention must, however, be made of Hillebrand and Lundell's *Applied Inorganic Analysis* (1929) and of Mitchell and Ward's *Modern Methods in Quantitative Chemical Analysis* (1932). In conclusion, the writer wishes to express his thanks: to Dr G. H. Jeffery, A.I.C., for reading the galley proofs and making numerous helpful suggestions; to Mr. A. S. Nickelson, B.Sc., for reading some of the galley proofs; to his laboratory steward, Mr. F. Mathie, for preparing a number of the diagrams, including most of those in Chapter VI, and for his assistance in other ways; to Messrs. A. Gallenkamp and Co., Ltd., of London, E.C.2, and to Messrs. Fisher Scientific Co, of Pittsburgh, Pa., for providing a number of diagrams and blocks; and to Mr. F. W. Clifford, F.L.A., Librarian to the Chemical Society, and his able assistants for their help in the task of searching the extensive literature.

Any suggestions for improving the book will be gratefully received by the author.

Arthur I. Vogel
Woolwich Polytechnic, London, SE18
June, 1939

Safety in the laboratory

There is always a potential hazard in a chemistry laboratory and, regrettably, accidents do occur on occasions. However, they can be minimised by adhering to the laboratory safety code, by **carrying out a hazard assessment before any experimentation and by taking all necessary safety precautions**. The following points should be borne in mind.

Protect eyes   Eye protection should be worn at all times when working in the laboratory. Special safety spectacles should be used by everyone as normal prescription spectacles are not usually made with safety-glass lenses.

Work safely   Do not play around in the laboratory and do not attempt unauthorised experiments.

No eating or drinking   Do not eat or drink in the laboratory, or use laboratory equipment for storing or holding food or drink, and do not touch your mouth or face with laboratory chemicals or glassware.

Avoid skin contact   Remember that many chemicals are corrosive or toxic even if in dilute solutions. Avoid getting chemicals on your skin, even solids, and wash off any contamination with large volumes of water. Immediately remove any clothing contaminated with corrosive substances. Safety and protection are more important than your appearance.

Wear protective clothing   Wear appropriate protective clothing and shoes, and avoid loose hair that might catch on moving equipment or dip into open solutions.

Use fume hoods   Only use toxic substances under fume hoods where there is adequate extraction.

Know safety procedures   Familiarise yourself with the locations of safety equipment and the safety procedures.

Units and reagent purity

Units

SI units have been used throughout this book, but as the litre (L) has been accepted as a special name for the cubic decimetre (dm^3), although this is not strictly speaking an SI term, we have considered it appropriate to use the litre throughout this book. Similarly we have chosen to use millilitres (mL) instead of cubic centimetres (cm^3). In accordance with established practice, the following prefixes are used for decimal multiples of units. Mass is the only exception: the prefixes are used with the gram (g) but the kilogram (kg) is the base unit.

Multiple	Prefix	Symbol	Multiple	Prefix	Symbol
10^{-1}	deci	d	10	deca	da
10^{-2}	centi	c	10^2	hecto	h
10^{-3}	milli	m	10^3	kilo	k
10^{-6}	micro	μ	10^6	mega	M
10^{-9}	nano	n	10^9	giga	G
10^{-12}	pico	p	10^{12}	tera	T
10^{-15}	femto	f	10^{15}	peta	P
10^{-18}	atto	a	10^{18}	exa	E

Concentrations of solutions are usually expressed as moles per litre; a molar solution (M) has one mole of solute per litre. In some instances the terms parts per million (ppm) and parts per billion (ppb) are still used for indicating trace levels of analytes. This is a dimensionless unit that can be subject to misinterpretation, and where it is employed it is essential that it is established if the levels indicate a weight/volume relationship, a volume/volume relationship or some other basis. Other units used in special cases are explained in the text.

Reagent purity

Unless otherwise stated, all reagents employed in the analytical procedures should be of appropriate analytical grade or spectroscopic grade. Similarly, where solutions are prepared in water this automatically means distilled or deionised water, from which all but very minor impurities will have been removed.

1

Chemical analysis

1.1 Introduction

Many different definitions exist for 'chemical analysis', but it may be reasonably stated as the application of a process or series of processes in order to identify and/or quantify a substance, the components of a solution or mixture, or the determination of the structures of chemical compounds.

This means that the scope of analytical chemistry is very broad and embraces a wide range of manual, chemical and instrumental techniques and procedures. Most people apply some form of chemical analysis almost every day, such as when smelling food to see if it has deteriorated or tasting substances to determine if they are sweet or sour. These constitute simple analytical processes, as compared with some of the more complex processes described in this volume that can be achieved with modern instruments. Note that it is not always necessary to apply advanced instrumental procedures to carry out accurate analyses and there may be many times when a simple, rapid analysis may actually be more desirable than a more complicated and time-consuming process. The objective and purpose of the analysis has to be sensibly assessed before selecting an appropriate procedure.

When a completely unknown sample is presented to an analyst, the first requirement is usually to ascertain what substances are present in it. This fundamental problem may sometimes be encountered in the modified form of deciding what impurities are present in a given sample, or perhaps of confirming that certain specified impurities are absent. The solution lies within the province of **qualitative analysis** and is outside the scope of this book.

Having ascertained the nature of the constituents of a given sample, the analyst is then frequently called upon to determine how much of each component, or of specified components, is present. Such determinations lie within the realm of **quantitative analysis**, and a variety of techniques are available to supply the required information.

1.2 Applications

With increasing demands for pure water, better food control and cleaner atmospheres, the analytical chemist has a greater and greater role to play within modern society. From the study of raw materials, such as crude oil and minerals to the finest quality scents and perfumes, the analytical chemist is called upon to play a part in determining composition, purity and quality. Manufacturing industries rely upon both qualitative and quantitative chemical analysis to ensure their raw materials meet certain specifications, and to check the quality of the final product. Raw materials are examined to ensure there are no unusual substances present which might upset the manufacturing process or appear as a harmful impurity in the final product. Furthermore, since the value of the raw material may be governed by the amount of the required ingredient it contains, a quantitative analysis is performed to establish the proportion of the essential component: this procedure is often called **assaying**. The final manufactured product is subject to **quality control** to ensure its essential components are present within a predetermined range of composition, whereas impurities do not exceed certain specified limits. The semiconductor industry is an example of an industry whose very existence depends on very accurate determination of substances present in extremely minute quantities.

The development of new products (which may be mixtures rather than pure materials, e.g. a polymer composition, or a metallic alloy) also requires the services of the analytical chemist. It will be necessary to ascertain the composition of the mixture which shows the optimum characteristics for the purpose for which the material has been developed.

Many industrial processes give rise to pollutants which can present a health problem. Quantitative analysis of air, water, and sometimes soil samples, must be carried out to determine the level of pollution, and to establish safe limits for pollutants.

In hospitals, chemical analysis is widely used to assist in the diagnosis of illness and in monitoring the condition of patients. In farming, the nature and level of fertiliser application is based on information obtained by analysing the soil to determine its content of the essential plant nutrients nitrogen, phosphorus and potassium, and the trace elements required for healthy plant growth.

Geological surveys require the services of analytical chemists to determine the composition of the numerous rock and soil samples collected in the field. One example is the qualitative and quantitative examination of moon rock brought back to Earth in 1969 by the first American astronauts to land on the moon.

Much government legislation can only be enforced by the work of analytical chemists, e.g. national and international agreements on water pollution and atmospheric pollution, food safety measures, regulations on substances hazardous to health, and laws governing the misuse of drugs.

When copper(II) sulphate is dissolved in distilled water, the copper is present in solution almost entirely as the hydrated copper ion $[Cu(H_2O)_6]^{2+}$. But if a natural water (spring water or river water) is substituted for the distilled water, then some of the copper ions will interact with various substances present in the natural water. These substances may include acids derived from vegetation (such as humic acids and fulvic acid), colloidal materials such as clay particles, carbonate ions (CO_3^{2-}) and hydrogencarbonate ions (HCO_3^-) derived from atmospheric carbon dioxide, and various other cations and anions leached from the rocks with which the water has been in contact. The copper ions which become adsorbed on colloidal particles, or those which form an organic complex, perhaps with fulvic acid, will no longer show the usual behaviour of hydrated copper(II) ions, so their biological and geological effects will be modified. To investigate these problems in natural waters, the analyst must devise procedures that determine the various copper-containing species in the solution and how the copper is distributed among them. These procedures are called '**speciation**'.

1.3   Stages of analysis

A complete chemical analysis, even for a single substance, involves a series of steps and procedures. Each one of them has to be carefully considered and assessed in order to minimise errors and to maintain accuracy and reproducibility. The steps are listed in

Table 1.1 *Stages in a chemical analysis*

Stages	Examples of procedure
1. Sampling	Depends on the size and physical nature of the sample
2. Preparation of analytical sample	Reduction of particle size, mixing for homogeneity, drying, determination of sample weight or volume
3. Dissolution of sample	Heating, ignition, fusion, use of solvent(s), dilution
4. Removal of interferences	Filtration, solvent extraction, ion exchange, chromatographic separation
5. Sample measurement and control of instrumental factors	Standardisation, calibration, optimisation, measurement of response; absorbance, emission signal, potential, current
6. Result(s)	Calculation of analytical result(s) and for the sample, statistical evaluation of data
7. Presentation of data	Printout, data plotting, storage (archiving)

Table 1.1 along with some of the procedures that may be employed. To obtain reliable results, correct sampling is paramount. This is especially true if the results obtained from a quantitative analysis of individual constituents of a given sample are used as the basis for calculating the composition and value of a bulk quantity of the commodity. Then it is essential to choose a sample that is representative of the entire material.

A homogeneous liquid presents few sampling problems, but a solid mixture is harder to sample. Several portions must be combined to ensure that a representative sample is finally selected for analysis. The analyst must therefore be acquainted with the standard sampling procedures for different types of material. Without doubt, sample preparation can be the most difficult step in the overall analytical process. This is especially true when dealing with solid samples which may require a number of stages before measurements can be made on the analyte. However, the direct conversion of a solid sample to a vapour by ablation methods (Section 16.9) or by treating the solid as a slurry has meant that inductively coupled plasma emission spectroscopy (ICP-OES) can be almost totally automatic.

Samples of liquids and gases are frequently amenable to automatic treatment; most readily automated are the instrumental procedures for measurement, calibration optimisation, statistical treatments, and data presentation and storage. If all operational stages in Table 1.1 take place without any human involvement, the analysis is considered fully automatic. Many autoanalysers used in clinical laboratories are totally automatic from the sampling stage right through to the presentation of data. Examples of automatic control are given throughout this text.

1.4   Selecting the method

Analysts are frequently spoilt for choice; several approaches can often be used to analyse a particular sample. To choose the most appropriate, analysts must be familiar with the practical details of the various techniques and the theoretical principles on which they are based. They must also understand the conditions under which each method is reliable, consider possible interferences which may arise, and find ways to circumvent any problems. Analysts will also be concerned with accuracy and precision, time and costing. The most accurate method for a certain determination may prove to be lengthy or it may require expensive reagents, and in the interests of economy it may be necessary to choose a method which, although somewhat less exact, yields results of sufficient accuracy in a reasonable time.

Important factors which must be taken into account when selecting an appropriate method of analysis include (a) the nature of the information which is sought, (b) the size of sample available and the proportion of the constituent to be determined, and (c) the purpose for which the analytical data is required.

The information sought may require very detailed data, or perhaps more general results. Chemical analyses may be classified into four types according to the data they generate:

Proximate analysis determines the amount of each element in a sample but not the compounds that are present.
Partial analysis determines selected constituents in the sample.
Trace constituent analysis is a type of partial analysis that determines specified components present in very minute quantity.
Complete analysis determines the proportion of each component in the sample.

On the basis of sample size, analytical methods are often classified as follows:

Macro for quantities of 0.1 g or more
Meso (semimicro) for quantities ranging from 10^{-2} g to 10^{-1} g
Micro for quantities in the range 10^{-3} g to 10^{-2} g
Submicro for samples in the range 10^{-4} g to 10^{-3} g
Ultramicro for quantities below 10^{-4} g
Trace for 10^2 to 10^4 $\mu g\,g^{-1}$ (100–10 000 parts per million)*
Microtrace for 10^{-1} to 10^2 $pg\,g^{-1}$ (10^{-7} to 10^{-4} ppm)
Nanotrace for 10^{-1} to 10^2 $fg\,g^{-1}$ (10^{-10} to 10^{-7} ppm)

The term 'semimicro' given as an alternative name for 'meso' is not very apt, referring as it does to samples larger than micro.

A major constituent accounts for 1% to 100% of the sample under investigation; a minor constituent is present in the range 0.01% to 1%; a trace constituent is present at a concentration of less than 0.01%.

When the sample weight is small (0.1–1.0 mg), the determination of a trace component at the 0.01% level may be called subtrace analysis. If the trace component is at the microtrace level, the analysis is called submicrotrace. With a still smaller sample (not larger than 0.1 mg) the determination of a component at the trace level is called ultratrace analysis, and with a component at the microtrace level, the analysis is called ultramicrotrace.

1.5   Searching the literature

The analytical chemist will often have to design a completely novel analysis and will have to seek information from previously published data. This may involve consulting multi-

---

* Concentrations are commonly quoted in parts per million (ppm), but this should be avoided as ppm is a dimensionless quantity.

volume reference works such as Kolthoff and Elving, *Treatise on analytical chemistry*; Wilson and Wilson, *Comprehensive analytical chemistry*; Fresenius and Jander, *Handbuch der analytischen Chemie*; a compendium of methods such as Meites, *Handbook of analytical chemistry*; or specialised monographs dealing with particular techniques or types of material. Details of recognised procedures for the analysis of many materials are published by various official bodies, such as the American Society for Testing and Materials (ASTM), the British Standards Institution and the European Commission. However, an initial search through the abstracting journals (e.g. *Analytical Abstracts, Chemical Abstracts*) will often provide additional information on recent developments and specific analytical procedures. Other general assessments of methods and results are available in review publications (e.g. *Annual Reports of the Chemical Society*), whereas current research is always available in journals specifically devoted to analytical chemistry (*The Analyst* and *Analytical Chemistry*).

1.6   Quantitative analysis

The main techniques employed in quantitative analysis are based on (a) the quantitative performance of suitable chemical reactions and either measuring the amount of reagent needed to complete the reaction, or ascertaining the amount of reaction product obtained; (b) appropriate electrical measurements (e.g. potentiometry); (c) the measurement of certain spectroscopic properties (e.g. absorption spectra); (d) the characteristic movement of a substance through a defined medium under controlled conditions. Sometimes two or more of these techniques may be used in combination to achieve both identification and quantification (e.g. gas chromatography may be linked to mass spectrometry).

The quantitative execution of chemical reactions is the basis of the traditional or 'classical' methods of chemical analysis: gravimetry, titrimetry and volumetry. In **gravimetric analysis** the substance being determined is converted into an insoluble precipitate which is collected and weighed; in the special case of **electrogravimetry**, electrolysis is carried out and the material deposited on one of the electrodes is weighed. Weight measurements or changes of energy are also important in thermal methods of analysis, where these features are recorded as a function of temperature. For example, it is possible to set up conditions where a precipitate from a gravimetric determination can be safely dried. Some common techniques record a parameter as a function of temperature or time. **Thermogravimetry (TG)** records the change in weight; **differential thermal analysis (DTA)** records the difference in temperature between a test substance and an inert reference material; **differential scanning calorimetry (DSC)** records the energy needed to establish a zero temperature difference between a test substance and a reference material.

In **titrimetric analysis** (sometimes called volumetric analysis) the substance to be determined is allowed to react with an appropriate reagent added as a standard solution, and the volume of solution needed for complete reaction is determined. The common titrimetric reactions are neutralisation (acid–base) reactions, complex-forming reactions, precipitation reactions and oxidation–reduction reactions. **Volumetry** measures the volume of gas evolved or absorbed in a chemical reaction.

Electrical methods of analysis (apart from electrogravimetry) involve the measurement of current, voltage or resistance in relation to the concentration of a certain species in solution. Electrical techniques include **voltammetry** (measurement of current at a micro-electrode at a specified voltage); **coulometry** (measurement of current and time needed to complete an electrochemical reaction or to generate sufficient material to react completely with a specified reagent); and **potentiometry** (measurement of the potential of an electrode in equilibrium with an ion to be determined).

5

Spectroscopic methods of analysis depend on measuring the amount of radiant energy of a particular wavelength absorbed by the sample, or measuring the amount of radiant energy of a particular wavelength emitted by the sample. Absorption methods are usually classified according to the wavelength involved as **visible, ultraviolet** or **infrared spectrophotometry**; visible spectrophotometry is sometimes called colorimetry. In addition to these techniques there is the increasing use of **nuclear magnetic resonance spectroscopy** in the quantitative analysis of organic compounds.

Atomic absorption spectroscopy involves atomising the specimen, often by spraying a solution of the sample into a flame, and then studying the absorption of radiation from an electric lamp producing the spectrum of the element to be determined. **Turbidimetric and nephelometric methods** measure the amount of light stopped or scattered by a suspension. They are not absorption methods in the strictest sense, but they deserve to be mentioned here.

Emission methods subject the sample to heat or electrical treatment so that atoms are raised to excited states causing them to emit energy; the intensity of this energy is then measured. Here are some of the common excitation techniques:

Emission spectroscopy subjects the sample to an inductively coupled plasma then examines the emitted light (which may extend into the ultraviolet region).

Flame photometry uses a solution of the sample injected into a flame.

Fluorimetry takes a suitable substance in solution (commonly a metal–fluorescent reagent complex) and excites it using visible or ultraviolet radiation.

Chromatographic and **electrophoretic** methods are essentially separative processes for mixtures of substances, but they are also adapted to identify components of mixtures. The nature of modern detector systems means that chromatography and electrophoresis can be used for reliable quantitative determinations.

In **mass spectrometry** the material under examination is vaporised using a high vacuum and the vapour is bombarded by a high-energy electron beam. Many of the vapour molecules undergo fragmentation and produce ions of varying size. These ions can be distinguished by accelerating them in an electric field, and then deflecting them in a magnetic field where they follow paths dictated by their mass/charge ratio (m/ze) to detection and recording equipment; each kind of ion gives a peak in the **mass spectrum**. Non-volatile inorganic materials can be examined by vaporising them using a high-voltage electric spark.

Mass spectrometry can be used for gas analysis, for the analysis of petroleum products, and in examining semiconductors for impurities. It is particularly useful for establishing the structure of organic compounds.

1.7   Special techniques

X-ray methods

X-rays are produced when high-speed electrons collide with a solid target (which can be the material under investigation). These X-rays are often known as **primary X-rays**. They arise because the electron beam may displace an electron from the inner electron shells of an atom in the target; the lost electron is then replaced by an electron from an outer shell, and energy is emitted as X-rays. It is possible to identify certain emission peaks which are characteristic of elements contained in the target. The wavelengths of the peaks can be related to the atomic number of the elements producing them, so they provide a means of identifying elements present in the target sample. Furthermore, under controlled conditions, the intensity of the peaks can be used to determine the amounts of the various elements present. This is the basis of **electron probe microanalysis**, in which a small target area of the sample is pinpointed for examination. This has important applications in metallurgical research, in the examination of geological samples, and in determining whether biological materials contain metallic elements.

When a beam of short-wavelength primary X-rays strikes a solid target, a similar mechanism as described above will cause the target material to emit X-rays at wavelengths characteristic of the atoms involved; this is called **secondary or fluorescence radiation**. The sample area can be large, and by examining the peak heights of the fluorescence radiation it is possible to get an indication of the sample composition. **X-ray fluorescence analysis** is a rapid process which finds application in metallurgical laboratories, in the processing of metallic ores, and in the cement industry.

Crystalline material will diffract a beam of X-rays, and X-ray powder diffractometry can be used to identify components of mixtures. These X-ray procedures are examples of **non-destructive analysis**.

Radioactivity

Methods based on radioactivity belong to the realm of radiochemistry and may involve measuring the intensity of the radiation from a naturally radioactive material; measuring radioactivity induced by exposing the sample to a **neutron source (activation analysis)**; or **isotope dilution** and **radioimmunoassay**. Typical applications are determining trace elements for investigating pollution problems; examining geological specimens; and quality control in semiconductor manufacturing.

Kinetic methods

Kinetic methods are based on increasing the speed of a reaction by adding a small amount of a catalyst; within limits, the rate of the catalysed reaction will be governed by the amount of catalyst present. If a calibration curve is prepared showing variation of reaction rate with amount of catalyst used, then reaction rate measurements will determine how much catalyst has been added in a certain instance. This provides a sensitive method for determining submicrogram amounts of appropriate organic substances. The method can also be adapted to determine the amount of a substance in solution by adding a catalyst which will destroy it completely, and measuring the concomitant change in the absorbance of the solution for visible or ultraviolet radiation. These procedures are applied in clinical chemistry.

Optical methods

Some optical methods are particularly appropriate for organic compounds. A **refractometer** can be used to measure the refractive index of liquids. This will often provide a means of identifying a pure compound, and in conjunction with a calibration curve, it can also be used to analyse a mixture of two liquids. The **optical rotation** of optically active compounds can be measured, and polarimetric measurements can be used to identify pure substances; it can also be used for quantitative analysis.

1.8 Instrumental methods

Many of the methods listed above, such as those which measure an electrical property, absorption of radiation or the intensity of an emission, require the use of a suitable instrument, polarograph, spectrophotometer, etc.; they are known as instrumental methods. Instrumental methods are usually much faster than purely chemical procedures, they are normally applicable at concentrations far too small to be determined by classical methods, and they find wide application in industry. In most cases a microcomputer can be interfaced to the instrument so that absorption curves, polarograms, titration curves, etc., can be plotted automatically, and in fact, by incorporating appropriate servomechanisms, the whole analytical process may sometimes be completely automated.

Despite the many advantages possessed by instrumental methods, their widespread adoption has not rendered the purely chemical or classical methods obsolete; the situation is influenced by four main factors:

1. The apparatus required for classical procedures is cheap and readily available in all laboratories, but many instruments are expensive and their use will only be justified if numerous samples have to be analysed, or when dealing with the determination of substances present in minute quantities (trace, subtrace or ultratrace analysis).
2. With instrumental methods it is necessary to carry out a calibration operation using a sample of material of known composition as reference substance.
3. Although an instrumental method is ideal for a large number of routine determinations, an occasional, non-routine analysis is often simpler by a classical method. Instrument calibration can be time-consuming.
4. To obtain accurate results with instrumental methods, the reagents still need careful weighing, measuring and preparation of standard solutions. Classical analysis provides the essential training and experience.

The good chemical analyst will always appreciate the value of developing the classical skills in order to maximise the quality of instrumental procedures.

1.9 Factors affecting the choice of analytical method

Analytical techniques have different degrees of sophistication, sensitivity and selectivity, as well as different cost and time requirements. An important task for the analyst is to select the best procedure for a given determination. This will require careful consideration of the following criteria:

(a) The type of analysis required: elemental or molecular, routine or occasional.
(b) Problems arising from the nature of the material to be investigated, e.g. radioactive substances, corrosive substances, substances affected by water.
(c) Possible interference from components of the material other than those of interest.
(d) The concentration range to be investigated.
(e) The accuracy required.
(f) The facilities available, particularly the instruments.
(g) The time required to complete the analysis; this will be particularly relevant when the results are required quickly for the control of a manufacturing process. This may mean that accuracy has to be a secondary consideration; it may require expensive instruments or in-line continuous analysis.
(h) The number of analyses of similar type which have to be performed; in other words, are there a limited number of determinations or are there many repetitive analyses?

(i)   Does the nature of the specimen, the kind of information sought, or the magnitude of the sample available indicate the use of non-destructive methods as opposed to the more commonly applied destructive methods? Destructive methods usually mean the sample is dissolved before analysis.

Many laboratory workers totally ignore the potential costs related to any proposed analytical work. The running costs alone of many modern instruments can be extremely high and need to be used regularly and efficiently to justify the initial expenditure. It is always worth bearing in mind that single or exploratory analyses may often be carried out much more cheaply and often more quickly by a traditional titrimetric or gravimetric procedure in which only limited sample preparation may be required. Many instrumental methods amply justify their use when there are multiple analyses. The cost of analysis does not only include the actual time the equipment is used, but also a proportion of the maintenance costs and the time the instruments may be standing idle.

1.10   Interferences

Whatever the method finally chosen for the required determination, it should ideally be a **specific method**; that is to say, it should be capable of measuring the amount of desired substance accurately, no matter what other substances may be present. In practice few analytical procedures attain this ideal, but many methods are **selective**; in other words, they can be used to determine any of a limited group of ions or molecules from a much wider selection of ions or molecules. Improved selectivity is frequently achieved by carrying out the analytical procedure under carefully controlled conditions. This is particularly the case with chromatographic separations and determinations. But other substances are often present, making it harder to obtain the desired measurements directly. Interfering substances may mean that extra procedures have to be carried out to remove the interference or to prevent it from taking part in the analytical process. Ionic procedures are used for inorganic substances, whereas solvent extraction and chromatographic procedures are better for organic substances. They may be divided into six categories.

Selective precipitation   Appropriate reagents may be added to convert interfering ions into precipitates which can be filtered off; careful pH control is often necessary to achieve a clean separation. Remember that precipitates tend to adsorb substances from solution, so take care to ensure that as little as possible of the substance to be determined is lost in this way.

Masking   A complexing agent is added. If the resultant complexes are sufficiently stable, they will fail to react with reagents added in a later step, perhaps a titrimetric procedure or a gravimetric precipitation method.

Selective oxidation (reduction)   The sample is treated with a selective oxidising or reducing agent which will react with some of the ions present; the resultant change in oxidation state will often facilitate separation. For example, to precipitate iron as hydroxide, the solution is always oxidised so that iron(III) hydroxide is precipitated; this precipitates at a lower pH than iron(II) hydroxide, which could be contaminated with the hydroxides of many bivalent metals.

Solvent extraction   When metal ions are converted into chelate compounds by treatment with suitable organic reagents, the resulting complexes are soluble in organic solvents and can be extracted from the aqueous solution. Many ion association complexes containing bulky ions which are largely organic in character (e.g. the tetraphenylarsonium ion $(C_6H_5)_4As^+$) are soluble in organic solvents, so they can be used to extract appropriate metals ions from aqueous solution. Solvent extraction, along with appropriate acids and bases, may also be used to separate organic compounds from each other before quantification.

Ion exchange   Insoluble ion exchange resins contain either anions or cations capable of exchanging with ions in solutions passed over them. They can be used to remove impurities from solutions or to enrich species under examination. Ion exchange is particularly valuable for concentrating low levels of ions in solution before carrying out quantitative analysis by methods of relatively low sensitivity.

Chromatography   Chromatography covers a whole of range of separation techniques in which chemicals in solution travel down columns or over surfaces by means of liquids or gases and are separated from each other due to their molecular characteristics. The processes involved are discussed in greater detail in Chapters 6 to 9, but note that collectively the forms of chromatography are now applicable to virtually all inorganic and organic materials, except very insoluble polymers; they are of immense value in obtaining quantitative analytical data. Typical applications include drug analysis in forensic work and high-quality food analysis.

1.11   Data handling

Once the best method of dealing with interferences has been decided on and the most appropriate method of determination chosen, the analysis should be carried out in duplicate and preferably in triplicate. All analytical results should be carefully recorded in a suitable notebook to provide a permanent record of the work. Besides this, many modern instruments are either computer-operated or interfaced with computers, so the results are not only displayed on a visual display unit but also presented graphically and/or in tabular form to serve as a detailed record. Further calculations on the analytical results may be necessary to present them in the form required. Many instruments are now able to treat the raw data and to relate it to calibration charts, action limits and statistical analyses.

Like all physical measurements, any results obtained are subject to a degree of uncertainty, and it is necessary to establish the magnitude of this uncertainty in order that meaningful results can be presented. It is therefore necessary to establish the **precision** of the results, by which we mean the extent to which they are reproducible. This is commonly expressed in terms of the numerical difference between a given experimental value and the mean value of all the experimental results. The **spread** or **range** in a set of results is the numerical difference between the highest and lowest results: this figure is also an indication of the precision of the measurements. However, the most important measures of precision are the standard deviation and the variance: these are discussed in Chapter 4.

The difference between the most probable analytical result and the true value for the sample is termed the **systematic error** in the analysis: it indicates the **accuracy** of the analysis.

1.12   Summary

Summarising, the following steps are necessary when confronted with an unfamiliar quantitative determination.

1.   Sampling.
2.   Literature survey and selection of possible methods of determination.
3.   Consideration of interferences and procedures for their removal.

After pooling the information obtained under headings (2) and (3), a selection will be made of a suitable method for analysis along with procedures for dealing with any interferences. Once this has been done the subsequent steps can be carried out.

4.   Dissolution of sample.
5.   Removal or suppression of interferences.
6.   Performance of the determination.
7.   Statistical analysis of the results.

The final step is the presentation of the results for further data processing or in the form of a report.

2

Solution reactions: fundamental theory

Many of the reactions of qualitative and quantitative chemical analysis take place in solution; the solvent is most commonly water but other liquids may also be used. It is therefore necessary to have a general knowledge of the conditions which exist in solutions and the factors which influence chemical reactions.

2.1 The law of mass action

Guldberg and Waage (1867) clearly stated the law of mass action (sometimes called the law of chemical equilibrium) in this form: 'The velocity of a chemical reaction is proportional to the product of the active masses of the reacting substances'. 'Active mass' was interpreted as concentration and expressed in moles per litre. By applying the law to homogeneous systems – systems in which all the reactants are present in one phase, e.g. in solution – we arrive at a mathematical expression for the condition of equilibrium in a reversible reaction.

Consider first the simple reversible reaction at constant temperature:

$$A + B \rightleftharpoons C + D$$

The rate of conversion of A and B is proportional to their concentrations, or

$$r_1 = k_1 \times [A] \times [B]$$

where k_1 is a constant known as the rate constant or rate coefficient, and the square brackets (see footnote on p. 45) denote the concentrations ($mol\,L^{-1}$) of the substances enclosed within the brackets.

Similarly, the rate of conversion of C and D is given by

$$r_2 = k_2 \times [C] \times [D]$$

At equilibrium, the two rates of conversion will be equal:

$$k_1 \times [A] \times [B] = k_2 \times [C] \times [D]$$

$$\text{or} \quad \frac{[C] \times [D]}{[A] \times [B]} = \frac{k_1}{k_2} = K$$

where K is the **equilibrium constant** of the reaction at the given temperature. The expression may be generalised. For a reversible reaction represented by

$$p_1 A_1 + p_2 A_2 + p_3 A_3 + \ldots \rightleftharpoons q_1 B_1 + q_2 B_2 + q_3 B_3 + \ldots$$

where p_1, p_2, p_3 and q_1, q_2, q_3 are the stoichiometric coefficients of the reacting species, the condition for equilibrium is given by the expression

$$\frac{[B_1]^{q_1} \times [B_2]^{q_2} \times [B_3]^{q_3} \ldots}{[A_1]^{p_1} \times [A_2]^{p_2} \times [A_3]^{p_3} \ldots} = K$$

This result may be expressed in words: when equilibrium is reached in a reversible re-action, at constant temperature, the product of the concentrations of the resultants (the substances on the right-hand side of the equation) divided by the product of the concentrations of the reactants (the substances on the left-hand side of the equation), each concentration being raised to a power equal to the stoichiometric coefficient of the substance concerned in the equation for the reaction, is constant.

The equilibrium constant of a reaction can be related to the changes in Gibbs function (ΔG), enthalpy (ΔH) and entropy (ΔS) which occur during the reaction:

$$\Delta G^{\ominus} = -RT \ln K^{\ominus} = -2.303RT \log_{10} K^{\ominus}$$

$$\frac{\mathrm{d}(\ln K^{\ominus})}{\mathrm{d}T} = \frac{\Delta H^{\ominus}}{RT^2}$$

$$\Delta G^{\ominus} = \Delta H^{\ominus} - T\Delta S^{\ominus}$$

In these expressions the superscript $\ominus$ indicates that the quantities concerned relate to a so-called 'standard state'. For the derivation and the significance of these expressions, a textbook of physical chemistry[1] should be consulted, but briefly a reaction will be spontaneous when ΔG is negative, it will be at equilibrium when ΔG is zero, and when ΔG is positive the reverse reaction will be spontaneous. It follows that a reaction is favoured when heat is produced, i.e. it is an exothermic reaction, so the enthalpy change ΔH is negative. It is also favoured by an increase in entropy, i.e. when ΔS is positive. A knowledge of the values of the equilibrium constants of certain selected systems can be of great value to the analyst when dealing with acid–base interactions, solubility equilibria, systems involving complex ions, oxidation–reduction systems, and many separation problems. But note that equilibrium constants do not give any indication of the rate of reaction. These matters are dealt with in detail in succeeding sections of this chapter, and in other pertinent chapters.

2.2 Activity and activity coefficient

In the deduction of the law of mass action it was assumed that the effective concentrations or active masses of the components could be expressed by the stoichiometric concentrations. According to thermodynamics, this is not strictly true. The rigorous equilibrium for a binary electrolyte is

$$AB \rightleftharpoons A^+ + B^-$$

$$\frac{a_{A^+} a_{B^-}}{a_{AB}} = K_t$$

where a_{A^+}, a_{B^-} and a_{AB} represent the **activities** of A^+, B^-, and AB respectively, and K_t is the true, or thermodynamic, **dissociation constant**. The concept of activity, a thermodynamic quantity, is due to G. N. Lewis. The quantity is related to the concentration by a factor called the activity coefficient:

activity = concentration $\times$ activity coefficient

Thus at any concentration

$$a_{A^+} = y_{A^+}[A^+] \qquad a_{B^-} = y_{B^-}[B^-] \qquad a_{AB} = y_{AB}[AB]$$

where y refers to the activity coefficients,* and the square brackets to the concentrations. Substituting in the above equation, we obtain

$$\frac{y_{A^+}[A^+] \times y_{B^-}[B^-]}{y_{AB}[AB]} = \frac{[A^+][B^-]}{[AB]} \times \frac{y_{A^+}y_{B^-}}{y_{AB}} = K_t$$

This is the rigorously correct expression for the law of mass action as applied to weak electrolytes.

The activity coefficient varies with the concentration. For ions it also varies with the ionic charge, and it is the same for all dilute solutions having the same **ionic strength**; ionic strength is a measure of the electrical field existing in the solution. The term 'ionic strength', designated by the symbol I, is defined as equal to one-half the sum of the products of the concentration of each ion multiplied by the square of its charge number, or $I = \frac{1}{2}\Sigma c_i z_i^2$, where c_i is the ionic concentration in moles per litre of solution and z_i is the charge number of the ion concerned. An example will make this clear. The ionic strength of $0.1\,M$ HNO_3 solution containing $0.2\,M$ $Ba(NO_3)_2$ is given by

$$0.5\{0.1 \text{ (for } H^+) + 0.1 \text{ (for } NO_3^-)$$
$$+ 0.2 \times 2^2 \text{ (for } Ba^{2+}) + 0.2 \times 2 \text{ (for } NO_3^-)\} = 0.5\{1.4\} = 0.7$$

It can be shown on the basis of the Debye–Hückel theory that for aqueous solutions at room temperature

$$\log y_i = \frac{-0.505 z_i^2 I^{0.5}}{1 + 3.3 \times 10^7 a I^{0.5}}$$

where y_i is the activity coefficient of the ion, z_i is the charge number of the ion concerned, I is the ionic strength of the solution, and a is the average 'effective diameter' of all the ions in the solution. For very dilute solutions ($I^{0.5} < 0.1$) the second term of the denominator is negligible and the equation reduces to

$$\log y_i = -0.505 z_i^2 I^{0.5}$$

For more concentrated solutions ($I^{0.5} > 0.3$) an additional term BI is added to the equation; B is an empirical constant. For a more detailed treatment of the Debye–Hückel theory a textbook of physical chemistry should be consulted.

2.3 Chemical equilibrium

If a mixture of hydrogen and iodine vapour is heated to a temperature of about $450\,^\circ C$ in a closed vessel, the two elements combine and hydrogen iodide is formed. It is found, however, that no matter how long the duration of the experiment, some hydrogen and iodine remain uncombined. If pure hydrogen iodide is heated in a closed vessel to a temperature of about $450\,^\circ C$, the substance decomposes to form hydrogen and iodine, but

---

* The symbol for the activity coefficient depends upon the method of expressing the concentration of the solution. The recommendations of the IUPAC Commission on Symbols, Terminology and Units (1969) are as follows: concentration in moles per litre (molarity), activity coefficient represented by y, concentration in moles per kilogram (molality), activity coefficient represented by γ, concentration expressed as mole fraction, activity coefficient represented by f.

no matter how prolonged the heating, some hydrogen iodide remains unchanged. This is an example of a **reversible reaction** in the gaseous phase:

$$H_2(g) + I_2(g) \rightleftharpoons 2HI(g)$$

An example of a reversible reaction in the liquid phase is afforded by the esterification reaction between ethanol and ethanoic acid forming ethyl ethanoate and water. But since ethyl ethanoate undergoes conversion to ethanoic acid and ethanol when heated with water, the esterification reaction never proceeds to completion:

$$C_2H_5OH + CH_3COOH \rightleftharpoons CH_3COOC_2H_5 + H_2O$$

It is found that after the elapse of a sufficient time interval, all reversible reactions reach a state of **chemical equilibrium**. In this state the composition of the equilibrium mixture remains constant, provided the temperature remains constant; some gaseous equilibria require constant temperature and constant pressure. Furthermore, provided the conditions (temperature and pressure) are maintained constant, the same state of equilibrium may be obtained from either direction of a given reversible reaction. In the equilibrium state, the two opposing reactions are taking place at the same rate, so the system is in a state of dynamic equilibrium.

Note that the composition of a given equilibrium can be altered by changing the conditions under which the system is maintained and it is necessary to consider the effect of changes in (a) the temperature, (b) pressure and (c) the concentration of the components. According to the **Le Chatelier–Braun principle**, if a constraint is applied to a system in equilibrium, the system will adjust itself so as to nullify the effect of the constraint. The effect of temperature, pressure and concentration can be considered in the light of this principle.

Temperature

The formation of ammonia from its elements

$$N_2(g) + 3H_2(g) \rightleftharpoons 2NH_3(g)$$

is a reversible process in which the forward reaction is accompanied by the evolution of heat (energy); it is an exothermic reaction. The reverse reaction absorbs heat; it is an endothermic reaction. If the temperature of an equilibrium mixture of nitrogen, hydrogen and ammonia is increased, the reaction which absorbs heat will be favoured, so ammonia is decomposed.

Pressure

Referring to the hydrogen iodide equilibrium system, the stoichiometric coefficients of the molecules on each side of the equation for the reaction are equal, and there is no change in volume when reaction occurs. Therefore, if the pressure of the system is doubled, thus halving the total volume, the two sides of the equation are equally affected, so the composition of the equilibrium mixture remains unchanged. In the nitrogen–hydrogen–ammonia equilibrium system there is a decrease in volume when ammonia is produced, hence an increase in pressure will favour the formation of ammonia. For equilibrium in the liquid phase, moderate changes in pressure have practically no effect on the volume, owing to the small compressibility of liquids, so moderate pressure changes do not affect the equilibrium.

Concentration of reagents

If hydrogen is added to the equilibrium mixture resulting from the thermal decomposition of hydrogen iodide, it is found that more hydrogen iodide is present when equilibrium is restored. In accordance with the Le Chatelier–Braun Principle, the system has reacted to remove some of the added hydrogen.

2.4 Factors affecting chemical reactions in solution

There are three main factors whose influence on chemical reactions in solution need to be considered: (a) the nature of the solvent; (b) temperature; and (c) the presence of catalysts.

Nature of the solvent

Reactions in aqueous solution generally proceed rapidly because they involve interaction between ions. Thus the precipitation of silver chloride from a chloride solution by the addition of silver nitrate solution can be formulated as

$$Ag^+ + Cl^- \rightleftharpoons AgCl(s)$$

Reactions between molecules in solution, e.g. the formation of ethyl ethanoate from ethanoic acid and ethanol, are generally comparatively slow. It is therefore convenient to classify solvents as **ionising solvents** if they tend to produce solutions in which the solute is ionised and as **non-ionising solvents** if they give solutions in which the solute is not ionised. Common ionising solvents include water, ethanoic acid, hydrogen chloride, ammonia, amines, bromine trifluoride and sulphur dioxide. Of these solvents, the first four are characterised by an ability to produce hydrogen ions:

Water $2H_2O \rightleftharpoons H_3O^+ + OH^-$

Ammonia $2NH_3 \rightleftharpoons NH_4^+ + NH_2^-$

These four solvents are called **protogenic solvents**, whereas bromine trifluoride and sulphur dioxide, which do not contain hydrogen, are **non-protonic solvents**. Non-ionising solvents include hydrocarbons, ethers, esters and higher alcohols; the lower alcohols, especially methanol and ethanol, do show slight ionising properties with appropriate solutes.

Temperature

Reaction rates increase rapidly with rising temperature, and in some analytical procedures it is necessary to heat the solution to ensure the required reaction takes place with sufficient rapidity. An example of this behaviour is the titration of acidified oxalate solutions with potassium permanganate solution. When potassium permanganate solution is added to a solution of an oxalate containing sulphuric acid at room temperature, reaction proceeds very slowly, and the solution sometimes acquires a brown tinge due to the formation of manganese(IV) oxide. If, however, the solution is heated to about 70 °C before adding any permanganate solution, the reaction becomes virtually instantaneous and no manganese(IV) oxide is produced.

Catalysts

The rates of some reactions can be greatly increased by the presence of a catalyst. This is a substance that alters the rate of a reaction without itself undergoing any net change. It follows that a small amount of the catalyst can influence the conversion of large quantities of the reactants. If the reaction under consideration is reversible, the catalyst affects both the forward and reverse reactions, and although the reaction is speeded up, the position of equilibrium is unchanged.

An example of catalytic action is provided by the titration of oxalates with potassium permanganate solution. It is found that even though the oxalate solution is heated, the first few drops of permanganate solution are only slowly decolorised; but as more permanganate solution is added, the decoloration becomes instantaneous. This is because the reaction between oxalate ions and permanganate ions is catalysed by the Mn^{2+} ions formed by the reduction of permanganate ions:

$$MnO_4^- + 8H^+ + 5e^- \rightarrow Mn^{2+} + 4H_2O$$

Other examples are the use of osmium(VIII) oxide (osmium tetroxide) as catalyst in the titration of solutions of arsenic(III) oxide with cerium(IV) sulphate solution, and the use of molybdate(VI) ions to catalyse the formation of iodine by the reaction of iodide ions with hydrogen peroxide. Certain reactions of various organic compounds are catalysed by several naturally occurring proteins known as enzymes.

The determination of trace quantities of many substances can be accomplished by examining the rate of a chemical reaction for which the substance to be determined acts as a catalyst. By comparing the observed rate of reaction with rates determined for the same reaction with known quantities of the same catalyst present, unknown concentrations can be calculated. Likewise a catalyst may be used to take a substance for which no suitable analytical reaction exists under the prevailing conditions, and convert it to a product which can be determined. Alternatively, the substance to be determined may be destroyed by adding a catalyst, and the resultant change in some measured property, e.g. the absorption of light, enables the amount of substance to be evaluated. Thus, uric acid in blood can be determined by measuring the absorption of ultraviolet radiation at a wavelength of 292 nm, but the absorption is not specific. The absorption meter reading is recorded, then the uric acid is destroyed by addition of the enzyme uricase. The absorption reading is repeated, and from the difference between the two results, the amount of uric acid present can be calculated.

2.5   The ionic product of water

Kohlrausch and Heydweiller (1894) found that the most highly purified water which can be obtained possesses a small but definite conductivity. Water must therefore be slightly ionised in accordance with the equation*

$$H_2O \rightleftharpoons H^+ + OH^-$$

Applying the law of mass action to this equation, we obtain for any given temperature that

$$\frac{a_{H^+}a_{OH^-}}{a_{H_2O}} = \frac{[H^+][OH^-]}{[H_2O]} \times \frac{y_{H^+}y_{OH^-}}{y_{H_2O}} = \text{a constant}$$

Since water is only slightly ionised, the ionic concentrations will be small, and their activity coefficients may be regarded as unity; the activity of the unionised molecules may also be taken as unity. The expression thus becomes

$$\frac{[H^+][OH^-]}{[H_2O]} = \text{a constant}$$

---

* Strictly speaking, the hydrogen ion H^+ exists in water as the hydronium ion H_3O^+ (Section 2.4). The electrolytic dissociation of water should therefore be written $2H_2O \rightleftharpoons H_3O^+ + OH^-$. For the sake of simplicity, the more familiar symbol H^+ will be retained.

In pure water or in dilute aqueous solutions, the concentration of the undissociated water may be considered constant. Hence

$$[H^+][OH^-] = K_w$$

where K_w is the ionic product of water. Note the assumption that the activity coefficients of the ions are unity and the activity coefficient of water is constant applies strictly to pure water and to very dilute solutions (ionic strength <0.01); in more concentrated solutions, i.e. in solutions of appreciable ionic strength, the activity coefficients of the ions are affected (Section 2.2), as is the activity of the unionised water. Then the ionic product of water will not be constant; it will depend upon the ionic strength of the solution. But it is difficult to determine the activity coefficients, except under specially selected conditions, so in practice the ionic product K_w, although not strictly constant, is employed.

The ionic product varies with the temperature, but under ordinary experimental conditions (at about 25 °C) its value may be taken as 1×10^{-14} with concentrations expressed in mol L^{-1}. This is sensibly constant in dilute aqueous solutions. If the product of $[H^+]$ and $[OH^-]$ in aqueous solution momentarily exceeds this value, the excess ions will immediately combine to form water. Similarly, if the product of the two ionic concentrations is momentarily less than 10^{-14}, more water molecules will dissociate until the equilibrium value is attained.

The hydrogen and hydroxide ion concentrations are equal in pure water, so $[H^+] = [OH^-]$ $= \sqrt{K_w} = 10^{-7}$ mol L^{-1} at about 25 °C. A solution in which the hydrogen and hydroxide ion concentrations are equal is called an **exactly neutral solution**. If $[H^+]$ is greater than 10^{-7}, the solution is **acid**, and if less than 10^{-7}, the solution is **alkaline** (or basic). It follows that at ordinary temperatures $[OH^-]$ is greater than 10^{-7} in alkaline solutions and less than 10^{-7} in acid solutions.

In all cases the reaction of the solution can be quantitatively expressed by the magnitude of the hydrogen ion (or hydronium ion) concentration, or less frequently, by the magnitude of the hydroxide ion concentration, since the following simple relations between $[H^+]$ and $[OH^-]$ exist:

$$[H^+] = \frac{K_w}{[OH^-]} \quad \text{and} \quad [OH^-] = \frac{K_w}{[H^+]}$$

The variation of K_w with temperature is shown in Table 2.1.

Table 2.1 *Ionic product of water at various temperatures*

Temp. (°C)	$K_w/10^{-14}$	Temp. (°C)	$K_w/10^{-14}$
0	0.12	35	2.09
5	0.19	40	2.92
10	0.29	45	4.02
15	0.45	50	5.47
20	0.68	55	7.30
25	1.01	60	9.61
30	1.47		

2.6   Electrolytic dissociation

2.6.1   Introduction

Aqueous solutions of many salts of the common 'strong acids' (hydrochloric, nitric and sulphuric) and of bases such as sodium hydroxide and potassium hydroxide are good conductors of electricity, whereas pure water shows only a very poor capability to conduct. The above solutes are therefore called electrolytes. On the other hand, certain solutes, e.g. ethane-1,2-diol (ethylene glycol) which is used as antifreeze, produce solutions that show a conducting capability only a little different from that of water; these solutes are called non-electrolytes. Most reactions of analytical importance occurring in aqueous solution involve electrolytes, and it is necessary to consider their nature.

Salts

The structure of numerous salts in the solid state has been investigated by means of X-rays and by other methods. They are composed of charged atoms or groups of atoms held together in a crystal lattice, and they are called ionic compounds. When these salts are dissolved in a solvent of high dielectric constant such as water, or heated to their melting point, the crystal forces are weakened and the substances dissociate into the pre-existing charged particles or ions, so the resultant liquids are good conductors of electricity; they are called **strong electrolytes**. But some salts, e.g. cyanides, thiocyanates, the halides of mercury and cadmium, and lead ethanoate, give solutions which show a significant electrical conductance, but it is not as great as for strong electrolytes of comparable concentration. Solutes showing this behaviour are called **weak electrolytes**; they are generally covalent compounds which undergo only limited ionisation when dissolved in water:

$$BA \rightleftharpoons B^+ + A^-$$

Acids and bases

An acid may be defined as a substance which, when dissolved in water, undergoes dissociation with the formation of hydrogen ions as the only positive ions:

$$HCl \rightarrow H^+ + Cl^-$$

$$HNO_3 \rightarrow H^+ + NO_3^-$$

Actually the hydrogen ion H^+ (or proton) does not exist in the free state in aqueous solution; each hydrogen ion combines with one molecule of water to form the hydronium ion H_3O^+. The hydronium ion is a hydrated proton, so the above reactions are more accurately written

$$HCl + H_2O \rightarrow H_3O^+ + Cl^-$$

$$HNO_3 + H_2O \rightarrow H_3O^+ + NO_3^-$$

The ionisation may be attributed to the great tendency of the free hydrogen ions H^+ to combine with water molecules to form hydronium ions. Hydrochloric acid and nitric acid are almost completely dissociated in aqueous solution, in accordance with the above equations; this is readily demonstrated by freezing point measurements and by other methods.

Polyprotic acids

Polyprotic acids ionise in stages. In sulphuric acid, one hydrogen atom is almost completely ionised:

$$H_2SO_4 + H_2O \rightarrow H_3O^+ + HSO_4^-$$

The second hydrogen atom is only partially ionised, except in very dilute solution:

$$HSO_4^- + H_2O \rightleftharpoons H_3O^+ + SO_4^{2-}$$

Phosphoric(V) acid also ionises in stages:

$$H_3PO_4 + H_2O \rightleftharpoons H_3O^+ + H_2PO_4^-$$

$$H_2PO_4^- + H_2O \rightleftharpoons H_3O^+ + HPO_4^{2-}$$

$$HPO_4^{2-} + H_2O \rightleftharpoons H_3O^+ + PO_4^{3-}$$

The successive stages of ionisation are known as the primary, secondary and tertiary ionisations. They do not take place to the same degree. The primary ionisation is always greater than the secondary, and the secondary very much greater than the tertiary.

Acids such as ethanoic acid (CH_3COOH) give an almost normal freezing point depression in aqueous solution; the extent of dissociation is correspondingly small. It is usual, therefore, to distinguish between acids which are completely or almost completely ionised in solution and those which are only slightly ionised. The former are called **strong acids** (e.g. hydrochloric, hydrobromic, hydriodic, iodic(V), nitric and perchloric [chloric(VII)] acids, primary ionisation of sulphuric acid), and the latter are called **weak acids** (e.g. nitrous acid, ethanoic acid, carbonic acid, boric acid, phosphorous (phosphoric(III)) acid, phosphoric(V) acid, hydrocyanic acid, hydrogen sulphide). There is, however, no sharp division between the two classes.

A base was originally defined as a substance which, when dissolved in water, undergoes dissociation with the formation of hydroxide ions OH^- as the only negative ions. Thus sodium hydroxide, potassium hydroxide and the hydroxides of certain bivalent metals are almost completely dissociated in aqueous solution:

$$NaOH \rightarrow Na^+ + OH^-$$

$$Ba(OH)_2 \rightarrow Ba^{2+} + 2OH^-$$

These are **strong bases**. Aqueous ammonia solution, however, is a weak base. Only a small concentration of hydroxide ions is produced in aqueous solution:

$$NH_3 + H_2O \rightleftharpoons NH_4^+ + OH^-$$

2.6.2 The Brønsted–Lowry theory of acid and bases

The simple concept given in the preceding paragraphs suffices for many of the requirements of quantitative inorganic analysis in aqueous solution. It is, however, desirable to have some knowledge of the general theory of acids and bases proposed independently by J. N. Brønsted and by T. M. Lowry in 1923, since this is applicable to all solvents. According to this theory, an acid is a species having a tendency to lose a proton, and a base is a species having a tendency to add on a proton. This may be represented as follows:

acid = proton + conjugate base

$$A \rightleftharpoons H^+ + B \qquad\qquad [2.1]$$

It must be emphasised that the symbol H^+ represents the proton and not the 'hydrogen ion' of variable nature existing in different solvents (OH_3^+, NH_4^+, $CH_3CO_2H_2^+$, $C_2H_5OH_2^+$, etc.); the definition is therefore independent of solvent. Reaction [2.1] represents a hypothetical scheme for defining A and B, not a reaction which can actually occur. Acids need not be neutral molecules (e.g. HCl, H_2SO_4, CH_3CO_2H), they may also be anions (e.g. HSO_4^-, $H_2PO_4^-$, $HOOCCOO^-$) and cations (e.g. NH_4^+, $C_6H_5NH_3^+$, $Fe(H_2O)_6^{3+}$). The same is true of bases, where the three classes can be illustrated by NH_3, $C_6H_5NH_2$, H_2O; CH_3COO^-, OH^-, HPO_4^{2-}, $OC_2H_5^-$; $Fe(H_2O)_5(OH)^{2+}$. Since the free proton cannot exist in solution in measurable concentration, reaction does not take place unless a base is added to accept the proton from the acid. By combining the equations $A_1 = B_1 + H^+$ and $B_2 + H^+ = A_2$, we obtain

$$A_1 + B_2 \rightleftharpoons A_2 + B_1 \qquad [2.2]$$

A_1–B_1 and A_2–B_2 are two conjugate acid–base pairs. This is the most important expression for reactions involving acids and bases; it represents the transfer of a proton from A_1 to B_2 or from A_2 to B_1. The stronger the acid A_1 and the weaker A_2, the more complete will be reaction [2.2]. The stronger acid loses its proton more readily than the weaker; similarly the stronger base accepts a proton more readily than the weaker base. It is evident that the base or acid conjugate to a strong acid or strong base is always weak, whereas the base or acid conjugate to a weak acid or weak base is always strong.

In aqueous solution a Brønsted–Lowry acid A,

$$A + H_2O \rightleftharpoons H_3O^+ + B$$

is strong when the above equilibrium is virtually complete to the right, so [A] is almost zero. A strong base is one for which [B], the equilibrium concentration of base other than hydroxide ion, is almost zero.

Acids may thus be arranged in series according to their relative combining tendencies with a base, which for aqueous solutions (in which we are largely interested) is water:

$$HCl + H_2O \rightleftharpoons H_3O^+ + Cl^-$$
$$\text{acid}_1 \quad \text{base}_2 \quad \text{acid}_2 \quad \text{base}_1$$

This process is essentially complete for all typical 'strong' (i.e. highly ionised) acids, such as HCl, HBr, HI, HNO_3, and $HClO_4$. In contrast with the 'strong' acids, the reactions of a typical 'weak' or slightly ionised acid, such as ethanoic acid or propanoic acid, proceeds only slightly to the right in the equation:

$$CH_3COOH + H_2O \rightleftharpoons H_3O^+ + CH_3COO^-$$
$$\text{acid}_1 \quad \text{base}_2 \quad \text{acid}_2 \quad \text{base}_1$$

The typical strong acid of the water system is the hydrated proton H_3O^+, and the role of the conjugate base is minor if it is a sufficiently weak base, e.g. Cl^-, Br^-, and ClO_4^-. The conjugate bases have strengths that vary inversely as the strengths of the respective acids. It can easily be shown that the basic ionisation constant of the conjugate base $K_{B, conj.}$ is equal to $K_w/K_{A, conj.}$, where K_w is the ionic product of water.

Reaction [2.2] includes reactions formerly described by a variety of names, such as dissociation, neutralisation, hydrolysis and buffer action (see below). One acid–base pair may involve the solvent (in water H_3O^+–H_2O or H_2O–OH^-), showing that ions such as H_3O^+ and OH^- are in principle only particular examples of an extended class of acids and bases, though they do occupy a particularly important place. It follows that the properties of an acid or base may be greatly influenced by the nature of the solvent employed.

Another definition of acids and bases is due to G. N. Lewis (1938). From the experimental viewpoint, Lewis regarded all substances which exhibit 'typical' acid–base properties (neutralisation, replacement, effect on indicators, catalysis), irrespective of their chemical nature and mode of action, as acids or bases. He related the properties of acids to the acceptance of electron pairs, and bases as donors of electron pairs, to form covalent bonds regardless of whether protons are involved. On the experimental side, Lewis's definition brings together a wide range of qualitative phenomena, e.g. solutions of BF_3, BCl_3, $AlCl_3$ or SO_2 in an inert solvent cause colour changes in indicators similar to those produced by hydrochloric acid, and these changes are reversed by bases so that titrations can be carried out. Compounds of the same type as BF_3 are usually described as **Lewis acids** or **electron acceptors**. The Lewis bases (e.g. ammonia, pyridine) are virtually identical with the Brønsted–Lowry bases. The great disadvantage of the Lewis definition of acids is that, unlike proton transfer reactions, it is incapable of general quantitative treatment.

The implications of the theory of the complete dissociation of strong electrolytes in aqueous solution were considered by Debye, Hückel and Onsager, and they succeeded in accounting quantitatively for the increasing molecular conductivity of a strong electrolyte producing singly charged ions with decreasing concentration of the solution over the concentration range 0 to $0.002\,M$. For full details, textbooks of physical chemistry must be consulted.[1]

It is important to realise that although complete dissociation occurs with strong electrolytes in aqueous solution, this does not mean the effective concentrations of the ions are identical with their molar concentrations in any solution of the electrolyte. If this were the case, the variation of the osmotic properties of the solution with dilution could not be accounted for. The variation of colligative, e.g. osmotic, properties with dilution is ascribed to changes in the activity of the ions; these changes depend upon the electrical forces between the ions. Using the Debye–Hückel theory for dilute solutions, it is possible to obtain expressions that describe the variation of activity and related quantities.

2.7 Acid–base equilibria in water

Consider the dissociation of a weak electrolyte, such as ethanoic acid, in dilute aqueous solution:

$$CH_3COOH + H_2O \rightleftharpoons H_3O^+ + CH_3COO^-$$

This will be written for simplicity in the conventional manner:

$$CH_3COOH \rightleftharpoons H^+ + CH_3COO^-$$

where H^+ represents the hydrated hydrogen ion. Applying the law of mass action, we have

$$\frac{[CH_3COO^-][H^+]}{[CH_3COOH]} = K$$

where K is the equilibrium constant at a particular temperature; it is usually known as the **ionisation constant** or the **dissociation constant**. If 1 mol of the electrolyte is dissolved in V litres of solution ($V = 1/c$, where c is the concentration in $mol\,L^{-1}$), and if α is the degree of ionisation at equilibrium, then the amount of unionised electrolyte will be $(1 - \alpha)\,mol$, and the amount of each of the ions will be $\alpha\,mol$. The concentration of unionised ethanoic acid will therefore be $(1 - \alpha)/V$ and the concentration of each of the ions α/V. Substituting in the equilibrium equation, we obtain the expression

$$\alpha^2/(1 - \alpha)V = K \quad \text{or} \quad \alpha^2 c/(1 - \alpha) = K$$

This is known as **Ostwald's dilution law**.

Interionic effects are not negligible even for weak acids, and the activity coefficient product must be introduced into the expression for the ionisation constant:

$$K = \frac{\alpha^2 c}{(1-\alpha)} \times \frac{y_{H^+} y_{A^-}}{y_{HA}} \quad \text{where} \quad A^- = CH_3COO^-$$

Reference must be made to textbooks of physical chemistry (Section 2.35) for details of the methods used to evaluate true dissociation constants of acids. From the viewpoint of quantitative analysis, sufficiently accurate values for the ionisation constants of weak monoprotic acids may be obtained by using the classical Ostwald dilution law expression: the resulting 'constant' is sometimes called the 'concentration dissociation constant'.

2.8 Strengths of acids and bases

Take the Brønsted–Lowry expression for acid–base equilibria (Section 2.6)

$$A_1 + B_2 \rightleftharpoons A_2 + B_1 \tag{2.2}$$

and apply the law of mass action; this leads to

$$K = \frac{[A_2][B_1]}{[A_1][B_2]} \tag{2.1}$$

where the constant K depends on the temperature and the nature of the solvent. This expression is strictly valid only for extremely dilute solutions; when ions are present the electrostatic forces between them have appreciable effects on the properties of their solutions and deviations are apparent from ideal laws (which are assumed in deriving the law of mass action using thermodynamic or kinetic methods). The deviations from the ideal laws are usually expressed in terms of activities or activity coefficients. For our purpose, the deviations due to interionic attractions and ionic activities will be regarded as small for small ionic concentrations, and the equations will be regarded as holding in the same form at higher concentrations, provided the total ionic concentration does not vary much in a given set of experiments.

To use equation (2.1) for measuring the strength of an acid, a standard acid–base pair, say A_2–B_2, must be chosen, and it is usually convenient to refer acid–base strength to the solvent. In water the acid–base pair H_3O^+–H_2O is taken as the standard. The equilibrium defining acids is therefore

$$A + H_2O \rightleftharpoons B + H_3O^+ \tag{2.3}$$

and the constant

$$K' = \frac{[B][H_3O^+]}{[A][H_2O]} \tag{2.2}$$

gives the strength of A; the strength of H_3O^+ is taken as unity. Reaction [2.3] represents what is usually described as the dissociation of the acid A in water. The constant K' is closely related to the dissociation constant of A in water as usually defined; it differs only by the inclusion of the term $[H_2O]$ in the denominator. The $[H_2O]$ term represents the 'concentration' of water molecules in liquid water ($55.5 \, mol \, L^{-1}$ on the ordinary volume concentration scale). When dealing with dilute solutions, the value of $[H_2O]$ may be regarded as constant and equation (2.2) may be expressed as

23

$$K_a = \frac{[\text{B}][\text{H}^+]}{[\text{A}]} \tag{2.3}$$

by writing H^+ for H_3O^+ and remembering that the hydrated proton is meant. This equation defines the strength of the acid A. If A is an uncharged molecule (e.g. a weak organic acid), B is the anion derived from it by the loss of a proton, and equation (2.3) is the usual expression for the ionisation constant. If A is an anion such as H_2PO_4^-, the dissociation constant $[\text{HPO}_4^{2-}][\text{H}^+]/[\text{H}_2\text{PO}_4^-]$ is usually called the second dissociation constant of phosphoric(V) acid. If A is a cation acid, e.g. the ammonium ion, which interacts with water as shown by the equation

$$\text{NH}_4^+ + \text{H}_2\text{O} \rightleftharpoons \text{NH}_3 + \text{H}_3\text{O}^+$$

the acid strength is given by $[\text{NH}_3][\text{H}^+]/[\text{NH}_4^+]$.

On this basis it is, in principle, unnecessary to treat the strength of bases separately from the strength of acids, since any protolytic reaction involving an acid must also involve its conjugate base. The properties of ammonia and various amines in water are readily understood on the Brønsted–Lowry concept:

$$\text{H}_2\text{O} \rightleftharpoons \text{H}^+ + \text{OH}^-$$

$$\text{NH}_3 + \text{H}^+ \rightleftharpoons \text{NH}_4^+$$

$$\text{NH}_3 + \text{H}_2\text{O} \rightleftharpoons \text{NH}_4^+ + \text{OH}^-$$

The basic dissociation constant K_b is given by

$$K_b = \frac{[\text{NH}_4^+][\text{OH}^-]}{[\text{NH}_3]} \tag{2.4}$$

Since $[\text{H}^+][\text{OH}^-] = K_w$ (the ionic product of water), we have

$$K_b = K_w/K_a$$

The values of K_a and K_b for different acids and bases vary through many powers of 10. It is often convenient to use the dissociation constant exponent pK defined by

$$pK = \log_{10}(1/K) = -\log_{10} K$$

The larger the value of pK_a, the weaker the acid and the stronger the base. For very weak or slightly ionised electrolyes, the expression $\alpha^2/(1 - \alpha)V = K$ reduces to $\alpha^2 = KV$ or $\alpha = (KV)^{1/2}$ since α may be neglected in comparison with unity. Hence for any two weak acids or bases at a given dilution V (in L), we have $\alpha_1 = (K_1 V)^{1/2}$ and $\alpha_2 = (K_2 V)^{1/2}$, or $\alpha_1/\alpha_2 = (K_1/K_2)^{1/2}$. Expressed in words, for any two weak or slightly dissociated electrolytes at equal dilutions, the degrees of dissociation are proportional to the square roots of their ionisation constants. Some values for the dissociation constants at $25\,^\circ\text{C}$ for weak acids and bases are collected in Appendix 7.

2.9 Dissociation of polyprotic acids

When a polyprotic acid is dissolved in water, the various hydrogen atoms undergo ionisation to different extents. For a diprotic acid H_2A, the primary and secondary dissociations can be represented by the equations

$$\text{H}_2\text{A} \rightleftharpoons \text{H}^+ + \text{HA}^-$$

$$\text{HA}^- \rightleftharpoons \text{H}^+ + \text{A}^{2-}$$

If the acid is a weak electrolyte, the law of mass action may be applied and the following expressions obtained:

$$[H^+][HA^-]/[H_2A] = K_1 \tag{2.5}$$

$$[H^+][A^{2-}]/[HA^-] = K_2 \tag{2.6}$$

where K_1 and K_2 are respectively known as the primary and secondary dissociation constants.

Each stage of the dissociation process has its own ionisation constant, and their magnitudes give a measure of the extent to which each ionisation has proceeded at any given concentration. The greater the value of K_1 relative to K_2, the smaller the secondary dissociation and the greater must be the dilution before K_2 becomes appreciable. It is therefore possible that a diprotic (or polyprotic) acid may behave as a monoprotic acid so far as dissociation is concerned. This is indeed characteristic of many polyprotic acids.

A triprotic acid H_3A, e.g. phosphoric(V) acid, will similarly yield three dissociation constants, K_1, K_2, and K_3, which may be derived in an analogous manner:

$$[H^+][H_2A^-]/[H_3A] = K_1 \tag{2.5'}$$

$$[H^+][HA^{2-}]/[H_2A^-] = K_2 \tag{2.6'}$$

$$[H^+][H^{3-}]/[HA^{2-}] = K_3 \tag{2.7}$$

The following numerical example provides a practical application of the theoretical principles.

Example 2.1

Calculate the concentrations of HS^- and S^{2-} in a saturated aqueous solution of hydrogen sulphide at 25 °C.

A saturated aqueous solution of hydrogen sulphide at 25 °C and atmospheric pressure is approximately $0.1\,M$, and for H_2S the primary and secondary dissociation constants may be taken as $1.0 \times 10^{-7}\,mol\,L^{-1}$ and $1.0 \times 10^{-14}\,mol\,L^{-1}$ respectively.

In the solution the following equilibria are involved:

$$H_2S + H_2O \rightleftharpoons HS^- + H_3O^+ \qquad K_1 = [H^+][HS^-]/[H_2S] \tag{2.4}$$

$$HS^- + H_2O \rightleftharpoons S^{2-} + H_3O^+ \qquad K_2 = [H^+][S^{2-}]/[HS^-] \tag{2.5}$$

$$H_2O \rightleftharpoons H^+ + OH^-$$

Electroneutrality requires that the total cation concentration must equal the total anion concentration, hence taking account of charge numbers,

$$[H^+] = [HS^-] + 2[S^{2-}] + [OH^-] \tag{2.8}$$

but since we are dealing with an acid solution, $[H^+] > 10^{-7} > [OH^-]$ and we can simplify equation (2.8) to read

$$[H^+] = [HS^-] + 2[S^{2-}] \tag{2.9}$$

The 0.1 mol H_2S is present partly as undissociated H_2S and partly as the ions HS^- and S^{2-}, and it follows that

$$[H_2S] + [HS^-] + [S^{2-}] = 0.1 \tag{2.10}$$

The very small value of K_2 indicates that the secondary dissociation and therefore $[S^{2-}]$ are extremely minute, and ignoring $[S^{2-}]$ in equation (2.9) we are left with the result

$$[H^+] \approx [HS^-]$$

Since K_1 is also small, $[H^+] \ll [H_2S]$ and so equation (2.10) can be reduced to $[H_2S] \approx 0.1$. Using these results with reaction [2.4] we find

$$[H^+]^2/0.1 = 1 \times 10^{-7} \qquad [H^+] = [HS^-] = 10 \times 10^{-4}\,\text{mol}\,L^{-1}$$

From reaction [2.5] it then follows that

$$(1.0 \times 10^{-4})[S^{2-}]/(1.0 \times 10^{-4}) = 1 \times 10^{-14}$$

$$\text{and} \quad [S^{2-}] = 1 \times 10^{-14}\,\text{mol}\,L^{-1}$$

2.10   The hydrogen ion exponent

For many purposes, especially when dealing with small concentrations, it is cumbersome to express concentrations of hydrogen and hydroxyl ions in terms of moles per litre ($\text{mol}\,L^{-1}$). A very convenient method was proposed by S. P. L. Sørensen (1909). He introduced the hydrogen ion exponent pH defined by the relationships

$$pH = \log_{10} 1/[H^+] = -\log_{10}[H^+] \quad \text{or} \quad [H^+] = 10^{-pH}$$

The quantity pH is thus the logarithm (to the base 10) of the reciprocal of the hydrogen ion concentration, or the logarithm of the hydrogen ion concentration with negative sign. This method has the advantage that, using the numbers between 0 and 14, it is possible to express all states of acidity and alkalinity from $1\,\text{mol}\,L^{-1}$ of hydrogen ions to $1\,\text{mol}\,L^{-1}$ of hydroxide ions. Thus a neutral solution with $[H^+] = 10^{-7}$ has a pH of 7; a solution with a hydrogen ion concentration of $1\,\text{mol}\,L^{-1}$ has a pH of 0 ($[H^+] = 10^0$); and a solution with a hydroxide ion concentration of $1\,\text{mol}\,L^{-1}$ has $[H^+] = K_w/[OH^-] = 10^{-14}/10^0 = 10^{-14}$, and possesses a pH of 14. A neutral solution is therefore one in which pH = 7, an acid solution one in which pH < 7, and an alkaline solution one in which pH > 7. An alternative definition for a neutral solution, applicable to all temperatures, is one in which the hydrogen ion and hydroxide ion concentrations are equal. In an acid solution the hydrogen ion concentration exceeds the hydroxide ion concentration, whereas in an alkaline or basic solution, the hydroxide ion concentration is greater.

Example 2.2

(i)   Find the pH of a solution in which $[H^+] = 4.0 \times 10^{-5}\,\text{mol}\,L^{-1}$.
(ii)   Find the hydrogen ion concentration corresponding to pH = 5.643.
(iii) Calculate the pH of a $0.01\,M$ solution of ethanoic acid in which the degree of dissociation is 12.5%.

(i) $pH = \log_{10} 1/[H^+] = \log 1 - \log [H^+]$

$$= \log 1 - \log 4.0 \times 10^{-5}$$

$$= 0 - (-4.398)$$

$$= 4.398$$

(ii) $pH = \log_{10} 1/[H^+] = \log 1 - \log [H^+] = 5.643$

so $\log [H^+] = -5.643$

By reference to a calculator we find $[H^+] = 2.28 \times 10^{-6}\,\mathrm{mol\,L^{-1}}$.

(iii)   The hydrogen ion concentration of the solution is

$$0.125 \times 0.01\,\mathrm{mol\,L^{-1}} = 1.25 \times 10^{-3}\,\mathrm{mol\,L^{-1}}$$
$$pH = \log_{10} 1/[H^+] = \log 1 - \log[H^+]$$
$$= 0 - (-2.903)$$
$$= 2.903$$

The hydroxide ion concentration may be expressed in a similar way:

$$pOH = -\log_{10}[OH^-] = \log_{10} 1/[OH^-] \quad \text{or} \quad [OH^-] = 10^{-pOH}$$

If we write the equation

$$[H^+][OH^-] = K_w = 10^{-14}$$

in the form

$$\log[H^+] + \log[OH^-] = \log K_w = -14$$
$$\text{then} \quad pH + pOH = pK_w = 14$$

This relationship should hold for all dilute solutions at about 25 °C. Figure 2.1 will serve as a useful mnemonic for the relation between $[H^+]$, pH, $[OH^-]$ and pOH in acid and alkaline solution.

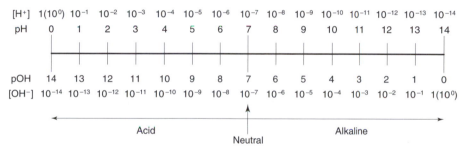

Figure 2.1   The pH scale

The logarithmic or exponential form is also useful for expressing other small quantities which arise in quantitative analysis. These include dissociation constants (Section 2.8), other ionic concentrations, and solubility products (Section 2.14):

(a)   For any acid with a dissociation constant K_a,

$$pK_a = \log(1/K_a) = -\log K_a$$

For any base with dissociation constant K_b,

$$pK_b = \log(1/K_b) = -\log K_b$$

(b)   For any ion I of concentration [I],

$$pI = \log 1/[I] = -\log[I]$$

Thus, for $[Na^+] = 8 \times 10^{-5}\,\mathrm{mol\,L^{-1}}$, pNa = 4.1.

(c)   For a salt with solubility product K_s,

$$pK_s = \log(1/K_s) = -\log K_s.$$

2.11   Buffer solutions

A solution of hydrochloric acid ($0.0001\,\mathrm{mol\,L^{-1}}$) should have a pH equal to 4, but the solution is extremely sensitive to traces of alkali from the glass of the containing vessel and to ammonia from the air. Likewise a solution of sodium hydroxide ($0.0001\,\mathrm{mol\,L^{-1}}$), which should have a pH of 10, is sensitive to traces of carbon dioxide from the atmosphere. Aqueous solutions of potassium chloride and aqueous solutions of ammonium ethanoate have a pH of about 7. Adding $1\,\mathrm{mL}$ of hydrochloric acid solution ($1\,\mathrm{mol\,L^{-1}}$) to $1\,\mathrm{L}$ of potassium chloride solution changes the pH from 7 to 3; adding $1\,\mathrm{mL}$ of hydrochloric acid solution ($1\,\mathrm{mol\,L^{-1}}$) hardly changes the pH of the ammonium ethanoate solution. The resistance of a solution to changes in hydrogen ion concentration upon the addition of small amounts of acid or alkali is called **buffer action**; a solution which exhibits buffer action is called a **buffer solution**. It is said to possess 'reserve acidity' and 'reserve alkalinity'. Buffer solutions usually consist of solutions containing a mixture of a weak acid HA and its sodium or potassium salt (A^-), or they consist of a weak base B and its salt (BH^+). A buffer, then, is usually a mixture of an acid and its conjugate base. In order to understand buffer action, consider first the equilibrium between a weak acid and its salt. The dissociation of a weak acid is given by

$$HA \rightleftharpoons H^+ + A^-$$

and its magnitude is controlled by the value of the dissociation constant K_a:

$$\frac{a_{H^+}a_{A^-}}{a_{HA}} = K_a \quad \text{or} \quad a_{H^+} = K_a\left(\frac{a_{HA}}{a_{A^-}}\right) \tag{2.11}$$

The expression may be approximated by writing concentrations for activities:

$$[H^+] = K_a\frac{[HA]}{[A^-]} \tag{2.12}$$

This equilibrium applies to a mixture of an acid HA and its salt, say MA. If the concentration of the acid is c_a and the concentration of the salt is c_s, then the concentration of the undissociated portion of the acid is $(c_a - [H^+])$. The solution is electrically neutral, hence $[A^-] = c_s + [H^+]$ (the salt is completely dissociated). Substituting these values into equation (2.12), we have

$$[H^+] = K_a\left(\frac{c_a - [H^+]}{c_s + [H^+]}\right) \tag{2.13}$$

This is a quadratic equation in $[H^+]$ and may be solved in the usual manner. It can, however, be simplified by introducing the following further approximations. In a mixture of a weak acid and its salt, the dissociation of the acid is repressed by the common ion effect, and $[H^+]$ may be taken as negligibly small by comparison with c_a and c_s. Equation (2.13) then reduces to

$$[H^+] = K_a\left(\frac{c_a}{c_s}\right) \quad \text{or} \quad [H^+] = K_a\frac{[\text{acid}]}{[\text{salt}]} \tag{2.14}$$

$$\text{or} \quad pH = pK_a + \log\frac{[\text{salt}]}{[\text{acid}]} \tag{2.15}$$

The equations can be readily expressed in a somewhat more general form when applied to a Brønsted–Lowry acid A and its conjugate base B:

$$A \rightleftharpoons H^+ + B$$

(e.g. CH_3COOH and CH_3COO^-). The expression for pH is

$$pH = pK_a + \log \frac{[B]}{[A]}$$

where $K_a = [H^+][B]/[A]$

Similarly, for a mixture of a weak base of dissociation constant K_b and its salt with a strong acid:

$$[OH^-] = K_b \frac{[base]}{[salt]} \tag{2.16}$$

$$\text{or} \quad pOH = pK_b + \log \frac{[salt]}{[base]} \tag{2.17}$$

Confining attention to the case in which the concentrations of the acid and its salt are equal, i.e. a half-neutralised acid, then $pH = pK_a$. Thus the pH of a half-neutralised solution of a weak acid is equal to the negative logarithm of the dissociation constant of the acid. For ethanoic acid, $K_a = 1.75 \times 10^{-5}\,mol\,L^{-1}$, $pK_a = 4.76$; a half-neutralised solution of say $0.1\,M$ ethanoic acid will have a pH of 4.76. If we add a small concentration of H^+ ions to it, they will combine with ethanoate ions to form undissociated ethanoic acid:

$$H^+ + CH_3COO^- \rightleftharpoons CH_3COOH$$

Similarly, if a small concentration of hydroxide ions is added, the hydroxide ions will combine with the hydrogen ions arising from the dissociation of the ethanoic acid and form water; the equilibrium will be disturbed, and more ethanoic acid will dissociate to replace the hydrogen ions removed in this way. In either case the concentration of the ethanoic acid and ethanoate ion (or salt) will not be appreciably changed. It follows from equation (2.15) that the pH of the solution will not be materially affected.

Example 2.3

Calculate the pH of the solution produced by adding $10\,mL$ of $1\,M$ hydrochloric acid to $1\,L$ of a solution which is $0.1\,M$ in ethanoic acid and $0.1\,M$ in sodium ethanoate ($K_a = 1.75 \times 10^{-5}\,mol\,L^{-1}$).

The pH of the ethanoic acid–sodium ethanoate buffer solution is given by the equation

$$pH = pK_a + \log \frac{[salt]}{[acid]} = 4.76 + 0.0 = 4.76$$

The hydrogen ions from the hydrochloric acid react with ethanoate ions, forming practically undissociated ethanoic acid; neglecting the change in volume from $1000\,mL$ to $1010\,mL$, we can say

$$[CH_3COO^-] = 0.1 - 0.001 = 0.09$$

$$[CH_3COOH] = 0.1 + 0.01 = 0.11$$

$$\text{and} \quad pH = 4.76 + \log 0.09/0.11 = 4.76 - 0.09 = 4.67$$

Thus the pH of the ethanoic acid–sodium ethanoate buffer solution is only altered by 0.09 pH unit on the addition of the hydrochloric acid. The same volume of hydrochloric acid added to 1 litre of water (pH = 7) would lead to a solution with pH = $-\log 0.01 = 2$; a change of 5 pH units. This illustrates how a buffer solution regulates pH.

A solution containing equal concentrations of acid and its salt, or a half-neutralised solution of the acid, has the maximum 'buffer capacity'. Other mixtures also possess considerable buffer capacity, but the pH will differ slightly from that of the half-neutralised acid. Thus, in a quarter-neutralised solution of acid, [acid] = 3 [salt]:

$$pH = pK_a + \log 1/3 = pK_a - 0.48$$

For a three-quarter-neutralised acid, [salt] = 3[acid]:

$$pH = pK_a + \log 3 = pK_a + 0.48$$

In general, we may say that the buffering capacity is maintained for acid : salt ratios in the range 1 : 10 to 10 : 1, and the approximate pH range of a weak acid buffer is

$$pH = pK_a \pm 1$$

The concentration of the acid is usually 0.05 to 0.2 mol L^{-1}. Similar remarks apply to weak bases. It is clear that the greater the concentrations of acid and conjugate base in a buffer solution, the greater the buffer capacity. A quantitative measure of buffer capacity is given by the number of moles of strong base required to change the pH of 1 litre of the solution by 1 pH unit.

The preparation of a buffer solution of a definite pH is a simple process once the acid (or base) of appropriate dissociation constant is found; small variations in pH are obtained by variations in the acid : salt ratio. One example is given in Table 2.2.

Before leaving the subject of buffer solutions, it is necessary to draw attention to a possible erroneous deduction from equation (2.15): the hydrogen ion concentration of a buffer solution depends only upon the ratio of the concentrations of acid and salt and upon K_a, not upon the actual concentrations; in other words, the pH of a buffer mixture should not change

Table 2.2 *pH of ethanoic acid–sodium ethanoate buffer mixtures*[a]

Ethanoic acid (x mL)	Sodium ethanoate (y mL)	pH
9.5	0.5	3.48
9.0	1.0	3.80
8.0	2.0	4.16
7.0	3.0	4.39
6.0	4.0	4.58
5.0	5.0	4.76
4.0	6.0	4.93
3.0	7.0	5.13
2.0	8.0	5.36
1.0	9.0	5.71
0.5	9.5	6.04

[a] 10 mL mixtures of x mL of 0.2 M ethanoic acid and y mL of 0.2 M sodium ethanoate

upon dilution with water. This interpretation is approximately true but not strictly true. In deducing equation (2.12), concentrations were substituted for activities, a step which is not entirely justifiable except in dilute solutions. The exact expression controlling buffer action is

$$a_{H^+} = K_a\left(\frac{a_{HA}}{a_{A^-}}\right) = K_a\left(\frac{c_a y_a}{c_s y_{A^-}}\right) \tag{2.18}$$

The activity coefficient y_a of the undissociated acid is approximately unity in dilute aqueous solution. Equation (2.18) thus becomes

$$a_{H^+} = K_a\frac{[\text{acid}]}{[\text{salt}]y_a} \tag{2.19}$$

$$\text{or} \quad \text{pH} = pK_a + \log\frac{[\text{salt}]}{[\text{acid}]} + \log y \tag{2.20}$$

This is known as the Henderson–Hasselbalch equation. If a buffer solution is diluted, the ionic concentrations are decreased, so as in Section 2.2, the ionic activity coefficients are increased. It follows from equation (2.20) that the pH is increased.

Buffer mixtures are not confined to mixtures of monoprotic acids or monoacid bases and their salts. We may employ a mixture of salts of a polyprotic acid, e.g. NaH_2PO_4 and Na_2HPO_4. The salt NaH_2PO_4 is completely dissociated:

$$NaH_2PO_4 \rightleftharpoons Na^+ + H_2PO_4^-$$

The ion $H_2PO_4^-$ acts as a monoprotic acid:

$$H_2PO_4^- \rightleftharpoons H^+ + HPO_4^{2-}$$

for which K ($\equiv K_2$ for phosphoric acid) is $6.2 \times 10^{-8}\,\text{mol L}^{-1}$. Adding the salt Na_2HPO_4 is analogous to adding ethanoate ions to a solution of ethanoic acid, since the tertiary ionisation of phosphoric acid ($HPO_4^{2-} = H^+ + PO_4^{3-}$) is small ($K_3 = 5 \times 10^{-13}\,\text{mol L}^{-1}$). The mixture of NaH_2PO_4 and Na_2HPO_4 is therefore an effective buffer over the range pH 7.2 ± 1.0 ($= pK \pm 1$). Note how this is a mixture of a Brønsted–Lowry acid and its conjugate base.

Buffer solutions find many applications in quantitative analysis, e.g. many precipitations are quantitative only under carefully controlled conditions of pH, as are many compleximetric titrations; numerous examples of their use will be found throughout the book.

2.12 The hydrolysis of salts

Salts may be divided into four main classes:

1. Those derived from strong acids and strong bases, e.g. potassium chloride.
2. Those derived from weak acids and strong bases, e.g. sodium ethanoate.
3. Those derived from strong acids and weak bases, e.g. ammonium chloride.
4. Those derived from weak acids and weak bases, e.g. ammonium methanoate or aluminium ethanoate.

When any salt from classes 2 to 4 is dissolved in water, the solution is not always neutral in reaction. Interaction may occur with the ions of water, and the resulting solution will be neutral, acid or alkaline according to the nature of the salt.

With an aqueous solution of a salt from class 1, the anions have no tendency to combine with the hydrogen ions of water, and the cations have no tendency to combine with the hydroxide ions of water; this is because the related acids and bases are strong electrolytes. The equilibrium between the hydrogen ions and the hydroxide ions in water

$$H_2O \rightleftharpoons H^+ + OH^- \qquad [2.6]$$

is therefore not disturbed and the solution remains neutral.

Consider, however, a salt MA derived from a weak acid HA and a strong base MOH (class 2). The salt is completely dissociated in aqueous solution:

$$MA \rightarrow M^+ + A^-$$

Very small concentrations of hydrogen ions and hydroxide ions, originating from the small but finite ionisation of water, will initially be present. HA is a weak acid, i.e. it is dissociated only to a small degree; the concentration of A^- ions which can exist in equilibrium with H^+ ions is therefore small. In order to maintain the equilibrium, the large initial concentration of A^- ions must be reduced by combination with H^+ ions to form undissociated HA:

$$H^+ + A^- \rightleftharpoons HA \qquad [2.7]$$

The hydrogen ions required for this reaction can be obtained only from the further dissociation of the water; this dissociation simultaneously produces an equivalent quantity of hydroxyl ions. The hydrogen ions are used in the formation of HA; consequently, the hydroxide ion concentration of the solution will increase and the solution will react alkaline.

It is usual in writing equations involving equilibria between completely dissociated and slightly dissociated or sparingly soluble substances to employ the ions of the completely dissociated substance and the molecules of the slightly dissociated substance. The reaction is therefore written

$$A^- + H_2O \rightleftharpoons OH^- + HA \qquad [2.8]$$

This equation can also be obtained by combining reactions [2.6] and [2.7] since both equilibria must coexist. This interaction between the ion (or ions) of a salt and water is called **hydrolysis**.

Consider now the salt of a strong acid and a weak base (class 3). Here the initial high concentration of cations M^+ will be reduced by combination with the hydroxide ions of water to form the slightly dissociated base MOH until the equilibrium

$$M^+ + OH^- \rightleftharpoons MOH$$

is attained. The hydrogen ion concentration of the solution will thus be increased, and the solution will react acid. The hydrolysis is here represented by

$$M^+ + H_2O \rightleftharpoons MOH + H^+$$

For salts of class 4, in which both the acid and the base are weak, two reactions will occur simultaneously

$$M^+ + H_2O \rightleftharpoons MOH + H^+ \qquad A^- + H_2O \rightleftharpoons HA + OH^-$$

The reaction of the solution will clearly depend upon the relative dissociation constants of the acid and the base. If they are equal in strength the solution will be neutral; if $K_a > K_b$ it will be acid; and if $K_b > K_a$ it will be alkaline.

Having considered all the possible cases, we are now in a position to give a more general definition of hydrolysis. Hydrolysis is the interaction between an ion (or ions) of a salt and water with the production of (a) a weak acid **or** a weak base, or (b) a weak acid **and** a weak base.

The phenomenon of salt hydrolysis may be regarded as a simple application of the general Brønsted–Lowry equation

$$A_1 + B_2 \rightleftharpoons A_2 + B_1$$

Thus the equation for the hydrolysis of ammonium salts

$$NH_4^+ + H_2O \rightleftharpoons NH_3 + H_3O^+$$

is really identical with the expression used to define the strength of the ammonium ion as a Brønsted–Lowry acid (Section 2.6), and the constant K_a for NH_4^+ is in fact what is usually termed the hydrolysis constant of an ammonium salt.

Hydrolysis of the sodium salt of a weak acid can be treated similarly. For a solution of sodium ethanoate

$$CH_3COO^- + H_2O \rightleftharpoons CH_3COOH + OH^-$$

the hydrolysis constant is

$$[CH_3COOH][OH^-]/[CH_3COO^-] = K_h = K_w/K_a$$

where K_a is the dissociation constant of ethanoic acid.

2.13   Degree of hydrolysis

Case 1: Salt of a weak acid and a strong base

The equilibrium in a solution of salt MA may be represented by

$$A^- + H_2O \rightleftharpoons OH^- + HA$$

Applying the law of mass action, we obtain

$$\frac{a_{OH^-} a_{HA}}{a_{A^-}} = \frac{[OH^-][HA]}{[A^-]} \times \frac{y_{OH^-} y_{HA}}{y_{A^-}} = K_h \tag{2.21}$$

where K_h is the hydrolysis constant. The solution is assumed to be dilute, so the activity of the unionised water may be taken as constant, and the approximation that the activity coefficient of the unionised acid is unity and that both ions have the same activity coefficient may be introduced. Equation (2.21) then reduces to

$$K_h = \frac{[OH^-][HA]}{[A^-]} \tag{2.22}$$

This is often written in the form

$$K_h = \frac{[base][acid]}{[unhydrolysed\ salt]}$$

The free strong base and the unhydrolysed salt are completely dissociated, and the acid is very slightly dissociated.

The degree of hydrolysis is the fraction of each mole of anion A^- hydrolysed at equilibrium. Let 1 mol of salt be dissolved in V L of solution, and let x be the degree of hydrolysis. The concentrations in $mol\,L^{-1}$ are

$$[HA] = [OH^-] = x/V \qquad [A^-] = (1 - x)/V$$

Substituting these values into equation (2.22):

$$K_h = \frac{[OH^-][HA]}{[A^-]} = \frac{(x/V)(x/V)}{(1-x)/V} = \frac{x^2}{(1-x)V}$$

This expression enables us to calculate the degree of hydrolysis at the dilution V; as V increases, the degree of hydrolysis x must increase. The two equilibria

$$H_2O \rightleftharpoons H^+ + OH^- \quad \text{and} \quad HA = H^+ + A^-$$

must coexist with the hydrolytic equilibrium:

$$A^- + H_2O \rightleftharpoons HA + OH^-$$

Hence the two relationships

$$[H^+][OH^-] = K_w \quad \text{and} \quad [H^+][A^-]/[HA] = K_a$$

must hold in the same solution as

$$[OH^-][HA]/[A^-] = K_h$$

But $\quad \dfrac{K_w}{K_a} = \dfrac{[H^+][OH^-][HA]}{[H^+][A^-]} = \dfrac{[OH^-][HA]}{[A^-]} = K_h$

therefore $\quad K_w/K_a = K_h$

or $\quad pK_h = pK_w - pK_a$

The hydrolysis constant is thus related to the ionic product of water and the ionisation constant of the acid. Since K_a varies slightly with temperature and K_w varies considerably with temperature, so K_h and consequently the degree of hydrolysis will be largely influenced by changes of temperature.

The hydrogen ion concentration of a solution of a hydrolysed salt can be readily calculated. The amounts of HA and OH^- ions formed as a result of hydrolysis are equal; so $[HA] = [OH^-]$ in a solution of the pure salt in water. If the concentration of the salt is $c \, \text{mol L}^{-1}$, then

$$\frac{[HA][OH^-]}{[A^-]} = \frac{[OH^-]^2}{c} = K_b = \frac{K_w}{K_a}$$

and $\quad [OH^-] = \sqrt{cK_w/K_a}$

or $\quad [H^+] = \sqrt{K_wK_a/c} \quad$ since $[H^+] = K_w/[OH^-]$

and $\quad pH = \frac{1}{2}pK_w + \frac{1}{2}pK_a + \frac{1}{2}\log c$

To be consistent we should use $pc = -\log c$, so the equation becomes

$$pH = \tfrac{1}{2}pK_w + \tfrac{1}{2}pK_a + \tfrac{1}{2}pc \tag{2.23}$$

Equation (2.23) can be employed for the calculation of the pH of a solution of a salt of a weak acid and a strong base. Thus the pH of a solution of sodium benzoate ($0.05 \, \text{mol L}^{-1}$) is given by

$$pH = 7.0 + 2.10 - \tfrac{1}{2}(1.30) = 8.45$$

since $K_a = 6.37 \times 10^{-5} \, \text{mol L}^{-1}$ ($pK_a = 4.20$) for benzoic acid. This type of calculation will provide useful information about the indicator to choose for the titration of a weak acid and a strong base (Section 10.34).

Example 2.4

Calculate (i) the hydrolysis constant, (ii) the degree of hydrolysis and (iii) the hydrogen ion concentration of a solution of sodium ethanoate (0.01 mol L^{-1}) at the laboratory temperature.

$$K_h = \frac{K_w}{K_a} = \frac{1.0 \times 10^{-14}}{1.75 \times 10^{-5}} = 5.7 \times 10^{-10}$$

The degree of hydrolysis x is given by

$$K_h = \frac{x^2}{(1 - x)V}$$

Substituting for K_h and V (= $1/c$), we obtain

$$5.7 \times 10^{-10} = \frac{0.01x^2}{(1 - x)}$$

Solving this quadratic equation, $x = 0.000\,238$ or 0.0238%.

If the solution were completely hydrolysed, the concentration of ethanoic acid produced would be 0.01 mol L^{-1}. But the degree of hydrolysis is 0.0238%, so the concentration of ethanoic acid is 2.38×10^{-6} mol L^{-1}. This is also equal to the hydroxide ion concentration produced, i.e. pOH = 5.62. Therefore

pH = 14.0 − 5.62 = 8.38

The pH may also be calculated from equation (2.23):

pH = $\frac{1}{2}$pK$_w$ + $\frac{1}{2}$pK$_a$ − $\frac{1}{2}$pc = 7.0 + 2.38 − $\frac{1}{2}$(2) = 8.38

Case 2: Salt of a strong acid and a weak base

The hydrolytic equilibrium is represented by

$$M^+ + H_2O \rightleftharpoons MOH + H^+$$

By applying the law of mass action along the lines of case 1, the following equations are obtained:

$$K_h = \frac{[H^+][MOH]}{[M^+]} = \frac{[acid][base]}{[unhydrolysed\ salt]} = \frac{K_w}{K_b} = \frac{x^2}{(1 - x)V}$$

K_b is the dissociation constant of the base. Furthermore, since [MOH] and [H$^+$] are equal, we have

$$K_h = \frac{[H^+][MOH]}{[M^+]} = \frac{[H^+]^2}{c} = \frac{K_w}{K_b}$$

$$[H^+] = \sqrt{cK_wK_b}$$

or   pH = $\frac{1}{2}$pK$_w$ − $\frac{1}{2}$pK$_b$ + $\frac{1}{2}$pc   (2.24)

Equation (2.24) may be used to calculate the pH for solutions of salts of strong acids and weak bases. Thus, the pH of a solution of ammonium chloride (0.2 mol L^{-1}) is

pH = 7.0 − 2.37 + $\frac{1}{2}$(0.70) = 4.98

since $K_b = 1.8 \times 10^{-5}$ mol L^{-1} (pK$_b$ = 4.74) for ammonia in water.

Case 3: Salt of a weak acid and a weak base

The hydrolytic equilibrium is expressed by the equation

$$M^+ + A^- + H_2O \rightleftharpoons MOH + HA$$

Applying the law of mass action and taking the activity of unionised water as unity, we have

$$K_h = \frac{a_{MOH}a_{HA}}{a_{M^+}a_{A^-}} = \frac{[MOH][HA]}{[M^+][A^-]} \times \frac{y_{MOH}y_{HA}}{y_{M^+}y_{A^-}}$$

Make the usual approximations; assume unity for the activity coefficients of the unionised molecules and, less justifiably, assume unity for the activity coefficients of the ions. Then the following approximate equation is obtained:

$$K_h = \frac{[MOH][HA]}{[M^+][A^-]} = \frac{[base][acid]}{[unhydrolysed\ salt]^2}$$

If x is the degree of hydrolysis of 1 mol of the salt dissolved in V L of solution, then the individual concentrations are

$$[MOH] = [HA] = x/V \qquad [M^+] = [A^-] = (1-x)/V$$

leading to the result

$$K_h = \frac{(x/V)(x/V)}{\{(1-x)/V\}\{(1-x)/V\}} = \frac{x^2}{(1-x)^2}$$

The degree of hydrolysis is independent of the solution concentration, hence the pH is also independent of the solution concentration.*

It may be readily shown that

$$K_h = K_b(K_w/K_a)$$

$$\text{or} \quad pK_h = pK_w - pK_a - K_b$$

This expression enables us to calculate the value of the degree of hydrolysis from the dissociation constants of the acid and the base.

The hydrogen ion concentration of the hydrolysed solution is calculated in the following manner:

$$[H^+] = K_a\frac{[HA]}{[A^-]} = K_a\left(\frac{x/V}{(1-x)V}\right) = K_a\left(\frac{x}{1-x}\right)$$

$$\text{but} \quad x/(1-x) = \sqrt{K_h}$$

$$\text{Hence} \quad [H^+] = K_a\sqrt{K_h} = \sqrt{K_wK_a/K_b}$$

$$\text{or} \quad pH = \tfrac{1}{2}pK_w + \tfrac{1}{2}pK_a - \tfrac{1}{2}pK_b \tag{2.25}$$

If the ionisation constants of the acid and the base are equal, i.e. $K_a = K_b$, pH $= \tfrac{1}{2}pK_w = 7.0$ and the solution is neutral, although hydrolysis may be considerable. If $K_a > K_b$ then pH < 7 and the solution is acid; but if $K_b > K_a$ then pH > 7 and the solution is alkaline.

---

* This applies only if the original assumptions about the activity coefficients are justified. In solutions of appreciable ionic strength, the activity coefficients of the ions will vary with the total ionic strength.

The pH of a solution of ammonium ethanoate is given by

$$pH = 7.0 + 2.38 - 2.37 = 7.1$$

i.e. the solution is approximately neutral. On the other hand, for a dilute solution of ammonium methanoate

$$pH = 7.0 + 1.88 - 2.37 = 6.51$$

since $K_a = 1.77 \times 10^{-4} \, mol \, L^{-1}$ ($pK_a = 3.75$) for methanoic acid, i.e. the solution has a slightly acid reaction.

2.14 Solubility product

For sparingly soluble salts (i.e. those for which the solubility is less than $0.01 \, mol \, L^{-1}$) it is an experimental fact that the mass action product of the concentrations of the ions is a constant at constant temperature. This product K_s is called the solubility product. For a binary electrolyte

$$AB \rightleftharpoons A^+ + B^-$$

$$K_{s(AB)} = [A^+][B^-]$$

And for an electrolyte A_pB_q, which ionises into pA^{q+} and qB^{p-} ions, then

$$A_pB_q = pA^{q+} + qB^{p-}$$

$$K_{s(A_pB_q)} = [A^{q+}]^p[B^{p-}]^q$$

Here is a plausible deduction of the solubility product relation. When excess of a sparingly soluble electrolyte, say silver chloride, is shaken up with water some of it passes into solution to form a saturated solution of the salt and the process appears to cease. The following equilibrium is actually present (the silver chloride is completely ionised in solution):

$$AgCl \, (s) \rightleftharpoons Ag^+ + Cl^-$$

The rate of the forward reaction depends only upon the temperature, and at any given temperature

$$r_1 = k_1$$

where k_1 is a constant. The rate of the reverse reaction is proportional to the activity of each of the reactants; hence at any given temperature

$$r_2 = k_2 a_{Ag^+} a_{Cl^-}$$

where k_2 is another constant. At equilibrium the two rates are equal, i.e.

$$k_1 = k_2 a_{Ag^+} a_{Cl^-}$$

or $a_{Ag^+} a_{Cl^-} = k_1/k_2 = K_{s(AgCl)}$

In the very dilute solutions with which we are concerned, the activities may be taken as practically equal to the concentrations, so $[Ag^+][Cl^-]$ = a constant.

Note that the solubility product relation applies with sufficient accuracy for purposes of quantitative analysis only to saturated solutions of slightly soluble electrolytes and with small additions of other salts. In the presence of moderate concentrations of salts, the ionic concentration, and therefore the ionic strength of the solution, will increase. In general, this will lower the activity coefficients of both ions, hence the ionic concentrations (and there-fore the solubility) must increase in order to maintain the solubility product constant. This effect is called the **salt effect**; it is most marked when the added electrolyte does not possess an ion in common with the sparingly soluble salt.

Two factors may come into play when a solution of a salt containing a common ion is added to a saturated solution of a slightly soluble salt. At moderate concentrations of the added salt, the solubility will generally decrease; at higher concentrations of the soluble salt, when the ionic strength of the solution increases considerably and the activity co-efficients of the ions decrease, the solubility may actually increase. This is one of the reasons why a very large excess of the precipitating agent is avoided in quantitative analysis.

The following examples illustrate the method of calculating solubility products from solubility data, and how to calculate solubilities from solubility products.

Example 2.5

The solubility of silver chloride is $0.0015 \, g \, L^{-1}$. Calculate the solubility product.

The relative molecular mass of silver chloride is 143.3. The solubility is therefore $0.0015/143.3 = 1.05 \times 10^{-5} \, mol \, L^{-1}$. In a saturated solution 1 mol of AgCl will give 1 mol each of Ag^+ and Cl^-. Hence $[Ag^+] = 1.05 \times 10^{-5} \, mol \, L^{-1}$ and $[Cl^-] = 1.05 \times 10^{-5} \, mol \, L^{-1}$.

$$K_{s(AgCl)} = [Ag^+][Cl^-] = (1.05 \times 10^{-5}) \times (1.05 \times 10^{-5})$$
$$= 1.11 \times 10^{-10} \, mol^2 \, L^{-2}$$

Example 2.6

Calculate the solubility product of silver chromate, given that its solubility is $2.5 \times 10^{-2} \, g \, L^{-1}$ and

$$Ag_2CrO_4 \rightleftharpoons 2Ag^+ + CrO_4^{2-}$$

The relative molecular mass of Ag_2CrO_4 is 331.7, hence

$$solubility = 2.5 \times 10^{-2}/331.7$$
$$= 7.5 \times 10^{-5} \, mol \, L^{-1}$$

Now 1 mol of Ag_2CrO_4 gives 2 mol of Ag^+ and 1 mol of CrO_4^{2-}, so

$$K_{s(Ag_2CrO_4)} = [Ag^+]^2[CrO_4^{2-}] = (2 \times 7.5 \times 10^{-5})^2 \times (7.5 \times 10^{-5})$$
$$= 1.7 \times 10^{-12} \, mol^3 \, L^{-3}$$

Example 2.7

The solubility product of magnesium hydroxide is $3.4 \times 10^{-11} \, mol^3 \, L^{-3}$. Calculate its solu-bility in grams per litre.

We have

$$Mg(OH)_2 \rightleftharpoons Mg^{2+} + 2OH^-$$

$$[Mg^{2+}][OH^-]^2 = 3.4 \times 10^{-11}$$

The relative molecular mass of magnesium hydroxide is 58.3. Each mole of magnesium hydroxide, when dissolved, yields 1 mol of magnesium ions and 2 mol of hydroxyl ions. If the solubility is s mol L^{-1}, $[Mg^{2+}] = s$ and $[OH^-] = 2s$. Substituting these values in the solubility product expression:

$$s \times (2s)^2 = 3.4 \times 10^{-11}$$

$$s = 2.0 \times 10^{-4} \, \text{mol} \, L^{-1}$$

$$= 2.0 \times 10^{-4} \times 58.3$$

$$= 1.2 \times 10^{-2} \, \text{g} \, L^{-1}$$

The great importance of the solubility product concept lies in its bearing upon precipitation from solution, which is one of the important operations of quantitative analysis. The solubility product is the ultimate value attained by the ionic concentration product when equilibrium has been established between the solid phase of a difficultly soluble salt and the solution. If the experimental conditions are such that the ionic concentration product is different from the solubility product, the system will attempt to adjust itself in such a manner that the ionic and solubility products are equal in value. Thus, for a given electrolyte, if the product of the concentrations of the ions in solution is arbitrarily made to exceed the solubility product, perhaps by the addition of a salt with a common ion, the adjustment of the system to equilibrium results in precipitation of the solid salt, provided supersaturation conditions are excluded. If the ionic concentration product is less than the solubility product or can arbitrarily be made so, perhaps by complex salt formation or by the formation of weak electrolytes, then a further quantity of solute can pass into solution until the solubility product is attained, or if this is not possible, until all the solute has dissolved.

2.15 Common ion effect

If a strong electrolyte such as a salt is added to a solution of a weak electrolyte so chosen because one of the ions into which it dissociates in solution is the same as one of the ions from the salt, the degree of dissociation of the weak electrolyte is decreased as a result of the common ion effect.

Example calculations are given below. In general, the effect is small; if the total concentration of the common ion is only slightly greater than the concentration which the original compound alone would furnish; but the effect is very great if the concentration of the common ion is very much increased, perhaps by the addition of a completely dissociated salt. A large common ion effect may be of considerable practical importance. Indeed, it provides a valuable method for controlling the concentration of the ions furnished by a weak electrolyte.

Example 2.8

Calculate the sulphide ion concentration in a $0.25\,M$ hydrochloric acid solution saturated with hydrogen sulphide.

The value $0.25\,M$ has been chosen since it is the concentration at which the sulphides of certain heavy metals are precipitated. The total concentration of hydrogen sulphide may be assumed to be approximately the same as in aqueous solution, i.e. $0.1\,M$; $[H^+]$ will equal the concentration of the completely dissociated HCl, i.e. $0.25\,M$, but $[S^{2-}]$ will be reduced below 1×10^{-14} (see Example 2.1). Using the equilibria [2.4] and [2.5], we find

$$[HS^-] = \frac{K_1[H_2S]}{[H^+]} = \frac{1.0 \times 10^{-7} \times 0.1}{0.25} = 4.0 \times 10^{-8}\,\text{mol}\,L^{-1}$$

$$[S^{2-}] = \frac{K_2[HS^-]}{[H^+]} = \frac{(1 \times 10^{-14})(4 \times 10^{-8})}{0.25} = 1.6 \times 10^{-21}\,\text{mol}\,L^{-1}$$

Thus, by changing the acidity from $1.0 \times 10^{-4}\,M$ (present in saturated H_2S water) to $0.25\,M$, the sulphide ion concentration is reduced from 1×10^{-14} to $1.6 \times 10^{-21}\,\text{mol}\,L^{-1}$.

Example 2.9

What effect has the addition of $0.1\,\text{mol}$ of anhydrous sodium ethanoate to $1\,L$ of $0.1\,M$ ethanoic acid upon the degree of dissociation of the acid?

The dissociation constant of ethanoic acid at $25\,°C$ is $1.75 \times 10^{-5}\,\text{mol}\,L^{-1}$ and the degree of ionisation α in $0.1\,M$ solution may be computed by solving the quadratic equation

$$\frac{[H^+][CH_3COO^-]}{[CH_3COOH]} = \frac{\alpha^2 c}{(1-\alpha)} = 1.75 \times 10^{-5}$$

For our purpose it is sufficiently accurate to neglect α in $(1-\alpha)$ since α is small, so

$$\alpha = \sqrt{K/c} = \sqrt{1.75 \times 10^{-4}} = 0.0132$$

Hence in $0.1\,M$ ethanoic acid, $[H^+] = 0.001\,32$, $[CH_3COO^-] = 0.001\,32$, and

$$[CH_3COOH] = 0.0987\,\text{mol}\,L^{-1}$$

The concentrations of sodium and ethanoate ions produced by the addition of the completely dissociated sodium ethanoate are

$$[Na^+] = 0.1 \qquad [CH_3COO^-] = 0.1\,\text{mol}\,L^{-1}$$

The ethanoate ions from the salt will tend to decrease the ionisation of the ethanoic acid, and consequently the ethanoate ion concentration derived from it. Hence we may write $[CH_3COO^-] = 0.1$ for the solution; and if α' is the new degree of ionisation then $[H^+] = \alpha'c = 0.1\alpha'$ and $[CH_3COOH] = (1 - \alpha')c = 0.1$, since α' is negligibly small.
    Substituting in the mass action equation:

$$\frac{[H^+][CH_3COO^-]}{[CH_3COOH]} = \frac{0.1\alpha' \times 0.1}{0.1} = 1.75 \times 10^{-5}$$

or $\quad \alpha' = 1.75 \times 10^{-4}$

$$[H^+] = \alpha'c = 1.75 \times 10^{-5}\,\text{mol}\,L^{-1}$$

The addition of a 0.1 mol of sodium ethanoate to a 0.1 M solution of ethanoic acid has decreased the degree of ionisation from 1.32% to 0.018%, and the hydrogen ion concentration from 0.001 32 to 0.000 018 mol L^{-1}.

Example 2.10

What effect has the addition of 0.5 mol of ammonium chloride to 1 L of 0.1 M aqueous ammonia solution upon the degree of dissociation of the base?
   (Dissociation constant of NH$_3$ in water = 1.8×10^{-5} mol L^{-1})

In 0.1 M ammonia solution $\alpha = \sqrt{1.8 \times 10^{-5}/0.1} = 0.0135$. Hence [OH$^-$] = 0.001 35, [NH$_4^+$] = 0.001 35, and [NH$_3$] = 0.0986 mol L^{-1}. Let α' be the degree of ionisation in the presence of the added ammonium chloride. Then [OH$^-$] = $\alpha'c = 0.1\alpha'$ and [NH$_3$] $(1 - \alpha')c = 0.1$, since α' may be taken as negligibly small. The addition of the completely ionised ammonium chloride will necessarily decrease [NH$_4^+$] derived from the base and increase [NH$_3$], and as a first approximation [NH$_4^+$] = 0.5.
   Substituting in the equation:

$$\frac{[NH_4^+][OH^-]}{[NH_3]} = \frac{0.5 \times 0.1\alpha'}{0.1} = 1.8 \times 10^{-5}$$

The addition of 0.5 mol of ammonium chloride to 1 L of a 0.1 M solution of aqueous ammonia has decreased the degree of ionisation from 1.35% to 0.0036%, and the hydroxide ion concentration from 0.001 35 to 0.000 0036 mol L^{-1}.

2.16 Common ion: quantitative effects

An important application of the solubility product principle is to the calculation of the solubility of sparingly soluble salts in solutions of salts with a common ion. Thus the solubility of a salt MA in the presence of a relatively large amount of the common M$^+$ ions,* supplied by a second salt MB, follows from the definition of solubility products:

$$[M^+][A^-] = K_{s(MA)}$$

$$[A^-] = K_{s(MA)}/[M^+]$$

The solubility of the salt is represented by the [A$^-$] which it furnishes in solution. It is clear that the addition of a common ion will **decrease** the solubility of the salt.

Example 2.11

Calculate the solubility of silver chloride in (a) 0.001 M and (b) 0.01 M sodium chloride solutions respectively ($K_{s(AgCl)} = 1.1 \times 10^{-10}$ mol^2 L^{-2}).

In a saturated solution of silver chloride [Cl$^-$] = $\sqrt{1.1 \times 10^{-10}} = 1.05 \times 10^{-5}$ mol L^{-1}; this may be neglected in comparison with the excess of Cl$^-$ ions added.

---

* This enables us to neglect the concentration of M$^+$ ions supplied by the sparingly soluble salt itself, and thus to simplify the calculation.

For (a) $[Cl^-] = 1 \times 10^{-3}$ $[Ag^+] = 1.1 \times 10^{-10}/1 \times 10^{-3}$

$$= 1.1 \times 10^{-7} \, mol \, L^{-1}$$

For (b) $[Cl^-] = 1 \times 10^{-2}$ $[Ag^+] = 1.1 \times 10^{-10}/1 \times 10^{-2}$

$$= 1.1 \times 10^{-8} \, mol \, L^{-1}$$

Thus the solubility is decreased 100 times in 0.001 M sodium chloride and 1000 times in 0.01 M sodium chloride. Similar results are obtained for 0.001 M and 0.01 M silver nitrate solutions.

Example 2.12

Calculate the solubilities of silver chromate in 0.001 M and 0.01 M silver nitrate solutions and 0.001 M and 0.01 M potassium chromate solutions (Ag_2CrO_4 has $K_s = 1.7 \times 10^{-12} \, mol^3 \, L^{-3}$ and its solubility in water $= 5.5 \times 10^{-5} \, mol \, L^{-1}$).

$$[Ag^+]^2[CrO_4^{2-}] = 1.7 \times 10^{-12}$$

$$[CrO_4^{2-}] = 1.7 \times 10^{-12}/[Ag^+]^2$$

For 0.001 M silver nitrate solution $[Ag^+] = 1 \times 10^{-3}$, so

$$[CrO_4^{2-}] = 1.7 \times 10^{-12}/1 \times 10^{-6} = 1.7 \times 10^{-6} \, mol \, L^{-1}$$

For 0.01 M silver nitrate solution $[Ag^+] = 1 \times 10^{-2}$, so

$$[CrO_4^{2-}] = 1.7 \times 10^{-12}/1 \times 10^{-4} = 1.7 \times 10^{-8} \, mol \, L^{-1}$$

The solubility product equation gives

$$[Ag^+] = \sqrt{1.7 \times 10^{-12}/[CrO_4^{2-}]}$$

For $[CrO_4^{2-}] = 0.001$ $[Ag^+] = \sqrt{1.7 \times 10^{-12}/1 \times 10^{-3}}$

$$= 4.1 \times 10^{-5} \, mol \, L^{-1}$$

For $[CrO_4^{2-}] = 0.01$ $[Ag^+] = \sqrt{1.7 \times 10^{-12}/1 \times 10^{-2}}$

$$= 1.3 \times 10^{-5} \, mol \, L^{-1}$$

This decrease in solubility by the common ion effect is of fundamental importance in gravimetric analysis. By the addition of a suitable excess of a precipitating agent, the solubility of a precipitate is usually decreased to so small a value that the loss from solubility influences is negligible. Consider the determination of silver as silver chloride. Here the chloride solution is added to the solution of the silver salt. If an exactly equivalent amount is added, the resultant saturated solution of silver chloride will contain $0.0015 \, g \, L^{-1}$ (Example 2.1). If 0.2 g of silver chloride is produced and the volume of the solution and washings is 500 mL, the loss owing to solubility will be 0.000 75 g or 0.38% of the weight of the salt; the analysis would then be 0.38% too low. By using an excess of the precipitant, perhaps to a concentration of 0.01 M, the solubility of the silver chloride is reduced to $1.5 \times 10^{-5} \, g \, L^{-1}$ (Example 2.4), and the loss will be $1.5 \times 10^{-5} \times 0.5 \times 100/0.2 = 0.0038\%$. Silver chloride is therefore very suitable for the quantitative determination of silver with high accuracy.

But note that as the concentration of the excess precipitant increases, so too does the ionic strength of the solution. This leads to a decrease in activity coefficient values, so to maintain the value of K_s **more** of the precipitate will dissolve. In other words, there is a limit to the amount of precipitant which can be safely added in excess. Also, addition of excess precipitant may sometimes result in the formation of soluble complexes, causing some precipitate to dissolve.

2.17   Fractional precipitation

The previous section used the solubility product principle in connection with the precipitation of one sparingly soluble salt. It is now necessary to examine the case where two slightly soluble salts may be formed. For simplicity, consider the situation which arises when a precipitating agent is added to a solution containing two anions, both of which form slightly soluble salts with the same cation, e.g. when silver nitrate solution is added to a solution containing both chloride and iodide ions. The questions which arise are: Which salt will be precipitated first, and how completely will the first salt be precipitated before the second ion begins to react with the reagent?

The solubility products of silver chloride and silver iodide are respectively $1.2 \times 10^{-10}\,mol^2\,L^{-2}$ and $1.7 \times 10^{-16}\,mol^2\,L^{-2}$, i.e.

$$[Ag^+][Cl^-] = 1.2 \times 10^{-10} \tag{2.26}$$

$$[Ag^+][I^-] = 1.7 \times 10^{-16} \tag{2.27}$$

Silver iodide is less soluble, its solubility product will be exceeded first, so it will be precipitated first. Silver chloride will be precipitated when the Ag^+ ion concentration is greater than

$$\frac{K_{s(AgCl)}}{[Cl^-]} = \frac{1.2 \times 10^{-10}}{[Cl^-]}$$

and then both salts will be precipitated simultaneously. When silver chloride begins to precipitate, silver ions will be in equilibrium with both salts, and equations (2.26) and (2.27) will be simultaneously satisfied, or

$$[Ag^+] = \frac{K_{s(AgI)}}{[I^-]} = \frac{K_{s(AgCl)}}{[Cl^-]} \tag{2.28}$$

and $\quad \dfrac{[I^-]}{[Cl^-]} = \dfrac{K_{s(AgI)}}{K_{s(AgCl)}} = \dfrac{1.7 \times 10^{-16}}{1.2 \times 10^{-10}} = 1.4 \times 10^{-6} \tag{2.29}$

Hence when the concentration of the iodide ion is about one-millionth part of the chloride ion concentration, silver chloride will be precipitated. If the initial concentration of both chloride and iodide ions is $0.1\,M$, then silver chloride will be precipitated when

$$[I^-] = 0.1 \times 1.4 \times 10^{-6} = 1.4 \times 10^{-7}\,M = 1.8 \times 10^{-5}\,g\,L^{-1}$$

Thus an almost complete separation is theoretically possible. The separation is feasible in practice if the point at which the iodide precipitation is complete can be detected. This may be done (a) by the use of an adsorption indicator (Section 10.93), or (b) by a potentiometric method with a silver electrode (Section 10.103).

For a mixture of bromide and iodide

$$\frac{[I^-]}{[Br^-]} = \frac{K_{s(AgI)}}{K_{s(AgBr)}} = \frac{1.7 \times 10^{-16}}{3.5 \times 10^{-13}} = \frac{1}{2.0 \times 10^3}$$

Precipitation of silver bromide will occur when the concentration of the bromide ion in the solution is 2.0×10^3 times the iodide concentration. The separation is therefore not so complete as in the case of chloride and iodide, but can nevertheless be effected with fair accuracy with the aid of adsorption indicators (Section 10.93).

2.18 Precipitate solubility: effect of acids

For sparingly soluble salts of a strong acid, the effect of the addition of an acid will be similar to that of any other indifferent electrolyte; but if the sparingly soluble salt MA is the salt of a weak acid HA, then acids will generally have a solvent effect upon it. If hydrochloric acid is added to an aqueous suspension of such a salt, the following equilibrium will be established:

$$M^+ + A^- + H^+ \rightleftharpoons HA + M^+$$

If the dissociation constant of the acid HA is very small, the anion A^- will be removed from the solution to form the undissociated acid HA. Consequently, more of the salt will pass into solution to replace the anions removed in this way, and this process will continue until equilibrium is established (i.e. until $[M^+][A^-]$ has become equal to the solubility product of MA), or if sufficient hydrochloric acid is present, until the sparingly soluble salt has dissolved completely. Similar reasoning may be applied to salts of acids, such as phosphoric(V) acid ($K_1 = 7.5 \times 10^{-3} \, mol \, L^{-1}$, $K_2 = 6.2 \times 10^{-8} \, mol \, L^{-1}$, $K_3 = 5 \times 10^{-13} \, mol \, L^{-1}$), oxalic acid ($K_1 = 5.9 \times 10^{-2} \, mol \, L^{-1}$; $K_2 = 6.4 \times 10^{-5} \, mol \, L^{-1}$) and arsenic(V) acid. Thus the solubility of, say, silver phosphate(V) in dilute nitric acid is due to the removal of the PO_4^{3-} ion as HPO_4^{2-} and/or $H_2PO_4^-$:

$$PO_4^{3-} + H^+ \rightleftharpoons HPO_4^{2-}; \quad HPO_4^{2-} + H^+ \rightleftharpoons H_2PO_4^-$$

With the salts of certain weak acids, such as carbonic, sulphurous and nitrous acids, an additional factor contributing to the increased solubility is the actual disappearance of the acid from solution, either spontaneously or on gentle warming. An explanation is thus provided for the well-known solubility of the sparingly soluble sulphites, carbonates, oxalates, phosphates(V), arsenites(III), arsenates(V), cyanides (with the exception of silver cyanide, which is actually a salt of the strong acid $H[Ag(CN)_2]$), fluorides, ethanoates and salts of other organic acids in strong acids.

The sparingly soluble sulphates (e.g. those of barium, strontium, and lead) also exhibit increased solubility in acids as a consequence of the weakness of the second-stage ionisation of sulphuric acid ($K_2 = 1.2 \times 10^{-2} \, mol \, L^{-1}$):

$$SO_4^{2-} + H^+ \rightleftharpoons HSO_4^-$$

But since K_2 is comparatively large, the solvent effect is relatively small; this is why in the quantitative separation of barium sulphate, precipitation may be carried out in slightly acid solution in order to obtain a more easily filterable precipitate and to reduce coprecipitation.

2.19 Precipitate solubility: effect of temperature

The solubility of the precipitates encountered in quantitative analysis increases with increasing temperature. With some substances the influence of temperature is small, but with others it is quite appreciable. Thus the solubilities of silver chloride at 10 and 100 °C are respectively 1.72 and $21.1 \, mg \, L^{-1}$, whereas the solubilities of barium sulphate at these two temperatures are respectively 2.2 and $3.9 \, mg \, L^{-1}$. In many instances the common ion effect reduces the solubility to a value so small that the temperature effect, which is otherwise appreciable, becomes very small. Wherever possible it is advantageous to filter while the solution is hot; the rate of filtration is increased, as is the solubility of foreign substances, thus rendering their removal from the precipitate more complete. The double phosphates of ammonium with magnesium, manganese or zinc, as well as lead sulphate and silver chloride, are usually filtered at the laboratory temperature to avoid solubility losses.

2.20   Precipitate solubility: effect of solvent

The solubility of most inorganic compounds is reduced by the addition of organic solvents such as methanol, ethanol, propan-1-ol and acetone. For example, the addition of about 20 vol% ethanol renders the solubility of lead sulphate practically negligible, thus permitting quantitative separation. Similarly, calcium sulphate separates quantitatively from 50 vol% ethanol. Other examples of the influence of solvents will be found in Chapter 11.

2.21   Complex ions

The increase in solubility of a precipitate upon adding excess precipitating agent is frequently due to the formation of a complex ion. A **complex ion** is formed by the union of a simple ion with other ions of opposite charge or with neutral molecules, as shown by the following examples.

When potassium cyanide solution is added to a solution of silver nitrate, a white precipitate of silver cyanide is first formed, because the solubility product of silver cyanide

$$[Ag^+][CN^-] = K_{s(AgCN)} \tag{2.30}$$

is exceeded. The reaction is expressed as

$$CN^- + Ag^+ \rightleftharpoons AgCN$$

The precipitate dissolves upon the addition of excess of potassium cyanide, producing the complex ion $[Ag(CN)_2]^-$:*

$$AgCN(s) + CN^- \text{ (excess)} \rightleftharpoons [Ag(CN)_2]^-$$

or $AgCN + KCN \rightleftharpoons K[Ag(CN)_2]$, a soluble complex salt. The $[Ag(CN)_2]^-$ complex ion dissociates to give silver ions, since the addition of sulphide ions yields a precipitate of silver sulphide (solubility product 1.6×10^{-49} mol^3 L^{-3}), and silver is deposited from the complex cyanide solution upon electrolysis. The complex ion thus dissociates in accordance with the equation

$$[Ag(CN)_2]^- \rightleftharpoons Ag^+ + 2CN^-$$

Applying the law of mass action, we obtain the dissociation constant of the complex ion

$$\frac{[Ag^+][CN^-]^2}{[\{Ag(CN)_2\}^-]} = K_{diss} \tag{2.31}$$

---

* Square brackets are commonly used for two purposes: (1) to denote concentrations and (2) to include the whole of a complex ion; curly brackets (braces) are sometimes used for the second purpose. With careful scrutiny there should be no confusion regarding the sense in which the square brackets are used; with complexes there will be no charge signs **inside** the brackets.

which has a value of $1.0 \times 10^{-21} \, mol^2 \, L^{-2}$ at the ordinary temperature. By inspection of this expression, and bearing in mind that excess cyanide ion is present, it is evident the silver ion concentration must be very small, so small in fact that the solubility product of silver cyanide is not exceeded.

The inverse of equation (2.31) gives us the stability constant or formation constant of the complex ion

$$K = \frac{[\{Ag(CN)_2\}^-]}{[Ag^+][CN^-]^2} = 10^{21} \, mol^{-2} \, L^2 \tag{2.32}$$

Consider now a somewhat different type of complex ion formation, namely the production of a complex ion with constituents other than the common ion present in the solution. This is exemplified by the solubility of silver chloride in ammonia solution. The reaction is

$$AgCl + 2NH_3 \rightleftharpoons [Ag(NH_3)_2]^+ + Cl^-$$

Here again electrolysis, or treatment with hydrogen sulphide, shows that silver ions are present in solution. The dissociation of the complex ion is represented by

$$[Ag(NH_3)_2]^+ \rightleftharpoons Ag^+ + 2NH_3$$

and the dissociation constant is given by

$$K_{diss} = \frac{[Ag^+][NH_3]^2}{[\{Ag(NH_3)_2\}^+]} = 6.8 \times 10^{-8} \, mol^2 \, L^{-2}$$

The stability constant $K = 1/K_{diss} = 1.5 \times 10^7 \, mol^{-2} \, L^2$

The magnitude of the dissociation constant clearly shows that only a very small silver ion concentration is produced by the dissociation of the complex ion.

The stability of complex ions varies within very wide limits. It is quantitatively expressed by means of the **stability constant**. The more stable the complex, the greater the stability constant, i.e. the smaller the tendency of the complex ion to dissociate into its constituent ions. When the complex ion is very stable, e.g. the hexacyanoferrate(II) ion $[Fe(CN)_6]^{4-}$, the ordinary ionic reactions of the components are not shown.

The application of complex ion formation in chemical separations depends upon the fact that one component may be transformed into a complex ion which no longer reacts with a given reagent, whereas another component does react. One example concerns the separation of cadmium and copper. Excess of potassium cyanide solution is added to the solution containing the two salts when the complex ions $(Cd(CN)_4]^{2-}$ and $[Cu(CN)_4]^{3-}$ are formed. Upon passing hydrogen sulphide into the solution containing excess of CN^- ions, a precipitate of cadmium sulphide is produced. Despite the higher solubility product of CdS ($1.4 \times 10^{-28} \, mol^2 \, L^{-2}$ as against $6.5 \times 10^{-45} \, mol^2 \, L^{-2}$ for copper(II) sulphide), the CdS is precipitated because the complex cyanocuprate(I) ion has a greater stability constant ($2 \times 10^{27} \, mol^{-4} \, L^4$ as compared with $7 \times 10^{10} \, mol^{-4} \, L^4$ for the cadmium complex).

2.22   Complexation

The processes of complex ion formation can be described by the general term **complexation**. A complexation reaction with a metal ion involves the replacement of one or more of the coordinated solvent molecules by other nucleophilic groups. The groups bound to the central ion are called ligands and in aqueous solution the reaction can be represented by the equation

$$M(H_2O)_n + L \rightleftharpoons M(H_2O)_{(n-1)}L + H_2O$$

Here the ligand (L) can be either a neutral molecule or a charged ion, and successive replacement of water molecules by other ligand groups can occur until the complex ML_n is formed; n is the coordination number of the metal ion and represents the maximum number of monodentate ligands that can be bound to it.

Ligands may be conveniently classified on the basis of the number of points of attachment to the metal ion. Thus simple ligands, such as halide ions or the molecules H_2O or NH_3, are **monodentate**, i.e. the ligand is bound to the metal ion at only one point by the donation of a lone pair of electrons to the metal. But when the ligand molecule or ion has two atoms, each of which has a lone pair of electrons, then the molecule has two donor atoms and it may be possible to form two coordinate bonds with the same metal ion; this is called a bidentate ligand. An example is the tris(ethylenediamine)cobalt(III) complex, $[Co(en)_3]^{3+}$. In this six-coordinate octahedral complex of cobalt(III), each of the bidentate ethylenediamine (1,2-diaminoethane) molecules is bound to the metal ion through the lone pair electrons of the two nitrogen atoms. This results in the formation of three five-membered rings, each including the metal ion; the process of ring formation is called **chelation**.

A **multidentate** ligand contains more than two coordinating atoms per molecule; for example, 1,2-diaminoethanetetra-acetic acid (ethylenediaminetetra-acetic acid, EDTA)* has two donor nitrogen atoms and four donor oxygen atoms in the molecule, so it can be hexadentate.

It has been assumed that the complex species does not contain more than one metal ion, but under appropriate conditions it is possible to form a binuclear complex, i.e. one containing two metal ions, or even a polynuclear complex, containing more than two metal ions. Thus interaction between Zn^{2+} and Cl^- ions may result in the formation of binuclear complexes, e.g. $[Zn_2Cl_6]^{2-}$, in addition to simple species such as $ZnCl_3^-$ and $ZnCl_4^{2-}$. The formation of bi- and polynuclear complexes will clearly be favoured by a high concentration of the metal ion; if the metal ion is present as a trace constituent of a solution, polynuclear complexes are unlikely to be formed.

2.23   Stability of complexes

The thermodynamic stability of a species is a measure of the extent to which this species will be formed from other species under certain conditions, provided that the system is allowed to reach equilibrium. Consider a metal ion M in solution together with a monodentate ligand L, then the system may be described by the following stepwise equilibria; for convenience, coordinated water molecules are not shown.

$$M + L \rightleftharpoons ML \qquad K_1 = [ML]/[M][L]$$
$$ML + L \rightleftharpoons ML_2 \qquad K_2 = [ML_2]/[ML][L]$$
$$ML_{(n-1)} + L \rightleftharpoons ML_n \qquad K_n = [ML_n]/[ML_{(n-1)}][L]$$

The equilibrium constants $K_1, K_2, \ldots, K_n$ are called **stepwise stability constants**. An alternative way of expressing the equilibria is as follows:

$$M + L \rightleftharpoons ML \qquad \beta_1 = [ML]/[M][L]$$
$$M + 2L \rightleftharpoons ML_2 \qquad \beta_2 = [ML_2]/[M][L]^2$$
$$M + nL \rightleftharpoons ML_n \qquad \beta_n = [ML_n]/[M][L]^n$$

* 1,2-Bis[bis(carboxymethyl)amino]ethane.

The equilibrium constants $\beta_1, \beta_2, \ldots, \beta_n$ are called the **overall stability constants** and are related to the stepwise stability constants by the general expression

$$\beta_n = K_1 K_2 \ldots K_n$$

In the above equilibria it has been assumed that no insoluble products are formed nor any polynuclear species.

   A knowledge of stability constant values is of considerable importance in analytical chemistry, since they provide information about the concentrations of the various complexes formed by a metal in specified equilibrium mixtures; this is invaluable in the study of complexometry and various analytical separation procedures such as solvent extraction, ion exchange and chromatography.[2,3]

2.24   Metal ion buffers

Consider the equation for complex formation

$$M + L \rightleftharpoons ML \qquad K = [ML]/[M][L]$$

and assume that ML is the only complex to be formed by the particular system. The equilibrium constant expression can be rearranged to give

$$[M] = (1/K)[ML]/[L]$$

$$\log[M] = \log(1/K) + \log\frac{[ML]}{[L]}$$

$$pM = \log K - \log\frac{[ML]}{[L]}$$

This shows that the pM value of the solution is fixed by the value of K and the ratio of complex ion concentration to that of the free ligand. If more of M is added to the solution, more complex will be formed and the value of pM will not change appreciably. Likewise, if M is removed from the solution by some reaction, some of the complex will dissociate to restore the value of pM. This recalls the behaviour of buffer solutions encountered with acids and bases (Section 2.11), and by analogy, the complex–ligand system may be called a **metal ion buffer**.

2.25   Complex stability: important factors

Complexing ability of metals

The relative complexing ability of metals is conveniently described in terms of the **Schwarzenbach classification**, which is broadly based upon the division of metals into class A and class B Lewis acids, i.e. electron acceptors. Class A metals are distinguished by an order of affinity (in aqueous solution) towards the halogens $F^- \gg Cl^- > Br^- > I^-$, and form their most stable complexes with the first member of each group of donor atoms in the periodic table (i.e. nitrogen, oxygen and fluorine). Class B metals coordinate much more readily with I^- than with F^- in aqueous solution, and form their most stable complexes with the second (or heavier) donor atom from each group (i.e. P, S, Cl). The Schwarzenbach classification defines three categories of metal ion acceptors:

Cations with noble gas configurations   The alkali metals, alkaline earths and aluminium belong to this group which exhibit class A acceptor properties. Electrostatic forces predominate in complex formation, so interactions between small ions of high charge are particularly strong and lead to stable complexes.

Cations with completely filled d subshells   Typical of this group are copper(I), silver(I) and gold(I), which exhibit class B acceptor properties. These ions have high polarising power and the bonds formed in their complexes have appreciable covalent character.

Transition metal ions with incomplete d subshells   In this group both class A and class B tendencies can be distinguished. The elements with class B characteristics form a roughly triangular group within the periodic table, with the apex at copper and the base extending from rhenium to bismuth. To the left of this group, elements in their higher oxidation states tend to exhibit class A properties; whereas to the right of the group, the higher oxidation states of a given element have a greater class B character.

The concept of **hard and soft acids and bases** is useful in characterising the behaviour of class A and class B acceptors. A soft base may be defined as one in which the donor atom is of high polarisability and low electronegativity, is easily oxidised, or is associated with vacant, low-lying orbitals. These terms describe, in different ways, a base in which the donor atom electrons are not tightly held, but are easily distorted or removed. Hard bases have the opposite properties, i.e. the donor atom is of low polarisability and high electronegativity, is difficult to reduce, and is associated with vacant orbitals of high energy which are inaccessible.

On this basis, class A acceptors prefer to bind to hard bases, e.g. with N, O and F donor atoms, whereas class B acceptors prefer to bind to the softer bases, e.g. P, As, S, Se, Cl, Br, I donor atoms. Examination of the class A acceptors shows them to have the following distinguishing features: small size, high positive oxidation state, and the absence of outer electrons which are easily excited to higher states. These are all factors which lead to low polarisability, and such acceptors are called hard acids. Class B acceptors have one or more of the following properties: low positive or zero oxidation state, large size, and several easily excited outer electrons (for metals these are the d electrons). These are all factors which lead to high polarisability, and class B acids may be called soft acids.

Here is a general principle for correlating the complexing ability of metals: hard acids tend to associate with hard bases and soft acids with soft bases. But do not regard it as exclusive; under appropriate conditions soft acids may complex with hard bases or hard acids with soft bases.

Characteristics of the ligand

Some characteristics of the ligand which are generally recognised as influencing the stability of its complexes are (i) its basic strength, (ii) its chelating properties (if any) and (iii) steric effects. From the viewpoint of analytical applications, the chelating effect is of paramount importance and therefore merits particular attention.

The term **chelate effect** refers to the fact that a chelated complex, i.e. one formed by a bidentate or a multidenate ligand, is more stable than the **corresponding** complex with monodentate ligands: the greater the number of points of attachment of ligand to the metal ion, the greater the stability of the complex. Thus the complexes formed by the nickel(II) ion with (a) the monodentate NH_3 molecule, (b) the bidentate ethylenediamine (1,2-diaminoethane), and (c) the hexadentate ligand 'penten' $\{(H_2N \cdot CH_2 \cdot CH_2)_2N \cdot CH_2 \cdot CH_2 \cdot N(CH_2 \cdot CH_2.NH_2)_2\}$ show an overall stability constant of 3.1×10^8 for the complex of ligand (a), which is increased by a factor of about 10^{10} for the complex of ligand (b), and is approximately 10 times greater still for the complex of ligand (c). The most common steric effect is inhibition of complex formation owing to the presence of a large group either attached to or in close proximity to the donor atom.

A further factor which must also be taken into consideration from the viewpoint of the analytical applications of complexes and of complex formation reactions is the rate of reaction. To be analytically useful, the reaction should usually be rapid. An important classification into labile and inert is based upon the rate at which complexes undergo substitution reactions. A **labile** complex completes nucleophilic substitution within the time required to mix the reagents. For example, when excess aqueous ammonia is added to an aqueous solution of copper(II) sulphate, the change in colour from pale blue to deep blue is instantaneous; the rapid replacement of water molecules by ammonia indicates that the Cu(II) ion forms kinetically labile complexes. An **inert** complex undergoes slow substitution reactions, i.e. reactions with half-times of the order of hours or even days at room temperature. Thus the Cr(III) ion forms kinetically inert complexes, so that the replacement of water molecules coordinated to Cr(III) by other ligands is a very slow process at room temperature.

Kinetic inertness or lability is influenced by many factors, but the following general observations form a convenient guide to the behaviour of the complexes of various elements:

1. Main group elements usually form labile complexes.
2. With the exception of Cr(III) and Co(III), most first-row transition elements form labile complexes.
3. Second- and third-row transition elements tend to form inert complexes.

For a full discussion of the topics introduced in this section, consult a textbook of inorganic chemistry[4] or a textbook dealing with complexes.[2]

2.26 Complexones

Formation of a single complex species rather than the stepwise production of several species will clearly simplify complexometric titrations and facilitate the detection of end points. Schwarzenbach[2] realised that the ethanoate ion is able to form ethanoato complexes of low stability with nearly all polyvalent cations, and that if this property could be reinforced by the chelate effect, then much stronger complexes would be formed by most metal cations. He found that the aminopolycarboxylic acids are excellent complexing agents; the most important of them is 1,2-diaminoethanetetra-acetic acid (ethylenediaminetetra-acetic acid). Formula [2.A] is preferred to formula [2.B] since it has been shown from measurements of the dissociation constants that two hydrogen atoms are probably held in the form of zwitterions. The values of pK are respectively pK_1 = 2.0, pK_2 = 2.7, pK_3 = 6.2, and pK_4 = 10.3 at 20 °C. These values suggest that it behaves as a dicarboxylic acid with two strongly acidic groups and that there are two ammonium protons; the first of them ionises in the pH region of about 6.3 and the second of them at a pH of about 11.5. Various trivial names are used for *ethylenediaminetetra-acetic* acid and its sodium salts, and these include Trilon B, Complexone III, Sequestrene, Versene and Chelation 3; the disodium salt is most widely employed in titrimetric analysis. To avoid the constant use of the long name, the abbreviation EDTA is used for the disodium salt.

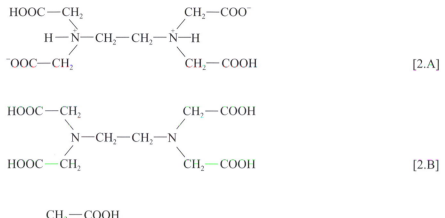

$$[2.A]$$

$$[2.B]$$

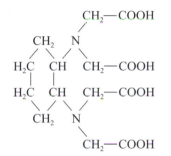

$$[2.C]$$

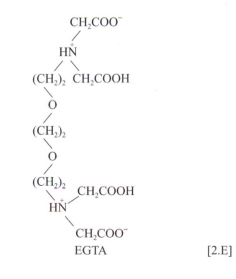

$$[2.D]$$

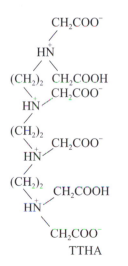

EGTA $\qquad$ [2.E]

TTHA $\qquad$ [2.F]

51

Other complexing agents (complexones) which are sometimes used include nitrilotriacetic acid [2.C], also called NITA, NTA or Complexone I ($pK_1 = 1.9$, $pK_2 = 2.5$ and $pK_3 = 9.7$); *trans*-1,2-diaminocyclohexane-N,N,N',N'-tetra-acetic acid (2.D], which should presumably be formulated as a zwitterion structure like [2.A] (the abbreviated name is CDTA, DCyTA, DCTA or Complexone IV); 2,2'-ethylenedioxybis{ethyliminodi(acetic acid)) [2.E], also known as ethylene glycolbis(2-aminoethyl ether)N,N,N',N'-tetra-acetic acid (EGTA); and triethylenetetramine-N,N,N',N'',N''',N'''-hexa-acetic acid (TTHA) [2.F]. CDTA often forms stronger metal complexes than EDTA and thus finds applications in analysis, but the metal complexes are formed rather more slowly than with EDTA, so that the end point of the titration tends to be drawn out with CDTA. EGTA finds analytical application mainly in the determination of calcium in a mixture of calcium and magnesium and is probably superior to EDTA in the calcium/magnesium water hardness titration (Section 10.76). TTHA forms $1:2$ complexes with many trivalent cations and with some divalent metals, and can be used for determining the components of mixtures of certain ions without the use of masking agents (Section 10.64).

However, EDTA has the widest general application in analysis because of its powerful complexing action and commercial availability. The spatial structure of its anion, which has six donor atoms, enables it to satisfy the coordination number of 6 frequently encountered among the metal ions and to form strainless five-membered rings on chelation. The resulting complexes have similar structures but differ from one another in the charge they carry.

To simplify the following discussion, EDTA is assigned the formula H_4Y; the disodium salt is therefore Na_2H_2Y and affords the complex-forming ion H_2Y^{2-} in aqueous solution; it reacts with all metals in a $1:1$ ratio. The reactions with cations, e.g. M^{2+}, may be written as

$$M^{2+} + H_2Y^{2-} \rightleftharpoons MY^{2-} + 2H^+ \qquad [2.9]$$

For other cations, the reactions may be expressed as

$$M^{3+} + H_2Y^{2-} \rightleftharpoons MY^- + 2H^+ \qquad [2.10]$$

$$M^{4+} + H_2Y^{2-} \rightleftharpoons MY + 2H^+ \qquad [2.11]$$

$$\text{or} \quad M^{n+} + H_2Y^{2-} \rightleftharpoons MY^{(n-4)+} + 2H^+ \qquad [2.12]$$

One mole of the complex-forming H_2Y^{2-} reacts in all cases with one mole of the metal ion, and in each case two moles of hydrogen ion are formed. Reaction [2.12] shows that the dissociation of the complex will be governed by the pH of the solution; lowering the pH will decrease the stability of the metal–EDTA complex. The more stable the complex, the lower the pH at which an EDTA titration of the metal ion in question may be carried out. Table 2.3 indicates minimum pH values for the existence of EDTA complexes of some

Table 2.3 *Stability with respect to pH of some metal–EDTA complexes*

Minimum pH at which complexes exist	Selected metals
1–3	Zr^{4+}; Hf^{4+}; Th^{4+}; Bi^{3+}; Fe^{3+}
4–6	Pb^{2+}; Cu^{2+}; Zn^{2+}; Co^{2+}; Ni^{2+}; Mn^{2+}; Fe^{2+}; Al^{3+}; Cd^{2+}; Sn^{2+}
8–10	Ca^{2+}; Sr^{2+}; Ba^{2+}; Mg^{2+}

selected metals. In general, EDTA complexes with metal ions of the charge number 2 are stable in alkaline or slightly acidic solution, whereas complexes with ions of charge numbers 3 or 4 may exist in solutions of much higher acidity.

2.27   Stability constants of EDTA complexes

The stability of a complex is characterised by the stability constant (or formation constant) K:

$$M^{n+} + Y^{4-} \rightleftharpoons (MY)^{(n-4)+} \qquad [2.13]$$

$$K = [(MY)^{(n-4)+}]/[M^{n+}][Y^{4-}] \qquad (2.33)$$

Some values for the stability constants (expressed as $\log K$) of metal–EDTA complexes are collected in Table 2.4; they apply to a medium of ionic strength $I = 0.1$ at 20 °C. In equation (2.33) only the fully ionised form of EDTA, i.e. the ion Y^{4-}, has been taken into account, but at low pH values the species HY^{3-}, H_2Y^{2-}, H_3Y^{-} and even undissociated H_4Y may well be present; in other words, only a part of the EDTA uncombined with metal may be present as Y^{4-}. Furthermore, in equation (2.33) the metal ion M^{n+} is assumed to be uncomplexed, i.e. in aqueous solution it is simply present as the hydrated ion. If, however, the solution also contains substances other than EDTA which can complex with the metal ion, then the whole of this ion uncombined with EDTA may no longer be present as the simple hydrated ion. Thus, in practice, the stability of metal–EDTA complexes may be altered (a) by variation in pH and (b) by the presence of other complexing agents. The stability constant of the EDTA complex will then be different from the value recorded for a specified pH in pure aqueous solution; the value recorded for the new conditions is called the **apparent** or **conditional stability constant**. Their effects need to be examined in some detail.

Table 2.4   *Stability constants (as log K) of metal–EDTA complexes*

Mg^{2+}	8.7	Zn^{2+}	16.7	La^{3+}	15.7
Ca^{2+}	10.7	Cd^{2+}	16.6	Lu^{3+}	20.0
Sr^{2+}	8.6	Hg^{2+}	21.9	Sc^{3+}	23.1
Ba^{2+}	7.8	Pb^{2+}	18.0	Ga^{3+}	20.5
Mn^{2+}	13.8	Al^{3+}	16.3	In^{3+}	24.9
Fe^{2+}	14.3	Fe^{3+}	25.1	Th^{4+}	23.2
Co^{2+}	16.3	Y^{3+}	18.2	Ag^{+}	7.3
Ni^{2+}	18.6	Cr^{3+}	24.0	Li^{+}	2.8
Cu^{2+}	18.8	Ce^{3+}	15.9	Na^{+}	1.7

pH effect

The apparent stability constant at a given pH may be calculated from the ratio K/α, where α is the ratio of the total uncombined EDTA (in all forms) to the form Y^{4-}. Thus K_H, the apparent stability constant for the metal–EDTA complex at a given pH, can be calculated from the expression

$$\log K_H = \log K - \log \alpha \qquad (2.34)$$

The factor α can be calculated from the known dissociation constants of EDTA, and since the proportions of the various ionic species derived from EDTA will depend upon the pH of the solution, α will also vary with pH. A plot of $\log \alpha$ against pH shows a variation of $\log \alpha = 18$ at pH $= 1$ to $\log \alpha = 0$ at pH $= 12$; such a curve is very useful for dealing with calculations of apparent stability constants. Thus, from Table 2.4 $\log K$ of the EDTA complex of the Pb^{2+} ion is 18.0, and from a graph of $\log \alpha$ against pH it is found that $\log \alpha = 7$ at a pH of 5.0. Hence, from equation (2.34), at a pH of 5.0 the lead–EDTA complex has an apparent stability constant given by

$$\log K_H = 18.0 - 7.0 = 11.0$$

Carrying out a similar calculation for the EDTA complex of the Mg^{2+} ion ($\log K = 8.7$), for the same pH (5.0), it is found that

$$\log K_H(Mg(II)\text{–}EDTA) = 8.7 - 7.0 = 1.7$$

These results imply that, at the specified pH, the magnesium complex is appreciably dissociated whereas the lead complex is stable; so titration of an Mg(II) solution with EDTA at this pH will be unsatisfactory, but titration of the lead solution under the same conditions will be quite feasible. In practice, for a metal ion to be titrated with EDTA at a stipulated pH, the value of $\log K_H$ should be greater than 8 when a metallochromic indicator is used.

The value of $\log \alpha$ is small at high pH values, hence the larger values of $\log K_H$ are found with increasing pH. However, increasing the pH of the solution will increase the tendency to form slightly soluble metallic hydroxides:

$$(MY)^{(n-4)+} + nOH^- \rightleftharpoons M(OH)_n + Y^{4-}$$

The extent of hydrolysis of $(MY)^{(n-4)+}$ depends upon the characteristics of the metal ion; it is largely controlled by the solubility product of the metallic hydroxide and, of course, the stability constant of the complex. Thus iron(III) is precipitated as hydroxide ($K_{sol} = 1 \times 10^{-36}$) in basic solution; but nickel(II), for which the relevant solubility product is 6.5×10^{-18}, remains complexed. Clearly the use of excess EDTA will tend to reduce the effect of hydrolysis in basic solutions. It follows that, for each metal ion, there exists an optimum pH which will give rise to a maximum value for the apparent stability constant.

The effect of other complexing agents

If another complexing agent (say NH_3) is also present in the solution then in equation (2.33) the concentration $[M^{n+}]$ will be reduced, owing to complexation of the metal ions with ammonia molecules. It is convenient to indicate this reduction in effective concentration by introducing a factor β, defined as the ratio of the sum of the concentrations of all forms of the metal ion not complexed with EDTA to the concentration of the simple (hydrated) ion. The apparent stability constant of the metal–EDTA complex, taking into account the effects of both pH and the presence of other complexing agents, is then given by

$$\log K_{HZ} = \log K - \log \alpha - \log \beta \tag{2.35}$$

2.28 Electrode potentials

When a metal is immersed in a solution containing its own ions, say, zinc in zinc sulphate solution, a potential difference is established between the metal and the solution. The potential difference E for an electrode reaction

$$M^{n+} + ne = M$$

is given by the expression

$$E = E^{\ominus} + \frac{RT}{nF} \ln a_{\text{M}^{n+}}$$

(2.36)

where R is the gas constant, T is the absolute temperature, F is the Faraday constant, n is the charge number of the ions, $a_{\text{M}^{n+}}$ is the activity of the ions in the solution, and $E^{\ominus}$ is a constant that depends on the metal. Equation (2.36) can be simplified by introducing the known values of R and F, and converting natural logarithms to base 10 by multiplying by 2.3026; it then becomes

$$E = E^{\ominus} + \frac{0.000\,1984\,T}{n} \log a_{\text{M}^{n+}}$$

For a temperature of 25 °C ($T = 298$ K)

$$E = E^{\ominus} + \frac{0.0591}{n} \log a_{\text{M}^{n+}}$$

(2.37)

For many purposes in quantitative analysis, it is sufficiently accurate to replace $a_{\text{M}^{n+}}$ by $c_{\text{M}^{n+}}$ the ion concentration (mol L^{-1}):

$$E = E^{\ominus} + \frac{0.0591}{n} \log c_{\text{M}^{n+}}$$

(2.38)

This is a form of the **Nernst equation**.

If in equation (2.38) $a_{\text{M}^{n+}}$ is made equal to unity, E is equal to $E^{\ominus}$. And $E^{\ominus}$ is called the **standard electrode potential** of the metal; both E and $E^{\ominus}$ are expressed in volts. In order to determine the potential difference between an electrode and a solution, it is necessary to have another electrode and solution of accurately known potential difference. The two electrodes can then be combined to form a voltaic cell, the e.m.f. of which can be directly measured. The e.m.f. of the cell is the difference of the electrode potentials at zero current; the value of the unknown potential can then be calculated. The primary reference electrode is the **normal** or **standard hydrogen electrode** (Section 13.15). This consists of a piece of platinum foil, coated electrolytically with platinum black, and immersed in a solution of hydrochloric acid containing hydrogen ions at unit activity. (This corresponds to 1.18 M hydrochloric acid at 25 °C.) Hydrogen gas at a pressure of one atmosphere is passed over the platinum foil through the side tube C (Figure 2.2) and escapes through the small holes B in the surrounding glass tube A. Because of the periodic formation of bubbles, the level of the liquid inside the tube fluctuates, and a part of the foil is alternately exposed to the solution and to hydrogen. The lower end of the foil is continuously immersed in the solution to avoid interruption of the electric current. Connection between the platinum foil and an external circuit is made with mercury in D. The platinum black has the property of adsorbing large quantities of atomic hydrogen. It permits the change from the gaseous to the ionic form, and the reverse process, to occur without hindrance; it therefore behaves as though it were composed entirely of hydrogen, i.e. as a hydrogen electrode. Under fixed conditions – hydrogen gas at atmospheric pressure and unit activity of hydrogen ions in the solution in contact with the electrode – the hydrogen electrode possesses a definite potential. By convention, the potential of the standard hydrogen electrode is equal to zero at all temperatures. The **standard electrode potential** of the metal may be determined by connecting the standard hydrogen electrode with a metal electrode consisting of a metal in contact with a solution of its ions of unit activity and measuring the cell e.m.f. The cell is usually written as

$$\text{Pt, H}_2 \,|\, \text{H}^+ \,(a = 1) \,\|\, \text{M}^{n+} \,(a = 1) \,|\, \text{M}$$

2 Solution reactions: fundamental theory

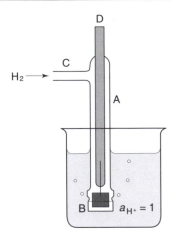

Figure 2.2 The standard hydrogen electrode

In this scheme, a single vertical line represents a metal–electrolyte boundary at which a potential difference is taken into account; the double vertical broken lines represent a liquid junction at which the potential is to be disregarded or is considered to be eliminated by a salt bridge.

When reference is made to the electrode potential of a zinc electrode, it means the e.m.f. of the cell

$$\text{Pt, } H_2 \,|\, H^+ \,(a = 1) \,||\, Zn^{2+} \,|\, Zn$$

or the e.m.f. of the half-cell $Zn^{2+} \,|\, Zn$. The cell reaction is

$$H_2 + Zn^{2+} \rightarrow 2H^+ \,(a = 1) + Zn$$

and the half-cell reaction is written as

$$Zn^{2+} + 2e = Zn$$

The electrode potential of the $Fe^{3+}, Fe^{2+} \,|\, Pt$ electrode is the e.m.f. of the cell

$$\text{Pt, } H_2 \,|\, H^+ \,(a = 1) \,||\, Fe^{3+}, Fe^{2+} \,|\, Pt$$

or the e.m.f. of the half-cell $Fe^{3+}, Fe^{2+} \,|\, Pt$. The cell reaction is

$$\tfrac{1}{2}H_2 + Fe^{3+} \rightarrow H^+ \,(a = 1) + Fe^{2+}$$

and the half-cell reaction is written

$$Fe^{3+} + e \rightleftharpoons Fe^{2+}$$

The convention is adopted of writing all half-cell reactions as reductions:

$$M^{n+} + ne \rightarrow M$$

$$\text{e.g.} \quad Zn^{2+} + 2e \rightarrow Zn \qquad (E^{\ominus} = -0.76\,\text{V})$$

When the activity of the ion M^{n+} is equal to unity (approximately true for a $1\,M$ solution), the electrode potential E is equal to the standard potential $E^{\ominus}$. Some important standard

Table 2.5 *Standard electrode potentials at 25 °C*

Electrode reaction	$E^{\ominus}$ (V)	Electrode reaction	$E^{\ominus}$ (V)
$Li^+ + e = Li$	−3.045	$Tl^+ + e = Tl$	−0.336
$K^+ + e = K$	−2.925	$Co^{2+} + 2e = Co$	−0.277
$Ba^{2+} + 2e = Ba$	−2.90	$Ni^{2+} + 2e = Ni$	−0.25
$Sr^{2+} + 2e = Sr$	−2.89	$Sn^{2+} + 2e = Sn$	−0.136
$Ca^{2+} + 2e = Ca$	−2.87	$Pb^{2+} + 2e = Pb$	−0.126
$Na^+ + e = Na$	−2.714	$2H^+ + 2e = H_2$	0.000
$Mg^{2+} + 2e = Mg$	−2.37	$Cu^{2+} + 2e = Cu$	+0.337
$Al^{3+} + 3e = Al$	−1.66	$Hg^{2+} + 2e = Hg$	+0.789
$Mn^{2+} + 2e = Mn$	−1.18	$Ag^+ + e = Ag$	+0.799
$Zn^{2+} + 2e = Zn$	−0.763	$Pd^{2+} + 2e = Pd$	+0.987
$Fe^{2+} + 2e = Fe$	−0.440	$Pt^{2+} + 2e = Pt$	+1.2
$Cd^{2+} + 2e = Cd$	−0.403	$Au^{3+} + 3e = Au$	+1.50

electrode potentials referred to the standard hydrogen electrode at 25 °C (in aqueous solution) are collected in Table 2.5.[5]

Note that the standard hydrogen electrode is rather difficult to manipulate. In practice, electrode potentials on the hydrogen scale are usually determined indirectly by measuring the e.m.f. of a cell formed from the electrode in question and a convenient reference electrode whose potential with respect to the hydrogen electrode is accurately known. The reference electrodes generally used are the calomel electrode and the silver–silver chloride electrode (Sections 13.16 and 13.17).

When metals are arranged in the order of their standard electrode potentials, the so-called electrochemical series of the metals is obtained. The greater the negative value of the potential, the greater the tendency of the metal to pass into the ionic state. A metal will normally displace any other metal below it in the series from solutions of its salts. Thus magnesium, aluminium, zinc or iron will displace copper from solutions of its salts; lead will displace copper, mercury or silver; copper will displace silver. The standard electrode potential is a quantitative measure of the readiness of the element to lose electrons. It is therefore a measure of the strength of the element as a reducing agent in aqueous solution; the more negative the potential of the element, the more powerful its action as a reductant.

Note that standard electrode potential values relate to an **equilibrium** condition between the metal electrode and the solution. Potentials determined under, or calculated for, such conditions are often known as 'reversible electrode potentials', and it must be remembered that the Nernst equation is only strictly applicable under these conditions.

2.29   Concentration cells

An electrode potential varies with the concentration of the ions in the solution. Hence two electrodes of the same metal, but immersed in solutions containing different concentrations of its ions, may form a cell. Such a cell is called a **concentration cell**. The e.m.f. of the cell will be the algebraic difference of the two potentials, if a salt bridge is inserted to eliminate the liquid–liquid junction potential. It may be calculated as follows.

At a temperature of 25 °C

$$E = \frac{0.0591}{n} \log c_1 + E^{\ominus} - \left(\frac{0.0591}{n} \log c_2 + E^{\ominus}\right)$$

$$= \frac{0.0591}{n} \log \frac{c_1}{c_2} \quad \text{where } c_1 > c_2$$

As an example, consider the cell

$$\text{Ag} \left| \begin{array}{c} \text{AgNO}_3 \text{ (aq)} \\ [\text{Ag}^+] = 0.004\,75\,M \\ \overleftarrow{E_2} \end{array} \right| \begin{array}{c} \text{AgNO}_3 \text{ (aq)} \\ [\text{Ag}^+] = 0.043\,M \\ \end{array} \left| \begin{array}{c} \text{Ag} \\ \overrightarrow{E_1} \end{array} \right.$$

Assuming there is no potential difference at the liquid junction, then

$$E = E_1 - E_2 = \frac{0.0591}{1} \log \frac{0.043}{0.004\,75} = 0.056\,\text{V}$$

2.30 Calculating the e.m.f. of a voltaic cell

An interesting application of electrode potentials is calculating the e.m.f. of a voltaic cell. One of the simplest of galvanic cells is the Daniell cell. It consists of a rod of zinc dipping into zinc sulphate solution and a strip of copper in copper sulphate solution; the two solutions are generally separated by placing one inside a porous pot and the other in the surrounding vessel. The cell may be represented as

$$\text{Zn} | \text{ZnSO (aq)} \| \text{CuSO (aq)} | \text{Cu}$$

At the zinc electrode, zinc ions pass into solution, leaving an equivalent negative charge on the metal. Copper ions are deposited at the copper electrode, rendering it positively charged. By completing the external circuit, the current (electrons) passes from the zinc to the copper. The chemical reactions in the cell are as follows:

$$\text{zinc electrode} \quad \text{Zn} \rightleftharpoons \text{Zn}^{2+} + 2e$$

$$\text{copper electrode} \quad \text{Cu}^{2+} + 2e \rightleftharpoons \text{Cu}$$

The net chemical reaction is

$$\text{Zn} + \text{Cu}^{2+} \rightleftharpoons \text{Zn}^{2+} + \text{Cu}$$

The potential difference at each electrode may be calculated by the formula given above, and the e.m.f. of the cell is the algebraic difference of the two potentials, the correct sign being applied to each.

As an example we may calculate the e.m.f. of the Daniell cell with molar concentrations of zinc ions and copper(II) ions:

$$E = E^{\ominus}_{(\text{Cu})} - E^{\ominus}_{(\text{Zn})} = +0.34 - (-0.76) = 1.10\,\text{V}$$

The small potential difference produced at the contact between the two solutions (the so-called liquid–junction potential) is neglected.

2.31 Oxidation–reduction cells

Reduction is accompanied by a gain of electrons, and oxidation by a loss of electrons. In a system containing both an oxidising agent and its reduction product, there will be an equilibrium between them and the electrons. If an inert electrode, such as platinum, is placed in a redox system, e.g. a system containing Fe(III) and Fe(II) ions, it will assume a definite potential indicative of the position of equilibrium. If the system tends to act as an oxidising agent ($Fe^{3+} \rightarrow Fe^{2+}$) it will take electrons from the platinum, leaving the platinum positively charged; but if the system has reducing properties ($Fe^{2+} \rightarrow Fe^{3+}$) then electrons will be given up to the metal, which will acquire a negative charge. The magnitude of the potential will thus be a measure of the oxidising or reducing properties of the system.

To obtain comparative values of the 'strengths' of oxidising agents, just as for the electrode potentials of the metals, it is necessary to measure the potential difference between the platinum and the solution relative to a reference standard and under standard experimental conditions. The primary standard is the standard or normal hydrogen electrode (Section 2.28) and its potential is taken as zero. The standard experimental conditions for the redox system are those in which the ratio of the activity of the oxidant to the activity of the reductant is unity. Thus for the Fe^{3+}–Fe^{2+} electrode, the redox cell would be

$$\text{Pt, H}_2 \left| \text{H}^+ (a=1) \right| \left| \frac{\text{Fe}^{3+}(a=1)}{\text{Fe}^{2+}(a=1)} \right| \text{Pt}$$

The potential measured in this way is called the **standard reduction potential**. A selection of standard reduction potentials is given in Table 2.6.

The standard potentials enable us to predict which ions will oxidise or reduce other ions at unit activity (or molar concentration). The most powerful oxidising agents occur at the top of the table, and the most powerful reducing agents occur at the foot of the table. Thus permanganate ion can oxidise Cl^-, Br^-, I^-, Fe^{2+} and $[Fe(CN)_6]^{4-}$; Fe^{3+} can oxidise H_3AsO_3 and I^- but not $Cr_2O_7^{2-}$ or Cl^-. Note that for many oxidants the pH of the medium is of great importance, since they are generally used in acidic media. Thus in measuring the standard potential of the MnO_4^-–Mn^{2+} system, $MnO_4^- + 8H^+ + 5e = Mn^{2+} + 4H_2O$, it is necessary to state that the hydrogen ion activity is unity; this leads to $E^\ominus = +1.52$ V. Similarly, the value of $E^\ominus$ for the $Cr_2O_7^{2-}$–Cr^{3+} system is +1.33 V. This means that the MnO_4^-–Mn^{2+} system is a better oxidising agent than the $Cr_2O_7^{2-}$–Cr^{3+} system. Since the standard potentials for Cl_2–$2Cl^-$ and Fe^{3+}–Fe^{2+} systems are +1.36 V and 0.77 V, respectively, permanganate and dichromate will oxidise Fe(II) ions but only permanganate will oxidise chloride ions; this explains why dichromate but not permanganate (except under very special conditions) can be used for the titration of Fe(II) in hydrochloric acid solution. Standard potentials do not give any information as to the speed of the reaction; sometimes a catalyst is necessary in order that the reaction may proceed with reasonable velocity.

Standard potentials are determined with full consideration of activity effects, and are really limiting values. They are rarely, if ever, observed directly in a potentiometric measurement. In practice, measured potentials determined under defined concentration conditions (formal potentials) are very useful for predicting the possibilities of redox processes. Further details are given in Section 10.127.

Table 2.6 *Standard reduction potentials at 25 °C*

Half-reaction	$E^{\ominus}$ (V)
$F_2 + 2e \rightleftharpoons 2F^-$	+2.65
$S_2O_8^{2-} + 2e \rightleftharpoons 2SO_4^{2-}$	+2.01
$Co^{3+} + e \rightleftharpoons Co^{2+}$	+1.82
$Pb^{4+} + 2e \rightleftharpoons Pb^{2+}$	+1.70
$MnO_4^- + 4H^+ + 3e \rightleftharpoons MnO_2 + 2H_2O$	+1.69
$Ce^{4+} + e \rightleftharpoons Ce^{3+}$ (nitrate medium)	+1.61
$BrO_3^- + 6H^+ + 5e \rightleftharpoons \frac{1}{2}Br_2 + 3H_2O$	+1.52
$MnO_4^- + 8H^+ + 5e \rightleftharpoons Mn^{2+} + 4H_2O$	+1.52
$Ce^{4+} + e \rightleftharpoons Ce^{3+}$ (sulphate medium)	+1.44
$Cl_2 + 2e \rightleftharpoons 2Cl^-$	+1.36
$Cr_2O_7^{2-} + 14H^+ + 6e \rightleftharpoons 2Cr^{3+} + 7H_2O$	+1.33
$Tl^{3+} + 2e \rightleftharpoons Tl^+$	+1.25
$MnO_2 + 4H^+ + 2e \rightleftharpoons Mn^{2+} + 2H_2O$	+1.23
$O_2 + 4H^+ + 4e \rightleftharpoons 2H_2O$	+1.23
$IO_3^- + 6H^+ + 5e \rightleftharpoons \frac{1}{2}I_2 + 3H_2O$	+1.20
$Br_2 + 2e \rightleftharpoons 2Br^-$	+1.07
$HNO_2 + H^+ + e \rightleftharpoons NO + H_2O$	+1.00
$NO_3^- + 4H^+ + 3e \rightleftharpoons NO + 2H_2O$	+0.96
$2Hg^{2+} + 2e \rightleftharpoons Hg_2^{2+}$	+0.92
$ClO^- + H_2O + 2e \rightleftharpoons Cl^- + 2OH^-$	+0.89
$Cu^{2+} + I^- + e \rightleftharpoons CuI$	+0.86
$Hg_2^{2+} + 2e \rightleftharpoons 2Hg$	+0.79
$Fe^{3+} + e \rightleftharpoons Fe^{2+}$	+0.77
$BrO^- + H_2O + 2e \rightleftharpoons Br^- + 2OH^-$	+0.76
$BrO_3^- + 3H_2O + 6e \rightleftharpoons Br^- + 6OH^-$	+0.61
$MnO_4^{2-} + 2H_2O + 2e \rightleftharpoons MnO_2 + 4OH^-$	+0.60
$MnO_4^- + e \rightleftharpoons MnO_4^{2-}$	+0.56
$H_3AsO_4 + 2H^+ + 2e \rightleftharpoons H_3AsO_3 + H_2O$	+0.56
$Cu^{2+} + Cl^- + e \rightleftharpoons CuCl$	+0.54
$I_2 + 2e \rightleftharpoons 2I^-$	+0.54
$IO^- + H_2O + 2e \rightleftharpoons I^- + 2OH^-$	+0.49
$[Fe(CN)_6]^{3-} + e \rightleftharpoons [Fe(CN)_6]^{4-}$	+0.36
$UO_2^{2+} + 4H^+ + 2e \rightleftharpoons U^{4+} + 2H_2O$	+0.33
$IO_3^- + 3H_2O + 6e \rightleftharpoons I^- + 6OH^-$	+0.26
$Cu^{2+} + e \rightleftharpoons Cu^+$	+0.15
$Sn^{4+} + 2e \rightleftharpoons Sn^{2+}$	+0.15
$TiO^{2+} + 2H^+ + e \rightleftharpoons Ti^{3+} + H_2O$	+0.10
$S_4O_6^{2-} + 2e \rightleftharpoons 2S_2O_3^{2-}$	+0.08
$2H^+ + 2e \rightleftharpoons H_2$	0.00
$V^{3+} + e \rightleftharpoons V^{2+}$	−0.26
$Cr^{3+} + e \rightleftharpoons Cr^{2+}$	−0.41
$Bi(OH)_3 + 3e \rightleftharpoons Bi + 3OH^-$	−0.44
$Fe(OH)_3 + e \rightleftharpoons Fe(OH)_2 + OH^-$	−0.56
$U^{4+} + e \rightleftharpoons U^{3+}$	−0.61
$AsO_4^{3-} + 3H_2O + 2e \rightleftharpoons H_2AsO_3^- + 4OH^-$	−0.67
$[Sn(OH)_6]^{2-} + 2e \rightleftharpoons [HSnO_2]^- + H_2O + 3OH^-$	−0.90
$[Zn(OH)_4]^{2-} + 2e \rightleftharpoons Zn + 4OH^-$	−1.22
$[H_2AlO_3]^- + H_2O + 3e \rightleftharpoons Al + 4OH^-$	−2.35

2.32 Calculating the standard reduction potential

A reversible oxidation – reduction redox system may be written in the form

oxidant + ne ⇌ reductant

or

ox + ne ⇌ red

(*oxidant* = substance in oxidised state, *reductant* = substance in reduced state). The electrode potential which is established when an inert or unattackable electrode is immersed in a solution containing both oxidant and reductant is given by the expression

$$E_T = E^\ominus + \frac{RT}{nF} \ln \frac{a_{ox}}{a_{red}}$$

where E_T is the observed potential of the redox electrode at temperature T relative to the standard or normal hydrogen electrode taken as zero potential, $E^\ominus$ is the standard reduction potential,* n is the number of electrons gained by the oxidant in being converted into the reductant, and a_{ox} and a_{red} are the activities of the oxidant and reductant respectively.

Since activities are often difficult to determine directly, they may be replaced by concentrations; although this introduces an error, it is usually of no great importance. The equation therefore becomes

$$E_T = E^\ominus + \frac{RT}{nF} \ln \frac{c_{ox}}{c_{red}}$$

Substituting the known values of R and F, and changing from natural to common logarithms, at a temperature of 25 °C ($T = 298$ K), we obtain

$$E_{25^\circ} = E^\ominus + \frac{0.0591}{n} \log \frac{[ox]}{[red]}$$

If the concentrations (or, more accurately, the activities) of the oxidant and reductant are equal then $E_{25^\circ} = E^\ominus$, i.e. the standard reduction potential. It follows from this expression that a tenfold change in the ratio of the concentrations of the oxidant to the reductant will produce a change in the potential of the system of $0.0591/n$ volts.

2.33 Equilibrium constants of redox reactions

The general equation for the reaction at an oxidation–reduction electrode may be written

$pA + qB + rC + \ldots + ne \rightleftharpoons sX + tY + uZ + \ldots$

The potential is given by

$$E = E^\ominus + \frac{RT}{nF} \ln \frac{a_A^p a_B^q a_C^r \ldots}{a_X^s a_Y^t a_Z^u \ldots}$$

where a refers to activities, and n to the number of electrons involved in the oxidation–reduction reaction. For a temperature of 25 °C this reduces to the following expression (concentrations are substituted for activities to permit ease of application in practice):

---

* $E^\ominus$ is the value of E_T at unit activities of the oxidant and reductant. If both activities are variable, e.g. Fe^{3+} and Fe^{2+}, $E^\ominus$ corresponds to an activity ratio of unity.

$$E = E^{\ominus} + \frac{0.0591}{n} \log \frac{c_A^p c_B^q c_C^r \dots}{c_X^s c_Y^t c_Z^u \dots}$$

It is of course possible to calculate the influence of the change in concentration of certain system constituents using this equation. Consider the permanganate reaction

$$MnO_4^- + 8H^+ + 5e \rightleftharpoons Mn^{2+} + 4H_2O$$

$$E = E^{\ominus} + \frac{0.0591}{5} \log \frac{[MnO_4^-][H^+]^8}{[Mn^{2+}]} \qquad \text{(at 25 °C)}$$

The concentration (or activity) of the water is taken as constant, since it is assumed the reaction takes place in dilute solution, and the concentration of the water does not change appreciably as the result of the reaction. The equation may be written in the form

$$E = E^{\ominus} + \frac{0.0591}{5} \log \frac{[MnO_4^-]}{[Mn^{2+}]} + \frac{0.0591}{5} \log [H^+]^8$$

This enables us to calculate the effect of change in the ratio $[MnO_4^-]/[Mn^{2+}]$ at any hydrogen ion concentration, other factors being maintained constant. In this system, however, difficulties are experienced in the calculation owing to the fact that the reduction products of the permanganate ion vary at different hydrogen ion concentrations. In other cases no such difficulties arise, and the calculation may be employed with confidence. Thus in the reaction

$$H_3AsO_4 + 2H^+ + 2e \rightleftharpoons H_3AsO_3 + H_2O$$

$$E = E^{\ominus} + \frac{0.0591}{2} \log \frac{[H_3AsO_4][H^+]^2}{[H_3AsO_3]} \qquad \text{(at 25 °C)}$$

$$\text{or} \quad E = E^{\ominus} + \frac{0.0591}{2} \log \frac{[H_3AsO_4]}{[H_3AsO_3]} + \frac{0.0591}{2} \log [H^+]^2$$

It is now possible to calculate the equilibrium constants of oxidation–reduction reactions, and thus to determine whether such reactions can find application in quantitative analysis. Consider first the simple reaction

$$Cl_2 + 2Fe^{2+} \rightleftharpoons 2Cl^- + 2Fe^{3+}$$

The equilibrium constant is given by

$$\frac{[Cl^-]^2[Fe^{3+}]^2}{[Cl_2][Fe^{2+}]^2} = K$$

The reaction may be regarded as taking place in a voltaic cell, where the two half-cells are a $Cl_2, 2Cl^-$ system and an Fe^{3+}, Fe^{2+} system. The reaction is allowed to proceed to equilibrium, and the total voltage or e.m.f. of the cell will then be zero, i.e. the potentials of the two electrodes will be equal:

$$E^{\ominus}_{Cl_2,2Cl^-} + \frac{0.0591}{2} \log \frac{[Cl_2]}{[Cl^-]^2} = E^{\ominus}_{Fe^{3+},Fe^{2+}} + \frac{0.0591}{1} \log \frac{[Fe^{3+}]}{[Fe^{2+}]}$$

Now $E^{\ominus}_{Cl_2,2Cl^-} = 1.36\,V$ and $E^{\ominus}_{Fe^{3+},Fe^{2+}} = 0.75\,V$, hence

$$\log \frac{[Fe^{3+}]^2[Cl^-]^2}{[Fe^{2+}]^2[Cl_2]} = \frac{0.61}{0.029\,65} = 20.67 = \log K$$

$$\text{or} \quad K = 4.7 \times 10^{20}$$

The large value of the equilibrium constant signifies that the reaction will proceed from left to right almost to completion, i.e. an iron(II) salt is almost completely oxidised by chlorine.

Consider now the more complex reaction

$$MnO_4^- + 5Fe^{2+} + 8H^+ \rightleftharpoons Mn^{2+} + 5Fe^{3+} + 4H_2O$$

The equilibrium constant K is given by

$$K = \frac{[Mn^{2+}][Fe^{3+}]^5}{[MnO_4^-][Fe^{2+}]^5[H^+]^8}$$

The term $4H_2O$ is omitted since the reaction is carried out in dilute solution and the water concentration may be assumed constant. The hydrogen ion concentration is taken as molar. The complete reaction may be divided into two half-cell reactions corresponding to the partial equations

$$MnO_4^- + 8H^+ + 5e \rightleftharpoons Mn^{2+} + 4H_2O \qquad [2.14]$$

$$\text{and} \quad Fe^{2+} \rightleftharpoons Fe^{3+} + e \qquad [2.15]$$

For [2.14] as an oxidation–reduction electrode, we have

$$E = E^{\ominus} + \frac{0.0591}{5} \log \frac{[MnO_4^-][H^+]^8}{[Mn^{2+}]}$$

$$= 1.52 + \frac{0.0591}{5} \log \frac{[MnO_4^-][H^+]^8}{[Mn^{2+}]}$$

The partial reaction [2.15] may be multiplied by 5 in order to balance [2.14] electrically:

$$5Fe^{2+} \rightleftharpoons 5Fe^{3+} + 5e \qquad [2.16]$$

For [2.16] as an oxidation–reduction electrode, we have

$$E = E^{\ominus} + \frac{0.0591}{5} \log \frac{[Fe^{3+}]^5}{[Fe^{2+}]^5} = 0.77 + \frac{0.0591}{5} \log \frac{[Fe^{3+}]^5}{[Fe^{2+}]^5}$$

Combining the two electrodes into a cell, the e.m.f. will be zero when equilibrium is attained, i.e.

$$1.52 + \frac{0.0591}{5} \log \frac{[MnO_4^-][H^+]^8}{[Mn^{2+}]} = 0.77 + \frac{0.0591}{5} \log \frac{[Fe^{3+}]^5}{[Fe^{2+}]^5}$$

$$\text{or} \quad \log \frac{[Mn^{2+}][Fe^{3+}]^5}{[MnO_4^-][Fe^{2+}]^5[H^+]^8} = \frac{5(1.52 - 0.77)}{0.0591} = 63.5$$

$$K = \frac{[Mn^{2+}][Fe^{3+}]^5}{[MnO_4^-][Fe^{3+}]^5[H^+]^8} = 3 \times 10^{63}$$

This result clearly indicates that the reaction proceeds virtually to completion. It is a simple matter to calculate the residual Fe(II) concentration in any particular case. Thus consider the titration of 10 mL of a 0.1 M solution of iron(II) ions with 0.02 M potassium permanganate in the presence of hydrogen ions, concentration 1 M. Let the volume of the solution at the equivalence point be 100 mL. Then $[Fe^{3+}] = 0.01\,M$, since it is known that the reaction is practically complete, $[Mn^{2+}] = \frac{1}{5}[Fe^{3+}] = 0.002\,M$, and $[Fe^{2+}] = x$. Let the excess of permanganate solution at the end point be one drop, or 0.05 mL; its concentration will be $0.05 \times 0.1/100 = 5 \times 10^{-5}\,M = [MnO_4^-]$. Substituting these values in the equation gives

$$K = \frac{(2 \times 10^{-3}) \times (1 \times 10^{-2})^5}{10^{-5} \times x^5 \times 1^8} = 3 \times 10^{63}$$

Standard reduction potentials may be employed to determine whether redox reactions are sufficiently complete for their possible use in quantitative analysis. But note that these calculations provide no information as to the speed of the reaction, upon which the application of that reaction will ultimately depend. This question must form the basis of a separate experimental study, which may include the investigation of the influence of temperature, variation of pH and the concentrations of the reactants, and the influence of catalysts. Thus, theoretically, potassium permanganate should quantitatively oxidise oxalic acid in aqueous solution. It is found, however, that the reaction is extremely slow at the ordinary temperature, but is more rapid at about 80 °C, and also increases in velocity when a little manganese(II) ion has been formed, which apparently acts as a catalyst.

It is of interest to consider the calculation of the equilibrium constant of the general redox reaction

$$a\,ox_1 + b\,red_2 \rightleftharpoons b\,ox_2 + a\,red_1$$

The complete reaction may be regarded as composed of two oxidation–reduction electrodes: $a\,ox_1$, $a\,red_1$ and $b\,ox_2$, $b\,red_2$ combined together into a cell; at equilibrium, the potentials of both electrodes are the same:

$$E_1 = E^\ominus + \frac{0.0591}{n} \log \frac{[ox_1]^a}{[red_1]^a}$$

$$E_2 = E^\ominus + \frac{0.0591}{n} \log \frac{[ox_2]^b}{[red_2]^b}$$

At equilibrium $E_1 = E_2$, so

$$E_1^\ominus + \frac{0.0591}{n} \log \frac{[ox_1]^a}{[red_1]^a} = E_2^\ominus + \frac{0.0591}{n} \log \frac{[ox_2]^b}{[red_2]^b}$$

$$\text{or} \quad \log \frac{[ox_2]^b[red_1]^a}{[red_2]^b[ox_1]^a} = \log K = \frac{n}{0.0591}(E_1^\ominus - E_2^\ominus)$$

This equation may be employed to calculate the equilibrium constant of any redox reaction, provided the two standard potentials $E_1^\ominus$ and $E_2^\ominus$ are known; from the value of K thus obtained, the feasibility of the reaction in analysis may be ascertained. It can readily be shown that the concentrations at the equivalence point – when equivalent quantities of the two substances ox_1 and red_2 are allowed to react – are given by

$$\frac{[red_1]}{[ox_1]} = \frac{[ox_2]}{[red_2]} = K^{1/(a+b)}$$

This expression enables the exact concentration at the equivalence point to be calculated in any redox reaction of the general type given above, and therefore the feasibility of a successful titration in quantitative analysis.

2.34   References

1. P W Atkins 1987 *Physical chemistry*, 3rd edn, Oxford University Press, Oxford
2. (a)   A Ringbom 1963 *Complexation in analytical chemistry*, Interscience, New York
    (b)   R Pribil 1982 *Applied complexometry*, Pergamon, Oxford
    (c)   G Schwarzenbach and H Flaschka 1969 *Complexometric titrations*, 2nd edn, Methuen, London
3. (a)   *Stability constants of metal ion complexes*, Special Publications 17 and 25, Chemical Society, London, 1964
    (b)   *Stability constants of metal ion complexes*, Part A, *Inorganic ligands*, E Högfeldt (ed) 1982; Part B, *Organic ligands*, D Perrin (ed) 1979; IUPAC and Pergamon, Oxford
4. (a)   F A Cotton and G Wilkinson 1988 *Advanced inorganic chemistry*, 5th edn, Interscience, Chichester
    (b)   N N Greenwood and E A Earnshaw 1995 *Chemistry of the elements*, 2nd edn, Pergamon, Oxford
5. A J Bard, R Parsons and J Jordan 1985 *Standard potentials in aqueous solution*, IUPAC and Marcel Dekker, New York

2.35   Bibliography

D R Crow 1979 *Principles and applications of electrochemistry*, 2nd edn, Chapman and Hall, London

Q Fernando and M D Ryan 1982 *Calculations in analytical chemistry*, Harcourt Brace Jovanovich, New York

F R Hartley, C Burgess and R M Alcock 1980 *Solution equilibria*, John Wiley, Chichester

S Kotrly and L Sucha 1985 *Handbook of chemical equilibria in analytical chemistry*, Ellis Horwood, Chichester

L Meites 1981 *An introduction to chemical equilibrium and kinetics*, Pergamon, Oxford

R W Ramette 1981 *Chemical equilibrium and analysis*, Addison-Wesley, London

J Robbins 1972 *Ions in solution*, Clarendon, Oxford

3

Common apparatus and basic techniques

3.1 Introduction

The development of advanced instrumentation and modern analytical procedures has tended to lead many people into believing that basic scientific techniques and simple apparatus are of less importance with respect to obtaining accurate, reproducible and reliable results. The importance of being able to handle simple quantitative equipment and of following well-established and set routines for cleanliness and orderly working cannot be emphasised strongly enough for the analyst seeking to maintain high standards of working. The following points should become second nature to the practising analyst:

1. Benches should always be kept clean and tidy and all spillages of both solids and liquids cleared away immediately.
2. All glassware must be scrupulously clean (Section 3.8). If it has been standing for any length of time, it must be rinsed with distilled or deionised water before use. The outsides of vessels may be dried with a lint-free glass cloth which is reserved exclusively for this purpose, and which is frequently laundered, but the cloth should not be used on the insides of the vessels.
3. Under no circumstances should the working surface of the bench become cluttered with apparatus. All the apparatus associated with some particular operation should be grouped together on the bench; this is most essential to avoid confusion when duplicate determinations are in progress. Apparatus for which no further immediate use is envisaged should be returned to the locker, but if it will be needed at a later stage, it may be placed at the back of the bench.
4. If a solution, precipitate, filtrate, etc., is set aside for subsequent treatment, the container must be labelled so that the contents can be readily identified, and the vessel must be suitably covered to prevent contamination of the contents by dust. For convenience, aluminium foil or flexible plastic film are most suitable, rather than traditional corks or rubber bungs. For temporary labelling, a felt-tip pen that will write directly on glass is preferable to wax pencils or adhesive labels. Wax pencil is difficult to wash off, and adhesive labels are mainly for long-term use.
5. Reagent bottles must never be allowed to accumulate on the bench; they must be replaced on the reagent shelves immediately after use.
6. It should be regarded as normal practice that all determinations are performed in duplicate.

7. A stiff covered notebook of A4 size must be provided for recording experimental observations as they are made.

Devote a double page to each determination, and give it a clear title and date. Use one page for the experimental observations, and the other for a brief description of the procedure followed, but with a full account of any special features associated with the determination. In most cases it will be convenient to divide the experimental page into two halves by a vertical line, then to halve the right-hand column with a second vertical line. The left-hand side of the page can then be used to indicate the observations, and the data for duplicate determinations can be recorded side by side in the two right-hand columns.

The record must conclude with the relevant calculations, so the equations for the principal chemical reactions should be shown together with a clear exposition of the procedure for calculating the result. Finally, appropriate comments should be made about the degree of accuracy and the level of precision.

Most modern laboratory instruments produce printed records of the analytical results in the form of spectra or chromatograms, and they are frequently interfaced or possess built-in computers that process the data and compare it automatically with standards and previous results stored in the computer memory. Any chart or printed data should be permanently attached within the laboratory notebook. These results should always be roughly checked to ensure they are consistent. Do not accept them blindly just because they have come from a computer.

Safety. Safety in the laboratory is essential at all times. You are responsible for the safety of any other person as well as your own. Many chemicals encountered in analysis are poisonous and must be carefully handled. The dangerous properties of concentrated acids and widely recognised poisons, such as potassium cyanide, are well known, but the dangers associated with halogenated solvents, benzene, mercury and many other chemicals are frequently overlooked. That is why it is essential to carry out a safety assessment of the chemicals and processes involved in any analysis before work is started for the first time.

Many operations involving chemical reactions are potentially dangerous, and recommended procedures must be carefully followed and obeyed. All laboratory workers should familiarise themselves with local safety requirements, which may include the compulsory wearing of lab coats and safety spectacles, and the position of first-aid equipment. Safety standards in laboratories have been greatly tightened in very many countries and established safety standards which are monitored by government officials frequently have to be observed. In the United Kingdom many of the operations of scientific laboratories are covered by the Health and Safety at Work Act 1974 coupled with the Control of Substances Hazardous to Health (COSHH) Regulations. Anyone working or training in a laboratory should become familiar with at least the basic controls and limitations specified under these statutes, especially the carrying out of COSHH assessments and following good working practices. In the United States similar standards on chemical health and safety are monitored by the US Occupational Safety and Health Administration, which works very closely with professional bodies and the chemical corporations.

All laboratory workers need to develop safety consciousness; study the relevant books devoted to laboratory and chemical hazards[1] and safe practices. Safety booklets may be obtained from many institutions and organisations dealing with these matters. The works in Section 3.39 contain further information.

Balances

3.2   The analytical balance

Most quantitative chemical processes depend at some stage upon the measurement of mass; it is by far the commonest procedure carried out by the analyst. Many chemical analyses are based upon the accurate determination of the mass of a sample, and the mass of a solid substance produced from it (gravimetric analysis), or upon ascertaining the volume of a carefully prepared standard solution (containing an accurately known mass of solute) which is required to react with the sample (titrimetric analysis). For the accurate measurement of mass in such operations, an analytical balance is employed; the operation is called weighing, and invariably reference is made to the weight of the object or material which is weighed.

The weight of an object is the force of attraction due to gravity that is exerted upon the object:

$$w = mg$$

where w is the weight of the object, m its mass and g is the acceleration due to gravity. Since the attraction due to gravity varies with altitude and with latitude, the weight of the object is variable, whereas its mass is constant. However, it has become the custom to employ the term 'weight' synonymously with mass, and it is in this sense that 'weight' is employed in quantitative analysis.

The analytical balance is one of the most important tools of the analytical chemist, and it has undergone radical changes. These changes have been prompted by the desire to produce an instrument which is more robust, less dependent upon the experience of the operator, less susceptible to the environment, and above all, one which will hasten the weighing operation. In meeting these requirements, the design of the balance has been fundamentally altered, and the traditional free-swinging, equal-arm, two-pan chemical balance together with its box of weights is now an uncommon sight.

An important development has been the replacement of the two-pan balance with its three knife-edges by a **two-knife single-pan balance**. In this instrument one balance pan and its suspension is replaced by a counterpoise, and dial-operated ring weights are suspended from a carrier attached to the remaining pan support (Figure 3.1). In this system all the weights are permanently in position on the carrier when the beam is at rest, and when an object to be weighed is placed upon the balance pan, weights must be removed from the carrier to compensate for the weight of the object. Weighing is completed by allowing the beam to assume its rest position, and then reading the displacement of the beam on an optical scale which is calibrated to read weights below 100 mg. Weighing is thus accomplished by **substitution**; many such manually operated balances are still in service in analytical laboratories.

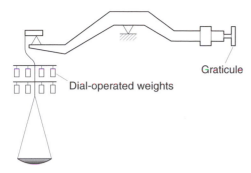

Graticule

Dial-operated weights

Figure 3.1   Two-knife single-pan balance

The standard modern instrument is the **electronic balance**, which provides convenience in weighing coupled with much greater freedom from mechanical failure, and greatly reduced sensitivity to vibration. It eliminates the operations of selecting and removing weights, smooth release of balance beam and pan support, noting the reading of weight dials and of an optical scale, returning the beam to rest, and replacing weights which have been removed. With an electronic balance, operation of a single on–off control permits the operator to read the weight of an object on the balance pan immediately from a digital display; most balances of this type can be coupled to a printer which gives a printed record of the weight. The majority of balances incorporate a **tare** facility which permits the weight of a container to be cancelled out, so that when material is added to the container, the weight recorded is simply the weight of material used. Many balances of this type incorporate a self-testing system which indicates the balance is functioning correctly each time it is switched on; and they also include a built-in weight calibration system. Operation of the calibration control leads to display of the weight of the standard incorporated within the balance, and thus indicates whether any correction is necessary. A more satisfactory calibration procedure is to check the balance readings against a series of calibrated analytical weights.

Electronic balances operate by applying an electromagnetic restoring force to the support for the balance pan; when an object is added to the balance pan, the resultant displacement of the support is cancelled out. The magnitude of the restoring force is governed by the value of the current flowing in the coils of the electromagnetic compensation system, and this is proportional to the weight placed on the balance pan; a microprocessor converts the value of the current into the digital display in grams.

The balance must be protected from draughts and from dust, and the balance pan is situated within an enclosure provided with glass doors which can be opened to provide access to the pan. The rest of the balance, including the electrical components, is contained in a closed compartment attached to the rear of the pan compartment.

Electronic balances are available to cover four weight ranges:

1.   Up to about 200 g and reading to 0.1 mg (macrobalance)
2.   Up to about 30 g and reading to 0.01 mg (semimicrobalance)
3.   Up to about 20 g and reading to 1 μg (microbalance)
4.   Up to 5 g and reading to 0.1 μg (ultramicrobalance)

Thus a wide variety of analytical balances is available; a typical electronic balance is shown in Figure 3.2.

Figure 3.2   Electronic balance (Courtesy Cherwell Laboratories, Bicester, Oxon)

3.3   Other balances

For many laboratory operations it is necessary to weigh objects or materials which are far heavier than the upper weight limit of a macroanalytical balance, or small amounts of material for which it is not necessary to weigh to this limit of sensitivity; this type of weighing is often called rough weighing. A wide range of electronic balances are available for rough weighing; here are some typical characteristics.

Maximum capacity (g)	Reading to (g)
500	0.01
5 000	0.1
16 000	1.0

With these top-pan balances it is not necessary to shield the balance pan from gentle draughts, and weighings can be accomplished very rapidly and with the usual facility of the results being recorded with a printer.

3.4   Weights and reference masses

With a modern balance it is not necessary to use a box of weights in the weighing process, but a set of weights is still desirable for checking its accuracy.

For scientific work, the fundamental standard of mass is the international prototype kilogram, a mass of platinum–iridium alloy made in 1887 and deposited at the International Bureau of Weights and Measures near Paris. Authentic copies of the standard are kept by the appropriate responsible authorities* in the various countries of the world; these copies are employed to compare secondary standards used in the calibration of weights for scientific work. The unit of mass that is almost universally employed in laboratory work is the **gram**, which may be defined as the one-thousandth part of the mass of the international prototype kilogram.

An ordinary set of analytical weights contains the following items:

Grams	100, 50, 30, 20, 10
Grams	5, 3, 2, 1
Milligrams	500, 300, 200, 100
Milligrams	50, 30, 20, 10

Notice the 5, 3, 2, 1 sequence. The weights from 1 g upwards are constructed from a non-magnetic nickel–chromium alloy (80% Ni, 20% Cr), or from austenitic stainless steel; plated brass is sometimes used but is less satisfactory. The fractional weights are made from the same alloys, or from a non-tarnishable metal such as gold or platinum. A pair of forceps are provided for handling the weights, and the weights themselves are stored in a box with suitably shaped compartments.

Analytical weights can be purchased which have been manufactured to class A standard; this is the only grade of laboratory weights officially recognised in the United Kingdom. The following tolerances are permitted in class A weights: 100 g (0.5 mg), 50 g (0.25 mg), 30 g (0.15 mg), 20 g (0.10 mg), 10 g to 100 mg (0.05 mg), 50 mg to 10 mg (0.02 mg). The National Bureau of Standards in Washington recognises the following classes of precision weights:

Class M For use as reference standards, for work of the highest precision, and where a high degree of constancy over a period of time is required.

Class S For use as working reference standards or as high-precision analytical weights.

Class S-1 Precision analytical weights for routine analytical work.

Class J Microweight standards for microbalances.

3.5 Care and use of analytical balances

No matter what type of analytical balance is employed, due attention must be paid to the manner in which it is used. The following remarks apply particularly to electronic balances:

1. Never exceed the stated maximum load of the balance.
2. Keep the balance clean. Remove dust from the pan and from the floor of the pan compartment with a camel-hair brush.
3. Objects to be weighed should never be handled with the fingers; always use tongs or a loop of clean paper.
4. Objects to be weighed should be allowed to attain the temperature of the balance before weighing, and if the object has been heated, sufficient time must be allowed for cooling. The time required to attain the temperature of the balance varies with the size, etc., of the object, but as a rule 30–40 min is sufficient.

---

* The National Physical Laboratory (NPL) in Great Britain, the National Bureau of Standards (NBS) in USA, etc.

5.   No chemicals or objects which might injure the balance pan should ever be placed directly on it. Substances must be weighed in suitable containers, such as small beakers, weighing bottles or crucibles, or on watch glasses. Liquids and volatile or hygroscopic solids must be weighed in tightly closed vessels, such as stoppered weighing bottles.

6.   Addition of chemicals to the receptacle must be done outside the balance case. It is good practice to weigh the chosen receptacle on the analytical balance, to transfer it to a rough balance, to add approximately the required amount of the necessary chemical, and then to return the receptacle to the analytical balance for reweighing, thus giving the exact weight of substance taken.

7.   Nothing must be left on the pan when the weighing has been completed. If any substance is spilled accidentally upon the pan or upon the floor of the balance compartment, it must be removed at once.

8.   Avoid exposing the balance to corrosive atmospheres.

The actual weighing process will include the following steps.

1.   Brush the balance pan lightly with a camel-hair brush to remove any dust.

2.   The object to be weighed must be at or close to room temperature. With the balance at rest, place the object on the pan and close the pan compartment case.

3.   Set the on–off control of the balance to the 'on' position, observe the value shown on the digital display and record it in the notebook. If the balance is linked to a printer, confirm that the printed result agrees with the digital display. Return the control to the 'off' position.

4.   When all weighings have been completed, remove the object which has been weighed, clear up any accidental spillages, and close the pan compartment.

These remarks apply particularly to analytical balances of the macrobalance range; microbalances and ultramicrobalances must be handled with special care, particularly with regard to the temperature of objects to be weighed.

3.6   Errors in weighing

Be aware of the potential sources of error that can occur in weighing, other than those arising from a defective balance. There are three sources of error that need to be considered in detail:

1.   Changes relating to the weighing vessel or the substance occurring between successive weighings.

2.   Buoyancy of the air and its effect upon the object, the container and any weights.

3.   Human errors in recording results.

Changes in the weighing vessel

Changes in the weight of the weighing vessel may arise from absorption or loss of moisture, electrification of the surface caused by rubbing, a difference in temperature between the weighing vessel and the balance case. These errors may be largely eliminated by wiping the vessel gently with a linen cloth, and allowing it to stand at least 30 min in proximity to the balance before weighing. The electrification – which may cause a comparatively large error, particularly if both the atmosphere and the cloth are dry – is slowly dissipated on standing; it may be removed by subjecting the vessel to the discharge from an antistatic gun. Hygroscopic, efflorescent and volatile substances must be weighed in completely closed vessels. Substances which have been heated in an air oven or ignited in a crucible are generally allowed to cool in a desiccator containing a suitable drying agent. The time of cooling in a desiccator cannot be exactly specified, since it will depend upon the temperature and the size of the crucible as well as the material it is made from. Platinum vessels require a shorter time than porcelain, glass or silica vessels. Before weighing, it has been customary to leave platinum crucibles in the desiccator for 20–25 min, and crucibles of other materials for 30–35 min. It is advisable to cover crucibles and other open vessels.

Buoyancy effects

Buoyancy occurs when an object is immersed in a fluid and its true weight is diminished by the weight of the fluid it displaces. This is the principle underpinning anything that floats. If the object and the weights have the same density, hence the same volume, no error will be introduced on this account. But if, as is usually the case, the density of the object is different from the density of the weights, the volumes of air displaced by each will be different. If the substance has a lower density than the weights, as is usual in analysis, the substance will displace a greater volume of air than the weights, and it will therefore weigh less in air than in a vacuum. Conversely, if a denser material (e.g. one of the precious metals) is weighed, the weight in a vacuum will be less than the apparent weight in air. Two examples will serve to illustrate these points.

Example 3.1

Consider the weighing of 1 L of water, first in vacuo, and then in air. It is assumed the flask containing the water is tared by an exactly similar flask, that the temperature of the air is 20 °C and the barometric pressure is 101 325 Pa (760 mm of mercury). The weight of 1 L of water in vacuo under these conditions is 998.23 g, but if weighed in air it appears to weigh less than this. The difference can be readily calculated. The weight of 1 L of air displaced by the water is 1.20 g. Assuming the weights to have a relative density of 8.0, they will displace 998.23/8.0 = 124.8 mL, or 124.8 × 1.20/1000 = 0.15 g of air. The net difference in weight will therefore be 1.20 − 0.15 = 1.05 g. Hence the weight in air of 1 L of water under the experimental conditions named is 998.23 − 1.05 = 997.18 g, a difference of 0.1% from the weight in vacuo.

Example 3.2

Consider the case of a solid, such as potassium chloride, under the conditions in Example 3.1. The relative density of potassium chloride is 1.99. If 2 g of the salt are weighed, the apparent loss in weight (= weight of air displaced) is 2 × 0.0012/1.99 = 0.0012 g. The apparent loss in weight for the weights is 2 × 0.0012/8.0 = 0.000 30 g. Hence 2 g of potassium chloride will weigh 0.0012 − 0.000 30 = 0.000 90 g less in air than in vacuo, a difference of 0.05%.

For most analytical purposes where it is desired to express the results in the form of a percentage, the ratio of the weights in air, so far as solids are concerned, will give a result that is practically the same as would be given by the weights in vacuo. Hence no buoyancy correction is necessary in these cases. However, where absolute weights are required, as in the calibration of graduated glassware, corrections for the buoyancy of the air must be made (Section 3.16). Although an electronic balance does not employ any weights, these remarks apply to weights recorded by the balance because the balance scale will have been established by reference to metal (stainless steel) weights used in air.

Now consider the general case. The weight of an object in vacuo is equal to the weight in air **plus** the weight of air displaced by the object **minus** the weight of air displaced by the weights. It can easily be shown that

$$W_v = W_a + d_a\left(\frac{W_v}{d_b} - \frac{W_a}{d_w}\right)$$

where
W_v = weight in vacuo
W_a = apparent weight in air
d_a = density of air
d_w = density of the weights
d_b = density of the body

The density of the air will depend on the humidity, the temperature and the pressure. For an average relative humidity (50%) and average conditions of temperature and pressure in a laboratory, the density of the air will rarely fall outside the limits 0.0011 and 0.0013 g mL^{-1}. It is therefore permissible for analytical purposes to take the weight of 1 mL of air as 0.0012 g.

Since the difference between W_v and W_a does not usually exceed 1–2 parts per thousand, we may write

$$W_v = W_a + d_a\left(\frac{W_a}{d_b} - \frac{W_a}{d_w}\right)$$

$$= W_a + W_a\left\{0.0012\left(\frac{1}{d_b} - \frac{1}{8.0}\right)\right\} = W_a + kW_a/1000$$

where

$$k = 1.20\left(\frac{1}{d_b} - \frac{1}{8.0}\right)$$

If a substance of density d_b weighs W_a grams in air, then $W_a k$ milligrams should be added to the weight in air to obtain the weight in vacuo. The correction is positive if the substance has a density lower than 8.0, and negative if the density of the substance is greater than 8.0.

Human errors

Many weighing errors are due to human mistakes made in checking the weights on the balance pan, or in reading the digital display on electronic balances. The correct reading of weights is best carried out by checking the weights as they are added to and removed from the balance. Any digital display should be read at least twice, paying very careful attention to the position of the decimal point.

Graduated glassware

3.7 Units of volume

For scientific purposes the convenient unit to employ for measuring reasonably large volumes of liquids is the cubic decimetre (dm^3) or for smaller volumes the cubic centimetre (cm^3). For many years the fundamental unit employed was the litre, based upon the volume occupied by one kilogram of water at 4 °C (the temperature of maximum density for water). The relationship between this litre and the cubic decimetre was established as

$$1 \text{ litre} = 1.000\,028\,dm^3 \quad \text{or} \quad 1 \text{ millilitre} = 1.000\,028\,cm^3$$

In 1964 the *Conférence Générale des Poids et des Mésures* (CGPM) decided to accept the term 'litre' as a special name for the cubic decimetre, and to discard the original definition. With this new definition of the litre (L), the millilitre (mL) and the cubic centimetre (cm^3) are identical.

3.8 Graduated apparatus

The most commonly used pieces of apparatus in titrimetric (volumetric) analysis are graduated flasks, burettes and pipettes. Graduated cylinders and weight pipettes are less widely employed. Each of them will be described in turn.

Graduated apparatus for quantitative analysis is generally made to specification limits, particularly with regard to the accuracy of calibration. In the United Kingdom the British Standards Institution (BSI) designates two grades of apparatus, class A and class B. The tolerance limits are closer for class A apparatus; this apparatus is intended for work of the highest accuracy. Class B apparatus is employed in routine work. In the United States there are specifications for only one grade of apparatus, equivalent to BSI class A; they are available from the National Bureau of Standards in Washington.

Most graduated glassware is manufactured in high-quality heat-resistant glass and will last for many years if carefully treated. Plastic graduated equipment is also available, but this is not usually suitable for accurate work as surfaces deteriorate rapidly and are often difficult to clean adequately.

Cleaning of glass apparatus

All such glassware must be perfectly clean and free from grease, otherwise the results will be unreliable. One test for cleanliness of glass apparatus is that on being filled with distilled water and the water withdrawn, only an unbroken film of water remains. If the water collects in drops, the vessel is dirty and must be cleaned. Various methods are available for cleaning glassware. Many commercially available detergents are suitable for this purpose, and some manufacturers market special formulations for cleaning laboratory glassware; some are claimed to be especially effective in removing contamination due to radioactive materials.

Teepol is a relatively mild and inexpensive detergent which may be used for cleaning glassware. The laboratory stock solution may consist of a 10% solution in distilled water. For cleaning a burette, 2 mL of the stock solution diluted with 50 mL of distilled water are poured into the burette and allowed to stand for 30–60 s; the detergent is then run off and the burette is rinsed three times with tap water followed by several times with distilled water. A 25 mL pipette may be similarly cleaned using 1 mL of the stock solution deionised with 25–30 mL of deionised water.

A method which is frequently used consists in filling the apparatus **carefully** with chromic acid cleaning mixture – a nearly saturated solution of powdered sodium dichromate or potassium dichromate in concentrated sulphuric acid – and allowing it to stand for several hours, preferably overnight. The acid is poured off then the apparatus is thoroughly rinsed with deionised water and allowed to drain until dry. Potassium dichromate is not very soluble in concentrated sulphuric acid (about $5 \, g \, L^{-1}$), whereas sodium dichromate ($Na_2Cr_2O_7 \cdot 2H_2O$) is much more soluble (about $70 \, g \, L^{-1}$); that is why sodium dichromate is usually preferred for the cleaning mixture, besides the fact it is much cheaper. From time to time it is advisable to filter the sodium dichromate–sulphuric acid mixture through glass wool placed in the apex of a glass funnel; this removes small particles or sludge that are often present and which may block the tips of burettes.

A very effective degreasing agent, claimed to be much quicker-acting than the cleaning mixture, is obtained by dissolving 100 g of potassium hydroxide in 50 mL of water, and after cooling, making up to 1 L with industrial methylated spirit.[2a] **Handle this with great care**.

3.9 Temperature standard

The capacity of a glass vessel varies with the temperature, so a temperature is defined at which the vessel's capacity is intended to be correct; in the UK it is 20 °C. A subsidiary standard temperature of 27 °C is accepted by the British Standards Institution, for use in tropical climates where the ambient temperature is consistently above 20 °C. The US Bureau of Standards, in compliance with the view held by some chemists that 25 °C more nearly approximates to the average laboratory temperature in the United States, will calibrate glass graduated apparatus marked either 20 °C or 25 °C.

Taking the coefficient of cubical expansion of soda glass as about $0.000\,030 \, °C^{-1}$ and of borosilicate glass as about $0.000\,010 \, °C^{-1}$, Table 3.1 gives the correction factors to obtain the capacity of a standard 1000 mL flask at temperatures other than 20 °C. Find the temperature in the temperature column, look along this row to the appropriate glass expansion column and add the value there to 1000 mL.

And when applying temperature corrections, make sure to consider the expansion of any liquids being measured. Table 3.1 gives the correction factors to obtain the volume occupied at 20 °C by a volume of water contained in a standard 1000 mL flask at some other

Table 3.1 *Temperature corrections (mL) for a 1 L graduated flask*

Temperature (°C)	Expansion of glass		Expansion of water	
	Soda glass	Borosilicate glass	Soda glass	Borosilicate glass
5	−0.39	−0.15	+1.37	+1.61
10	−0.26	−0.10	+1.24	+1.40
15	−0.13	−0.05	+0.77	+0.84
20	0.00	0.00	0.00	0.00
25	+0.13	+0.05	−1.03	−1.11
30	+0.26	+0.10	−2.31	−2.46

temperature. They are used in the same way as the correction factors for glass expansion. And notice, that they are considerably greater. For dilute (e.g. 0.1 M) aqueous solution the corrections can be regarded as approximately the same as for water, but with more concentrated solutions the correction increases, and for non-aqueous solutions the corrections can be quite large.[2b]

3.10 Graduated flasks

A graduated flask, also known as a volumetric flask, is a flat-bottomed, pear-shaped vessel with a long narrow neck. A thin line etched around the neck indicates the volume that it holds at a certain definite temperature, usually 20 °C (both the capacity and temperature are clearly marked on the flask); the flask is then said to be graduated **to contain**. Flasks with one mark are always taken **to contain** the volume specified. A flask may also be marked **to deliver** a specified volume of liquid under certain definite conditions, but these flasks are not suitable for exact work and are not widely used. Vessels intended to contain definite volumes of liquid are marked C or TC or In, whereas those intended to deliver definite volumes are marked D or TD or Ex.

The mark extends completely around the neck, in order to avoid errors due to parallax when making the final adjustment; the lower edge of the meniscus of the liquid should be tangential to the graduation mark, and both the front and the back of the mark should be seen as a single line. The neck is made narrow so that a small change in volume will have a large effect upon the height of the meniscus; the error in adjustment of the meniscus is accordingly small.

The flasks should be fabricated in accordance with BS 5898 (1980)* and the opening should be ground to standard (interchangeable) specifications and fitted with an interchangeable glass or plastic (commonly polypropylene) stopper. They should conform to either class A or class B specification BS 1792 (1982); examples of permitted tolerances for class B are as follows:

Flask size (mL)	5	25	100	250	1000
Tolerance (mL)	0.04	0.06	0.15	0.30	0.80

For class A flasks the tolerances are approximately halved; these flasks may be purchased with a works calibration certificate, or with a British Standard Test (BST) certificate.

Graduated flasks are available in the following capacities: 1, 2, 5, 10, 20, 50, 100, 200, 250, 500, 1000, 2000 and 5000 mL. They are employed in making up standard solutions to a given volume; they can also be used for obtaining, with the aid of pipettes, aliquot portions of a solution of the substance to be analysed.

3.11 Pipettes

There are three kinds of pipette:

Transfer pipettes have one mark and deliver a constant volume of liquid under certain specified conditions.

---

* Many modern British Standards are closely lined to the specifications laid down by the International Standardisation Organisation based in Geneva; in the above example the relevant reference is to ISO 384-1978.

Graduated or measuring pipettes have graduated stems which are employed to deliver various small volumes as required.

Syringe pipettes have fixed or variable volume and are usually employed for dispensing large numbers of identical volumes very quickly.

Transfer pipettes

Transfer pipettes consist of a long glass tube with large central cylindrical bulb; a calibration mark is etched around the upper (suction) tube, and the lower (delivery) tube is drawn out to a fine tip. The graduated or measuring pipette is usually intended for the delivery of predetermined variable volumes of liquid; it does not find wide use in accurate work, for which a burette is generally preferred. Transfer pipettes are constructed with capacities of 1, 2, 5, 10, 20, 25, 50 and 100 mL; those of 10, 25 and 50 mL capacity are most frequently employed in macro work. They should conform to BS 1583 (1986); ISO 684-1984 and should carry a colour code ring at the suction end to identify the capacity, BS 5898 (1980). As a safety measure, an additional bulb is often incorporated above the graduation mark. They may be fabricated from soda-lime or Pyrex glass, and some high-grade pipettes are manufactured in Corex glass (Corning Glass Works, USA). This is glass which has been subjected to an ion exchange process that strengthens the glass and also leads to a greater surface hardness, thus giving a product which is resistant to scratching and chipping. Pipettes are available to class A and class B specifications. Here are some typical class B tolerances. For class A they are approximately halved.

Pipette capacity (mL)	5	10	25	50	100
Tolerance (mL)	0.01	0.04	0.06	0.08	0.12

The filling of pipettes should never be carried out by mouth suction, and the pipette should never be placed to the lips, irrespective of which liquids are being measured.

To use transfer pipettes, a suitable **pipette filler** is first attached to the upper or suction tube. These devices are obtainable in various forms; a simple version consists of a rubber or plastic bulb fitted with glass ball valves which can be operated between finger and thumb. The valves control the entry and expulsion of air from the bulb and thus the flow of liquid into and out of the pipette.

Before measuring out the volume of liquid required, rinse the pipette with a small amount of the liquid then discard it. The pipette should then be filled with the liquid to about 1–2 cm above the graduation mark. Any adhering liquid is removed from the outside of the lower stem by wiping with a piece of filter paper, and then by careful manipulation of the filler, the liquid is allowed to run out slowly until the bottom of the meniscus just reaches the graduation mark; the pipette must be held vertically and with the graduation mark at eye level. Any drops adhering to the tip are removed by stroking against a glass surface. The liquid is then allowed to run into the receiving vessel, the tip of the pipette touching the wall of the vessel. When the continuous discharge has ceased, the jet is held in contact with the side of the vessel for a **draining time** of 15 s. At the end of the draining time, the tip of the pipette is removed from contact with the wall of the receptacle; the liquid remaining in the jet of the pipette must not be removed either by blowing or by other means.

A pipette will not deliver constant volumes of liquid if discharged too rapidly. The orifice size must produce an outflow time of about 20 s for a 10 mL pipette, 30 s for a 25 mL pipette and 35 s for a 50 mL pipette.

Graduated pipettes

Graduated pipettes consist of straight, fairly narrow tubes with no central bulb, and are also constructed to a standard specification, BS 6696 (1986); they are likewise colour-coded in accordance with ISO 1769. Three different types are available:

Type 1 delivers a measured volume from a top zero to a selected graduation mark.
Type 2 delivers a measured volume from a selected graduation mark to the jet, i.e. the zero is at the jet.
Type 3 is calibrated to **contain** a given capacity from the jet to a selected graduation mark, and thus to **remove** a selected volume of solution.

For type 2 pipettes the final drop of liquid remaining in the tip must be expelled, which is contrary to the usual procedure. These pipettes are therefore distinguished by a white or sandblasted ring near the top of the pipette.

Syringe pipettes or micropipettes

Syringe pipettes or micropipettes are now very common in laboratories and are used particularly for dispensing toxic solutions and large numbers of repeat volumes for multiple analyses. They may be of fixed or variable volumes. They have a push-button design in which the syringe is operated by pressing a button on the top of the pipette; the plunger travels between two fixed stops and a reliable constant volume of liquid is delivered. Syringe pipettes are fitted with disposable plastic tips (usually of polythene or polypropylene) which are not wetted by aqueous solutions, thus helping to ensure constancy of the volume of liquid delivered. The liquid is contained entirely within the plastic tip and so, by replacing the tip, the same pipette can be employed for different solutions. They are available for delivering volumes of 1 μL to 10 mL, and the delivery is reproducible to within about 1%.

The dispensing of volumes smaller than 1 μL is usually carried out using special needle syringes of the type employed for gas chromatography (Chapter 9). Micrometer pipettes are also available for the dropwise dispensing of solutions. They are fitted with a micrometer control that operates the plunger of the syringe, which has a stainless steel needle tip. The small volume of liquid delivered at any time is measured accurately by the micrometer scale.

3.12   Burettes

Burettes are long graduated cylindrical tubes of uniform bore terminating at the lower end in a glass or polytetrafluoroethylene (PTFE) stopcock and a jet. The PTFE taps have the great advantage that they do not require lubrication.

It is sometimes advantageous to employ a burette with an extended jet which is bent twice at right angles so that the tip of the jet is displaced by some 7.5–10 cm from the body of the burette. Insertion of the tip of the burette into complicated assemblies of apparatus is thus facilitated; and if heated solutions have to be titrated, the body of the burette is kept away from the source of heat. Burettes fitted with two-way stopcocks are useful for attachment to reservoirs of stock solutions.

As with other graduated glassware, burettes are produced to both class A and class B specifications in accordance with BS 846 (1985) or ISO 385 (1984), and class A burettes may be purchased with BST certificates. All class A and some class B burettes have graduation marks which completely encircle the burette; this is a very important for avoiding parallax errors when taking readings. Here are some typical values for class A tolerances. For class B they are approximately doubled.

Total capacity (mL)	5	10	50	100
Tolerance (mL)	0.02	0.02	0.05	0.10

In addition to the volume requirements, limits are also imposed on the length of the graduated part of the burette and on the drainage time.

When in use, a burette must be firmly supported on a stand, and various types of burette holder are available for this purpose. The use of an ordinary laboratory clamp is not recommended; the ideal type of holder permits the burette to be read without removing it from the stand.

Lubricants for glass stopcocks

The object of lubricating the stopcock of a burette is to prevent sticking or 'freezing' and to ensure smoothness in action. The simplest lubricant is pure vaseline, but this is rather soft, and unless used sparingly, portions of the grease may readily become trapped at the point where the jet is joined to the barrel of the stopcock, eventually blocking the jet. Commercial lubricants for stopcocks are available from laboratory suppliers. Silicone-based lubricants should not be used as they tend to creep and cause contamination of the inside of the burette.

To lubricate the stopcock, the plug is removed from the barrel and two thin streaks of lubricant are applied to the length of the plug on lines roughly midway between the ends of the bore of the plug. Upon replacing in the barrel and turning the tap a few times, a uniform thin film of grease is distributed round the ground joint. A spring or some other form of retainer may then be attached to the key to lessen the chance of it becoming dislodged when in use.

Using a burette

The burette is thoroughly cleaned using one of the cleaning agents described in Section 3.8 and is then well rinsed with distilled water. The plug of the stopcock is removed from the barrel, and after wiping the plug and the inside of the barrel dry, the stopcock is lubricated as described in the preceding paragraph. Using a small funnel, about 10 mL of the solution to be used are introduced into the burette, and then after removing the funnel, the burette is tilted and rotated so that the solution flows over the whole of the internal surface; the liquid is then discharged through the stopcock. After repeating the rinsing process, the burette is clamped **vertically** in the burette holder and then filled with the solution to a little above the zero mark. The funnel is removed, and the liquid discharged through the stopcock until the lowest point of the liquid meniscus is just below the zero mark; the jet is inspected to ensure that all air bubbles have been removed and that it is completely full of liquid.

To read the position of the meniscus, the eye must be at the same level as the meniscus, in order to avoid errors due to parallax. In the best type of burette, the graduations are carried completely round the tube for each millilitre (mL) and halfway round for the other graduation marks; parallax is thus easily avoided. To assist in reading the position of the meniscus, use a piece of white paper or cardboard, its lower half blackened by painting with dull black paint or by pasting on a piece of dull black paper. When this is placed so that the sharp dividing line is 1–2 mm below the meniscus, the bottom of the meniscus appears to be darkened and is sharply outlined against the white background; the level of the liquid can then be accurately read. A variety of burette readers are available from laboratory supply houses, and a home-made device which is claimed to be particularly effective has been described by Woodward and Redman.[2c] For ordinary purposes, readings are made to 0.05 mL; for precision work, readings should be made to 0.01–0.02 mL using a lens to assist in estimating the subdivisions.

To deliver liquid from a burette into a conical flask or other similar receptacle, place the fingers of the left hand behind the burette and the thumb in front, and hold the tap on the right-hand side between the thumb and the fore and middle fingers. In this way there is no tendency to pull the plug out of the barrel of the stopcock, and the operation is under complete control. Any drop adhering to the jet after the liquid has been discharged is removed by bringing the side of the receiving vessel into contact with the jet. During the delivery of the liquid, the flask may be gently rotated with the right hand to ensure the added liquid is well mixed with any existing contents of the flask.

3.13   Weight burettes

Weight burettes are used for work demanding the highest possible accuracy in transferring various quantities of liquids. As the name implies, they are weighed before and after a transfer of liquid. A very useful form is shown in Figure 3.3(a). There are two ground-glass caps; the lower cap is closed and the upper cap has a capillary opening; this means there is negligible loss by evaporation. For hygroscopic liquids, a small ground-glass cap is fitted to the top of the capillary tube. The burette is roughly graduated in 5 mL intervals. The titre thus obtained is in terms of weight loss of the burette, and for this reason the titrants are prepared on a weight/weight basis rather than a weight/volume basis. The errors associated with a volumetric burette, such as drainage, reading and change in temperature, are avoided, and weight burettes are especially useful when dealing with non-aqueous solutions or with viscous liquids.

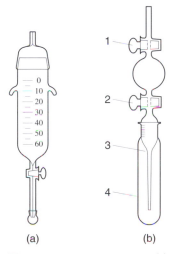

Figure 3.3   Weight burettes: (a) a very useful form, (b) the Lunge–Rey pipette

An alternative form of weight burette due to Redman[2d] consists of a glass bulb, flattened on one side so it will stand on a balance pan. Above the flattened side is the stopcock-controlled discharge jet, and a filling orifice which is closed with a glass stopper. The stopper and short neck into which it fits are pierced with holes; air can be admitted by aligning the holes, allowing the contents of the burette to be discharged through the delivery jet.

The **Lunge–Rey pipette** is shown in Figure 3.3(b). There is a small central bulb (5–10 mL capacity) closed by two stopcocks (1, 2); the pipette (3) below the stopcock has a capacity of about 2 mL, and is fitted with a ground-on test-tube (4). This pipette is especially useful for weighing out corrosive and fuming liquids.

3.14   Piston burettes

In piston burettes the delivery of the liquid is controlled by movement of a tightly fitting plunger within a graduated tube of uniform bore. They are particularly useful when the piston is coupled to a motor drive, forming the basis of automatic titrators. Piston burettes can provide automatic plotting of titration curves; they also allow a variable rate of delivery as the end point is approached, so there is no danger of overshooting.

3.15   Graduated (measuring) cylinders

Cylinders are graduated in capacities from 2 to 2000 mL. Since the surface area of the liquid is much greater than in a graduated flask, the accuracy is not very high. This means graduated cylinders cannot be used for work demanding even a moderate degree of accuracy, but they are okay for rough measurements.

3.16   Calibration of graduated apparatus

For most analytical purposes, graduated apparatus manufactured to class A standard will prove satisfactory, but for work of the highest accuracy it is advisable to calibrate all apparatus for which a **recent** test certificate is unavailable. The calibration procedure involves determining the weight of water contained in or delivered by the particular piece of apparatus. The temperature of the water is observed, and from the known density of water at that temperature, the volume of water can be calculated. Tables giving density values are usually based on weights in vacuo (Section 3.6), but the data given in Table 3.2 is based on weighings in air with stainless steel weights, and these values can be used to calculate the

Table 3.2   *Volume of 1 g of water at various temperatures*

Temp. (°C)	Volume (mL)	Temp. (°C)	Volume (mL)
10.00	1.0013	22.00	1.0033
12.00	1.0015	24.00	1.0037
14.00	1.0017	26.00	1.0044
16.00	1.0021	28.00	1.0047
18.00	1.0023	30.00	1.0053
20.00	1.0027		

relevant volume directly from the observed weight of water. Plot a graph to find the volume of 1 g of water at the calibration temperature. Fuller tables are given in BS 6696 (1986).[3]

In all calibration operations the apparatus to be calibrated must be carefully cleaned and allowed to stand adjacent to the balance which is to be employed, together with a supply of distilled or deionised water, so they assume the temperature of the room. Flasks will also need to be dried, and this can be accomplished by rinsing twice with a little acetone then blowing a current of air through the flask to remove the acetone.

Graduated flask

After allowing it to stand in the balance room for an hour, the clean dry flask is stoppered and weighed. A small filter funnel, the stem of which has been drawn out so it reaches below the graduation mark of the flask, is then inserted into the neck and deionised (distilled) water, which has also been standing in the balance room for an hour, is added slowly until the mark is reached. The funnel is then carefully removed, taking care not to wet the neck of the flask above the mark, and then, using a dropping tube, water is added dropwise until the meniscus stands on the graduation mark. The stopper is replaced, the flask reweighed, and the temperature of the water noted. The true volume of the water filling the flask to the graduation mark can be calculated with the aid of Table 3.2.

Pipette

The pipette is filled with the distilled water which has been standing in the balance room for at least an hour, to a short distance above the mark. Water is run out until the meniscus is exactly on the mark, and the outflow is then stopped. The drop adhering to the jet is removed by bringing the surface of some water contained in a beaker in contact with the jet, and then removing it without jerking. The pipette is then allowed to discharge into a clean, weighed stoppered flask (or a large weighing bottle) and held so the jet of the pipette is in contact with the side of the vessel (it will be necessary to incline slightly either the pipette or the vessel). The pipette is allowed to drain for 15 s after the outflow has ceased, the jet still being in contact with the side of the vessel. At the end of the draining time, the receiving vessel is removed from contact with the tip of the pipette, thus removing any drop adhering to the outside of the pipette and ensuring the drop remaining in the end is always of the same size. To determine the instant at which the outflow ceases, the motion of the water surface down the delivery tube of the pipette is observed, and the delivery time is considered to be complete when the meniscus comes to rest slightly above the end of the delivery tube. The draining time of 15 s is counted from this moment. The receiving vessel is weighed, and the temperature of the water noted. The capacity of the pipette is then calculated with the aid of Table 3.2. At least two determinations should be made.

Burette

If it is necessary to calibrate a burette, it is essential to establish that it is satisfactory with regard to leakage and delivery time, before undertaking the actual calibration process. To test for leakage, the plug is removed from the barrel of the stopcock and both parts of the stopcock are carefully cleaned of all grease; after wetting well with deionised water, the stopcock is reassembled. The burette is placed in the holder, filled with deionised water, adjusted to the zero mark, and any drop of water adhering to the jet removed with a piece of filter paper. The burette is then allowed to stand for 20 min, and if the meniscus has not fallen by more than one scale division, the burette may be regarded as satisfactory as far as leakage is concerned.

To test the delivery time, again separate the components of the stopcock, dry, grease and reassemble, then fill the burette to the zero mark with deionised water, and place in the holder. Adjust the position of the burette so the jet comes inside the neck of a conical flask standing on the base of the burette stand, but does not touch the side of the flask. Open the stopcock fully, and note the time taken for the meniscus to reach the lowest graduation mark of the burette: this should agree closely with the time marked on the burette, and it must fall within the limits laid down by BS 846 (1985).

If the burette passes these two tests, the calibration may begin. Fill the burette with the deionised water which has been allowed to stand in the balance room to acquire room temperature: ideally, this should be as near to 20 °C as possible. Weigh a clean, dry stoppered flask of about 100 mL capacity, then, after adjusting the burette to the zero mark and removing any drop adhering to the jet, place the flask in position under the jet, open the stopcock fully and allow water to flow into the flask. As the meniscus approaches the desired calibration point on the burette, reduce the rate of flow until eventually it is discharging dropwise, and adjust the meniscus exactly to the required mark. Do not wait for drainage, but remove any drop adhering to the jet by touching the neck of the flask against the jet, then restopper and reweigh the flask. Repeat this procedure for each graduation to be tested; for a 50 mL burette, this will usually be every 5 mL. Note the temperature of the water, and then, using Table 3.2, the volume delivered at each point is calculated from the weight of water collected. The results are most conveniently used by plotting a calibration curve for the burette.

Water for laboratory use

3.17   Purified water

From the earliest days of quantitative chemical measurements it has been recognised that some form of purification is required for water used in analytical operations, and with increasingly lower limits of detection being attained in instrumental methods of analysis, correspondingly higher standards of purity are imposed upon the water used for preparing solutions. Standards have now been laid down for water to be used in laboratories;[4] they prescribe limits for non-volatile residue, for residue remaining after ignition, for pH and for conductance. British Standard 3978 (1987) (ISO 3696-1987) recognises three different grades of water, as summarised in Table 3.3:

Grade 3 is suitable for ordinary analytical purposes and may be prepared by single distillation of tap water, by deionisation, or by reverse osmosis (see below).
Grade 2 is suitable for more sensitive analytical procedures, such as atomic absorption spectroscopy and the determination of substances present in trace quantities. Water of this quality can be prepared by redistillation of grade 3 distilled water, by distillation of deionised water, or by distillation of reverse osmosis water.
Grade 1 is suitable for the most stringent requirements, including high-performance liquid chromatography and the determination of substances present in ultratrace amounts. It is obtained by subjecting grade 2 water to reverse osmosis or deionisation, followed by filtration through a membrane filter of pore size 0.2 μm to remove particulate matter. Alternatively, grade 2 water may be redistilled in an apparatus constructed from fused silica.

Table 3.3 *Standards for water to be used in analytical operations*

Parameter	Grade of water		
	1	2	3
pH at 25°C	a	a	5.0–7.5
Electrical conductance at 25°C (mS m^{-1})	0.01	0.1	0.5
Oxidisable matter ≡ amount of oxygen (mg L^{-1})	b	0.08	0.4
Absorbance at 254 nm, 1 cm cell	0.001	0.01	c
Residue after evaporation (mg kg^{-1})	c	1	2
SiO$_2$ content (mg L^{-1})	0.01	0.02	c

[a] pH measurements in highly purified water are difficult; results are of doubtful significance.
[b] Not applicable.
[c] Not specified.

For many years the sole method of water purification available was by distillation, and distilled water of varying quality was universally employed for laboratory purposes. A modern water-still is usually made of glass and is heated electrically. The current may be interrupted if there is a failure in the cooling water supply or the boiler-feed supply; the current is also cut off when the receiver is full. The equipment should be regularly checked and cleaned to ensure a constant quality of water.

Deionised water

Deionised water, often of greater purity than laboratory distilled water, is produced by allowing tap water to pass through a mixture of ion exchange resins. A strong acid resin is used to replace any cations from the water by hydrogen ions, and a strong base resin (OH$^-$ form) is used to remove any anions. In commercial deionisers the quality of the water produced is monitored by a conductance meter. The resins are usually supplied in an interchangeable cartridge, so that maintenance is reduced to a minimum. A mixed-bed ion exchange column fed with distilled water is capable of producing water with the very low conductance of about $2.0 \times 10^{-6} \, \Omega^{-1} \, cm^{-1}$ (2.0 μS cm^{-1}), but in spite of this very low conductance, the water may contain traces of organic impurities which can be detected by means of a spectrofluorimeter. For most purposes, however, the traces of organic material present in deionised water can be ignored, and it may be used in most situations where distilled water is acceptable.

An alternative method of purifying water is by **reverse osmosis**. Under normal conditions, if an aqueous solution is separated by a semipermeable membrane from pure water, osmosis will lead to water entering the solution to dilute it. But if sufficient pressure is applied to the solution, i.e. a pressure in excess of its osmotic pressure, then water will flow through the membrane from the solution; the process of reverse osmosis is taking place. This principle has been adapted as a method of purifying tap water. The tap water, at a pressure of 3–5 atm (300–500 kPa) is passed through a tube containing the semipermeable membrane. The permeate usually still contains traces of inorganic material and is therefore not suitable for operations requiring very pure water, but it will serve for many laboratory purposes, and is very suitable for further purification by ion exchange treatment. The water produced by reverse osmosis is passed first through a bed of activated charcoal which removes organic contaminants, and is then passed through a mixed-bed ion exchange column and the resultant effluent is finally filtered through a submicron filter membrane to remove any last traces of colloidal organic particles.

The **high-purity water** thus produced typically has a conductance of about $0.5 \times 10^{-6}\,\Omega^{-1}\,cm^{-1}$ ($0.5\,\mu S\,cm^{-1}$) and is suitable for use under the most stringent requirements. It will meet the purity required for trace element determinations and for operations such as ion chromatography. But remember that it can readily become contaminated from the vessels in which it is stored, and also by exposure to the atmosphere. For organic determinations the water should be stored in containers made of resistant glass (e.g. Pyrex), or ideally of fused silica, whereas for inorganic determinations the water is best stored in containers made from polythene or from polypropylene.

3.18   Wash bottles

Traditional glass wash bottles are 500–750 mL flat-bottomed flasks fitted with tubes and jets (Figure 3.4a); they are very rarely used except where nothing else is available. However, a very wide range of **polythene** wash bottles purpose-made for the job are readily available and inexpensive. They usually have ~250 mL capacity, a hard plastic cap and a plastic jet; the jet is inserted through the cap (Figure 3.4b). The bottles are easy to hold in the hand and provide a controllable jet of liquid when squeezed with a gentle pressure. They are virtually unbreakable, but should be kept away from Bunsen flames and hot surfaces; they can be used with a wide variety of solutions and solvents, but seldom with hot liquids.

Polythene wash bottles used with liquids other than water should normally be designated for use only with a specific solvent or solution. This is because the components of some wash solutions may be adsorbed into the polythene and then contaminate any other liquid if the wash bottle is used for another purpose. Repeated fillings and rinsings may be required to clean a bottle before it can be used for a different liquid.

Note   **Wash bottles should always be clearly labelled and colour-coded if they are used for different liquids, but should not be used for concentrated acids or highly corrosive substances.**

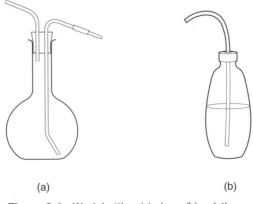

(a)   (b)

Figure 3.4   Wash bottles: (a) glass, (b) polythene

General apparatus

3.19 Glassware, ceramics, plastic ware

Glassware

To avoid introducing impurities during analysis, use apparatus made from resistance glass. A borosilicate glass is preferred for most purposes. Resistance glass is very slightly affected by all solutions, but attack by acid solutions is generally less than by pure water or alkaline solutions; hence alkaline solutions should be acidified, whenever possible, if they must be kept in glass for any length of time. Attention should also be given to watch, clock, and cover glasses; they should also be made from resistance glass. As a rule, glassware should not be heated with a naked flame; a wire gauze should be interposed between the flame and the glass vessel. For special purposes, Corning Vycor glass (96% silica) may be used. It has great resistance to heat and equally great resistance to thermal shock, and is unusually stable to acids (except hydrofluoric acid), water and various solutions.

Beakers The most satisfactory beakers for general use are beakers with a spout. The advantages of this form are convenience of pouring; the spout forms a convenient place at which a stirring rod may protrude from a covered beaker, and it forms an outlet for steam or escaping gas when the beaker is covered with an ordinary clockglass. The size of a beaker must be selected with due regard to the volume of the liquid it is to contain. The most useful sizes are from 250 to 600 mL.

Conical flasks Conical flasks of 150–500 mL capacity are used extensively in titrations, for mixing liquids and for heating solutions.

Funnels Funnels should enclose an angle of 60°. The most useful sizes for quantitative analysis have diameters of 5.5, 7 and 9 cm. The stem should have an internal diameter of about 4 mm and should not be more than 15 cm long. For filling burettes and transferring solids to graduated flasks, a short-stem, wide-necked funnel is useful.

Porcelain apparatus

Porcelain is generally employed for operations in which hot liquids are to remain in contact with the vessel for prolonged periods. Compared with glass, it is usually considered to be more resistant to solutions, particularly alkaline solutions, although this will depend primarily upon the quality of the glaze. Shallow porcelain basins with lips are employed for evaporations. Casseroles are lipped, flat-bottomed porcelain dishes provided with handles; they are more convenient to use than dishes.

   Porcelain crucibles are very often used for igniting precipitates and heating small quantities of solids because of their cheapness and their ability to withstand high temperatures without appreciable change. Some reactions, such as fusion with sodium carbonate or other alkaline substances, and also evaporations with hydrofluoric acid, cannot be carried out in porcelain crucibles owing to the resultant chemical attack. A slight attack of the porcelain also takes place with pyrosulphate fusions.

Fused-silica apparatus

Two varieties of silica apparatus are available commercially, translucent and transparent. Translucent apparatus is much cheaper and can usually be employed instead of the transparent variety. Silica ware has several advantages: it has a great resistance to heat shock because of its very small coefficient of expansion; it is not attacked by acids at a high temperature, except by hydrofluoric acid and phosphoric acid; and it is more resistant than porcelain to pyrosulphate fusions. And the chief disadvantages: it is attacked by alkaline solutions, particularly fused alkalis and carbonates; it is more brittle than ordinary glass; and it requires a much longer time for heating and cooling than platinum apparatus. Corning Vycor apparatus (96% silica glass) possesses most of the merits of fused silica and is transparent.

Plastic apparatus

Plastic materials are widely used for a variety of items of common laboratory equipment such as aspirators, beakers, bottles, Buchner funnels and flasks, centrifuge tubes, conical flasks, filter crucibles, filter funnels, measuring cylinders, scoops, spatulas, stoppers, tubing and weighing bottles; plastic items are often cheaper than their glass counterparts, and are frequently less fragile. Although inert towards many chemicals, there are some limitations on the use of plastic apparatus, including the generally rather low maximum temperature to which it may be exposed. Salient properties of the commonly used plastic materials are summarised in Table 3.4.

Teflon is extremely inert; it is so lacking in reactivity that it is used as the liner in pressure digestion vessels in which substances are decomposed by heating with hydrofluoric acid, or with concentrated nitric acid (Section 3.31).

Table 3.4   *Plastics used for laboratory apparatus*

Material	Appearance[a]	Highest temperature (°C)	Acids[b]		Alkalis[b]		Attacking organic solvents[c]
			Weak	Strong	Weak	Strong	
Polythene (LD)[d]	TL	80–90	R	R*	V	R	1, 2
Polythene (HD)[d]	TL–O	100–110	V	R*	V	V	2
Polypropylene	T–TL	120–130	V	R*	V	V	2
TPX (polymethylpentene)	T	170–180	V	R*	V	V	1, 2
Polystyrene	T	85	V	R*	V	V	most
PTFE (Teflon)	O	250–300	V	V	V	V	V
Polycarbonate	T	120–130	R	A	F	A	most
PVC (polyvinyl chloride)	T–O	50–70	R	R*	R	R	2, 3, 4
Nylon	TL–O	120	R	A	R	F	V

[a] O = opaque; T = transparent; TL = translucent.
[b] A = attacked; F = fairly resistant; R = resistant; R* = generally resistant but attacked by oxidising mixtures; V = very resistant.
[c] 1 = hydrocarbons; 2 = chlorohydrocarbons; 3 = ketones; 4 = cyclic ethers; V = very resistant.
[d] LD = low density; HD = high density.

3.20 Metal apparatus

Platinum

Platinum is used mainly for crucibles, dishes and electrodes; it has a very high melting point (1773 °C), but the pure metal is too soft for general use, and is therefore always hardened with small quantities of rhodium, iridium or gold. These alloys are slightly volatile at temperatures above 1100 °C, but retain most of the advantageous properties of pure platinum, such as resistance to most chemical reagents, including molten alkali carbonates and hydrofluoric acid (the exceptions are dealt with below), excellent conductivity of heat, and extremely small adsorption of water vapour. A 25 mL platinum crucible has an area of 80–100 cm^2, so the error due to volatility may be appreciable if the crucible is made of an alloy of high iridium content. The magnitude of this loss will be evident from Table 3.5, which gives the approximate loss in weight of crucibles expressed in mg per 100 cm^2 per hour at the temperature indicated. An alloy consisting of 95% platinum and 5% gold is known as a 'non-wetting' alloy and fusion samples are readily removed from crucibles composed of this alloy; removal is assisted by keeping the crucible tilted while the melt is solidifying. Crucibles made of this alloy are used in preparing samples for X-ray fluorescence investigation.

Table 3.5 *Weight loss of platinum crucibles*

Temp. (°C)	Pure Pt	99%Pt–1%Ir	97.5%Pt–2.5%Ir
900	0.00	0.00	0.00
1000	0.08	0.30	0.57
1200	0.81	1.2	2.5

A recent development is the introduction of ZGS (zirconia grain stabilised) platinum. This is produced by adding a small amount of zirconia (zirconium(IV) oxide) to molten platinum, which leads to modification of the microstructure of the solid material with increased hot strength and greater resistance to chemical attack. Whereas the recommended operating temperature for pure platinum is 1400 °C, the ZGS material can be used up to 1650 °C. Apparatus can also be constructed from TRIM, which consists of palladium coated with ZGS platinum; this permits stouter apparatus with the corrosion resistance of ZGS platinum at an appreciably cheaper price.

Use a platinum triangle to support platinum crucibles during heating, or a silica triangle if platinum is not available; avoid Nichrome and other metal triangles. Pipeclay triangles usually contain enough iron to damage the platinum. Hot platinum crucibles must always be handled with platinum-tipped crucible tongs as unprotected brass or iron tongs produce stains on the crucible. Platinum vessels must not be exposed to a luminous flame, nor should they be allowed to come into contact with the inner cone of a gas flame; this disintegrates the surface of the metal, causing it to become brittle, probably through formation of a platinum carbide.

At high temperatures platinum permits the flame gases to diffuse through it, and this may cause the reduction of some substances not otherwise affected. Hence if a covered crucible is heated by a gas flame, there is a reducing atmosphere in the crucible; but with an open crucible, diffusion into the air is so rapid this effect is not appreciable. If iron(III) oxide is heated in a covered crucible, it is partly reduced to metallic iron, which alloys with the platinum; sodium sulphate is similarly partly reduced to the sulphide. So, in the ignition of iron compounds or sulphates, it is advisable to place the crucible in a slanting position with free access of air.

There are several procedures where platinum apparatus may be used without significant loss:

1. Fusions with (a) sodium carbonate or fusion mixture, (b) borax and lithium metaborate, (c) alkali bifluorides, and (d) alkali hydrogensulphates (slight attack in the last case above 700 °C, which is diminished by the addition of ammonium sulphate).
2. Evaporations with (a) hydrofluoric acid, (b) hydrochloric acid in the absence of oxidising agents which yield chlorine, and (c) concentrated sulphuric acid (a slight attack may occur).
3. Ignition of (a) barium sulphate and sulphates of metals which are not readily reducible, (b) the carbonates, oxalates, etc., of calcium, barium and strontium, and (c) oxides which are not readily reducible, e.g. CaO, SrO, Al_2O_3, Cr_2O_3, Mn_3O_4, TiO_2, ZrO_2, ThO_2, MoO_3, and WO_3. (BaO, or compounds which yield BaO on heating, attack platinum.)

Platinum is **attacked** under the following conditions, and these procedures must not be conducted in platinum vessels:

1. Heating with the following liquids: (a) aqua regia, (b) hydrochloric acid and oxidising agents, (c) liquid mixtures which evolve bromine or iodine, and (d) concentrated phosphoric acid (slight, but appreciable, action after prolonged heating).
2. Heating with the following solids, their fusions, or vapours: (a) oxides, peroxides, hydroxides, nitrates, nitrites, sulphides, cyanides, hexacyanoferrate(III), and hexacyanoferrate(II) of the alkali and alkaline earth metals (except oxides and hydroxides of calcium and strontium): (b) molten lead, silver, copper, zinc, bismuth, tin, or gold, or mixtures which form these metals upon reduction; (c) phosphorus, arsenic, antimony, or silicon, or mixtures which form these elements upon reduction, particularly phosphates, arsenates, and silicates in the presence of reducing agents; (d) sulphur (slight action), selenium, and tellurium; (e) volatile halides (including iron(III) chloride), especially those which decompose readily; (f) all sulphides or mixtures containing sulphur and a carbonate or hydroxide; and (g) substances of unknown composition. Also heating in an atmosphere containing chlorine, sulphur dioxide or ammonia, whereby the surface is rendered porous.

Solid carbon, however produced, presents a hazard. It may be burnt off at low temperatures, with free access to air, without harm to the crucible, but it should never be ignited strongly. Precipitates in filter paper should be treated in a similar manner; strong ignition is only permissible after **all** the carbon has been removed. Ashing in the presence of carbonaceous matter should not be conducted in a platinum crucible, since metallic elements which may be present will attack the platinum under reducing conditions.

Cleaning and preservation of platinum ware

All platinum apparatus (crucibles, dishes, etc.) should be kept clean, polished and in proper shape. If a platinum crucible becomes stained, a little sodium carbonate should be fused in the crucible, the molten solid poured out on to a dry stone or iron slab, the residual solid dissolved out with water, and the vessel then digested with concentrated hydrochloric acid; this treatment may be repeated, if necessary. If fusion with sodium carbonate is without effect, potassium hydrogensulphate may be substituted; a slight attack of the platinum will occur. Disodium tetraborate may also be used. Sometimes the use of hydrofluoric acid (**care**) or potassium hydrogenfluoride may be necessary. Iron stains may be removed by heating the covered crucible with a gram or two of pure ammonium chloride and applying the full heat of a burner for 2–3 min.

All platinum vessels must be handled with care to prevent deformation and denting. Platinum crucibles must on no account be squeezed with the object of loosening the solidified cake after a fusion. Boxwood formers can be purchased for crucibles and dishes; these are invaluable for reshaping dented or deformed platinum ware.

Other metals

Platinum-clad stainless steel laboratory ware   This is available for evaporating solutions of corrosive chemicals. These vessels have all the corrosion resistance of platinum up to about 550 °C. The main features are (1) much lower cost than similar apparatus of platinum; (2) an overall thickness about four times that of similar all-platinum apparatus, leading to greater mechanical strength; and (3) lower susceptibility to damage by handling with tongs, etc.

Silver apparatus   The chief uses of silver crucibles and dishes in the laboratory are in the evaporation of alkaline solutions and for fusions with caustic alkalis; during fusions the silver is slightly attacked. Gold vessels (m.p. 1050 °C) are more resistant than silver to fused alkalis. Silver melts at 960 °C, so take care when heating it over a bare flame.

Nickel ware   Nickel crucibles and dishes are used for fusions with alkalis and with sodium peroxide (**care**). In the peroxide fusion a little nickel is absorbed into the mass, but this is usually not objectionable. No metal entirely withstands the action of fused sodium peroxide. Nickel oxidises in air, hence nickel apparatus cannot be used for operations involving weighing.

Iron ware   Iron crucibles may be substituted for those of nickel in sodium peroxide fusions. They are not so durable, but are much cheaper.

Stainless steel ware   Beakers, crucibles, dishes, funnels, etc., made of stainless steel are available commercially and have obvious uses in the laboratory. They will not rust, are tough, strong, and highly resistant to denting and scratching.

Metal crucibles and beaker tongs   Crucibles, evaporating basins and beakers which have been heated need to be handled with suitable tongs. Crucible tongs should be made of solid nickel, nickel steel or another rustless ferroalloy. For handling hot platinum crucibles or dishes, platinum-tipped tongs must be used. Beaker tongs are available for handling beakers of 100–2000 mL capacity. An adjustable screw with locknut limits the jaw span of the tongs, allowing them to be adjusted for the container size.

3.21   Heating apparatus

Burners

The ordinary Bunsen burner is still widely employed for providing moderately high temperatures, and more specialised burners such as Meker Burners can reach slightly higher temperatures for ignition purposes. The maximum temperature is reached when the air regulator is adjusted to admit slightly more air than required to produce a non-luminous flame; but too much air will give a noisy flame, unsuitable for combustion or ignition in quantitative analysis. Owing to the different combustion characteristics and calorific values of the common gaseous fuels – natural gas, liquefied petroleum (bottled) gas – burners do vary slightly in their dimensions, including jet size and aeration controls. Unless it is an 'all gases' type, which can be adjusted, choose the appropriate burner to obtain maximum efficiency from the available gas supply.

Hotplates

Electrically heated hotplates are available in a very wide range of shapes and sizes with controls varying from simple 'low, medium, high' to very advanced thermostats and temperature monitoring. They should satisfy all standard safety requirements, with totally enclosed wiring protected from possible chemical spillages. The best hotplates incorporate a magnetic stirrer; they are valuable for getting substances into solution rapidly before dilution to standard volumes. Low-temperature heating can always be carried out on steam baths, which are usually available with thermostatically controlled heater units.

Electric ovens

The most convenient oven is an electrically heated, thermostatically controlled drying oven having a temperature range from room temperature to about 250–300 °C; the temperature can be controlled to within ±1–2 °C. Electric ovens are mainly used for drying precipitates or solids at comparatively low temperatures; they have virtually superseded the steam oven.

Microwave ovens

Microwave ovens are now being used very extensively for drying and heating. They are particularly valuable when determining moisture contents of materials, as water is removed very rapidly on exposure to microwave radiation. They also give greatly reduced drying times for precipitates.

Muffle furnaces

An electrically heated furnace of muffle form should be available in every well-equipped laboratory. The maximum temperature should be about 1200 °C. If possible, a thermocouple and indicating pyrometer should be provided; otherwise the ammeter in the circuit should be calibrated, and a chart constructed showing ammeter and corresponding temperature readings. Gas-heated muffle furnaces may give temperatures up to about 1200 °C.

Air baths

An electric oven should not be used for drying solids and precipitates at temperatures up to 250 °C in which acid or other corrosive vapours are evolved. An air bath may be constructed from a cylindrical metal (copper, iron or nickel) vessel, its bottom pierced with numerous holes. A silica triangle, legs appropriately bent, is inserted inside the bath for supporting an evaporating dish, crucible, etc. The whole set-up is heated by a Bunsen flame, shielded from draughts. The insulating layer of air prevents bumping by reducing the rate at which heat reaches the contents of the inner dish or crucible. An air bath of similar construction but with special heat-resistant glass sides may also be used; this gives visibility inside the air bath.

Infrared lamps and heaters

Powerful infrared lamps with concentrating reflectors are available commercially and are useful for evaporating solutions and drying even relatively large quantities of solid materials. If the lamps are mounted above the liquid to be heated, evaporation will occur rapidly, usually without spattering. Specially designed infrared units can be used with a number of dishes simultaneously, and the heat can be directed to both the top and bottom of the containers simultaneously. Take care when handling the lamps; they can become extremely hot and are fragile immediately after use and before they have cooled down.

Immersion heaters

An immersion heater, consisting of a radiant heater encased in a silica sheath, is useful for the direct heating of most acids and other liquids (except hydrofluoric acid and concentrated caustic alkalis). Infrared radiation passes through the silica sheath with little absorption, so a large proportion of heat is transferred to the liquid by radiation. The heater is almost unaffected by violent thermal shock due to the low coefficient of thermal expansion of the silica.

Heating mantles

Heating mantles consist of a flexible 'knitted' fibreglass sheath that fits snugly around a flask and contains an electrical heating element which operates at black heat. The mantle may be supported in an aluminium case which stands on the bench, but for use with suspended vessels the mantle is supplied without a case. Electric power is supplied to the heating element through a control unit which may be either a continuously variable transformer or a thyristor controller, so the operating temperature of the mantle can be smoothly adjusted. Heating mantles are particularly designed for the heating of flasks and find wide application in distillation operations. For details of the distillation procedure and description of the apparatus employed, consult a textbook of practical organic chemistry.[5]

3.22   Desiccators and dry boxes

Desiccators

A desiccator is a covered glass container designed for the storage of objects in a dry atmosphere; inside the base is a drying agent, such as anhydrous calcium chloride (largely used in elementary work), silica gel, activated alumina or anhydrous calcium sulphate (Drierite). Silica gel, alumina and calcium sulphate can be obtained which have been impregnated with a cobalt salt so that they are self-indicating; the colour changes from blue to pink when the desiccant is exhausted. The spent material can be regenerated by heating

Table 3.6   *Comparative efficiency of drying agents*

Drying agent	Residual water per litre of air (mg)	Drying agent	Residual water per litre of air (mg)
$CaCl_2$	1.5	Al_2O_3	0.005
NaOH (sticks)	0.8	$CaSO_4$	0.005
H_2SO_4 (95%)	0.3	Molecular sieve	0.004
Silica gel	0.03	H_2SO_4	0.003
KOH (sticks)	0.014	$Mg(ClO_4)_2$	0.002
		P_2O_5	0.000 02

in an electric oven at 150–180 °C (silica gel); 200–300 °C (activated alumina); 230–250 °C (Drierite); it is therefore convenient to place these drying agents in a shallow dish situated at the bottom of the desiccator, allowing easy removal for baking as required.

The action of desiccants can be considered from two viewpoints. The amount of moisture that remains in the closed space of the desiccator is related to the vapour pressure of the inexhausted desiccant, i.e. the vapour pressure measures the extent to which the desiccant can remove moisture, and therefore measures its efficiency. A second factor is the weight of water that can be removed per unit weight of desiccant, i.e. the drying capacity. In general, substances that form hydrates have higher vapour pressures but they also have greater drying capacities. Remember that a substance cannot be dried by a desiccant which has a vapour pressure greater than the vapour pressure of the substance itself.

The relative efficiencies of various drying agents will be evident from the data presented in Table 3.6. These values were determined by aspirating properly conditioned air through U-tubes charged with the desiccants; they are applicable, strictly, to the use of these desiccants in absorption tubes, but the figures may reasonably be applied as a guide when selecting desiccants for desiccators. It would appear that a hygroscopic material such as ignited alumina should not be allowed to cool in a covered vessel over 'anhydrous' calcium chloride; anhydrous magnesium perchlorate or phosphorus pentoxide would be satisfactory.

The normal (or Scheibler) desiccator is provided with a porcelain plate having apertures to support crucibles, etc. The plate is supported on a constriction situated roughly halfway up the wall of the desiccator. For small desiccators, it is possible to use a silica triangle with the wire ends suitably bent. The ground edge of the desiccator should be lightly coated with white vaseline or a special grease in order to make it airtight.

There is, however, controversy regarding the effectiveness of desiccators. If the lid is briefly removed from a desiccator, it may take as long as 2 h to remove the atmospheric moisture thus introduced, and to re-establish the dry atmosphere; during this period a hygroscopic substance may actually gain in weight while in the desiccator. It is therefore advisable that any substance which is to be weighed should be kept in a vessel with a lid as tightly fitting as possible while it is in the desiccator.

Cooling of hot vessels within a desiccator is also an important problem. A crucible which has been strongly ignited and immediately transferred to a desiccator may not have attained room temperature even after 1 h. The situation can be improved by allowing the crucible to cool for a few minutes before transferring to the desiccator, and then a further cooling time of 20–25 min is usually adequate. The inclusion in the desiccator of a metal

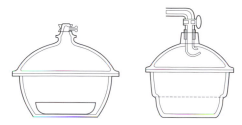

Figure 3.5   Vacuum desiccators

block (e.g. aluminium), upon which the crucible may be stood, also helps to ensure that temperature equilibrium is reached.

When a hot object, such as a crucible, is placed in a desiccator, about 5–10 s should elapse for the air to become heated and expand before putting the cover in place. When reopening, the cover should be slid away very gradually in order to prevent any sudden inrush of air, which might blow material out of the crucible. Air rushes in to fill the partial vacuum created when the expanded gas content of the desiccator cools down.

A desiccator is frequently employed for the thorough drying of solids. Its efficient operation depends upon the condition of the desiccant, which should be renewed at frequent intervals, particularly if its drying capacity is low. A vacuum desiccator is advisable when dealing with large quantities of solid.

Convenient types of **vacuum desiccator** are illustrated in Figure 3.5. Large surfaces of the solid can be exposed; the desiccator may be evacuated, so drying is much more rapid than in the ordinary Scheibler type. These desiccators are made of heavy glass, plastics or even metal, and they are designed to withstand reduced pressure; nevertheless, no desiccator should be evacuated unless it is surrounded by an **adequate guard** in the form of a stout wire cage.

For most purposes the vacuum produced by an efficient water pump (20–30 mm mercury) will suffice; a guard tube containing desiccant should be inserted between the pump and the desiccator. The sample to be dried should be covered with a watch or clock glass, so that no solid is lost during removal or admission of air. Air must be admitted slowly into an exhausted desiccator; if the substance is very hygroscopic, a drying train should be attached to the stopcock. In order to maintain a satisfactory vacuum within the desiccator, the flanges on both the lid and the base must be well lubricated with vaseline or other suitable grease. In some desiccators an elastomer ring is incorporated in a groove in the flange of the lower component of the desiccator; when the pressure is reduced, the ring is compressed by the lid of the desiccator, and an airtight seal is produced without the need for any grease. The same desiccants are used as with an ordinary desiccator.

Dry boxes

Dry boxes (glove boxes) are used for manipulating materials that are very sensitive to atmospheric moisture (or to oxygen); they consist of a plastic or metal box with glass or clear plastic windows on the upper side and on the sidewalls. Rubber or plastic gloves are fitted using airtight seals through the front side of the box, and by placing the hands and forearms into the gloves, manipulations may be carried out inside the box. One end of the box is fitted with an airlock so that apparatus and materials can be introduced into the box without disturbing the atmosphere inside. A tray of desiccant placed inside the box maintains a dry atmosphere, but to counter the unavoidable leakages it is usual to supply a slow current of dry air to the box; inlet and outlet taps are provided to control this operation. If the box is flushed out before use with an inert gas (e.g. nitrogen), and a slow stream of the gas is maintained while the box is in use, materials which are sensitive to oxygen can be safely handled. A detailed discussion of their construction and use is given in the literature.[6]

3.23 Stirring apparatus

Stirring rods

Stirring rods are made from glass rod 3–5 mm in diameter, cut into suitable lengths. Both ends should be rounded by heating in the Bunsen or blowpipe flame. The length of the stirring rod should be suitable for the size and shape of the vessel, e.g. a spouted beaker requires a stirring rod that projects 3–5 cm beyond the lip when in a resting position. Glass stirring rods should not be used for stirring viscous liquids as they can cause serious hand injuries if they break. A short piece of Teflon or rubber tubing (or a rubber cap) fitted tightly over one end of a stirring rod of convenient size provides the so-called **policeman**, used for detaching particles of a precipitate adhering to the side of a vessel which cannot be removed by a stream of water from a wash bottle; it should not, as a rule, be employed for stirring, nor should it be allowed to remain in a solution.

Boiling rods

Boiling liquids and liquids in which a gas, such as hydrogen sulphide or sulphur dioxide, has to be removed by boiling can be prevented from superheating and 'bumping' by the use of a boiling rod (Figure 3.6). This consists of a piece of glass tubing closed at one end and sealed approximately 1 cm from the open end; the open end is immersed in the liquid. When the rod is removed, the liquid in the open end must be shaken out and the rod rinsed with a jet of water from a wash bottle. This device should not be used in solutions which contain a precipitate.

Stirring may be conveniently effected with the so-called **magnetic stirrer**. A rotating magnet induces a variable-speed stirring action within closed or open vessels. The actual stirrer is a small cylinder of iron sealed in Pyrex glass, polythene or Teflon, which is caused to rotate by the rotating magnet.

The usual glass paddle stirrer is also widely used. It works in conjunction with an electric motor controlled by a transformer or solid-state speed device. The stirrer may be connected directly to the motor shaft or to a spindle actuated by a gearbox which forms an integral part of the motor housing; it is possible to obtain a wide variation in stirrer speed.

Under some circumstances, e.g. the dissolution of a sparingly soluble solid, it may be better to use a **mechanical shaker**. They range from wrist action shakers which accommodate small or medium-sized flasks, to powerful shakers which can take large bottles and give their contents a vigorous agitation.

Figure 3.6 Boiling rod

3.24   Filtration apparatus

The simplest filtration apparatus is a filter funnel fitted with a filter paper. To promote rapid filtration, the funnel angle should be as close as possible to 60° and the funnel stem should be 15 cm long. Filter papers are made in varying grades of porosity, so choose one appropriate to the filtration (Section 3.35).

In the majority of quantitative determinations involving the collection and weighing of a precipitate, it is convenient to collect the precipitate in a crucible for direct weighing, and various forms of **filter crucible** have been devised for this purpose. Sintered-glass crucibles are made of resistance glass and have a porous disc of sintered ground glass fused into the body of the crucible. The filter disc is made in varying porosities, 0 is the coarsest and 5 the finest; the range of pore diameter for the various grades is as follows.

Porosity	0	1	2	3	4	5
Pore diameter (μm)	200–250	100–120	40–50	20–30	5–10	1–2

Porosity 3 is suitable for precipitates of moderate particle size, and porosity 4 for fine precipitates such as barium sulphate. These crucibles should not be heated above about 200 °C. Silica crucibles of similar pattern are also available; although expensive, they have certain advantages in thermal stability.

Filter crucibles with a porous filter base are available in porcelain (porosity 4), in silica (porosities 1, 2, 3, 4) and in alumina (coarse, medium and fine porosities); they can be heated to much higher temperatures than sintered crucibles. Nevertheless, the heating must be gradual otherwise the crucible may crack at the join between porous base and glazed side.

Large quantities of material are usually filtered through a **Buchner funnel**, or one of the modified funnels in Figure 3.7. In the ordinary porcelain funnel and the 'slit sieve' glass funnel (Figure 3.7(a) and (b)) two good quality filter papers are placed on the plate; the transparent glass funnel is preferable since it is easy to see whether the funnel is perfectly clean. In the Pyrex funnel with a sintered glass plate (Figure 3.7(c)) no filter paper is required, so strongly acidic and weakly alkaline solutions can be readily filtered with this funnel. In all cases the funnel of appropriate size is fitted into a filter flask (Figure 3.7(d)) and the filtration conducted under the diminished pressure provided by a filter pump or vacuum line. One of the disadvantages of the porcelain Buchner funnel is that, being of one-piece construction, the filter plate cannot be removed for thorough cleaning and it is

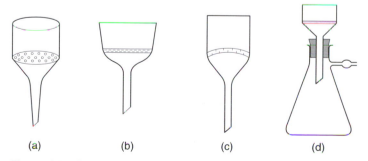

(a)   (b)   (c)   (d)

Figure 3.7   Buchner funnels: (a) ordinary porcelain, (b) 'slit sieve' glass, (c) Pyrex, (d) Buchner funnel fitted to a filter flask

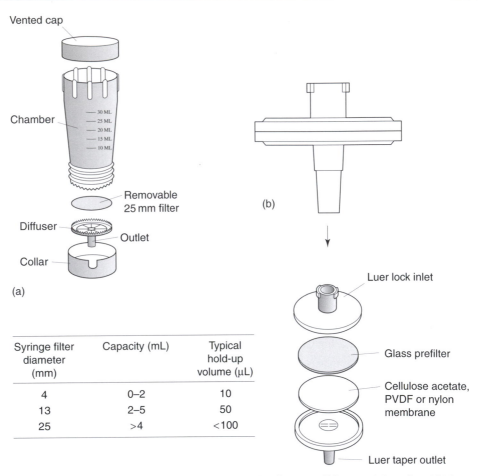

Syringe filter diameter (mm)	Capacity (mL)	Typical hold-up volume (µL)
4	0–2	10
13	2–5	50
25	>4	<100

Figure 3.8 (a) Disposable filter funnel: construction (courtesy Whatman International Ltd, Maidstone, Kent). (b) Construction of a syringe type filter.

difficult to see whether the whole of the plate is clean on both sides. In a modern polythene version, the funnel is made in two sections which can be unscrewed for inspecting both sides of the plate.

Also available are special disposable filter funnels with removable filter plates and capacities of up to 30 mL; they are particularly useful in filtering small quantities of radioactive materials and proteins (Figure 3.8(a)). The choice of filter should achieve a compromise between separation time and separation quality. For many filtrations, single-use cartridges made of plastic are more convenient than traditional filter funnels and folded papers. But these throwaway items come in such a huge range of sizes, shapes and compositions that choosing the appropriate cartridge may seem quite daunting to start with. However, if each parameter in turn is compared against the proposed experiment, the choice is normally fairly straightforward. The most widely used form is a sealed flat filter in a plastic housing fitted with male and female Luer lock (syringe) fittings on either side of the filter (Figure 3.8(b)).

The pore size of the filter is normally considered first; it can range from $1.0\,\mu m$ down to less than $0.1\,\mu m$ (since these membranes are made of plastic materials, the pore size is remarkably constant across the membrane). For many chemical separations, a pore size of $0.45\,\mu m$ is often chosen as a good compromise between effective filtration and speed. However, where very finely divided materials are anticipated or where biological activity must be minimised, then $0.2\,\mu m$ filters are used. Once the pore size has been determined, the actual material for the filter and its housing need to be specified, along with the size of the filter.

For many aqueous solutions, a cellulose acetate membrane (CAM) or glass fibre–polypropylene membrane can be used, but for non-aqueous or aggressive solutions, the membrane and its housing can be constructed of a wide range of other materials such as PTFE or polysulphone plastics. Equally the size of the filter can be optimised for the volume of solution to be filtered, ranging from 3 mm discs which will filter up to 2 mL, to 25 mm discs which will filter 100 mL of liquid in a reasonable time.

Separation of solid from liquid is sometimes better achieved by using a **centrifuge**. A small, electrically driven centrifuge is a useful piece of equipment for an analytical laboratory; it may be used for removing mother liquor from recrystallised salts, for collecting precipitates that are difficult to filter, and for washing certain precipitates by decantation. It is particularly useful when small quantities of solids are involved; centrifuging, followed by decantation and recentrifuging, avoids transference losses and yields the solid phase in a compact form. Another valuable application is the separation of two immiscible phases.

3.25   Weighing bottles

Most chemicals are weighed **by difference** through placing the material inside a stoppered weighing bottle which is then weighed. The requisite amount of substance is shaken out into a suitable vessel (beaker or flask), and the weight of substance taken is determined by reweighing the weighing bottle. In this way, the substance dispensed receives the minimum exposure to the atmosphere during the actual weighing process, important if the material is hygroscopic.

The most convenient form of weighing bottle is fitted with an external cap and made of glass, polythene or polycarbonate. A weighing bottle with an **internally** fitting stopper is not recommended; there is always the danger that small particles may lodge at the upper end of the bottle and be lost when the stopper is pressed into place.

If the substance is unaffected by exposure to the air, it may be weighed on a watchglass, or in a disposable plastic container. The weighing funnel (Figure 3.9) is very useful, particularly when the solid is to be transferred to a flask. Having weighed the solid into the scoop-shaped end, flattened so it will stand on the balance pan, the narrow end is inserted into the neck of the flask and the solid washed into the flask with a stream of water from a wash bottle.

Woodward and Redman[2c] have described a specially designed weighing bottle which will accommodate a small platinum crucible; when a substance has been ignited in the crucible, the crucible is transferred to the weighing bottle and subsequently weighed inside. This obviates the need for a desiccator. If the substance to be weighed is a liquid, it is placed in a weighing bottle fitted with a cap carrying a dropping tube.

Figure 3.9   Weighing funnel

Reagents and standard solutions

3.26 Reagents

The purest reagents available should be used for quantitative analysis; wherever possible, use reagents of analytical quality. In Great Britain AnalaR chemicals from BDH Ltd (Merck) conform to the specifications given in the handbook *AnalaR standards for laboratory chemicals*.[7] In the United States the American Chemical Society Committee on Analytical Reagents has established standards for certain reagents,[8] and manufacturers supply reagents labelled 'conforms to ACS specifications'. In addition, certain manufacturers market chemicals of high purity, and each package of these analysed chemicals has a label giving the manufacturer's limits of certain impurities.

With the increasingly lower limits of detection being achieved in various types of instrumental analysis, there is an ever growing demand for reagents of correspondingly improved specification, and some manufacturers are now offering a range of specially purified reagents such as the BDH Aristar chemicals.

In some instances, where a reagent of the requisite purity is not available, it may be advisable to weigh out a suitable portion of the appropriate **pure** metal (e.g. the Johnson Matthey Specpure range), and to dissolve this in the appropriate acid.

The label on a bottle is not an infallible guarantee of the contents. Purity may be compromised for several reasons:

(a) Some impurities may not have been tested for by the manufacturer.
(b) The reagent may have been contaminated after its receipt from the manufacturers; the stopper may have been left open for some time, exposing the contents to the laboratory atmosphere, or there may have been accidental return of an unused portion of reagent to the bottle.
(c) A solid reagent may not be sufficiently dry, perhaps due to insufficient drying by the manufacturers, perhaps due to leakage through the stoppers, or perhaps a combination of the two.

However, if the analytical reagents are purchased from a reputable manufacturer, if no bottle is open for longer than absolutely necessary, and if no reagent is returned to the bottle, then the likelihood of these errors is considerably reduced. Liquid reagents should be poured from the bottle; a pipette should never be inserted into the reagent bottle. Particular care should be taken to avoid contamination of the stopper of the reagent bottle. When a liquid is poured from a bottle, the stopper should never be placed on the shelf or on the working bench; it may be placed upon a clean watchglass, and many chemists cultivate the habit of holding the stopper between the thumb and fingers of one hand. The stopper should be returned to the bottle immediately after the reagent has been removed, and all reagent bottles should be kept scrupulously clean, particularly round the neck or mouth of the bottle.

If there is any doubt as to the purity of the reagents used, they should be tested by standard methods for the impurities that might cause errors in the determinations. Where a chemical required for quantitative analysis is not available in the form of analytical reagents, the purest commercially available products should be purified by known methods (see below). The exact mode of drying, if required, will vary with the reagent; details are given for specific reagents in the text.

3.27   Purification of substances

If a reagent of adequate purity for a particular determination is not available, then the purest available product must be purified; for inorganic compounds this is most commonly done by recrystallisation from water. Using a conical flask, a known weight of the solid is dissolved in a volume of water sufficient to give a saturated or nearly saturated solution at the boiling point. The hot solution is filtered through a fluted filter paper placed in a short-stemmed funnel, and the filtrate collected in a beaker; this process will remove any insoluble material present. If the substance crystallises out in the funnel, it should be filtered through a heated or jacketed funnel. The clear hot filtrate is cooled rapidly by immersion in a dish of cold water or in a mixture of ice and water, according to the solubility of the solid; the solution should be constantly stirred in order to promote the formation of small crystals and to prevent the trapping of mother liquor. The solid is then separated from the mother liquor by filtration, using one of the funnels shown in Figure 3.7. When all the liquid has been filtered, the solid is pressed down on the funnel with a wide glass stopper, sucked as dry as possible, then washed with small portions of the original solvent to remove the adhering mother liquor. The recrystallised solid is dried in an oven above the laboratory temperature with the exclusion of dust. The dried solid is preserved in glass-stoppered bottles. When the solid is removed from the funnel, take great care to avoid introducing fibres from the filter paper, or small particles of glass from the glass filter disc.

Some inorganic solids are either too soluble, or the solubility does not vary sufficiently with temperature, in a given solvent for direct crystallisation to be practicable. In many cases the solid can be precipitated from a concentrated aqueous solution by the addition of a liquid, miscible with water, in which it is less soluble. Many inorganic compounds are almost insoluble in ethanol, so ethanol is generally used. Take care that the amount of ethanol or other solvent is not so large that the impurities are also precipitated. Potassium hydrogencarbonate and antimony potassium tartrate may be purified by this method.

Many organic compounds can be purified by recrystallisation from suitable organic solvents, and here again, precipitation by the addition of another solvent in which the required compound is insoluble, may be effective. Liquids can be purified by fractional distillation.

Sublimation

Sublimation is used to separate volatile substances from non-volatile impurities. Iodine, arsenic(III) oxide, ammonium chloride and a number of organic compounds can be purified in this way. The material to be purified is gently heated in a porcelain dish, and the vapour produced is condensed on a flask which is kept cool by circulating cold water inside it.

Zone refining

Zone refining was originally developed to refine certain metals; it is applicable to all substances of reasonably low melting point which are stable at the melting temperature. In a zone refining apparatus, the substance to be purified is packed into a column of glass or stainless steel, which may vary in length from 15 cm (semimicro apparatus) to 1 m. An electric ring heater which heats a narrow band of the column is allowed to fall slowly by a motor-controlled drive, from the top to the bottom of the column. The heater is set to produce a molten zone of material at a temperature 2–3 °C above the melting point of the substance, and the substance travels slowly down the tube with the heater. Since impurities normally lower the melting point of a substance, the impurities tend to flow down the column in step with the heater, becoming concentrated in the lower part of the tube. The process is usually repeated several times (the apparatus may be programmed to reproduce automatically a given number of cycles) until the required degree of purification has been achieved.

3.28   Standard solutions

In any analytical laboratory it is essential to maintain stocks of various reagents in solution. Standard solutions of accurately known concentration need to be stored correctly. They may be classified into four types:

1. Reagent solutions which are of approximate concentration.
2. Standard solutions which have a known concentration of some chemical.
3. Standard reference solutions which have a known concentration of a primary standard substance (Section 10.6) (IUPAC primary standard solutions).
4. Standard titrimetric solutions which have a known concentration (determined either by weighing or by standardisation) of a substance other than a primary standard (IUPAC secondary standard solutions).

For **reagent solutions** (item 1) it is usually sufficient to weigh out approximately the amount of material required, using a watchglass or a plastic weighing container, and then to add this to the required volume of solvent which has been measured with a measuring cylinder.

To prepare a **standard solution** use the following procedure. A short-stemmed funnel is inserted into the neck of a graduated flask of the appropriate size. A suitable amount of the chemical is placed in a weighing bottle which is weighed, and the required amount of substance is then transferred from the weighing bottle to the funnel, taking care that no particles are lost. After the weighing bottle has been reweighed, the substance in the funnel is washed down with a stream of the liquid. The funnel is thoroughly washed, inside and out, and then removed from the flask; the contents of the flask are dissolved, if necessary, by shaking or swirling the liquid, and then made up to the mark. For the final adjustment of volume, use a dropping tube drawn out to form a very fine jet.

If a watchglass is employed for weighing out the sample, the contents are transferred as completely as possible to the funnel, and then a wash bottle is used to remove the last traces of the substance from the watchglass. If the weighing scoop (Figure 3.9) is used, then a funnel is not needed, provided the flask is big enough to allow the end of the scoop easily into its neck.

If the substance is not readily soluble in water, it is advisable to add the material from the weighing bottle or the watchglass to a beaker, followed by deionised water; the beaker and its contents are then heated gently with stirring until the solid has dissolved. After allowing the resulting concentrated solution to cool a little, it is transferred through the short-stemmed funnel to the graduated flask, the beaker is rinsed thoroughly with several portions of deionised water, added to the contents of the flask, and finally the solution is made up to the mark. To ensure the solution is at room temperature, it may be necessary to allow the flask to stand for a while before making the final adjustment to the mark. **Never heat the graduated flask**.

It is also possible to prepare the standard solution by using one of the commercial volumetric solutions supplied in sealed ampoules; all they require is dilution in a graduated flask to produce a standard solution.

Solutions which are comparatively stable and unaffected by exposure to air may be stored in 1 L or 2.5 L bottles. For work requiring the highest accuracy, the bottles should be Pyrex, or other resistance glass, and fitted with ground-glass stoppers; this considerably reduces the solvent action of the solution. For alkaline solutions, a plastic stopper is preferable to a glass stopper, and a polythene container may often replace glass vessels. But some solutions, e.g. iodine and silver nitrate, can be stored only in glass containers; and for iodine and silver nitrate the bottle should be made of dark (brown) glass. Solutions of EDTA (Section 10.66) are best stored in polythene containers.

Bottles for storing standard solutions should initially be clean and dry. They should be rinsed with a small amount of the standard solution then allowed to drain before the bulk of the solution is poured in and the bottle stoppered. Bottles which have been washed and are still wet with water will require rinsing and draining with at least three small volumes of the standard solution, and they should be well drained between each rinse; then they can be filled with the standard solution. Immediately after the solution has been transferred to the stock bottle, it should be labelled with the name of the solution, its concentration, the date of preparation and the initials of the person who prepared the solution, together with any other relevant data. Unless the bottle is completely filled, internal evaporation and condensation will cause drops of water to form on the upper part of the inside of the vessel. For this reason, the bottle must be thoroughly shaken before removing the stopper. For expressing concentrations of reagents, the molar system is universally applicable, i.e. the number of moles of solute present in 1 L of solution. Concentrations are not usually expressed now in normalities, although they are still used by some people and their relationship is explained briefly in Appendix 18.

Some standard solutions are likely to be affected by air (e.g. alkali hydroxides which absorb carbon dioxide; iron(II) and titanium(III) salts which are readily oxidised). Ideally, they should be kept under an inert atmosphere, such as nitrogen, in bottles fitted with automatic dispensers or burettes. A simple apparatus for storage and use of standard solutions is shown in Figure 3.10. The solution is contained in the storage bottle (A), and the

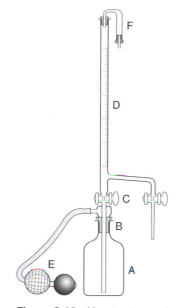

Figure 3.10 How to store a standard solution: A = storage bottle, B = ground-glass joint, C = burette tap, D = burette, E = small bellows, F = small guard tube

50 mL burette is fitted into this by means of a ground-glass joint (B). To fill the burette (D), the tap (C) is opened and the liquid pumped in using the small bellows (E). A small guard tube (F) is filled with soda-lime or Carbosorb when caustic alkali is contained in the storage bottle. Bottles with a capacity up to 2 L are provided with standard ground-glass joints; large bottles, up to 15 L capacity, can also be obtained. With both of these storage vessels, for strongly alkaline solutions, the ground-glass joints should be replaced by rubber bungs or rubber tubing. Many fixed volumes of standard solution for multiple analyses can be dispensed by automatic syringe pipettes of the type used in automatic analysers.

Some basic techniques

3.29 Preparing substances for analysis

In many instances the analyst is presented with the problem of selecting a representative sample from a large quantity of available material. Sometimes this may also mean that a large bulky material has to be broken into smaller more uniform pieces in order to obtain a sample suitable for laboratory work. This is considered in Sections 5.3 and 5.4. Before analysis most samples are dried at 105–110 °C to remove moisture.

3.30 Weighing the sample

Section 3.5 explains the operation of a chemical balance, and Sections 3.25 and 3.22 cover weighing bottles and desiccators respectively. The material, prepared as above, is usually transferred to a weighing bottle which is stoppered and stored in a desiccator. Samples of appropriate size are withdrawn from the weighing bottle as required; the bottle is weighed before and after the withdrawal, so the weight of substance is obtained by difference.

3.31 Dissolving the sample

Most organic substances can be readily dissolved in a suitable organic solvent, and some are directly soluble in water or can be dissolved in aqueous solutions of acids or alkalis. Many inorganic substances can be dissolved directly in water or in dilute acids, but materials such as minerals, refractories and alloys must usually be tested with a variety of reagents in order to discover a suitable solvent; the preliminary qualitative analysis will have revealed the best procedure to adopt. Each case must be considered on its merits, but it is worth considering how to dissolve a sample in water or in acids, and how to treat insoluble substances.

For a substance which dissolves readily, the sample is weighed out into a beaker, and the beaker immediately covered with a clockglass of suitable size with its convex side facing downwards. The beaker should have a spout in order to provide an outlet for the escape of steam or gas. The solvent is added by pouring it carefully down a glass rod, the lower end of which rests against the wall of the beaker, with the clockglass slightly displaced. If a gas is evolved during addition of the solvent (e.g. acids with carbonates, metals, alloys), the beaker must be kept covered as far as possible during the addition. The reagent is then best added by pipette or by funnel with a bent stem inserted beneath the clockglass at the spout of the beaker; loss by spurting or as spray is thus prevented. When the evolution of gas has ceased and the substance has completely dissolved, the underside of the clockglass is well rinsed with a stream of water from a wash bottle, care being taken that the washings fall on to the side of the beaker and not directly into the solution. If warming is necessary, it is usually best to carry out the dissolution in a conical flask with a small funnel in the mouth; loss of liquid by spurting is thus prevented and the escape of gas is not hindered. When using volatile solvents, the flask should be fitted with a reflux condenser.

It may often be necessary to reduce the volume of the solution, or sometimes to evaporate completely to dryness. Wide, shallow vessels are most suitable, since a large surface is thus exposed and evaporation is thereby accelerated. Pyrex evaporating dishes, porcelain basins or casseroles, and silica or platinum basins may be employed; the material selected will depend on the extent of attack by the hot liquid, and on the constituents being determined in the subsequent analysis. Evaporations should be carried out on the steam bath or on a low-temperature hotplate; slow evaporation is preferable to vigorous boiling, since boiling may lead to some mechanical loss in spite of the precautions mentioned below. During evaporations, the vessel must be covered by a Pyrex clockglass of slightly larger diameter than the vessel, and supported either on a large all-glass triangle or three small U-rods of Pyrex glass hanging over the rim of the container. At the end of the evaporation the sides of the vessel, the lower side of the clockglass and the triangle and glass hooks (if employed) should be rinsed with distilled water into the vessel.

For evaporation at the boiling point, use a conical flask with a short Pyrex funnel in the mouth or a round-bottomed flask inclined at an angle of about 45°; with the round-bottomed flask the drops of liquid, etc., thrown up by ebullition or effervescence, will be retained by striking the inside of the flask, whereas gas and vapour will escape freely. When organic solvents are used the flask should be fitted with a swan neck and a condenser, so the solvent is recovered, or a rotary evaporator may be used. Consider the possibility of losses during the concentration procedure; for example, boric acid, halogen acids and nitric acid are lost from boiling aqueous solutions.

Substances which are insoluble (or only slightly soluble) in water can often be dissolved in an appropriate acid, but the possible loss of gaseous products must be borne in mind. The respective evolution of carbon dioxide, hydrogen sulphide and sulphur dioxide from carbonates, sulphides and sulphites will be immediately apparent; less obvious are losses of boron and silicon as the corresponding fluorides during evaporations with hydrofluoric acid, or loss of halogen by the treatment of halides with a strong oxidising agent such as nitric acid. The following more powerful reagents may be used to dissolve difficult materials.

Concentrated acids   Concentrated hydrochloric acid will dissolve many metals (generally those situated above hydrogen in the electrochemical series), as well as many metallic oxides. Hot concentrated nitric acid dissolves most metals, but antimony, tin and tungsten are converted to slightly soluble acids, providing a separation of these elements from other components of alloys. Hot concentrated sulphuric acid dissolves many substances, and many organic materials are charred and then oxidised by this treatment.

Aqua regia   Aqua regia is 75 vol% hydrochloric acid and 25 vol% nitric acid. Largely due to its oxidising character, it is a very potent solvent; and the effectiveness of the acid is frequently increased by adding other oxidants, e.g. bromine and hydrogen peroxide.

Hydrofluoric acid   Hydrofluoric acid is mainly used for the decomposition of silicates; excess hydrofluoric acid is removed by evaporation with sulphuric acid leaving a residue of metallic sulphates. Complexes of fluoride ions with many metallic cations are very stable, so the normal properties of the cation may not be exhibited. It is therefore essential to ensure complete removal of fluoride, and to achieve this it may be necessary to repeat the evaporation with sulphuric acid two or three times. Hydrofluoric acid causes serious and painful skin burns. **Use hydrofluoric acid with great care**.

Perchloric acid   Perchloric acid attacks stainless steels and a number of iron alloys that do not dissolve in other acids. A mixture of perchloric and nitric acids is valuable as an oxidising solvent for many organic materials to produce a solution of inorganic constituents of the sample. For safety, the substance should be treated first with concentrated nitric acid, the mixture heated, and then careful additions of small quantities of perchloric acid can be made until the oxidation is complete. Even then, the mixture should not be evaporated, because the nitric acid evaporates first allowing the perchloric acid to reach dangerously high concentrations. When using a mixture of 60 vol% nitric acid, 20 vol% perchloric acid and 20 vol% sulphuric acid, the perchloric acid is also evaporated to leave a sulphuric acid solution of the components for analysis. The organic part of the material under investigation is destroyed and the process is known as **wet ashing**. Hot concentrated perchloric acid gives explosive reactions with organic materials or easily oxidised inorganic compounds; if frequent reactions and evaporations involving perchloric acid are to be performed, it is wise to use a fume cupboard free from combustible organic materials. **Use perchloric acid with great care**.

Fusion reagents   Fusion reagents, commonly known as **fluxes**, are used to solubilise substances that are not soluble in normal solvents or acids. Typical fluxes include anhydrous sodium carbonate, either alone or mixed with potassium nitrate or sodium peroxide; potassium pyrosulphate or sodium pyrosulphate; sodium peroxide; sodium hydroxide or potassium hydroxide. Anhydrous lithium metaborate is especially suitable for materials containing silica;[9] when the resulting fused mass is dissolved in dilute acids, no separation of silica takes place as it does when a sodium carbonate melt is similarly treated. Here are some other advantages claimed for lithium metaborate:

1.   No gases are evolved during the fusion or during the dissolution of the melt, hence there is no danger of losses due to spitting.
2.   Fusions with lithium metaborate are usually quicker (15 min will often suffice) and can be performed at a lower temperature than with other fluxes.
3.   The loss of platinum from the crucible is less during a lithium metaborate fusion than with a sodium carbonate fusion.
4.   Many elements can be determined directly in the acid solution of the melt without the need for tedious separations.

The flux employed will depend on the nature of the insoluble substance. Acidic materials are attacked by basic fluxes (carbonates, hydroxides and metaborates), whereas basic materials are attacked by acidic fluxes (pyroborates, pyrosulphates and acid fluorides). In some instances an oxidising medium is useful, in which case sodium peroxide or sodium carbonate mixed with sodium peroxide or potassium nitrate may be used. The fusion vessel must be carefully chosen; platinum crucibles are employed for sodium carbonate, lithium metaborate and potassium pyrosulphate; nickel or silver crucibles for sodium hydroxide or potassium hydroxide; nickel, gold, silver or iron crucibles for sodium carbonate and/or sodium peroxide; nickel crucibles for sodium carbonate and potassium nitrate (platinum is slightly attacked).

To prepare samples for X-ray fluorescence spectroscopy, lithium metaborate is the preferred flux because lithium does not give rise to interfering X-ray emissions. The fusion may be carried out in platinum crucibles or in crucibles made from specially prepared graphite; these graphite crucibles can also be used for the vacuum fusion of metal samples for the analysis of occluded gases.

To carry out the fusion, a layer of flux is placed at the bottom of the crucible, then an intimate mixture of the flux and the finely divided substance added; the crucible should be not more than about half-full, and should generally be kept covered during the whole process. The crucible is very gradually heated at first, and the temperature slowly raised to the required temperature. The final temperature should not be higher than is actually necessary; any possible further attack of the flux upon the crucible is thus avoided. When the fusion, which usually takes 30–60 min, has been completed, the crucible is grasped by means of the crucible tongs and gently rotated and tilted so the molten material distributes itself around the walls of the container and solidifies there as a thin layer. This procedure greatly facilitates the subsequent detachment and solution of the fused mass. When cold, the crucible is placed in a casserole, porcelain dish, platinum basin or Pyrex beaker (according to the nature of the flux) and covered with water. Acid is added, if necessary, the vessel is covered with a clockglass, and the temperature is raised to 95–100 °C and maintained until solution is achieved.

Many of the substances which require fusion treatment to render them soluble will in fact dissolve in mineral acids if the digestion with acid is carried out under pressure, and consequently at higher temperatures than those normally achieved. Such drastic treatment requires a container capable of withstanding the requisite pressure, and also resistant to chemical attack; these conditions are met in **acid digestion vessels** (bombs). Acid digestion vessels comprise a stainless steel pressure vessel (capacity 50 mL) with a screw-on lid and fitted with a Teflon liner. They may be heated to 150–180 °C and will withstand pressures of 80–90 atm (8–9 MPa); under these conditions it is possible to decompose refractory materials in about 45 min. Apart from the time saving and the cost saving – there is no need for expensive platinum ware – other advantages are that no losses can occur during the treatment, and the resulting solution is free from the heavy loading of alkali metals which follows the usual fusion procedures. Digestions of this type can now be carried out in microwave ovens using vessels specially constructed from Teflon. A full discussion of decomposition techniques is given in the literature.[9]

3.32 Decomposing organic compounds

Analysis of organic compounds for elements such as halogens, phosphorus or sulphur is achieved by combustion of the organic material in an atmosphere of oxygen; the inorganic constituents are thus converted to forms which can be determined by titrimetric or spectrophotometric procedures. The method was developed by Schöniger[10,11] and is usually known as the Schöniger oxygen flask method. A number of reviews have been published[12,13] giving considerable detail on all aspects of the procedure.

In outline the procedure consists of carefully weighing about 5–10 mg of sample on to a shaped piece of paper (Figure 3.11(b)) which is folded in such a way that the tail (wick) is free. This is then placed in a platinum basket or carrier suspended from the ground-glass stopper of a 500 mL or 1 L flask. The flask, containing a few millilitres of absorbing solution (e.g. aqueous sodium hydroxide), is filled with oxygen then sealed using the stopper with the platinum basket attached.

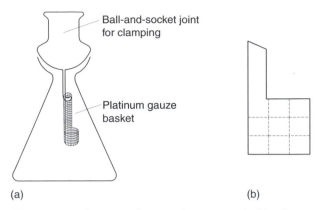

Figure 3.11   Decomposing organic compounds: (a) safety oxygen flask, (b) paper shape for wrapping samples

The wick of the sample paper can either be ignited before the stopper is placed in the neck of the flask, or better still ignited by remote electrical control or an infrared lamp. Combustion is rapid and usually complete within 5–10 s. After standing for a few minutes until any combustion cloud has disappeared, the flask is shaken for 2–3 min to ensure complete absorption has taken place. The solution can then be treated by a method appropriate to the element being determined.

Organic sulphur is converted to sulphur trioxide and sulphur dioxide by the combustion, absorbed in hydrogen peroxide, and the sulphur determined as sulphate. The combustion products of organic halides are usually absorbed in sodium hydroxide containing some hydrogen peroxide. The resulting solutions may be analysed by a range of available procedures. For chlorides the method most commonly used is argentimetric potentiometric titration,[14] whereas for bromides it is mercurimetric titration.[15]

Phosphorus from organophosphorus compounds, which are combusted to give mainly orthophosphate, can be absorbed by either sulphuric acid or nitric acid and readily determined spectrophotometrically either by the molybdenum blue method or as the phosphovanadomolybdate (Section 17.27). Procedures have also been devised for determining metallic constituents. Mercury is absorbed in nitric acid and titrated with sodium diethyldithiocarbamate, and zinc is absorbed in hydrochloric acid and determined by an EDTA titration (Section 10.88).

The simplest method for decomposing an organic sample is to heat it in an open crucible until all carbonaceous matter has been oxidised, leaving a residue of inorganic components, usually as oxide. The residue can then be dissolved in dilute acid to give a solution which can be analysed by appropriate procedures. This technique is known as **dry ashing**; it is obviously inapplicable when the inorganic component is volatile. Under these conditions the wet ashing procedure described under perchloric acid must be used. A full discussion is given in the literature.[16]

3.33   Precipitation

The conditions for precipitation of inorganic substances are given in Section 11.2. Precipitations are usually carried out in resistance-glass beakers, and the solution of the precipitant is added slowly (e.g. by pipette, burette or tap funnel) and with efficient stirring of the suitably diluted solution. The addition must always be made without splashing; this is best achieved by allowing the solution of the reagent to flow down the side of the beaker or precipitating vessel. Only a moderate excess of the reagent is generally required; a very large excess may lead to increasing solubility (Section 2.14) or contamination of the precipitate. After the precipitate has settled, a few drops of the precipitant should always be added to determine whether further precipitation occurs. As a general rule, precipitates are not filtered off immediately after they have been formed; unless they are definitely colloidal, such as iron(III) hydroxide, most precipitates require more or less digestion to complete the precipitation and make all particles of filterable size. Sometimes digestion is carried out by setting the beaker aside and leaving the precipitate in contact with the mother liquor at room temperature for 12–24 h; and where a higher temperature is permissible, digestion is usually effected near the boiling point of the solution. Hotplates, water baths, or even a low flame if no bumping occurs, are used to heat the mixture; in all cases the beaker should be covered with a clockglass, convex down. If the solubility of the precipitate is appreciable, the solution may need to attain room temperature before filtration.

3.34   Filtration

Filtration is the separation of the precipitate from the mother liquor; the object is to get the precipitate and the filtering medium quantitatively free from the solution. The systems employed for filtration are (1) filter paper; (2) porous fritted plates made of resistance glass, e.g. Pyrex (sintered-glass filtering crucibles), of silica (Vitreosil filtering crucibles), or made of porcelain (porcelain filtering crucibles); see Section 3.24. The choice of the filtering medium will be controlled by the nature of the precipitate (filter paper is especially suitable for gelatinous precipitates) and the cost. The limitations of the various filtering media are given below.

3.35   Filter papers

Quantitative filter papers must have a very small ash content; this is achieved during manufacture by washing with hydrochloric and hydrofluoric acids. The sizes generally chosen are circles of 7.0, 9.0, 11.0 and 12.5 cm diameter; 9.0 and 11.0 cm are most widely used. The ash of an 11 cm circle should not exceed 0.0001 g; if the ash exceeds this value, it should be deducted from the weight of the ignited residue. Manufacturers give values for the average ash per paper; this may also be determined by igniting several filter papers in a crucible. Quantitative filter paper is made in various degrees of porosity. The texture of the filter paper must allow rapid filtration while retaining the smallest particles of precipitate. Three textures are generally made, one for very fine precipitates, one for the average precipitate containing medium-sized particles, and one for gelatinous precipitates and coarse particles. The speed of filtration depends on the porosity of the paper.

'Hardened' filter papers are made by further treatment of quantitative filter papers with acid; they have an extremely small ash, a much greater mechanical strength when wet, and are more resistant to acids and alkalis. They should be used in all quantitative work. Table 3.7 gives the characteristics for the Whatman series of hardened ashless filter papers. The size of the filter paper selected for a particular operation is determined by the bulk of the precipitate, not by the volume of the liquid to be filtered. The entire precipitate should occupy about one-third of the filter's capacity at the end of the filtration. The funnel should match the filter paper in size; the folded paper should extend to within 1–2 cm of the top of the funnel, but never closer than 1 cm.

Table 3.7   *Whatman quantitative filter papers*

Filter paper	Hardened ashless		
Number	540	541	542
Speed	medium	fast	slow
Particle size retention	medium	coarse	fine
Ash (%)	0.008	0.008	0.008

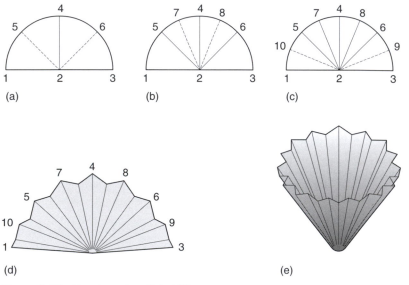

Figure 3.12   How to make a fluted filter paper

To promote rapid filtration, use a funnel with an angle as near as possible to 60° and a stem of length about 15 cm. The filter paper must be carefully fitted into the funnel so the upper portion beds tightly against the glass. To prepare the filter paper for use, the dry paper is usually folded exactly in half and exactly again in quarters. The folded paper is then opened so that a 60° cone is formed with three thicknesses of paper on one side and a single thickness on the other; the paper is then adjusted to fit the funnel. The paper is placed in the funnel, moistened thoroughly with water, pressed down tightly to the sides of the funnel, then filled with water. If the paper fits properly, the stem of the funnel will remain filled with liquid during the filtration. Sometimes filtration may be achieved more rapidly using a fluted filter paper (Figure 3.12).

To carry out a filtration, the funnel containing the properly fitted paper is placed in a funnel stand (or supported vertically in some other way) and a clean beaker is placed so the stem of the funnel just touches the side; this will prevent splashing. The liquid to be filtered is then poured down a glass rod into the filter, directing the liquid against the side of the filter, not into the apex; the lower end of the stirring rod should be very close to the filter paper, but not quite touching, on the side having three thicknesses of paper. The paper is never filled completely with the solution; the level of the liquid should not rise closer than within 5–10 mm of the top of the paper. A precipitate which tends to remain in the bottom of the beaker should be removed by holding the glass rod across the beaker, tilting the beaker, and directing a jet of water from a wash bottle so the precipitate is rinsed into the filter funnel. This procedure may also be used to transfer the last traces of the precipitate in the beaker to the filter. Any precipitate which adheres firmly to the side of the beaker or to the stirring rod may be removed with a rubber-tipped rod or policeman (Section 3.23).

Filtration by suction is rarely necessary; with gelatinous and some finely divided precipitates, the suction will draw the particles into the pores of the paper, and the speed of filtration will actually be reduced rather than increased.

3.36   Crucibles with permanent porous plates

Section 3.24 has already mentioned crucibles with porous plates. In use they are supported in a special holder, known as a crucible adaptor, by means of a wide rubber tube (Figure 3.13); the bottom of the crucible should be quite free from the side of the funnel and free from the rubber gasket; this ensures the filtrate does not come into contact with the rubber. The adaptor passes through a one-holed rubber bung into a large filter flask of about 750 mL capacity. The tip of the funnel must project below the side arm of the filter flask so there is no risk that the liquid may be sucked out of the filter flask. The filter flask should be coupled with another flask of similar capacity which is connected to a water filter pump; if the water in the pump should suck back, it will first enter the empty flask and the filtrate will not be contaminated. It is advisable also to have some sort of pressure regulator to limit the maximum pressure under which filtration is conducted. A simple method is to insert a glass tap in the second filter flask, as in Figure 3.13; alternatively, a glass T-piece may be introduced between the receiver and the pump, and one arm closed either by a glass tap or by a piece of heavy rubber tubing (pressure tubing) carrying a screw clip.

When the apparatus is assembled, the crucible is half-filled with distilled water, then gentle suction is applied to draw the water through the crucible. When the water has passed through, suction is maintained for 1–2 min to remove as much water as possible from the filter plate. The crucible is then placed on a small ignition dish, saucer or watchglass and dried to constant weight at the same temperature as will be used in drying the precipitate. For temperatures up to about 250 °C, use a thermostatically controlled electric oven. For higher temperatures, the crucible may be heated in an electrically heated muffle furnace. In all cases the crucible is allowed to cool in a desiccator before weighing.

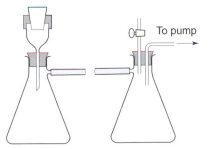

Figure 3.13   How to support crucibles using a wide rubber tube

When transferring a precipitate into the crucible, the same procedure is employed as described in Section 3.35. Take care that the liquid level in the crucible is never less than 1 cm from the top of the crucible.

With both sintered-glass and porous-base crucibles, avoid filtering materials that may clog the filter plate. A new crucible should be washed with concentrated hydrochloric acid then with distilled water. The crucibles are chemically inert and are resistant to all solutions which do not attack silica; they are attacked by hydrofluoric acid, fluorides and strongly alkaline solutions.

Crucibles fitted with permanent porous plates are cleaned by shaking out as much of the solid as possible, then dissolving out the remainder of the solid with a suitable solvent. A hot $0.1\,M$ solution of the tetrasodium salt of the ethylenediaminetetra-acetic acid is an excellent solvent for many of the precipitates encountered in analysis, except metallic sulphides and hexacyanoferrates(III). These include barium sulphate, calcium oxalate, calcium phosphate, calcium oxide, lead carbonate, lead iodate, lead oxalate and ammonium magnesium phosphate. The crucible may be completely immersed in the hot reagent, or the hot reagent may be drawn by suction through the crucible.

3.37   Washing precipitates

Most precipitates are produced in the presence of one or more soluble compounds. Since the soluble compounds are frequently not volatile at the drying temperature of the precipitate, it is necessary to wash the precipitate to remove impurities as completely as possible. The minimum volume of the washing liquid required to remove the objectionable matter should be used, since no precipitate is absolutely insoluble. Qualitative tests for removal of the impurities should be made on small volumes of the filtered washing solution. It is better to wash with a number of small portions of the washing liquid, which are well drained between each washing, than with one or two large portions, or by adding fresh portions of the washing liquid while solution still remains on the filter (Section 11.2).

The ideal washing liquid should comply as far as possible with the following conditions:

1. It should have no solvent action on the precipitate, but dissolve foreign substances easily.
2. It should have no dispersive action on the precipitate.
3. It should form no volatile or insoluble product with the precipitate.
4. It should be easily volatile at the drying temperature of the precipitate.
5. It should contain no substance which is likely to interfere with subsequent determinations in the filtrate.

In general, pure water should not be used unless it is certain it will not dissolve appreciable amounts of the precipitate. If the precipitate is appreciably soluble in water, a common ion is usually added, since any electrolyte is less soluble in a dilute solution containing one of its ions than it is in pure water (Section 2.16); calcium oxalate may be washed with dilute ammonium oxalate solution. If the precipitate tends to become colloidal and pass through the filter paper (this is frequently observed with gelatinous or flocculent precipitates), a wash solution containing an electrolyte must be employed (Section 11.2). The nature of the electrolyte is immaterial, provide it has no action on the precipitate during washing and is volatilised during the final heating. Ammonium salts are usually selected for this purpose; ammonium nitrate solution is used for washing iron(III) hydroxide.

Sometimes it is possible to select a solution which will both reduce the solubility of the precipitate and prevent peptisation; for example, dilute nitric acid is used with silver chloride. Some precipitates tend to oxidise during washing; then the precipitate cannot be allowed to run dry, and a special washing solution which reconverts the oxidised compounds into the original condition must be employed, e.g. acidified hydrogen sulphide water for copper sulphide. Gelatinous precipitates, like aluminium hydroxide, require more washing than crystalline precipitates, such as calcium oxalate.

In most cases, particularly if the precipitate settles rapidly or is gelatinous, **washing by decantation** may be carried out. As much as possible of the liquid above the precipitate is transferred to the prepared filter (either filter paper or filter crucible), observing the usual precautions, and taking care to minimise any disturbance of the precipitate. Some 20–50 mL of a suitable wash liquid is added to the residue in the beaker, the solid stirred up and allowed to settle. If the solubility of the precipitate allows, the solution should be heated, since the rate of filtration will be higher. When the supernatant liquid is clear, as much as possible of the liquid is decanted through the filtering medium. This process is repeated three to five times (or as many times as is necessary) before the precipitate is transferred to the filter.

The main bulk of the precipitate is first transferred by mixing with the wash solution and pouring off the suspension, the process being repeated until most of the solid has been removed from the beaker. Precipitate adhering to the sides and bottom of the beaker is then transferred to the filter with the aid of a wash bottle as described in Section 3.34, using a policeman if necessary to transfer the last traces of precipitate. Finally, a wash bottle is used to wash the precipitate down to the bottom of the filter paper or to the plate of the filter crucible.

In all cases, tests for the completeness of washing must be made by collecting a small sample of the washing solution after it is estimated that most of the impurities have been removed, and applying an appropriate qualitative test. Where filtration is carried out under suction, a small test tube is placed under the crucible adaptor.

3.38   Drying and igniting precipitates

After a precipitate has been filtered and washed, it must be brought to a constant composition before it can be weighed. The further treatment, drying or igniting the precipitate, will depend on the nature of the precipitate and the nature of the filtering medium. The choice of drying or igniting depends on the temperature to which the precipitate is heated. In general, drying applies when the temperature is below 250 °C (the maximum temperature which is readily reached in the usual thermostatically controlled, electric drying oven), and ignition applies from 250 to 1200 °C. Precipitates for drying should be collected on filter paper, or in sintered-glass or porcelain filtering crucibles. Precipitates for igniting are collected on filter paper, porcelain filtering crucibles, or silica filtering crucibles. Ignition is simply effected by placing in a special ignition dish and heating with the appropriate burner; alternatively, these crucibles (indeed, any type of crucible) may be placed in an electrically heated muffle furnace equipped with a pyrometer and a means for controlling the temperature.

Thermogravimetry (TG) provides information on the temperature range to which a precipitate should be heated for a particular composition.[17,18] In general, TG curves seem to suggest that, in the past, precipitates were heated for too long and at too high a temperature. But remember that the TG curve is sometimes influenced by the experimental conditions of precipitation, and even if a horizontal curve is not obtained, it is possible that a suitable weighing form may be available over a certain temperature range. Nevertheless, TG curves do provide valuable data concerning the range of temperature over which a precipitate has a constant composition under the conditions that the analysis was made; at the very least, they provide a guide for the temperature at which a precipitate should be dried and heated for quantitative work, but due regard must be paid to the general chemical properties of the weighing form.

Although precipitates which require ignition will usually be collected in porcelain or silica filtering crucibles, there may be times when filter paper is used, then the ignition method will be somewhat different. The exact technique depends on whether or not the precipitate may be safely ignited in contact with the filter paper. Remember that some precipitates, e.g. barium sulphate, may be reduced or changed in contact with filter paper or its decomposition products.

Incineration of the filter paper in the presence of the precipitate

A silica crucible is first ignited to constant weight (i.e. to within 0.0002 g) at the same temperature as the precipitate ultimately reaches. The well-drained filter paper and precipitate are carefully detached from the funnel; the filter paper is folded so as to enclose the precipitate completely taking care not to tear the paper. The packet is placed point-down in the weighed crucible, which is supported on a pipeclay, or better, a silica triangle resting on a ring stand. The crucible is slightly inclined, and partially covered with the lid, which should rest partly on the triangle. A **very small flame** is then placed under the crucible lid; drying thus proceeds quickly and without undue risk.

When the moisture has been expelled, the flame is increased slightly so as to carbonise the paper **slowly**. The paper should not be allowed to inflame, as this may cause a mechanical expulsion of fine particles of the precipitate, owing to the rapid escape of the products of combustion. If it does catch fire, the flame should be extinguished by momentarily placing the cover on the mouth of the crucible with the aid of crucible tongs. When the paper has completely carbonised and vapours are no longer evolved, the flame is moved to the back (bottom) of the crucible and the carbon slowly burned off while the flame is gradually increased.* After all the carbon has been burned away, the crucible is covered completely (if desired, this may be done with the crucible in a vertical position) and heated to the required temperature by means of a Bunsen burner. Usually it takes about 20 min to char the paper, and 30–60 min to complete the ignition

When the ignition is ended, the flame is removed and, after 1–2 min, the crucible and lid are placed in a desiccator containing a suitable desiccant (Section 3.22) and allowed to cool for 25–30 min. The crucible and lid are then weighed. The crucible and contents are then ignited at the same temperature for 10–20 min, allowed to cool in a desiccator as before, and weighed again. The ignition is repeated until constant weight is attained. Crucibles should always be handled with clean crucible tongs and preferably with platinum-tipped tongs. Remember that 'heating to constant weight' has no real significance unless the periods of heating, cooling of the **covered** crucible, and weighing are duplicated.

Incineration of the filter paper apart from the precipitate

This method is used whenever the ignited substance is reduced by the burning paper; for example, barium sulphate, lead sulphate, bismuth oxide and copper oxide. The funnel containing the precipitate is covered by a piece of qualitative filter paper, made secure by

---

* If the carbon on the lid is oxidised only slowly, the cover may be heated separately in a flame. It should be held in clean crucible tongs.

crumpling its edges over the rim of the funnel so they engage the outer conical portion of the funnel. The funnel is placed in a drying oven maintained at 100–105 °C, for 1–2 h or until completely dry. A sheet of glazed paper about 25 cm square (white or black, to contrast with the colour of the precipitate) is placed on the bench away from all draughts. The dried filter paper is removed from the funnel, then as much as possible of the precipitate is removed and allowed to drop onto a clockglass resting on the glazed paper. This may be done by very gently rubbing the sides of the filter paper together, when the bulk of the precipitate becomes detached and drops onto the clockglass. Any small particles of the precipitate which may have fallen onto the glazed paper are brushed into the clockglass with a small camel-hair brush.

The clockglass containing the precipitate is covered with a larger clockglass or with a beaker. The filter paper is then carefully folded and placed inside a weighed porcelain or silica crucible. The crucible is placed on a triangle and the filter paper incinerated as above. The crucible is allowed to cool, and the filter ash subjected to a suitable chemical treatment in order to convert any reduced or changed material into the form finally desired. The cold crucible is then placed on the glazed paper and the main part of the precipitate carefully transferred from the clockglass to the crucible. A small camel-hair brush will assist in the transfer. Finally, the precipitate is brought to constant weight by heating to the necessary temperature as explained in the previous section.

3.39 References

1. S G Luxon (ed) 1992 *Hazards in the chemical laboratory*, 5th edn, Royal Society of Chemistry, London
2. C Woodward and H N Redman 1973 *High precision titrimetry*, Society for Analytical Chemistry, London: (a) p. 5; (b) p. 14; (c) p. 11; (d) p. 10
3. Graduated apparatus: British Standards and their ISO equivalents

(a)	Graduated flasks	BS 5898: 1980 (1994)	ISO 384: 1978
		BS 1792: 1982 (1993)	ISO 1042: 1983
(b)	Pipettes	BS 1583: 1986 (1993)	ISO 648: 1984
(c)	Burettes	BS 846: 1985 (1993)	ISO 385/1, 385/2
(d)	Calibration procedures	BS 6696: 1986 (1992)	ISO 4787: 1984

4. BS 3978 (1987) *Water for laboratory use* British Standards Institution, London
5. B S Furniss, A J Hannaford, V Rogers, P W G Smith and A R Tatchell 1978 *Vogel's practical organic chemistry*, 4th edn, Longman, London
6. D F Shriver 1969 *The manipulation of air-sensitive materials*, McGraw-Hill, New York
7. D J Bucknell (ed) 1984 *'AnalaR' standards for laboratory chemicals*, 8th edn, BDH Chemicals, Poole
8. *Reagent Chemicals*, 7th edn, American Chemical Society, Washington DC, (1986)
9. R Bock 1979 *Handbook of decomposition methods in chemistry*, International Textbook Company, Glasgow (translated by I Marr)
10. W Schöniger 1955 *Mikrochim, Acta*, 123
11. W Schöniger 1956 *Mikrochim, Acta*, 869
12. A M G Macdonald 1965 In C N Reilley (ed) *Advances in analytical chemistry and instrumentation*, Volume 4, Interscience, New York
13. A M G Macdonald 1961 *Analyst*, **86**; 3
14. Analytical Methods Committee 1963 *Analyst*, **88**; 415
15. R C Denney and P A Smith 1974 *Analyst*, **99**; 166

16. T Gorsuch 1970 *The destruction of organic matter*, Pergamon, Oxford
17. C Duval 1963 *Inorganic thermogravimetric analysis*, Elsevier, Amsterdam
18. *Wilson and Wilson Comprehensive Analytical Chemistry*, Volume XII, *Thermal analysis*, Elsevier, Amsterdam: Part A, J Paulik and F Paulik (eds) 1981; Part D, J Seslik (ed) 1984

3.40   Bibliography

J A Beran 1994 *Chemistry in the laboratory*, 2nd edn, John Wiley, New York

G Christian 1994 *Analytical Chemistry*, 5th edn, John Wiley, New York

N T Freeman and J Whitehead 1982 *Introduction to safety in the chemical laboratory*, Academic Press, London

I M Kolthoff and P J Elving 1978 *Treatise on analytical chemistry*, Part 1, *Theory and practice*, Volume 1, 2nd edn, Interscience, New York

S Kotrly and L Sucha 1985 *Handbook of chemical equilibria in analytical chemistry*, Ellis Horwood, Chichester

R E Lawn, M Thompson and R Walker 1997 *Proficiency testing in analytical chemistry*, Royal Society of Chemistry, London

National Bureau of Standards 1979 *Handbook 44: specifications, tolerances and other technical requirements for weighing and measuring as adopted by the National Conference on Weights and Measures*, NBS, Washington DC

P J Potts 1987 *A handbook of silicate rock analysis*, 2nd edn, Blackie, Glasgow

Royal Society of Chemistry 1989–92 *Chemical safety data sheets*, Volumes 1–5, Royal Society of Chemistry, London

Royal Society of Chemistry 1996 *COSHH in laboratories*, 2nd edn, Royal Society of Chemistry, London

Royal Society of Chemistry *Laboratory Hazards Bulletin*, Royal Society of Chemistry, London

D A Skoog and J J Leary 1992 *Principles of instrumental analysis*, 4th edn, Saunders, New York

G Weiss (ed) 1986 *Hazardous chemicals data book*, 2nd edn, Noyes, Trenton NJ

A Weissberger (ed) 1986 *Techniques of Chemistry*, Volume 2, *Organic solvents: physical properties and methods of purification*, 4th edn, John Wiley, New York

Wilson and Wilson Comprehensive Analytical Chemistry, Volumes 1A (1959) and 1B (1960), Elsevier, Amsterdam

4

Statistics: introduction to chemometrics

4.1 Limitations of analytical methods

The function of the analyst is to obtain a result as near to the true value as possible by the correct application of the analytical procedure employed. The level of confidence that analysts may enjoy in their results will be very small unless they have knowledge of the accuracy and precision of the method used as well as being aware of the sources of error which may be introduced. Quantitative analysis is not simply a case of taking a sample, carrying out a single determination and then claiming that the value obtained is irrefutable. It also requires a sound knowledge of the chemistry involved, of the possibilities of interferences from other ions, elements and compounds as well as the statistical distribution of values.

The aims of this chapter may be summarised as follows:

1. To explain some of the terms used and to describe the classical statistical procedures that can be applied to analytical results.
2. To give an introduction in methods for the design and optimisation of measurement procedures.
3. To outline the scope that chemometrics may provide to enhance the information obtained from analytical data.

4.2 Classification of errors

Systematic (determinate) errors

These are errors which can be avoided, or whose magnitude can be determined. The most important of them are operational and personal errors, instrumental and reagent errors, errors of method.

Operational and personal errors These are due to factors for which the individual analyst is responsible and are not connected with the method or procedure; they form part of the 'personal equation' of an observer. The errors are mostly physical in nature and occur when sound analytical technique is not followed. Examples are incomplete drying of analytical samples before weighing; mechanical loss of materials during sample dissolution from effervescence or from bumping; incorrect technique involving the transfer of solutions; lack of reproducibility in solvent extraction methods. Sample treatment before measurement is often the source of major error in chemical analysis. Lack of care in this stage will too often give rise to meaningless results. Personal errors may arise from the constitutional

117

inability of an individual to make certain observations accurately. Thus, some persons are unable to judge colour changes sharply in visual titrations, which may result in a slight overstepping of the end point. Other personal decisions include the estimation of a value between two scale divisions of a burette or a meter.

Instrumental and reagent errors   These arise from the faulty construction of balances, the use of uncalibrated or improperly calibrated weights, graduated glassware and other instruments; the attack of reagents upon glassware, porcelain, etc., resulting in the introduction of foreign materials; and the use of reagents containing impurities.

Errors of method   These are the most serious errors because often they can be difficult to detect. Examples include a pH meter that has been wrongly standardised; background absorption in atomic absorption spectroscopy; faulty detector response in chromatographic and spectroscopic methods. Errors in classical analysis include solubility of precipitates and the decomposition or volatilisation on ignition of weighing forms in gravimetry. Errors may arise in titrimetry if there are differences between the observed end point and the stoichiometric equivalence point of a reaction. One of the most common errors in methodology is due to 'matrix effects' when there is a difference between the bulk composition of the analyte sample solution and the composition of the standard solutions used to establish the calibration graph. (Section 15.15). It is especially important in trace analysis to ensure the solvent has a high state of purity and certainly should be free from any traces of the analyte whose concentration is to be measured.

Random (indeterminate) errors

These errors manifest themselves by the slight variations that occur in successive measurements made by the same observer with the greatest care under as nearly identical conditions as possible. They are due to causes over which the analyst has no control, and which in general are so intangible they are incapable of analysis. If a **sufficiently large number of observations** is taken, these errors lie on a curve of the form shown in Figure 4.1 (Section 4.9). An inspection of this error curve shows (a) small errors occur more frequently than large ones; and (b) positive and negative errors of the same numerical magnitude are equally likely to occur.

4.3   Accuracy

The accuracy of a determination may be defined as the concordance between it and the true or most probable value. It follows, therefore, that systematic errors cause a constant error (either too high or too low) and thus affect the accuracy of a result. For analytical methods there are two possible ways of determining the accuracy; the so-called absolute method and the comparative method.

Absolute method

A synthetic sample containing known amounts of the constituents in question is used. Known amounts of a constituent can be obtained by weighing out pure elements or compounds of known stoichiometric composition. These substances, primary standards, may be available commercially or they may be prepared by the analyst and subjected to rigorous purification by recrystallisation, etc. The substances must be of known purity. The test of the accuracy of the method under consideration is carried out by taking varying amounts of the constituent and proceeding according to specified instructions. The amount of the constituent must be varied, because the determinate errors in the procedure may be a function of the amount used. The difference between the mean of an adequate number of results and the amount of the constituent actually present is a measure of the accuracy of the method in the absence of foreign substances.

The constituent in question will usually have to be determined in the presence of other substances, and it will therefore be necessary to know their effect upon the determination. This will require testing the influence of a large number of elements, each in varying amounts – a major undertaking. The scope of such tests may be limited by considering the determination of the component in a specified range of concentration in a material whose composition is more or less fixed with respect to the elements which may be present and their relative amounts. It is desirable, however, to study the effect of as many foreign elements as feasible. In practice it is frequently found that separations will be required before a determination can be made in the presence of varying elements; the accuracy of the method is likely to be largely controlled by the separations involved.

Comparative method

Sometimes, as in the analysis of a mineral, it may be impossible to prepare solid synthetic samples of the desired composition. It is then necessary to resort to standard samples of the material in question (mineral, ore, alloy, etc.) in which the content of the constituent sought has been determined by one or more supposedly 'accurate' methods of analysis. This comparative method, involving secondary standards, is obviously not altogether satisfactory from the theoretical standpoint, but is nevertheless very useful in applied analysis. Standard samples can be obtained from various sources (Section 4.5).

If several fundamentally different methods of analysis for a given constituent are available, e.g. gravimetric, titrimetric, spectrophotometric, the agreement between at least two methods of essentially different character can usually be accepted as indicating the absence of an appreciable systematic error in either (a systematic error is one which can be evaluated experimentally or theoretically).

4.4   Precision

Precision may be defined as the concordance of a series of measurements of the same quantity. Accuracy expresses the correctness of a measurement, and precision the 'reproducibility' of a measurement (the definition of precision will be modified later). Precision always accompanies accuracy, but a high degree of precision does not imply accuracy. This may be illustrated by the following example.

A substance was known to contain 49.10% ± 0.02% of constituent A. The results obtained by two analysts using the same substance and the same analytical method were as follows.

ANALYST 1 %A 49.01; 49.25; 49.08; 49.14

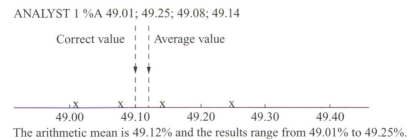

The arithmetic mean is 49.12% and the results range from 49.01% to 49.25%.

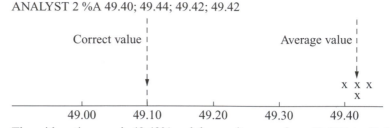

ANALYST 2 %A 49.40; 49.44; 49.42; 49.42

The arithmetic mean is 49.42% and the results range from 49.40% to 49.44%.

We can summarise the results of the analyses as follows:

(a)   The values obtained by Analyst 1 are accurate (very close to the correct result), but the precision is inferior to the results given by Analyst 2. The values obtained by Analyst 2 are very precise but are not accurate.

(b)   The results of Analyst 1 occur on both sides of the mean value and could be attributed to random errors. It is apparent that there is a constant (systematic) error present in the results of Analyst 2.

Precision was previously described as the reproducibility of a measurement. However, the modern analyst makes a distinction between the terms 'reproducible' and 'repeatable'. On further consideration:

(c)   If Analyst 2 had made the determinations on the same day in rapid succession, then this would be defined as 'repeatable' analysis. However, if the determinations had been made on separate days when laboratory conditions may vary, this set of results would be defined as 'reproducible'.

Thus, there is a distinction between a within-run precision (repeatability) and a between-run precision (reproducibility).

4.5   How to reduce systematic errors

Calibration of apparatus and application of corrections   All instruments (weights, flasks, burettes, pipettes, etc.) should be calibrated, and the appropriate corrections applied to the original measurements. In some cases where an error cannot be eliminated, it is possible to apply a correction for the effect that it produces; thus an impurity in a weighted precipitate may be determined and its weight deducted. To reduce systematic errors, recalibrate instruments frequently.

Running a blank determination   This consists in carrying out a separate determination, the sample being omitted, under exactly the same experimental conditions as employed in the actual analysis of the sample. The object is to find out the effect of the impurities introduced through the reagents and vessels, or to determine the excess of standard solution necessary to establish the end point under the conditions met with in the titration of the unknown sample. A large blank correction is undesirable, because the exact value then becomes uncertain and the precision of the analysis is reduced.

Running a control determination   This consists in carrying out a determination on a standard substance under conditions as close as possible to the experimental conditions.

The quantity of the standard substance should contain the same weight of the constituent as is contained in the unknown sample. The weight of the constituent in the unknown sample can then be calculated from the relation

$$\frac{\text{result found for standard}}{\text{result found for unknown}} = \frac{\text{weight of constituent in standard}}{x}$$

where x is the weight of the constituent in the unknown.

Note that standard samples which have been analysed by a number of skilled analysts are commercially available. These include certain primary standards (sodium oxalate, potassium hydrogenphthalate, arsenic(III) oxide, and benzoic acid) and ores, ceramic materials, irons, steels, steelmaking alloys, and non-ferrous alloys. Many of them are also available as BCS Certified Reference Materials (CRM) supplied by the Bureau of Analysed Samples Ltd, Newham Hall, Middlesborough, UK, who also supply EURONORM Certified Reference Materials (ERCM), the composition of which is specified on the basis of results obtained by a number of laboratories within the EU. BCS Reference Materials are obtainable from the Community Bureau of Reference, Brussels, Belgium. In the United States similar reference materials are supplied by the National Bureau of Standards.

Use of independent methods of analysis   In some instances the accuracy of a result may be established by carrying out the analysis in an entirely different manner. Thus in water hardness, the calcium and magnesium concentrations determined by atomic absorption (Section 15.20) may be compared with the results obtained by EDTA titration (Section 10.76). Another example that may be mentioned is the determination of the strength of a hydrochloric solution both by titration with a standard solution of a strong base and by precipitation and weighing as silver chloride. If the results obtained by the two radically different methods are concordant, it is highly probable that the values are correct within small limits of error.

Running parallel determinations   These serve as a check on the result of a single determination and indicate only the precision of the analysis. The values obtained for constituents which are present in not too small an amount should not vary among themselves by more than three parts per thousand. If larger variations are shown, the determinations must be repeated until satisfactory concordance is obtained. Duplicate, and at most triplicate, determinations should suffice. Note that good agreement between duplicate and triplicate determinations does not justify the conclusion that the result is correct; a constant error may be present. The agreement merely shows that the accidental errors, or variations of the determinate errors, are the same or nearly the same in the parallel determinations.

Standard addition   A known amount of the constituent being determined is added to the sample, which is then analysed for the total amount of constituent present. The difference between the analytical results for samples with and without the added constituent gives the recovery of the amount of added constituent. If the recovery is satisfactory, our confidence in the accuracy of the procedure is enhanced. The method is usually applied to physicochemical procedures such as polarography and spectrophotometry. Alternatively, a series of standard additions can be performed and the concentration of the analyte is determined by graphical extrapolation. For full details the reader is referred to Section 4.20.

Internal standards   This procedure is of particular value in chromatographic determinations. It involves adding a fixed amount of a reference material (the internals standard) to a series of known concentrations of the material to be measured. The ratios of the physical value (peak size) of the internal standard and the series of known concentrations are plotted against the concentration values. This should give a straight line. Any unknown concentration can then be determined by adding the same quantity of internal standard and finding where the ratio obtained falls on the concentration scale.

Amplification methods   In determinations in which a very small amount of material is to be measured, this may be beyond the limits of the apparatus available. In these circumstances if the small amount of material can be reacted in such a way that every molecule produces two or more molecules of some other measurable material, the resultant amplification may then bring the quantity to be determined within the scope of the apparatus or method available.

Isotopic dilution   A known amount of the element being determined, containing a radioactive isotope, is mixed with the sample and the element is isolated in a pure form (usually as a compound), which is weighed or otherwise determined. The radioactivity of the isolated material is measured and compared with the radioactivity of the added element; the weight of the element in the sample can then be calculated.

4.6   Significant figures

The term 'digit' denotes any one of the ten numerals, including the zero. A significant figure is a digit which denotes the amount of the quantity in the place in which it stands. The digit 0 is a significant figure except when it is the first figure in a number. Thus in the quantities 1.2680 g and 1.0062 g the zero is significant, but in the quantity 0.0025 kg the zeros are not significant figures; they serve only to locate the decimal point and can be omitted by proper choice of units, i.e. 2.5 g. The first two numbers contain five significant figures, but 0.0025 contains only two significant figures.

Observed quantities should be recorded with one uncertain figure retained. Thus in most analyses weights are determined to the nearest tenth of a milligram, e.g. 2.1546 g. This means that the weight is less than 2.1547 g and more than 2.1545 g. A weight of 2.150 g would signify that it had been determined to the nearest milligram, and that the weight is nearer to 2.150 g than it is to either 2.151 g or 2.149 g. The digits of a number which are needed to express the precision of the measurement from which the number was derived are known as significant figures.

There are a number of rules for computations with which the student should be familiar.

1. Retain as many significant figures in a result or in any data as will give only one uncertain figure. Thus a volume which is known to be between 20.5 mL and 20.7 mL should be written as 20.6 mL, but not as 20.60 mL, since 20.60 mL would indicate that the value lies between 20.59 mL and 20.61 mL. If a weight, to the nearest 0.1 mg, is 5.2600 g it should not be written as 5.260 g or 5.26 g, since 5.26 g indicates an accuracy of a centigram and 5.260 g indicates an accuracy of a milligram.
2. In rounding off quantities to the correct number of significant figures, add 1 to the last figure retained if the following figure (which has been rejected) is 5 or over. Thus the average of 0.2628, 0.2623, and 0.2626 is 0.2626 (0.2625_7).
3. In addition or subtraction there should be in each number only as many significant figures as there are in the least accurately known number. Thus the addition

$$168.11 + 7.045 + 0.6832$$

should be written

$$168.11 + 7.05 + 0.68 = 175.84$$

The sum or difference of two of more quantities cannot be more precise than the quantity having the largest uncertainty.

4. In multiplication or division, retain in each factor one more significant figure than is contained in the factor having the largest uncertainty. The percentage precision of a product or quotient cannot be greater than the percentage precision of the least precise factor entering into the calculation. Thus the multiplication

$$1.26 \times 1.236 \times 0.6834 \times 24.8652$$

should be carried out using the values

$$1.26 \times 1.236 \times 0.683 \times 24.87$$

and the result expressed to three significant figures. When using a calculator, it is more useful to retain all the digits which are then 'rounded off' with the final answer.

4.7 Calculators and computers

In addition to the normal arithmetic functions, a suitable calculator for statistical work should enable the user to evaluate the mean and standard deviation (Section 4.8), linear regression and correlation coefficient (Section 4.16). The results obtained by the use of the calculator must be carefully scrutinised to ascertain the number of significant figures to be retained, and should always be checked against a 'rough' arithmetical calculation to ensure there are no gross computational errors. Computers are used for processing large amounts of data. Although computer programming is outside the scope of this book, note that many standard programs now exist.

The computer may also be interfaced with most types of electronic equipment used in the laboratory. This facilitates the collection and processing of the data, which may be stored on floppy or hard disks for later use. There is a large amount of commercial software available for performing the calculations described later in this chapter.

4.8 Mean and standard deviation

When a quantity is measured with the greatest exactness of which the instrument, method and observer are capable, it is found that the results of successive determinations differ among themselves to a greater or lesser extent. The average value is accepted as the most probable. This may not always be the true value. In some cases the difference may be small, in others it may be large; the reliability of the result depends upon the magnitude of this difference. It is therefore of interest to enquire briefly into the factors which affect and control the trustworthiness of chemical analysis.

The absolute error of a determination is the difference between the observed or measured value and the true value of the quantity measured. It is a measure of the accuracy of the measurement.

The relative error is the absolute error divided by the true value; it is usually expressed in terms of percentage or in parts per thousand. The true or absolute value of a quantity cannot be established experimentally, so the observed result must be compared with the most probable value. With pure substances the quantity will ultimately depend upon the relative atomic mass of the constituent elements. Determinations of the relative atomic mass have been made with the utmost care, and the accuracy obtained usually far exceeds that attained in ordinary quantitative analysis; the analyst must accordingly accept their reliability. With natural or industrial products, we must accept provisionally the results obtained by analysts of repute using carefully tested methods. If several analysts determine the same constituent in the same sample by different methods, the most probable value, which is usually the average, can be deduced from their results. In both cases the establishment of the most probable value involves the application of statistical methods and the concept of precision.

In analytical chemistry one of the most common statistical terms employed is the **standard deviation** of a population of observations. This is also called the root mean square deviation as it is the square root of the mean of the sum of the squares of the differences between the values and the mean of those values (this is expressed mathematically below) and is of particular value in connection with the normal distribution.

If we consider a series of n observations arranged in ascending order of magnitude,

$$x_1, x_2, x_3, \ldots, x_{n-1}, x_n$$

the arithmetic mean (often simply called the mean) is given by

$$\bar{x} = \frac{x_1 + x_2 + x_3 + \ldots + x_{n-1} + x_n}{n}$$

The spread of the values is measured most efficiently by the standard deviations, defined by

$$s = \sqrt{\frac{(x_1 - \bar{x})^2 + (x_2 - \bar{x})^2 + \ldots + (x_n - \bar{x})^2}{n-1}}$$

In this equation the denominator is $(n - 1)$ rather than n when the number of values is small.

The equation may also be written as

$$s = \sqrt{\frac{\Sigma (x - \bar{x})^2}{n-1}}$$

The square of the standard deviation is called the **variance**. A further measure of precision, known as the relative standard deviation (RSD), is given by

$$\text{RSD} = \frac{s}{\bar{x}}$$

This measure is often expressed as a percentage, known as the **coefficient of variation** (CV):

$$\text{CV} = \frac{s \times 100}{\bar{x}}$$

Example 4.1

Analyses of a sample of iron ore gave the following percentage values for the iron content: 7.08, 7.21, 7.12, 7.09, 7.16, 7.14, 7.07, 7.14, 7.18, 7.11. Calculate the mean, standard deviation and coefficient of variation for the values.

Results (x)	$x - \bar{x}$	$(x - \bar{x})^2$
7.08	−0.05	0.0025
7.21	0.08	0.0064
7.12	−0.01	0.0001
7.09	−0.04	0.0016
7.16	0.03	0.0009
7.14	0.01	0.0001
7.07	−0.06	0.0036
7.14	0.01	0.0001
7.18	0.05	0.0025
7.11	−0.02	0.0004

$\Sigma x = 71.30$ $\Sigma(x - \bar{x})^2 = 0.0182$

Mean $\bar{x}$ 7.13%

$$s = \sqrt{\frac{0.0182}{9}}$$

$$= \sqrt{0.0020}$$

$$= \pm 0.045\%$$

$$CV = \frac{0.045 \times 100}{7.13} = 0.63\%$$

The mean of several readings $\bar{x}$ will make a more reliable estimate of the true mean μ than is given by one observation. The greater the number of measurements (n), the closer to the true mean will be the sample average. The standard error of the mean s_x is given by

$$s_x = \frac{s}{\sqrt{n}}$$

In Example 4.1 we have

$$s_x = \pm\frac{0.045}{\sqrt{10}} = \pm 0.014$$

and if 100 measurements were made

$$s_x = \pm\frac{0.045}{\sqrt{100}} = \pm 0.0045$$

Hence the **precision** of a measurement may be improved by increasing the number of measurements.

4.9 Distribution of random errors

Section 4.8 showed that the spread of a series of results obtained from a given set of measurements can be ascertained from the value of the standard deviation. However, this term does not indicate how the results are distributed.

125

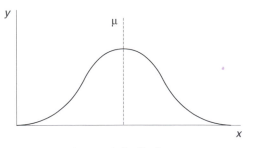

Figure 4.1 A normal distribution

If a large number of replicate readings, at least 50, are taken of a continuous variable, e.g. a titrimetric end point, the results obtained will usually be distributed about the mean in a roughly symmetrical manner. The mathematical model that best satisfies such a distribution of random errors is called the normal (or Gaussian) distribution. This is a bell-shaped curve that is symmetrical about the mean, as shown in Figure 4.1.

The curve satisfies the equation

$$\frac{1}{\sigma\sqrt{2\pi}}\exp[-(x-\mu)^2/2\sigma]$$

It is important to know that the Greek letters σ and μ refer to the standard deviation and mean respectively of a total population, whereas the Roman letters s and $\bar{x}$ are used for samples of populations, irrespective of the values of the population mean and the population standard deviation.

With this type of distribution about 68% of all values will fall within one standard deviation on either side of the mean, 95% will fall within two standard deviations, and 99.7% within three standard deviations.

For Example 4.1 on the analysis of an iron ore sample, the standard deviation is ±0.045%. If the assumption is made that the results are normally distributed, then 68% (approximately 7 out of 10 results) will be between ±0.045% and 95% will be between ±0.090% of the mean value. It follows that there will be a 5% probability (1 in 20 chance) of a result differing from the mean by more than ±0.090%, and a 1 in 40 chance of the result being 0.090% **higher** than the mean.

4.10 Reliability of results

Statistical figures obtained from a set of results are of limited value by themselves. Analysis of the results can be considered in two main categories: (a) the reliability of the results; and (b) comparison of the results with the true value or with other sets of data (Section 4.12).

A most important consideration is to be able to arrive at a sensible decision as to whether certain results may be rejected. It must be stressed that values should be rejected only when a suitable statistical test has been applied, or when there is an obvious chemical or instrumental reason that could justify exclusion of a result. Too frequently, however, there is a strong temptation to remove what may appear to be a 'bad' result without any sound justification. Consider the following example.

Example 4.2

The following values were obtained for the determination of cadmium in a sample of dust: 4.3, 4.1, 4.0, 3.2 µg g^{-1}. Should the value 3.2 be rejected?

The Q-test may be applied to solve this problem.

$$Q = \frac{|\text{questionable value} - \text{nearest value}|}{\text{largest value} - \text{smallest value}}$$

$$Q = \frac{|3.2 - 4.0|}{4.3 - 3.2} = \frac{0.8}{1.1} = 0.727$$

If the calculated value of Q exceeds the critical value given in the Q-table (Appendix 14), the questionable value may be rejected.

In this example Q calculated is 0.727 and Q critical is 0.831 for a sample size of 4. Hence the result 3.2 µg g^{-1} should be retained. But if three additional measurements were made, with the results (µg g^{-1}):

4.3, 4.1, 4.0, 3.2, 4.2, 3.9, 4.0

then

$$Q = \frac{|3.2 - 3.9|}{4.3 - 3.2} = \frac{0.7}{1.1} = 0.636$$

The value of Q critical is 0.570 for a sample size of 7, so it is justified to reject the value 3.2 µg g^{-1}. Note that the value Q has no regard to algebraic sign.

4.11   Confidence interval

When a small number of observations are made, the value of the standard deviation s does not by itself give a measure of how close the sample mean $\bar{x}$ might be to the true mean. But it is possible to calculate a confidence interval to estimate the range within which the true mean may be found. The limits of this confidence interval, known as the confidence limits, are given by the expression

$$\text{confidence limits of } \mu \text{ for } n \text{ replicate measurements} = \bar{x} \pm \frac{ts}{\sqrt{n}} \tag{4.1}$$

where t is a parameter that depends upon the number of degrees of freedom v (Section 4.12) and the confidence level required. A table of the values of t at different confidence levels and degrees of freedom v is given in Appendix 11.

Example 4.3

The mean $\bar{x}$ of four determinations of the copper content of a sample of an alloy was 8.27% with a standard deviation $s = 0.17\%$. Calculate the 95% confidence limit for the true value.

From the t-tables, the value of t for the 95% confidence level with $(n-1)$, i.e. 3 degrees of freedom, is 3.18.

Hence from equation (4.1) the 95% confidence level is

$$95\% \text{ (CL) for } \mu = 8.27 \pm \frac{3.18 \times 0.17}{\sqrt{4}}$$

$$= 8.27\% \pm 0.27\%$$

Thus, there is 95% confidence that the true value of the copper content of the alloy lies in the range 8.00% to 8.54%.

If the number of determinations in the above example had been 12, then the reader may wish to confirm that

$$95\%(\text{CL}) \text{ for } \mu = 8.27 \pm \frac{2.20 \times 0.17}{\sqrt{12}}$$

$$= 8.27\% \pm 0.11\%$$

Hence, on increasing the number of replicate determinations, the values of t and $s/\sqrt{n}$ decrease with the result that the confidence interval is smaller. But there is often a limit to the number of replicate analyses that can be sensibly performed. A method for estimating the optimum number of replicate determinations is given in Section 4.15.

4.12   Comparison of results

The comparison of the values obtained from a set of results with either the true value or with other sets of data makes it possible to determine whether the analytical procedure has been accurate and/or precise, or if it is superior to another method. There are two common methods for comparing results, Student's t-test and the variance ratio test (F-test).

These methods require a knowledge of the number of **degrees of freedom**. In statistical terms, this is the number of independent values necessary to determine the statistical quantity. Thus a sample of n values has n degrees of freedom, whereas the sum $\sum (x - \bar{x})^2$ is considered to have $n - 1$ degrees of freedom, as for any defined value of $\bar{x}$ only $n - 1$ values can be freely assigned, the nth being automatically defined from the other values.

Student's t-test

This is a test used for small samples; its purpose is to compare the mean from a sample with some standard value and to express some level of confidence in the significance of the comparison. It is also used to test the difference between the means of two sets of data $\bar{x}_1$ and $\bar{x}_2$.

The value of t is obtained from the equation

$$t = \frac{(\bar{x} - \mu)\sqrt{n}}{s} \tag{4.2}$$

where μ is the true value.

It is then related to a set of t-tables (Appendix 11) in which the probability P of the t-value falling within certain limits is expressed, either as a percentage or as a function of unity, relative to the number of degrees of freedom.

Example 4.4   *t*-test when the true mean is known

If the mean of the 12 determinations is $\bar{x} = 8.37$, and the true value is $\mu = 7.91$, say whether or not this result is significant if the standard deviation is 0.17.

From equation (4.2) we have

$$t = \frac{(8.37 - 7.91)\sqrt{12}}{0.17} = 9.4$$

From *t*-tables for 11 degrees of freedom (one less than the 12 degrees used in the calculation)

for $P = 0.10$ (10%)	0.05 (5%)	0.01 (1%)
$t = 1.80$	2.20	3.11

and as the calculated value for *t* is 9.4 the result is highly significant. The *t*-table tells us that the probability of obtaining the difference of 0.46 between the experimental and true result is less than 1 in 100. This implies that some particular bias exists in the laboratory procedure.

Had the calculated value for *t* been less than 1.80 then there would have been no significance in the results and no apparent bias in the laboratory procedure, as the tables would have indicated a probability of greater than 1 in 10 of obtaining that value. Note that these values refer to what is known as a double-sided, or two-tailed, distribution because it concerns probabilities of values both less than and greater than the mean. In some calculations an analyst may only be interested in one of these two cases, and under these conditions the *t*-test becomes single-tailed, so the probability from the tables is halved.

F-test

This is used to compare the precisions of two sets of data, e.g. the results of two different analytical methods or the results from two different laboratories. It is calculated from the equation

$$F = \frac{s_A^2}{s_B^2} \tag{4.3}$$

The larger value of *s* is always used as the numerator, so the value of *F* is always greater than unity. The value obtained for *F* is then checked for its significance against values in the *F*-table calculated from an *F*-distribution (Appendix 12) corresponding to the numbers of degrees of freedom for the two sets of data.

Example 4.5   *F*-test comparison of precisions

The standard deviation from one set of 11 determinations was $s_A = 0.210$, and the standard deviation from another 13 determinations was $s_B = 0.641$. Is there any significant difference between the precision of these two sets of results?

From equation (4.3) we have

$$F = \frac{(0.641)^2}{(0.210)^2} = \frac{0.411}{0.044} = 9.4$$

for

$$P = 0.10 \quad 0.05 \quad 0.01$$

$$F = 2.28 \quad 2.91 \quad 4.71$$

The value 2.28 corresponds to 10% probability, the value 2.91 to 5% probability and the value 4.71 to 1% probability.

Under these conditions there is less than 1 chance in 100 that these precisions are similar. To put it another way, the difference between the two sets of data is highly significant.

Had the value of F turned out to be less than 2.28, it would have been possible to say there was no significant difference between the precisions, at the 10% level.

4.13   Comparing the means of two samples

When a new analytical method is being developed it is usual practice to compare the values of the mean and precision of the new (test) method with those of an established (reference) procedure.

The value of t when comparing two sample means $\bar{x}_1$ and $\bar{x}_2$ is given by the expression

$$t = \frac{\bar{x}_1 - \bar{x}_2}{s_p \sqrt{1/n_1 + 1/n_2}} \tag{4.4}$$

where s_p, the pooled standard deviation, is calculated from the two sample standard deviations s_1 and s_2, as follows:

$$s_p = \sqrt{\frac{(n_1 - 1)s_1^2 + (n_2 - 1)s_2^2}{n_1 + n_2 - 2}} \tag{4.5}$$

Note that there must not be a significant difference between the precisions of the methods. Hence the F-test (Section 4.12) is applied before using the t-test in equation (4.5).

Example 4.6   Comparison of two sets of data

The following results were obtained in a comparison between a new method and a standard method for the determination of the percentage nickel in a special steel.

	New method	Standard method
Mean	$\bar{x}_1 = 7.85\%$	$\bar{x}_2 = 8.03\%$
Standard deviation	$s_1 = \pm0.130\%$	$s_2 = \pm0.095\%$
Number of samples	$n_1 = 5$	$n_2 = 6$

Test at the 5% probability value if the new method mean is significantly different from the standard reference mean.

The _F_-test must be applied to establish there is no significant difference between the precisions of the two methods:

$$F = \frac{s_A^2}{s_B^2} = \frac{(0.130)^2}{(0.095)^2} = 1.87$$

The _F_-value ($P = 5\%$) from the tables (Appendix 12) for 4 and 5 degrees of freedom for s_A and s_B is 5.19.

The calculated value of _F_ (1.87) is less than the tabulated value, so the methods have comparable precisions (standard deviations) and the _t_-test can be used with confidence.

From equation (4.5) the pooled standard deviation s_P is given by

$$s_P = \sqrt{\frac{(5-1) \times 0.0169 + (6-1) \times 0.0090}{9}} = \pm 0.112$$

and from equation (4.4) we have

$$t = \frac{7.85 - 8.03}{0.112\sqrt{1/5 + 1/6}} = \frac{0.18}{0.112 \times 0.605} = 2.66$$

At the 5% level, the tabulated value of _t_ for ($n_1 + n_2 - 2$), i.e. 9 degrees of freedom, is 2.26.

Since $t_{calculated} = 2.66 > t_{tabulated} = 2.26$ there is a significant difference, at the specified probability, between the mean results of the two methods.

4.14 Paired _t_-test

Another method of validating a new procedure is to compare the results using samples of varying composition with the values obtained by an accepted method. The calculation is best illustrated by an example.

Example 4.7 The _t_-test using samples of different composition (the paired _t_-test)

Two different methods, A and B, were used for the analysis of five different iron compounds. The percentage iron content in each of the five samples is tabulated for methods A and B.

	1	2	3	4	5
Method A	17.6	6.8	14.2	20.5	9.7
Method B	17.9	7.1	13.8	20.3	10.2

In this example it would not be correct to attempt the calculation by the method described previously (Section 4.13).

In this case the differences _d_ between each pair of results are calculated and $\bar{d}$, the mean of the difference, is obtained. The standard deviation s_d of the differences is then evaluated. The results are tabulated as follows.

Method A	Method B	d	$d - \bar{d}$	$(d - \bar{d})^2$
17.6	17.9	+0.3	0.2	0.04
6.8	7.1	+0.3	0.2	0.04
14.2	13.8	−0.4	0.5	0.25
20.5	20.3	−0.2	−0.3	0.09
9.7	10.2	+0.5	0.4	0.16
		$\Sigma d = 0.5$		$\Sigma (d - \bar{d})^2 = 0.58$
		$\bar{d} = 0.1$		

$$s_d = \sqrt{\frac{0.58}{4}} = \pm 0.38$$

Then t is calculated from the equation

$$t = \frac{\bar{d}\sqrt{n}}{s_d} = \frac{0.10\sqrt{5}}{0.38} = 0.58_9$$

The tabulated value of t is 2.78 ($P = 0.05$) and since the calculated value is less than this, there is no significant difference between the methods.

4.15   The number of replicate determinations

To avoid unnecessary time and expenditure, an analyst needs some guide to the number of repetitive determinations needed to obtain a suitably reliable result from the determinations performed. The larger the number the greater the reliability; but after a certain number of determinations, any improvement in precision and accuracy is very small.

   Although rather complicated statistical methods exist for establishing the number of parallel determinations, a reasonably good assessment can be made by establishing the variation of the value for the absolute error Δ obtained for an increasing number of determinations:

$$\Delta = \frac{ts}{\sqrt{n}}$$

The value for t is taken from the 95% confidence limit column of the t-tables for $n - 1$ degrees of freedom.

   The values for Δ are used to calculate the reliability interval L from the equation

$$L(\%) = \frac{100\Delta}{z}$$

where z is the approximate percentage level of the unknown being determined. The number of replicate analyses is assessed from the magnitude of the change in L with the number of determinations.

Example 4.8

Ascertain the number of replicate analyses desirable (a) for the determination of approximately 2% Cl⁻ in a material if the standard deviation for determinations is 0.051, (b) for approximately 20% Cl⁻ if the standard deviation of determinations is 0.093.

(a) For 2% Cl⁻:

Number of determinations	$\Delta = \dfrac{ts}{\sqrt{n}}$	$L = \dfrac{100\Delta}{z}$	Difference (%)
2	$12.7 \times 0.051 \times 0.71 = 0.4599$	22.99	
3	$4.3 \times 0.051 \times 0.58 = 0.1272$	6.36	16.63
4	$3.2 \times 0.051 \times 0.50 = 0.0816$	4.08	2.28
5	$2.8 \times 0.051 \times 0.45 = 0.0642$	3.21	0.87
6	$2.6 \times 0.051 \times 0.41 = 0.0544$	2.72	0.49

(b) For 20% Cl⁻:

Number of determinations	$\Delta = \dfrac{ts}{\sqrt{n}}$	$L = \dfrac{100\Delta}{z}$	Difference (%)
2	$12.7 \times 0.093 \times 0.71 = 0.838$	4.19	
3	$4.3 \times 0.093 \times 0.58 = 0.232$	1.16	3.03
4	$3.2 \times 0.093 \times 0.50 = 0.148$	0.74	0.42
5	$2.8 \times 0.093 \times 0.45 = 0.117$	0.59	0.15
6	$2.6 \times 0.093 \times 0.41 = 0.099$	0.49	0.10

In (a) the reliability interval is greatly improved by carrying out a third analysis. This is less the case with (b) as the reliability interval is already narrow. In this second case no substantial improvement is gained by carrying out more than two analyses.

Shewell[1] has discussed other factors which influence the value of parallel determinations.

4.16   Correlation and regression

When using instrumental methods it is often necessary to carry out a calibration procedure by using a series of samples (standards) each having a known concentration of the analyte to be determined. A **calibration curve** is constructed by measuring the instrumental signal for each standard and plotting this response against concentration (Sections 17.14 and 17.16). Provided the same experimental conditions are used for the measurement of the standards and for the test (unknown) sample, the concentration of the test sample may be determined from the calibration curve by graphical interpolation.

Two statistical procedures should be applied to a calibration curve:

(a)   Test whether the graph is linear or in the form of a curve.
(b)   Find the best straight line (or curve) through the data points.

Correlation coefficient

To establish whether there is a linear relationship between two variables x_1 and y_1, use Pearson's correlation coefficient r:

$$r = \frac{n\Sigma x_1 y_1 - \Sigma x_1 \Sigma y_1}{\{[n\Sigma x_1^2 - (\Sigma x_1)^2][n\Sigma y_1^2 - (\Sigma y_1)^2]\}^{1/2}} \tag{4.6}$$

where n is the number of data points.

The value of r must lie between $+1$ and -1; the nearer it is to ±1, the greater the probability that a definite linear relationship exists between the variables x and y; values close to $+1$ indicate positive correlation and values close to -1 indicate negative correlation. Values of r that tend towards zero indicate that x and y are not linearly related (they may be related in a non-linear fashion).

Although the correlation coefficient r would easily be calculated with the aid of a modern calculator or computer package, the following example will show how the value of r can be obtained.

Example 4.9

Quinine may be determined by measuring the fluorescence intensity in $1\,M$ H_2SO_4 solution (Section 17.3). Standard solutions of quinine gave the following fluorescence values. Calculate the correlation coefficient r.

Concentration of quinine x_1 ($\mu g\,mL^{-1}$)	0.00	0.10	0.20	0.30	0.40
Fluorescence intensity y_1 (arb. units)	0.00	5.20	9.90	15.30	19.10

The terms in equation (4.6) are found from the following tabulated data.

x_1	y_1	x_1^2	y_1^2	x_1y_1
0.00	0.00	0.00	0.00	0.00
0.10	5.20	0.01	27.04	0.52
0.20	9.90	0.04	98.01	1.98
0.30	15.30	0.09	234.09	4.59
0.40	19.10	0.16	364.81	7.64
$\Sigma x_1 = 1.00$	$\Sigma y_1 = 49.5$	$\Sigma x_1^2 = 0.30$	$\Sigma y_1^2 = 723.95$	$\Sigma x_1y_1 = 14.73$

Therefore

$$(\Sigma x_1)^2 = 1.000 \qquad (\Sigma y_1)^2 = 2450.25 \qquad n = 5$$

Substituting the above values in equation (4.6) we have

$$r = \frac{5 \times 14.73 - 1.00 \times 49.5}{\{(5 \times 0.30 - 1.000)(5 \times 723.95 - 2450.25)\}^{1/2}} = \frac{24.15}{\sqrt{584.75}} = 0.9987$$

Hence there is a very strong indication that a linear relation exists between fluorescence intensity and concentration (over the given range of concentration).

But note that a value of r close to either $+1$ or -1 does not necessarily confirm there is a linear relationship between the variables. It is sound practice first to plot the calibration curve on graph paper and ascertain by visual inspection whether the data points could be described by a straight line or whether they may fit a smooth curve.

The significance of the value of r is determined from a set of tables (Appendix 14). Consider the following example using five data points $(x_1 y_1)$. From the table the value of r at 5% significance is 0.878. If the value of r is greater than 0.878 or less than -0.878 (if there is negative correlation), then the chance this value could have occurred from random data points is less than 5%. The conclusion can therefore be drawn that it is likely x_1 and y_1 are linearly related. With the value of $r = 0.998_7$ obtained in Example 4.9 there is confirmation of the statement that the linear relation between fluorescence intensity and concentration is highly likely.

Note   A value of the correlation coefficient r of near $+1$ or -1 does **not** confirm a linear relationship. Many non-linear plots will give a high positive value of the correlation coefficient. A scatter diagram should always be plotted first to ensure the calibration curve is linear.

4.17   Linear regression

Once a linear relationship has been shown to have a high probability by the value of the correlation coefficient r, then the best straight line through the data points has to be estimated. This can often be done by visual inspection of the calibration graph, but in many cases it is far more sensible to evaluate the best straight line by linear regression (the method of least squares).

The equation of a straight line is

$$y = bx + a$$

where y, the **dependent** variable, is plotted as a result of changing x, the **independent** variable. For example, a calibration curve (Section 15.15) in atomic absorption spectroscopy would be obtained from the measured values of absorbance (y-axis) which are determined by using known concentrations of metal standards (x-axis).

To obtain the regression line 'y on x', the slope b of the line and the intercept a on the y-axis are given by the following equations:

$$b = \frac{n\sum x_1 y_1 - \sum x_1 \sum y_1}{n\sum x_1^2 - (\sum x_1)^2} \tag{4.7}$$

$$\text{and} \quad a = \bar{y} - \bar{x} \tag{4.8}$$

where $\bar{x}$ is the mean of all values of x_1, and $\bar{y}$ is the mean of all values of y_1.

Example 4.10

Calculate by the least squares method the equation of the best straight line for the calibration curve given in Example 4.9.

From Example 4.9 the following values have been determined:

$$\sum x_1 = 1.00 \qquad \sum y_1 = 49.5 \qquad \sum x_1^2 = 0.30 \qquad \sum x_1 y_1 = 14.73 \qquad (\sum x_1)^2 = 1.000$$

The number of points is $n = 5$ and the values of $\bar{x}$ and $\bar{y}$ are

$$\bar{x} = \frac{\sum x_1}{n} = \frac{1.00}{5} = 0.2$$

and

$$\bar{y} = \frac{\Sigma y_1}{n} = \frac{49.5}{5} = 9.9$$

By substituting the values in equations (4.7) and (4.8) we have

$$b = \frac{5 \times 14.73 - 1.00 \times 49.5}{(5 \times 0.30) - (1.00)^2} = \frac{24.15}{0.5} = 48.3$$

and

$$a = 9.9 - (48.3 \times 0.2) = 0.24$$

So the equation of the straight line is

$$y = 48.3x + 0.24$$

If the fluorescence intensity of the test solution containing quinine was found to be 16.1, then an estimate of the concentration of quinine ($x \, \mu g \, mL^{-1}$) in this unknown could be

$$16.10 = 48.3x + 0.24$$

$$x = \frac{15.86}{48.30} = 0.32_8 \, \mu g \, mL^{-1}$$

4.18   Errors in the slope and the intercept

The determination of errors in the slope b and the intercept a of the regression line may be calculated in the following manner. First, the term $S_{y/x}$ must be evaluated from the following equation:

$$S_{y/x} = \sqrt{\Sigma(y_1 - \hat{y})^2 / (n - 2)} \tag{4.9}$$

The $\hat{y}$ values are obtained from the calculated regression line for given values of x. Thus from the equation of the line obtained in Example 4.10, $y = 48.3x + 0.24$; when $x = 0.10 \, mg \, L^{-1}$ then $\hat{y} = 48.3 \times 0.10 + 0.24 = 5.07$.

Once the value $S_{y/x}$ has been obtained, both the standard deviations of the slope S_b, and the intercept S_a can be calculated from the following equations:

$$S_b = S_{y/x} / \sqrt{\Sigma(x_1 - \bar{x})^2} \tag{4.10}$$

$$S_a = S_{y/x} \sqrt{\Sigma x_1^2 / n \Sigma(x_1 - \bar{x})^2} \tag{4.11}$$

Example 4.11

Calculate the standard deviations and the 95% confidence limits for the slope and intercept of $y = 48.3x + 0.24$, the regression line obtained in Example 4.10.

From the data determined in Examples 4.9 and 4.10, we can tabulate values as follows.

x_1	x_1^2	$(x_1 - \bar{x})^2$	y_1	$\hat{y}$	$(y_1 - \hat{y})^2$
0.00	0.00	0.04	0.00	0.24	0.0576
0.10	0.01	0.01	5.20	5.07	0.0169
0.20	0.04	0.00	9.90	9.90	0.0000
0.30	0.09	0.01	15.30	14.73	0.3249
0.40	0.16	0.04	19.10	19.56	0.2116
$\Sigma x_1 = 1.00$	$\Sigma x_1^2 = 0.30$	$\Sigma(x_1 - \bar{x})^2 = 0.10$	$\Sigma y_1 = 49.50$		$\Sigma(y_1 - \hat{y})^2 = 0.6110$
$\bar{x}_1 = 0.20$			$\bar{y}_1 = 9.9$		

Hence from equation (4.9)

$$S_{y/x} = \sqrt{\frac{0.611}{3}} = 0.4513$$

The value of S_b, the standard deviation of the slope b, is obtained by substituting values in equation (4.10):

$$S_b = \frac{0.4513}{\sqrt{0.10}} = \frac{0.4513}{0.3162} = \pm 1.427$$

The 95% confidence limits for the slope are given by $b \pm ts$ where $t = 3.18$ at 95% for $n - 2$, i.e. 3 degrees of freedom.

Thus the 95% confidence limits for the slope b are given by

$$b = 48.3 \pm 3.18 \times 1.427 = 48.3 \pm 4.54$$

From equation (4.11) the standard deviation of the intercept a is

$$S_a = 0.4513 \left[\frac{0.30}{5 \times 0.1} \right]^{1/2} = 0.3496$$

The 95% confidence limits for the intercept a are

$$a = 0.24 \pm 3.18 \times 0.35 = 0.24 \pm 1.11$$

4.19   Error in the estimate of concentration

The estimation of the error in the concentration determined by the use of the regression line involves the following expression:

$$S_{xc} = \frac{S_{y/x}}{b} \left[1 + \frac{1}{n} + \frac{(y_0 - \bar{y})^2}{b^2 \Sigma(x - \bar{x})^2} \right]^{1/2} \tag{4.12}$$

where y_0 is the value of y from which the concentration x_c is to be evaluated and where S_{xc} is the standard deviation of x_c.

If we refer to Example 4.10, the concentration of quinine was estimated at $0.32_8 \, \mu g \, mL^{-1}$ when the fluorescence intensity was 16.1.

Example 4.12

Using the data given in Examples 4.10 and 4.11, calculate the S_{xc} value given the estimate of concentration $x_c = 0.32\,\mu\text{g}\,\text{mL}^{-1}$.

The value of S_{xc} can be obtained by substituting the appropriate values in equation (4.12):

$$S_{xc} = \frac{0.4513}{48.3}\left[1 + \frac{1}{5} + \frac{(16.1 - 9.9)^2}{48.3^2 \times 0.1^2}\right]^{1/2}$$

$$= 0.009\,34\left[1.20 + \frac{38.44}{23.33}\right]^{1/2}$$

$$= \pm 0.016$$

Thus, the 95% confidence limits (given $t = 3.18$ with 3 degrees of freedom) are

$$x_c \pm 0.016 t_3 = 0.32_8 \pm 0.05\,\mu\text{g}\,\text{mL}^{-1}$$

Close scrutiny of equation (4.12) reveals that the value of $y_0 - \bar{y}$ is lowest near the centre of the calibration plot. You may wish to confirm that in this example a fluorescence intensity of 9.9 gives a concentration estimate

$$x_c = 0.20_0\,\mu\text{g}\,\text{mL}^{-1} \quad \text{and} \quad S_{xc} = \pm 0.01_0$$

and 95% confidence limits of $0.20_0 \pm 0.03\,\mu\text{g}\,\text{mL}^{-1}$.

The value of the confidence limits may also be reduced by increasing the number n of calibration points, for not only does it lower S_{xc}, but the value of t is reduced with an increasing number of degrees of freedom.

If the concentration of an analyte is measured by graphical interpolation, the error will be lowest near the centre of the calibration plot. Values near the extremities of the plot are subject to the greatest error.

Never assume that the calibration plot can be extended without the relevant measure of appropriate standards.

4.20   Standard additions

Consider the determination of strontium in river water by flame emission spectroscopy. The calibration procedure described in Section 4.16 may result in serious errors if the strontium standards contain only a pure strontium salt dissolved in water. The strontium emission signal obtained from the river water sample may be enhanced or reduced by the presence of other components. This so-called matrix effect would be eliminated if the standards had the same bulk composition as the unknown sample solution. In many cases, however, matrix matching is virtually impossible to achieve.

The technique of standard additions is widely applied to spectroscopic and electrochemical methods in order to overcome matrix effects. In our example, equal volumes of the river water sample are taken and different known amounts of a strontium solution are added to each sample solution, except one, then all are diluted to the same volume.

The emission signals are measured for every solution and the results are plotted on a graph with the dependent variable on the y-axis and the amount of strontium added on the

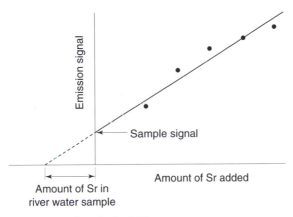

Figure 4.2 Standard additions

x-axis. Extrapolation of the straight line thus obtained to the point where the *x*-axis is cut (i.e. when $y = 0$) provides a measure of strontium in the river water (Figure 4.2).

Example 4.13

In the determination of strontium in river water by flame emission spectrometry using the method of standard additions, the following results were obtained.

Sr standard added ($\mu g\,mL^{-1}$)	0.0	10.0	15.0	20.0	25.0	30.0
Emission signal	2.3	4.4	5.3	6.1	7.5	8.7

Determine the strontium concentration in the river water and evaluate the 95% confidence limits for this value.

By application of equations (4.7) and (4.8) we obtain the linear regression line

$$y = 0.210x + 2.22$$

Hence the strontium concentration in the river water obtained from the extrapolated value x_E when $y = 0$ is $10.6\,\mu g\,mL^{-1}$.

To find the confidence limits for the above value, initially use a modified form of equation (4.12) to calculate S_{xE}, the standard deviation of the extrapolated value:

$$S_{xE} = \frac{S_{y/x}}{b}\left[\frac{1}{n} + \frac{\bar{y}^2}{b^2\sum(x_1 - \bar{x})^2}\right]^{1/2} \tag{4.13}$$

From equation (4.9) we can show that $S_{y/x} = 0.2041$. The value of $\bar{y}^2$ is readily shown to be 32.68 and $\sum(x_1 - \bar{x})^2 = 583.4$. Thus S_{xE}, the error of the extrapolated value x_E, is found using equation (4.13) to be ± 1.166.

The 95% confidence limits of x_E can be found from the relationship $x_E \pm tS_{xE}$. Hence they are $10.6 \pm 2.78 \times 1.166 = 10.6 \pm 3.24\,\mu g\,mL^{-1}$.

The standard additions method must be used with care. It is absolutely imperative that the response (absorbance or emission) is linearly related to the concentration. Often at higher concentrations a linear relationship is not valid. Therefore the concentration of the added substance must not be too high.

Furthermore, any extrapolation method is less precise than a graphical interpolation, so an extrapolation method should only be used when there is no alternative method. Whenever possible, the results obtained from an analysis involving standard additions should be compared to an established (referee) method. It is also sound practice to obtain the 95% confidence limits (described in Example 4.13) for the result obtained by extrapolation, which will serve to modify the optimistic results obtained by rigid adherence to this method.

4.21   Non-linear regression

By examination of a scatter diagram, it is often possible to see whether or not there is a linear relationship between the variables. It is usually the policy of the analytical chemist, by ensuring the appropriate experimental conditions, to try to achieve a linear relationship if at all possible. There are, however, situations where the calibration curve is non-linear over the whole concentration range. Then a curve-fitting process may be used to attempt to fit the variables to an appropriate polynomial expression of the form $y = a + bx + cx^2 + \ldots$. Computer programs can be used to obtain the best-fit curve by an iterative process; this curve can then be assessed by polynomial regression in a similar way to the least squares method for linear regression. The works listed in Section 4.38 give more information on this subject.

4.22   Comparison of more than two means

The comparison of more than two means is a situation that often arises in analytical chemistry. It may be useful to compare (a) the mean results obtained from different spectrophotometers all using the same analytical sample; (b) the performance of a number of analysts using the same titration method. Assume that three analysts, using the same solutions, each perform four replicate titrations. In this case there are two possible sources of error: the random error associated with replicate measurements, and the variation that may arise between the individual analysts. These variations may be calculated and their effects estimated by a statistical method known as the analysis of variance (ANOVA), where the **square of the standard deviation** s^2 is called the **variance** V. Thus $F = s_1^2/s_2^2$ where $s_1^2 > s_2^2$, and may be written as $F = V_1/V_2$ where $V_1 > V_2$.

Example 4.14

Three analysts were each asked to perform four replicate titrations using the same solutions. The titres (mL) are given below.

Analyst A	Analyst B	Analyst C
22.53	22.48	22.57
22.60	22.40	22.62
22.54	22.48	22.61
22.62	22.43	22.65

To simplify the calculation it is sound practice to subtract a common number, e.g. 22.50, from each value. The sum of each column is then determined. This will have no effect on the final values.

Analyst A	Analyst B	Analyst C
0.03	−0.02	0.07
0.10	−0.10	0.12
0.04	−0.02	0.11
0.12	−0.07	0.15
Sum = 0.29	−0.21	0.45

The following steps have to be made in the calculation:

(a)   The grand total

$$T = 0.29 - 0.21 + 0.45$$

$$= 0.53$$

(b)   The correction factor (CF)

$$CF = \frac{T^2}{N} = \frac{(0.53)^2}{12} = 0.0234$$

where N is the total number of results.

(c)   The total sum of squares is obtained by squaring each result, summing the totals of each column and then subtracting the correction factor (CF).

Analyst A	Analyst B	Analyst C
0.0009	0.0004	0.0049
0.0100	0.0100	0.0144
0.0016	0.0004	0.0121
0.0144	0.0049	0.0225
Sum = 0.0269	0.0157	0.0539

$$\text{Total sum of squares} = (0.0269 + 0.0157 + 0.0539) - CF$$

$$= 0.0965 - 0.0234 = 0.0731$$

(d)   To obtain the between-treatments (analysts) sum of squares, take the sum of the squares of each individual column and divide it by the number of results in each column, then subtract the correction factor:

$$\text{Between sum of squares} = \tfrac{1}{4}(0.29^2 - 0.21^2 + 0.45^2) - 0.0234$$

$$= 0.0593$$

(e)   To obtain the within-sample sum of squares, subtract the between sum of squares from the total sum of squares.

$$0.0731 - 0.0593 = 0.0138$$

(f)   The number of degrees of freedom v is obtained as follows:

The total number of degrees of freedom $= N - 1 = 11$

The between-treatments degrees of freedom $= C - 1 = 2$

The within-sample degrees of freedom $= (N - 1) - (C - 1) = 9$

where C is the number of columns (here the number of analysts).

(g)   A table of analysis of variance (ANOVA table) may now be set up.

Source of variation	Sum of squares	v	Mean square
Between analysts	0.0593	2	0.0593/2 = 0.0297
Within titrations	0.0138	9	0.0138/9 = 0.001 53
Total	0.0731	11	

(h)   The F-test is used to compare the two mean squares:

$$F_{2,9} = \frac{0.0297}{0.00153} = 19.41$$

From the F-tables (Appendix 12), the value of F at the 1% level for the given degrees of freedom is 8.02. The calculated result (19.41) is higher than 8.02, hence there is a significant difference in the results obtained by the three analysts. Having ascertained in this example there is a significant difference between the three analysts, the next stage would be to determine whether the mean result is different from the others, or whether all the means are significantly different from each other.

The procedure adopted to answer these questions for the example given above is as follows:

(a)   Calculate the titration means for each analyst. The mean titration values are $\bar{x}(A) = 22.57$ mL, $\bar{x}(B) = 22.45$ mL and $\bar{x}(C) = 22.61$ mL.

(b)   Calculate the quantity defined as the 'least significant difference', which is given by $s\sqrt{2/n}\ t_{0.05}$ where s is the square root of the residual mean square, i.e. the within-titration mean square. Hence $s = \sqrt{0.00153}$; n is the number of results in each column (here it is 4); t is the 5% value from the t-tables (Appendix 11), with the same number of degrees of freedom as that for the residual term, i.e. the within-titration value. In this example the number of degrees of freedom is 9, so the least significant difference is given by

$$\sqrt{0.00153} \times \sqrt{2/4} \times 2.26 = 0.06\ \text{mL}$$

If the titration means are arranged in increasing order, then $\bar{x}(B) < \bar{x}(A) < \bar{x}(C)$, and $\bar{x}(C) - \bar{x}(B)$ and $\bar{x}(A) - \bar{x}(B)$ are both greater than 0.06, whereas $\bar{x}(C) - \bar{x}(A)$ is less than 0.06. Hence there is no significant difference between analysts A and C, but the results of analyst B are significantly different from those of both A and C.

Note that in this example the performance of only one variable, the three analysts, is investigated and thus this technique is called a one-way ANOVA, If **two** variables, e.g. the three analysts with four **different** titration methods, were to be studied, this would require the use of a two-way ANOVA. An example of two-way ANOVA is given in Section 4.24.

Notes

1. If a negative value is found in the total sum of squares, the between sum of squares and the within sum of squares, then there has been a calculation error.
2. In an ANOVA problem it is always advisable to use the largest number of significant figures and round off at the end of the calculation. There is a danger that if numbers are rounded off prematurely, meaningless zeros will appear and negate the subsequent calculations.

4.23   Experimental design

In Example 4.14 only one variable was studied – the analyst variable. In analytical chemistry there are many situations where more than one variable has to be considered. For example, in colorimetric analysis the chosen wavelength, the solution temperature, the solution pH and the standing time of the solution before measurement are four variables that could affect the absorbance (the response) of the experiment. Variables that could affect the experimental result are called **factors**. Factors which can have specific values, e.g. pH and wavelength, are termed **controlled factors**. However, factors such as the variation of ambient laboratory conditions over a period of time are clearly **uncontrolled factors**. Experimental design can help to achieve three objectives:

1. To identify factors that may affect the experimental result.
2. To ensure the effects of uncontrolled factors are kept to a minimum.
3. To use statistical methods for interpreting the results thus obtained.

An example of randomisation

Four aliquots of three different samples A, B and C are each analysed for their tin content. There may be a tendency to begin by analysing the four aliquots taken from sample A, then the four from sample B and finally the four from sample C. But there could be several uncontrolled factors that might affect the results obtained by such a procedure. There can be instrument drift, the loss of reagent strength and operator fatigue, and in this case all three could, over time, have a dramatic effect on the four analyses of the last sample C. Other uncontrolled factors such as laboratory temperature and pressure variation over a number of days could also give rise to systematic errors. In order to overcome the effects of uncontrolled factors, the technique of **randomisation** is used. Randomisation is achieved by mixing up the order in which the experiments are carried out. One method for determining the order of experiments is to use random number tables, or random number functions generated by most electronic calculators. Thus, instead of the order of experiments being 1 to 4 for sample A, then 5 to 8 for B and 9 to 12 for C, the following random number experimental sequence may be generated.

09	04	02	06	01	03	08	10	12	05	07	11
C	A	A	B	A	A	B	C	C	B	B	C

This gives a random order of treatment, which minimises the effect of uncontrolled factors. But what happens if it is only possible to perform three analyses on each day? The above order would then produce the following schedule.

Day 1	C	A	A
Day 2	B	A	A
Day 3	B	C	C
Day 4	B	B	C

Sample A would be analysed on the first two days, which would not necessarily minimise any uncontrolled factors evident over a longer period of time.

A more satisfactory design would be one where each sample is analysed **once** each day. The order for each day is randomised.

Day 1	C	A	B
Day 2	A	C	B
Day 3	B	A	C
Day 4	B	C	A

Such a diagram is called a **randomised block design**, and the analyses performed on each day are called a **block**. This design allows us to separate three sources of variation:

1. Between blocks (different days)
2. Between treatments (samples A, B and C)
3. Random (due to indeterminate errors)

In this case a two-way analysis of variance (two-way ANOVA) is used, then the between-blocks variance and the between-treatments variance are compared with the estimated random error variance.

4.24   Two-way analysis of variance

Example 4.15

Four standard solutions, each calculated to contain 5.00% (by weight) of copper(II) were prepared. Three methods of titration, each with a different end point determination were used for the analysis of each solution. The results (wt%) for the copper(II) content are given below. The sequence of the experiments was generated in a random manner.

Solution	Method A	Method B	Method C
1	5.08	5.17	5.09
2	5.02	5.15	5.15
3	5.06	5.22	5.10
4	5.00	5.13	5.05

Determine whether there are significant differences at the 5% level between (i) the concentrations of copper(II) in the different solutions and (ii) the results obtained by the different titrimetric methods.

There are two variables involved, the different copper standard solutions and the different titration methods. Hence a two-way ANOVA is used to solve the problem. Note that the steps in the solution are similar to the steps in Example 4.14.

In order to simplify the problem, a constant such as e.g. 5.00 can be subtracted from each value. The sum of each column and of each row is then determined.

Solution	A	B	C	Row sum
1	0.08	0.17	0.09	0.34
2	0.02	0.15	0.15	0.32
3	0.06	0.22	0.10	0.38
4	0.00	0.13	0.05	0.18
Column sum	0.16	0.67	0.39	1.22 = T

(a)   The grand total T is obtained by adding up the column sums, or by adding up the row sums. In this example $T = 1.22$.

(b)   The correction factor is

$$\text{CF} = \frac{T^2}{N} = \frac{(1.22)^2}{12} = 0.1240$$

(c)   The total sum of squares, obtained as in Example 4.14, is

$$0.1702 - 0.1240 = 0.0461$$

(d)   To obtain the between-treatments (titration method) sum of squares, take the sum of the squares of each individual column and divide it by the number of results in each column, then subtract the correction factor:

$$\text{Between-treatments sum of squares} = \tfrac{1}{4}(0.16^2 + 0.67^2 + 0.39^2) - 0.1240$$

$$= 0.0327$$

(e)   To obtain the between-solutions sum of squares, take the sum of squares of each individual row and divide it by the number of results in each row, then subtract the correction factor:

$$\text{Between-solutions sum of squares} = \tfrac{1}{3}(0.34^2 + 0.32^2 + 0.38^2 + 0.18^2) - 0.1240$$

$$= 0.0075$$

(f)   To obtain the residual (random) sum of squares, take the total sum of squares and subtract the between-treatments sum of squares and the between-solutions sum of squares:

$$\text{Residual sum of squares} = 0.0461 - (0.0327 + 0.0075) = 0.0059$$

(g)   The number of degrees of freedom v is obtained as follows:

The total number of degrees of freedom $= N - 1 = 11$

The between-treatments degrees of freedom $= C - 1 = 2$

The between-solutions degrees of freedom $= R - 1 = 3$

The residual degrees of freedom (v) are

$$(N - 1) - [C - 1 + R - 1] = 11 - (2 + 3) = 6$$

where

N = total number of experiments (12)

C = number of columns (3)

R = number of rows (4)

A two-way ANOVA table may now be drawn up.

Source of variation	Sum of squares	v	Mean square
Between treatments	0.0327	2	0.0164
Between solutions	0.0075	3	0.0025
Residual	0.0059	6	0.000 98
Total	0.0461	11	

(i)   The F-test is used to compare the between-treatments mean square with the residual mean square:

$$F_{2,6} = \frac{0.0164}{0.000\,98} = 16.73$$

From the F-tables (Appendix 12) the value of F at the 5% level is 5.14 for (2, 6) degrees of freedom). The calculated result 16.73 is much higher than 5.14, hence there is a significant difference between the titrimetric end points.

(ii) Compare the between-solutions mean square with the residual mean square:

$$F_{3,6} = \frac{0.0025}{0.000\,98} = 2.55$$

From the F-tables, the value of F at 5% = 4.76 for (3, 6 degrees of freedom). Since 2.55, the calculated value, is less than 4.76, the value from the tables, there is no significant difference between the standard solutions used.

A special design, the **Latin square**, can deal with two factors that may affect the treatments. In this design each treatment appears once in each row and once in each column; it must therefore be a square design. The previous example could be extended to include three different analysts to perform the three titration methods, but in this case, only **three** different standard solutions could be used. This design would then allow the between-treatments, between-solutions and also the between-analysts variations to be compared with the random error variation.

4.25   Chemometrics and experimental design

Chemometrics may be defined as the application of mathematical and statistical methods to design and/or optimise measurement procedures and to provide chemical information by analysing relevant data. For a successful design the following procedure should be adopted:

1.   Consider what result(s) are required and what variables (factors) may affect the outcome of the experiment whose results have to be evaluated.

2.   Select an appropriate design to solve the problem.
3.   Carry out the experimental work.
4.   Examine the data thus obtained, separating the important variables from those of little significance and deciding if other techniques may produce useful information.

It may be necessary to repeat these steps with an alternative approach to the design. Step 2 is selection of an experimental design. Designs can be divided into two broad classes:

Simultaneous designs mean that all experiments are completed before the analysis of the data is carried out.

Sequential designs mean that the results from the previous experiments determine the conditions for the next experiment.

4.26   Factorial design

The most common form of simultaneous design is the **factorial experimental design**. In a factorial design a set of levels is decided for each factor (variable) to be studied and then a series of experiments is carried out, perhaps once or maybe several times, with each of the possible combinations of the factors.

Note how this approach is completely different from univariate search – varying one factor at a time while holding the remaining factors constant. Univariate search is frquently more time-consuming and less efficient than the factorial design; most significantly, it does not take into account that **variables can interact**. Before discussing a specific example of a factorial design experiment in detail, a number of terms should be defined:

A **factor** is a variable believed to affect the experimental result. There are two types of factor. A **quantitative factor** may be assigned different magnitudes; examples are pH, temperature and reagent concentration. A **qualitative factor** may be distinguished by its presence or absence, e.g. whether or not there is a catalyst. In a factorial design each factor has two or more **levels**. For a quantitative factor such as pH, these levels could be 2, 3, 4 or 5. For a qualitative factor such as the use of an inert atmosphere, these levels may be absence (low level) or presence (high level).

The combination of factor levels used in an experiment is called the **treatment combination**. In an experimental factorial design involving three factors at two levels, there are 2^3 treatment combinations. The number of treatment combinations when there are three levels and two factors = 3^2. Thus, the integer gives the number of levels and the power gives the number of factors.

A **response** is the result observed from a given treatment; it could be the titrimetric end point or it could be an instrument response, e.g. absorbance, fluorescence emission or the signal-to-noise ratio. An **interaction** occurs when the effects of two or more factors are not additive. Details are given in Sections 4.27 and 4.28.

Example 4.16

In an atomic absorption experiment the absorption of a calcium solution ($10\,\mu g\,mL^{-1}$) using the resonance line at 422.7 nm was measured at two different levels for each of the following factors: (A) flame height, (B) lamp current (C) fuel ratio (acetylene/air). The levels selected for each factor are shown below.

	Low Level (−)	High Level (+)
A flame height (mm)	15	25
B lamp current (mA)	2	3
C fuel ratio (C_2H_2/air)	4/9	5/9

Since there are three factors at two levels, there will be $2^3 = 8$ treatment combinations. The order of the treatment combinations was randomised and two replicate measurements were made on each treatment combination.

The results obtained by a graduate student are presented in Table 4.1. In order to simplify the calculation, the response (absorbance) was rounded to 2 significant figures and then multiplied by 100. Table 4.1 gives the average value for each pair of replicate measurements. If possible, replicate measurements should be made of each treatment combination, allowing an estimation of interaction effects.

Table 4.1 *Table of levels*

Lamp current	Fuel ratio low		Fuel ratio high	
	Flame = 15 mm	Flame = 25 mm	Flame = 15 mm	Flame = 25 mm
2 mA	(1) 54	a 28	c 69	ac 52
3 mA	b 50	ab 23	bc 63	abc 47

Each treatment combination is coded in the following manner. Lower case letter a denotes that factor A is at a high level; its absence indicates that factor A is at a low level. The label (1) signifies that all factors are at a low level. Thus c denotes that factor C (the fuel ratio) is high but factors A and B (flame height and lamp current) are low. Similarly, ab shows that factors A and B are high and factor C is low. Finally, abc means that all three factors are high.

In order to calculate the magnitude of the factors and their interactions, a **table of signs** can be constructed (Table 4.2). A plus sign (+) indicates the factor is high, a minus sign (−) means it is low. The signs of the interaction terms are obtained by the algebraic product of the signs of the factors involved in the interaction. Thus, for treatment combination (1) the sign of the interaction term of factors A and B, i.e. AB, is obtained from the product of a − sign for factor A and a − sign for factor B, resulting in a + sign for the interaction AB. The reader can confirm from Table 4.2 that the sign of the interaction BC for the treatment combination ab is a −, and the sign of the triple interaction ABC for treatment combination ac is also a − (the product of +, − and +).

Using Table 4.2 it is now possible to calculate the effects of the factors and the interactions as follows. The effect of factor A (the flame height) is calculated by subtracting the mean of the responses measured at low level from the mean of the responses obtained at high level. From Table 4.2 the effect of factor A using the coded treatment combinations is

$$A = \tfrac{1}{4}(a + ab + ac + abc) - \tfrac{1}{4}((1) + b + c + bc)$$

↑   ↑

Treatment combinations **where A is at high level**   Treatment combinations **where A is at low level**

Table 4.2 *Table of signs*

Treatment combination	Factors and Interactions							Response (absorbance × 100)
	A	B	AB	C	AC	BC	ABC	
(1)	−	−	+	−	+	+	−	54
a	+	−	−	−	−	+	+	28
b	−	+	−	−	+	−	+	50
ab	+	+	+	−	−	−	−	23
c	−	−	+	+	−	−	+	69
ac	+	−	−	+	+	−	−	52
bc	−	+	−	+	−	+	−	63
abc	+	+	+	+	+	+	+	47

$$\text{Therefore} \quad 4A = (28 + 23 + 52 + 47) - (54 + 50 + 69 + 63)$$

$$= -86$$

$$A = -\frac{86}{4} = -21.5$$

The effect of factor B (the lamp current) is derived in a similar manner.

$$B = \tfrac{1}{4}(b + ab + bc + abc) - \tfrac{1}{4}((1) + a + c + ac)$$

$$= \tfrac{1}{4}(50 + 23 + 63 + 47) - \tfrac{1}{4}(54 + 28 - 69 - 52)$$

$$= -\frac{20}{4} = -5$$

Using the Table 4.2 it is possible to confirm that the effect of factor C (the fuel ratio) = 19 and that the effect of the interaction BC = 0.5.

The realisation that factors may interact is possibly the most important statement in this chapter.

4.27 Yates' method

However, a more elegant way to calculate the effects of factors and their interactions is the **Yates method**. In this method the treatment combinations are listed in the systematic way shown in Table 4.3. The appropriate response (absorbance) to each treatment combination is placed in the Response column. Column (i) is derived from the response column as follows: the first entry is given by the sum of the first two absorbance values in the response column (54 + 28); the second entry is the sum of the second pair of absorbances (50 + 23); the third and fourth entries are the sum of the third and then the fourth pairs respectively, i.e. (69 + 52) and (63 + 47). This completes the top half of column (i). The lower half is derived from the response column by taking the differences of the same pairs, **the first from the second** in every case. Thus, the first entry in the lower half of column (i) is (28 − 54), the second (23 − 50), the third (52 − 69) and the fourth (47 − 63). Column (ii)

Table 4.3   *The Yates method*

Treatment combination	Response	(i)	(ii)	(iii)	Sum of squares (mean square)
none	54	82	151	386 = total	
a	28	73	231	−86 = 4A	1849
b	50	121	−53	−20 = 4B	50
ab	23	110	−33	0 = 4AB	0
c	69	−26	−9	76 = 4C	1444
ac	52	−27	−11	20 = 4AC	100
bc	63	−17	−1	−2 = 4BC	1
abc	47	−16	1	2 = 4ABC	1
Total	386				

is derived in exactly the same manner by summing and differencing the pairs of values given in column (i). Column (iii) is obtained from column (ii) in the same way.

Note   The number of summing and differencing operations is equal to the number of factors.

The sum of squares can be calculated from the estimated effects by the expression

$$\text{sum of squares} = \frac{N}{4} \times (\text{estimated effect})^2$$

where N is the total number of experiments (here 16 because each treatment combination was measured twice). So

$$\text{sum of squares of factor A} = \frac{16}{4} \times 21.5^2 = 1849$$

Note   As a useful check to the Yates calculation, the top value in column (iii) (386) should be the same as the sum of the values in the response column (386).

The results in column (iii) show that factor A (the flame height) is the most important, followed by factor C (the fuel ratio). Factor B (the lamp current) and the interaction AC may be significant (see later).

   The negative values for the effects of factor A and factor B show that the response is lowered going from the low level to the high level. The positive value for the effect of factor C indicates the response is raised when going from low level to high level. Thus, it is likely that operating at low flame height, low lamp current and high fuel ratio (treatment combination c) would produce the best response. This is in fact the case; treatment combination c has the highest response (69).

   The sum of squares can give an estimate of those factors and interactions that are significant. The mean square is compared with the error mean square. When a replicated factorial design is used, the error (residual) mean square can be calculated by the method described below. The individual values of the replicate experiments given in Example 4.16 are shown as columns (i) and (ii) in Table 4.4.

   The error (residual) mean square is calculated as the sum of the square of each value in column (i) and the square of each value in column (ii) minus the sum of the square of

Table 4.4 *Replicate values*

Treatment combination	(i)	(ii)	(iii) = (i) + (ii)	Mean squares from Table 4.3
none	53	55	108	
a	27	28	55	$A = 1849$
b	49	51	100	$B = 100$
ab	23	22	45	$AB = 0$
c	70	68	138	$C = 1444$
ac	51	53	104	$AC = 100$
bc	63	63	126	$BC = 1$
abc	48	46	94	$ABC = 1$

each value in column (iii) divided by the number of replicate measurements in each treatment combination (here it is 2). Thus, the residual mean square is given by

$$53^2 + 55^2 + 27^2 + 28^2 + 49^2 + 51^2 + \ldots + 63^2 + 63^2 + 48^2 + 46^2$$
$$- \tfrac{1}{2}(108^2 + 55^2 - 100^2 + \ldots + 126^2 + 94^2)$$
$$= 40\,654 - \tfrac{1}{2}(81\,286) = 11$$

Thus the error mean square = 11 with 8 degrees of freedom.

To test for significance, the mean square each with 1 degree of freedom is compared with the residual mean square. The interaction ABC and the interactions AB and BC are not significant for the resulting F values are less than unity:

$$F_{1,8} = \frac{1.0}{11} = 0.0909 \quad \text{for BC and ABC}$$

Now test for the interaction AC and the factor B (the lamp current) which have the same value for the mean square (100):

$$F_{1,8} = \frac{100}{11} = 9.09$$

From the F-tables, the 5% value of $F_{1,8} = 5.32$ and the 1% value = 11.3. Hence the lamp current B and the interaction of the flame height and fuel ratio AB are significant at the 5% level but not at the 1% level. Obviously, both the flame height A and the fuel ratio C are highly significant both with F-values greater than 100.

In order to ascertain whether the absorption signal could be further enhanced, it may be expedient to perform a 2^2 factorial design with the suggested settings given below, and keeping factor B (the lamp current) at the low value of 2 mA.

	Low level (−)	High level (+)
A flame height (mm)	8	15
C fuel ratio[a]	5/9	6/9

[a] When selecting fuel/air ratios, ensure they lie within the instrument manufacturer's guidelines for safety.

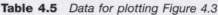

Table 4.5   *Data for plotting Figure 4.3*

Point on Figure 4.3	Factors	Treatment combinations	Response values		Mean response
R	A low C low	(1) and b	54	50	52
P	A high C low	a and ab	28	23	25.5
S	A low C high	c and bc	69	63	66
Q	A high C high	ac and abc	52	47	49.5

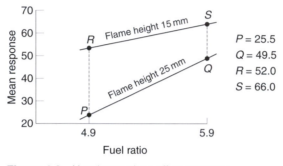

Figure 4.3   How interactions affect response

4.28   Interaction effect: alternative calculation

The interaction AC was shown to be significant at the 5% level by the *F*-test. An alternative approach can be adopted to show if there is any appreciable interaction. Consider the effect of changing the flame height (factor A) from a low level (15 mm) to a high level (25 mm) with the fuel ratio (factor C) at a low level (4/9). Then change the flame height from 15 mm to 25 mm with the fuel ratio at 5/9, a high level. The values taken from Example 4.16 are presented in Table 4.5.

This data can be represented graphically as shown in Figure 4.3. The lines *PQ* and *RS* are not parallel (the distance *QS* = 14, the distance *PR* = 24). Hence the effects of the two factors are **not** additive, which indicates interaction between factor A and factor C. The reader can confirm that for the interaction term AB the effects are additive and the resulting graph produces two parallel lines, hence no interaction. This graphical method is particularly useful in showing interaction effects when more than two levels are used in the factorial design.

4.29   Factorial design: critical appraisal

A major disadvantage of the factorial design is that an increase in the number of factors is accompanied by a dramatic rise in the number of experiments. For example, in an unreplicated design with five factors at two levels there are $2^5 = 32$ experiments. A four-factor, three-level design requires $3^4 = 81$ experiments. However, the number of experiments may be reduced without a substantial loss of information by the use of **fractional factorial**

designs. A quarter fractional design of five factors at two levels has $\frac{1}{4}(2^5) = 8$ experiments. In fractional factorial designs, with four or more factors, the higher interaction terms can be ignored so that only the main effects and two-factor interaction effects are evaluated.

An **incomplete factorial design** which measures only the main effects (no interactions) was devised by Plackett and Burman.[2] The number of experiments is reduced greatly but useful data is generated. The matrix of the proposed experiments (the table of signs) is presented below:

Experiment no.	Factors							Response
	A	B	C	D	E	F	G	
1	+	+	+	−	+	−	−	R_1
2	−	+	+	+	−	+	−	R_2
3	−	−	+	+	+	−	+	R_3
4	+	−	−	+	+	+	−	R_4
5	−	+	−	−	+	+	+	R_5
6	+	−	+	−	−	+	+	R_6
7	+	+	−	+	−	−	+	R_7
8	−	−	−	−	−	−	−	R_8

In this type of design for n factors there are $n + 1$ experiments. The effects of each factor are determined in the same way as for the complete factorial design described in Section 4.26. Thus, the effect of changing factor C from low level to high level is given by

$$C = \tfrac{1}{4}(R_1 + R_2 + R_3 + R_6) - \tfrac{1}{4}(R_4 + R_5 + R_7 + R_8)$$

This design is often called the **ruggedness test** and has application in validating methods that could be adopted as routine practice in laboratories. A method is shown to be rugged if it is reproducible without hidden pitfalls. Factors which have a large effect on the results are thus identified and can be subjected to a more rigorous investigation before the method is validated. In collaborative trials this design is particularly useful in giving prior warning to participating laboratories of the factors that require careful control.

Another important consideration with factorial designs is that a sensible choice of the levels should be made. If the levels are set too close together or too far apart, they may give a response difference that is not significant, whereas the factor effect may be significant. In Figure 4.4(a) and (b) the response difference is small, but in Figure 4.4(c) a

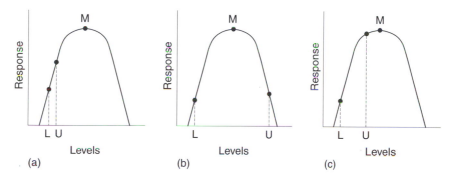

Figure 4.4 Choosing levels: in (a) and (b) the upper and lower levels (U and L) give a relatively small difference in response; in (c) the difference is much larger, so (c) is a better choice (M = maximum response)

153

better choice of levels has resulted in a much larger response difference. A better design would use three levels, although this would increase the number of experiments.

4.30   Optimisation methods

As the demand increases for analytical methods to detect and determine even smaller amounts of trace materials, it follows that those factors which affect the response of instruments should be held at those settings which give the maximum value. The traditional univariate method of varying one factor at a time while holding remaining factors constant often fails to realise optimum conditions. This failure is best illustrated with the aid of a contour diagram (Figure 4.5) where the contour lines represent equal response. The highest point X is regarded as the optimum measure.

The levels of the two factors A and B are given on the x and y axes respectively. If the level of factor A is held at P_1, then by varying the level of factor B, the optimum value would be found at C. The value of factor B is then held at P_2 and the level of factor A is varied, which is most likely again to reach an optimum value at C. This false maximum is nowhere near the optimum value X, which could only have been reached if factor A had been held at P_3.

A series of factorial design experiments may be used for optimisation by applying the method of **steepest ascent**. This process is laborious and most difficult if a large number of factors are involved. Undoubtedly, the most widely used optimisation method is a sequential experimental design called **simplex optimisation**.

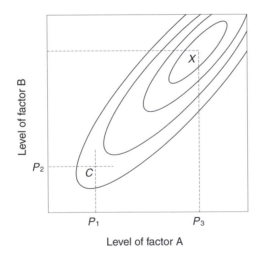

Figure 4.5   Contour diagram: highest point X is deemed the optimum measure

4.31   Sequential simplex optimisation

A **simplex** is a geometric figure defined by a number of points equal to one more than the number of factors. If two factors are chosen, the simplex will be a triangle. For three factors, the simplex will be a tetrahedron. More than three factors can be considered, but such simplexes cannot be visualised in three-dimensional space. The fundamental idea is

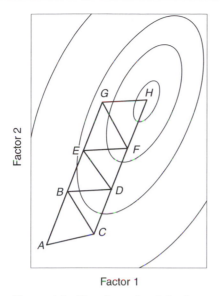

Factor 1

Figure 4.6 Two-dimensional simplex: contours represent lines of equal response

to reach the optimum in a sequential fashion with the minimum number of experiments. The underlying philosophy is elegantly described by Betteridge as 'to climb a mountain without a map, where the objective is to get to the top'. For easy visualisation, a two-dimensional simplex with two factors will be used to illustrate the optimisation procedure. Figure 4.6 shows a map where the contours represent lines of equal response.

The aim of the simplex procedure is to force the simplexes away from regions of poor response towards the region of optimum response. This is achieved by a series of 'moves' to the simplex *FGH* where the vertex *H* has reached the optimum. In order to achieve this objective, there are a series of rules to be followed when using the simplex method:[3]

Rule 1 A move is made after each response is observed.

Rule 2 A new simplex is formed by rejecting the point with the worst response in the current simplex and replacing it with its mirror image across the line defined by the remaining points. The new point generated usually shows a better response than at least one of the two points remaining. If the new point has the worst response of the simplex thus formed, a continuation of rule 2 would lead to an oscillation between simplexes, thus causing the optimisation procedure to be halted. This situation leads to the next rule.

Rule 3 If the reflected point has the least desirable response on the new simplex, the next to worst response is rejected in the original simplex and its mirror image is used to form the new simplex.

Rule 4 If a point falls outside the boundaries, reject it and proceed with rules 2 and 3. This rule will be explained later with the aid of an example.

Rule 5 If one point is retained in $n + 1$ simplexes, where n = the number of factors, as the response then it is considered the optimum.

The method outlined above has no provision for accelerated moves to reach the optimum and can sometimes lead to a false maximum.

155

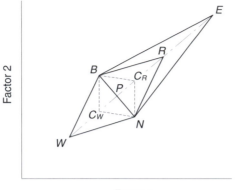

Figure 4.7 Operations on a two-factor modified simplex

The basic simplex was modified by Nelder and Mead.[4] This was achieved by the addition of two new operations, expansion and contraction, to the basic simplex operation of reflection. This modified simplex allows a more rapid and more precise location of the optimum response. The movements of the simplexes are governed by the same rules listed above, but there are additional tests to decide which operation to use. The operations available for a two-factor modified simplex are shown in Figure 4.7. *BNW* is the initial simplex where B = best response, N = next best response and W = worst response:

Reflection is achieved by extension of the line *WP* to the point R where $R = P + (P - W)$.
Expansion takes place to the point E where $E = P + \alpha(P - W)$; usually $\alpha = 2$.
Contraction may be to a new vertex that lies closer to R than W, then

$$C_R = P + \beta(P - W); \text{ usually } \beta = \tfrac{1}{2}$$

or it may be to a new vertex that lies closer to W than R, then

$$C_W = P - \beta(P - W)$$

Here is an example that shows how it works in practice. Not every stage of the calculation is reproduced, but there is sufficient detail to elucidate the method.

Example 4.17

Follow a modified simplex procedure to obtain the optimum response for measuring a calcium solution (containing $3 \, \text{mg L}^{-1}$ of Ca) by atomic absorption spectroscopy.

Define the quantity to be optimised   The absorbance of the calcium resonance line at 422.7 nm is selected as the response. For most spectroscopic techniques the response could be absorbance, emission or the signal-to-noise ratio. For chromatographic methods the response is not so easily defined (Section 4.31).

Select the factors   The factors selected are lamp current, flame height, fuel flow rate and nebuliser uptake.

Identify the system constraints The following system constraints (boundary conditions) are applied to each factor.

	Low (x_n)	High
Lamp current (mA)	1	10
Flame height (mm)	0	25
Fuel flow rate	2	8
Nebulizer uptake (mL min^{-1})	0	9.2

Too high a lamp current would appreciably shorten the useful life of a hollow cathode lamp, so an upper limit of 10 mA is chosen. The whole range of the flame height setting on the instrument is used. The acetylene flow rate is set within the limits on the grounds of safety. The maximum value for the nebuliser uptake is found to be 9.2 mL min^{-1}. Any values generated by the simplex operations that lie outside the boundary conditions may be identified.

Locate the initial simplex A matrix (Figure 4.8) described by Yabro and Deming[5] is used to design the initial simplex. The term S_n is known as the **step size** of the nth factor. It is calculated by subtracting the lowest value (denoted by x_n in the matrix) from the highest value. The step size of the lamp current is therefore $(10 - 1) = 9$ mA. Both p_n and q_n can be calculated for each factor from equations (4.14) and (4.15). By adding them to the relevant x_n in the matrix, the initial simplex can be formed. Vertex 1 is obtained from the lowest values selected for each factor:

Vertex 1	Lamp current 1 mA	Flame height 0 mm	Fuel rate 2	Nebuliser uptake 0 mL min^{-1}

Factors

Vertex	1	2	3	4	...	$n-1$	n
1	x_1	x_2	x_3	x_4	...	x_{n-1}	x_n
2	p_1+x_1	q_2+x_2	q_3+x_3	q_4+x_4	...	$q_{n-1}+x_{n-1}$	q_n+x_n
3	q_1+x_1	q_2+x_2	q_3+x_3	q_4+x_4	...	$q_{n-1}+x_{n-1}$	q_n+x_n
4	q_1+x_1	q_2+x_2	q_3+x_3	q_4+x_4	...	$q_{n-1}+x_{n-1}$	q_n+x_n
⋮	⋮	⋮	⋮	⋮	⋮	⋮	⋮
n	q_1+x_1	q_2+x_2	q_3+x_3	q_4+x_4	...	$p_{n-1}+x_{n-1}$	q_n+x_n
$n+1$	q_1+x_1	q_2+x_2	q_3+x_3	q_4+x_4	...	$q_{n-1}+x_{n-1}$	p_n+x_n

where

$$p_n = \frac{S_n}{n\sqrt{2}} [\sqrt{n+1} + (n-1)] \qquad (4.14)$$

$$q_n = \frac{S_n}{n\sqrt{2}} [\sqrt{n+1} - 1)] \qquad (4.15)$$

Figure 4.8 Initial simplex matrix

Vertex 2 is obtained as follows. From Figure 4.8 the first factor, the lamp current, is given by

$$(p_1 + x_1) \quad \text{where} \quad p_1 = \frac{S_1}{n\sqrt{2}}[\sqrt{n+1} + (n-1)]$$

where S_1 is the step size and n is the number of factors; here $S_1 = 10.0 - 1.0$, $x_1 = 1.0$ and $n = 4$, so

$$p_1 = \frac{10.0 - 1.0}{4\sqrt{2}}(\sqrt{4+1} + 4 - 1) = 8.33 \text{ mA}$$

and $\quad p_1 + x_1 = 8.33 + 1.0 = 9.33 \text{ mA}$

The second factor, the flame height, is given by

$$q_2 + x_2 \quad \text{where} \quad q_2 = \frac{25}{4\sqrt{2}}(\sqrt{4+1} - 1) = 5.46 \text{ mm}$$

and $\quad q_2 + x_2 = 5.46 + 0.0 = 5.46 \text{ mm}$

The third factor, the fuel rate, is given by

$$q_3 + x_3 \quad \text{where} \quad q_3 = \frac{8.0 - 2.0}{4\sqrt{2}}(\sqrt{5} - 1) = 1.31$$

and $\quad q_3 + x_3 = 1.31 + 2.0 = 3.31$

The fourth factor, the nebuliser uptake, is given by

$$q_4 + x_4 = 2.01 \text{ mL min}^{-1}$$

So vertex 2 is as follows:

Vertex 2	Lamp current	Flame height	Fuel rate	Nebuliser uptake
	9.33 mA	5.46 mm	3.31	2.01 mL mm^{-1}

The instrument is adjusted so these values (or values as close as possible) are set on the instrument and the absorbance measured is 0.103. Vertices 3, 4 and 5 are calculated in the same manner with the aid of the initial simplex matrix. The absorbance of each vertex is measured. The results obtained are recorded on a simplex worksheet (Figure 4.9).

Search for the optimum response   The vertex with the worst response (vertex 1) is rejected and the next vertex (vertex 6) is obtained using the lower part of the simplex worksheet (Figure 4.9). The first row of the lower section has the symbol Σ on the left-hand side. The values of Σ are simply formed from the sum of the remaining four vertices. So for lamp current

$\Sigma = 9.33 + 2.97 + 2.97 + 2.97 = 18.24$

Row 2 $P = \Sigma/n = 18.24/4 = 4.56$

Row 3 $P - W$ where W is the value of the rejected vertex

$P - W = 4.56 - 1.0 = 3.56$

Row 4 $\frac{1}{2}(P - W) = 1.78$

Row 5 $R = P + (P - W) = 8.12$ (reflection)

Last row $E = R + (P - W) = 11.68$ (expansion)

Vertex	Lamp current (mA)	Flame height (mm)	Fuel rate	Nebuliser uptake (mL min^{-1})	Absorbance
1	1.0	0.0	2.0	0.0	0.000
2	9.33	5.46	3.31	2.01	0.103
3	2.97	23.14	3.31	2.01	0.020
4	2.97	5.46	7.55	2.01	0.077
5	2.97	5.46	3.31	8.52	0.190
Sum Σ	18.24	39.52	17.48	14.55	
$P = \Sigma/n$	4.56	9.88	4.37	3.64	
$P - W$	3.56	9.88	2.37	3.64	
$\frac{1}{2}(P - W)$	1.78	4.94	1.19	1.82	
$R = P + (P - W)$	8.12	19.76	6.74	7.28	0.202
$C_R = P + \frac{1}{2}(P - W)$					
$C_W = P - \frac{1}{2}(P - W)$					
$E = R + (P - W)$	11.68				

Figure 4.9 Simplex worksheet

The value 11.68 is outside the boundary conditions, hence E is rejected in favour of R. Figure 4.9 shows that all four factors lie within the boundary conditions. So vertex 6 uses the instrument settings shown in the R row. This gives an absorbance of 0.202.

Vertex 7 is generated in exactly the same way; it gives a value in the R row that lies outside the boundary conditions, so the C_W row is used. Sequential steps are carried out in the same manner until vertex 21. And the response at vertex 15 may then be confirmed as the highest absorbance (0.235) for five successive simplexes. So, according to rule 5, this is the optimum response. The optimum response should be anticipated as the absorbance values are all close together for vertices 17 to 21 but less than for vertex 15.

Vertex	15	17	18	19	20	21
Absorbance	0.235	0.214	0.215	0.215	0.216	0.221

Once the optimum response has been located, a univariate search may be undertaken. Each factor is varied in turn while the remaining three are held constant at the values given by vertex 15. If, when varied, each factor still gives a maximum at the value determined by vertex 15, this will confirm that the simplex has not settled on a false maximum.

In order to determine the effects of the four factors, a factorial design could be used around the region of maximum response. A design using three levels would be preferable (although more lengthy) because of the difficulty that arises with a two-level factorial (Section 4.28).

4.32   Simplex optimisation: critical appraisal

Unlike factorial design, the number of experiments in the simplex method does not increase markedly with the number of factors. In order to accelerate the process of locating the optimum, a **supermodified simplex** has been developed. With a supermodified simplex it is possible to vary the scale factors α and β in the expansion and contraction operations; the modified simplex is usually restricted to $\alpha = 2$, $\beta = \pm\frac{1}{2}$. Among the most useful applications is where an instrument is interfaced with a microcomputer, enabling the simplex procedure to be automated to achieve optimisation.

There are, of course, many instances when the attainment of the optimum signal is not absolutely necessary and therefore it is claimed that the simplex method is not required. However, if the analyte sample is small in size and/or expensive, a considerable amount may be wasted using the univariate approach. The repeatability of results by different workers using the generated optimum conditions could be investigated subsequently. For example, after the simplex optimisation of an HPLC experiment, the significance of the four factors used was investigated by factorial design. Two analysts then checked the reproducibility of response using the F-test (Section 4.12) and the t-test between two means (Section 4.13). Finally, a two-way ANOVA (Section 4.24) was used to investigate if there was a variation using three different samples and three analysts.

The definition of the quantity to be measured (the response) is not always straight-forward. In gas chromatography the response must involve peak size, peak separation and retention time. For details of the optimisation of chromatographic procedures, consult the items listed in Section 4.38. In atomic spectroscopic techniques the signal/noise ratio may be a preferred response. Interference effects, or measuring the signal towards the limits of the detector range, can result in both emission and absorbance signals having an appreciably noisy background. The simplex method gives no information on both the effects of factors and on interactions. It is sometimes possible for the simplex to locate a false optimum, but a univariate search will confirm whether the optimum has been established. Furthermore, the sharpness of the response peak around the optimum region can serve as a measure of any critical variation in the level of a factor.

4.33 Treatment of multivariate data

The conversion of multivariate data into useful information is one of the most important areas of chemometrics, including **pattern recognition** and **principal component analysis**. Only an introduction will be given in this section with the aid of examples from one branch of pattern recognition, namely **cluster analysis**. In classical analysis a number of replicate experiments yielded just one item of information, e.g. the end point of a titration. With sophisticated instrumentation, however, a single experiment can give rise to a large amount of multivariate data. The intensities and frequencies of an infrared spectrum are a case in point. It is reasonable to state that our understanding of the information supplied by many modern instruments is often limited. A complete knowledge of the fingerprint region of an infrared spectrum is frequently unattainable. Chemometric methods can assist in the interpretation of data thus obtained. The country of origin of an oil-spill may be deduced from a chemometric appraisal of the frequencies and intensities obtained from infrared spectra.

Multivariate data exist in multidimensional space, which is clearly impossible to visualise above three dimensions. The main objective of pattern recognition is to reduce the dimensions of the data set. The two-dimensional patterns thus generated can be easily detected and may be classified by the human eye. The process of classification is indeed important in abstracting useful information. There are two main approaches to classification, supervised and unsupervised. Supervised methods require a test data set; this involves having certain samples of known origin and classification which are first analysed to set up a model. Unsupervised methods require no prior test set. Cluster analysis falls into this category.

There are many variants of cluster analysis, but only one type will be described with aid of examples. For more detail on cluster analysis, the reader is referred to the works in Section 4.38.

Example 4.18   Agglomerative hierarchical cluster analysis

Four different compounds were examined by thin layer chromatography using three different stationary phases. From the tabulated R_f values, multiplied by 100 and rounded off, determine which stationary phases are the most similar.

Compound	Stationary phase		
	A	B	C
1	90	70	70
2	70	50	60
3	60	40	30
4	50	30	40

The first stage in the calculation is to construct a **dissimilarity matrix**. This may be achieved in a variety of ways; one of the most common uses the **Euclidean distance** (Figure 4.10). In two-dimensional space the Euclidean distance may be determined using Pythagoras's theorem.
   In three dimensions;

$$d_{AB} = \sqrt{(\Delta x)^2 + (\Delta y)^2 + (\Delta z)^2}$$

This concept can be extended into n-dimensional space, where

$$d_{AB} = \sqrt{(\Delta x_1)^2 + (\Delta x_2)^2 + (\Delta x_3)^2 + \ldots + (\Delta x_n)^2}$$

The term d_{AB} is called the dissimilarity of AB. As the distance d_{AB} increases then the further apart the points A and B become, hence they become more dissimilar.
   In this example

$$d_{AB} = \sqrt{\underbrace{(90-70)^2}_{\Delta x_1} + \underbrace{(70-50)^2}_{\Delta x_2} + \underbrace{(60-40)^2}_{\Delta x_3} + \underbrace{(50-30)^2}_{\Delta x_4}} = \sqrt{1600}$$

so $d_{AB} = 40(R_f \times 100)$

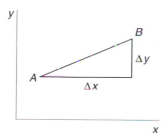

Figure 4.10   Euclidean distance: $d_{AB} = [(\Delta x)^2 + (\Delta y)^2]^{1/2}$

similarly

$$d_{AC} = \sqrt{(90-70)^2 + (70-60)^2 + (60-30)^2 + (50-60)^2} = \sqrt{1500}$$

so $d_{AC} = 38.7(R_f \times 100)$

and

$$d_{BC} = \sqrt{(70-70)^2 + (50-60)^2 + (40-30)^2 + (30-40)^2} = \sqrt{300}$$

so $d_{BC} = 17.3(R_f \times 100)$

The dissimilarity matrix can now be constructed as follows:

	A	B	C
A	0	40.0	38.7
B	40.0	0	17.3
C	38.7	17.3	0

The next stage is to identify the most similar stationary phases and combine them by taking the mean of the distances to form a cluster, the **linkage algorithm**. The distance BC is the shortest, so the phases B and C are combined (hence agglomerative) to form the cluster B^*.

Since $\dfrac{d_{AB} + d_{AC}}{2} = \dfrac{40.0 + 38.7}{2} = 39.3$

the new matrix may be written

	A	B*
A	0	39.3
B*	39.3	0

This linkage process is repeated until a 2×2 matrix has been formed (as is the case here).

Finally, a **dendogram** is constructed as a visual display of this information (Figure 4.11). Phases B and C are similar and they are both different from A, therefore BC can form a cluster.

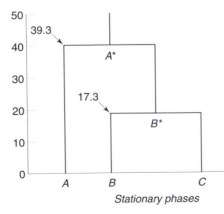

Figure 4.11 Dendrogram: phases B and C are similar and they are both different from A, therefore BC can form a cluster

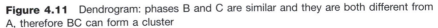

Instead of the Euclidean distance, the correlation coefficient values, or the **Manhattan distance**, can be used to construct the dissimilarity matrix:

(a) The correlation coefficient is calculated from the plots of A versus B, A versus C, and B versus C (Section 4.16) and then these values are put into the dissimilarity matrix. Note that the matrix diagonal is given the value 1.000.

(b) The Manhattan distance is given by the sum of distances of each variable, thus in Example 4.18 the sum of the distance AB is given by

$$(90 - 70) + (70 - 50) + (60 - 40) + (50 - 30) = 80$$

and the sum of distance $AC = 70$, and the sum of distance $BC = 30$.

Hence the matrix using the Manhattan distance may be written as follows:

	A	B	C
A	0	80	70
B	80	0	30
C	70	30	0

Quite often more than one clustering technique is used to test whether the same number of clusters are obtained. A more detailed example is given below of hierarchical agglomerative clustering to demonstrate how useful information can be extracted from a fairly incomprehensible data set. Even this example with 40 data points is small compared with the very large sets frequently encountered.

Example 4.19

The following data, measured at eight different wavelengths, was obtained from the electronic absorption spectra of five different plant extracts. Determine whether the extracts can be formed into groups using an agglomerative clustering technique.

Extract	Absorbance × 100							
	λ_1	λ_2	λ_3	λ_4	λ_5	λ_6	λ_7	λ_8
1	22	4	12	6	50	8	7	1
2	16	10	9	1	45	13	11	1
3	11	37	29	16	8	34	39	0
4	10	4	7	4	27	6	5	1
5	4	17	16	7	3	17	21	0

The dissimilarity matrix is obtained as in Example 4.18:

	1	2	3	4	5
1	0.00	13.11*	71.16	26.65	54.74
2	13.11*	0.00	63.04	22.22	46.45
3	71.16	63.04	0.00	63.40	36.57
4	26.65	22.22	63.40	0.00	35.54
5	54.74	46.45	36.57	34.77	0.00

163

The most similar extracts are 1 and 2; the linkage algorithm is then applied to give

	1, 2	3	4	5
1* ≡ 1, 2	0.00	**67.10**	**24.44***	**50.60**
3	67.10	0.00	63.40	46.45
4	24.44	63.40	0.00	35.34
5	50.60	46.45	35.34	0.00

The value 67.10 is obtained from the distance 1*3, i.e. the mean of 71.16 and 63.04. Similarly, $\frac{1}{2}(26.65 + 22.22) = 24.44$ and $\frac{1}{2}(54.74 + 46.45) = 50.60$.

Continuing the linkage method with wavelength 4 now being the most similar (24.44), we obtain

	1, 2, 4	3	5
1, 2, 4	0.00	65.25	42.97*
3	65.25	0.00	46.45
5	42.97	46.45	0.00

where $\frac{1}{2}(67.10 + 63.40) = 65.25$
and $\frac{1}{2}(50.60 + 35.34) = 42.97$

which finally reduces to

	1, 2, 4, 5	3
1, 2, 4, 5	0.00	55.85
3	55.85	0.00

The dendogram is shown in Figure 4.12.

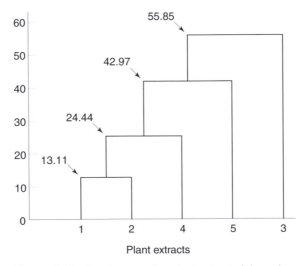

Figure 4.12 Dendrogram: the plant extracts join up in one large cluster, 1 and 2 first, then 4 then 5 then 3

The plant extracts join up in one large cluster. The most similar extracts 1 and 2 first, followed by 4 then 5 then 3. Extract number 3 may be treated as a potential outlier from the main cluster.

By use of the Manhattan distance (Example 4.18) instead of the Euclidean distance, the reader can confirm that the same overall result is obtained, although the difference between groups 1, 2, 4, 5 and groups 1, 2, 4, 3 was very small.

4.34   Factor analysis

Factor analysis is one of the most widely used chemometric techniques. Large data sets are analysed using this method. For example, McCue and Malinowski[6] have used the technique to investigate the infrared spectra of multicomponent mixtures. A series of ten mixtures were prepared from four strongly overlapping components (the three isomeric xylenes and ethylbenzene), together with two additional mixtures containing chloroform as an impurity to test the robustness of the method. An FTIR instrument was used to record the spectra of mixtures and of the pure components. The resulting data was then treated by factor analysis.

The first stage involves the preparation of a data matrix and subsequent data reduction to determine the number of factors. How this is achieved may be explained with the aid of graphical representation. A series of data points are plotted on a graph (Figure 4.13).

Each point can be identified by a pair of coordinates defining its position relative to two perpendicular axes. The first axis (factor 1), found in a manner similar to linear regression (Section 4.17) is where this axis passes through the highest concentration of data points. Hence the largest variance in the data is accounted for.

A second axis (factor 2) is found, perpendicular to the first, which in this simple example accounts for all the residual variance in the data. These two new axes allow the position of each data point to be identified (the dashed lines in Figure 4.13). These sets of figures, when all the data points have been identified, make up the **abstract matrices**. The axes represent factors involved in producing the data; the **abstract axes** are related to the true factors by rotation about the origin. When many factors are involved, this process is carried out until all the variance of the data has been accounted for. The importance of a factor is indicated by the size of its **eigenvalue**. Factors, called **eigenvectors**, are produced

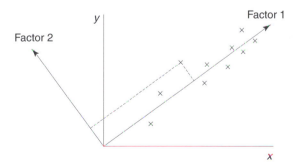

Figure 4.13   Factor analysis: the (x, y) coordinate frame is rotated about the origin to a new coordinate frame where the perpendicular axes are the eigenvectors representing factor 1 and factor 2

by **principal component analysis** in decreasing order of importance. The eigenvalue of each eigenvector is related to the amount of variance. When this value becomes small, it can be attributed to experimental (random) error. The eigenvectors caused by experimental error may be removed, resulting in more coherent and reliable data.

The next stage in factor analysis is to transform the abstract data into individual real factors. One method of conversion of the abstract solution into a real solution is called **target transformation (TT)**. This allows real factors to be tested individually. In target testing, a vector is defined which is believed to be a factor of the data set. Now the data described by two factors lies in a plane, and if the test vector lies outside the plane it is not a true factor. Target testing projects the test vector onto the plane of the data, producing the **predicted vector**. Comparison is then made between the test vector and the predicated vectors. If the test vector is a true factor in the data set, then the values for the test vector and the predicted vector must be the same.

A further aim of factor analysis is to quantify how much of a component is contained in a mixture. In two-dimensional cases, experimental measurements should lie in the plane corresponding to two factors. The position of a point in this plane is related directly to the relative proportion of the components in a mixture. Thus, McCue and Malinowski[6] were able to identify the components of the mixtures of the three isomeric xylenes and ethylbenzene by testing for each of the compounds present in a mixture with the spectra of the pure components. This was accomplished without any information on the remaining constituents.

Target combination was performed using the molar absorptivities calculated from the spectra of the pure components. The concentrations of each component in the mixtures agreed well with the true values. The presence of chloroform, used as a contaminant, was noted by a marked increase in the error term after the abstract factor analysis. The contaminant was then listed using the spectrum of pure chloroform. Its presence was confirmed by the similarity between the test vector and the predicted vector.

These results illustrate the power and usefulness of the target testing procedure, bearing in mind the lack of prior information about the identity and number of components in the system.

4.35 Quick statistics

In analytical chemistry, quite frequently a few replicate measurements are made. It is usual practice for a gravimetric determination to be performed in duplicate; in titrimetry the mean of three replicate titrations is the normal approach. It is certainly relevant to ask if this mean is normally distributed, the assumption that has been made with statistical tests that use the mean and standard deviation.

There are statistical tests available that are considered distribution-free, called **non-parametric** methods. In the majority of examples, calculations involving non-parametric methods are very simple, so they are amenable to a quick evaluation. Instead of the mean, the **median** is used as the measure of central tendency. The number of observations is n, and the observations are arranged in ascending order. If n is odd the median is the value of observation $\frac{1}{2}(n + 1)$. If n is even the median is the average of observation $\frac{1}{2}n$ and observation $\frac{1}{2}(n + 1)$. The **range** is the difference between the highest observation and the lowest observation in a data set; it is used as a measure of dispersion rather than the standard deviation. For the following titration results (mL):

$$10.00 \quad 10.05 \quad 10.07 \quad 10.25$$

the median is $\frac{1}{2}(10.05 + 10.07) = 10.06\,\text{mL}$ and the range is $(10.25 - 10.00) = 0.25\,\text{mL}$.

Some tests use the range as a measure of dispersion, though strictly they are not non-parametric since the arithmetic mean $\bar{x}$ is used. Range tests can be used instead of the tests described in Sections 4.12 and 4.13:

(a) The t-test used to compare the experimental mean $\bar{x}$ with the true or known mean μ (Section 4.12) may be replaced with a range test using the statistic T_1, where T_1 is given by

$$T_1 = \frac{|\bar{x} - V|}{R} \qquad (4.16)$$

where V is the known mean and R is the range.

(b) Instead of the t-test for the comparison of two means $\bar{x}_1$ and $\bar{x}_2$ (Section 4.13), the statistic T_d may be used, where T_d is given by

$$T_d = \frac{2|\bar{x}_1 - \bar{x}_2|}{R_1 + R_2} \qquad (4.17)$$

Similarly, an alternative to the F-test (Section 4.13) based on the range uses the statistic F_R, where F_R is given by

$$F_R = \frac{R_1}{R_2} \quad \text{where} \quad R_1 > R_2 \qquad (4.18)$$

Example 4.20

In a new method used for the determination of calcium in tap water, a single sample was analysed four times; the results ($\mathrm{mg\,L^{-1}}$) were

104.5   106.0   103.9   105.1

These values were compared with a standard method which gave

106.2   105.8   106.3   105.6

Use a test based on ranges to show whether the methods differ significantly in (i) their precisions and (ii) their means.

Precision   From equation (4.18)

$$F_R = \frac{R_1}{R_2} = \frac{2.1}{0.7} = 3.0$$

From tables (Appendix 17) the calculated value of F_R (3.0) is less than the critical value of F_R (4.0). Hence the precisions are not significantly different.

Means   From equation (4.17)

$$T_d = \frac{2|\bar{x}_1 - \bar{x}_2|}{R_1 + R_2}$$

where $\bar{x}_1$ (mean of new method) $= 104.87\ \mathrm{mg\,L^{-1}}$

and $\bar{x}_2$ (mean of standard method) $= 105.97\ \mathrm{mg\,L^{-1}}$

so $T_d = \dfrac{2\,|104.87 - 105.97|}{2.1 + 0.7} = \dfrac{2.2}{2.8} = 0.786$

The calculated value of T_d (0.786) is less than the tabulated value of T_d (0.81) from Appendix 16, so the means are not significantly different.

The paired t-test (Section 4.14) may also be evaluated by an alternative non-parametric method, the **Wilcoxon signed rank test**. This method is best explained using an example.

Example 4.21

Each of ten different samples of canned fruit juice was divided into two portions; one portion was sent to laboratory 1 and the other to laboratory 2. The laboratories determined the amount of tin (mg^{-1}) in each sample. The results were as follows. Is there evidence of a systematic difference between the two laboratories?

Sample	A	B	C	D	E	F	G	H	I	J
Laboratory 1 ($mg\,L^{-1}$)	51.7	82.1	73.3	35.7	65.9	95.3	21.9	16.2	45.1	103.6
Laboratory 2 ($mg\,L^{-1}$)	50.9	81.9	73.4	35.4	64.8	94.8	22.3	15.0	44.2	103.1

The stages of the Wilcoxon signed rank test are as follows:

1.   Calculate the signed differences between each sample:

A	B	C	D	E	F	G	H	I	J
+0.8	+0.2	−0.1	+0.3	+1.1	+0.5	−0.4	+1.2	+0.9	+0.5

2.   Arrange them in increasing numerical order, ignoring the sign:

$$-0.1 \quad 0.2 \quad 0.3 \quad -0.4 \quad 0.5 \quad 0.5 \quad 0.8 \quad 0.9 \quad 1.1 \quad 1.2$$

3.   The results are now 'ranked'. The lowest is given rank 1 and in this example, with ten results, the highest is given rank 10. With the tied values (0.5), at rank 5, they average between 5 and 6 to give 5.5. The ranking, keeping the + and − signs, becomes

$$-1 \quad +2 \quad +3 \quad -4 \quad +5.5 \quad +5.5 \quad +7 \quad +8 \quad +9 \quad +10$$

4.   The positive ranks sum to 50 and the negative ranks sum to −5. The lower rank sum, 5, irrespective of sign, is taken as the test value.

From the table of Wilcoxon signed rank test (Appendix 15) the value for ten pairs = 8. In tests of this type, if the calculated value of the lower rank sum (5) is **less than or equal to** the tabulated value (8) then there is a significant difference between the laboratories.

This brief introduction to non-parametric and rapid methods demonstrates the relative simplicity of the calculations. There are many non-parametric methods that can be of use to the practising analytical chemist. The reader is referred to Section 4.38, where some useful books on the subject are listed.

4.36   The value of chemometrics

Correctly used, the methods described in this chapter are an invaluable asset to the practising analytical chemist. It has only been possible in this chapter to give an introduction to the many statistical and chemometric methods now available. The main purpose of this chapter has been to make the reader aware of the increasing potential of the rapidly expanding subject of chemometrics. The selection of topics to be included has therefore been difficult and inevitably there have been omissions which may appear regrettable to practising chemometricians.

The approach throughout this chapter has been to use specific examples which serve to illustrate the scope of the subject when applied to the treatment and interpretation of analytical data. Clearly, there is a danger that some basic concepts may thus be overlooked and the reader is strongly advised to become more fully conversant with chemometric methods by obtaining a selection of the excellent explanatory texts now available.

Note that signal processing methods are classified as a branch of chemometrics and are included in this book. Information will be found on Fourier transform spectroscopy (Section 18.5) and on derivative spectroscopy (Section 17.13).

There is now a large amount of computer software easily obtainable for both classical statistics and chemometric methods. Packages are available for all the techniques covered in this chapter. A detailed listing of available software is not included. The rapid development of more sophisticated computer programs would soon make a current list obsolete.

Never forget the chemistry in chemometrics! If the answer is chemical nonsense, the method was incorrectly applied or the design was faulty.

4.37   References

1. C T Shewell 1959 *Anal. Chem.*, **31** (5); 21A
2. R L Plackett and J P Burman 1946 *Biometrika*, **33**; 385
3. S N Deming and S L Morgan 1973 *Anal. Chem.*, **45**; 278A
4. J A Nelder and R Mead 1965 *Comput J.*, **7**; 308
5. L A Yabro and S N Deming 1974 *Anal. Chim. Acta*, **73**; 391
6. M McCue and E R Malinowski 1981 *Anal. Chim. Acta*, **133**; 125

4.38   Bibliography

M J Adams 1995 *Chemometrics in analytical spectroscopy*, Royal Society of Chemistry, Cambridge

K R Beebe, R J Pell and M B Seasholtz 1998 *Chemometrics: a practical guide*, Wiley, Chichester

R G Brereton 1990 *Chemometrics*, Ellis Horwood, Chichester

C Chatfield 1996 *Statistics for technology*, 3rd edn, Chapman and Hall, London

S N Deming and S Morgan 1993 *Experimental design: a chemometric approach*, 2nd edn, Elsevier, Amsterdam

D L Massart, B G M Vandeginste, S N Deming, Y Michotte and L Kaufman 1998 *Chemometrics: a textbook*, Elsevier, Amsterdam

J C Miller and J N Miller 1993 *Statistics for analytical chemistry*, 3rd edn, Wiley, Chichester

E Morgan 1995 *Chemometrics: experimental design*, ACOL–Wiley, Chichester

P Sprent 1993 *Applied nonparametric statistical methods*, 2nd edn, Chapman and Hall, London

5

Sampling

5.1 Introduction

5.1.1 Sampling techniques

Analytical measurements are used for a wide range of purposes: to monitor and regulate the composition of raw materials used in trade, to control or optimise manufacturing processes, to monitor impurities and by-products, to ensure compliance with legal requirements for both maximum and minimum composition, to ensure that food and drink are wholesome, to safeguard personal health and safety in the workplace and to maintain a safe working environment, and to monitor and protect the general environment.

It has been estimated[1] that in the developed world approximately 3% of GDP is used for analysis and 1 billion analytical measurements are made per year in the United Kingdom alone (although as many as 10% of them are not of adequate quality).

At first sight, analysis asks simple questions such as, What is in the sample? What concentrations are present? What health risks do these materials present? Unfortunately, it is often very difficult to answer these questions in a simple way and arrive at a correct answer unless a **representative sample** can be obtained from a complex matrix. Thus sampling represents the first, and often the hardest, step in the analytical protocol, although this is not always explicitly realised by the customer who requests the analysis, or even by all analysts. However, the awareness of this problem is growing, as shown by a recent quote from the Laboratory of the Government Chemist: 'Poor measurements are, at best, expensive and inconvenient but at worst can be dangerous or hazardous to health. . . . There is no point in having first rate analysts, expensive instruments or state of the art sensors if the sample is not representative or has altered prior to analysis.'[2]

This chapter will discuss ways of collecting representative samples from a range of different materials in the gas, liquid and solid phases, and it will consider how correct and sensible answers can be obtained for both industrial and environmental monitoring studies. Although the exact protocols used will differ according to the problem, some general comments can be made which apply equally to gas, liquid and solid sampling.

5.1.2 Sampling statistics

Firstly, although a commonsense approach to sampling is always required, the use of statistical techniques can usually make the analyst's task quicker, and also provide a more reliable answer to the questions posed. A detailed consideration of the statistics of sampling

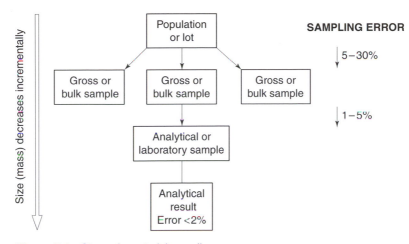

Figure 5.1   Stages in material sampling

will be postponed until the section on solid sampling but note that, like all scientific statistical techniques, it is the **number** of analyte particles which is important rather than the amount (mass) of the sample. The following relationship applies in all cases:

$$n \propto \frac{1}{R^2}$$

where R is the relative standard deviation and n is the number of particles. The consequence of this simple relationship means that different approaches are used for each of the three phases.

For gases and liquids the particle size is very small, e.g. 1 mL of a pure gas at STP contains ~2.5×10^{19} particles and 1 mL of liquid, ~2.5×10^{16}; but for solids the requirements of homogeneity depend upon taking sufficient analyte particles to be representative. Thus, while finely divided materials can often give a gross sample weight of a few grams or less, for bulky materials the gross sample weight may have to be measured in tons.

Since modern methods of analysis often require considerably less than 1 mg of material in order to perform a number of analyses, this small amount (the analytical sample) which is ideally an exact but miniature replica of the entire mass of material (the population) is often obtained by first collecting a gross sample which is large enough to be representative of the population and then subsequently reducing it down to the analytical sample (Figure 5.1). It cannot be stressed too highly that the reduction of sampling error during this sequence is often the hardest step in any analytical procedure and relies on the analyst having as much information about the system as possible.

The second statistical term that is important in this area is the **level of confidence**. Generally it is impossible to provide an answer with absolute confidence. For example, the statement 'all eggs from a particular farm are free from salmonella' requires that all the eggs are sampled (and thus destroyed). A lower confidence level, of say 95%, is often used and allows a very much smaller number of samples to be taken, but for many sampling situations even this level of confidence is too high. Normally it is good practice to quote both the numerical value determined and the level of uncertainty in that value, so there is an awareness of the uncertainties in the measurement.

5.1.3   Variability in the sample

Variability, as used in sampling, is the difference in analytical content or composition of the matrix with respect to additional external factors; it is **not** simply the different numerical values that may be produced as result of random fluctuations in the analytical technique or even the different values which **may** be produced as a result of poor sampling technique, but it is due to real changes in the sample composition. Two main causes of variability will be encountered in this chapter:

Variability with position is fairly obvious in the case of solid sampling, where analyte concentrations can vary by several orders of magnitude over quite small distances; but this problem is also encountered in liquid and gas sampling, especially where large populations, or gross samples, are considered and the normal mixing effects seen in gases and liquids in the laboratory do not occur. Although all gases are 'mutually miscible with each other' it is not sensible to suggest that the composition of the atmosphere is the same across the globe or even at different altitudes. Equally, the composition of large bodies of liquid such as the oceans will vary considerably with position.

Variability with time may occur in a random or semirandom fashion, or with some pronounced cyclic nature. This chronological variation may be due to natural factors, 'it sometimes rains', or perhaps where many environmental problems have a diurnal variation; it may also be due to some periodic variation in a manufacturing process, causing the value to fluctuate. In each case it is important to take samples in such a way as to avoid the 'coincidence of periodicity' in the sampling process.

5.1.4   Sample stability

Since it is rare for an analysis to be performed at the instant the sample is taken, an apparent variability can sometimes be observed where the sample is changing in composition during the time between collection and determination. For accurate analysis, the material must not be allowed to change significantly with respect to the bulk with time. Although some samples are stable, even they require general precautions to avoid gain or loss of weight due to water; and many other samples need some form of preservation to maintain their integrity.

5.1.5   Regulation and legislation

Analysis should always be done with a purpose, often this is to monitor substances in either the industrial or general environment to ensure that defined (safe) levels are not exceeded. The regulations or laws which define these levels have been introduced progressively worldwide, and form a complex framework within which the analyst must work. Often the precise way that samples are collected and analysed is stipulated within the regulation, and although other methods **may** in the analyst's opinion be just as suitable, they can only be used after a thorough evaluation against the standard method. If the analysis is to form the basis of legal proceedings, it is normally better to use a standard method which can be demonstrated reliable. The specific legislation concerning different sample types will be introduced for gas, liquid and solid samples in each of the relevant sections.

5.1.6   The terminology of sampling

As with most branches of science, the language, particularly the abbreviations used in sampling protocols, can be somewhat daunting at first. Unfortunately, there is no short cut and the jargon must be learned if a proper appreciation of the problems and their solutions are to be obtained.

5.2   Gases and vapours

Since the error R is inversely proportional to the number of sample particles taken, we will start by looking at the analysis of gases and vapours, where the particle size is the molecule (or occasionally the atom), hence large numbers of particles can be sampled easily. This section will concentrate mainly on **open** systems where the gas is not contained, since this constitutes a more difficult sampling problem than for gases in **closed** systems such as pipes or cylinders for which simplified protocols are available, normally based on the assumption that the bulk sample is homogeneous or nearly so.

Normally the sampling of gases and vapours in an industrial environment (workplace monitoring), where levels may be quite high, is considered separately from analysis in the environment (environmental monitoring), where it is hoped that gaseous pollutants are at low concentrations. Certainly different legislation and guidelines do apply in each area, but progressively the distinctions between the two areas in terms of concentration levels found, or expected, are decreasing and the techniques traditionally used for one area are being used in another. Although this section begins by looking at workplace monitoring, many of the techniques and protocols are beginning to be used in more general situations.

Workers in industry are protected against excessive exposure to hazardous materials, including gases and vapours, by a wide range of legislation and guidelines which vary according to the country in which the industry is situated,[3] although there is a progressive move towards harmonisation of this legislation, particularly in the developed world. In each case three key factors have to be considered:

1. The intrinsic hazard posed by the gas; logically, possible exposure to HCN is more serious than exposure to say EtOH vapour.
2. The level or concentration of the gas; for many, but not all gaseous chemicals there is a dose-related response, where it is assumed that the higher the level, the more serious the effect.
3. The duration of exposure; exposure to low concentrations for long periods of time might not have the same effect as short-term exposure to high levels.

Note   The words 'hazard' and 'risk' are often used rather loosely in general conversation, and often appear to be descriptions of the same thing. However, scientifically, a hazard is intrinsic to the substance or situation being described; thus concentrated acids are always hazards. Risk, on the other hand, varies with the action being performed; thus there is little risk of harm from concentrated acids in properly sealed bottles, but if they are placed in unlabelled, open vessels, the risk is increased dramatically.

Gas sampling protocols have to consider all three factors if a sensible estimate of actual risk to the worker is to be obtained, and in the United Kingdom the legislation protecting the worker addresses each factor. For the first, and in many respects the simplest factor, the Health and Safety Executive (HSE) publishes an annual list of occupational exposure limits known as EH40/X, where X is the year of publication; these limits form part of the Control

Table 5.1 *Occupational exposure standards for some common gases and vapours found in industrial environments*

	8 h average[a]		10 min exposure[b]	
	ppm	mg m^{-3}	ppm	mg m^{-3}
Benzene	5	16	–	–
Carbon tetrachloride	2	12.6	–	–
Chlorobenzene	50	230	–	–
Chloroform	2	9.8	–	–
1,1-Dichloroethane	200	810	400	1620
Dichloromethane[c]	100	350	300	1050
Formaldehyde[c]	2	2.5	2	2.5
Hydrogen cyanide[c]	–	–	10	10
Mercury (metal)	–	0.05	–	0.15
Methyl cellosolve	5	16	–	–
Nitrotoluene	5	30	10	60
Phenol	5	19	10	38
Styrene[c]	100	420	250	1050
Toluene	50	188	150	560
Vinyl chloride[c]		3 ppm over 1 year		

[a] Time-weighted average (TWA).
[b] Short-term exposure limit (STEL).
[c] Maximum exposure limit (MEL).

Of Substances Hazardous to Health Regulations 1988 or COSHH Regulations.[4] It is in fact two lists; the smaller list contains about 30 substances assigned maximum exposure limits (MELs); these are the gases and vapours considered to be the most harmful. Exposure of workers to levels in excess of the published value constitutes a breach of the regulations. In simple terms the manager responsible for that process is breaking the law and is liable to prosecution.

The larger list consists of occupational exposure standards (OESs) and contains recommendations for maximum exposure to a very wide range of gases and vapours likely to be found in an industrial environment (Table 5.1). In simple terms this is a list of 'safe' working levels for these materials. Since the list is published annually, individual levels can be adjusted up or down in the light of new information regarding the substance and its hazard.

Although specific to the United Kingdom, OES levels are normally similar or identical to threshold limit values (TLVs), their American equivalents. In each case the list specifies not only maximum levels, but also the time period over which the measurements are made. Normally two reference periods are used; a short-term exposure level (STEL), commonly 10 minutes, and a longer time-weighted average (TWA) figure of 8 hours, corresponding to an average shift in industry. Lower limits are generally set for long-term exposure than for short. It is desirable to monitor for a given gas or vapour using both short and long time periods, but each uses different protocols.

Sampling techniques for gases and vapours in the workplace fall into four general categories:

1. Gas sampling vessels
2. Static sensors
3. Entrapment
4. Real-time analysis

Each method has advantages in certain situations, but no one method is satisfactory in all cases, especially as different (statutory) requirements are based on different sampling protocols which specify different maximum levels and different reference periods.

5.2.1 Gas sampling vessels

Where a simple 'grab' sample of the atmosphere in a workshop or laboratory is all that is required, then glass or metal vessels of between 0.01 and 10 L (Figure 5.2), either evacuated prior to use or flushed with gas, are simple and reliable for sampling high concentrations of gases, but they are bulky and can suffer 'memory' effects if re-used. Gas sampling syringes, which can be fitted with a simple valve, are a modification of this approach. Inflatable plastic bags are less bulky in storage but can be just as inconvenient in use as rigid vessels since, when inflated, they have volumes of up to 100 L. In order to prevent loss of sample via permeation through the walls, or adsorption and loss on the inner skin, the bags are normally produced with a triple-layer construction: Mylar/aluminium/polythene. This increases the cost, so they are often reused a number of times with the consequent

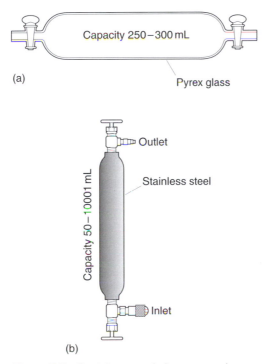

Figure 5.2 Sampling vessels for gases and vapours: (a) glass, (b) metal

risk of contamination from a previous sampling experiment. In each case, although simple to use, the device can only give an indication of the gas levels at the time it was filled, and long-term measurements are not possible. Also, because the total volume of gas collected is limited, they are not suitable for low-level concentrations.

5.2.2 Static sensors

Two forms of static sensors are generally found in industrial monitoring situations. **Adsorbent sensors** take the form of badges containing a small amount of adsorbent material which can be exposed to the atmosphere via a semipermeable membrane. Normally they are purchased in sealed foil containers and simply need to be removed from the foil and clipped to the clothing when adsorption of gases or vapours in the air takes place by a diffusion process (Figure 5.3). Although simple and cheap in use, they suffer from the problem that since the analyte only reaches the adsorbent via diffusion, they tend to require long sampling times in order to achieve acceptable sensitivity. They can, however, be worn with little inconvenience by workers who may be in an environment where they are exposed to long-term, low-level exposure, and as such are widely used for routine monitoring of OES levels over an 8 hour period. A number of these 'badge monitors' are available where a colour change is observed at specified levels so that simple visual inspection is all that is required. (A domestic version of these devices can be purchased which monitors levels of CO in the home.)

The main disadvantage of these devices is that, since they rely on a chemical reaction to produce a colour change, they are specific to one substance or group of substances and different badges are required if several gases are being monitored. To overcome this difficulty, general-purpose adsorption badges, generally containing activated carbon or another powerful adsorbent, can be used, where the total exposure to gases or vapours is measured at the end of the sampling period by desorbing the analyte and then measuring the total gases, often by gas chromatography. This type of sensor suffers from the problem that it is post-collection analysis and may only indicate a problem when it is too late, and

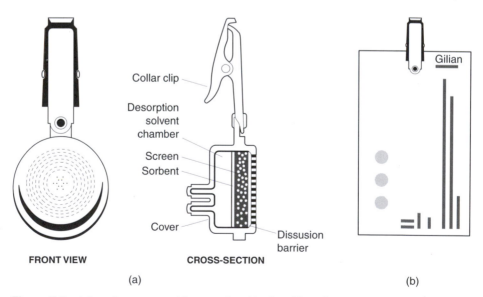

Figure 5.3 Adsorption sensors: (a) conventional badge, (b) credit card

expensive equipment is required for desorption and analysis. It is not necessary to contain the adsorbent in a badge form and several commercial adsorption-by-diffusion systems take the form of glass or stainless steel tubes containing the adsorbent bed plus a membrane; however, since this configuration can be used for entrapment as well, a fuller description will be given in Section 5.2.3.

Specialist (electrochemical) sensors can be used for a particular gas or vapour. Possibly the most widely known and certainly the most widely used is the 'Breathalyser' carried by police officers for measuring alcohol in breath. Since generally they are capable of giving an instantaneous response, a description of these sensors will be given in Section 5.2.4.

5.2.3   Entrapment

Although, strictly speaking, the static adsorbent sensors described above are entrapment systems, since they use an adsorption process to collect the analyte of interest, the word 'entrapment' is normally applied to active collection devices, where the atmosphere is pumped mechanically through a trap which either adsorbs the analytes or reacts directly with them. Currently, pumped systems are the most widely used method for gas sampling prior to analysis, since by adjusting the pump speed and the sampling time, different volumes of gas can be sampled; this means it can be used to provide TWA measurements at quite low levels, and by choosing the correct type of trap, a wide range of sample types can be determined.

Liquid traps

Liquid traps have been in use for a long time and are still used for some types of monitoring programmes such as long-term evaluation of gases in an environment, where the simplicity, specificity and sensitivity are well established. Gas is drawn through a bubbler or impinger by a pump, and simple but reliable chemistry is employed to trap the substance of interest. As an example, SO_2 can be trapped at low levels by passing the gas through a dilute solution of hydrogen peroxide, when the following reaction takes place:

$$SO_2 + H_2O_2 \rightarrow H_2SO_4$$

The acid produced is stable, so measurements can be done over long periods (30 days or more) then the total amount of acid produced can be determined, either by titration or pH measurement. Liquid traps tend not to be as useful for personal monitoring because it is difficult to design apparatus containing liquids which can be readily worn on the clothing without the risk of spillage, but at least one system is commercially available (Figure 5.4) based on diffusion of gas through a membrane and into a small volume of liquid.

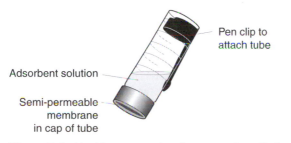

Adsorbent solution

Semi-permeable membrane in cap of tube

Pen clip to attach tube

Figure 5.4   Liquid sensor system for personal monitoring of gases and vapours

Solid adsorbents

By using a pump to force the gas sample through a bed of adsorbent, either the sensitivity can be increased considerably over that of static diffusion sensors, or the sampling period required to trap a given mass of analyte can be reduced.

The simplest of these systems, applicable where high levels of analyte are expected, uses a simple hand-operated bellows-type pump which forces a few litres of gas through a tube packed with adsorbent material. One widely used device of this type uses an adsorbent bed of granular silica gel onto which is coated a reagent or reagents that react specifically with the analyte to produce a colour. The contamination level is then determined directly by measuring the length of the coloured stain in the tube. This principle was used in the original Breathalyser tubes, where ethanol in the breath was oxidised by acid dichromate adsorbed on the silica to give a green chromate stain. These devices are widely used in industry, particularly where an immediate indication of possible contamination, and thus risk, is required. Although simple in use and relatively accurate, they can only be used for short-term monitoring of high levels of gaseous emissions, and since they rely on specific chemistry to produce the colour change, different tubes have to be used for each different gas.

Where low levels of gases or vapours need to be measured in either the industrial or general environment, possibly over an extended time period (remember that much industrial legislation is based upon an 8 hour sampling period), the gas can be pumped through a bed of adsorbent material using a 'constant-volume' pump. These battery-operated pumps are available in a form which allows them to be worn by the worker with minimal discomfort for the whole of an 8 hour shift, but which will suck air from the immediate environment at a flow of 0.20 to $2 \, \text{L} \, \text{min}^{-1}$ through a bed of adsorbent material contained in a sampling tube. These 'personal monitors' are currently the best way of obtaining a representative TWA sample of exposure to gases and vapours in an industrial environment. It is critical that the pump can provide a constant flow rate, since the volume of gas collected is normally simply calculated as

$$\text{sampling period} \times \text{flow rate} = \text{total volume}$$

Typically at a flow of $0.2 \, \text{L} \, \text{min}^{-1}$ for 8 hours, 9.6 L of gas, or $0.096 \, \text{m}^3$ will have been collected. The choice of adsorbent and the way it is contained in the tube can vary considerably, and a huge number of commercially available configurations can be purchased for different sampling protocols. However, two problems should immediately be obvious; the adsorbent must **effectively** adsorb the material of interest, **and** allow complete desorption of the material prior to analysis. Secondly, since the sampling period is of finite length and the analysis is only performed after collection, these devices give post-collection analysis. Generally, two types of tube are used, glass and stainless steel. Glass tubes are normally about 4 mm internal diameter and 70 mm long and are typically packed with 100 mg of adsorbent, held in place by glass/quartz wool plugs or porous polymer plugs. They are sealed at each end until just before use, when the ends are broken and the tube connected by flexible tubing to the pump.

They are often known either as ORBO, or NIOSH tubes because a lot of the early evaluation of them was done by the National Institute of Occupational Safety and Health Administration in the United States. These tubes are designed to be used only once, and although they are relatively cheap when purchased commercially, they can readily be made in the laboratory using glass dropping pipette tubing and a Bunsen burner. Several absorbent materials are available,[5] but charcoal is the most common for disposable tubes, since it is a powerful adsorbent for most non-polar organics. Silica gel is used for more polar

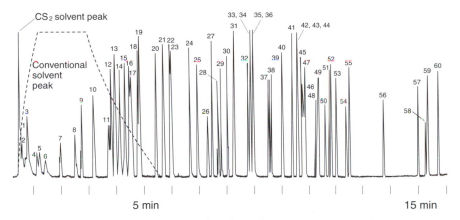

Figure 5.5   GC trace of volatile organics after solvent extraction with CS_2: sampling tube 100 mg 20–30 mesh Carbotrap; desorption 100 μL CS_2; column 30 m DBS (ID = 0.22 mm); temperature programme 1 min at 10 °C then to 80 °C at 5 °C min^{-1}

organics and some inorganics but it absorbs water, unlike charcoal, and care is required that absorption efficiency is not impaired by moisture in the air. Once the tube has been used to adsorb analytes from a measured volume of air, they must be desorbed and quant-itatively analysed. Since most organic gases and vapours are strongly adsorbed onto char-coal (which is why it is used in the first place), it is important to be able to desorb the gases, quantitatively and without change. Normally this is done using a solvent desorption process, which is more reliable than simple thermal desorption from the charcoal.

In principle a number of organic solvents could be used, but in practice carbon disulph-ide is often the preferred solvent. The choice of a toxic, highly flammable and extremely noxious-smelling solvent may, at first, seem a strange one, but there are good reasons for this preference. Normally the final analytical technique used to separate, identify and quantify the substances which have been collected is gas chromatography, and in many cases the detector used is a flame ionisation detector. It is found that carbon disulphide gives a very small response at an FID and thus it will be less likely to mask the important first few minutes of the chromatogram in which most of the sample compounds will elute (Figure 5.5). Secondly the solvent is available in pure form at reasonable cost. This is again important, since in most cases only a few micrograms of a particular analyte are collected, and any impurities in the solvent which eluted in the region of interest would readily mask the peak due to sample.

The advantage of carbon disulphide giving a very small response at an FID detector is not important if other detection systems such as GC/MS or even HPLC analysis are used, but the importance of using the purest solvent available in these experiments cannot be overemphasised. The simplest way of extracting the analyte is to break the glass sampling tube in the middle, tip the carbon granules into a small sample bottle and add 0.25–0.50 mL of solvent and then agitate for a few minutes, preferably by partially immersing the bottle in an ultrasonic bath. A portion of the solvent can then be injected directly into the gas chromatograph. The technique is simple and with care reliable results can be obtained, although producing a calibration line for quantitative analysis can be quite a lengthy pro-cess, since this should be done by adding increasing, known quantities of the analyte in question to blank tubes and then carrying them through the complete extraction process.

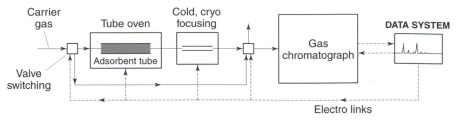

Figure 5.6   Automated thermal desorption/GC system

For small sample numbers the simplicity of the equipment is attractive, but the process is labour-intensive, and for larger sample numbers, perhaps where several dozen workers are monitored every day, alternative systems are gaining favour because they can be readily automated. In these systems the tubes, still about 4 mm diameter and 70 mm long, are made of stainless steel and are open at each end, although they are fitted with push-on caps which prevent gas adsorption until required. Typically 100 or 500 mg of adsorbent is held in the tube using small stainless steel gauze frits or screens. The adsorbent is normally a synthetic porous polymer such as Porapak or Tenax, originally developed as packing materials for gas chromatography. To collect the sample, the caps are removed and air is pumped, at constant flow for a defined time, through the tube as before. Once the sampling period is over, the analytes are desorbed from the tube and analysed in one automated process. The tubes are heated in a flowing gas stream to thermally remove the adsorbed sample, then they are passed straight to the inlet of a gas chromatograph. Several systems are available in which up to 50 tubes can be loaded into a sample carousel and then the complete process of desorption and chromatographic analysis can be carried out unattended, overnight if required (Figure 5.6).

With adsorbent porous polymers, thermal desorption effectively removes all adsorbed gases and vapours, and this confers two advantages:

(a)   All of the sample is injected into the gas chromatograph as one 'slug', giving much higher sensitivity than the same mass of analyte desorbed into a relatively large volume of solvent.

(b)   The tubes are cleaned by the thermal process, and can thus be reused many times without loss of performance, but the adsorption capacity of these materials is much lower than for charcoal.

This capability for high throughput of samples with high sensitivity and enhanced precision is often essential in large- or medium-scale sampling programmes and the high initial cost is soon offset by much lower running costs. One problem which can occur with either type of sampling tube is breakthrough, where the adsorption capacity of the adsorbent bed is exceeded, either by high concentrations of the material under study or by other substances, e.g. water vapour, which are adsorbed on to the bed and displace the analyte. In this case, if solvent extraction is used, two compartment tubes may be employed where the first 100 mg section is separated from a second, backup, section of 50 mg by a small urethane foam plug (Figure 5.7). The two sections are analysed separately; if any analyte is determined in

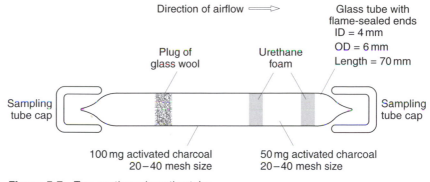

Figure 5.7   Two-section adsorption tube

the backup section then breakthrough has occurred and the experiments must be repeated. This two-compartment tube is unsuitable where automatic thermal desorption is used, but two single-section tubes can be used in series and then each analysed separately. A comprehensive review of methods recommended in the United States is available from the Environmental Protection Agency (EPA).[6]

Diffusive samplers

Rather than use active pump sampling to collect volatiles in tube samplers, it has been established by a series of studies that passive diffusion can also be used. The process relies on Fick's first law of diffusion, where it is found that the mass uptake of material onto a given bed of adsorbent is proportional to both the concentration of that material in the gas phase, and to the time of exposure. In use, tubes identical to those used for pumped sampling have a clip attached to them so they may be worn on the lab coat, rather like a ballpoint pen, and the sample molecules are able to diffuse (at a constant rate) into the tube and onto the adsorbent. The absolute sensitivity is lower than for active pumping, but for many purposes, especially long-term monitoring, it has been shown[7] that the results obtained are just as reliable, with accuracy and precision at least as good as pumped systems. In the United Kingdom and the Netherlands, but not yet in the United States, diffusion samplers have been adopted as recommended (although not exclusive) methods for industrial hygiene monitoring to check compliance with OES standards.

For long-term or large-scale analysis programmes, entrapment techniques offer very high sensitivity and good precision for a very wide range of analytes in industrial hygiene monitoring, but due to the apparatus used, they tend to be costly to set up and they are always post-collection analysis, which is sometimes not desirable. They are also beginning to be used for environmental monitoring, even where the lower expected level of analytes and the open nature of the system has presented problems in the past. For example, NO_2 can be monitored using diffusion sampling; an acrylic tube containing a wire gauze impregnated with triethanolamine is used for monthly average measurements in several cities. The NO_2 reacts specifically, and quantitatively, to form a nitrite which is then determined colorimetrically.

5.2.4 Real-time analysis

A number of specialist instruments are becoming available for instant or near-instant analysis. Normally these devices are specific for one component or are programmable to determine one component from a range of detectable materials. Some of these devices are extremely sensitive and can be used for real-time monitoring of environmental systems, while others are much less sensitive (and cheaper) and are designed as hazard warning monitors in an industrial environment. Since these devices are based on a range of technologies, it is best to group them into four sub-types:

1. Electrochemical sensors
2. Infrared spectroscopy
3. Chromatography
4. Specialist sensors

Electrochemical (amperometric) sensors

A number of specific sensors are available for the determination of a range of commonly encountered gases in the industrial environment. These use a gas phase electrolytic cell to reduce or oxidise the material of interest, generating in the process a response which is related to the concentration of analyte. They normally contain two polarised electrodes immersed in a common electrolyte, which may be in the form of a gel, and separated from the environment by a gas-permeable membrane. As a particular gas permeates through the membrane, a redox reaction occurs at the electrodes, generating a current which is proportional to the gas concentration in the region of the sensor. The response of these devices is instantaneous and they are often used as hazard warnings. Devices of this type are most useful in situations where a single gas is used, but where excessive levels must be avoided for safe working conditions.

Typical applications would include the monitoring of Cl_2 in the water industry, HCl or ammonia in chemical process plant, or even oxygen levels, where the alarm would be triggered by low rather than high levels. Although these sensors are available for several gases of environmental as well as industrial interest, such as NO_x and SO_2, the sensitivity is currently too low to be of use for general environmental monitoring. As an example, one commercially available SO_2 monitor is quoted to have a range of 0–199 ppm SO_2 with a resolution of 0.1 ppm. In an industrial environment, where the OES standard for SO_2 is 2 ppm with a STEL ceiling of 5 ppm, this is perfectly satisfactory; but it is no use for environmental monitoring, where a level of $100 \, \mu g \, m^{-3}$ (equivalent to about 0.03 ppm) is considered high. They are, however, both physically small, cheap and robust, making them ideal 'personal monitors' or hazard warning devices.

Note Although for most scientific applications the terms parts per million (ppm) and parts per billion (ppb) are not considered to be acceptable – the more precise terms of $\mu g \, g^{-1}$ or $\mu g \, kg^{-1}$ being used instead – they continue to be used in the field of gas sampling. This is because, unlike the condensed phase of liquid or solid solutions where $1 \, \mu g \, g^{-1}$ (1 ppm) can be readily derived from measured parameters (weight or volume), in the gas phase the relationship between $\mu g \, g^{-1}$ and ppm is more complicated and relies on a knowledge of the molar mass of the determined gas. Table 5.2 gives the conversion between ppm and $\mu g \, g^{-1}$ for the more environmentally important gases.

Table 5.2 *Conversion factors for some common gaseous pollutants*[a]

Gas	Molar mass	UK urban value[b]		TLV[c]		US NAAQS[d]	
		mg m^{-3}	ppm	mg m^{-3}	ppm	mg m^{-3}	ppm
SO_2	64	0.08	0.03		2	0.365	0.14 (24 h)
NO_2	46	0.376	0.2		1	0.1	0.053 (annual)
CO	28	0.17	0.15		35	10	9 (1 h)
NO	30	0.37	0.30		25		
O_3	48	0.196	0.1		0.1	0.235	0.12 (1 h)
CO_2	44	635	353		10 000		

[a] For a gas at 25 °C and 1 atm (100 kPa)

$$\text{concentration (mg m}^{-3}) = \frac{\text{concentration (ppm)} \times \text{molar mass (g)}}{24.45}$$

1 ppm = 1 molecule per million molecules, but since the volume occupied by these molecules depends on temperature T and pressure P, it is preferable to use mass/volume with units mg m^{-3} or µg m^{-3}; mass/volume is independent of T and P.
[b] These are typical values.
[c] TLV = threshold limit value.
[d] NAAQS = national ambient air quality standard.

Infrared spectroscopy

Since many species of concern absorb IR radiation strongly at a characteristic wavelength, many analytes may be determined successfully even at low concentrations by using a form of portable IR spectrometer. The simplest of them is shown in Figure 5.8(a). In this device the complete range of IR radiation from a simple source is passed alternately through a reference cell, containing a non-absorbing gas, normally dry nitrogen, and a sample cell containing the analyte. A two-compartment detector cell, in which both compartments are filled with the substance to be determined (in this example CO), is used to measure the difference in transmitted energy between the sample and reference cells. Although each instrument is specific only for one substance, different devices can be constructed for any IR-absorbing gas by simply changing the detector.

A variant on this design, (Figure 5.8(b)) is to use a single cell and detector, but pass the IR radiation through a gas filter wheel comprising one section with nitrogen gas and a second section with CO. As the wheel rotates, the radiation is chopped and a modulated signal is produced. When the source radiation passes through the CO section, complete absorption occurs at the wavelength(s) at which this gas absorbs, hence no further adsorption is seen as the radiation passes through the sample cell. When the nitrogen section is in the beam, this allows all the source intensity to pass through the sample cell, where adsorption proportional to the concentration of CO in the sample occurs. The difference

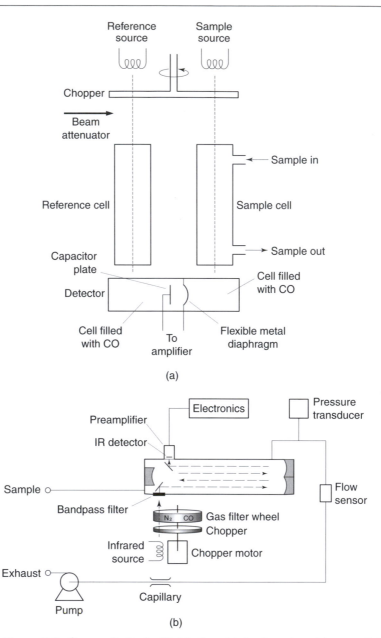

(a)

(b)

Figure 5.8 Gas monitoring by IR detectors: (a) simple system, (b) sequential system for several different gases

between the two signals gives the concentration of CO in the sample. Different gases may again be measured, this time by changing the gas in the gas filter wheel.

Often in these instruments, a mirror system inside the sample cell causes the radiation to pass many times along the length of the cell, considerably increasing the sensitivity.

These instruments are relatively sensitive and widely used for 'process gas' monitoring; however, a more sophisticated (and expensive) device is available known as Miran™ (miniature infrared analyser).[8] This too is a single-beam system, but a single wavelength or a narrow band of wavelengths can be selected by means of a three-segment variable filter anywhere in the region 2.5–14.5 μm; this passes through a single cell containing the sample gas.

By using a series of reflecting mirrors within the cell compartment, the beam of radiation can be made to traverse the cell many times (effectively increasing the path length of the cell to a maximum of 20 m) before reaching a detector which measures the amount of absorption by the sample. Since the Beer law is obeyed, the absorption can be directly related to concentration. In fact, although the Miran is normally used as a non-dispersive instrument, i.e. at one wavelength, it can even be used to give a scanned spectrum of a gas atmosphere. This enables identification of any contamination prior to quantitative determination at a selected characteristic wavelength, and over one hundred different gases can be determined, many of them with detection limits of less than 1 ppm.

A more recent introduction to this type of gas monitor is to use photoacoustic spectroscopy; and it is claimed this gives even higher sensitivity than IR techniques.[9] Here an infrared source is allowed to pass through a narrow bandpass filter, a mechanical chopper and then into the gas sample held in a sealed cell. If any absorption of radiation at this wavelength occurs, then the temperature of the cell rises and the gas tends to expand. Since the radiation is chopped at about 1000 Hz, this alternate expansion and contraction of the gas actually corresponds to a sound wave, and its intensity may be measured using a conventional microphone and high-gain audio amplifier, giving lower detection limits than is possible using normal IR detectors. Both systems can be used only for gases and vapours which have infrared adsorption bands in the measured region, so they are not suitable for symmetric molecules. Nevertheless, it is the method of choice for many inorganics, including CO_2.

With the advent of reliable Fourier transform infrared systems, the technique can also be used for remote sensing applications. By simply transmitting a collimated infrared beam across a site to a distant reflector and then back to a detector, the presence of infrared-active materials in a path of up to 1000 m can be determined and quantified.

Although not uniquely operating in the infrared region, the LIDAR system (light detection and ranging) deserves to be mentioned. A high-power laser beam is pulsed at two frequencies, one absorbed by the species of interest and the other not. The beam can be directed at a target source such as a distant stack, or even a cloud, and from the reflected signal received at the detector it is possible to estimate both the concentration of analyte and the distance of the sample from the source. Normally these systems are large and extremely costly, but for a number of environmental monitoring tasks they are invaluable.

Chromatography

In the laboratory, gas chromatography (GC) is probably the first choice for the analysis of most gases and vapours; the sample is already in the correct form for introduction into the system; it is capable of separating a number of quite similar compounds from each other, and even with a routine, non-specific detector such as a flame ionisation detector (FID), it is capable of high sensitivity. However, conversion of the standard laboratory GC into a portable instrument is not particularly easy, since this would require a portable source of electrical **power** (portable electronic devices are usually low power), to heat up the column oven and portable sources of up to three different gases for routine operation. Although fully portable GC systems are available, it is becoming more common to simplify the system so the restrictions of power and gas supplies are reduced or removed. If a solid-state detector is used rather than an FID, then the only gases required are the carrier gas, usually nitrogen, which can be stored in a single small cylinder. By reducing the length of the column, it is often possible to obtain separation of common gases from each other at ambient temperature; this means a column oven is no longer required and the power consumption of the system is considerably reduced; in fact, in some systems the column is removed altogether and a small pump is used to introduce sample directly into the detector, thus removing the need for compressed gases completely.

Normally a system of this type could not give separation and possible identification of different species in the sample, but would only give an indication of total organic vapour, which in itself is a valuable measurement. However, by using selective detectors such as a photoionisation detector (PID), or the newer chemical sensitive transistor detectors, then some selectivity of response can be built in to the system. Depending upon the exact need, a gas chromatograph can be purchased which will either give a total organic vapour response (leak detectors), selectivity of one compound from several, or from a background containing organics, or even complete separation and display of a normal chromatogram with the retention time of each peak shown. Each is capable of high sensitivity and can give a response within a few seconds.

Special application monitors

Due to the increasing interest in low-level gaseous pollutants in the environment, and the need to determine them reliably over extended periods of time, a number of urban monitoring networks[10] have been developed in the United Kingdom and the European Union. The aim of these systems is to monitor a selected number of environmental urban pollutants so that long-term trends in levels can be seen and the effect of reduction measures can be evaluated. Because these measurements often need to be performed continuously for periods of up to a month with little or no maintenance, specialist monitors have been developed which can measure the major environmental pollutant gases SO_x, NO_x, O_3 and peroxyacetylnitrate (PAN). In many cases these specialist monitors are now the recommended systems for routine, continuous monitoring of environmental urban pollutants. Although they all use established chemistry in their operation, they have generally used this chemistry in a novel way to produce particular advantages for long-term continuous monitoring. The instruments in Section 5.2.5, together with NDIR for CO_2 and CO (Chapter 18) give an integrated suite of devices for monitoring many of the environmentally significant gases.

5.2.5 Some special application monitors

Sulphur dioxide SO_2

The most widely used system for this pollutant employs gas phase molecular fluorescence in which sulphur dioxide in the air is caused to fluoresce by irradiation with an intense source at about 215 nm, and detected at 240–420 nm by a photomultiplier at 90° to the source. Since, as in all forms of molecular fluorescence, the intensity of the emitted radiation is proportional to both the concentration of the absorbing species (the sample) and the intensity of the source, if an intense source can be used, low limits of detection result. The source used in these sulphur dioxide monitors is in fact a zinc hollow cathode lamp as used in atomic absorption spectroscopy (AAS). Virtually all of the radiation intensity of

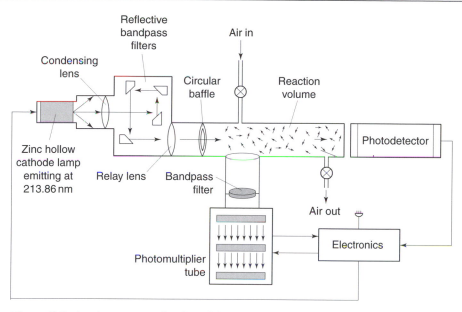

Figure 5.9 Luminescent monitor for sulphur dioxide

this lamp is at one wavelength (213.86 nm), so it is effectively more intense even than a laser source at this one wavelength (Figure 5.9). It is claimed there are virtually no interferences other than a small quenching effect due to water, which can easily be removed by drying the gas before measurement. Detection limits in the low ppb range are claimed, with linear response from 1 ppb to 100 ppm.[11]

Since the monitor is mechanically quite simple and requires no chemical reagents, it can be left unattended for long periods without maintenance, and the use of standard solid-state electronics means it is also very stable once calibrated. Calibration of the monitor, as for all gas monitors, is often quite difficult since it is much harder to prepare and handle a gas mixture with known, low concentrations of analyte than it is for a solution. Probably the best way to obtain gas mixtures of known concentration for calibration purposes is to use **permeation tubes**, where gas diffuses through a semipermeable membrane at a known rate for a given temperature. The tubes themselves are normally quite simple, consisting of a PTFE-walled tube filled with liquid SO_2, but they need to be housed in a thermostatic enclosure through which a constant flow of gas is maintained. If the total volume of gas passing through the apparatus in a given time is measured, and the loss in weight of the tube in the same time is determined, then the concentration of analyte in the gas flow can be calculated.

Ozone O_3

Normally a direct absorbance method is used for ozone, since it has a strong absorbance at 254 nm. The absorbance at 254 nm of an air sample is measured in a cell with a fairly long path length (0.5 m); the same air is then pumped through a catalytic converter to remove the ozone but to leave behind any other stable gases which also absorb at this wavelength. The absorbance is then measured again. By using the difference in the two measurements, which corresponds only to the ozone absorbance, and by applying the Beer

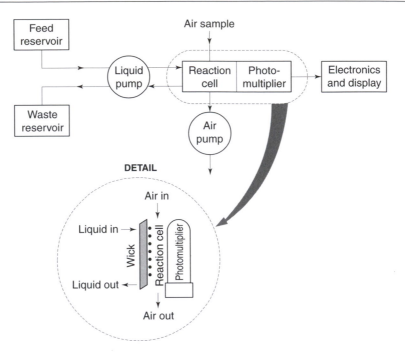

Figure 5.10 Gas/liquid chemiluminescent monitor for ozone

law, a direct linear response is obtained with a limit of detection (LOD) of 1 ppb. Again, since this monitor requires no reagents and only a source of power, it can be completely automated, and left in situ for months at a time if required.

An alternative and more sensitive method for ozone monitoring relies upon the luminescence produced when air containing ozone is allowed to come into contact with a surface coated with eosin dye. Using a small pump, the air sample is drawn through the instrument into a lightproof compartment, and flows across a fabric wick continuously wetted by a solution of eosin. The chemiluminescence so produced is measured by a standard photomultiplier focused on the centre of the wick. Since the reaction is highly specific and localised on the surface of the wick, very high sensitivity is claimed of the order of 0.1 ppb. Compact and lightweight, the instrument (Figure 5.10) is highly automated with all the pumping and measurement functions internally controlled, but the eosin reagent must be replaced at infrequent intervals.

Oxides of nitrogen

Although at least five well-defined compounds containing only nitrogen and oxygen are known, three are commonly found in the atmosphere of polluted towns and cities. Nitrous oxide, N_2O, or laughing gas, is formed naturally during microbiological processes and is present even in unpolluted air at up to 0.25 ppm; it often reaches levels ten times this value in polluted atmospheres. The gas is relatively unreactive and does not take part in reactions at low level, but is photolytically decomposed in the stratosphere in a sequence of reactions to produce nitric oxide. The two gases which are of most concern from an environmental viewpoint are nitrogen oxide (NO) and nitrogen dioxide (NO_2); nitrogen dioxide is sometimes called dinitrogen tetroxide (N_2O_4). They are collectively known as NO_x. At normal laboratory concentrations, NO is very quickly oxidised by molecular oxygen to give the dioxide,

$$2NO + O_2 \rightleftharpoons 2NO_2$$

giving the familiar brown fumes seen when concentrated nitric acid reacts with many substances. However, this is a second-order reaction with respect to NO, and reaction occurs very slowly at concentrations below about 500 ppb of NO. Even at polluted atmospheric levels, the two gases exist in equilibrium with each other, and the ratio of the two gases depends on the levels of ozone present:

$$NO + O_3 \rightarrow NO_2 + O_2$$

This reaction is reversed by high levels of radiation below about 435 nm, i.e. sunshine, and it is this capability of the brown NO_2 gas to absorb energy from sunlight at ground level which is the driving force behind a complex sequence of reactions. These involve, among others, hydrocarbons (from vehicle exhausts) and ozone, and they eventually produce **photochemical smog**, extremely damaging to health in the inner cities. Each gas may be monitored separately, but because of the equilibrium which exists between them it is better to monitor both.

Nitrogen oxide NO   When nitric oxide and excess ozone are allowed to react in the gas phase, metastable nitrogen dioxide is produced, which then immediately emits a photon of radiation to decay to normal nitrogen dioxide in a chemiluminescent reaction:

$$NO + O_3 \rightarrow NO_2^* \rightarrow NO_2 + h\nu$$

The photon can be detected between 600 and 2000 nm, $\lambda_{max} = 1200$ nm. This reaction may be used to monitor NO at low levels in air, using a system which measures the amount of emitted light (Figure 5.11). An air sample is pumped into an enclosed lightproof enclosure and an excess of ozone (produced electrically within the apparatus) is admitted to produce

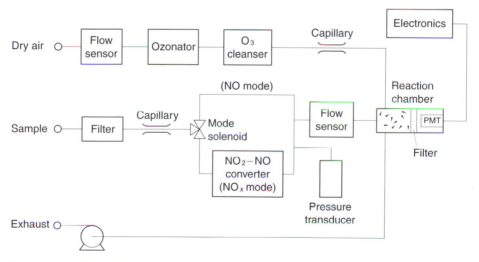

Figure 5.11   Chemiluminescent monitor for nitrogen oxides: PMT = photomultiplier tube

the reaction. It is assumed that the sampled air has low ozone levels, otherwise the reaction would proceed in the air itself. Although specific to NO, the air sample may be first passed through a catalytic cell which converts NO_2 into NO, thus enabling measurements to be made either of total NO_x, or each of the species separately. The monitor requires only a source of power to operate and does not rely on any chemical reagents. The LOD is claimed to be about 5 ppm. The same reaction can be used in a modified system to monitor O_3, but this time it requires a source of NO which is added in excess.

Nitrogen dioxide NO_2   Nitrogen dioxide is possibly the pollutant arousing most current interest, since not only does it play an important part in the complex chemistry of photochemical smog, it has been directly implicated in the increased incidence of asthma and other respiratory problems, especially in the young. Using a system that is almost identical to the ozone set-up, nitrogen dioxide in air may also be measured using condensed phase chemiluminescence. In this case the reaction at the surface of the wick is between nitrogen dioxide and luminol reagent. An LOD of 5 parts per trillion is claimed with virtually no interference from other species. The advantage of this system is that it responds directly and instantly (1 s response) to nitrogen dioxide rather than to NO.

Peroxyacetyl nitrate (PAN)

PAN is one of the principal hazardous components of photochemical smog, formed by reaction between NO_x, O_3, hydrocarbons and $h\nu$; it is also thought to be the main hazardous component of smog, causing smarting of the eyes and respiratory problems. Until recently it has been difficult to monitor, since it is present in very low concentrations and it is extremely reactive. But it may now be determined at levels as low as 30 parts per trillion, by using a gas chromatographic separation of PAN from other air pollutants and then catalytically converting it to NO_2, which is then determined using the system shown below (Figure 5.12). The chromatographic step is integrated into the equipment, but this slows the response down to about 10 measurements per hour.

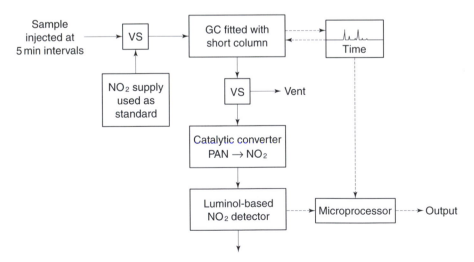

Figure 5.12   Block diagram of a PAN monitor: VS = valve switching

Mercury

Although not an environmental pollutant, mercury vapour is often monitored using similar specialist technology and measurements of mercury levels are important in industrial situations. A direct absorbance method for mercury is often used in which air is pumped into a gas cell with a long path length (0.7 m) and the absorbance at 254 nm is measured. A zero reference for a particular atmosphere is obtained before or immediately after measurement by passing the air through a charcoal filter; this removes interference created by organic materials which absorb at this wavelength. A sensitivity of $1\,\mu g\,m^{-3}$ is claimed, which is low enough for most purposes.

An alternative and much more portable system is based upon the adsorption of mercury onto a gold foil or wire. This causes a change in electrical resistance of the gold, which may be determined and equated to mercury concentration. At the end of the measurement cycle, the foil or wire is heated to sublime the mercury and reactivate the system. Although much less sensitive than the previous system, with a detection limit of about $1\,mg\,m^{-3}$, the device is much smaller and cheaper, and it may also be used in time-weighted average measurements for occupational exposure standards, because the gold foil automatically traps all the sample as air is pumped over it.

5.3 Liquids

The literature of liquid sampling is extensive, and for many specific cases it is possible to use or modify an existing sampling protocol for use, but this section attempts to give as broad an overview as possible of techniques for collection and sampling of liquids. Since much of the liquid sampling which is performed is concerned with drinking water or its precursors, most of the following will also be related to drinking-water supplies, although the techniques can often be used for other liquids as well. A number of parameters are important when considering how to set up a sampling protocol, including the requirements of any legislation or guidelines. Water is essential for life, so strict control over its purity is both necessary and desirable, but it is by no means uniform, and the legislation and its implementation, involving water and water supplies, varies widely from country to country. Although by no means complete, Table 5.3 gives some of the more important acts and directives which govern the sampling techniques used in the United Kingdom, the United States and the European Union. These regulations, directives and recommendations concerning water composition are complex and they continue to increase in number as our concern grows for improving the environment.

Somewhere in excess of 700 organic compounds are found in water supplies, and different administrations take widely differing views about how they should be monitored and reduced. Increasingly the trend worldwide is to set up controls which incorporate maximum allowed values of named substances, giving rise to lists of 'dangerous' chemicals. This process was first started in the United States by the Environmental Protection Agency (EPA), which in the early 1970s introduced a priority pollutant list containing 129 pollutants with **maximum concentration levels (MCLs)** for each.

This list has since been modified and expanded, and in Europe similar systems apply; there are red, grey and black lists where each substance has a **maximum allowable concentration (MAC)** or prescribed concentration value (PCV). This mechanical approach has a number of advantages, not least the ease of interpreting breaches of the regulations; but it also has a number of disadvantages such as the inability to respond to threats from new pollutant materials and the lack of scientific judgement as to whether a chemical, or group

Table 5.3   *Major items of water legislation*

United Kingdom	United States	European Union
1973 The Water Act is designed to provide 'wholesome water'; it is amended in 1989; the National Rivers Authority (NRA) is set up	1970 The Environmental Protection Agency (EPA) is set up by President Nixon	1976 Dangerous Substances Directive (76/464/EEC); 129 Black List substances are to be eliminated from water (actually only 17 in full agreed list) and a Grey List of substances are to be reduced
1974 Control of Pollution Act (COPA) 'the polluter pays' does not refer to specific compounds	1972 The Federal Water Pollution Control Act (Clean Water Act) introduces the priority pollutant list of 129 compounds in 13 classes; EPA Series 600 methods are set up for industrial waters; the Act is amended in 1977, 1981 and 1987	
1990 The Environmental Protection Act is to be comprehensive legislation on all aspects of pollution; not yet fully implemented		1980 Directive on quality of water for human consumption (80/778/EEC) sets maximum admissible concentrations (MACs) and minimum required concentrations (MRCs)
1990 Third North Sea Conference (The Hague) introduces Red List substances, strictly for estuary waters	1974 The Safe Drinking Water Act (SDWA) establishes national standards for drinking water; it introduces maximum contaminant levels (MCLs) and recommended maximum contaminant levels (RCMLs); the Act is amended in 1977, 1981 and 1987	MAC The interpretation of what is meant by an MAC, particularly average levels, has caused a great deal of debate; the UK sets environmental quality objectives (EQOs) for each body of water (i.e. after the pollution has entered and mixed with the water); the rest of Europe uses fixed emission standards at source
1991 Water Industry Act		
1995 The Environment Act is a major piece of legislation; eventually it deals with all aspects of the environment, not just water		
1996 The Environment Agency comes into existence on 1 April, created by the Environment Act of 1995	1980 The Comprehensive Environmental Response, Compensation and Liability Act (CERCLA) 'Super Fund' is mainly concerned with pollution from waste sites	
	1986 Super Fund Amendments and Reauthorisation Act (SARA)	

of chemicals, in a given environment does or does not pose a real threat. But however the analytical results are finally interpreted, they should as far as possible be representative of the whole and a true indication of the levels of any determined material.

Just as for gases, even a small volume of water will contain a large number of particles, normally molecules or ions, and as such, the **sample taken** will be amenable to statistical methods. But the problem remains that this sample must be representative of the whole, and **variability** is again a factor that must be carefully evaluated.

Note   The term 'variability' in this context means that the analyte values depend upon external factors, independent of the sampling process. Variation once the sample has been taken may still occur due to decomposition, adsorption, etc., but this is, or should be, under the control of the analyst.

Although in the laboratory it is assumed that liquids are homogeneous, at least for a single phase, for large volumes the sample sites must be chosen carefully. It would not be sensible to assume that the North Sea is homogeneous, with constant concentrations of analytes across its breadth and depth. But even for quite small sample volumes, positional variation can be important due to factors such as density variation, laminar flow at surfaces, or lack of equilibrium in the bulk. Variability with time and meteorological conditions (rain) impose further constraint upon the sampling method. If an interface exists in the liquid due to the mutual immiscibility of two or more phases, then the problems of obtaining a representative sample can be severe, and the analyst would normally choose to sample, and subsequently analyse, each layer as a separate problem.

Since the range of analyte concentration can vary from the order of a few percent of the bulk, to traces at ppb level or even less, the volume of sample required can also vary considerably and often preconcentration techniques are employed in order to obtain measurable amounts of analyte. The stability of the collected sample can also cause problems since in many cases, without proper preservation or stabilisation of the sample, changes will occur between collection and analysis. Finally it is important to be aware of the reason for doing the analysis. In many cases the exact form of the question being posed – What is the average value? What is the maximum value? Does the sample contain X? Or is the analyte concentration below defined limits? – can determine both the exact form of sample collection and subsequent analysis.

Several approaches are used to obtain a gross liquid sample; they depend to a large extent upon the type of system to be sampled:

1.   Small static systems
2.   Flowing contained volumes
3.   Open flowing systems
4.   Large static volumes

The protocol also depends upon whether it is desirable to obtain a number of **discrete** samples which will be analysed separately, or to bulk the individual samples as taken to give a **composite** sample. Discrete sampling is the more commonly employed method, although if a large number of samples are taken, which should always be in duplicate, then problems of bulk and the need for a large number of analyses can make this a very expensive way of obtaining analytical results. If only an average value of analyte is required, it is often better to design and set up a composite sampling plan which will drastically reduce both the volume of sample required and the number of analyses needed. Composite sampling will not, however, be able to produce data that indicates variation in position or time, and often a combination of discrete and composite sampling is required. In either case make sure to observe cleanliness in all stages of the sampling protocol; this avoids contamination and false values. Remember the saying: Keep it simple, keep it clean.

5.3.1   Discrete sampling

Providing the analyst is aware of the possible variability of the bulk material (positional and time variation) and assuming the sample is accessible, discrete sampling can be performed with very simple apparatus such as clean glass or polythene bottles of 50–1000 mL capacity. But if samples are to be taken at various depths within the bulk material, then more complicated sampling vessels are required. A number of designs are available, but the two most common are the Knudson bottle and the bomb sampler (Figure 5.13).

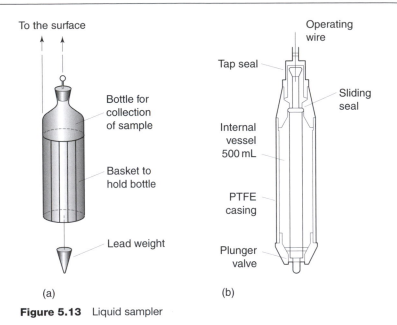

Figure 5.13   Liquid sampler

Both of these devices can be used for sampling at various depths; generally they are simple and reliable sampling vessels, provided they are cleaned thoroughly between each use to prevent cross-contamination. But there are problems with such a simple approach; if a large number of samples (in duplicate) are taken, then storage, transport and handling rapidly become major components in the overall timescale and cost for the process. If a large number of samples have to be treated in some way before analysis, this will increase the requirement for chemical reagents. The time interval between collection and analysis also tends to become undesirably long. Often, if the original purpose of the analysis is examined carefully, it is not necessary to take large numbers of discrete samples. Suppose the question posed is, What is the average value? Then it is a waste of time collecting and analysing large numbers of individual samples, only to take the numerical average of all of the results. In this case it is sensible to use composite sampling instead.

5.3.2   Composite sampling

In essence composite sampling is performed by taking an average sample from the bulk prior to final analysis. The simplest approach is to simply bulk a number of discrete samples in the laboratory prior to analysis, and then take a representative sample from this bulked volume. Although this reduces the total analysis time (and cost), it still requires manual collection, storage and transport of a number of discrete samples. So generally it is better, if possible, to bulk the individual aliquots at the point of collection rather than later in the laboratory.

For long-term monitoring of a bulk liquid, such as a lake or reservoir, it is possible to use a pumped sampling system where samples are drawn from various fixed positions within the bulk by using a system of interlinked sample outlets, joined by tubing to a central pump and holding vessel (Figure 5.14). This gives a **positional composite**. Although a positional composite is often the desired result, it does assume that analyte

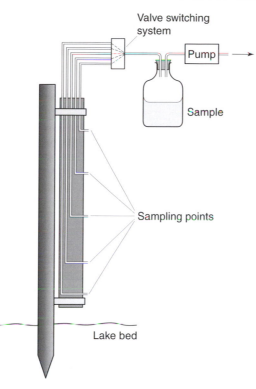

Valve switching
system

Pump

Sample

Sampling points

Lake bed

Figure 5.14   Fixed-position composite sampler

levels will vary only with position, not with some other variable such as time. But often the level of analyte does change with time, either randomly (say for effluents in a sewer) or in some periodic way (perhaps hourly, daily or even seasonally). The composite sample needs to reflect these changes, and this can be achieved using a **timed composite**.

Timed composites can be collected either by continuously extracting a small volume with a pump, or by using a more sophisticated programmable collection device which can be left unattended for extended periods and which collects samples at regular time intervals, or even when triggered by excessive flow (Figure 5.15). These automatic collectors can either produce a single composite sample or can collect up to 24 separate samples in individual containers, thus giving some discrimination to the collection.

5.3.3   Sample pretreatment

Whichever form of sample collection is employed, it is usually necessary to carry out some form of pretreatment in order to stabilise the sample and prevent composition changes between collection and analysis. The length of time that can be safely allowed between collection and analysis, the **hold time**, is very variable; it depends on the analyte and the pretreatment. For many samples, especially those collected in the 'environment', the first stage should be some form of **filtration** to remove unwanted 'solid material' from the sample. This may seem to be a simple task, but in reality it can make a significant difference to both the hold time, and the final analytical result, since the inclusion or removal of

Hinged lid and stainless steel handle can be locked in the vertical position to render sampler tamper-proof

Removable rechargeable battery

Quick-release hose adaptor

Sample tract made from chemically inert glass, PTFE, medical grade silicone rubber and stainless steel; plastic chambers are also available

Detachable programmer with LCD display and four-button controls; operates in eight European languages

Moulded polyethylene casing; diagram shows large container module, other sizes are available

PVC inlet hose (ID = 9.5 mm) with sinker weight and optional filter; Teflon hoses are also available

Figure 5.15   Automated composite sampler

small particles, including micro-organisms, will significantly affect the values of a number of common analyte species. As examples, it is found that levels of nitrate and nitrite can be greatly changed on standing if certain forms of bacteria are present or that the concentrations of many cations, especially heavy metals, will appear to increase on standing, simply because finely divided solid material slowly dissolves.

For simple filtration of solid material from solution, filter papers and gravity filtration can be used, but it is more common to use either glass fibre filter discs for fast but coarse filtering, or synthetic membrane filters of closely defined pore size to remove smaller particle sizes. The most commonly used are cellulose acetate membranes (CAM) with pore sizes of 2.0–0.2 μm. Membranes with pore size 0.2 μm will filter nearly all insoluble components, including microbiological species, but they are very slow even when used under pressure or vacuum. Many chemists use a slightly coarser membrane (0.45 μm) which gives a compromise between speed of filtration and retained species. For small volumes of sample, disposable syringe-tip filters with a preformed Luer fitting, which can be pushed firmly onto a glass syringe, are fast and reliable, although if used in large numbers they are expensive.

Whichever form of filtration is employed, it is important to be aware of the possibility of inadvertently separating or losing some of the sample which may be present in the form of more than one species. This ability of a number of cations to exist in several different forms (speciation) is widely known; for example, aluminium can coexist in aqueous solutions as ions, complexes, colloidial particles and adsorbed onto other solid particles in the solution. But it is only now being recognised that this can lead to quite different analytical results for the same sample using different analytical methods.

The next stabilisation step after filtration is often some form of chemical or physical **preservation**; this can help to increase the hold time of the sample. A number of methods are used, either alone or in combination, depending on the analyte to be determined and the technique used for analysis. For analysis of cations and some anions, acidification with nitric acid is a common approach since this not only tends to keep the required species in solution but also slows or stops biological changes. Typically the addition of 1 mL of conc. HNO_3 to each 100 mL sample will be all that is required. Storing the sample at low temperature, either 4 °C or –10 °C, normally slows down any chemical or biological changes, and the exclusion of light by wrapping the sample vessel in foil is generally a good idea, especially for low levels of organic species such as pesticide residues. Samples containing low levels of volatile organic species are more difficult to preserve; they should be collected in completely filled glass containers with no air gap at the top, and analysed as soon as practicable. It is generally much more difficult to maintain the integrity of biologically active samples, which should either be analysed immediately or very carefully preserved.

5.3.4   Field sampling and analysis

In recognition that preservation and storage can alter the composition of materials, a number of techniques are now being developed for the analysis of environmental and ecological samples, where the normal process of collection, transport and analysis is avoided by actually performing the analysis at the sample collection point.

Portable test equipment for determining parameters such as temperature, pH and conductivity have been available for some time, as have simple colour tests for the more important anions and cations, but a number of more sophisticated and sensitive spectroscopic devices are becoming available,[12] which allow the more or less specific determination of a wide range of anions and cations and sometimes even organic species. Although nearly all these 'test kits' are based on standard procedures used in the laboratory for the UV/visible spectrometric analysis of materials, the reagents have been carefully packaged either in the form of tablets or sealed sachets so that the chemistry can be performed easily and reliably in the field. The intensity of the coloured solutions resulting from the test are then generally measured either against a colour chart, or on a small battery-operated visible photometer operated at a defined wavelength, and often directly calibrated in the desired concentration units for the sought analyte.

Optically matched light-Emitting Diode (LED) source/detector combinations and stable solid-state electronics allow these instruments to be small, relatively cheap and extremely stable. Combined chemical/biochemical systems have increased the number of species that can be determined and enable even lower minimum levels to be determined; some important organic pollutants such as pesticides or explosive residues can be specifically quantified using modifications of standard immunoassays, especially enzyme-linked immunosorbent assays (ELISAs). The biochemistry of these tests is somewhat complicated, and relies on binding or competitive binding of the substrate to an immobilised antibody, but the end result is the production of a coloured solution which is inversely proportional to analyte concentration.[13]

Since they effectively take the laboratory to the sample, rather than the other way around, these techniques largely overcome the problems of sample deterioration on storage, but they tend to be relatively slow and they must still be used with regard to proper sampling protocols. Also, since they tend to be specific for one analyte, they are not as versatile as traditional laboratory-based spectroscopic methods or as sensitive as much of the instrumentation found in the laboratory. The decision whether to use field test equipment, or to preserve and transport the samples back to the laboratory for analysis, still rests with the trained analyst and is based on experience, availability of equipment and the requirements of the original request for analysis.

5.3.5 Sample preconcentration

In many cases water samples can be analysed for the important anions or cations at the expected level of pollution with minimum pretreatment or concentration, especially if the analysis is done in a well-equipped laboratory. For these substances, detection limits of $\mu g\,L^{-1}$ (ppb) are easily attained, and below these values very few inorganic substances are thought to present any health risk. However, for the analysis of some organic species, which often need to be determined at very low levels $(ng\,L^{-1})$, some form of preconcentration is usually required. Five general methods are available to the analyst.

Direct injection

Direct injection is not strictly a preconcentration method since the water sample is injected directly into a gas chromatograph. Until recently the technique was only used for samples containing relatively high levels of non-polar materials, because although GC can normally accommodate small sample sizes, of the order of $0.1–2.5\,\mu L$, it is not very sensitive and analyte concentrations generally have to be at least in the ppm region for satisfactory determination. Also, although non-polar columns (for non-polar analytes) are relatively tolerant to large amounts of water, it is not desirable to inject water onto polar columns since they are easily degraded and will only have a short life. However, newer techniques of injecting large sample volumes with carefully controlled temperature programmes make the process more attractive for some sample types. (A simple calculation shows that in order to present the detector of a GC with 10 pg of material, $1–2\,mL$ of aqueous solution at a concentration of $10\,ng\,L^{-1}$ are required.) Some analytes may also be determined by direct injection onto a reverse-phase HPLC system, but again sensitivity is often a problem with this technique.

Headspace analysis

For volatiles or semivolatiles, an extremely attractive preconcentration method may be to partition organic material between liquid and a gas space above the liquid by simply heating a partly filled but sealed container containing the sample. In its simplest form, a small (5–10 mL) screw-capped bottle fitted with a rubber septum cap is partially filled with water sample and immersed in a water bath at about $45\,°C$. After a short interval, during which the volatiles enter the gas phase, a sample is taken using a gas-tight syringe of about 1 mL capacity and injected directly into a GC. Although it can be a very useful technique for the GC analysis of volatiles in water, it may only be used for volatile species; for highest precision, a fully automated system (which is expensive) should be used. Headspace analysis is not only used for water samples, but finds widespread application in the food and beverages industries, where it is ideally suited for analysis of volatiles such as alcohol, and flavour and fragrance components.[14] Because it is a concentrating technique, it may be used even for samples containing low levels of analyte.

Purge and trap

In many respects this is similar to the techniques used for trapping organic volatiles in gases, except that a first stage of purging the organics from the liquid phase is employed. In this technique an inert gas such as nitrogen or helium is bubbled through the water sample contained in a large vessel. The purge gas passes through a sintered glass frit below the liquid surface to produce a stream of very fine bubbles, which remove the organic components as they pass through the liquid. The gas, plus volatiles, is then passed into a small adsorbent tube very similar to those used for gas sampling and the volatiles are adsorbed. The adsorbent material used in these tubes must be chosen so it is not affected by the large amounts of water vapour that are also pumped through the tube, since as much as 20 mg of water will be adsorbed compared to about 10 μg of each organic component. At the end of a suitable sampling period, the adsorbent tube is connected to a gas chromatograph and the volatile organics are thermally desorbed into the system. When used correctly, the technique is capable of giving reliable results even at very low analyte concentrations, but care must be exercised to ensure the system does not give unrepresentative values. The system can only be used for volatile species; for high reliability it is desirable to use a commercial system in which each component has been optimised for the analysis.

Solvent extraction

This is the oldest, and possibly the most familiar of the extraction techniques. It is also based on selective partition between two phases, but this time both are liquid. Although still widely used in the organic chemistry laboratory for 'working up' materials, its use for preconcentration of organics from water is now much less common. A small amount of an immiscible solvent is added to a large volume of water, shaken then allowed to stand; in principle, as the phases separate, the solvent containing a large proportion of the organics can be removed and analysed. However, if one is trying to extract material present in the low ppm or even ppb level in the water, then solvents of very high purity have to be used in order to avoid contaminating the sample. The method is also labour-intensive and increasingly concern is expressed about the safe disposal of the used solvents at the end of the procedure. Thus, although the method can be can be used for the extraction and preconcentration of a wide range of non-volatile or semivolatile species from water using only routine laboratory equipment, its use is decreasing in most situations because solvents of the required purity tend to be expensive, and also can cause problems with proper disposal after use.

Solid phase extraction (SPE)

Solvent extraction has largely been replaced by solid phase extraction (SPE). This is often the best way to extract and preconcentrate organics at low or very low levels from aqueous systems. Although widely used for water analysis, SPE is also extensively used for extraction of material from biological systems, e.g. drugs of abuse in urine, or blood analysis. The most widely used SPE systems use a small bed of a specific adsorbent material contained in the barrel of a disposable hypodermic syringe. In use (Figure 5.16) the adsorbent bed is first preconditioned by passing through it a small volume of organic solvent, often alcohol, then the aqueous sample is sucked or pumped through the bed where the organics are retained.

The last step involves stripping the organics from the bed using another small volume of organic solvent. A large range of adsorbent materials (mostly based on HPLC packings) are commercially available in bed sizes to handle between 2 and 2000 mL of sample.[15] For relatively clean waters, the technique is very fast and easy to use, and by correct choice of adsorbent bed, can extract almost all organics or inorganics from aqueous samples with minimum contamination and with concentration ratios approaching 1000 in favourable cases. However, where very large volumes of water containing low levels of analyte need to be analysed, or where the water contains appreciable amounts of solids, it can take an unacceptably long time to pass the liquid through the adsorbent bed, even when using a

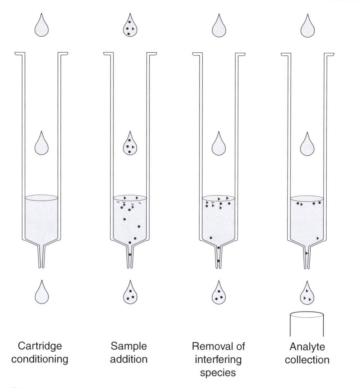

Cartridge conditioning Sample addition Removal of interfering species Analyte collection

Figure 5.16 Using solid phase extraction cartridges: (▲) analyte, (●) interferences

vacuum. In these cases it is possible to use SPE discs, in which the adsorbent material is trapped in the matrix of a thin filtration disc. The chemistry of the system is the same as for SPE cartridges but large volumes of liquid can be processed more quickly.

Solid phase microextraction (SPME)

The most recent of the solid phase techniques, and in many ways the most promising, is solid phase microextraction. Here a thin silica capillary fibre about 5 cm long and coated on its outside surface with a layer of stationary phase between 10 and 100 μm thick is attached to a syringe-like holder which can be adjusted to cover the capillary with a stainless steel protective sheath (Figure 5.17). In use the silica capillary is immersed into

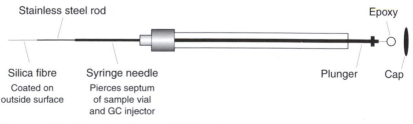

Stainless steel rod Epoxy

Silica fibre Syringe needle Plunger Cap
Coated on outside surface Pierces septum of sample vial and GC injector

Figure 5.17 Microextraction: an SPME device

a small volume of stirred water sample for 2–15 minutes, during which organic material present in the water partitions into the stationary phase. The capillary is then withdrawn into its protective sheath, and the sheath and its capillary are inserted through the injection septum of a GC and into the heated injector zone. Any organic material absorbed on the fibre is then rapidly thermally desorbed into the GC to give a normal chromatogram.

Since no organic solvents are required at all, the technique is much less prone to contamination problems, and it gives rise to simpler chromatograms because there is no solvent peak, which in other methods has the potential to obscure early eluting components of the sample. The action of inserting the fibre into the heated injection zone completely desorbs any organics from the stationary phase, so it is cleaned and ready for reuse. A number of thicknesses and coatings are already available for different applications, although for general use poly(dimethylsiloxane) is the most popular, combining good absorption properties with high mechanical stability of the film; this permits it to be used many times. Because the design of the apparatus allows it to replace the normal syringe in conventional autosamplers, which can, with modified timing parameters, be used both to extract and inject the sample, giving a completely automated system that extracts and injects a number of samples with minimal operator time.

The technique is used to extract organics directly from solution and as such it may be used for a wide range of substances from organic solvents through to non-volatile, polar materials like pesticide residues, but its use can be modified by allowing the fibre only to come into contact with the vapour above a sample of water. In that case only the volatile components in the water sample are adsorbed, giving rise to a simplified chromatogram in the analysis stage. Applications are also being developed where the fibre, after immersion into the sample, is injected into a modified HPLC injector, where a small amount of the eluting solvent strips the sample and then passes it directly to the head of the HPLC column. This allows analysis of compounds which are not amenable to GC analysis. Although very new and relatively untested, it is claimed that trace levels of most organics, down to $ng\,L^{-1}$, can be quickly determined (15 min per sample) using only small volumes of aqueous sample without the requirement of organic solvents and their possible complications.

5.4   Solids

The sampling of solids presents analysts with their most challenging task. For gases and liquids, although both spatial and chronological variation may occur, at a particular location and time most gas and liquid samples are homogeneous and contain a large number of particles (this is assumed every time one takes an aliquot of liquid from a graduated flask in the laboratory). This is generally not true in solid sampling, where small variations in position can be highly significant and the particle size can be in excess of 10 cm in diameter. The sampling of coal in a trainload of material and the analysis of rock samples from a mountainside are typical, if extreme, examples. Generally then, the sampling of solids requires an even more rigorous and systematic approach than for gases and liquids. The collection of a representative sample from the population presents the analyst with an extremely difficult task, and it is worth re-emphasising that the sampling error is often the greatest source of error in the overall sampling procedure.

For solid samples, variation with time is generally less important than in the gas or liquid phase, but variation with position is normally of much more concern, since analyte concentrations can vary by orders of magnitude over even small distances. But the greatest potential problem with solids is the size of individual particles. It will be shown below that sampling errors are inversely proportional to the number of particles taken. For gases or liquids, where the particle is an ion or molecule, even quite small amounts of sample contain large numbers of particles (10^{16} particles per millilitre for liquids); for solids, where the particle may be as much as several kilograms in mass, the problem is much more significant and a systematic approach to sampling is always required to reduce the sampling error as much as possible. Generally a **sampling plan** is devised which will enable the analyst to calculate how many samples should be taken from the population and the size of each of these samples.

The language of solid sampling varies from text to text and it is perhaps best to define and explain the terms which will be used in this section. The whole of the material to be analysed is known as the **population** or the **lot**; this may be a single, more or less homogeneous mass, or it may contain a number of individual units. From this is drawn the **gross** sample or **bulk** sample. This is obtained by taking **increments** from the lot in such a way that, when combined, they should be representative of the lot, but much smaller; however, in the case of certain solids this may still be as much as a tonne in mass. The **analytical** sample or **laboratory** sample, which is smaller still, is taken from this prior to analysis (Figure 5.1).

Note Analysts often refer to the **size** of the sample as the mass or volume, but statisticians mean the **number** of samples taken.

For convenience, sample types are often categorised as follows:

1. Bulk samples
2. Stratified (segregated) samples
3. Discrete units (packets, drums, bottles etc.)

Since a statistical approach is to be developed here, only bulk samples will be considered in detail and the reader is referred to more specialised treatments of the other categories. The overall sampling procedure involves reducing the amount of material progressively by taking increments in such a way that at any stage the composition is as far as possible representative of the original, i.e. the avoidance of bias. The exact way that this is performed will depend crucially on the nature of the sample, and for many materials, especially manufactured goods or industrial raw materials, the procedures have been well established to achieve this objective. However, in this section, some general principles and guidelines will be discussed.

Bulk samples are those which may contain several components and particle sizes, but which do not have an observable boundary between one part of the sample and another. Examples include a trainload of coal or metal ore, the soil of a factory yard which has been contaminated with waste material, or a single packet of washing powder. A simplified discussion of the statistics of sampling from a bulk material is given below, but it is important to realise that, even for moderately complex samples, the statistical approach becomes very complex.

5.4.1 Statistics of sampling

Errors in scientific measurements are classified as determinate (systematic) errors, which give a measure of accuracy, and indeterminate (random) errors, which are a measure of precision. For normal (laboratory) statistical manipulations it is often assumed that the determinate errors are reduced to a low value and the remaining uncertainty in the result is due to random errors, which are quantified by calculating the standard deviation. However, the standard deviation of a method is composed of the standard deviations of **each step** in the analysis; thus, when analysing real samples

$$s_o = \sqrt{s_a^2 + s_s^2} \quad \text{or} \quad V_o = s_a^2 + s_s^2 \quad \text{or} \quad V_o = (V_a + V_s) \tag{5.1}$$

Where s_o is the overall standard deviation, s_a the analytical standard deviation and s_s the sampling standard deviation. Often it is better to use the square of each of these terms, the variance, V since in a multistep process, variances are additive whereas standard deviations are not.

The analytical standard deviation s_a is considered first, and it can be separated into two discrete components:

Same-sample variation This is the variation in instrument or signal response for repetitive measurements of the same sample. It is normally the smallest of all the variances, although often it is the one which the analyst is (mistakenly) most concerned with.

Different-sample variation This is the variation in response obtained for repeat measurements of different analytical samples drawn from a single gross sample. It is the **within-lot** variance and it depends on the relative composition of the sample and the homogeneity of the particles. The value can be reduced to an acceptable level by repeat analysis of aliquots of the gross sample, but it will not on its own give a reliable estimate of the value or variability of the population.

The sampling variation s_s, or **between-sample** variation, measures the variability between different parts of the population. For bulk samples this can be large, and it can only be reduced to acceptable values by taking large numbers of bulk samples from the population. The relative significance of each of these terms can be estimated, once they have been evaluated, by using analysis of variance (ANOVA), but 'typical' figures are shown in Figure 5.1. Substitution of likely values into these equations shows that if the sampling error is high, it is a waste of time and money to use a very precise analytical technique to obtain answers, since the overall error will largely determined by the sampling error. Provided the analytical technique has a precision three times better than the sampling precision, there is little point in trying to reduce the analytical precision further. It is often better to take a larger number of samples (which improves the sampling precision) and to analyse them with a method of moderate precision than to use a high-precision technique on a small number of samples. The use of composite samples can also greatly increase the overall precision of the experiment under suitable circumstances.

Example 5.1

For a particular analysis, it has been determined that the sampling error is 6% ($s_s = 0.06$). A particular analytical technique can give a precision of 1%, what is the overall precision and is it worth considering a slower technique which can give a precision of 0.2%?

From equation (5.1) we have

$$s_o = \sqrt{s_s^2 + s_a^2} = \sqrt{(0.06)^2 + (0.01)^2} = 0.0608 \approx 6.1\%$$

For 0.2% analytical precision

$$s_s = \sqrt{(0.06)^2 + (0.002)^2} = 0.060$$

No, it is not worth using a technique of higher precision.

Example 5.2

For $s_s = 15\%$ and $s_a = 5\%$ calculate s_o? If s_a is reduced to 1% what is the total precision?

When attempting to obtain a representative sample from a bulk solid, extreme care must be exercised in order to reduce the sampling errors to a minimum, otherwise the whole time and cost of the analysis will have been wasted. However, since taking an excessive number of samples or samples which are too large will rapidly make the economics of the process non-viable, it is desirable to develop a **sampling plan** that is designed to provide a satisfactory analytical answer while maintaining realistic values for the **number** and **size** of the samples. Without exception a sampling plan must be devised which tries to ensure a sufficient number of random samples are taken from the bulk so that the composition can be determined to a given degree of accuracy.

Note   The word 'random' in statistics means without bias; it does not mean haphazard.

Often it is necessary to use an **iterative** approach where a preliminary set of experiments give the analyst some information about the population. This information is then used to devise a second set of experiments, and so on, until the desired sampling precision is obtained. Since the answer obtained depends upon **how much** sample is taken from the bulk and **how many** samples are taken, each of these terms will be considered further.

5.4.2   Minimum sample size

In order to calculate the mass of each gross sample taken, it is first necessary to decide upon the acceptable sampling variation. This value, which is chosen by the analyst, is normally expressed as a percentage variation; it is very similar to the confidence level used in simple statistics and, in fact, it is the sampling relative standard deviation R. Note carefully that this acceptable sampling variation, measured as the sampling precision or sampling relative standard deviation, is chosen by the analyst; whereas the analytical standard deviation, obtained in the laboratory, is calculated from a number of repeat determinations. Once R is chosen by the analyst, it is fairly easy to calculate, for a given sample type, how much material is required to achieve this value.

Considering the simple case of a homogeneous bulk material containing only two components A and B, and assuming a normal distribution (i.e. binomial statistics), it may be shown that the variance V for a given component decreases as the number of particles taken for analysis increases:

$$n \propto 1/V \quad \text{or} \quad n \propto 1/s^2 \tag{5.2}$$

Equally the variance will depend upon the composition of the mixture. (Common sense indicates that the smaller the composition of the analyte in the matrix, the greater the error in determination.) This is best expressed by considering the proportion of one component, say A, compared to the total number of particles (remember that statistics is concerned primarily with numbers). Thus the probability p of extracting (analysing) a single particle A from a mixture of A and B is

$$p = n_a/(n_a + n_b)$$

where n_a is the number of type A particles in the total population $(n_a + n_b)$. Thus for n particles removed

$$\text{probability} = np \tag{5.3}$$

For solid B there will be nq particles where $q = n_b/(n_a + n_b)$; and by definition $q = 1 - p$. The standard deviation s for either type of particle is $s = \sqrt{npq}$, thus the relative standard deviation R for analyte A is given by

$$R = \frac{\text{standard deviation}}{\text{number of particles withdrawn}} \tag{5.4}$$

$$\text{i.e.} \quad R = s/np = \sqrt{npq}/np = \sqrt{q/np} \tag{5.5}$$

This can be rearranged to give

$$n = (1 - p)/pR^2 \quad \text{(remember } q = 1 - p) \tag{5.6}$$

The calculation above assumes that the proportion (concentration) of A in the mixture is known, which is the whole object of the exercise in the first place; so in reality a preliminary series of experiments is conducted to allow approximate calculation of this value for insertion into the equations.

These equations calculate how many particles are required to give a desired value of sampling variability R, but normally we are concerned with mass. The mass of the particles can therefore be calculated as follows. Since the mass of n particles is

$$m = n(4/3)\pi r^3 \rho \tag{5.7}$$

where
r = particle radius
ρ = particle density

Substituting for n from equation (5.6) gives

$$m = \frac{(1 - p)(4/3)\pi r^3 \rho}{pR^2} \tag{5.8}$$

Thus the particle size of the material will have a profound effect on the mass of material taken, since $m \propto r^3$. For example, 10^4 particles with $r = 0.5$ mm (1 mm diameter) and density 2 g mL^{-1} (i.e. common salt) will weigh 10.5 g. The same number of particles of 5 mm diameter will weigh 1308 g, i.e. 1.3 kg. Equally, if the particle size could be reduced to say 80/120 mesh size (~150 μm), then less than 40 mg would contain the desired 10^4 particles. Equation (5.8) may be rearranged, and the constants grouped together to give

$$mR^2 = K_s \tag{5.9}$$

where m is the mass of material taken and K_s is a constant known as the sampling constant. It is numerically identical to the mass of material required to give a sampling relative standard deviation R of 1%. Although having the virtue of simplicity, the sampling constant includes in it both the density and composition of the analyte under test, which are not normally known prior to the test sequence, unless the same type of material is analysed regularly. Its use is therefore limited to routine analysis of the same or similar sample types, but where different levels of variance may need to be calculated.

Note Although reducing particle size mechanically will dramatically reduce the mass of material required for a given sampling variance, it is not normally possible at this stage, since the whole procedure is designed to calculate the mass of material taken from the population or bulk, which may well be simply too large to modify. At later stages in the process, when considering subsets of this sample (the gross or analytical samples), it is often possible, and even desirable, to reduce particles size appreciably, and various crushing, grinding and milling systems are available which can reduce particle size from several centimetres (small rocks) to fine, homogeneous powders. But for these calculations, unless the whole of the material is treated identically, the process becomes invalid.

Once these equations have been used to calculate the minimum mass of material required to give a specified sampling relative standard deviation or variability, the minimum number of samples required for a given confidence is also calculated.

5.4.3 Minimum number of samples

Once the minimum mass for each of the gross samples has been calculated, it is necessary to calculate the minimum number of samples that are required to be taken from the population in order to maintain the desired sampling variation. This is really asking how many samples are required such that the analyst is certain the answer lies between say ±5% of the mean value, i.e. 95% confidence.

Obviously an equation is needed that includes both n (the number of samples) and a term expressing the confidence which the analyst requires. This is available in the form of the Student t expression, familiar from simple statistics:

$$\mu = \bar{x} \pm ts/\sqrt{n} \qquad (5.10)$$

and with a simple rearrangement it may be used to calculate the **allowable error in measurement** of a number of samples, or the number of samples required to give a specified error. This is done by expressing the error as the difference between the true result and the mean value:

$$\mu - \bar{x} = E = tR/\sqrt{n} \quad \text{or} \quad n = t^2R^2/E^2 \qquad (5.11)$$

where the general form of standard deviation s has been written as R, the relative standard deviation of the sampling operation; E is the error. Note carefully that E is the sampling error, chosen by and acceptable to the analyst; it determines the minimum number of samples taken.

For simplicity, since chemists often use 95% confidence levels, which give $t = 1.96$ for large numbers of samples, this can be rounded up to $t = 2$, giving

$$n = 4R^2/E^2 \qquad (5.12)$$

Take care with dimensions. When R is expressed as a fraction, then E must be expressed as a fraction. When R is expressed as a percentage, then E must be expressed as a percentage.

The smaller the standard deviation required in the overall process, the larger the total number of samples required to achieve it. Thus, with a sampling standard deviation of 7% and an allowable error of 5%, equation (5.12) gives $n = 8$. If the sampling standard deviation increases to 10% then, for the same error, $n = 16$. Normally these calculations are carried out in an iterative way, i.e. the result from the first calculation is used to refine the values for a second calculation, and so on, until the answer approaches a constant value.

Example 5.3

From a preliminary determination of s_s on a single lot from the bulk, a value of 9.9% was calculated for the sampling relative standard deviation of analyte. How many samples need to be taken if the analyst wishes to have 95% confidence that the measured value of analyte does not vary by more than 5% about the measured mean?

Assuming a value of $t = 2$ (for 95% confidence) and substituting into equation (5.12) gives

$$n = 2^2 \times (9.9)^2/(5)^2 = 16 \text{ samples}$$

If R were expressed as 0.099 and the error E as 0.05 then the calculation would be

$$n = 2^2 \times (0.099)^2/(0.05)^2 = 16 \text{ samples}$$

The calculation is then repeated but this time using the value $t = 2.13$ (the value of t at 95% confidence for 16 samples) to give a revised estimate of n. This gives $n = 17.78$. A third calculation using the value for $t = 2.11$ (t_{18}) gives $n = 17.45$, and so on, until a constant value of n is obtained. For this problem it may be confirmed with only four calculations that $n = 18$ samples.

If one assumes that the sampling has been performed without bias, then the error term E, which has been used so far, and is defined as the difference between the true value and the mean, can be called the accuracy A. This gives an approximate formula

$$n = 4V_s/A^2 \tag{5.13}$$

where V_s is the sampling variance. This simplified form of the general equation is often used for calculating the number of samples n to achieve a specified accuracy.

5.4.4 The significance of statistical calculations

All the calculations in Section 5.4.3 are based on the assumption that sampling errors are related to the number of particles taken. Although the error is also related to a number of other factors such as the concentration (composition) and complexity of the sample, there is an inverse square relationship between the standard deviation and the number of particles taken. This is why the current chapter started by looking at gases and then liquids; in each case even a small physical sample contains sufficient particles to make that part of the sampling process error-free compared to the sampling of solids, where the number of particles taken becomes important. Although the calculations are based on number of particles, in reality we tend to measure not the number but the mass of the sample. By looking at the equation (5.7), we can see there is a cubic relationship between the mass and the particle diameter.

Thus, where possible, reducing the particle size has a dramatic effect on the mass required for a given number of particles.

5.4.5 The taking of increments

Having calculated the mass and the number of samples required for analysis, we then require a method or methods for obtaining these samples from the bulk – without bias. For a homogeneous bulk material, the total material is normally considered as a number of

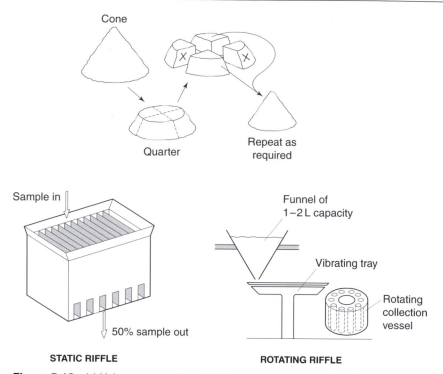

Figure 5.18   (a) Using a cone and quarter for sample reduction. (b) Riffles used for more rapid sample reduction

identically sized sections (a matrix), and the required number of samples is taken from within that matrix with the aid of random number tables. The resulting samples are then either analysed separately, or they may be combined and treated as a new bulk, with identical characteristics to the original lot, but smaller.

Note   The reduction in size of a bulk sample without introduction of bias is possible at this stage using a number of established techniques. The simplest, and often the most practical, is the cone and quarter; this involves arranging the total material into a pile (cone) and then drawing two vertical lines through the pile at right angles to each other, producing four identical quarters (Figure 5.18(a)). Two opposite quarters are then combined and retained, whereas the other two are discarded. Assuming the particle size of the material is constant, or nearly constant, the mass has been halved, without bias. The process may be repeated several times until the original mass has been reduced to manageable proportions for laboratory manipulation. A number of mechanical devices, known as riffles (Figure 5.18(b)) can also be used to reduce gross samples to manageable proportions. Although they are capable of taking up to 16 identical samples in one cycle, from one mass of material, they need to be kept clean if they are not to introduce possible contamination from one sample to another, and for some designs this is difficult to achieve.

Segregated or stratified samples, which exhibit visible divisions into markedly different zones, are best treated by considering each layer or division as a different bulk material, and sampled separately using a different plan if necessary. Alternatively, a composite sample may be obtained from the stratified bulk by taking increments from each zone and combining them in such a way that the original make-up of the sample is preserved. Thus, if the bulk appears to have three zones of volume $1:5:20$, then the increments should be in this ratio. The difficulty with this approach is that although the number and size of increments should reflect the composition of the original bulk, this is what the analyst is trying to find in the first place!

In all cases a sampling plan must be devised, and the analyst must comply with it. Modifications and substitutions introduced during the collection of materials produce a sample which will probably not represent the composition of the original, and as such will be a waste of time and money to collect and analyse. The word 'random' in the context of sample collection means without bias; it does not mean haphazard. This does not mean that no modifications to the sampling protocol are allowed, only that any changes to the original plan are introduced as a result of data produced from the material in the preliminary analytical stages. This is the process of iteration which can be applied not only to refine calculations, but also to improve the accuracy and precision of the sampling process, and thus ultimately the final answer. Iteration is very time-consuming and costly, but it is often essential in order to obtain reliable results from a complex matrix.

5.4.6   Sampling of discrete units

Often the bulk material for analysis is composed of a number of discrete units: drums, bottles, packets or tablets; in this situation the total variance of the system is now composed of three separate components:

$$\text{total variance} = \text{variance of analytical operation}$$
$$+ \text{ variance within a single unit} + \text{variance between units}$$

The first of these terms is determined by the choice of analytical method and all the earlier comments apply in this situation as well. The second and third terms are independent of each other and may be determined separately, although the same cyclic (iterative) plan can be used to optimise the number and size of samples from each unit. Provided the preliminary sampling plan reveals no appreciable differentiation between units (as is often the case with mass-produced articles) then individual units may be combined into a composite sample to reduce the effect of the third term.

5.4.7   General comments

The variation of measured analyte concentration often consists of two distinct components:

Between-sample variation is the variance between different parts of the population; this is often large and can only be reduced to acceptable values by taking large numbers of samples from different parts of the bulk to give a representative bulk sample.

Within-sample variation measures the variance of the analyte in a single gross or bulk sample. This depends on the relative composition of the sample and the homogeneity of the particles, which can sometimes be improved by reducing particle size (by grinding, etc.). This value can be reduced to an acceptable level by repeat analysis of aliquots of a gross sample, but it will only give the value for that bulk sample; it will not, on its own, give a reliable estimate of the value or variability of the population.

In each type of sample the final sampling plan will depend on prior knowledge of the system and upon experience gained by the analyst in the early stages of sampling. Remember that random sampling does not mean the indiscriminate collection of samples; it means rigid adherence to a carefully conceived sampling plan designed to minimise uncertainty and bias, thereby giving the best estimate possible of the analyte composition.

5.5   References

1. Laboratory of the Government Chemist 1996 *VAM Bulletin*, **15** (Autumn); 13–14
2. Laboratory of the Government Chemist, July 1993
3. C C Lee 1995 *Sampling, analysis and Monitoring methods: a guide to EPA requirements*, Government Institutes Inc., Rockville MD
4. HSE Books, PO Box 1999, Sudbury, Suffolk, UK
5. SKB, Blandford Forum, Dorset, UK
6. US Environmental Protection Agency 1997 *Compendium of methods for the determination of toxic organic compounds in ambient air*, EPA 625/r-96/010B
7. Health and Safety Executive 1995 *Methods for determination of hazardous substances* (MDHS Nos 27, 70, 80), HSE, Sheffield
8. The Foxboro Co., Foxboro MA 02035, USA
9. Bruel & Kjaer, DK-2850, Naerum, Denmark
10. R M Harrison 1994 *Chemistry in Britain*, **30** (12); 987–1000
11. Thermo Electron Ltd, Warrington, Cheshire, UK
12. Palintest Ltd, Gateshead, Tyne & Wear, UK
13. Phillip Harris Scientific, Litchfield, Staffs, UK
14. Perkin-Elmer Corp., Norwalk CT
15. Jones Chromatography Ltd, Hengoed, Mid-Glamorgan, UK

5.6   Bibliography

T R Crompton 1992 *Comprehensive water analysis*, Elsevier, London

N T Crosby 1995 *General principles of good sampling practice*, Royal Society of Chemistry, London

Health and Safety Executive 1998 *Occupational exposure limits 1998: guidance note EH40/98*, HSE Books, Sudbury

L H Keith 1996 *Compilation of EPA's sampling and analysis methods*, CRC, Lewis, London

L H Keith 1996 *Principles of environmental sampling* 2nd edn, American Chemical Society, Washington DC

C L Paul Thomas and H Schofield 1995 *Sampling source book*, Butterworth-Heinemann, Oxford

R Perry and R M Harrison 1993 *Handbook of air pollution analysis*, Chapman and Hall, London

P Pradyot 1997 *Handbook of Environmental Analysis* (chemical pollutants in air, water, soil and solid wastes), CRC, Lewis, London

R Reeve 1993 *Environmental analysis*, ACOL–Wiley, Chichester

C L Rose 1997 *Provisional environmental sampling protocols*, AEA Technology, National Environmental Technology Centre

6

Separation

6.1 Introduction

The validity of results from chemical analyses depends initially upon the quality and the integrity of the sample that has been obtained. The problems associated with sampling have been dealt with in Chapter 5, and in the majority of cases the sample to be analysed will contain more than one component and may require some separation and/or concentration to be carried out prior to analysis. Even pure single substances used as standards rarely exceed 99.99% purity, which still leaves the possibility of $10\,\mu g\,g^{-1}$ of another material being present. Normally the sample is considered to consist of the substance(s) to be analysed, **the analyte**, and the rest of the material, **the matrix**.

Some analytical techniques are, by their very nature, either selective, or even specific, and can be used for detection or determination of the desired analyte without a prior separation, but in these cases the analyst must be satisfied that matrix effects are not significant under the conditions of the experiment, so that the true level of analyte is not over- or underestimated. Even for these cases the sample is often subjected to some form of pretreatment prior to the actual determination, either to change the phase, or to stabilise the sample prior to analysis. Equally the analyst is now expected to identify and quantify substances at much lower levels than previously possible. Determining concentrations at $\mu g\,kg^{-1}$ (ppb) levels is now often routine in many laboratories, and for some techniques which look at specific substances it is possible to measure, or at least identify, materials at $ng\,kg^{-1}$ (parts per trillion) levels. Theoretically, provided there is sufficient sample and a reliable preconcentration technique, even if the final analytical technique is not ultrasensitive, low levels in the sample can still be determined by multiplying the measured value by the concentration factor.

Remember the distinction between low concentrations and small masses, which is often not clearly explained by instrument manufacturers when extolling the virtues of their newest analytical instrument. For example, the technique of inductively coupled plasma spectroscopy (ICP) is often quoted as having detection limits (for suitable elements) in the $\mu g\,kg^{-1}$ range; but since it is possible that up to 20 mL of sample is required to obtain a reliable reading, then the amount, or mass, of analyte determined is of the order of 20 ng. GC analysis, on the other hand, often relies on analyte concentrations in the $\mu g\,g^{-1}$ range for reliable results; but since the sample size injected can be less than $0.1\,\mu L$, the mass determined is 0.1 ng.

6.2   Separation techniques

For the purposes of this chapter, separation techniques can be considered to fall into two main groupings:

Bulk separations Large-scale separations of one component from another, typically they include filtration, temperature-dependent effects, (distillation, evaporation and drying), solubility effects (solvent extraction, crystallisation and precipitation), ion exchange, dialysis and lyophilisation. Many of these techniques fall into the category of standard laboratory manipulation, and have been mentioned elsewhere in this book, but it will be useful to draw them together here so that, for a given experiment, each can be considered on its merits.

Instrumental separations The most common instrumental separations are chromatography, (GC, HPLC, TLC, supercritical fluid chromatography (SFC)) and the electrophoretic group of separation techniques, especially capillary electrophoresis (CE). For these techniques the amounts of analyte involved are often of the order of micrograms or less and are too small to be physically observed. In fact, one way of distinguishing between the two is on the basis of the simple test. Can the analyte be seen?

In each case a number of different mechanisms can be employed to perform the separation, and for the purposes of this chapter they can be grouped into either physical methods or chemical methods. Often, as we shall see, bulk separations rely mainly on physical mechanisms, and instrumental separations mainly on chemical mechanisms, but there are numerous exceptions to this general guide; sometimes the distinction is somewhat blurred as to the mechanisms employed. For example, even when used on a large scale as bulk separation techniques, solvent extraction or ion exchange still rely on solvent–solute or chemical interactions. A number of detailed works on separation processes are available if a more rigorous approach is required, e.g. the text by Seader and Henley.[1]

6.2.1   Filtration

At first sight, filtration is a very simple process of separating solid material from solution using some form of filter medium, and for many laboratory tasks this is still the case. However, when the task is analysed more carefully, it is not nearly so simple as first imagined. The distinction between solid and liquid, although theoretically clear, is actually getting more difficult to define in real terms. Particles of say 0.1 mm or above, which are not only visible but often settle to the bottom of a solution, are solid and can be separated using simple filter papers in a funnel. But are particles of say 0.3 μm diameter, solid or liquid? If ordinary filter papers are used, then these particles pass through the paper and are thus considered liquid for the remainder of the experiment.

Clearly the separation of liquid from solid will depend upon the pore size of the filtration medium, and one would be tempted always to use the finest grade of membrane available, which can even retain some colloids or viruses. But this approach can lead to severe difficulties with the separation step, since it is found that as the pore size of the membrane decreases, the speed of transfer through the membrane decreases very rapidly, so gravity alone is not sufficient to allow sensible separation times. Either positive pressure must be applied to the top of the filter vessel, or more commonly, a vacuum is used to suck the solution through the filter.

6.2.2   Temperature-dependent effects

Temperature-dependent effects fall into the category of general laboratory manipulation, but some special precautions are often necessary when they are used for analytical separations. Distillation is sometimes used for analytical separations, although not widely; but provided the apparatus is of a suitable scale for the sample (and it is clean), then it is a straightforward way of separating substances. Evaporation and drying are really often two sides of the same process – removal of a liquid from a solid.

Where the analyte is a liquid phase that needs to be separated from a solid matrix, a fairly simple apparatus consisting of a heat source and some form of condenser is often all that is required, although a vacuum system such as a rotary evaporator can often help, and occasionally the process is carried out in an inert atmosphere such as nitrogen. However, if quantitative measurements need to be performed, then collection of all of the liquid phase without modifying the solid matrix as it is heated can often cause significant problems. For example, evaluation of water content in solids is by no means a trivial task. Most solid samples contain water in some form or another, and accurate quantitative measurements either require that the amount of water is known, or that the solid is dried.

The drying process will depend to a large extent on the form of water present in the sample. Water can be non-essential water, which is retained in the matrix by physical forces such as adsorption or is just held between the particles of solid, i.e. it is wet. Alternatively the water may form a closer and more defined part of the material, e.g. part of a solid hydrate such as $CaSO_4 \cdot 2H_2O$. Most solid samples, even those without any of these forms of water, are in some form of equilibrium with the moisture in the atmosphere and their weight will change as the humidity of their surroundings alters, although this is often only by small amounts.

Generally, for the most demanding analysis, it is still worth standardising the moisture content of the sample. Although by no means foolproof, the most commonly accepted method is to dry to constant weight at 105 °C or some other similar temperature. This at least means that samples can be compared with each other in a standardised way. However, for many samples where the sample is biological in some form (and this will include food analysis), the process of heating the material will alter its composition, making this unsuitable as a method of pretreatment. Here more specialised methods are necessary, such as homogenisation of the sample in a high-speed blender to produce a slurry, or occasionally freeze drying. In all cases where the amount of analyte in a given amount of sample (matrix) is quoted, it should be clear on what basis the total weight of sample was measured.

The process of headspace analysis relies on evaporation or at least on differential volatility of different components in the sample. Here a portion of sample, either solid or liquid, is placed in a small sealed vessel which is then heated to constant temperature. Volatile components in the sample are thus driven from the condensed phase into the gas phase above the sample (the vessel is only partially filled with sample), from where they can be extracted, usually by means of a syringe passed through a septum in the lid of the vessel, and then analysed directly.

Headspace analysis is finding wide application in the analysis of volatiles in complex samples in such diverse applications as organic material in river sediment, and the characterisation of aroma components in food and drink, especially alcoholic beverages.[2] For many determinations it has great attraction, since when coupled to gas chromatography it can give qualitative fingerprints of complex samples very readily. However, although very simple to set up in the laboratory, only requiring a number of small septum-capped bottles and a water bath, headspace analysis can give variable quantitative results if the conditions of the experiment are not closely controlled. Then it is better to use a completely automated system that fits directly onto the inlet of a GC.

6.3 Solvent extraction

6.3.1 Solvent extraction

Solvent extraction of solid materials is still widely performed, either by simple mixing of liquid and solid (often aided by immersing the flask in an ultrasonic bath) followed by filtration or centrifugation, or by using some form of continuous extraction apparatus such as a Sohxlet apparatus. It is especially appropriate where quantitative measurements are required, and this is described elsewhere in the text. However, problems with the use of solvents (see below) mean that, where possible, other forms of pretreatment should be considered, such as headspace analysis or supercritical fluid extraction (SFE); these alternative methods are now gaining favour, especially if only qualitative data is required.

The extraction of an analyte, or group of analytes from one liquid phase into another, **liquid–liquid extraction**, is still probably the most widely employed technique for bulk separations where appreciable quantities of analyte are involved. (For smaller quantities of analyte, SPE and its related techniques are preferred.) Liquid–liquid extraction involves the partitioning of the analyte between two immiscible liquid phases, normally by shaking them in a simple separatory funnel. In most cases one of the phases is aqueous, perhaps pure water or perhaps an acidic, basic or buffered solution; the other phase is an organic solvent (Table 6.1). The analyte or analytes may initially be present in either phase. The selectivity of separation and its efficiency are controlled by the choice of these two phases, as are a number of other factors discussed below.

Table 6.1 *Solvents used in liquid–liquid extraction*

Aqueous phase	Organic phase
Pure water	*Chlorinated solvents*
Acidic solution (pH 0–6)	Dichloromethane
Basic solution (pH 8–14)	Chloroform
High ionic strength (salting out)	*Hydrocarbons*
Complexing agents	Aliphatics: C_5 (pentane) and above
Ion pair reagents	Aromatics: toluene and xylenes
Chiral complexing agents	Alcohols: C_6 and above are immiscible with water
	Esters
	Ketones: C_6 and above
	Ethers: diethyl and above

Aqueous phase

If the aqueous phase is the extracting solvent, i.e. it does not initially contain the analyte, then it is generally possible to ensure this is of adequate purity to avoid contaminating the final sample, although if modifiers such as acids, bases, buffers or complexing agents are added, they too must be of high purity. On the other hand, if the sample is contained in a dirty water sample and is to be extracted from it, the analyst is wise to consider what effect, if any, the other components may have on the extraction (matrix effects). And if the analyte is initially present in the aqueous phase, then in order to efficiently extract it into an organic phase, it will probably need some modification in order to make it 'dissolve' preferentially into the organic layer, i.e. it will need to be made less hydrophilic and more hydrophobic.

Organic phase

At this stage, choice of the second solvent is determined by a simple criterion: it must be immiscible with water so that two distinct layers are formed. Strictly speaking, although no two liquids in contact are completely immiscible (small amounts of each dissolve in the other), for practical purposes the solubility of one solvent in the other should not exceed about 10%. This in itself precludes some of the more commonly used solvents such as acetone and the lower alcohols, but it allows many others. The solvent density should be different from the aqueous layer so that it forms a clear layer at the bottom of the separating funnel, or if less dense than the aqueous layer, at the top. The problems of emulsion formation between the two liquids are often exacerbated when the densities are similar, especially if the liquids contain either surfactant or fatty material.

If the organic phase is to be the extracting phase, which will be carried forward to the next stage of the process, it is an advantage if it is volatile so it can be removed by evaporation, but again it must be as pure as possible to avoid contamination of the final analytical sample. Since relatively large amounts of organic solvent are used in these processes, remember to consider the toxicity and the environmental consequences of solvent disposal at the end of the experiment. If there is a choice between the less toxic $CHCl_3$ and the more toxic CCl_4, choose the $CHCl_3$. Several solvents, such as benzene and CH_2Cl_2, may be favoured on chemical grounds, but they must nowadays be excluded on toxicity grounds. Finally the polarity of the solvent must be considered, since generally speaking the more polar the solute, the more polar the extracting solvent needs to be: 'like attracts like' is a useful phrase to remember for many separation processes involving partition.

6.3.2 Partition: the theory of extraction

Nearly all liquid–liquid extractions rely on the process of partition, where the analyte is distributed between two liquid phases in contact. The Nernst distribution law states that if a substance which is soluble in both liquid phases is allowed to remain in contact with both, then at equilibrium the ratio of activities of the material in each phase is a constant known as the partition coefficient:

$$K = \frac{a_{s1}}{a_{s2}}$$

In dilute solution the activities can be replaced by solubilities S_o and S_{aq}, then the constant K is written K_D and is known as the distribution constant:

$$K_D = \frac{S_o}{S_{aq}}$$

The law, as stated, is not thermodynamically rigorous and takes no account of the activities of the various species. For this reason it would be expected to apply only in very dilute solutions, where the ratio of the activities approaches unity. Nor does this simple form apply when any of the distributing species undergo dissociation or association in either phase. However, in the practical application of solvent extraction we are interested primarily in the fraction of the total solute in each phase, regardless of association, dissociation or other interactions with dissolved species. It is thus more convenient to introduce the distribution ratio D, where

$$D = \frac{C_o}{C_{aq}}$$

where C represents the concentration. The value of D depends only upon the nature of each solvent, the solute and the temperature. This form of the Nernst distribution law is more useful for liquid–liquid extraction. It may be shown that if V mL of aqueous solution containing x_{aq} g of solute is extracted n times with v mL portions of organic solvent, then the weight of solute remaining in the aqueous layer is

$$x_n = x_{aq}\left(\frac{DV}{DV + v}\right)^n$$

and the weight extracted will be $x_{aq} - x_n$. Note that the equation assumes extraction with n portions of solvent, rather than with the equivalent volume used once. The advantage of using several small portions of organic solvent to extract an analyte from another phase is best seen by example.

Example 6.1

A 50 mL volume of water containing 0.1 g of analyte is to undergo liquid–liquid extraction by shaking with 25 mL of an organic phase. It is known that the distribution ratio for the analyte is 1/85, i.e. it is 85 times more soluble in the organic layer than in water. Compare the result of (a) one extraction using all 25 mL of organic phase and (b) three extractions each using 8.33 mL of organic phase.

(a)   For one extraction using all 25 mL of organic phase

$$\text{weight remaining} = \frac{1/85 \times 50 \times 0.1}{1/85 \times 50 + 25} = 0.0023 \text{ g}$$

$$\text{weight extracted} = 0.1 - 0.0023 \quad = 0.0977 \text{ g}$$

(b)   For three extractions each using 8.33 mL of organic phase (and combining the extracts)

$$\text{weight extracted} = 0.1 - \left[0.1\left(\frac{1/85 \times 50}{50/85 + 8.33}\right)\right]^3 = 0.099\ 999\ 713 \text{ g}$$

i.e. virtually complete extraction.

An alternative expression is

$$E_\% = \frac{100D}{D + (V_{aq}/v)}$$

where $E_\%$ is the percentage extraction.

If the solution contains two (or more) solutes, A and B, and both tend to be extracted, then a separation coefficient or factor β can be defined as

$$\beta = \frac{[A_o]/[B_o]}{[A_{aq}]/[B_{aq}]} = \frac{[A_o]/[A_{aq}]}{[B_o]/[B_{aq}]} = \frac{D_a}{D_b}$$

Thus if $D_a = 10$ and $D_b = 0.1$, a single extraction will remove 90.9% of A, but only 9.1% of B (ratio 10 : 1). A second extraction of the aqueous phase will bring the total amount of A extracted to 99.2% and of B to 17.4% (ratio 5.7 : 1). If B were a contaminant, then more complete extraction of A would involve proportionally greater extraction of contaminant.

The best situation is where the distribution factor for A is large and the distribution factor for B is small. For many separations involving organic species, a knowledge of the distribution coefficients of analyte in various aqueous–organic pairs is sufficient to enable a sensible choice of extraction conditions to be made, provided the restrictions of health and environmental concern are also taken into account. However, for some organic species and most inorganics, it is possible to modify the extraction conditions so that analyte extraction into the desired phase is maximised, and possibly extraction of contaminants is reduced. For example, if the organic species is acidic (HA) or contains acidic groupings, then by keeping the pH of the aqueous solution low, the organic will exist mainly as HA which will extract into the organic layer readily; but by increasing the pH of the aqueous solution and thus converting the organic into A^-, it will tend to stay in the aqueous layer. Conversely, for basic materials such as amines a high pH will favour solution in the organic phase, whereas a low pH (BH^+) will favour retention in the aqueous layer. Remember that the more ionic (polar) the material, the greater the tendency to pass into the more polar (water) layer.

6.3.3   Factors favouring solvent extraction of inorganic species

Where the analyte is an inorganic ion, measures have to be taken to modify the species in order to enable efficient extraction into an organic phase. The most obvious way to do this is to neutralise the charge and produce a neutral species. This can be done in one of two main ways: (a) formation of chelated complexes using suitable organic ligands, or (b) formation of ion-associated complexes. In each case the object is to change the small hydrated metal or anion into a species which is more soluble in the organic layer. Many metal ions may be complexed to an organic ligand, making them soluble in organic solvents:

$$M^{n+} + nL^- \rightarrow ML_n$$

where L is the ligand which complexes to metal ion M^{n+}. A quantitative treatment of this process can be made on the basis of the following assumptions: (a) the reagent and the metal complex exist as simple unassociated molecules in both phases; (b) solvation plays no significant part in the extraction process; and (c) the solutes are uncharged molecules and their concentrations are generally so low that the behaviour of their solutions departs little from ideality. The dissociation of the ligand (chelating agent) HL in the aqueous phase is represented by the equation

$$HL \rightleftharpoons H^+ + L^-$$

The various equilibria involved in the solvent extraction process are expressed in terms of the following thermodynamic constants:

Dissociation constant of complex $\quad K_c = [M^{n+}]_w[L^-]^n_w/[ML_n]_w$

Dissociation constant of reagent $\quad K_r = [H^+]_w[L^-]_w/[HL]_w$

Partition coefficient of complex $\quad p_c = [ML_n]_w/[ML_n]_o$

Partition coefficient of reagent $\quad p_r = [HL]_w/[HL]_o$

where the subscripts c and r refer to complex and reagent, and w and o to aqueous and organic phase respectively.

The distribution ratio, i.e. the ratio of the amount of metal extracted as complex into the organic phase to that remaining in all forms in the aqueous phase, is given by

$$D = [ML_n]_o/\{[ML_n]_w + [M^{n+}]_w\}$$

which reduces to

$$D = K[HL]^n_o/[H^+]^n_w \quad \text{where} \quad K = (K_r p_r)^n/K_c p_c$$

If the reagent concentration remains virtually constant, then

$$D = K^*/[H^+]^n_w \quad \text{where} \quad K^* = K[HL]^n_o$$

and the percentage of solute extracted, E, is given by

$$\log E - \log(100 - E) = \log D = \log K^* + n(\text{pH})$$

Thus the distribution of the metal in a given system of this type is a function of the pH alone. The equation represents a family of sigmoid curves when E is plotted against pH, with the position of each along the pH axis depending only on the magnitude of K^* and the slope of each uniquely depending upon n. Figure 6.1 shows some theoretical extraction curves for divalent metals, and it illustrates how their position depends upon the magnitude of K^*. Figure 6.2 illustrates how the slope depends upon n. A tenfold change in reagent concentration is exactly offset by a tenfold change in hydrogen ion concentration, i.e. by a single pH unit; this change of pH is much easier to produce in practice. If $\text{pH}_{1/2}$ is defined as the pH value at 50% extraction ($E_\% = 50$) we see from the above equation that

$$\text{pH}_{1/2} = -\frac{1}{n}\log K^*$$

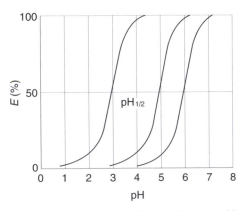

Figure 6.1 Percentage E of metal extracted into organic phase as a function of pH

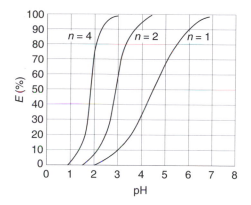

Figure 6.2   Variation of slope as *n* changes

Table 6.2   *Optimum pH for extraction of metal dithizonates into chloroform*

Metal ion	Cu(II)	Hg(II)	Ag	Sn(II)	Co	Ni, Zn	Pb
Optimum pH of extraction	1	1–2	1–2	6–9	7–9	8	8.5–11

The difference in $pH_{1/2}$ values of two metal ions in a specific system is a measure of the ease of separation of the two ions. If the $pH_{1/2}$ values are sufficiently far apart, then excellent separation can be achieved by controlling the pH of extraction. It is often helpful to plot the extraction curves of metal chelates.

If one takes as the criterion of a successful single-stage separation of two metals by pH control a 99% extraction of one with a maximum of 1% extraction of the other, for bivalent metals a difference of two pH units would be necessary between the two $pH_{1/2}$ values; the difference is less for tervalent metals. Some figures for the extraction of metal dithizonates in chloroform are given in Table 6.2. If the pH is controlled by a buffer solution, then those metals with $pH_{1/2}$ values in this region, together with all metals having smaller $pH_{1/2}$ values, will be extracted.

The $pH_{1/2}$ values may be altered (and the selectivity of the extraction thus increased) by using a competitive complexing agent or a masking agent. In the separation of mercury and copper by extraction with dithizone in carbon tetrachloride at pH 2, the addition of EDTA forms a water-soluble complex which completely masks the copper but does not affect the mercury extraction. Cyanides raise the $pH_{1/2}$ values of mercury, copper, zinc and cadmium in dithizone extraction with carbon tetrachloride.

6.3.4   Ion association complexes

Instead of forming neutral metal chelates for solvent extraction, the species of analytical interest may be associated with oppositely charged ions to form a neutral extractable species. Such complexes may form clusters with increasing concentration which are larger than just simple ion pairs, particularly in organic solvents of low dielectric constant. Three types of ion association complex may be recognised. The first two represent extraction systems involving coordinately unsolvated large ions, and in this important respect they differ from the third type.

Complexes formed from a reagent yielding a large organic ion

The tetraphenylarsonium, $(C_6H_5)_4As^+$, and tetrabutylammonium, $(n\text{-}C_4H_9)_4N^+$ ions form large ion aggregates or clusters with suitable oppositely charged ions, e.g. the perrhenate ion, ReO_4^-. These large and bulky ions do not have a primary hydration shell and cause disruption of the hydrogen-bonded water structure; the larger the ion the greater the amount of disruption and the greater the tendency for the ion association species to be pushed into the organic phase. These large ion extraction systems lack specificity, since any relatively large unhydrated univalent cation will extract any such large univalent anion. On the other hand, because of their greater hydration energy, polyvalent ions are not so easily extracted and good separations are possible between MnO_4^-, ReO_4^- or TcO_4^- and CrO_4^{2-}, MoO_4^{2-} or WO_4^{2-}.

Complexes involving an ionic chelate complex of a metal ion

Chelating agents having two uncharged donor atoms, such as 1,10-phenanthroline, form cationic chelate complexes which are large and hydrocarbon-like. Tris(phenanthroline) iron(II) perchlorate extracts fairly well into chloroform, and extraction is virtually complete using large anions such as long-chain alkyl sulphonate ions in place of ClO_4^-. The determination of anionic detergents using ferroin has also been described.[3] Dagnall and West[4] have described the formation and extraction of a blue ternary complex, Ag(I)–1,10-phenanthroline-bromopyrogallol red (BPR), as the basis of a highly sensitive spectrophotometric procedure for the determination of traces of silver. The reaction mechanism for the formation of the blue complex in aqueous solution was investigated by photometric and potentiometric methods, and these studies led to the conclusion that the complex is an ion association system, $[Ag(phen)_2]_2BPR^{2-}$, i.e. involving a cationic chelate complex of a metal ion (Ag^+) associated with an anionic counterion derived from the dyestuff (BPR).

Complexes in which solvent molecules are directly involved

Most of the solvents (ethers, esters, ketones and alcohols) which participate in this way contain donor oxygen atoms, and the coordinating ability of the solvent is of vital significance. The coordinated solvent molecules facilitate the solvent extraction of salts such as chlorides and nitrates by contributing both to the size of the cation and the resemblance of the complex to the solvent. A class of solvents which show very marked solvating properties for inorganic compounds are the esters of phosphoric(V) acid (orthophosphoric acid). The functional group in these molecules is the semipolar phosphoryl group, $\rightarrow P^+\!\!-\!\!O^-$, which has a basic oxygen atom with good steric availability. A typical compound is tri-n-butyl phosphate (TBP), which has been widely used in solvent extraction on both the laboratory and industrial scale; of particular note is the use of TBP for the extraction of uranyl nitrate and its separation from fission products.

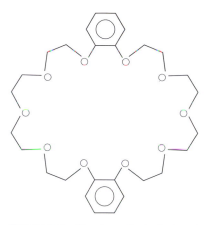

Figure 6.3   Structure of a typical crown ether

The mode of extraction in these 'oxonium' systems may be illustrated by considering the ether extraction of iron(III) from strong hydrochloric acid solution. In the aqueous phase, chloride ions replace the water molecules coordinated to the Fe^{3+} ion, yielding the tetrahedral $FeCl_4^-$ ion. It is recognised that the hydrated hydronium ion, $H_3O^+(H_2O)_3$ or $H_9O_4^+$ normally pairs with the complex halo-anions, but in the presence of the organic solvent, solvent molecules enter the aqueous phase and compete with water for positions in the solvation shell of the proton. On this basis the primary species extracted into the ether (R_2O) phase is considered to be $[H_3O(R_2O)_3^+, FeCl_4^-]$, although aggregation of this species may occur in solvents of low dielectric constant. The principle of ion pair formation has long been used for the extraction of many metal ions, but not the alkali metals, due to the lack of complexing agents forming stable complexes with them.

A significant development of recent years has been the application of the so-called **crown ethers**, which form stable complexes with a number of metal ions, particularly the alkali metal ions. These crown ethers are macrocyclic compounds containing 9–60 atoms, including 3–20 oxygen atoms, in the ring. Complexation is considered to result mainly from electrostatic ion–dipole attraction between the metal ion situated in the cavity of the ring and the oxygen atoms surrounding it (Figure 6.3). The ion pair extraction of rare earth ions with 15-crown-5 as a complex-forming reagent has been described in the literature.[5] The most commonly used crown ether is 18-crown-6, particularly for dissolution of biological compounds for mass spectral analysis.

6.3.5   Some extraction reagents specifically used for inorganic ions

Due to advances in instrumentation such as AAS, ICP and ICP/MS, it is now less common to have to separate or concentrate metals in solution before analysis. Nevertheless, for some purposes the techniques of liquid–liquid extraction remain useful, especially if colorimetric analysis is to be used in the final step. Bulk separation of one metal in solution from another is also required where interference in a reaction would be a problem. Table 6.3 provides a brief list of common chelating and extracting reagents, some of which are dealt with in more detail in Chapter 11. The text by Hiraoka,[6] or the handbook of Ueno *et al.*[7] should be consulted for more detailed accounts of organic analytical reagents.

Table 6.3 *Common reagents for solvent extraction*

Acetylacetone (pentane-2,4-dione)	$CH_3COCHCOCH_3$	Chelates with 60 metals
Thenoyltrifluoroacetone (TTA)	$C_6H_4SCOCH_2COCF_3$	Useful for lanthanides and actinides
8-Hydroxyquinoline (oxine)	C_9H_6ON	See Section 11.3
Dimethylglyoxime	$C_4H_8O_2N_2$	See Section 11.3
I-Nitroso-2-naphthol	$C_{10}H_7O_2N$	For Co and Fe extraction
Cupferron (Ammonium salt of *N*-nitroso-*N*-phenylhydroxylamine)	$C_6H_9O_2N_3$	See Section 11.3
Diphenylthiocarbazone (dithizone)	$C_6H_5N=NCSNHNHC_6H_5$	Selective reagent for Pb, Zn, Cd, Ag, Hg, Cu, Pd
Sodium diethyldithiocarbamate	$\{(C_2H_5)_2NCSS\}^-Na^+$	Extraction reagent for more than 20 heavy metals
Ammonium pyrrolidine dithiocarbamate (APDC)	$C_5H_2N_2S_2$	For heavy metals giving complexes soluble in organic solvents
Tri-*n*-butyl phosphate	$(n\text{-}C_4H_9)_3PO_4$	Particularly useful for Fe, Ce, U, Tl
Tri-*n*-octylphosphine oxide (TOPO)	$(n\text{-}C_8H_{17})_3PO$	Used for U, Cr, Zr, Fe, Mo, Sn
Cetyltrimethylammonium bromide (CTMB)	$CH_3(CH_2)_{15}N(CH_3)^+Br^-$	Acts as a surfactant to give high ligand-to-metal ratios of many metals

6.3.6 Determination of copper as the diethyldithiocarbamate complex

Discussion Sodium diethyldithiocarbamate [6.A] reacts with a weakly acidic or ammoni-acal solution of copper(II) in low concentration to produce a brown colloidal suspension of the copper(II) diethyldithiocarbamate. The suspension may be extracted with an organic solvent (chloroform, carbon tetrachloride or butyl acetate) and the coloured extract ana-lysed spectrophotometrically at 560 nm (butyl acetate) or 435 nm (chloroform or carbon tetrachloride).

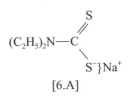

[6.A]

Many of the heavy metals give slightly soluble products (some white, some coloured) with the reagent, most of which are soluble in the organic solvents mentioned. The selectivity of the reagent may be improved by the use of masking agents, particularly EDTA. The reagent decomposes rapidly in solutions of low pH.

Procedure   Dissolve 0.0393 g of pure copper(II) sulphate pentahydrate in 1 L of water in a graduated flask. Pipette 10.0 mL of this solution (containing about 100 μg Cu) into a beaker, add 5.0 mL of 25% aqueous citric acid solution, render slightly alkaline with dilute ammonia solution and boil off the excess of ammonia; alternatively, adjust to pH 8.5 using a pH meter. Add 15.0 mL of 4% EDTA solution and cool to room temperature. Transfer to a separatory funnel, add 10 mL of 0.2% aqueous sodium diethyldithiocarbamate solution, and shake for 45 s. A yellow-brown colour develops in the solution. Pipette 20 mL of butyl acetate (ethanoate) into the funnel and shake for 30 s. The organic layer acquires a yellow colour. Cool, shake for 15 s and allow the phases to separate. Remove the lower aqueous layer; add 20 mL of 5% v/v sulphuric acid, shake for 15 s, cool and separate the organic phase. Determine the absorbance at 560 nm in 1.0 cm absorption cells against a blank. All the copper is removed in one extraction. Repeat the experiment in the presence of 1 mg of iron(III); no interference can be detected.

6.3.7   Determination of copper as the neocuproin complex

Discussion   Neocuproin (2,9-dimethyl-1,10-phenanthroline) can, under certain conditions, behave as an almost specific reagent for copper(I). The complex is soluble in chloroform and absorbs at 457 nm. It may be applied to the determination of copper in cast iron, alloy steels, lead/tin solder, and various metals.

Procedure   To 10.0 mL of the solution containing up to 200 μg of copper in a separatory funnel, add 5.0 mL of 10% hydroxylammonium chloride solution to reduce Cu(II) to Cu(I), and 10 mL of a 30% sodium citrate solution to complex any other metals which may be present. Add ammonia solution until the pH is about 4, followed by 10 mL of a 0.1% solution of neocuproin in absolute ethanol. Shake for about 30 s with 10 mL of chloroform and allow the layers to separate. Repeat the extraction with a further 5 mL of chloroform. Measure the absorbance at 457 nm against a blank on the reagents which have been treated similarly to the sample.

6.3.8   Determination of iron as the 8-hydroxyquinolate

Discussion   Iron(III) (50–200 μg) can be extracted from aqueous solution with a 1% solution of 8-hydroxyquinoline in chloroform by double extraction when the pH of the aqueous solution is between 2 and 10. At a pH of 2–2.5 nickel, cobalt, cerium(III) and aluminium do not interfere. Iron(III) oxinate is dark-coloured in chloroform and absorbs at 470 nm.

Procedure   Weigh out 0.0226 g of hydrated ammonium iron(III) sulphate, and dissolve it in 1 L of water in a graduated flask; 50 mL of this solution contains 100 μg of iron. Place 50.0 mL of the solution in a 100 mL separatory funnel, add 10 mL of a 1% oxine (analytical grade) solution in chloroform and shake for 1 min. Separate the chloroform layer. Transfer a portion of the chloroform layer to a 1.0 cm absorption cell. Determine the absorbance at 470 nm in a spectrophotometer, using the solvent as a blank or reference. Repeat the extraction with a further 10 mL of 1% oxine solution in chloroform, and measure the absorbance to confirm that all the iron was extracted. Repeat the experiment using 50.0 mL of the iron(III) solution in the presence of 100 μg of aluminium ion and 100 μg of nickel ion at pH 2.0. Measure the absorbance. Confirm that an effective separation has been achieved.

Typical results   Absorbance after first extraction 0.605; after second extraction 0.004; in presence of 100 μg Al and 100 μg Ni the absorbance obtained is 0.602.

6.3.9   Determination of lead by the dithizone method

Warning   **This experiment is not recommended for elementary students or those having little experience of analytical work.**

Discussion   Diphenylthiocarbazone (dithizone) behaves in solution as a tautomeric mixture of [6.B] and [6.C]:

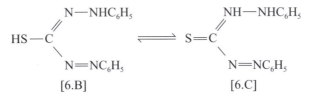

[6.B]   [6.C]

It functions as a monoprotic acid ($pK_a = 4.7$) up to a pH of about 12; the acid proton is that of the thiol group in [6.B].

Primary metal dithizonates are formed according to the reaction

$$M^{n-} + nH_2Dz \rightleftharpoons M(HDz)_n + nH^-$$

Some metals, notably copper, silver, gold, mercury, bismuth and palladium, form a second complex (which we may term secondary dithizonates) at a higher pH range or with a deficiency of the reagent:

$$2M(HDz)_n \rightleftharpoons M_2Dz_n + nH_2Dz$$

In general, the primary dithizonates are of greater analytical utility than the secondary dithizonates. Dithizone is a violet-black solid which is insoluble in water, soluble in dilute ammonia solution, and also soluble in chloroform and in carbon tetrachloride to yield green solutions. It is an excellent reagent for the determination of small (microgram) quantities of many metals, and can be made selective for certain metals by resorting to one or both of the following devices:

(a)   Adjusting the pH of the solution to be extracted. Thus from acid solution (0.1–$0.5\,M$) silver, mercury, copper and palladium can be separated from other metals; bismuth can be extracted from a weakly acidic medium; lead and zinc can be extracted from a neutral or faintly alkaline medium; cadmium can be extracted from a strongly basic solution containing citrate or tartrate.

(b)   Adding a complex-forming agent or masking agent, e.g. cyanide, thiocyanate, thiosulphate, or EDTA.

Note that dithizone is an extremely sensitive reagent and is applicable to quantities of metals of the order of micrograms. Only the purest dithizone may be used, since the reagent tends to oxidise to diphenylthiocarbadiazone, $S{=}C(N{=}NC_6H_5)_2$; this does not react with metals, is insoluble in ammonia solution, and dissolves in organic solvents to give yellow or brown solutions. Reagents for use in dithizone methods of analysis must be of the highest purity. Deionised water and redistilled acids are recommended; ammonia solution should be prepared by passing ammonia gas into water. Weakly basic and neutral solutions can frequently be freed from reacting heavy metals by extracting them with a fairly concentrated solution of dithizone in chloroform until a green extract is obtained. Pyrex vessels should be rinsed with dilute acid before use. Blanks must always be run. Only one example of the use of dithizone in solvent extraction will be given in order to illustrate the general technique involved.

Procedure   Dissolve 0.0079 g of pure lead nitrate in 1 L of water in a graduated flask. To 10.0 mL of this solution (containing about 50 µg of lead), contained in a 250 mL separatory funnel, add 75 mL of ammonia–cyanide–sulphite mixture (note 1), adjust the pH of the solution to 9.5 (pH meter) by the cautious addition of hydrochloric acid (care), then add 7.5 mL of a 0.005% solution of dithizone in chloroform (note 2) followed by 17.5 mL of chloroform. Shake for 1 min and allow the phases to separate. Determine the absorbance at 510 nm against a blank solution in a 1.0 cm absorption cell. A further extraction of the same solution gives zero absorption, indicating the complete extraction of the lead. Almost the same absorbance is obtained in the presence of 100 µg of copper ion and 100 µg of zinc ion.

Notes

1. This solution is prepared by diluting 35 mL of concentrated ammonia solution (density 0.88 g cm^{-3}) and 3.0 mL of 10% potassium cyanide solution (caution) to 100 mL and then dissolving 0.15 g of sodium sulphite in the solution.
2. One millilitre of this solution is equivalent to about 20 µg of lead. The solution should be freshly prepared using the analytical grade reagent, ideally taken from a new or recently opened reagent bottle.

6.4   Crystallisation and precipitation

Both techniques are covered in Chapter 11, as they form integral parts of normal laboratory manipulation, but it is worth mentioning them here as sometimes their value as preconcentration techniques can be overlooked. Crystallisation has long been one of the prime methods used by organic chemists to clean up impure compounds during reaction or synthetic sequences. In the majority of cases, if conditions are chosen carefully, then crystalline material of high purity can be isolated from solutions containing many other components.

Although not of great use to the analyst when dealing with low concentrations of analyte in dilute solution, crystallisation or the slightly cruder process of salting out material by addition of an organic solvent to an aqueous phase can sometimes be a useful first step in obtaining analyte of sufficient purity for analysis. Equally, precipitation of predominantly inorganic species by addition of simple selective reagents such as H_2S or modification of pH enables the analyst to identify and even quantify very small amounts of material in a variety of matrices.

Sadly the days of 'schematic qualitative and quantitative analysis' have passed and many analysts have a much less rigorous training in this area than one would like. However, even in the modern laboratory, where limits of detection of metals and anions using sophisticated instrumentation can be less than ng kg^{-1} in many cases, it is often worth looking at these older schematic methods, based upon simple but well-understood chemistry, to determine whether complications may arise in any of the manipulative steps now used prior to analysis.

6.5   Ion exchange separations

6.5.1   Introduction

Since ion exchange processes are often used in bulk separations as well as in analytical chromatography, a reasonably full description of the process is given here. The term 'ion exchange' is generally understood to mean the exchange of ions of like sign between a solution and a solid, highly insoluble body in contact with it. The solid (ion exchanger) must contain ions of its own, and for the exchange to proceed sufficiently rapidly and extensively to be of practical value, the solid must have an open, permeable molecular structure so that ions and solvent molecules can move freely in and out.

Many substances, both natural (e.g. certain clay minerals) and artificial, have ion-exchanging properties, but for analytical work synthetic organic ion exchangers are chiefly of interest, although some inorganic materials, e.g. zirconyl phosphate and ammonium 12-molybdophosphate, also possess useful ion exchange capabilities and have specialised applications. All ion exchangers of value in analysis have several properties in common: they are almost insoluble in water and in organic solvents, and they contain active or counterions that will exchange reversibly with other ions in a surrounding solution without any appreciable physical change occurring in the material.

The ion exchanger is complex and is in fact polymeric. The polymer carries an electric charge that is exactly neutralised by the charges on the counterions. These active ions are cations in a cation exchanger and anions in an anion exchanger. Thus a **cation exchanger** consists of a polymeric anion and active cations, whereas an **anion exchanger** is a polymeric cation with active anions. A widely used cation exchange resin is obtained by the copolymerisation of styrene [6.D] and a small proportion of divinylbenzene [6.E], followed by sulphonation; it may be represented as [6.F]:

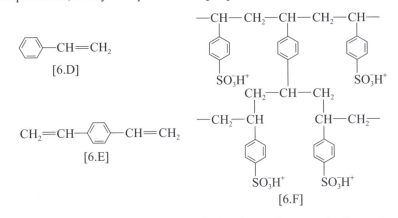

The formula enables us to visualise a typical cation exchange resin. It consists of a polymeric skeleton, held together by linkings crossing from one polymer chain to the next; the ion exchange groups are carried on this skeleton. The physical properties are largely determined by the degree of cross-linking. This cannot be determined directly in the resin itself; it is often specified as the mol% of the cross-linking agent in the mixture polymerised. Thus 'polystyrene sulphonic acid, 5% DVB' refers to a resin containing nominally 1 mole in 20 of divinylbenzene; the true degree of cross-linking probably differs somewhat from the nominal value, but the nominal value remains useful for grading resins. Highly cross-linked resins are generally more brittle, harder and more impervious than the lightly cross-linked materials; the preference of a resin for one ion over another is influenced by the degree of cross-linking. The solid granules of resin swell when placed in water to give a gel structure, but the swelling is limited by the cross-linking.

In our example the divinylbenzene units 'weld' the polystyrene chains together and prevent it from swelling indefinitely and dispersing into solution. The resulting structure is a vast sponge-like network with negatively charged sulphonate ions attached firmly to the framework. These fixed negative charges are balanced by an equivalent number of cations: hydrogen ions in the hydrogen form of the resin and sodium ions in the sodium form of the resin, etc. These ions move freely within the water-filled pores and are sometimes called mobile ions; they are the ions which exchange with other ions.

When a cation exchanger containing mobile ions C^+ is brought into contact with a solution containing cations B^+, the B^+ diffuse into the resin structure and cations C^+ diffuse out until equilibrium is attained. The solid and the solution then contain both cations C^+ and B^+ in numbers that depend on the position of equilibrium. The same mechanism operates for the exchange of anions in an anion exchanger.

Anion exchangers are also cross-linked polymers with high molecular weight. Their basic character is due to the presence of amino, substituted amino, or quaternary ammonium groups. The polymers containing quaternary ammonium groups are strong bases; those with amino or substituted amino groups possess weak basic properties. A widely used anion exchange resin is prepared by copolymerisation of styrene and a little divinylbenzene, followed by chloromethylation (introduction of the –CH$_2$Cl grouping, say, in the free *para* position) and interaction with a base such as trimethylamine. Structure [6.G] shows a hypothetical formulation of such a polystyrene anion exchange resin.

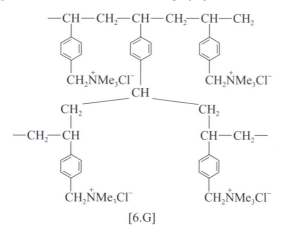

[6.G]

A useful resin should satisfy four fundamental requirements:

1. It must be sufficiently cross-linked to have only a negligible solubility.
2. It must be sufficiently hydrophilic to permit diffusion of ions through the structure at a finite and usable rate.
3. It must contain a sufficient number of accessible ionic exchange groups and it must be chemically stable.
4. When swollen it must be denser than water.

A relatively new polymerisation technique yields a cross-linked ion exchange resin having a truly macroporous structure quite different from that of the conventional homogeneous

Table 6.4   *Comparable ion exchange materials*

Type	Duolite	Rohm & Haas	Dow Chemical	Bio-Rad Labs	pH range
Strong acid cation exchangers	Duolite C20 Duolite C255 Duolite C26S[a]	Amberlite IR-120 Amberlite IRC-200[a] AMB 200	Dowex 50WX	AG50W AGMP-50[a]	0–14
Weak acid cation exchangers	Duolite C433 Duolite C464[a]	Amberlite IRC-84 Amberlite IRC-50[a]		Bio-Rex 70[a]	5–14
Strong base anion exchangers	Duolite A113 Duolite A116 Duolite A162[a]	Amberlite IRA-400 Amberlite IRA-938[a] Amberlite IRA-900[a]	Dowex 1X Dowex 2	AGI AGMP-1[a]	0–14
Weak base anion exchangers	Duolite A303 Duolite A378[a]	Amberlite IRA-67 Amberlite IRA-68 Amberlite 93[a]	Dowex 3	AG4-X4A	0–9
Chelating resins	Duolite ES466[a]	Amberlite 718[a]		AG501-X8 Chelex 100	depends on ions involved
Mixed bed	Duolite MB6113	MB1		Biorex MSZ 501	

[a] Macroporous/macroreticular resins.

gels already described. An average pore diameter of 130 nm is not unusual, and the introduction of these **macroreticular** resins (e.g. the Amberlyst resins developed by Rohm & Haas) has extended the scope of the ion exchange technique. Thus, the large pore size allows the more complete removal of high-molecular-weight ions than is the case with the gel-type resins. Macroporous resins are also well suited for non-aqueous ion exchange applications. A review of the properties of these macroporous resins has recently appeared.[8]

New types of ion exchange resins have also been developed to meet the specific needs of high-performance liquid chromatography (HPLC). They include pellicular resins and microparticle packings (e.g. the Aminex-type resins produced by Bio-Rad). Some of the commercially available ion exchange resins are listed in Table 6.4. These resins, produced by different manufacturers, are often interchangeable and similar types will generally behave in a similar manner. Comprehensive lists of ion exchange resins and their properties are available from the major suppliers.

Silica-based ion exchange packings are used for HPLC. Their preparation is similar to that for bonded-phase packings, then the ion exchange groups are subsequently introduced into the organic backbone. The small particle size (10–15 μm diameter) and narrow distribution provide high column efficiencies, and typical applications include high-resolution analysis of amino acids, peptides, proteins, nucleotides, etc. These silica-based packings are preferred when column efficiency is the main criterion, but resin microparticle packings should be used when capacity is the main requirement.

Action of ion exchange resins

Cation exchange resins contain free cations which can be exchanged for cations in solution (soln). These will be represented by $(Res.A^-)B^+$, where Res. is the basic polymer of the resin, A^- is the anion attached to the polymeric framework, and B^+ is the active or mobile cation. Thus a sulphonated polystyrene resin in the hydrogen form would be written as $(Res.SO_3^-)H^+$. A similar nomenclature will be employed for anion exchange resins, e.g. $(Res.NMe_3^+)Cl^-$. Here is the equilibrium for a cation exchange resion:

$$(Res.A^-)B^+ + C^+ \text{ (soln)} \rightleftharpoons (Res.A^-)C^+ + B^+ \text{ (soln)}$$

If the experimental conditions are such that the equilibrium is completely displaced from left to right, the cation C^+ is completely fixed on the cation exchanger. If the solution contains several cations (C^+, D^+, and E^+) the exchanger may show different affinities for them, thus making separations possible. A typical example is the displacement of sodium ions in a sulphonate resin by calcium ions:

$$2(Res.SO_3^-)Na^+ + Ca^{2+} \text{ (soln)} \rightleftharpoons (Res.SO_3^-)_2Ca^{2+} + 2Na^+ \text{ (soln)}$$

The reaction is reversible. By passing a solution containing sodium ions through the product, the calcium ions may be removed from the resin and the original sodium form regenerated. Similarly, by passing a solution of a neutral salt through the hydrogen form of a sulphonic resin, an equivalent quantity of the corresponding acid is produced by the following typical reaction:

$$(Res.SO_3^-)H^+ + Na^+Cl^- \text{ (soln)} \rightleftharpoons (Res.SO_3^-)Na^+ + H^+Cl^- \text{ (soln)}$$

For the strongly acidic cation exchange resins, such as the cross-linked polystyrene sulphonic acid resins, the exchange capacity is virtually independent of the pH of the solution. For weak acid cation exchangers, such as those containing the carboxylate group, ionisation occurs to an appreciable extent only in alkaline solution, i.e. in their salt form; consequently, the carboxylic resins have very little action in solutions below pH 7. These carboxylic exchangers in the hydrogen form will absorb strong bases from solution:

$$(Res.COO^-)H^+ + Na^+OH^- \text{ (soln)} \rightleftharpoons (Res.COO^-)Na^+ + H_2O$$

but will have little action upon, say, sodium chloride; hydrolysis of the salt form of the resin occurs so that the base may not be completely absorbed even if an excess of resin is present.

Strongly basic anion exchange resins, e.g. a cross-linked polystyrene containing quaternary ammonium groups, are largely ionised in both the hydroxide and the salt forms. These resins are similar to the sulphonate cation exchange resins in their activity, and their action is largely independent of pH. Weakly basic ion exchange resins contain little of the hydroxide form in basic solution. The equilibrium

$$(Res.NMe_2) + H_2O \rightleftharpoons (Res.NHMe_2)^+OH^-$$

is mainly to the left and the resin is largely in the amine form. This may also be expressed by stating that in alkaline solutions the free base $Res.NHMe_2 \cdot OH$ is very little ionised. In acidic solutions, however, they behave like the strongly basic ion exchange resins, yielding the highly ionised salt form:

$$(Res.NMe_2) + H^+Cl^- \rightleftharpoons (Res.NHMe_2^+)Cl^-$$

They can be used in acid solution for the exchange of anions, e.g.

$$(Res.NHMe_2^+)Cl^- + NO_3^- \ (soln) \rightleftharpoons (Res.NHMe_2^+)NO_3^- + Cl^- \ (soln)$$

Basic resins in the salt form are readily regenerated with alkali.

Ion exchange equilibria

The ion exchange process, involving the replacement of the exchangeable ions A_r of the resin by ions of like charge B_s from a solution, may be written

$$A_r + B_s \rightleftharpoons B_r + A_s$$

The process is reversible, and for ions of like charge the selectivity coefficient K is defined by

$$K_A^B = \frac{[B]_r[A]_s}{[A]_r[B]_s}$$

where the terms in brackets represent the concentrations of ions A and B in either the resin or solution phase. The values of selectivity coefficients are obtained experimentally and provide a guide to the relative affinities of ions for a particular resin. Thus if $K_A^B > 1$ the resin shows a preference for ion B, whereas if $K_A^B < 1$ its preference is for ion A; this applies to both anion and cation exchanges.

Table 6.5 summarises, for singly charged ions, the relative selectivities of strongly acid and strongly basic polystyrene resins with about 8% DVB. Note that the relative selectivities for certain ions may vary with a change in the extent of cross-linking of the resin. For example, with a 10% DVB resin the relative selectivity values for Li^+ and Cs^+ ions are 1.00 and 4.15, respectively.

The extent to which one ion is absorbed in preference to another is of fundamental importance; it will determine the readiness with which two or more substances, which form ions of like charge, can be separated by ion exchange and also the ease with which the ions can subsequently be removed from the resin:

(a) At low aqueous concentrations and at ordinary temperatures the extent of exchange increases with increasing charge of the exchanging ion:

$$Na^+ < Ca^{2+} < Al^{3+} < Th^{4+}$$

Table 6.5 *Relative selectivities of polystyrene – 8% DVB resins for singly charged ions*

Cation	Relative selectivity	Anion	Relative selectivity
Li^+	1.00	F^-	0.09
H^+	1.26	OH^-	0.09
Na^+	1.88	Cl^-	1.00
NH_4^+	2.22	Br^-	2.80
K^+	2.63	NO_3^-	3.80
Rb^+	2.89	I^-	8.70
Cs^+	2.91	ClO_4^-	10.0

(b)   Under similar conditions and constant charge, for singly charged ions the extent of exchange increases with decrease in size of the hydrated cation:

$$Li^+ < H^+ < Na^+ < NH_4^+ < K^+ < Rb^+ < Cs^+$$

while for doubly charged ions the ionic size is an important factor but the incomplete dissociation of their salts also plays a part:

$$Cd^{2+} < Be^{2+} < Mn^{2+} < Mg^{2+} = Zn^{2+} < Cu^{2+}$$
$$= Ni^{2+} < Co^{2+} < Ca^{2+} < Sr^{2+} < Pb^{2+} < Ba^{2+}$$

(c)   With strongly basic anion exchange resins the extent of exchange for singly charged anions varies with the size of the hydrated ion in a similar manner to their cation counterparts. In dilute solution, multicharged anions are generally absorbed preferentially.

(d)   When a cation in solution is being exchanged for an ion of different charge, the relative affinity of the ion of higher charge increases in direct proportion to the dilution.

Thus, to exchange an ion of higher charge on the exchanger for one of lower charge in solution, exchange will be favoured by increasing the concentration; but if the ion of lower charge is in the exchanger and the ion of higher charge is in solution, exchange will be favoured by high dilutions.

   The absorption of ions will depend upon the nature of the functional groups in the resin. Here the term 'absorption' is used whenever ions or other solutes are taken up by an ion exchanger. It does not imply any specific types of forces responsible for this uptake. It will also depend upon the degree of cross-linking; as the degree of cross-linking is increased, resins become more selective towards ions of different sizes (the volume of the ion is assumed to include the water of hydration); the ion with the smaller hydrated volume will usually be absorbed preferentially.

Exchange of organic ions

Although similar principles apply to the exchange of organic ions, the following features must also be taken into consideration:

1.   The sizes of organic ions differ to a much greater extent than the sizes of inorganic ions and may exceed a hundred or even a thousand times the average size of inorganic ions.
2.   Many organic compounds are only slightly soluble in water, so non-aqueous ion exchange has an important role in operations with organic substances.

Clearly the application of macroreticular (macroporous) ion exchange resins will often be advantageous in the separation of organic species.

Ion exchange capacity

The total ion exchange capacity of a resin depends upon the total number of ion-active groups per unit weight of material, and the greater the number of ions, the greater the capacity. The **total ion exchange capacity** is usually expressed as millimoles per gram of exchanger. The capacities of the weakly acidic and weakly basic ion exchangers are functions of pH; the weakly acid exchangers reach moderately constant values at pH > ~9 and the weakly basic exchangers at pH < ~5. Here are the values of the total exchange capacities, expressed as $mmol\,g^{-1}$ of dry resin, for a few typical resins: Amberlite IR-120

231

(Na$^+$ form) = 4.4, Amberlite IRC-50 (H$^+$ form) = 10.0, Duolite A113 (Cl$^-$ form) = 4.0, Amberlite IRA-67 = 5.6. The total exchange capacity expressed as mmol mL^{-1} of the wet resin is about one-third to one-half the value for the dry resin. These figures are useful in estimating very approximately the quantity of resin required in a determination; an adequate excess must be employed, since the breakthrough capacity is often much less than the total capacity of the resin. In most cases a 100% excess is satisfactory.

The exchange capacity of a cation exchange resin may be measured in the laboratory by determining the number of millimoles of sodium ion absorbed by 1 g of the dry resin in the hydrogen form. Similarly, the exchange capacity of a strongly basic anion exchange resin is evaluated by measuring the amount of chloride ion taken up by 1 g of dry resin in the hydroxide form. Note that large ions may not be absorbed by a medium cross-linked resin, so its effective capacity is seriously reduced. A resin with larger pores should be used for such ions.

6.5.2   Changing the ionic form

Strongly acidic cation exchangers (polystyrene sulphonic acid resins), such as Duolite C255 and Amberlite IR-120, are usually marketed in the sodium form, and if it is necessary to convert them into the hydrogen form (also available), they need to be treated with 2 M or 10% hydrochloric acid. This is a simple task and the suppliers normally give full experimental details.

Weakly acidic cation exchangers (e.g. polymethylacrylic acid resins), such as Duolite C433 and Amberlite IRC-50, are usually supplied in the hydrogen form. They are readily changed into the sodium form if required by treatment with 1 M sodium hydroxide; an increase in volume of 80–100% may be expected. The swelling is reversible and does not appear to cause any damage to the bead structure. Below a pH of about 3.5 the hydrogen form exists almost entirely in the little ionised carboxylic acid form. Exchange with metal ions will occur in solution only when they are associated in solution with anions of weak acids, i.e. pH > ~4. The exhausted resin is more easily regenerated than the strongly acidic exchangers; about 1.5 bed volumes of 1 M hydrochloric acid will usually suffice.

Strongly basic anion exchangers (polystyrene quaternary ammonium resins), such as Duolite A113 and Amberlite IRA-400, are usually supplied in the chloride form. For conversion into the hydroxide form, treatment with 1 M sodium hydroxide is employed; the volume depends upon the extent of conversion desired, but two bed volumes are satisfactory for most purposes. To rinse the alkali from the resin, use deionised water free from carbon dioxide; this avoids converting the resin into the carbonate form; about 2 L of deionised water will suffice for 1 g of resin. An increase in volume of about 20% occurs in the conversion of the resin from the chloride to the hydroxide form. Weakly basic anion exchangers (polystyrene tertiaryamine resins), such as Duolite A303 and Amberlite IRA-67, are generally supplied in the free base (hydroxide) form. The salt form may be prepared by treating the resin with the appropriate acid (e.g. 1 M hydrochloric acid) and rinsing with water to remove the excess.

6.5.3   Experimental techniques

The simplest apparatus for ion exchange work in analysis consists of a burette provided with a glass wool plug or sintered glass disc (porosity 0 or 1) at the lower end. Another simple column is shown in Figure 6.4(a). Here the ion exchange resin is supported on a glass wool plug or sintered glass disc. A glass wool pad may be placed at the top of the

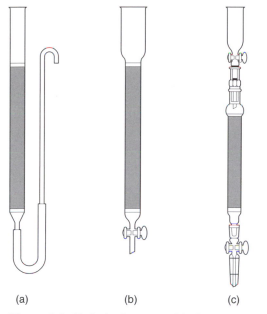

(a)   (b)   (c)

Figure 6.4   Typical columns used for low-pressure ion exchange chromatography

bed of resin and the eluting agent is added from a tap funnel supported above the column. The siphon overflow tube, attached to the column by a short length of PVC tubing, ensures the level of the liquid does not fall below the top of the resin bed, so the bed is always wholly immersed in the liquid. The ratio of the height of the column to the diameter is not very critical but it is usually 10 : 1 or 20 : 1. Another form of column is depicted in Figure 6.4(b) (not to scale); a convenient size is 30 cm long with lower portion about 10 mm internal diameter and upper portion about 25 mm internal diameter. A commercially available column, fitted with groundglass joints is illustrated in Figure 6.4(c).

The ion exchange resin should be of small particle size, so as to provide a large surface of contact, but it should not be so fine as to produce a very slow flow rate. For much laboratory work 50–100 mesh or 100–200 mesh materials are satisfactory. In all cases the diameter of the resin bed should be less than one-tenth the diameter of the column. Resins of medium and high cross-linking rarely show any further changes in volume, and only if subjected to large changes of ionic strength will any appreciable volume change occur. Resins of low cross-linking may change in volume appreciably, even with small variations of ionic strength. This may produce channelling and possible blocking of the column, hence these resins have limitations on their use. To obtain satisfactory separations, it is essential that the solutions should pass through the column in a uniform manner.

The resin particles should be packed uniformly in the column; the resin bed should be free from air bubbles so that there is no channelling. To prepare a well-packed column, a supply of exchange resin of narrow size range is desirable. An ion exchange resin swells if the dry solid is immersed in water; no attempt should therefore be made to set up a column by pouring the dry resin into a tube and then adding water, since the expansion will probably shatter the tube. The resin should be stirred with water in an open beaker for several minutes, any fine particles removed by decantation, and the resin slurry transferred in portions to the tube previously filled with water. The tube may be tapped gently to prevent the formation of air bubbles. To ensure the removal of entrained air bubbles, or any remaining fine particles, and to ensure an even distribution of resin granules, it is advisable to backwash the resin column before use, i.e. a stream of good quality distilled water or of deionised water is run up through the bed from the bottom at a sufficient flow rate to loosen and suspend the exchanger granules. The enlarged upper portion of the exchange tube shown in Figure 6.4(b) or (c) will hold the resin suspension during washing.

If a tube of uniform bore is used, the volume of resin employed must be suitably adjusted or else a tube attached by a rubber bung to the top of the column; the tube dips into an open filter flask, the side arm of which acts as the overflow and is connected by rubber tubing to waste. When the wash water is clear, the flow of water is stopped and the resin is allowed to settle in the tube. The excess of water is drained off; the water level must never fall below the surface of the resin else channelling will occur, leading to incomplete contact between the resin and solutions used in subsequent operations. The apparatus with a side-arm outlet (Figure 6.4(a)) has an advantage in this respect because its outlet is above the surface of the resin, so the resin will not run dry even if left unattended.

Ion exchange resins (standard grades) as received from the manufacturers may contain unwanted ionic impurities and sometimes traces of water-soluble intermediates or incompletely polymerised material; these must be washed out before use. This is best done by passing $2M$ hydrochloric acid and $2M$ sodium hydroxide alternately through the column, with distilled water rinsings in between, and then washing with water until the effluent is neutral and salt-free. Analytical grade and/or chromatographic grade ion exchange resins that have undergone this preliminary washing are available commercially. For analytical work, use the analytical grade or chromatographic grade of resin with a particle size of 100–200 mesh. For student work, use the standard grade of resin with particle size 50–100 or 15–50 mesh; it is less expensive and generally proves satisfactory. But the standard grade of resin must be conditioned before use.

Cation exchange resins must be soaked in a beaker in about twice the volume of $2M$ hydrochloric acid for 30–60 min with occasional stirring; the fine particles are removed by decantation or by backwashing a column with distilled or deionised water until the supernatant liquid is clear. Anion exchange resins may be washed with water in a beaker until the colour of the decanted wash liquid reaches a minimum intensity; they may then be transferred to a wide glass column and cycled between $1M$ hydrochloric acid and $1M$ alkali. Sodium hydroxide is used for strongly basic resins, and ammonia (preferably) or sodium carbonate for weakly basic resins. For all resins the final treatment must be with a solution leading to the resin in the desired ionic form.

A 50 mL or 100 mL burette, with a Pyrex glass wool plug or sintered-glass disc at the lower end, can generally be used for the determinations described below; alternatively, the column with side arm (Figure 6.4(a)) is equally convenient for student use. Reference will be made to the Amberlite resins; the equivalent Duolite resin or another equivalent resin (Table 6.4) may also be used.

6.5.4 Separation of zinc and magnesium on an anion exchanger

Theory Several metal ions (e.g. Fe, Al, Zn, Co, Mn ions) can be absorbed from hydrochloric acid solutions on anion exchange resins owing to the formation of negatively charged chloro complexes. Each metal is absorbed over a well-defined range of pH, and this property can be used as the basis of a method of separation. Zinc is absorbed from $2M$ acid, whereas magnesium (and aluminium) are not; thus a separation is effected by passing a mixture of zinc and magnesium through a column of anion exchange resin. The zinc is subsequently eluted with dilute nitric acid.

Procedure Prepare a column of the anion exchange resin using about 15 g of Amberlite IRA-400 in the chloride form. The column should be made up in $2 M$ hydrochloric acid. Prepare separate standard solutions of zinc ions (about $2.5 \, mg \, mL^{-1}$) and magnesium ions (about $1.5 \, mg \, mL^{-1}$) by dissolving accurately weighed quantities of zinc shot and magnesium in $2 M$ hydrochloric acid and diluting each to volume in a 250 mL graduated flask. Pipette 10.0 mL of the zinc ion solution and 10.0 mL of the magnesium ion solution into a small separatory funnel supported in the top of the ion exchange column, and mix the solutions. Allow the mixed solution to flow through the column at a rate of about $5 \, mL \, min^{-1}$. Wash the funnel and column with 50 mL of $2 M$ hydrochloric acid; do not permit the level of the liquid to fall below the top of the resin column. Collect all the effluent in a conical flask; this contains all the magnesium. Now change the receiver. Elute the zinc with 30 mL of water, followed by 80 mL of ~$0.25 M$ nitric acid.

Determine the magnesium and the zinc in the respective eluates by neutralisation with sodium hydroxide solution, followed by titration with standard EDTA solution using a buffer solution of pH = 10 and solochrome black indicator. The following weights were obtained in a typical experiment:

Zn taken = 25.62 mg   found = 25.60 mg

Mg taken = 14.95 mg   found = 14.89 mg

Magnesium may conveniently be determined by atomic absorption spectroscopy, if a smaller amount (~4 mg) is used for the separation. Collect the magnesium effluent in a 1 L graduated flask, dilute to the mark with deionised water and aspirate the solution into the flame of an atomic absorption spectrometer. Calibrate the instrument using standard magnesium solutions covering the range $2-8 \, \mu g \, g^{-1}$.

6.5.5 Separation of chloride and bromide on an anion exchanger

Theory The anion exchange resin, originally in the chloride form, is converted into the nitrate form by washing with sodium nitrate solution. A concentrated solution of the chloride and bromide mixture is introduced at the top of the column. The halide ions exchange rapidly with the nitrate ions in the resin, forming a band at the top of the column. Separation is possible because the sodium nitrate solution elutes the chloride ions from the band more rapidly than the bromide ions. The progress of halide elution is followed by titrating fractions of the effluents with standard silver nitrate solution.

Procedure Prepare an anion exchange column using about 40 g of Amberlite IRA-400 (chloride form). The ion exchange tube may be 16 cm long and about 12 mm internal diameter. Wash the column with $0.6 M$ sodium nitrate until the effluent contains no chloride ion (silver nitrate test) and then wash with 50 mL of $0.3 M$ sodium nitrate. Weigh out accurately about 0.10 g of analytical grade sodium chloride and about 0.20 g of potassium bromide, dissolve the mixture in about 2.0 mL of water and transfer quantitatively to the top of the column with the aid of $0.3 M$ sodium nitrate. Pass $0.3 M$ sodium nitrate through the column at a flow rate of about $1 \, mL \, min^{-1}$ and collect the effluent in 10 mL fractions. Transfer each fraction in turn to a conical flask, dilute with an equal volume of water, add 2 drops of $0.2 M$ potassium chromate solution and titrate with standard $0.02 M$ silver nitrate.

Before commencing the elution, titrate 10.00 mL of the 0.3 *M* sodium nitrate with the standard silver nitrate solution, and retain the product of this blank titration for comparing with the colour in the titrations of the eluates. When the titre of the eluate falls almost to zero (i.e. nearly equal to the blank titration), ~150 mL of effluent, elute the column with 0.6 *M* sodium nitrate. Titrate as before until no more bromide is detected (titre almost zero). A new blank titration must be made with 10.0 mL of the 0.6 *M* sodium nitrate.

Plot a graph of the total effluent collected against the concentration of halide in each fraction (mmol L^{-1}). The sum of the titres using 0.3 *M* sodium nitrate eluant (less blank for each titration) corresponds to the chloride, and the parallel figure with 0.6 *M* sodium nitrate corresponds to the bromide recovery. A typical experiment gave the following results:

Weight of sodium chloride used	= 0.1012 g	equivalent to 61.37 mg Cl^-
Weight of potassium bromide used	= 0.1934 g	equivalent to 129.87 mg Br^-
Concentration of silver nitrate solution	= 0.019 36 *M*	
Cl^- total titres (less blanks)	= 89.54 mL	equivalent to 61.47 mg
Br^- total titres (less blanks)	= 83.65 mL	equivalent to 129.4 mg

6.5.6   Determination of the total cation concentration in water

Theory   This is a rapid procedure for determing the total cations present in water, particularly water used for industrial ion exchange plants, but it may be used for all samples of water, including tap water. When water containing dissolved ionised solids is passed through a cation exchanger in the hydrogen form, all cations are removed and replaced by hydrogen ions. Any alkalinity present in the water is destroyed, and the neutral salts present in solution are converted into the corresponding mineral acids. The effluent is titrated with 0.02 *M* sodium hydroxide using screened methyl orange as indicator.

Procedure   Prepare a 25–30 cm column of Amberlite IR-120 in a 14–16 mm chromatographic tube. Pass 250 mL of 2 *M* hydrochloric acid through the tube over about 30 min; rinse the column with distilled water until the effluent is just alkaline to screened methyl orange or until a 10 mL portion of the effluent does not require more than one drop of 0.02 *M* sodium hydroxide to give an alkaline reaction to bromothymol blue indicator. The resin is now ready for use; the level of the water should never be permitted to drop below the upper surface of the resin. Pass 50.0 mL of the sample of water under test through the column at a rate of 3–4 mL min^{-1} and discard the effluent. Now pass two 100.0 mL portions through the column at the same rate, collect the effluents separately and titrate each with standard 0.02 *M* sodium hydroxide using screened methyl orange as indicator. After the determination has been completed, pass 100–150 mL of distilled water through the column.

From the results of the titration calculate the number of millimoles of calcium present in the water. It may be expressed, if desired, as the equivalent mineral acidity (EMA) in terms of mg $CaCO_3$ per litre of water, (i.e. μg g^{-1} of $CaCO_3$). In general, if the titre is *A* mL of sodium hydroxide of molarity *B* for an aliquot volume of *V* mL, the EMA is given by

$$\left(\frac{AB}{V}\right) \times 50 \times 1000$$

Commercial samples of water are frequently alkaline due to the presence of hydrogen carbonates, carbonates or hydroxides. The alkalinity is determined by titrating a 100.0 mL sample with $0.02\,M$ hydrochloric acid using screened methyl orange as indicator (or to pH 3.8). To obtain the total cation content in terms of $CaCO_3$, the total methyl orange alkalinity is added to the EMA.

6.6   Dialysis and lyophilisation

Until recently these two techniques were rarely used by chemists, but were left to the biologists or life scientists to use. However, the traditional boundaries of these scientific disciplines are now much less well defined, with many problems originating in the life sciences but requiring analytical solutions. Lyophilisation is the process of removing water from a frozen sample by application of a vacuum. It is often called freeze drying, though this is slightly less precise. The technique can be extremely useful for removing water from inorganic and organic samples, provided the analyte is not appreciably volatile under the conditions used. Semi-automated machines have been developed, mainly for biological applications; they can handle volumes as high as $1000\,mL$ per sample.

Because water is removed at ambient temperature, fewer changes are likely to occur in the sample than in normal distillation systems, and the vacuum in which the sample is placed means there is minimal contamination. The matrix remaining at the end of the process, often only a small amount of dry powder, is generally quite stable with respect to time, and the technique can be a convenient way of preprocessing samples which cannot be analysed immediately. But when analysing simple organic materials of low molar mass, it is advisable to check that material has not been lost during the initial vacuum process.

Dialysis is another technique which is more familiar to the biologist than to the chemist. Normally a semipermeable membrane (cellulose acetate or similar with pore sizes of 1–5 nm) is placed between two aqueous phases containing different concentrations of metal ions; over a period of time, which may be as long as 48 hours, the **small** species (ions) transfer across the membrane to equalise the concentration. In biological applications one of the aqueous phases contains both ionic species and large biomolecules, and the other can be pure water. The equilibrium that is set up effectively reduces the ionic concentrations in the first phase without loss or change of the larger molecules, usually proteins. This is the process of desalting protein solutions. However, since many colloidal substances also have particle sizes of approximately 1–5 nm, the technique can be of great value for separation or preconcentration of certain difficult colloidal solutions which are inorganic in nature. It can also be used to clean up samples before some forms of instrumental analysis; for example, it is often used to deproteinate samples before HPLC since proteins may cause difficulties with the separation process and can irreversibly clog the column.

6.7   Instrumental separations

At the beginning of this chapter, separations were classified into two groups according to the amount of analyte: bulk separations and instrumental separations. Although most of the material in this section acts as a precursor for Chapters 7 to 9, which deal with instrumental separations, the processes described here are increasingly being used in simple apparatus to allow both separation and preconcentration before chromatography or one of the other instrumental techniques. All of the processes described below are concerned with the differential affinity of related analytes towards two different phases in which they are in mutual contact.

In all cases we are concerned with, one phase is kept stationary and the other is allowed to move (the mobile phase). The separation thus occurs at the boundary between a mobile phase and a stationary phase. Until fairly recently, for most chemical analysts, the preceding sentence would have been a reasonable definition of chromatography, but with the increasing use of electrophoretic methods, which also use separation of analytes at a phase boundary, the definition of 'chromatography' has become less clear. However, if we are to consider separations of related materials from each other at a phase boundary, in which one phase remains fixed and the other moves, whatever we decide to call it, we must firstly consider what is actually going on there. Once that is done, the skill of the analyst lies in using their chemical knowledge to achieve the required separation most effectively. This is seldom straightforward and there is often no right or wrong answer, with several techniques having the potential to separate a given mixture.

The choice is made on the basis of sensitivity, speed or even what is available; it doesn't have to give the greatest separation of components. And it's here that the analyst's experience is so very important. But there are a number of guidelines which, used with caution, will help to simplify the process. Small organic molecules that are volatile and uncharged, these tend to be separated by gas chromatography; whereas molecules with an electric charge, or which can be given an electric charge, would probably be separated using electrophoresis; this is especially true of large molecules. Equally, although chromatography and electrophoresis have tended to develop along different lines, producing their own separate vocabularies and often not even appearing in the same texts, they share the same simple mechanism of differential migration of analyte through a stationary phase. Thus, to a large extent, they share the same problems and the same limitations. This means that, independent of the choice of technique, it is only possible to obtain reliable, reproducible and efficient separations if the analyst controls the chemistry and physics of the mobile and stationary phases. Small changes in either the chemical composition, or even the temperature, may cause large changes in the observed separation.

6.7.1   Mechanisms of separation at a phase boundary

All five mechanisms in this section may be used in the liquid phase, whereas in the gas phase (gas chromatography) only adsorption and partition are possible. It has been estimated that only about 20% of known chemicals are sufficiently stable and sufficiently volatile for separation by gas chromatography. In theory the remaining 80% can be separated by an appropriate form of liquid chromatography, but in practice some may be more readily separated by electrophoresis.

Adsorption

Adsorption is possibly the most widely recognised, and often least chromatographically used of these mechanisms, although adsorption from the gas or liquid phase onto the surface of a solid has many scientific and commercial applications. For example, the adsorption of gases and vapours onto activated charcoal is used in many homes in the familiar fume extraction system fitted above the hob, and charcoal is used in many industrial processes to remove coloured impurities from solutions, e.g. in sugar production. Other widely used adsorbent materials are silica or silica gel, which is in fact a highly hydrated form of silicon dioxide ($SiO_2 \cdot xH_2O$) with a very large surface area and the related aluminium oxide ($Al_2O_3 \cdot xH_2O$) known as alumina.

The problem with adsorption as a mechanism of separation is that the sorption processes tend to be so strong that, once adsorbed, the analytes are difficult to desorb again; this makes chromatography difficult or unreliable in some circumstances, although it is used in both liquid and gas chromatography. However, particularly in gas chromatography, deactivation of silica surfaces such as injection ports or columns is essential to limit adsorption and allow an alternative mechanism to cause separation.

Partition

The theory of partition is discussed in Section 6.3.2, which considers liquid–liquid separations on a bulk scale, but the mechanism is also widely used in chromatographic separations, and as employed here, it is **relative solubility** of solute in two immiscible phases. In gas chromatography this involves the partitioning of the solute between the mobile gas phase and a stationary liquid phase, either supported on small particles in packed columns, or bonded onto the inner walls of a capillary column. Partition is the most widely used of the separating mechanisms in GC. And relative solubility between a mobile liquid phase and an immobile stationary phase is also widely employed in HPLC. The stationary phase is normally bonded to a support material to prevent problems of dissolution in the mobile phase. (Chapter 8). A simple rule of thumb operates for partition chromatography: like separates like. Non-polar materials dissolve in and are separated by non-polar phases, and polar materials require stationary phases that are even more polar.

Affinity

Affinity is the most recently employed and the most selective of the mechanisms. It uses highly specific interactions between the immobile phase and certain solute molecules. These interactions are usually enzyme or antibody–antigen reactions and may give very high selectivity for one type of molecule, such as proteins, in complex mixtures. It is known that antibodies are extremely specific in their reactions with antigens, and this can be used in affinity chromatography where an antibody, immobilised on a stationary phase (by covalently binding to it), can react with one protein (antigen) from a mixture containing several hundred similar proteins, binding it to the column. Once the column has been washed to remove all other proteins, then by changing the ionic strength of the eluent, the desired substance can be mobilised again and collected.[9]

The mechanism is sometimes described as the lock and key effect due to the high specificity between the immobile phase and the analyte. This mechanism, which is only used in the liquid phase, has already greatly simplified the separation and determination of a number of mixtures in the life sciences which, until recently, were considered too difficult to separate. The basic principles of immunoaffinity separations used in biotechnology have recently been reviewed.[10] Since the stationary phases contains a biologically active material, they tend to be rather expensive to produce and much less tolerant to adverse conditions than conventional stationary phases, so extreme care is needed in their use and storage if they are to provide a reasonable working life.

Ion exchange

Ion exchange has already been discussed as a bulk separation processes (Section 6.5), but it can also be used in liquid chromatography. Ion exchange chromatography can only occur in the liquid phase, where ions of one polarity in the mobile phase are temporarily bound to immobilised anions or cations on the stationary phase, or ion exchange resin, as it is called. These ions can then be selectively displaced by elution with a buffer of increasing ionic strength.

Uses of ion exchange as a bulk medium to separate two or more ions was discussed earlier in this chapter, but quite complex mixtures of ions or ionic species, especially some biologically significant substances such as amino acids, have been separated using conventional-sized columns of 10 mm diameter or more and allowing elution to occur by gravity. The advantage of this technique is that relatively large amounts of material (> 1 mg) can be separated and isolated before further experimentation. However, ion exchange techniques can be adapted to make them suitable for HPLC. Here the particle size of the synthetic resin is chosen to be small enough to produce efficient columns, or sometimes the ion exchange groups are attached to the surface of silica particles in the same way as bonded-phase partition columns are made; this produces a rigid particle with ion exchange properties. Normally a conductivity detector is also used, since the species separated using HPLC–ion exchange are not readily determined using a UV detector. Although the processes involved are intrinsically the same as in Section 6.5, this high-performance version of ion exchange is now more usually known as **ion chromatography**. Ion chromatography may be used for separation of either anions or cations, and for both small and large ions, but its major analytical use at present is for the separation and determination of the common anions (F^-, Cl^-, Br^-, I^-, NO_2^-, NO_3^-, SO_4^{2-}, PO_4^{3-}, etc.) found in aqueous solutions, particularly natural waters, or for cations and anions in food.[11]

Gel permeation

Gel permeation is a simple mechanism for separating species according to physical size; it is known by several other names, including gel filtration, molecular exclusion chromatography and molecular sieving, now outdated but perhaps the most descriptive of all. The stationary phase is a polymeric material known as a gel. Among the most widely used series of gels are the Sephadex materials, produced by inducing varying degrees of cross-linking to a dextran (polymeric carbohydrate) structure (Figure 6.5). By altering the amount of cross-linking, pores of different sizes are introduced into the structure and a 'sieve' is produced which can separate molecules according to their size.

The Sephadex range can be used for separating biological molecules such as peptides and proteins over a very wide range; Sephadex G-10 is used for molecules up to relative molecular mass (RMM) 700, and G-200 gives separations of molecules having RMM 750 000. Sephadex gels are fairly polar and will absorb water (swelling considerably in the process), but gels based on cross-linked polacrylamides (Bio-Gel) or polystyrene (Styragel) are completely inert to water and may be used in non-aqueous systems using common organic solvents such as toluene or methylene chloride. Section 8.2.4 considers the use of size exclusion gels in liquid chromatography for separation of large molecules, but size exclusion also finds application for the separation of small molecules, notably gases in GC, when the stationary phases are often called molecular sieves and are based upon inorganic matrices with holes just large enough to separate small gaseous molecules from each other.

In all cases of size exclusion chromatography or gel permeation, the large molecules are eluted first followed by progressively smaller molecules. This is because small molecules can penetrate more deeply into the spaces or holes in the gel, hence they are more strongly retained, whereas the large molecules cannot penetrate the gel structure. (This is where the sieve analogy fails, because it is the smaller particles which pass more readily through a sieve.) For the separation of macromolecules, a comparison has been made between gel permeation (larger molecules elute first) and electrophoresis (smaller molecules travel faster); this study provides more details.[12]

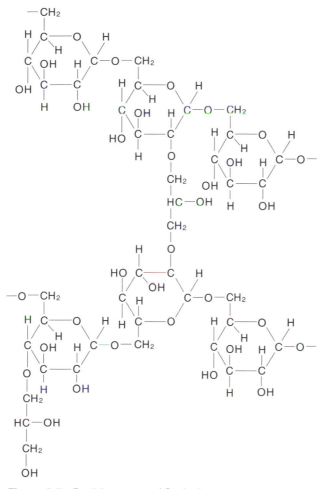

Figure 6.5   Partial structure of Sephadex

All five mechanisms of separation may be employed in the liquid phase, whereas in the gas phase, (gas chromatography), only adsorption and partition are possible.

6.8   Solid phase extraction

Solid phase extraction (SPE) is a relatively new technique. In Section 5.3.5 it was compared to other processes for preconcentration of water samples. Although very simple in concept (and in use), SPE and its variants have quickly become standard techniques in the analytical laboratory for pretreatment of liquid samples, where SPE products are available in a wide range of sizes, and containing stationary phases that use four of the above mechanisms of separation (affinity SPE is not yet available). However, precisely because it is so versatile, and can be used for such a wide range of sample types, it can at times appear to be rather a complicated task to chose the appropriate SPE protocol. The

241

manufacturers of SPE discs and cartridges all supply free application briefs or more detailed guides and bibliographies on the use of their products,[13] and the literature also describes a large number of uses.[14]

Try to remember that, at its centre, SPE relies upon using the separation mechanisms of Section 6.7 in a simple column (the cartridge); this should greatly simplify the task of choosing an appropriate cartridge from the many that are available. When beginning to develop a new method of analysis, or updating an existing one, if the sample is a liquid (it does not even need to be aqueous), and the analyte or analytes are present in low or very low concentrations in a complex matrix, then it is almost certainly worth considering the use of SPE, both as a means of separating the analyte(s) from the matrix and as a means of concentrating the analyte prior to analysis.

6.9   Comparing separation efficiencies

It is essential to be able to make comparisons between separative systems and between the various materials employed as stationary and mobile phases in chromatographic columns. In all forms of column chromatography the efficiency of columns is usually expressed in terms of the plate count N of the column, and can be related theoretically to the plate theory of separation first introduced by Martin and Synge to explain efficiency of separation in fractional distillation.[15]

The number of theoretical plates for a column measured for a given solute is

$$N = 16(t_r/w)^2 \quad \text{or} \quad N = 5.54(t_r/w_{1/2})^2$$

where each term is indicated on Figure 6.6. The plate count can be increased for a given separation by using longer and longer columns (note the square relationship, i.e. quadrupling the length only doubles the plate count).

For comparison it is more useful to modify N to become independent of length, such that $L/N = H$ where H is the plate height for the column. The ratio L/N used to be known as the height equivalent to a theoretical plate (HETP). It is often expressed in metres per plate, and its reciprocal N/L is given in plates per metre. The following comparison between a typical high-performance capillary GC column and an HPLC column produces some interesting figures.

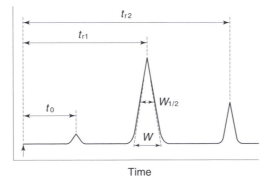

Time

Figure 6.6   Idealised chromatogram of a two-component system

Property	Capillary GC	5 µm HPLC
Column length L (m)	100	0.125
Number of plates N	300 000	6000
N/L (plates/m)	3000	48 000
Plate height H (mm)	0.34	0.02

It shows clearly that HPLC columns are less efficient than typical GC columns in terms of total plate count, but much more efficient in terms of N/L.

The three terms α, k and N can be related to each other by the resolution equation

$$R_s = 0.25\left(\frac{k}{\alpha}\right)\left(\frac{\alpha-1}{k+1}\right)N^{1/2}$$

where α is the relative retention for the two solutes ($\alpha = k_1/k_2$), k is the average capacity factor for the two solutes, N is the average number of plates.

Figure 6.7 shows how they influence the resolution. By definition $\alpha \geq 1$, so the plot of $(\alpha - 1)/\alpha$ has a limiting value of 1; notice that the resolution is most sensitive to changes of α close to 1, say between 1 and 2. Similarly a plot of $k/(1 + k)$ takes a limiting value of 1, but values of k between 1 and 10 can be used successfully. Both α and k are related only to the interactions between the solutes and the two phases, and they can only be modified by altering the nature or composition of the phases, or the temperature. Resolution is related to $N^{1/2}$, which in turn is a measure of the overall performance of the column, so for well-made (efficient) columns N is probably going to give the least effect on resolution.

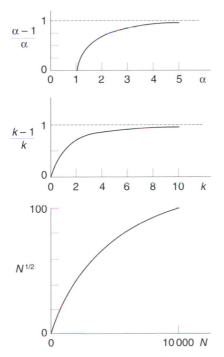

Figure 6.7   How α, k and N affect resolution

6.10 Kinetic factors: rate theory

The equations in Section 6.9 have made no assumptions about the speed of the interactions at the phase boundary; for GC this is just acceptable since diffusion in the gas phase (into and out of the stationary phase) is relatively rapid, but in HPLC these processes are about three orders of magnitude slower than in the gas phase. Since equilibrium is not achieved infinitely quickly, this leads to a broadening of bands and consequent loss of resolution. There are three kinetically controlled mechanisms which need to be considered.

Mass transfer effects are possibly the most obvious to visualise; they are the result of the finite time required for solute molecules to move from the mobile phase into the stationary phase and back again. The larger the particle size of the stationary phase and the more viscous the mobile phase, the greater this effect becomes. And the greater the mobile phase velocity, the greater the band broadening due to this effect.

Longitudinal diffusion occurs as the plug of solute is carried through the column by the mobile phase. The zone tends to expand in a longitudinal direction due to diffusion of solute to the edges of the zone, where the solute concentration is lowest. This will be more pronounced the longer the solute remains in the column, so it is reduced by using high flow rates and short columns. Longitudinal diffusion is the most important source of dispersion in gas chromatography, where long columns, slow flow rates and rapid gaseous diffusion all emphasise its effects.

Flow dispersion is the term used to describe the multiple flow paths that solute molecules must travel as they pass down the packed column. The different molecules travel through the column via different routes; if each were moving at the same velocity then they would take different times to travel the length of the column and this would produce band broadening. In fact, the process is more complicated than this simple explanation, and flow dispersion depends on the particle size and the size distribution, as would be expected, but also on the width of the column (strictly the ratio of particle size to column diameter).

For narrow columns, wall effects become significant, and most columns operate with a ratio of between 10 and 100, although radial dispersion is relatively limited in preparative columns, where this ratio is very high (> 1000), and the wall effects are small. In each case the broadening is independent of mobile phase velocity through the column (Figure 6.8).

Mass transfer, flow dispersion and longitudinal diffusion effects react differently to changes in the mobile phase velocity, but the overall effect can be described by the van Deemter equation:

$$H = A + (B/v) + Cv$$

where
H = the plate height, a measure of efficiency
v = the linear velocity of the mobile phase through the column
A = the flow dispersion term
B = the longitudinal diffusion term
C = the mass transfer term

Figure 6.9 plots the van Deemter equation for gaseous flow. Notice that efficiency strongly depends on the particle size; it also depends on flow rate but since the curve after the minimum is relatively flat, one can often use higher than optimum flow rates, reducing the analysis time without a great sacrifice in resolution, especially for the smallest particle sizes.

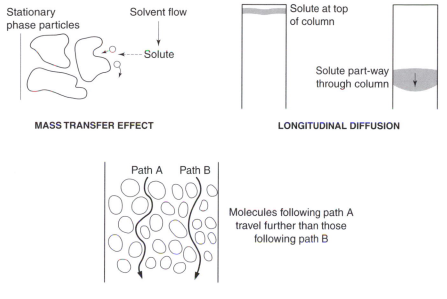

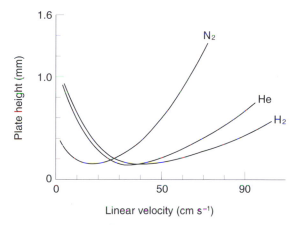

Figure 6.8   Kinetically controlled mechanisms: mass transfer, longitudinal diffusion and flow dispersion

Figure 6.9   A van Deemter plot for N_2, He and H_2

However, when using capillary columns with their much smaller mass of stationary phase and much lower flow rates, Figure 6.9 shows that hydrogen or helium will give better efficiency than nitrogen, except at very low linear velocities of carrier. The reason for this improvement in resolution is that diffusion of solute vapour is more rapid in these gases than in nitrogen. Since a 'normal' capillary column is operated at linear velocities of 20– 50 cm s^{-1}, helium or hydrogen are more efficient carrier gases than nitrogen.

It is quite surprising to discover that when a column is temperature-programmed the flow normally decreases as the temperature increases; this is because gas viscosity

245

increases with temperature, so if the column is operated at constant pressure, as is normally the case for capillary columns, the amount of gas passing through the column decreases with increasing temperature. Figure 6.9 shows that hydrogen has a flatter curve in the normal operating region than helium. This means the resolution obtained using hydrogen is almost constant as flow decreases, whereas for helium and nitrogen the resolution is quite markedly dependent on flow.

From a theoretical standpoint, therefore, hydrogen should be the carrier of choice for capillary chromatography, but other considerations are also important; for example, the purity of the carrier must be considered as even quite small amounts of oxygen will react at high temperatures with the stationary phase, thereby degrading the column performance.

6.11   Separations by electrophoresis

Traditional electrophoresis involves the movement of a colloidal or dissolved substance relative to a buffer solution as a result of an applied field. This leads to migration of the particles or ions towards either the cathode or the anode, depending on the effective charge of the particles. The term 'zone electrophoresis' has been applied to those systems in which ionic mobilities are studied on strips of paper, cellulose acetate or acrylamide. These systems have been used extensively over many years for studying biological and biochemical systems, especially separation of proteins. The more recent application of DNA fingerprinting in forensic science laboratories[16] and for parental identification is typical of the value of modern electrophoretic methods.

More recently the big expansion in electrophoresis has been through the use of capillaries, work begun by Mikkers *et al.*,[17] who used capillary tubes with 200 μm internal diameter to achieve rapid separations. This has been advanced even further by Jorgenson and Lukacs,[18] who reduced the capillary diameter to 75 μm using Pyrex tubes and led to the development of commercial instruments. By using capillaries, as distinct from flat bead systems, it was possible to minimise zone spreading, improve heat dissipation, shorten separation times and increase efficiencies so they were comparable with values for high-performance liquid chromatography. Because of these factors, **capillary electrophoresis (CE)**, also called capillary zone electrophoresis (CZE), has rapidly developed into a widely used process, and modern equipment can achieve the separation of complex steroid or drug mixtures within 10 to 15 minutes.[19]

6.12   The theory of electrophoresis

The accelerating force F_i on a charged particle i under the influence of a constant electric field E is given by

$$F_i = z_i e E \tag{6.1}$$

where z_i is the charge number on charged particle i and e is the charge on an electron. The particle will continue to accelerate until the accelerating force is equal to the resistance within the medium (the drag force F_d). Hence at this stage

$$F_i = F_d \tag{6.2}$$

Once this point is reached, the ions will continue to move with a velocity v_i proportional to the electric field intensity. So

$$v_i \propto E$$
$$v_i = u_i E \tag{6.3}$$

where u_i is the electrophoretic mobility for particle i under the given conditions. Values of u for different substances under specified conditions have been tabulated.[16] The drag force F_d acting on a moving particle in a viscous medium is proportional to the viscosity η of the medium and the velocity of the particle. Hence

$$F_d = k\eta v_i \qquad (6.4)$$

But according to Stokes' law, the drag force on a spherical particle moving through a viscous medium is expressed as

$$F_d = 6\pi\eta r_i v_i$$

where r_i is the particle radius. Hence

$$z_i e E = 6\pi\eta r_i v_i$$

and

$$v_i = \frac{z_i e E}{6\pi r_i \eta} = u_i E$$

So the electrophoretic mobility is given by

$$u_i = \frac{z_i e}{6\pi r_i \eta}$$

Clearly the mobilities of ions depend upon the nature of the electrolyte solution, the concentration and the temperature, so the standard values[20] for u are given as limiting ionic mobilities at 298 K in very dilute solutions.

Note that both anions and cations are responsible for the charge transfer rate (i.e. the electric current passing through the system) and that all ions contribute to the current density in proportion to the charge they carry, their concentrations in solution and their velocities. The cations and anions will move in opposite directions from each other, with the cations being attracted to the negatively charged cathode and the anions moving towards the positively charged anode. In CE the system employed is one in which diffusion of the zones of individual ionic species is minimised. Modern instrumentation is designed to achieve the maximum separation in the minimum time with as much flexibility as possible.

6.13   Instrumentation for CE

The essential CE system consists of a fused silica capillary filled with an aqueous buffer electrolyte; the two ends dip into containers of the electrolyte, one holding the anode, the other holding the cathode. The layout is shown schematically in Figure 6.10. The sample is introduced by inserting the anode end of the capillary into the sample vial and then applying an electric field to the sample vial, leading to electrokinetic injection, or by using pressure on the vial to produce hydrodynamic injection. Once a few nanolitres of the sample are in the capillary, it is replaced in the anode electrolyte container and the high-voltage power supply is used to bring about the separation. Autosamplers are now commonly available for handling large numbers of samples one after the other. The nature and consistency of the buffer solution must be constantly monitored as it is likely to be affected by the migration and accumulation of solutes. CE development has occurred very rapidly, but most features in design of the components are now fairly standard. The various parts of the instrument are now described in more detail.

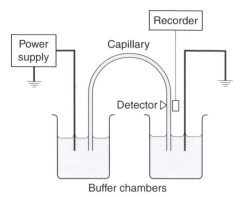

Figure 6.10   Equipment for capillary electrophoresis

6.14   Capillaries

Although glass and Teflon capillary tubes have been used, fused silica capillaries 30–100 cm long are now virtually standard equipment. They possess an internal diameter of 50–100 μm and external diameters of 200–400 μm. They are coated externally with polyimide for protection. Sometimes the capillary may have an internally bonded chemical surface, such as glycerol, in order to reduce interaction between the moving molecules and the capillary walls.

6.15   The applied field

Capillary electrophoresis needs high-strength electric fields to achieve rapid and efficient separation. The high-voltage power supply needs to provide voltages between 20 and 100 kV. At constant voltage a typical current would be between 50 and 200 μA. In the ideal system it should be possible to operate either under constant-current or constant-voltage conditions and to reverse the polarity of the system.

6.16   The detector

Any CE instrument is only as good as its detector, and detectors continue to attract a great deal of study. The detector should aim to avoid causing any effect that might increase band broadening or the dead volume. Because of this, the most commonly used systems are on-column systems where the UV absorbance or fluorescence is measured while the material flows through a short length of the capillary. The standard approach for detection is to remove a short stretch of the polyimide plastic covering from around the capillary tube and to use it as the detector cell for a beam of light on one side of the capillary passing through to a detector on the other side. The usual absorption laws (Section 17.2) then apply, and the signal from the photodetector can be coupled to a recorder. As not all substances absorb at a single fixed wavelength, the system should be able to operate at a range of wavelengths or it should incorporate a photodiode array (Section 17.5) to obtain complete spectral data as the zones pass through.

   Detection may also be carried out using laser-induced fluorescence, benefiting from the way in which laser beams can be accurately directed to very specific sites in the capillary sample window. The fluorescence is observed at right angles to the laser beam and is passed through a filter to the photomultiplier. Other detection systems use the properties of chemiluminescence, thermo-optical absorbance and conductivity. CE may also be coupled with mass spectrometry, forming a highly sensitive and specific detector system.[21] In the CE/MS detector the capillary outlet is directed into an electrospray ionisation interface with the mass spectrometer; this is achieved by passing the capillary outlet into a stainless steel sheath which completes the electrical circuit for the electrophoresis system.[22] Detection limits for CE detectors are normally in the range of 10^{-16} to 10^{-20} mol.

6.17   Applications

Capillary electrophoresis is still a rapidly expanding field; new developments and applications are regularly published for both qualitative and quantitative analyses. As the use of migration times cannot be totally conclusive for qualitative analysis, it is in quantitative analysis that the greatest application occurs, based upon integration of detector signals and comparisons with the necessary standards. However, the great value of CE is its very reliable application to the analysis of inorganic ions and to organic anions.[23] As with other separative procedures, samples can be spiked with reference standards. But in qualitative analysis, errors of peak assignments occur due to migration variation that arises from applied field fluctuations, changes in the capillary wall lining, pH alterations and progressive modifications in the buffer composition. To help overcome these problems, it is common to incorporate a migration time marker in the sample as a reference point for the calculations of relative migration times.

In quantitative analysis it is common to use the same calibration techniques as other analytical procedures: internal and external standards and standard addition methods (Section 9.5). They greatly improve the level of reproducibility in CE quantitative analysis. But always take great care with respect to maintaining the quality of the capillary and the purity of the buffer system.

A great deal of work has been carried out on amino acid separations using dansyl derivatives,[24] and there are now many applications, qualitative and quantitative, to important biochemical compounds such as vitamins, peptides, nucleosides and nucleotides.[25] An increasing area of application is in drug analysis,[26] where the speed of separation and the use of minute sample volumes is invaluable. Capillary electrophoresis has quickly established itself as a valuable laboratory procedure with increasing areas of application.

6.18   References

1. J D Seader and E J Henley 1998 *Separation process principles*, John Wiley, Chichester
2. R A Kjonaas, J L Soller and L A McCoy 1997 *J. Chem. Ed.*, **9**; 1104–5
3. D Sicilia, S Rubio and D Ferezbendito 1994 *Anal. Chim. Acta*, **298** (3); 405–13
4. R M Dagnall and T S West 1964 *Talanta*, **11**; 1627
5. J S Shih 1992 *J. Chinese Chem. Soc.*, **39** (6); 551–9
6. M Hiraoka (ed) 1992 *Crown ethers and analogous compounds*, Elsevier, Amsterdam
7. K Ueno 1992 *Handbook of organic analytical reagents*, 2nd edn, CRC Press, Boca Raton FL
8. I M Abrams and J R Millar 1997 *Reactive and Functional Polymers*, **35** (1/2); 7–22
9. M Leonard 1997 *J. Chrom. B.*, **699** (1/2); 3–27
10. M L Yarmush *et al*. 1992 *Biotechnol. Adv.*, **10** (3); 413–46
11. P L Buldini, S Cavlli and A Trifiro 1997 *J. Chrom. A.*, **789** (1/2); 529–48

12.   J L Viovy and J Lesec 1994 *Adv. Polym. Sci.*, **114**; 1–41
13.   P D McDonald and E S P Bouvier (ed) 1995 *Solid phase extraction applications: guide and bibliography*, Waters, Milford MA
14.   L A Berrueta, B Gallo and F Vicente 1995 *Chromatographia*, **40** (7/8); 474–83
15.   A J P Martin and R L M Synge 1941 *Biochem. J.*, **35**; 1358
16.   J Robertson, A M Ross and L A Burgoyne (eds) 1990 *DNA in forensic science*, Ellis Horwood, New York
17.   F E P Mikkers, F M Everaerts and T P E M Verheggen 1979 *J. Chromatogr.*, **169**; 1–10 and 11–20
18.   J W Jorgenson and K D Lukacs 1981 *Anal. Chem.*, **53**; 1298–1302
19.   W G Kuhr and C A Monnig 1992 *Anal. Chem.*, **64**; 389R–407R
20.   J Pospíchal, P Gebauer and P Boček 1989 *Chem. Rev.*, **89**; 419–30
21.   M Parker 1994 *Lab Products Technol.*, June; 2
22.   H R Udseth, J A Loo and R D Smith 1989 *Anal. Chem.*, **61**; 228–32
23.   F Foret, M Deml and P Boček 1985 *J. Chromatogr.*, **320**; 159–65
24.   F Foret, L Křivánková and P Boček 1993 *Capillary zone electrophoresis*, VCH, Weinheim
25.   N H H Heegard and F A Robey 1994 *Int. Chromatogr. Lab.*, **21**; 2–8
26.   J A Walker *et al.* 1996 *J. Foren. Sci.*, **41**; 824–9

6.19   Bibliography

Anon 1981 *Ion exchange resins*, 6th edn, BDH Chemicals, Poole, UK

G W Gokel 1994 *Crown ethers and cryptands*, Royal Society of Chemistry, Cambridge

N A Guzman (ed) *Journal of capillary electrophoresis*, ISC Technological Publications, Shelton CT

C E Harland 1994 *Ion exchange theory and practice*, 2nd edn, Royal Society of Chemistry, Cambridge

J Korkisch 1989 *Handbook of ion exchange resins: their application to inorganic chemistry*, CRC Press, Boca Raton FL

A S Lindsay 1992 *High performance liquid chromatography*, 2nd edn, John Wiley, Chichester

P D McDonald and E S P Bouvier (eds) 1995 *Solid phase extraction: applications guide and bibliography*, 6th edn, Waters, Milford, MA

J Rydberg, C Musikas and G R Chopin (eds) 1992 *Principles and practice of solvent extraction*, Marcel Dekker, New York

J D Seader and C J Henley 1998 *Separation process principles*, John Wiley, Chichester

P Sewell 1987 *Chromatographic separations*, John Wiley, Chichester

A G Sharpe 1992 *Inorganic chemistry*, 3rd edn, Longman, Harlow

G Svehla 1992 *Vogel's qualitative inorganic analysis*, 7th edn, Longman, Harlow

P A Williams, A Dyer and M J Hudson 1997 *Progress in ion exchange: advances and applications*, Royal Society of Chemistry, Cambridge

W S Winston Ho and K K Sirkar (eds) 1992 *Membrane handbook*, Van Nostrand Reinhold, New York

C Wu (ed) 1995 *Handbook of size exclusion chromatography*, Marcel Dekker, New York

M Zief and L Crane 1988 *Chromatographic chiral separations*, Marcel Dekker, New York

7

Thin-layer chromatography

7.1 Introduction

Developments that have taken place in the field of thin-layer chromatography (TLC) have elevated it from the level of a semiquantitative analytical procedure to one in which highly reliable quantitative results can be obtained. This means it has become as much of an instrumental technique as other forms of chromatography. There is no doubt that laboratories seeking to minimise analytical costs, and laboratories which are less well equipped with advanced instrumentation, find that TLC has much to commend it for pharmaceutical and environmental analyses. TLC has several advantages over other forms of chromatography:

(a) Sample preparation is usually relatively simple.
(b) Samples may be directly compared, often as they are running.
(c) Parallel development of related and unrelated samples can be carried out simultaneously.
(d) A range of detection procedures can be applied, often to the same plate.
(e) The separation can be followed throughout the whole process and stopped when desired or when the solvent systems are changed.
(f) Solvents and other reagents are required in very small volumes.

Although the basic approach to TLC remains the same, the introduction of instrumentation for sample application, development, densitometry and recording have greatly extended the scope of the procedure.[1]

The important difference between TLC and other forms of chromatography is one of practical technique rather than physical phenomenon (adsorption, partition, etc.). Thus in TLC the stationary phase consists of a thin layer of sorbent (e.g. silica gel, cellulose powder or alumina) coated on an inert, rigid backing material such as a glass plate, aluminium foil or plastic foil. As a result, the separation process takes place on a flat and essentially two-dimensional surface. The technique of paper chromatography has been almost entirely superseded by TLC in analytical laboratories, although it is still useful for demonstrating the general principles behind chromatographic separations with simple mixtures of dyes. The more recent performance improvements, in terms of separations and quantitative measurements, are discussed in Section 7.6 on high-performance TLC and in Section 7.5 on two-dimensional TLC.

Forensic chemistry has greatly benefited from TLC, where it is extensively employed for initial screening of a wide variety of drugs.[2,3] Of particular value are the specialist applications in studying the composition of fibre-tip pens,[4] and typewriter[5] and computer ribbons.[6]

7.2 The technique of TLC

Preparation of the plate

In TLC a variety of coating materials are available, but silica gel is most frequently used. A slurry of the adsorbent (silica gel, cellulose powder, etc.) is spread uniformly over the plate by means of one of the commercial forms of spreader, the recommended thickness of adsorbent layer being 150–250 μm. After air drying overnight, or oven drying at 80–90 °C for about 30 min, it is ready for use. Ready-to-use thin layers (i.e. precoated plates or plastic sheets) are commercially available; the chief advantage of plastic sheets is that they can be cut to any size or shape required, but unless they are supported, they do bend in the chromatographic tank. Two points of practical importance:

1. TLC plates should never be handled or touched on the surface, but carefully held only by the edges. This will avoid possible contamination due to perspiration.
2. Precleansing of the plate is advisable in order to remove extraneous material that might be contained in the layer. This may be carried out by running the development solvent to the top of the plate, and then redrying it before use.

Sample application

The origin line, to which the sample solution is applied, is usually located 2.0–2.5 cm from the bottom of the plate. The accuracy and precision with which the sample 'spots' are applied is very important when quantitative analysis is required. Volumes of 1, 2 or 5 μL are applied using an appropriate measuring instrument, e.g. a syringe or micropipette (the micropipette is a calibrated capillary tube fitted with a small rubber teat). Care must be taken to avoid disturbing the surface of the adsorbent as this causes distorted shapes of the spots on the subsequently developed chromatogram and so hinders quantitative measurement. To assist in the positioning of the sample spots, plastic raised cover sheets are available with regularly cut holes for appropriate location of sites.

In carrying out semiquantitative and quantitative analyses, remember that losses of sample may occur if the applied spots are dried using a current of air. The use of mixtures in low boiling solvents aids natural atmospheric drying and helps to ensure the spots remain compact (less than 2–3 mm diameter). Figure 7.1 illustrates the type of arrangement that might be expected with a TLC plate being used to compare a number of samples and standards.

Development of plates

The chromatogram is usually developed by the ascending technique in which the plate is immersed in the developing solvent to a depth of 0.5 cm (redistilled or chromatographic grade solvent should be used). The tank or chamber is preferably lined with sheets of filter paper which dip into the solvent in the base of the chamber; this ensures the chamber is saturated with solvent vapour (Figure 7.2). Development is allowed to proceed until the solvent front has travelled the required distance (usually 10–15 cm), the plate is then removed from the chamber and the solvent front immediately marked with a pointed object. The plate is allowed to dry in a fume cupboard or in an oven; the drying conditions should take into account the heat and light sensitivity of the separated compounds.

The positions of the separated solutes can be located by various methods. Coloured substances can be seen directly when viewed against the stationary phase, whereas colourless

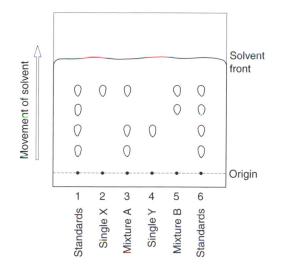

Figure 7.1   A thin-layer plate: standards are run alongside the mixtures

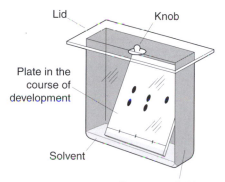

Figure 7.2   Developing the plate (Reprinted, with permission, from D. Abbott and R. S. Andrews, 1965, *An introduction to chromatography*, Longman, London)

species may usually be detected by spraying the plate with an appropriate reagent that produces coloured areas in the regions which they occupy. Spray reagents for locating solutes should only be used in a fume cupboard. Some compounds may be located without spraying if they fluoresce under ultraviolet light. Alternatively, if the adsorbent used for the TLC plate contains a fluorescing material, the solutes can be observed as dark spots on a fluorescent background when viewed under ultraviolet light. When locating zones by this method, **protect the eyes** by wearing special goggles or spectacles. The spots located by this method can be delineated by marking with a needle.

Measurement and identification of solutes

The success of TLC depends upon the different affinities of the solutes for the TLC plate and the developing solvent(s), and the order in which substances separate from each

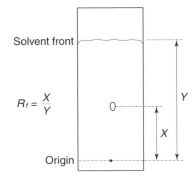

Figure 7.3 How to measure R_f values

other varies depending upon the various combinations that are applied. Under constant conditions of temperature, solvent system and adsorbent, any individual solute (drug, dye, steroid, etc.) will move by a constant ratio with respect to the solvent front. This is known as the R_f value (relative front or retardation factor)[7] where

$$R_f = \frac{\text{distance compound has moved from origin}}{\text{distance of solvent front from origin}}$$

This is shown in Figure 7.3. The calculation produces results which are always decimal values less than unity. Because of this, and to avoid decimal points, it is very common to use what have become known as hR_f values, in which a whole number value is obtained by simply multiplying the R_f value by 100. Thus an R_f value of 0.57 becomes an hR_f value of 57.

To assist in the comparison of results and identification of solutes by TLC, libraries of R_f and hR_f values have been compiled. They have been of particular value for toxicological studies,[8] where very detailed standardised TLC systems have been specified.

Quantitative evaluation

Methods for the quantitative measurement of separated solutes on a thin-layer chromatogram can be divided into two categories. In the more generally used in situ methods, quantitation is based on measurement of the photodensity of the spots directly on the thin-layer plate, preferably using a densitometer. The densitometer scans the individual spots by reflectance or absorption of a light beam; the scan is usually along the line of development of the plate. The difference in intensity of the reflected (or transmitted) light between the adsorbent and the solute spots is observed as a series of peaks plotted by a chart recorder. The areas of the peaks correspond to the quantities of the substances in the various spots. This type of procedure requires comparison with spots obtained using known amounts of standard mixtures, which must be chromatographed on the same plate as the sample. Improvements in the design of densitometers have considerably increased the reliability of quantitative TLC determinations.

A cheaper procedure is to remove the separated components by scraping off the relevant portion of the adsorbent after visualisation by a non-destructive technique. The component is conveniently extracted by placing the adsorbent in a centrifuge tube and adding a suitable solvent to dissolve the solute. When the solute has dissolved, the tube is spun in a centrifuge; then the supernatant liquid is removed and analysed by an appropriate quantitative technique, e.g. ultraviolet, visible or fluorescence spectrometry or gas–liquid chromato-graphy. Alternatively the solute may be extracted by transferring the adsorbent on to a short column of silica gel, supported by a sinter filter, and eluting with the solvent. Again the extract is analysed by a suitable quantitative technique. In each case it is necessary to obtain a calibration curve for known quantities of the solute in the chosen solvent.

To obtain the best results in any of these quantitative TLC methods, the spots being used should have $R_f = 0.3$–0.7; spots with $R_f < 0.3$ tend to be too concentrated whereas those with $R_f > 0.7$ are too diffuse.

7.3   Stationary phases

The range of substances now available for use as stationary phases in TLC extends far beyond the traditional silica gel and alumina. The introduction of microcrystalline cellulose has meant that TLC has almost totally replaced paper chromatography. Silica gel is now available in numerous particle sizes and mixed with a variety of binders, such as calcium sulphate, which may also incorporate fluorescent indicators. Similar combinations occur with alumina, which can be dried to give various levels of activation. So-called reversed-phase materials are also available for use with aqueous eluents. These consist of long-chain hydrocarbons (e.g. C_{10} to C_{18}) which have been bonded to the silica gel via the hydroxyl groups. Their use has extended the range of solvent systems that can now be applied for development of TLC.

7.4   Mobile phases

In selecting a mobile phase for development of TLC plates, it is important to ensure the solvent system does not react chemically with the substances in the mixture under examination. In many instances this will preclude the use of mixtures containing substances such as ethanoic acid (acetic acid). Similar considerations need to be given to mixtures containing ammonia. Carcinogenic solvents, e.g. benzene, or environmentally dangerous solvents, e.g. dichloromethane, should always be avoided. Solvent systems range from non-polar single solvents, e.g. hexane, through to polar solvent mixtures of ethanol and organic acids. With polar stationary phases it is normal to use solvents of medium to low polarity, whereas reversed-phase systems tend to require more polar solvents, e.g. acetonitrile or butanol.

7.5   Two-dimensional TLC

In many TLC processes, and even with several solvent steps, the separation of mixtures along a single line does not always give a clear resolution of the individual components. However, this problem can often be overcome by using two-dimensional TLC, even for closely related compounds. In this procedure the separation is carried out using a square TLC plate; the spot of the mixture of solutes is placed near to one corner of the plate before being developed using the first solvent system (Figure 7.4(a)). When complete (Figure 7.4(b)) the chromatogram is removed from the development tank, dried and rotated through 90° in order that the separated row of spots becomes the starting line (Figure 7.4(c)). The chromatogram is then rerun using the second solvent system. The end result (Figure 7.4(d)) should be an effective separation covering the broad area of the square chromatographic plate. One great advantage of two-dimensional TLC is that it is possible to obtain good separations of closely related materials even on cut squares of chromatographic plates measuring only $10\,cm \times 10\,cm$.

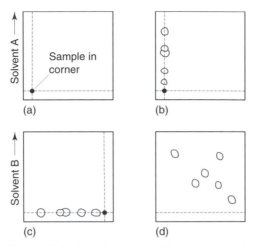

Figure 7.4 Two-dimensional thin-layer chromatography: (a) first development, (b) results of first separation, (c) plate rotated 90° for second development, (d) final chromatogram (Reprinted, with permission, from R. C. Denney, 1982, *A dictionary of chromatography*, 2nd edn, Macmillan, London)

7.6   High-performance thin-layer chromatography (HPTLC)

Developments which have taken place in the quality of TLC adsorbents and in the procedures for sample application have led to such improvements in performance of TLC separations that the expression 'high-performance thin-layer chromatography (HPTLC)' has been used for separations in which high resolution are achieved. The main features which have led to HPTLC are summarised below, but remember that the smaller particle size of the adsorbents also leads to slower development rates, hence the elution distances are made shorter and smaller solute quantities are used.

Quality of the adsorbent layer

Layers for HPTLC are prepared using specially purified silica gel with average particle diameter of 3–5 μm and a narrow particle size distribution. The silica gel may be modified if necessary, e.g. chemically bonded layers are available commercially as reverse-phase plates. Layers prepared using these improved adsorbents give up to about 5000 theoretical plates and so provide a much improved performance over conventional TLC; this enables more difficult separations to be effected using HPTLC, and also enables separations to be achieved in much shorter times.

Methods of sample application

Due to the lower sample capacity of the HPTLC layer, the amount of sample applied to the layer is reduced. Typical sample volumes are 100–200 nL, which give starting spots of only 1.0–1.5 mm diameter; after developing the plate for a distance of 3–6 cm, compact separated spots are obtained giving detection limits about 10 times better than in conventional TLC. A further advantage is that the compact starting spots allow an increase in the number of samples which may be applied to the HPTLC plate.

The introduction of the sample into the adsorbent layer is a critical process in HPTLC. For most quantitative work, a convenient spotting device is a platinum–iridium capillary of fixed volume (100 or 200 nL), sealed into a glass support capillary of larger bore. The capillary tip is polished to provide a smooth, planar surface of small area (~0.05 mm^2), which when used with a mechanical applicator minimises damage to the surface of the plate; spotting by manual procedures invariably damages the surface.

TLC can be a highly labour-intensive system of analysis as it has several processing stages, including sample preparation, often followed by solvent extraction and concentration, sample spotting, development and spot exposure by either fluorescence or chemical sprays. And quantitative analysis requires further processing by scraping and dissolving or scanning. Because of this, there have been major efforts in recent years to shorten procedures and to speed up the average time required for each sample. This has been achieved by using solid phase extraction columns to concentrate samples, followed by automatic spotting devices to deal with multiple samples in a short period of time and provide highly reproducible and quantitative spot application. These systems are particularly valuable for analyses where there is a constant flow of fairly repetitive samples.

Scanning densitometers

Commercial instruments for in situ quantitative analysis based on direct photometric measurements have played an important role in modern TLC. Both double- and single-beam instruments are available. They are of particular value in HPLC where the quality and surface homogeneity of the plates are generally very good. The densitometers scan the individual spots by reflectance or absorption of a light beam. In reflectance densitometry the chromatogram is scanned by the moving beam of light and the intensity of the reflected light from the surface is measured. The differences in the light intensity between the adsorbent and any spots are observed as a series of peaks in the scan display. The areas of the peaks correspond to the quantities of the materials in the spots. Alternative procedures produce a similar record by measuring the light transmitted through the plate. With photodensitometers the transmitted or reflected radiation is used to produce a photograph of the chromatogram revealing dark and light zones for the areas of the separated compounds. The standard deviation for quantitative TLC determinations by densitometry is better than 5%.

Experimental section

> ! Safety. Before carrying out any experiments in this section, pay full
> ■ attention to any safety warnings and make sure you adhere to national
> laboratory and safety regulations.

7.7   Separation and recovery of dyes

Introduction   This experiment illustrates the technique required to recover a number of pure substances after their separation by TLC. By using standard concentrations of the individual dyes, it is also possible to extend the experiment to demonstrate the quantitative determination of the recovered compounds.

Apparatus
Prepared silica gel plates
Chromatographic tank (Figure 7.2)
Micropipette or microcapillary

Chemicals
Indicator solutions (~0.1% aq): bromophenol blue, Congo red, phenol red
Mixture (M) of above three indicator solutions
Developing solvent: butan-1-ol: ethanol: $0.2\,M$ ammonia (60 : 20 : 20 by volume)
Chromatographic grade solvents should be used.

Procedure   Pour the developing solvent into the chromatographic tank to a depth of about 0.5 cm and replace the lid. Take a prepared plate and carefully spot 5 μL of each indicator on the origin line using a micropipette (follow the advice on sample application in Section 7.2). Allow to dry, slide the plate into the tank and develop the chromatogram by the ascending solvent for about 1 h. Remove the plate, mark the solvent front and dry the plate in an oven at 60 °C for about 15 min. Evaluate the R_f value for each of the indicators.

Take a second prepared plate and spot three separate 5 μL of mixture M on to the origin line using a micropipette. Place the dry plate into the tank, replace the lid and allow the chromatogram to run for about 1 h. Remove the plate, mark the solvent front and dry the plate at 60 °C for about 15 min. Identify the separated components on the basis of their R_f values.

Carefully scrape the separated bromophenol blue spots on to a sheet of clean, smooth-surfaced paper using a narrow spatula (this is easier if two grooves are made down to the glass on either side of the spots). Pour the blue powder into a small centrifuge tube, add 2 mL of ethanol, 5 drops of 0.880 ammonia solution, and stir briskly until the dye is completely extracted. Centrifuge and remove the supernatant blue solution from the residual white powder. Repeat this procedure with the separated Congo red and phenol red spots.

An alternative elution technique is to transfer the powder (e.g. for bromophenol blue) to a glass column fitted with a glass wool plug or glass sinter, and elute the dye with ethanol containing a little ammonia. The eluted solution, made up to a fixed volume in a small graduated flask, may be used for colorimetric or spectrophotometric analysis of the recovered dye (Chapter 17). A calibration curve must be constructed for each of the individual compounds.

7.8   Separation of carbohydrates

Introduction   This experiment is applicable to a range of carbohydrates,[9] but those chosen here will produce well-defined spots with good separation from each other. Quantitative evaluations can be readily carried out by preparing calibration plots using ranges of standard solutions of the individual carbohydrates.

Apparatus
Kieselgel G or Keiselguhr G preprepared plates
Chromatographic tank (Figure 7.2)
Micropipette or microcapillary

Chemicals
Solution of carbohydrates: fructose, glucose, lactose, maltose and sucrose (approximately 0.1% of each component) and for quantitative determinations a range of concentrations of the individual carbohydrates from 0.01% to 0.2% w/v in deionised water
Developing solvent: acetone : butan-1-ol : water (50 : 40 : 10 by volume)
Spray reagent: equal volumes of 20% aqueous trichloroethanoic acid and 0.2% ethanolic solution of 1,3-dihydroxynaphthalene.

Procedure   Use the micropipette or microcapillary to apply the mixture, pure substances and quantitative standards to the line; apply them 1.0–1.5 cm from the edge of a square (20 cm × 20 cm) thin-layer plate. Place in the development chamber previously equilibrated with the solvent mixture. Allow the plate to develop until the solvent has travelled 12–15 cm up the plate. Remove from the tank and immediately mark the solvent front, then allow the plate to dry before spraying with the reagent. The carbohydrates should separate with increasing R_f value in the order fructose, lactose, glucose, maltose and sucrose.

7.9   Separation of artificial colourants in confectionery

Introduction   A number of sweets (e.g. Smarties and M&M's Peanuts) are manufactured with hard outer coatings in a variety of colours. These are water-soluble dyes which are readily dissolved and can be identified using TLC. The eight most common colours used for this purpose are Allura Red, Brilliant Blue FCF, Carmoisine, Cochineal, Patent Blue V, Ponceau 4R, Quinoline Yellow and Sunset Yellow FCF. The experiment is straightforward to carry out.

Apparatus
Standard 20 cm × 20 cm silica gel coated TLC plates
Chromatographic tank (Figure 7.2)
Micropipettes or microcapillaries
100 mL glass beakers

Chemicals
A standard set of the eight colours as individual 1% solutions in water, and as a mixture
Developing solvent: propanol–ammonia (4 : 1)

Procedure   Take 5 g of each of the different coloured sweets and place in individual beakers with 10 mL of deionised water. Leave to soak for 5 min with gentle swirling by hand once or twice. Avoid being too vigorous or the sweets will disintegrate. Decant the separate solutions into fresh beakers and allow them to settle and clear, if necessary decant again. Concentrate the solutions on a water bath until each volume is less than 1 mL.
   Apply 20 μL of each extract to the TLC plate alongside spots of all the standard colours and/or the mixture of colours. Run the chromatogram with the propanol–ammonia development solvent for as long as possible. When development is complete, remove the plate and mark the solvent front. When it has dried, the individual colours should be clearly identifiable by comparison with the standards and measurement of R_f values.

7.10   References

1.   I Ojanpera 1997 *Bull. Int. Assoc. Forensic Toxicologists*, **XXVIII** (2); 5
2.   T A Gough (ed) 1991 *The analysis of drugs of abuse*, John Wiley, Chichester
3.   M D Cole and B Caddy 1995 *The analysis of drugs of abuse: an instruction manual*, Ellis Horwood, New York
4.   O P Jasuja and A K Singla 1990 *Indian J. Forensic Sci.*, **4** (4); 167–70
5.   R L Brunelle *et al.* 1977 *J. Forensic Sci.*, **22** (4); 807–14
6.   N Kaw, O P Jasuja and A K Singla 1992 *Forensic Sci. Int.*, **53** (1); 51–60
7.   R C Denney 1982 *A dictionary of chromatography*, 2nd edn, Macmillan, London, p. 161
8.   A C Moffat (ed) 1987 *Thin-layer chromatographic R_f values of toxicologically relevant substances on standardized systems*, Deutsche Forschungsgemeinschaft, Bonn, and the International Association of Forensic Toxicologists, Newmarket
9.   R S Kirk and R Sawyer 1991 *Pearson's composition and analysis of foods*, 9th edn, Longman, Harlow, pp. 184–5

7.11   Bibliography

B Fried and J Sherma 1994 *Thin-layer chromatography: techniques and applications*, 3rd edn, Marcel Dekker, New York

R Hamilton and S Hamilton 1987 *Thin layer chromatography*, ACOL–Wiley, Chichester

E J Shellard (ed) 1968 *Quantitative paper and thin-layer chromatography*, Academic Press, London

J Sherma and B Fried 1996 *Handbook of thin-layer chromatography*, 2nd edn, Marcel Dekker, New York

J C Touchstone 1992 *Practice of thin layer chromatography*, 3rd edn, John Wiley, Chichester

8

Liquid chromatography

8.1 Introduction

The term 'liquid chromatography' has been used to cover a range of chromatographic systems, including liquid–solid, liquid–liquid, ion exchange and size exclusion chromatography, all of which employ a mobile liquid phase. Classical liquid column chromatography is associated with the use of wide-diameter glass columns filled with a finely divided stationary phase through which the mobile phase percolates under gravity. These systems have led to very successful separations of complex mixtures, although they are frequently slow and it can be tedious to perform the chemical or spectroscopic examinations of fractions. High-performance systems have meant that liquid chromatography has overtaken gas chromatography as HPLC now provides the following features:

1. High resolving power
2. Speedy separation
3. Continuous monitoring of the column effluent
4. Accurate quantitative measurement
5. Repetitive and reproducible analysis using the same column
6. Automation of the analytical procedure and data handling

High-performance liquid chromatography is in some respects more versatile than gas chromatography since it is not limited to volatile and thermally stable samples, and the choice of mobile and stationary phases is wider.

The first instrumental liquid chromatograph was constructed by Csaba Horvath[1] at Yale University in 1964, and was described as a high-pressure liquid chromatograph (HPLC). But Horvath himself called the process 'high performance liquid chromatography' – the term that is now more widely used.

Although the technique has undergone extensive development in the last three decades, the original chromatograph which operated at pressures up to 1000 psi (6897 kPa or 68.97 bar), with an aqueous buffer mobile phase passing through columns of 1 mm internal diameter and with an ultraviolet (UV) spectrometer as detector, is remarkably similar to systems in use today. The first mixtures to be separated by Horvath's group were nucleic acid components associated with thyroid function.[2] This may help to explain the tremendous growth of the technique since 1964, because in its many variants, HPLC allows researchers to study molecules that are difficult to separate by any other means, particularly biomolecules.

8.2   Types of liquid chromatography

8.2.1   Liquid–solid chromatography (LSC)

Liquid–solid chromatography, often called adsorption chromatography, is based on inter-actions between the solute and fixed active sites on a finely divided solid adsorbent used as the stationary phase. The adsorbent may be packed in a column or spread on a plate (as in TLC); it is generally an active solid with a high surface area, e.g. alumina, charcoal or silica gel; silica gel is the most widely used. A practical consideration is that highly active adsorbents may give rise to irreversible solute adsorption; silica gel, which is slightly acidic, may strongly retain basic compounds, whereas alumina (non-acid washed) is basic and should not be used for the chromatography of base-sensitive compounds. Adsorbents of varying particle size, perhaps down to 5 μm for HPLC, may be purchased commercially.

The role of the solvent in LSC is clearly vital since mobile phase (solvent) molecules compete with solute molecules for polar adsorption sites. The stronger the interaction between the mobile phase and the stationary phase, the weaker the solute adsorption, and vice versa. The classification of solvents according to their strength of adsorption is called an eluotropic series,[3] which may be used as a guide to find the optimum solvent strength for a particular separation; a trial-and-error approach may, however, be required and this is done more rapidly by TLC than by using a column technique. Solvent purity is very important in LSC since water and other polar impurities may significantly affect column performance, and the presence of UV-active impurities is undesirable when using UV-type detectors.

In general, the compounds best separated by LSC are those which are soluble in organic solvents and are non-ionic. Water-soluble non-ionic compounds are better separated using either reverse-phase or bonded-phase chromatography.

8.2.2   Liquid–liquid (partition) chromatography (LLC)

Liquid–liquid chromatography is similar in principle to solvent extraction (Chapter 6); it is based upon the distribution of solute molecules between two immiscible liquid phases according to their relative solubilities. The separating medium consists of a finely divided inert support (e.g. silica gel, kieselguhr) holding a fixed (stationary) liquid phase, and separation is achieved by passing a mobile phase over the stationary phase. The stationary phase may be in the form of a packed column, a thin layer on glass, or a paper strip.

It is convenient to divide LLC into two categories, based on the relative polarities of the stationary and mobile phases. The term 'normal LLC' is used when the stationary phase is polar and the mobile phase is **non-polar**. In this case the solute elution order is based on the principle that non-polar solutes prefer the mobile phase and elute earlier, whereas polar solutes prefer the stationary phase and elute later. In **reverse-phase chromatography (RPC)** the stationary phase is non-polar and the mobile phase is polar; the solute elution order is commonly the reverse of that observed in normal LLC, i.e. polar compounds elute earlier and non-polar compounds elute later. This is a popular mode of operation due to its versatility and scope; the almost universal application of RPC arises because nearly all organic molecules have hydrophobic regions in their structure and are therefore capable of interacting with the non-polar stationary phase. Since the mobile phase in RPC is polar, and commonly contains water, the method is particularly suited to the separation of polar substances which are either insoluble in organic solvents or bind too strongly to solid adsorb-ents (LSC) for successful elution. Table 8.1 shows some typical stationary and mobile phases which are used in normal and reverse-phase chromatography.

Table 8.1 *Typical stationary and mobile phases for normal and reverse-phase chromatography*

Stationary phases	Mobile phases
Normal	
β,β′-Oxydipropionitrile	Saturated hydrocarbons, e.g. hexane, heptane;
Carbowax (400, 600, 750, etc.)	aromatic solvents, e.g. toluene, xylene; saturated
Glycols (ethylene, diethylene)	hydrocarbons mixed with up to 10% dioxan,
Cyanoethylsilicone	methanol, ethanol, chloroform, methylene chloride (dichloromethane)
Reverse-phase	
Squalane	Water and alcohol–water mixtures; acetonitrile and
Zipax-HCP	acetonitrile–water mixtures
Cyanoethylsilicone	

Although the stationary and mobile phases in LLC are chosen to have as little solubility in one another as possible, even slight solubility of the stationary phase in the mobile phase may result in the slow removal of the stationary phase as the mobile phase flows over the column support. For this reason the mobile phase must be presaturated with stationary phase before entering the column. This is conveniently done by using a precolumn before the chromatographic column; the precolumn should contain a large-particle packing (e.g. 30–60 mesh silica gel) coated with a high percentage (30–40%) of the stationary phase to be used in the chromatographic column. As the mobile phase passes through the precolumn, it becomes saturated with stationary phase before entering the chromatographic column.

The support materials for the stationary phase can be relatively inactive supports, e.g. glass beads, or adsorbents similar to those used in LSC. It is important, however, that the support surface should not interact with the solute, as this can produce a mixed mechanism (partition and adsorption) rather than true partition. This complicates the chromatographic process and may give non-reproducible separations. For this reason, high loadings of liquid phase are required to cover the active sites when using porous adsorbents with a high surface area.

To overcome some of the problems associated with conventional LLC, such as loss of stationary phase from the support material, the stationary phase may be chemically bonded to the support material. This form of liquid chromatography, in which both monomeric and polymeric phases have been bonded to a wide range of support materials, is called bonded-phase chromatography.

Silylation reactions have been widely used to prepare bonded phases. The silanol groups ($-\overset{|}{\underset{|}{Si}}-OH$) at the surface of silica gel are reacted with substituted chlorosilanes. A typical example is the reaction of silica with a dimethylchlorosilane which produces a monomeric bonded phase, since each molecule of the silylating agent can react with only one silanol group:

$$-\overset{|}{\underset{|}{Si}}-OH + Cl-\overset{CH_3}{\underset{CH_3}{Si}}-R \longrightarrow -\overset{|}{\underset{|}{Si}}-O-\overset{CH_3}{\underset{CH_3}{Si}}-R + HCl$$

263

The use of di- or trichlorosilanes in the presence of moisture can cause a polymeric layer to be formed at the silica surface, i.e. a polymeric bonded phase. But monomeric bonded phases are preferred since they are easier to manufacture reproducibly than the polymeric type. The nature of the main chromatographic interaction can be varied by changing the characteristics of the functional group R; in analytical HPLC the most important bonded phase is the non-polar C-18 type in which the modifying group R is an octadecyl hydrocarbon chain. Unreacted silanol groups are capable of adsorbing polar molecules and will therefore affect the chromatographic properties of the bonded phase, sometimes producing undesirable effects such as tailing in RPC. These effects can be minimised by the process of end-capping, in which these silanol groups are rendered inactive by reaction with trimethylchlorosilane:

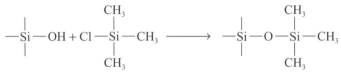

An important property of these siloxane phases is their stability under the conditions used in most chromatographic separations; the siloxane bonds are attacked only in very acidic (pH < 2) or basic (pH > 9) conditions. A large number of commercial bonded-phase packings are available in particle sizes suitable for HPLC.[4]

8.2.3   Ion chromatography

Ion chromatography is simply a high-performance version of ion exchange chromatography, which was first developed in the early 1940s in order to separate and concentrate rare earth and transuranic ions for development of the first nuclear weapons. It can be performed on standard HPLC equipment rather than using a dedicated ion chromatograph, but it may be difficult to choose a solvent system with the correct elution characteristics and UV absorption. Both cations and anions can be separated on appropriate forms of ion exchange resins (Section 6.5). However, for HPLC, conventional ion exchange resins caused problems as slow diffusion through the resin bed and the relatively low strength of the material led to compression of the column bed. These difficulties were solved by bonding the resin material to the surface of either moderately large (30–50 μm) glass beads to form a pellicular packing, or by coating the surface of a rigid microparticle to give a type of reverse-phase packing.[5] Columns packed with this material are capable of giving high resolution of both cationic and anionic systems. Once held on the column, the ions of interest are eluted using a buffered mobile phase of increasing ionic strength.

Conventional HPLC detectors, e.g. UV detectors, may be used for organic ions e.g. amino acids; but until recently there was no satisfactory system for detecting inorganic ions. However, the importance of certain anions and certain metal ions stimulated research, and several techniques have now been developed. The most widely used relies on the change of conductance in the eluent as ionic analyte is displaced from the column. It is a change of conductance (1/resistance) not conductivity (1/resistivity) that is actually measured, although the term 'conductivity detection' is normally used.

An alternative to conductivity detection is the indirect UV detection, which may be used for a number of systems containing simple ions. Here the eluting solvent contains a constant concentration of a strongly absorbing material, often potassium hydrogen phthalate or sodium benzoate, which produces a high absorbance signal at an ordinary UV detector. The solute ions are transparent at the detection wavelength; as they pass through the detector, they decrease the absorbance and negative peaks are observed. By reversing the polarity of the recording device, the negative peaks can be displayed as the more usual positive peaks.

8.2.4   Exclusion chromatography

The mechanism used for separating solute molecules according to their size, or more accurately their solution volume, is generally known as exclusion or size exclusion chromatography (SEC). Separations specifically performed in aqueous solvents on water-soluble species are described as gel filtration chromatography (GFC), whereas those performed in non-aqueous solvents are known as gel permeation chromatography (GPC). In each case the mechanism is the same and relies only on the relative pore sizes of the stationary phase compared to the size (volume) of the solute molecules with no chemical interaction between the solute and stationary phase.

The stationary phases used in GPC are porous materials with a closely controlled pore size; the primary mechanism for retention of solute molecules is the different penetration (or permeation) by each solute molecule into the interior of the gel particles. Molecules whose size is too great will be effectively barred from certain openings into the gel network and will therefore pass through the column chiefly by the interstitial liquid volume. Smaller molecules are better able to penetrate into the interior of the gel particles, depending on their size and the distribution of pore sizes available to them; smaller molecules are more strongly retained.

The materials originally used as stationary phases for GPC were the xerogels of the polyacrylamide (Bio-Gel) and cross-linked dextran (Sephadex) type. However, these semi-rigid gels are unable to withstand the high pressures used in HPLC, and modern stationary phases consist of microparticles of styrene–divinylbenzene copolymers (Ultrastyragel, manufactured by Waters Associates), silica or porous glass.

The extensive analytical applications of exclusion chromatography cover both organic and inorganic materials.[6] Although there have been many applications of exclusion chromatography to simple inorganic and organic molecules, it has mainly been applied to studies of complex biochemical or highly polymerised molecules.

The total volume V_t of a size exclusion column is composed of three parts:

1. About 20% of the total is occupied by the solid parts of the packing.
2. Some 40% of the total is the pore volume V_p of the packing; solvent and small molecules can completely permeate these pores and will pass through them as they travel down the column.
3. The remaining 40% is the void volume V_v, the spaces between the individual particles.

Large molecules, which cannot enter the pore volume, only travel through this region. For a standard 3000 mm × 7.8 mm column with a total volume of 15 mL, this means that all very large molecules will elute in about 6 mL and all very small molecules will elute in about 12 mL, with intermediate sizes eluting somewhere in between. Note that in this form of chromatography the capacity factor k ranges from 0 (for large molecules) to 1 (for small molecules) and the total usable elution volume, which is between V_p and V_v, is only about half the volume of the column. This means that if no other retarding factors are present, the total elution volume for a given system is known, but it also means that the maximum number of components which can be reasonably separated is only of the order of 12.

We have used the words 'large' and 'small' without quantification because these general concepts can be applied to a wide range of molecular sizes; the plot of log molecular mass

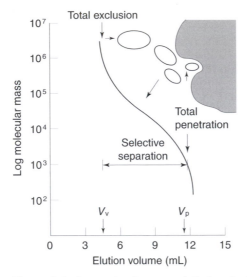

Figure 8.1   Log molecular mass plotted against elution volume

versus elution volume (Figure 8.1) can be drawn for all exclusion packings. However, a given packing material with a defined pore size can only effectively separate molecules over a small range of masses. Normally this fractionation will be over 1.5–2.5 orders of magnitude in terms of mass. Packing materials are available in a range of pore sizes from $50\,\text{Å}$ to $10^6\,\text{Å}$ (angstroms are conventionally used in this area).

The quoted pore size is not the actual size of hole in the packing, but the minimum mass of a polystyrene standard that is not retained, i.e. it is the exclusion limit of the packing. In exclusion chromatography it is fairly common to couple columns together in order to improve the experimental separation. Two or more columns of the same exclusion limit can be joined to increase the plate count, hence the resolution, without modifying the mass range; or a number of columns of different pore size can be coupled to extend the range of masses separated. Many commercially available columns are packed using a given solvent during manufacture and should be used and stored with this solvent in the column. Eluents commonly used are toluene, tetrahydrofuran (THF), dichloromethane, chloroform, dimethylformamide (DMF) and dimethyl sulphoxide (DMSO).

Although exclusion chromatography has a wide range of application, many of its uses can be categorised into two types:

(a)   The separation of large molecules or groups of molecules from smaller ones.
(b)   The separation and characterisation of polymeric materials (both natural and synthetic).

Separation of large molecules finds wide application in the biosciences, where separation of biopolymers from smaller molecules is often the first step in a multistage analysis, but it can also be used in more traditional chemical analysis; for example, pesticides can be determined in a complex matrix (chicken fat) if they are first separated from the matrix by exclusion chromatography and then reinjected on a reverse-phase column for identification and quantification.

Characterisation of polymers is often a problem due to the lack of suitable materials for calibration of the system. Most commonly encountered substances of high molecular mass are not single substances (monodisperse) but consist of a range of closely related compounds differing in mass by small amounts. This molecular weight distribution or polydispersivity of synthetic polymers is extremely important and can often be responsible for the bulk properties such as flexibility, melt viscosity or stability of the material. The most common approach to these systems is to determine the weight average molecular weight

$$\overline{M}_w = \sum_i M_i^2 N_i \Big/ \sum_i M_i N_i$$

and the number average molecular weight

$$\overline{M}_n = \sum_i M_i N_i \Big/ \sum_i N_i$$

then the dispersivity is given by

$$\text{dispersivity} = \overline{M}_w \Big/ \overline{M}_n$$

These terms cannot be evaluated using GPC unless the elution volume of the column has been calibrated against accurate absolute masses. One way is to use suitable monodisperse standards of a similar type to the sample, but it is becoming more common to attempt a 'universal' calibration using the Mark–Houwink relationship; this is a linear relationship between the solution volume (hydrodynamic radius) and the logarithm of the product of the molecular mass M and the intrinsic viscosity v of the substance:

$$\ln vM = \text{elution volume} \quad \text{or} \quad v = KM^{\alpha}$$

where K and α are constants which depend only on the pore size of the material not the sample being determined, although they do show some temperature and solvent dependence. This calculation can best be performed by eluting monodisperse standards of say polystyrene and using a dual detection system comprising a refractive index detector and a differential viscometer. The refractive index detector is a mass-sensitive system that determines concentration c, and the differential viscometer which measures specific viscosity. Since $v_{sp} = vc$ both K and α may be determined.[7] Once the relationship between elution volume and molecular mass is established using polystyrene standards, a combination of column and mobile phase can be used for other sample types using other detectors, e.g. spectroscopic detectors operating in the ultraviolet or infrared regions, which may be more sensitive than refractive index detection.

Choice of column chromatography system

To select the most appropriate column type, understand the physical characteristics of the sample and the type of information required. Figure 8.2 is a general guide to selecting chromatographic methods for separating compounds of relative molecular mass (RMM) < 2000; for RMM > 2000 the method of choice would be size exclusion or gel permeation chromatography. But a prediction of the correct chromatographic system to be used for a given sample cannot be made with certainty; it must usually be confirmed by experiment. For a complex sample, no single method may be completely adequate for the separation, and a combination of techniques may be required. Computer-aided methods for optimisation of separation conditions in HPLC have been described.[8]

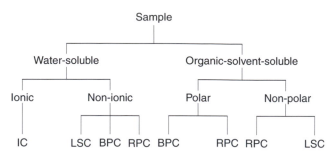

Figure 8.2 How to choose a column chromatography system: IC = ion chromatography, LSC = liquid–solid chromatography, BPC = bonded-phase chromatography, RPC = reverse-phase chromatography

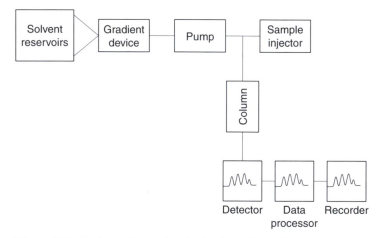

Figure 8.3 Features of a modern liquid chromatograph

8.3 Mobile phase, sample injection, column design

Figure 8.3 shows the essential features of a modern liquid chromatograph:

1. Solvent delivery system

    (a) pump
    (b) pressure controls
    (c) flow controls
    (d) inlet filter

2. Sample injection system
3. Column
4. Detector
5. Data control and display

Consult the literature for detailed descriptions of the extensive range of instrumentation that is now available.[9] In any case, manufacturers' instructions must be consulted for details of the mode of operation of particular instruments.

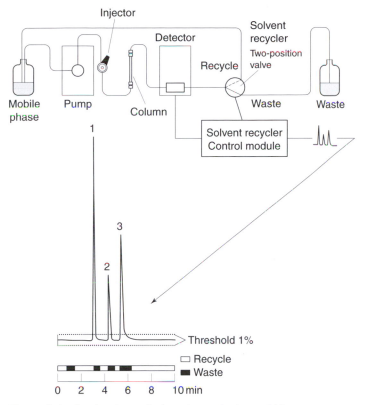

Figure 8.4   A solvent recycler is an increasingly good idea

8.3.1   The mobile phase

A successful chromatographic separation depends upon differences in the interaction of the solutes with the mobile phase and the stationary phase, and in liquid chromatography the choice and variation of the mobile phase is of critical importance in achieving optimum efficiency. However, before considering the theoretical aspects of solvent choice, here are a few general comments. HPLC grade solvents tend to be costly. To ensure consistent performance, the solvent needs to contain no more than trace amounts of other materials, including water for organic solvents. And if the system uses a UV detector then even small traces of absorbing species will be unacceptable; they must be exhaustively removed. Particulates in the solvent are also highly undesirable; with prolonged use they will lead to wear in the pump and injector and cause blockage of the column.

Although normal laboratory grade solvents could be suitably purified, this is a time-consuming step and many laboratories prefer to purchase HPLC grade reagents, including water. HPLC grade reagents can be used directly without further purification, although they may need to be degassed immediately before use. These solvents are an appreciable fraction of the running cost of an HPLC laboratory, and increasingly the use of solvent recylers is to be recommended (Figure 8.4). Recyclers are computer-controllable valving systems fitted beyond the detector; they ensure that solvent which does not contain absorbing analyte is returned to the reservoir and only solvent which contains absorbing species is

Table 8.2 *Properties of common solvents*

	Polarity index	Eluent strength	Refractive index	UV cut-off (nm)	Viscosity at 20°C (mN s m^{-2})	Boiling point (°C)
Fluoroalkanes	<–2	–0.25	1.27–1.29	200	0.4–2.0	50–174
Cyclohexane	0.04	–0.2	1.423	200	0.90	81
n-Hexane	0.1	0.01	1.372	195	0.33	69
1-Chlorobutane	1.0	0.26	1.400	220	0.42	78
Carbon tetrachloride	1.6	0.18	1.459	265	0.97	76.8
2-Propylether	2.4	0.28	1.367	210	0.38	68
Methylbenzene	2.4	0.29	1.494	280	0.59	110
Chlorobenzene	2.7	0.30	1.523	290	0.80	132
Diethylether	2.8	0.38	1.352	205	0.23	35
Dichloromethane	3.1	0.42	1.421	233	0.44	40
Tetrahydrofuran	4.0	0.45	1.404	238	0.55	66
Chloroform	4.1	0.40	1.444	245	0.58	61.2
Ethanol	4.3	0.88	1.359	207	1.20	78.5
Ethylethanoate	4.4	0.58	1.370	255	0.45	77
1,4-Dioxane	4.8	0.56	1.420	216	1.54	101.3
Methanol	5.1	0.95	1.326	210	0.60	64.7
2-Propanone	5.1	0.56	1.357	330	0.33	56.2
Acetonitrile	5.8	0.65	1.342	190	0.37	82.0
Nitromethane	6.0	0.64	1.380	380	0.70	101
Ethanoic acid	6.0	high	1.370	230	1.28	117.9
Ethylene glycol	6.9	1.11	1.429	235	19.9	182
Dimethylsulphoxide	7.2	0.62	1.476	270	2.0	189
Water	10.2	high	1.333	190	1.002	100

vented to waste. The manufacturers of recyclers claim the capital cost of the system can quickly be recovered in greatly reduced solvent usage and solvent disposal.

The choice of a suitable **mobile phase** is vital in HPLC and it is appropriate to refer to the factors influencing this choice. Thus, the eluting power of the mobile phase is determined by its overall polarity, the polarity of the stationary phase and the nature of the sample components. For normal-phase separations, eluting power increases with increasing polarity of the solvent; for reverse-phase separations, eluting power decreases with increasing solvent polarity. The popularity of reverse-phase chromatography is partly due to the types of material commonly encountered in LC, but also because there are several advantages in using polar mobile phases such as water and methanol. Compared with their non-polar counterparts, polar mobile phases are cheaper, generally purer, less affected by small amounts of impurities and less toxic.

Currently there is no one system which adequately ranks all of the common solvents in terms of polarity, but probably the best available is the Snyder classification which has been applied to 81 solvents.[3] Each solvent is given a polarity index P on the basis of gas chromatographic data (Table 8.2), and then three further parameters are evaluated: proton acceptor, proton donor and strong dipole parameter. Using this information, solvents are split into eight groups with similar characteristics. This is more satisfactory than the older and simpler eluotropic series based on a single term ε^0 known as the solvent strength parameter, although these values are also listed in Table 8.2 for comparison. The polarity index ranges from -2 to 10.2 and in theory any value between these limits can be obtained by mixing two or more solvents together.

Other properties of solvents which need to be considered are boiling point, viscosity (lower viscosity generally gives greater chromatographic efficiency), detector compatibility, flammability and toxicity. Many of the common solvents used in HPLC are flammable and some are toxic, so it is advisable for HPLC instrumentation to be used in a well-ventilated laboratory, if possible under an extraction duct or hood.

Special grades of solvents are available for HPLC which have been carefully purified to remove UV-absorbing impurities and any particulate matter. If, however, other grades of solvent are used, purification may be required since impurities present would affect the detector if strongly UV-absorbing, or influence the separation if of higher polarity than the solvent (e.g. traces of water or ethanol, commonly added as a stabiliser, in chloroform). It is also important to remove dissolved air or suspended air bubbles, a major cause of practical problems affecting the pump and detector. These problems may be avoided by degassing the mobile phase before use; this can be accomplished by placing the mobile phase under vacuum, or by heating and ultrasonic stirring. Compilations of useful solvents for HPLC are available.[10]

There is no simple solution to choosing the most suitable phase for a given separation other than 'like separates like', and the analyst must be guided by the literature and information supplied by the manufacturer.

8.3.2   Optimisation of the mobile phase

Since mobile phase composition is the parameter which the chromatographer can vary most easily in order to optimise a separation in LC, considerable effort has gone into ways of making this choice as effective as possible. For relatively simple separations, it is often possible to choose a single solvent mixture for the whole analysis. This isocratic operation is simpler than gradient elution methods (described later). However, the required solvent system will probably contain more than one component, perhaps as many as four. The best way of defining these complex solvent systems is to draw a solvent triangle for ternary systems and a tetrahedron for quaternary compositions (Figure 8.5). The composition of any mixture is then given by a point within the triangle or tetrahedron. The problem is to find the point which gives the best separation for a given mixture. A number of methods

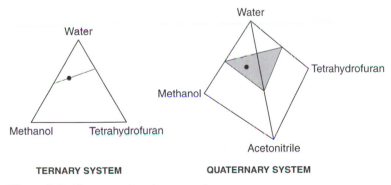

Figure 8.5   Ternary and quaternary systems

have evolved, based on simplex algorithms, in which a sequence of experiments are performed and the composition is adjusted until an optimum set of conditions are found.

These sequential or iterative approaches can require a large number of chromatographic experiments to reach the desired optimum, and under certain circumstances they may be replaced by predictive methods in which the results from a restricted number of experiments are processed by a computer program. The first of these experiments is usually a gradient elution experiment from 0 to 100% of each of the solvent pairs; this defines either a line (for ternary systems) or a plane (for quaternary systems) in which the optimum composition will lie. Commercial software is available for this approach, and the whole process can be automated if used on suitable systems fitted with automatic injection. These optimisation systems have been applied mainly to reverse-phase chromatography, and the most common solvent combinations are water (or buffer) W, methanol M, acetonitrile A and tetrahydrofuran T. If solvent optimisation methods are available, they can often replace gradient elution techniques for repetitive analysis; gradient elution is generally slower than isocratic operation if the re-equilibration time after each gradient run is taken into account.

8.3.3   Gradient elution

Gradient elution can frequently separate solutes that cannot be separated by isocratic operation, even with mixed solvents. It is especially effective where the sample components differ widely in polarity.

For gradient elution using a low-pressure mixing system, the solvents from separate reservoirs are fed to a mixing chamber and the mixed solvent is then pumped to the column; modern instruments use time-proportioning electrovalves regulated by a microprocessor.[11] Not only can this increase the resolution for each chromatogram, it can also reduce the run-time and increase the sensitivity, as peaks which would have excessive retention times and possibly show tailing are made to elute quicker as the gradient develops.

Many commercial liquid chromatographs have liquid-handling systems that can mix two, or occasionally more, solvents in a progressive manner from 0 to 100% of one component, so that a solvent composition profile is generated with respect to time (Figure 8.6). The profile, which may be convex, concave or linear, has a similar effect to temperature programming in GC, enabling separation of closely eluting peaks in one part of the chromatogram without changing the resolution in other parts. But there are several constraints on this technique. If one of the solvents gives an appreciable response at the detector, say absorption at a UV detector, then the generation of a solvent gradient will also introduce a baseline drift in response. The column will also need time to regenerate to the starting solvent composition each time a fresh gradient run is started, and ideally a blank gradient is run between samples to prevent the occurrence of artefact peaks which can be observed. This can make gradient elutions seem slower than the published retention times. The selection of the correct solvent profile can also be a time-consuming step, again showing the similarity to temperature programming in GC. Gradients of changing buffer strength in an aqueous system, or mainly aqueous system, may also be used in ion chromatography.

8.3.4   Solvent delivery and sample injection

Although a variety of different pumping systems have been used in HPLC, the most popular type is the small volume, constant flow, reciprocating pump which can provide accurately controlled flow rates of $1-15 \, mL \, min^{-1}$ against a column back pressure of up to 7250 psi (50 MPa). The reason for their popularity can be understood when the performance

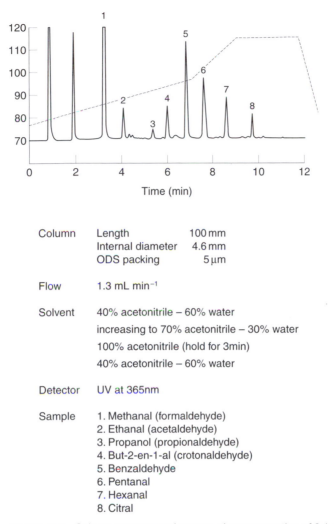

Column	Length	100 mm
	Internal diameter	4.6 mm
	ODS packing	5 μm

Flow	1.3 mL min⁻¹

Flow   1.3 mL min^{-1}

Solvent   40% acetonitrile – 60% water

increasing to 70% acetonitrile – 30% water

100% acetonitrile (hold for 3min)

40% acetonitrile – 60% water

Detector   UV at 365nm

Sample   1. Methanal (formaldehyde)
2. Ethanal (acetaldehyde)
3. Propanol (propionaldehyde)
4. But-2-en-1-al (crotonaldehyde)
5. Benzaldehyde
6. Pentanal
7. Hexanal
8. Citral

Figure 8.6   Solvent-programmed reverse-phase separation of 2,4-dinitrophenylhydrazine derivatives of aldehydes

characteristics are considered in more detail. The heart of the pump is the pump head containing a piston and a solvent chamber, which can be of very small volume (10–50 μL) containing two check valves mounted in line with each other (Figure 8.7).

As the main piston is withdrawn from the chamber, valve 1 opens, allowing mobile phase to enter the pump head, and valve 2 is shut, preventing liquid from being drawn out of the column. As the piston enters the chamber on the return stroke, valve 1 is closed, preventing a backflow into the solvent reservoir, and valve 2 opens to allow mobile phase to be pumped into the column. This process is repeated at such a rate that the desired flow is produced through the column. (For a head with volume 10 μL this would correspond to 100 strokes per minute.) A single-headed pump will provide a pulsed flow, since mobile phase is only forced through the column on alternate strokes of the piston, and this can cause flow noise at the detector.

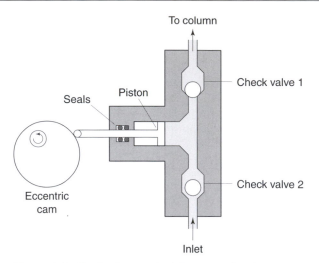

Figure 8.7 Dual-valve constant-flow pump head

Flow noise can be reduced by using a pump with a constant, very rapid stroke cycle, say 20 strokes per second, so the detector cannot respond rapidly enough to the fluctuations; the flow rate is then controlled by altering the length of the piston stroke. But the more common way of reducing flow noise is to use a twin-headed pump in which the operation of each head is 180° out of phase with the other; then as one chamber is filling, the other is delivering solvent to the column. This type of pump is very reliable and trouble-free; it can be constructed of relatively inert materials so the mobile phase only comes into contact with stainless steel, ruby (the ball valves are made of this) and inert seals or valve seats. Also, because it has a small volume, it can easily be used for solvent programming.

The range of flow rates that can be accurately delivered by a given piston pump is becoming of more concern, especially for the low flow rates required for narrow bore and capillary chromatography. These systems require flow rates of between 1 and $500\,\mu L\,min^{-1}$, where piston pumps are not reliable, so pressurised gas control has been used, but pressurised gas cannot give constant flow. Membrane pumping systems are a further alternative. A more radical approach, currently under development, uses the electro-osmotic effect of the mobile phase to provide small accurately controlled flows.[12]

Sample injection system

Introduction of the sample is generally achieved in one of two ways, either by using syringe injection or through a sampling valve. Septum injectors allow sample introduction by a high-pressure syringe through a self-sealing elastomer septum. One of the problems associated with septum injectors is the leaching effect of the mobile phase in contact with the septum, which may give rise to ghost peaks. In general, syringe injection for HPLC is more troublesome than in gas chromatography.

Although the problems associated with septum injectors can be eliminated by using stop-flow septumless injection, currently the most widely used devices in commercial chromatographs are the microvolume sampling valves (Figure 8.8) which enable samples to be introduced reproducibly into pressurised columns without significantly interrupting the flow of the mobile phase. The sample is loaded at atmospheric pressure into an external

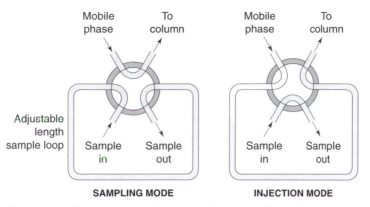

Figure 8.8   The sample loop in a microvolume sampling valve

loop in the valve and introduced into the mobile phase by an appropriate rotation of the valve. The volume of sample introduced, ranging from $2\,\mu L$ to over $100\,\mu L$, may be varied by changing the volume of the sample loop or by using special variable-volume sample valves. Automatic sample injectors are also available which allow unattended (e.g. overnight) operation of the instrument. Valve injection is preferred for quantitative work because of its higher precision compared to syringe injection.

8.3.5   The column

The columns most commonly used are made from precision-bore polished stainless steel tubing; typical dimensions are 10–30 cm long and 4 or 5 mm internal diameter. The stationary phase or packing is retained at each end by thin stainless steel frits with a mesh of $2\,\mu m$ or less.

The packings used in modern HPLC consist of small, rigid particles having a narrow particle size distribution. The types of packing may conveniently be divided into the following general categories:

(a)   Porous, polymeric beads based on styrene–divinylbenzene copolymers. These are used for ion chromatography and exclusion chromatography (Section 8.2), but have been replaced for many analytical applications by silica-based packings which are more efficient and mechanically stable.

(b)   Porous-layer beads (diameter $30–55\,\mu m$) consisting of a thin shell ($1–3\,\mu m$) of silica, or modified silica or other material, on an inert spherical core (e.g. glass beads). These pellicular packings are still used for some ion exchange applications, but their general use in HPLC has declined with the development of totally porous microparticulate packings.

(c)   Totally porous silica particles (diameter $< 10\,\mu m$, with narrow particle size range) are now the basis of the most commercially important column packings for analytical HPLC. Compared with the porous-layer beads, totally porous silica particles give considerable improvements in column efficiency, sample capacity and speed of analysis.

The development of bonded phases (Section 8.2) for liquid–liquid chromatography on silica gel columns is of major importance. For example, the widely used C-18 type permits the separation of moderately polar mixtures and is used for the analysis of pharmaceuticals, drugs and pesticides.

The procedure chosen for column packing depends chiefly on the mechanical strength of the packing and its particle size. Particles of diameter > 20 µm can usually be dry-packed, whereas particles with diameters < 20 µm are packed using slurry techniques in which the particles are suspended in a suitable solvent and the suspension (or slurry) driven into the column under pressure. The essential features for successful slurry packing of columns are summarised in the literature.[13] But many analysts will prefer to purchase the commercially available HPLC columns, for which the appropriate catalogues should be consulted.

The length of the column can affect not only the resolution of a given separation – the longer the column the greater the plate number – but also the speed of separation, and it is often the separation speed which is most significant in choosing a given length. Standard lengths vary from manufacturer to manufacturer, but the most common values are 300, 250, 150, 125, 100 and 75 mm; the longer columns are described as standard columns and the shorter columns as high-speed columns.

Effect of column length

To achieve a given efficiency, say 12 500 plates, a 250 mm column containing 5 µm packing could be used. The same efficiency could be obtained by using a 150 mm column packed with 3 µm particles. (Halving the size of the particle doubles the efficiency.) For the same linear velocity through the column, the speed of analysis will be increased by 250/150; but since smaller particles have a higher optimum linear velocity, this increase in separation speed is in fact $250/150 \times 5/3 = 3$ times faster for the same resolution. The column packed with the finer particles is more expensive than the standard 5 µm packing.

Effect of diameter

A wide range of analytical column diameters is commercially available, ranging from 4.6 mm internal diameter to less than 0.2 mm (200 µm). If preparative LC columns are considered, they may be as wide as 50 mm or more. We will only consider analytical columns. Generally, the wider the column the higher the loading capacity, hence the greater the sample size that may be injected. The converse is of course also true, but other factors occur as the column diameter decreases which make narrow bore columns more attractive, at least in theory. The reason for this terminology can best be explained with an example.

Consider two columns with diameters 4.6 mm and 2.1 mm (often known as narrow bore), each packed with the same material, and thus having the same plate number and efficiency. In order to maintain the same linear velocity in the two columns, the flow rate in the narrower column will be reduced by almost a factor of 5 (4.8 to be precise). This means that less solvent is used for a given separation, and thus less solvent has to be disposed of, considerably reducing the operating costs of the system. A second, and often important advantage, is that a given substance elutes in a smaller volume of solvent, so that for most detectors, which have a concentration-dependent response, the peak height will be increased by a factor of almost 5 and the detection limit will be reduced by the same factor. The advantages of using narrow bore columns have led to considerable research into the use of columns as small as 0.1 mm (100 µm), but small diameters place severe constraints on the other components in the system, since they require very low flow rates and minimal dead volumes.

Although no general consensus exists for describing different columns, it has been suggested[14] that the term 'microbore' is used for diameters between 2.0 and 0.5 mm and the term 'microcolumn' for diameters below 0.5 mm. Microcolumns are further subdivided into packed capillary (40–300 μm) and open tubular (3–50 μm). Open tubular (OT) columns are in fact constructed of coated fused silica and are mechanically identical to capillary GC columns. At the time of writing, few commercial systems are capable of using either micro-bore or microcolumns, possibly because the system has to be carefully designed to minim-ise dead volumes, and the pumping system and detector must be able to operate at flow rates as low as $1\,\mu L\,min^{-1}$. Although there are a number of mechanical difficulties in the use of these columns, if they could be overcome then it should be possible to perform separations which require in excess of 100 000 plates in less than 30 min, using very small amounts of mobile phase and at high sensitivity.

Finally, the useful life of an analytical column is increased by introducing a **guard column**. This is a short column which is placed between the injector and the HPLC column to protect the HPLC column from damage or loss of efficiency caused by particulate matter or strongly adsorbed substances in samples or solvents. It may also be used to saturate the eluting solvent with soluble stationary phase. Guard columns may be packed with micro-particulate stationary phases or with porous-layer beads; beads are cheaper and easier to pack than microparticulates, but they have lower capacities and therefore need to be changed more frequently.

8.4   Choosing a detector

The function of the detector in HPLC is to monitor the mobile phase as it emerges from the column. The detection process in liquid chromatography has presented more prob-lems than in gas chromatography; there is no equivalent to the universal flame ionisation detector of gas chromatography for use in liquid chromatography. Suitable detectors can be broadly divided into the following two classes:

Bulk property detectors measure the difference in some physical property of the solute in the mobile phase compared to the mobile phase alone, e.g. refractive index and conductivity detectors. Although generally universal in application, they tend to have poor sensitivity and limited range. They are usually affected by even small changes in the composition of the mobile phase, which precludes their use with gradient elution.

Solute property detectors may be spectrophotometric, fluorescence and electrochemical detectors. They respond to a particular physical or chemical property of the solute, and ideally they are independent of the mobile phase. In practice, however, complete inde-pendence of the mobile phase is rarely achieved, but the signal discrimination is usually sufficient to permit operation with solvent changes, e.g. gradient elution. They generally provide high sensitivity (about 1 in 10^9 is attainable with UV and fluorescence detectors) and a wide linear response range but, as a consequence of their more selective natures, more than one detector may be required to meet the demands of an analytical problem. Some commercially available detectors have a number of different detection modes built into a single unit, e.g. the Perkin-Elmer '3D' system, which combines UV absorption, fluorescence and conductimetric detection.

Some of the important characteristics required of a detector are as follows.

Sensitivity is often expressed as the noise equivalent concentration, i.e. the solute concen-tration C_n which produces a signal equal to the detector noise level. The lower the value of C_n for a particular solute, the more sensitive the detector for that solute.

Linear range is the concentration range over which a detector's response is directly pro-portional to the concentration of solute. Quantitative analysis is more difficult outside the linear range of concentration.

Table 8.3 *Detectors used in high-performance liquid chromatography*

Name	Approximate limit of detection ($\mu g \, mL^{-1}$)	Gradient	Application
UV/visible absorbance	10^{-4}	Y	Selective, versatile
Fluorescence	10^{-5}	Y	Selective, limited number of compounds
Chemiluminescence	2×10^{-7}	Y	Selective, restricted group of compounds
Laser-induced fluorescence	low	Y	Selective, limited range of compounds
Low-angle laser light scattering (LALLS)	10	N	Species with high molecular weight
FT-IR	1	Y	Selective, versatile
Conductivity	10^{-2}	Y/N	Ions and ionisable species
Amperometric	10^{-5}	N	Selective, oxidisable or reducible species
Refractive index	10^{-2}	N	Universal detector
Mass spectrometry	10^{-5}	Y	Universal detector
Optical activity	10^{-4}	N	Chiral centres
ICP-MS	2.5×10^{-3}	Y	Metallic species

Type of response, i.e. whether the detector is universal or selective. A universal detector will sense all the constituents of the sample, whereas a selective detector will only respond to certain components. Although the response of the detector will not be independent of the operating conditions, e.g. column temperature or flow rate, it is advantageous if the response does not change too much when there are small changes of these conditions.

A great deal of effort has been put into the design and construction of detectors suitable for HPLC. One review[15] listed no fewer than 30 different types of detection system based upon almost all the conventional spectroscopic techniques, including NMR and ESR, plus a number of non-spectroscopic systems. In each case claims were made for increased sensitivity, selectivity or compatibility with microbore or capillary columns, since it is in these areas that constraints on detector design are still most apparent. The majority of HPLC detectors are spectroscopic (Table 8.3), with electrochemical systems forming the next largest group, followed by a number of special techniques for specific applications.

8.4.1   Ultraviolet detectors

Most widely used of the HPLC detectors, the UV absorption detector works by measuring how much UV/visible light is absorbed as effluent from the column is passed through a small flow cell held in the radiation beam. It is characterised by high sensitivity (detection limit of about $1 \times 10^{-9} \, g \, mL^{-1}$ for highly absorbing compounds), and since it is a solute property detector, it is relatively insensitive to changes of temperature and flow rate. The UV detector is generally suitable for gradient elution work since many of the solvents used

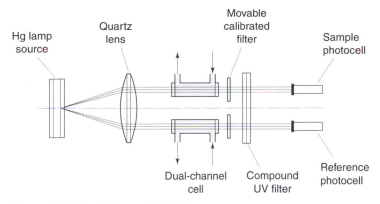

Figure 8.9   Double-beam UV detector

in HPLC do not absorb to any significant extent at the wavelengths used for monitoring the column effluent. The presence of air bubbles in the mobile phase can greatly impair the detector signal, causing spikes on the chromatogram; this effect can be minimised by degassing the mobile phase prior to use, e.g. by ultrasonic vibration. Both single- and double-beam instruments are commercially available (Figure 8.9). Although the original detectors were single- or dual-wavelength instruments (254 and/or 280 nm), some manufacturers now supply variable-wavelength detectors covering the range 210–800 nm so that more selective detection is possible.

Although single selective wavelength UV detectors are used extensively in HPLC, they are rapidly being displaced by photodiode arrays and charge transfer systems which can record the whole of a spectrum many times a second. In these detectors, polychromatic light is passed through the HPLC flow cell and the emerging radiation is diffracted by a grating to fall on the array. This may consist of a series of over a thousand semiconductor sensors, photodiodes or semiconductor charge transfer devices. Each unit of the detector receives a different narrow wavelength band and is scanned several times per second by a microprocessor. The resulting spectrum from the collected signals can be presented on a visual display unit, and the changes in the strength and nature of the spectrum are recorded as separated compounds pass through the detector cell. An important feature of the multi-channel detector is that it can be programmed to give changes in detection wavelength at specified points in the chromatogram; this facility can be used to 'clean up' a chromatogram, e.g. by discriminating against interfering peaks due to compounds in the sample which are not of interest to the analyst.

8.4.2   Luminescence detectors

The fluorescence of a number of compounds may be used in conventional fluorescence detectors; these detectors are both selective and sensitive for materials which naturally fluoresce or materials which can be made fluorescent by post-column derivatisation. Since the emission intensity in this technique is directly proportional to excitation intensity, a number of systems are beginning to use laser sources rather than the older xenon lamps; these systems are called laser-induced fluorescence (LIF) detectors. One system uses a frequency-doubled argon ion laser operated at 257 nm. The extremely high intensity and spatial resolution of lasers mean they are well suited to microbore or capillary systems, and can provide high selectivity and extremely low detection limits for suitable compounds. Several authors have quoted detection limits of between 3×10^{-18} and 10×10^{-18} mol for systems such as nucleotides. The application of fluorescence detectors has been extended by means of pre- and post-column derivatisation of non-fluorescent or weakly fluorescing compounds (Section 8.7).

Two factors limit fluorescence detection: the compound of interest must be fluorescent, and the detector signal will be swamped by any fluorescent impurities in the sample or the solvent. This has led to the development of chemiluminescent detection in which the excitation energy is supplied by chemical rather than spectroscopic means. Most systems use post-column mixing of the chromatographic eluent with a luminol or peroxyoxalate solution; this is then passed through a commercial fluorescence detector which has the source switched off, or blanked out with a shutter. However, some have used an immobilised reagent system. Detection limits for suitable materials are an order of magnitude lower than for fluorescence, and the selectivity is high. Most systems studied to date have been of biological or pharmaceutical interest[17] but trace metals, including cobalt, have also been determined.

8.4.3 Other spectroscopic detectors

Under the general heading of spectroscopic detectors, one should include mass spectroscopy, for which a number of interfaces have already been described. The potential analytical power of LC/MS is so high that many combinations of interface and analyser are available commercially, and it is often difficult to decide upon the best combination for a problem given that, whichever system is chosen, some compromise has to be made. Although not commercially available at present, several promising applications of direct coupling of microbore or capillary columns have been reported[18] and it is probable they will replace the more complicated systems, as has occurred in GC/MS.

8.4.4 Refractive index detectors

Refractive index detectors are based on the change of refractive index of the eluant from the column with respect to pure mobile phase. Although widely used, they suffer from several disadvantages: lack of high sensitivity, lack of suitability for gradient elution, and the need for strict temperature control ($\pm 0.001\,^\circ$C) to operate at their highest sensitivity. A pulseless pump, or a reciprocating pump equipped with a pulse dampener, must also be employed. The effect of these limitations may to some extent be overcome by the use of differential systems in which the column eluant is compared with a reference flow of pure mobile phase. The two chief types are deflection refractometers and Fresnel refractometers.

Deflection refractometers

Deflection refractometers measure the deflection of a beam of monochromatic light by a double prism in which the reference and sample cells are separated by a diagonal glass divide (Figure 8.10). When both cells contain solvent of the same composition, no deflection of the light beam occurs; but if the composition of the column's mobile phase is changed because of the presence of a solute, then the altered refractive index causes the beam to be deflected. The magnitude of this deflection depends on the concentration of the solute in the mobile phase.

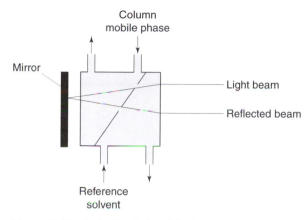

Figure 8.10   Refractive index detector

Fresnel refractometers

Fresnel refractometers measure the change in the fractions of reflected and transmitted light at a glass–liquid interface as the refractive index of the liquid changes. In this detector both the column's mobile phase and a reference flow of solvent are passed through small cells on the back surface of a prism. When the two liquids are identical there is no difference between the two beams reaching the photocell, but when the mobile phase containing solute passes through the cell there is a change in the amount of light transmitted to the photocell, and a signal is produced. The smaller cell volume (about 3 µL) in this detector makes it more suitable for high-efficiency columns, but for sensitive operation the cell windows must be kept scrupulously clean.

8.4.5   Electrochemical detectors

Although detectors are commercially available for monitoring a number of electron transfer reactions – amperometric, voltammetric or polarographic, potentiometric or electrode, and coulometric – they have failed to achieve widespread popularity. Polarographic detectors are probably the most widely used (Figure 8.11); they can measure nanoampere currents at the glassy carbon working electrode corresponding to as little as 5 pg of reducible material. However, although over 500 different compounds have been detected, ranging from drugs of abuse to phenols in water, polarographic detectors are very sensitive to flow rate and temperature; they also require an aqueous solvent, free from oxygen. Hence polarographic detectors are not as widely used as spectroscopic detectors. The surface of the glassy carbon electrode must also be kept highly polished to ensure a sufficiently fast response time for chromatographic detection.

In contrast to detectors based upon oxidation or reduction, those which are based on an equilibrium in the system, in particular conductivity, have found wide application in ion chromatography. Due to their mode of operation, these detectors are simple, have small internal volumes, and can be considered universal for the species which are determined (mainly anions, but also cations). But the sensitivity is determined mainly by processes which occur prior to the detector.

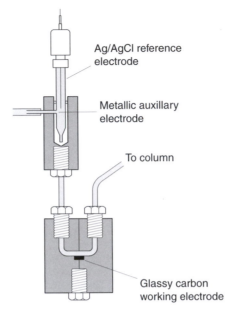

Ag/AgCl reference
electrode

Metallic auxillary
electrode

To column

Glassy carbon
working electrode

Figure 8.11   Electrochemical detector

8.5   Column efficiency

The many factors that affect the separation efficiency led to the development of the so-called plate theory, first introduced into chromatography by Martin and Synge. And plate theory led to the more sophisticated treatment afforded by rate theory, involving the various aspects of the van Deemter equation. The ideas behind these theoretical concepts are considered in more detail in Chapter 9, but also look at Section 6.9 and Figure 6.6, which explain the equation for calculating the number of theoretical plates in a column as well as solute retention times t_r, peak widths w and half-widths $w_{1/2}$.

In all forms of chromatography, the final width of a band depends not only on chemical effects within the column, but also on the design and construction of the total system. Any dead volume in the system can have serious effects on resolution. This is most pronounced in HPLC, where small flow rates and complex plumbing can produce severe broadening unless the system is carefully planned. Take special care to follow any advice for minimising dead volumes. And make sure the separating mechanism, the column and the mobile phase are matched to the sample.

The capacity factor for a given solute is directly related to the nature of the solute, the stationary phase, the mobile phase and the temperature. It is a measure of the efficiency of the separation process occurring at the phase boundary. For two solutes, separation of the peaks will require different values of k' for each component, and a separation or selectivity factor α can be defined as

$$\alpha = k'_2/k'_1 = \frac{t_{r2} - t_0}{t_{r1} - t_0}$$

By convention the peaks are chosen so that $\alpha > 1$. But the two terms α and k' are not sufficient to define the system on their own; besides the interactions between solute and the two phases (solvent efficiency), other mechanisms always lead to peak broadening. The amount of broadening can be measured by the resolution R_s between two adjacent peaks, which in chromatography is defined as

$$R_s = \frac{t_{r2} - t_{r1}}{0.5(w_1 + w_2)}$$

With this definition, baseline separation occurs at $R_s = 1.5$.

A few example calculations based on HPLC systems will show how important it is to use a well-designed total system, especially for modern, small columns. Assume the total peak width w_t consists of the width w_c due to column dispersion and the width w_e due to extra column effects. Since w_c and w_e are sources of error in measurement, they are treated as variances, i.e.

$$w_t^2 = w_c^2 + w_e^2$$

Example 8.1

For a column of 25 cm × 4.6 mm packed with 5 µm packing and assuming 70% of the volume is solvent, we have

$$\text{total volume} \quad = 250 \times (4.6)^2/2 \times \pi = 4.15 \text{ mL}$$

$$\text{solvent volume} = 4.15 \times 0.7 = 2.9 \text{ mL}$$

Assuming a plate count N of 12 500, which is typical for this type of column, and that an unresolved peak would take one solvent volume to elute.

$$\text{From} \quad N = 16 t_r^2/w \quad \text{where} \quad t_r = 2.9 \text{ mL}$$

$$w = t_r \times 16^{1/2}/N = 104 \text{ µL}$$

For w_t to be no more than 110% of w_c, i.e. 10% increase in column widths due to extra column effects, then

$$114^2 = 104^2 + w_e^2 \quad \text{thus} \quad w_e = 47 \text{ µL}$$

Since normal injection volumes are 20 µL, we have

$$\text{detector volume} \quad = 5\text{–}20 \text{ µL}$$

$$\text{connector volume} = 20 \text{ µL (assuming 10 cm of 0.5 mm i.d. s.s.)}$$

$$\text{total volume} \quad = 45\text{–}60 \text{ µL}$$

Thus the increase in peak width due to extra column effects is seldom more than 10% in these systems.

Example 8.2

For a 25 cm × 2.1 mm column packed with 3 µm packing (same packing density)

$$\text{total volume} \quad = 866 \text{ µL}$$

$$\text{solvent volume} = 606 \text{ µL}$$

For $N = 30\,000$ (note the increased plate number) this corresponds to $w_c = 14\,\mu L$, hence for a maximum of 10% increase due to w_e this gives $w_e = 6.4\,\mu L$. For this type of column, not only must the sample volume be kept small ($1\,\mu L$), but all other sources of broadening such as detector volume and connecting tubing must be carefully considered, since less than $7\,\mu L$ of dead volume will give a 10% increase in peak width.

Detectors of less than $5\,\mu L$ volume are becoming available, as is low-volume connection tubing

Internal diameter (mm)	0.5	0.25	0.12
Volume (µL) of a 10 cm length	19.63	4.5	1.1

Similar considerations apply to GC systems but because of the relatively narrow column bore, especially in capillary GC, the most likely problems are dead or unswept volumes in the injector or detector. Careful design of these components is again essential for maximum resolution. Care is also needed with the design of injection systems if relatively large volumes of sample are to be injected. For example, $1\,\mu L$ of a typical solvent may expand in the injector to $1000\,\mu L$ of gas, which could seriously affect resolution in a column with flow rates of only $1-2\,mL\,min^{-1}$.

8.6   Chiral chromatography

A number of biologically important substances can exist as enantiomer pairs or optical isomers – there are two forms of the material, with identical chemical and physical behaviour, but differing properties in an asymmetrical environment. Often it is found that one enantiomer is inactive or even harmful compared to the other in medical applications. Possibly the most infamous example of this to date has been the thalidomide tragedy, where the other enantiomeric form of a beneficial sedative drug gave rise to severe malformation in the foetus of pregnant women taking the drug to reduce morning sickness. It is thus increasingly important that chiral analysis (separation and quantification of two enantiomers) is performed on medicinal and agricultural chemicals. The first stationary phases capable of performing chiral separations were produced in the early 1980s by Professor W. H. Pirkle and colleagues[16] and are commonly known as Pirkle packings. They are constructed by bonding a short asymmetric organic chain to the surface of conventional silica packings in a similar way to the production of bonded phases. Although effective at separating the optical isomers of a number of organic compounds, they are used as normal-phase materials with THF–hexane or isopropanol–hexane as the eluting solvent. They are therefore not very effective for separating hydrophilic materials.

Other packings based on larger optically active materials, notably the cyclodextrins, can be used in normal and reversed modes of separation, but the most important group of chiral stationary phases are those based upon natural protein molecules that are inherently asymmetric. Probably the most widely used of these is immobilised human serum albumin (HSA), which may be bonded to silica and used as a reversed-phase material to separate water-soluble substances by utilising the affinity of only one enantiomer to the stationary phase.[17] This is therefore a sophisticated form of affinity chromatography, whereby optically active materials can be separated from each other.

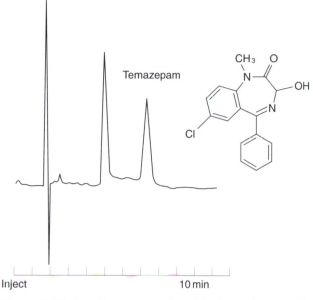

Figure 8.12   Enantiomer separation of the benzodiazepine Temazepam: column (4.6 mm × 150 mm) of chiral HSA; mobile phase 0.05 M phosphate buffer (pH 6.9) and acetonitrile in 9:1 mixture; flow rate 1.5 mL min^{-1}; UV detector set at 254 nm

Like all proteins, HSA is sensitive to temperature, ionic strength, organic modifiers and particularly pH, and these (expensive) chiral columns must be handled and used with care. Although stable at pH 6–8, separations are normally carried out with a low concentration of phosphate buffer (0.05 M) at pH 7 and using either 1-propanol or acetonitrile up to a maximum of 10–15% as the organic modifier. These columns are finding wide application in chiral separation of neutral and acidic materials, particularly drugs, and a typical separation of a diazepam drug is shown in Figure 8.12. For more basic biological materials, the more acidic alpha-1-acid glycoprotein can give better separations. A number of manufacturers now supply these columns specifically tailored for a particular application. Apart from the analytical application of this type of column, it is found that more fundamental data about the binding of drugs to HSA may be obtained, and this direct measurement of affinity can be related to the pharmacokinetic and pharmacological behaviour of new compounds more quickly than using the traditional techniques of equilibrium dialysis or ultracentrifugation.

8.7   Derivatisation

In liquid chromatography, in contrast to gas chromatography (Section 9.2), derivatives are almost invariably prepared to enhance the response of a particular detector to the substance of analytical interest. For example, with compounds lacking an ultraviolet chromophore in the 254 nm region but having a reactive functional group, derivatisation provides a means of introducing into the molecule a chromophore suitable for its detection. Derivative preparation can be carried out either before the separation (precolumn derivatisation) or afterwards (post-column derivatisation). The most commonly used techniques are pre-column off-line and post-column on-line derivatisation.

Precolumn off-line derivatisation requires no modification to the instrument and, compared with the post-column techniques, imposes fewer limitations on the reaction conditions. Disadvantages are that the presence of excess reagent and by-products may interfere with the separation, and the group introduced into the molecules may change the chromatographic properties of the sample.

Post-column on-line derivatisation is carried out in a special reactor situated between the column and detector. A feature of this technique is that the derivatisation reaction need not go to completion provided it can be made reproducible. But the reaction needs to be fairly rapid at moderate temperatures and there should be no detector response to any excess reagent present. An advantage of post-column derivatisation is that ideally the separation and detection processes can be optimised separately. A problem which may arise, however, is that the most suitable eluant for the chromatographic separation rarely provides an ideal reaction medium for derivatisation; this is particularly true for electrochemical detectors which operate correctly only within a limited range of pH, ionic strength and aqueous solvent composition.

Reagents which form a derivative that strongly absorbs UV/visible radiation are called **chromatags**; an example is the reagent ninhydrin, commonly used to obtain derivatives of amino acids which show absorption at about 570 nm. Derivatisation for fluorescence detectors is based on the reaction of non-fluorescent reagent molecules with solutes to form fluorescent species. These non-fluorescent reagents are called **fluorotags**. The reagent dansyl chloride [8.A] is used to obtain fluorescent derivatives of proteins, amines and phenolic compounds; its excitation and emission wavelengths are 335–365 nm and 520 nm, respectively.

[8.A]

Consult a relevant textbook[18] for a more comprehensive account of chemical derivatisation in liquid chromatography, and also the items listed in Section 8.10.

8.8 Quantitative analysis

Quantitative analysis by HPLC ideally requires a linear relationship between the magnitude of the signal and the concentration of any particular solute in the sample; the signal is measured by either the corresponding peak area or the peak height. Peak area measurements are preferred when the column flow can be controlled precisely, since peak area is relatively independent of mobile phase composition. Most HPLC systems are supplied with computerised data handling systems, and using standard concentrations to prepare calibration curves, it is possible to obtain an automatic set of results for the concentrations of individual components in samples matched with their corresponding retention times. But remember that the detector response is likely to differ for the various components of a mixture, even if they are chemically similar, and it may be necessary to know the relative response factors for the detector system (Section 8.4).

Experimental section

> ! **Safety.** Before carrying out any experiments in this section, pay full
> ■ attention to any safety warnings and make sure you adhere to national
> laboratory and safety regulations.

8.9   Aspirin, phenacetin and caffeine in a mixture

High-performance liquid chromatography is used for the separation and quantitative analysis of a wide variety of mixtures, especially those in which the components are insufficiently volatile and/or thermally stable to be separated by gas chromatography. This is illustrated by the following method which may be used for the quantitative determination of aspirin and caffeine in the common analgesic tablets, using phenacetin as internal standard; where APC tablets are available the phenacetin can also be determined by this procedure.

Sample mixture   A suitable sample mixture is obtained by weighing out accurately about 0.601 g of aspirin, 0.076 g of phenacetin and 0.092 g of caffeine. Dissolve the mixture in 10 mL absolute ethanol, add 10 mL of 0.5 M ammonium formate solution and dilute to 100 mL with deionised water.

Solvent (mobile phase)   Ammonium formate $(0.05\,M)$ in 10 vol% ethanol–water at pH 4.8. Use a flow rate of 2 mL min^{-1} with inlet pressure of about 117 bar (11.7 MPa).

Column   Dimensions 15.0 cm × 4.6 mm, packed with a 5 μm silica SCX (strong cation exchanger) bonded phase.

Detector   UV absorbance at 244 nm (or 275 nm).

Procedure   Inject 1 μL of the sample solution and obtain a chromatogram. Under the given conditions the compounds are separated in about 3 min; the elution sequence is (1) aspirin, (2) phenacetin, (3) caffeine. Record the peak areas and express each peak area as a percentage of the total peak area. Compare these results with the known composition of the mixture; discrepancies arise because of different detector response to the same amount of each substance.

Determine the response factors r for the detector relative to phenacetin (= 1) as internal standard by carrying out three runs, using 1 μL injection, and obtaining the average value of r:

$$r = \frac{(\text{peak area of compound})/(\text{mass of compound})}{(\text{peak area of standard})/(\text{mass of standard})}$$

To correct the peak areas initially obtained, divide by the appropriate response factor and normalise the corrected values. Compare this result with the known composition of the mixture.

8.10   References

1. J V Mortimer 1967 Liquid chromatography discussion group session. In A B Little-wood (ed) *Gas chromatography 1966 (Rome symposium)*, Institute of Petroleum, London, p. 414
2. C S Horvath and S R Lipsky 1966 *Nature*, **211**; 748
3. L R Snyder 1978 *J. Chromatogr. Sci.*, **16**; 223
4. R E Majors 1975 *Am. Lab.*, **7**; 13
5. Millipore UK Ltd, Waters Chromatography Division, Croxley Green, Hertfordshire, England
6. D M W Anderson, I C M Dea and A Hendrie 1971 *Sel. Ann. Rev. Anal. Sci.*, **1**; 1
7. P K Dutta *et al.* 1991 *J. Chromatogr.*, **113**; 536
8. J C Berridge 1985 *Anal. Proc.*, **22**; 323
9. J F K Huber (ed) 1976 *Instrumentation for high performance liquid chromatography*, Elsevier, Amsterdam
10. R P W Scott 1976 *Contemporary liquid chromatography*, John Wiley, New York
11. P A Bristow 1976 *Anal. Chem.*, **48**; 237
12. W D Pfeffer and E S Yeung 1990 *Anal. Chem.*, **62**; 2178
13. J H Knox 1977 *J. Chromatogr. Sci.*, **15**; 353
14. J G Dorsey *et al.* 1990 *Anal. Chem.*, **62** (12); 324R
15. T J Bahowick *et al.* 1992 *Anal. Chem.*, **64** (12); 257R
16. W H Pirkle and T C Pochapsky 1989 *Chem. Rev.*, **89**; 347
17. J Hermansson 1984 *J. Chromatogr.*, **316**; 537
18. J F Lawrence and R W Frei 1976 *Chemical derivatisation in liquid chromatography*, Elsevier, Amsterdam

8.11   Bibliography

T A Berger 1995 *Packed column super fluid chromatography*, Royal Society of Chemistry, London

W D Conway and R J Petroski (eds) 1995 *Modern countercurrent chromatography*, American Chemical Society, Washington DC

S Lindsay 1992 *High performance liquid chromatography*, ACOL–Wiley, Chichester

M C McMaster 1994 *HPLC: a practical user's guide*, VCH, New York and Cambridge

V Meyer 1988 *Practical high performance liquid chromatography*, Wiley, New York

G Patonary 1992 *HPLC detection – newer methods*, VCH, New York

R M Smith 1988 *Supercritical fluid chromatography*, Royal Society of Chemistry, London

9

Gas chromatography

9.1 Introduction

Gas chromatography separates a mixture into its constituents by passing a moving gas phase over a stationary sorbent. It is similar to liquid–liquid chromatography except that the mobile liquid phase is replaced by a moving gas phase. Only two possibilities exist for the stationary phase; it can be a solid or a liquid. This immediately limits the separation mechanisms to adsorption or partition, both of which are extensively employed in gas chromatography. Originally two types of gas chromatography were described, gas–liquid chromatography (GLC) and gas–solid chromatography (GSC). This terminology has been superseded by the simpler and more satisfactory term **gas chromatography (GC)**.

The first true GC experiments were performed by Martin and James in 1951 on the lower fatty acids.[1] These early experiments used partition as the separation mechanism, and were described by Martin and coworkers as gas liquid partition chromatography (GLPC). The rapid development of the technique was due to the fact that much of the theory had been developed by Martin and Synge a decade earlier when partition chromatography in the liquid phase was first described.[2] Partition in the gas phase was also immediately seen by a number of scientists to have enormous potential to solve the separation problems of many systems, and this work was rapidly developed by researchers at ICI, British Petroleum and Shell Research Laboratories. Although the first commercial gas chromatograph was not available until 1955, it is now one of the most useful and widely available separation techniques in the analytical laboratory.

Gas chromatography has been refined so that it is possible to separate very complex mixtures containing up to 200 related compounds using either partition or adsorption, with very small sample sizes, but it does have inherent limitations. The sample must be able to exist in the gas phase, so it may only be applied to volatile materials, although this includes substances which have an appreciable vapour pressure at temperatures up to 400 °C. The requirement for volatility of the sample means that non-polar materials are generally easier to handle than polar materials, and ionic materials cannot pass through a gas chromatograph. This limits the technique to about 20% of the known chemicals.

This chapter deals with gas chromatography and some of its applications in the field of quantitative chemical analysis. But it begins by describing the apparatus (Figure 9.1) and explaining some of the basic principles. A comprehensive account is beyond the scope of this book; more detailed accounts are given in the works listed in Section 9.11.

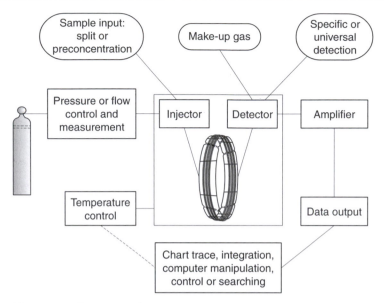

Figure 9.1 Typical gas chromatograph

9.2 Apparatus

9.2.1 A supply of carrier gas from a high-pressure cylinder

The carrier gas is either helium, nitrogen, hydrogen or argon; the choice depends on factors such as availability, purity, consumption and the type of detector employed. Thus helium is preferred when thermal conductivity detectors are employed because of its high thermal conductivity relative to the vapours of most organic compounds. Associated with this high-pressure supply of carrier gas are the attendant pressure regulators and flowmeters to control and monitor the carrier gas flow; the operating efficiency of the apparatus is very dependent on the maintenance of a constant flow of carrier gas. Two important safety considerations:

1. Free-standing gas cylinders must always be supported by means of clamps or chains.
2. Waste gases, especially hydrogen, must be vented through an extraction hood.

Because of safety and storage problems associated with gas cylinders, a number of gases for chromatography (e.g. air, hydrogen and nitrogen) are available from benchtop generators providing high-purity gases with flow rates of $300-900\,\text{mL min}^{-1}$.

9.2.2 Sample injection system and derivatisation

Numerous devices have been developed for introducing the sample, but the major applications involve liquid samples that are introduced using a microsyringe with hypodermic needle. The needle is inserted through a self-sealing silicone rubber septum and the sample injected smoothly into a heated metal block at the head of the column. Manipulation of the syringe may be regarded as an art developed with practice, and the aim must be to introduce the sample in a reproducible manner. The temperature of the sample port should be

Table 9.1 *Injection systems used in capillary gas chromatography*

Name	Sample volume (µL)	Concentration	Application
Split	0.1–2	high	general-purpose
Splitless	0.5–10	low–trace	dilute solutions
Cold on column	1.0–500	low–trace	'delicate' low concentrations
Programmable temperature vaporising (PTV)	1.0–50	low–high	thermally labile species
Thermal desorption	not applicable	low	trapped gases and vapours
Pyrolysis	10 µg to 1 mg	high	polymers, biopolymers

such that the liquid is rapidly vaporised but without either decomposing or fractionating the sample; a useful rule of thumb is to set the sample port temperature approximately to the boiling point of the least volatile component. For greatest efficiency, the smallest possible sample size (1–10 µL) consistent with detector sensitivity should be used. Gaseous samples of 0.5–10 mL can be injected in a similar fashion, provided a gas-tight syringe capable of withstanding the back pressure at the head of the column is used.

Unfortunately, sample injection into capillary systems is much more difficult, due to the small sample loading required on the column and the low flow rates used.[3] A number of different injection systems are in current use (Table 9.1) and the sampling method must be chosen to match the sample characteristics. The most widely used method of placing samples onto the column is probably **split injection**, and for many sample types it is perfectly adequate. For this purpose 0.1–1.0 µL of sample is injected into a glass-lined heated injection port, where it vaporises and mixes with carrier gas; a preset proportion, normally between 1% and 10%, is allowed to pass into the column and the rest is vented to the atmosphere via a variable needle valve (Figure 9.2). This avoids problems with overloading, but the technique only works if concentrations of material in the sample are high (> 0.01 vol%). It has the disadvantage that carrier gas is being continuously vented to the laboratory.

For samples containing low concentrations of material, **splitless injection** is often the most appropriate. In this system a larger volume of between 0.5 and 5 µL of a dilute solution in a volatile solvent is injected relatively slowly into the same injector as above, but with the split vent closed and the column temperature set at 20–25 °C lower than the boiling point of the solvent. Under these conditions, a plug of solvent condenses in the first few millimetres of column and the sample molecules are concentrated in this region in a process known as solvent trapping. A predetermined time after injection, normally 40–60 s, the split vent is opened to purge away any remaining solvent in the injector, otherwise it would cause tailing and broadening of peaks, and the column temperature is raised to its operational value. The solvent and any low-boiling solute molecules are eluted rapidly as a large peak and then the higher-boiling solutes are eluted as a normal chromatogram. The quality of the data obtained by this technique critically depends on the timing of events (injection time, split opening time and beginning of temperature ramp) and the temperatures that are used in each stage of the process. Splitless injection is satisfactory for trace analysis on samples containing solute at ppm levels, but by its very nature it can only

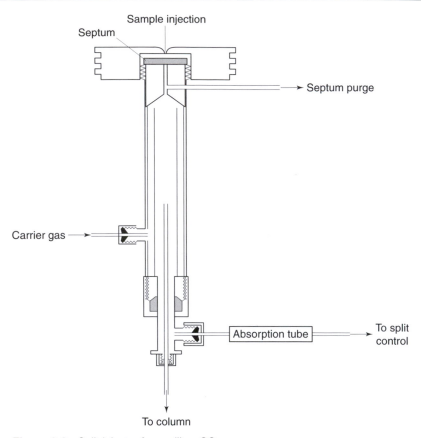

Figure 9.2   Split injector for capillary GC

reliably perform separations if the solute molecules have boiling points at least $100-150\,°C$ above the boiling point of the solvent, and this can be a severe limitation for some types of analysis.

Potentially the most universal method of introducing sample to a capillary column is to revert back to the **direct on-column method**, and this is commonly employed for packed systems. But two problems have to be overcome. Firstly the total mass of material that a 0.25 mm column can handle without overloading is of the order of 20 ng per component; this corresponds to $0.05\,\mu L$ of pure liquid. Secondly the device for introducing the sample onto the column, i.e. the syringe needle, must be smaller than the internal diameter of the column ($0.25\,\mu m$). Special syringes are available which can deliver these very small volumes. The needle is a short length of quartz capillary tube of smaller diameter than the column into which it is to be inserted. It cannot be used with a normal septum inlet, since the needle could not penetrate the rubber disc without breakage, and special pneumatic injection ports have been developed which allow insertion of the needle directly onto the head of the column without disrupting the carrier gas flow. Some versions of these septum-less injectors are cooled and thus subject the sample to the minimum thermal stress and discrimination of any current injector, but the range of sample volumes and concentrations is still somewhat limited, so their acceptance has been only gradual.

For compounds of high molecular mass there is a major problem of involatility in the application of GC. This difficulty may be overcome by breaking the large molecules into smaller, more volatile fragments which may then be analysed. This process is known as **pyrolysis gas chromatography (PGC)**. PGC is a technique in which a non-volatile sample is pyrolysed under rigidly controlled conditions, usually in the absence of oxygen, and the decomposition products separated in the gas chromatographic column. The resulting chromatogram (pyrogram) is used for both qualitative and quantitative analysis of the sample. If the quantitative analysis is very complex, complete identification of the pyrolysis fragments may not be possible, so the pyrogram may simply be used to 'fingerprint' the sample. PGC has been applied to a wide variety of samples, but its major use has been in polymer analysis for the investigation of both synthetic and naturally occurring polymers. The various PGC systems can generally be classified into two distinct types:

Static-mode (furnace) reactors typically consist of a quartz reactor tube and a Pregl type of combustion furnace. Solid samples are placed in the reactor tube and the system is closed. The furnace is then placed over the combustion tube and the sample heated to the pyrolysis temperature. In this type of pyrolysis system, the time required to reach the necessary temperature is much longer (up to 30 s) than in dynamic pyrolysis, resulting in a greater number of secondary reactions. The static-mode system usually has a larger sample capacity, and this is an important advantage.

Dynamic (filament) reactors accept samples placed on the tip of a filament or wire igniter (platinum and Nichrome wires have been used) which is then sealed in a reactor chamber; in PGC the reactor chamber is typically the injection port of the gas chromatograph. As the carrier gas passes over the sample, a d.c. charge is applied, and the sample is heated rapidly to the pyrolysis temperature. As the sample decomposes, the pyrolysis products are carried away into a cooler area (reducing the possibility of secondary reactions) before entering the GC column.

Pyrolysis prior to GC/MS (Section 19.18) is also an elegant way of analysing certain molecules which cannot be chromatographed by conventional methods.

For GC analysis of highly volatile substances in solution or solid–liquid matrices (e.g. ethanol in blood samples) it is often desirable to use the procedure of **headspace analysis**. The sample, usually in a sealed vial, is heated to a fixed temperature to reach equilibrium with the atmosphere above. This creates the headspace, which is then sampled in the normal manner and is injected directly into the GC using a gas-tight syringe. This type of experiment can be performed simply by placing the sample in a septum-capped bottle immersed in a water bath to maintain a constant temperature, or using automatic devices which heat the sample at constant temperature for fixed time intervals before injecting the gaseous sample. This is extremely useful for aroma analysis of fresh foods[4] and beverages, where a combination of headspace and trapping techniques is often employed, and a number of static and dynamic systems have been described.

But many samples are unsuitable for direct injection into a gas chromatograph; perhaps they have high polarity, low volatility or thermal instability. In this respect the versatility and application of gas chromatography has been greatly extended by the formation of volatile derivatives, especially silylation reagents. The term **silylation** is normally taken to mean the introduction of the trimethylsilyl, $—Si(CH_3)_3$, or similar group in place of active hydrogen atoms in the substance under investigation. A considerable number of these reagents are now available,[5] including some special silylating agents which give improved detector response, usually by incorporating a functional group suitable for a selective detector system. Reagents containing chlorine and bromine atoms in the silyl group are used particularly for preparing derivatives injected on to gas chromatographs fitted with electron capture detectors. Derivatisation can also give enhanced resolution from other components in a mixture and improved peak shape for quantitative analysis.

Although inorganic compounds are generally not so volatile as organic compounds, gas chromatography has been applied in the study of certain inorganic compounds which possess the requisite properties. If gas chromatography is to be used for metal separation and quantitative analysis, the types of compounds which can be used are limited to those that can be readily formed in virtually quantitative and easily reproducible yield. Together with the requirements of sufficient volatility and thermal stability necessary for successful gas chromatography, this makes neutral metal chelates the most favourable compounds for use in metal analysis. β-Diketone ligands, e.g. acetylacetone and the fluorinated derivatives, trifluoroacetylacetone (TFA) and hexafluoroacetylacetone (HFA) form stable, volatile chelates with aluminium, beryllium, chromium(III) and a number of other metal ions; it is thus possible to chromatograph a wide range of metals as their β-diketone chelates.

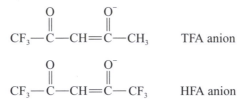

The number of reported applications to analytical determinations at the trace level appear to be few, probably the best-known example is the determination of beryllium in various samples. The method generally involves the formation of the volatile beryllium trifluoro-acetylacetonate chelate, its solvent extraction into benzene with subsequent separation and analysis by gas chromatography.[6]

Various types of derivatisation have now been developed for both gas and liquid chromatography. Consult the literature for more detailed information on choosing a suitable derivative for a particular analytical problem.[7,8]

9.2.3 The column

The actual separation of sample components is effected in the column; the nature of the solid support, the type and amount of liquid phase, the method of packing, the column length and the temperature are important factors in obtaining the desired resolution. The column is enclosed in a thermostatically controlled oven so that its temperature is held constant to within 0.5 °C, thus ensuring reproducible conditions. The operating temperature may range from ambient to over 400 °C and for isothermal operation is kept constant during the separation process.

Packed columns

The earliest experiments were performed using packed columns which were tubes up to 5 m in length, 2–4 mm internal diameter, and made of glass, metal (aluminium, stainless steel or copper) or high-temperature plastic (PTFE). The packing consisted of a support which was an inert material, often diatomaceous earth which had been washed and deactivated with acid treatment and then sieved into a close range of particle sizes (mesh sizes of 60–120 are most common corresponding to a range of diameters from 250–125 μm). The role of the support was to hold the liquid phase in a form through which gas could be passed.

A large number of liquid phases have been employed, with more than 400 still listed in supply catalogues.

The chosen phase was coated uniformly onto the support at a level of 1–15%, then the coated support was packed evenly into the chosen column. This produced a bewildering range of packed columns, each one believed to be the best at separating a specified mixture. The decision about which phase to use was often difficult, but most phases can be grouped into non-polar, intermediate polarity and polar then the idea 'like separates like' proves to be a useful guide. Liquid phases can be broadly classified as follows:

1. Non-polar hydrocarbon-type liquid phases, e.g. paraffin oil (Nujol), squalane, Apiezon L grease and silicone-gum rubber; silicone-gum rubber is used for high-temperature work (upper limit ~400 °C).
2. Compounds of intermediate polarity which possess a polar or polarisable group attached to a large non-polar skeleton, e.g. esters of high-molecular-weight alcohols such as dinonyl phthalate.
3. Polar compounds containing a relatively large proportion of polar groups, e.g. the carbowaxes (polyglycols).
4. Hydrogen-bonding class, i.e. polar liquid phases such as glycol, glycerol and hydroxy-acids, which possess an appreciable number of hydrogen atoms available for hydrogen bonding.

Capillary columns

Fortunately the use of packed columns has decreased markedly in the last decade with the introduction of open tubular or capillary columns. Capillary columns rely on interaction between the mixture and the stationary phase which is coated as a thin coherent film on the inside surface of a long thin (capillary) tube. Made of glass or stainless steel, they were used by a few development laboratories from the late 1950s but they really only became viable for routine use with the introduction of fused silica or quartz columns in 1979. The equipment to produce these long, thin quartz tubes is costly so, unlike the earlier packed columns, capillaries are almost invariably purchased rather than fabricated in the chromatography lab.

The initial high cost of capillary columns compared with a home-packed columns meant their introduction was slower than one might have hoped, but nowadays the vast majority of partition gas chromatography is performed on capillary columns rather than packed columns. Since capillary columns must be purchased, it is important to choose them correctly. The majority of separations can be adequately made on a much smaller range of columns. Many laboratories perform >90% of their work on six or fewer different columns.

The earliest quartz capillaries were constructed by dynamically coating the inside surface of a fused silica tube 50 m in length, inside diameter 0.22 mm, with a thin film of stationary phase identical to that used in packed columns, e.g. silicone oil (OV 1). The outer surface of the tube was protected from the moment of fabrication with a coating of tough, high-temperature plastic normally based on a polyimide material. This produced a relatively strong and durable column with much higher resolving capability than any packed equivalent. But these early columns were not without problems, especially the constraints on plumbing them into the total system and their relatively short life. As the use of capillary columns becomes more widespread, a wider range is becoming available (Table 9.2) and the first-time purchaser is again confronted with a bewildering variety; but by considering the type of separation, a sensible choice can be made. In these capillary

Table 9.2 *Variable parameters in capillary columns*

Length (m)	5	10	15	25	30	50	60	100
Internal diameter (mm)	0.1	0.2		0.25	0.32		0.53	0.75
Phase	non-polar		intermediate		special		highly polar	
Maximum temperature (°C)	200							>450
Film thickness (µm)	0.2			1.5		3.0		5
Coating	polymide plastic						aluminium	
Maximum loading (ng)	10							1000
Efficiency (plates/m)	5000							20 000

columns the stationary phase is coated on the inner wall of the tube; two basic types are available:

Wall-coated open tubular (WCOT) columns have the stationary phase coated directly on to the inner wall of the tubing.
Support-coated open tubular (SCOT) columns have a finely divided layer of solid support material deposited on the inner wall, on to which the stationary phase is then coated.

Rules of thumb for column selection

Like separates like The same general concept of like separating like is still true for capillary columns, but since their resolution is very high, it is less critical to select exactly the correct phase, as most non-polar materials can be separated on a non-polar phase, and many polar materials on a general polar phase.

Long column, long elution but high resolution The longer the column the higher the resolution, but the longer a given material will take to elute (the cost also increases with length). For many applications the 'standard' 25 m column is satisfactory, but columns of 10 m or less can be used for rapid separations of fairly simple mixtures. (Note that these short columns can give the same separation as the best packed columns but in a much shorter time). For more demanding separation, longer columns may be used. The longest column so far produced is 2100 m (1.3 miles) with an estimated plate count of over 2 million, but normally the longest length commercially available is of the order of 100 m.

Small diameter, high efficiency Column efficiency increases markedly with decrease in internal diameter, so by using tubing with an internal diameter of 100 µm (0.1 mm) the highest resolution is obtained, but at the expense of severely restricting the amount of material that can be placed on the column as well as placing more demands on the design and construction of the whole system, since the small flow rates of carrier gas through these columns mean that any upswept zones will cause severe peak broadening. Table 9.2 shows that a range of columns is available with diameters from 0.1 to 0.53 mm, the so-called megabore columns. For certain applications such as interfacing to another technique, e.g. FT-IR, there are advantages to using wider-bore columns, which can tolerate a greater mass of sample loading than the more conventional 0.2 or 0.3 mm column, but for normal GC applications there appears to be no good reason to use wide-bore columns.

Film thickness   Yet another variable parameter is the film thickness. In earlier columns the stationary phase simply coated the inner surface of the tube and was then held in place by capillary action. The film thicknesses were about 1 μm because anything much greater meant the film became unstable, giving rise to pooling and excessive column bleed when in use. It is now possible to bond the phase chemically to the silica by carrying out a chemical reaction between the phase and the wall after coating. A similar approach is to cause the phase to polymerise and cross-link, either by a radical process which can be initiated after coating, or by exposing the column to high-energy radiation after coating.

PLOT columns

Until recently capillary columns were restricted to partition as the separation mechanism, since only liquid stationary phases could be introduced into the column, but it is now possible to purchase porous layer open tubular (PLOT) columns in which a very thin layer (5–50 μm) of solid adsorbent is deposited evenly on the inside wall. These solid adsorbents are identical to the adsorbents used for packed column adsorption chromatography, except the particle size is much smaller, and they can be used for the same applications. Columns containing molecular sieve, alumina, carbosieve (a type of carbon molecular sieve) and the Porapak range of cross-linked polymers are currently available, and they give extremely efficient separation of the permanent gases or the C_1–C_5 hydrocarbons.[5] However, the mechanical stability of the coating is not as good as for the liquid phases and more care is required in their use.

9.2.4   The detector

The function of the detector, which is situated at the exit of the separation column, is to sense and measure the small amounts of the separated components present in the carrier gas stream leaving the column. The output from the detector is fed to a device which produces a trace called a **chromatogram**. The choice of detector will depend on factors such as the concentration level to be measured and the nature of the separated components. The detectors most widely used in gas chromatography are the thermal conductivity, flame ionisation and electron capture detectors, and a brief description of these will be given. Table 9.3 lists a number of commonly used detectors; each has different sensitivity, selectivity and compatibility with modern, low-volume gas chromatographs.

Sensitivity   This is usually defined as the detector response (mV) per unit concentration of analyte ($mg\,mL^{-1}$). It is closely related to the limit of detection (LOD) since high sensitivity often gives a low limit of detection. But since the limit of detection is generally defined as the amount (or concentration) of analyte which produces a signal equal to twice the baseline noise, the limit of detection will be raised if the detector produces excessive noise. The sensitivity also determines the slope of the calibration graph (slope increases with increasing sensitivity) and therefore influences the precision of the analysis.

Linearity   The linear range of a detector refers to the concentration range over which the signal is directly proportional to the amount (or concentration) of analyte. Linearity in detector response will give linearity of the calibration graph and allows the graph to be drawn with more certainty. With a convex calibration curve, the precision is reduced at the higher concentrations where the slope of the curve is much less. A large linear range is a

Table 9.3 *Detectors used in gas chromatography*

Detector	Abbreviation	Dynamic range	Sensitivity	Application
Hot-wire	HWD	10^4 to 10^5	10^{-8} g mL^{-1}	Universal
Thermal conductivity	TCD	10^4 to 10^5	10^{-8} g mL^{-1}	Universal
Flame ionisation	FID	10^7	2 pg s^{-1}	Organics
Electron capture	ECD	10^2 to 10^3	0.01 pg s^{-1}	Electrophilics (halogens)
Flame photometric	FPD	10^3 to 10^4	1–10 pg s^{-1}	sulphur, phosphorus
Alkali flame	AFD[a]	10^4 to 10^5	0.05 pg s^{-1}	nitrogen, phosphorus
Photoionisation	PID	10^7	1 pg	Organics
Atomic emission	AED	2×10^4	1–100 pg s^{-1}	Elements
Mass spectrometric	MS	$>10^5$	1–100 pg	Structure
Fourier transform infrared	FT-IR	10^4	0.5–50 ng	Structure

[a] Sometimes abbreviated to NPD.

great advantage, but detectors with a small linear range may still be used because of their other qualities, although they will need to be recalibrated over a number of different concentration ranges.

Stability An important characteristic of a detector is the extent to which the signal output remains constant with time, assuming there is a constant input. Lack of stability can be exhibited in two ways: by **baseline noise** or by **drift**, both of which will limit the sensitivity of the detector. Baseline noise, caused by a rapid random variation in detector output, makes it difficult to measure small peaks against the fluctuating background. Baseline drift, a slow systematic variation in output, produces a sloping baseline which in severe cases may even go off scale during the analysis. Drift is often due to factors external to the detector, such as temperature change or column bleed, and so is controllable, whereas noise is usually due to poor contacts within the detector and imposes a more fundamental limit on its performance.

Universal or selective response A universal detector will respond to all the components present in a mixture. In contrast, a selective detector senses only certain components in a sample; this can be advantageous if the detector responds only to components of interest, thus giving a considerably simplified chromatogram and avoiding interference.

Hot-wire detector (HWD)

The hot-wire detector, also known as the thermal conductivity or katherometer detector, is the oldest GC detector; due to its inherently large volume, low sensitivity and contamination problems, it was long dismissed as unsuitable for capillary systems. However, two of its operating characteristics are extremely suitable for capillary column use. The sensitivity is inversely proportional to flow rate, increasing dramatically on going from flows of the

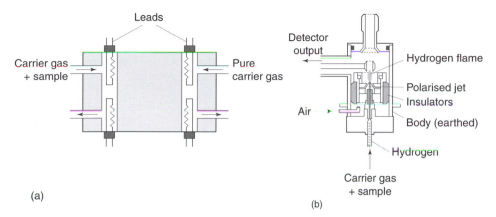

Leads

Carrier gas
+ sample

Pure
carrier gas

Detector
output

Hydrogen flame

Polarised jet
Insulators

Air

Body (earthed)

Hydrogen

Carrier gas
+ sample

(a)

(b)

Figure 9.3 Detectors: (a) thermal conductivity, (b) flame ionisation

order of $50\,mL\,min^{-1}$ down to $1\,mL\,min^{-1}$. And maximum sensitivity is obtained if the carrier gas is either hydrogen or helium (commonly used for capillary systems), since these gases have the greatest difference in thermal conductivity from most organic species. Hot-wire detectors employ a heated metal filament or a thermistor (a semiconductor of fused metal oxides) to sense changes in the thermal conductivity of the carrier gas stream. Helium and hydrogen are the best carrier gases to use in conjunction with this type of detector since their thermal conductivities are much higher than any other gases; on safety grounds, helium is preferred because of its inertness.

In the detector two pairs of matched filaments are arranged in a Wheatstone bridge circuit; two filaments in opposite arms of the bridge are surrounded by the carrier gas only, whereas the other two filaments are surrounded by the effluent from the chromatographic column. This type of thermal conductivity cell is illustrated in Figure 9.3(a) with two gas channels through the cell, a sample channel and a reference channel. When pure carrier gas passes over both the reference and sample filaments the bridge is balanced, but when a vapour emerges from the column the rate of cooling of the sample filaments changes and the bridge becomes unbalanced. The extent of this imbalance is a measure of the concentration of vapour in the carrier gas at that instant, and the out-of-balance signal is fed to a recorder, producing the chromatogram. The differential technique used is thus based on the measurement of the difference in thermal conductivity between the carrier gas and the carrier gas/sample mixture.

Most major manufacturers will supply a hot-wire detector that incorporates a number of mechanical and electronic improvements to enhance its performance. They have a small internal volume and are capable of high-temperature operation, although normally the detector is operated at the lowest available temperature, because the sensitivity is proportional to the temperature difference between the heated filament and the block. The sensitivity also increases roughly with the square of the filament current, so this is maintained at a high value. With the older designs of detector this often meant the chromatographer had to decide either to use a high filament temperature, which gave higher sensitivity but short working lifetimes, or to use a lower and safer temperature in order to prevent the filament from burning out. With advanced electronic control and signal processing, the detector can be operated in a constant (maximum) current mode, rather than the simpler but less sensitive constant-voltage mode; this also improves the linearity of response.

Flame ionisation detector

The basis of the flame ionisation detector (FID) is that the effluent from the column is mixed with hydrogen and burned in air to produce a flame which has sufficient energy to ionise solute molecules having low ionisation potentials. The ions produced are collected at electrodes and the resulting ion current measured; the burner jet is the negative electrode whereas the anode is usually a wire or grid extending into the tip of the flame. This is shown in Figure 9.3(b).

The combustion of mixtures of hydrogen and air produces very few ions so that, with only the carrier gas and hydrogen burning, an essentially constant signal is obtained. But when carbon-containing compounds are present, ionisation occurs and there is a large increase in the electrical conductivity of the flame. Because the sample is destroyed in the flame, a stream-splitting device is employed when further examination of the eluate is necessary; this device is inserted between the column and detector and allows the bulk of the sample to bypass the detector.

Widely applicable, the FID is very nearly a universal detector for gas chromatography of organic compounds, and coupled with its high sensitivity, stability, fast response and wide linear response range ($\sim 10^7$), this has made it the most popular detector in current use.[9]

The FID is mass sensitive rather than concentration sensitive, so the response is un-affected by changes in flow through the detector and the high sensitivity of approximately $2 \times 10^{-12} \, \mathrm{g \, s^{-1}}$ is maintained even at low flow rates. Sensitivity is normally expressed as mass per unit time, so the effects of peak broadening can be eliminated from the term. Thus a peak which is 2 s wide and results from 10^{-8} g of material will be as easy to see as a peak which is 20 s wide and contains 10 times as much analyte. This high sensitivity, combined with the large linear range and immunity from contamination, has meant that the FID is still the most widely used detector for routine capillary GC. But the low response factors seen for certain oxygenated compounds, such as alcohols and carbonyls, and for many halogenated or nitrogen compounds still apply when used for capillary work.

By a relatively simple modification to the interior of the FID, it may be made much more sensitive to either nitrogen- or phosphorus-containing compounds. Often known as the alkali flame detector (AFD), this modified FID has a small electrically heated bead of rubidium silicate positioned between the flame and the collector electrode, and a potential of about 200 V between the bead and the collector. In use the normal hydrogen flame is operated at a low fuel/air ratio to suppress the normal ionisation of hydrocarbons, and the flame impinges upon the bead, which is maintained at 600–800 °C. A small plasma is thought to be formed close to the surface of the bead, allowing large numbers of ions to be produced from nitrogen- and phosphorus-containing compounds while suppressing the response for carbon.

Although a satisfactory explanation for the exact mechanism of this effect is still not available, the detector does work and can show sensitivity enhancements of ×50 for nitro-gen compounds (nitrogen carrier cannot be used) and ×500–1000 for phosphorus com-pounds relative to the normal FID response, making it the most sensitive of the routine detectors under these conditions. But this detector is not as straightforward to use as the normal FID since the response to either nitrogen or phosphorus depends on the exact operating conditions used, particularly the temperature of the bead, which in turn depends partly on the size of the hydrogen flame impinging on it. This means that day-to-day reproducibility is difficult to maintain and the linear range is not as great as for the unmodified detector, but it is extremely useful for the sensitive detection and quantification of a number of compounds containing nitrogen or phosphorus, e.g. the newer pesticides.

A third general type of flame detector is the flame photometric detector (FPD), also based on the standard FID. In this system phosphorus- or sulphur-containing compounds are burned in a hydrogen/oxygen flame, where they produce molecular emission at 536 nm for phosphorus or 394 nm for sulphur. After passing through a narrowband optical filter, this emission is detected by a conventional photomultiplier tube mounted at right angles to the flame axis. The detector, which can be equipped with two separate optical sensors for simultaneous determination of sulphur and phosphorus, is very sensitive to these two elements relative to hydrocarbons, giving a high degree of selectivity. A major problem with this type of detector, is that the response is not linear at any concentration; it is in fact approximately a square law response which depends on the exact nature of the compound being detected. Some chromatographs have a linearisation function programmed into the system to provide a root factor which may be user controlled between 1.5 and 2.5. Despite its problems, this detector is extremely useful, particularly for sulphur analysis at low concentrations in a number of environmentally important problems.

The FID, the AFD and the FPD, each may be built to operate at up to 400 °C, so they can be used with high-temperature columns and contamination by condensation is reduced.

Electron capture detector

Most ionisation detectors are based on measuring the increase in current (above the background current from ionised carrier gas) that occurs when a more readily ionised molecule appears in the gas stream. The electron capture detector (ECD) differs from other ionisation detectors in that it exploits the recombination phenomenon based on electron capture by compounds having an affinity for free electrons; the detector thus measures a decrease rather than an increase in current.

A β-ray source (commonly a foil containing ^{3}H or ^{63}Ni) is used to generate 'slow' electrons by ionisation of the carrier gas (nitrogen preferred) flowing through the detector. These slow electrons migrate to the anode under a fixed potential and give rise to a steady baseline current. When an electron-capturing gas (i.e. eluate molecules) emerges from the column and reacts with an electron, the net result is the replacement of an electron by a negative ion of much greater mass and there is a corresponding reduction in current flow. The response of the detector is clearly related to the electron affinity of the eluate molecules; it is particularly sensitive to compounds containing halogens and sulphur, anhydrides, conjugated carbonyls, nitrites, nitrates and organometallic compounds.

A number of commercial capillary ECDs are now available where the internal volume has been reduced to as little as 200 μL, so they can respond even to very narrow (low-volume) peaks. The resulting detector has the same advantages as earlier detectors of high selectivity and sensitivity (5×10^{-15} g s^{-1}) for halogenated compounds, and it still suffers from the same susceptibility to contamination, so it must be used and maintained with care if reliable results are to be obtained. Maximum sensitivity for this type of detector is only obtained if the carrier is either nitrogen or a mixture of nitrogen and methane, and for capillary operation using hydrogen or helium it is preferable to introduce a make-up supply of nitrogen to the column eluent prior to entry into the detector, even though this decreases the resolution of the system.

Although not directly in response to the requirements of capillary operation, the electronic performance of the detector has also been drastically altered in the newer designs. The polarisation voltage between the radioactive foil and the collector may be modulated in a number of different ways, providing either constant frequency pulses or constant (average) current. This gives a linear working range of nearly four orders of magnitude, not as efficient as an FID but still enough for a wide range of concentrations to be determined fairly easily. Many of the earlier applications of ECDs were for the determination of persistent (organochlorine) pesticide residues in food, water or environmental samples. These compounds are now found less often because the second- and third-generation pesticides, based on phosphorus or nitrogen structures, have replaced the organochlorines. These newer materials can be determined using AFD or FPD detectors. But the ECD should still be regarded as a medium-sensitivity universal detector, for both organic and inorganic species, and as such it can be used for the determination of water and other low-molecular-weight species such as H_2S, CO_2, N_2, and O_2 which are not easily detected with other systems.

Photoionisation detector (PID)

The photoionisation detector (PID) is one of the most recent introductions, so it will be described in a little more detail. It works as an ionisation detector similar to the FID or ECD, and the response results from the collection and amplification of ions at a positively charged collector electrode using a conventional high-impedance amplifier (Figure 9.4). As organic solute molecules elute from the column, ions are produced by irradiating the eluent with light from a high-intensity ultraviolet lamp which produces photons in the range 9.5–11.7 eV, depending upon the wavelength of the radiation (106 nm corresponds to 9.5 eV and 149 nm to 11.7 eV). Since the bond energies of most organic species fall within this range, the PID can either be used as a universal detector for organics, or it can cause selective

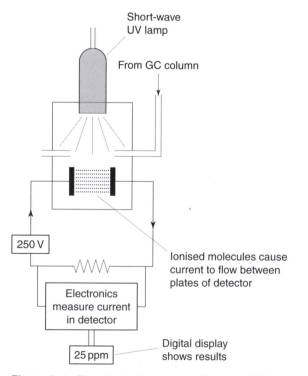

Figure 9.4 Photoionisation detector (Courtesy ELE International Ltd, Hemel Hempstead, Herts)

ionisation of only some types of molecules within the sample which have low ionisation energies. This detector has only been commercially available for a comparitively short time and there are still a number of problems with its long-term use, especially with regard to contamination of the lamp window, and the lifetime of the lamp. But it has considerable potential as an alternative to the FID, giving similar sensitivity and universal response. Unlike the normal FID, the PID does not require ancillary gas supplies other than carrier gas, which gives it the potential for portable operation. A number of portable organic monitors using the PID and a simple pump to force air, acting as both carrier and sample, through a short column are available for gas and vapour determination on-site.

Atomic emission detector (AED)

Strictly perhaps a hyphenated device, the atomic emission detector (AED) looks set to find widespread application. Eluent from the end of a capillary column, using helium as carrier gas, is passed into a water-cooled microwave cavity where a helium plasma is produced. The very high temperature of the plasma is sufficient to decompose the sample into its constituent atoms, which then emit their characteristic atomic emission spectra. The resulting radiation is focused then dispersed by a grating onto a movable diode array detector in a similar manner to conventional plasma spectroscopy. This system allows the detection of elements (other than the helium used as carrier) at very high sensitivity with typical detection limits quoted as $pg\,g^{-1}$. A number of systems allow up to six elements to be detected simultaneously, provided the spectral lines of the chosen elements are all in a similar region of the spectrum, giving a series of elemental chromatographic traces.

Since the diode array detector can be set to monitor a selected range of wavelengths across the UV/visible region, a large number of elements within a given sample can be detected by reinjecting the sample and observing a different portion of the spectrum. This unique specificity for elements can be extremely useful when determining all of the heteroatoms such as the halogens, phosphorus, sulphur, nitrogen and oxygen which may be seen by other detectors. It may also be used to determine other potentially important elements such as silicon, the heavy metals (Pb, Hg,) tin, arsenic, copper and iron, or it may be used as an organic detector by monitoring carbon and hydrogen. It may even selectively detect isotopes such as ^{13}C or deuterium, allowing the possibility of isotope ratio measurements.

The AED not only allows the detection of sample types which are not possible using conventional detectors, but it has a wide linear response and can give quantitative data on each element fairly readily. The software accompanying the detector can also use the response of the diode array to monitor any column bleed or background; it then subtracts this signal from the sample response to give a background-corrected trace. This detector is already in widespread use for such diverse sample types as hydrocarbons, pharmaceuticals, drugs and especially environmental substances.[10] Without doubt, the range of applications will grow, even though the purchase price is currently high and the system requires careful maintenance.

Coupled GC detectors

Gas chromatography is widely coupled to spectroscopic instruments to give hyphenated systems which can combine the separation of chromatography with the identification capability of spectroscopy. These coupled systems are described in Chapter 19, but there is also a pronounced trend towards constructing the spectroscopic section of the combination as a **dedicated** chromatographic detector. This is found for the AED described above.

9.3 Programmed temperature

Gas chromatograms are usually obtained with the column kept at a constant temperature. Two important disadvantages result from this isothermal mode of operation:

1. Early peaks are sharp and closely spaced (i.e. resolution is relatively poor in this region of the chromatogram), whereas late peaks tend to be low, broad and widely spaced (i.e. resolution is excessive).
2. Compounds of high boiling point are often undetected, particularly in the study of mixtures of unknown composition and wide boiling point range; the solubilities of the higher-boiling substances in the stationary phase are so large they are almost completely immobilised at the inlet to the column, especially where the column is operated at a relatively low temperature.

These consequences of isothermal operation may be largely avoided by using the technique of programmed temperature gas chromatography (PTGC) in which the temperature of the whole column is raised during the sample analysis. A temperature programme consists of a series of changes in column temperatures which may be conveniently selected by a microprocessor controller. The programme commonly consists of an initial isothermal period, a linear temperature rise segment, and a final isothermal period at the temperature which has been reached, but may vary according to the separation to be effected. The rate of temperature rise, which may vary over a wide range, is a compromise between the need for a slow rate of change to obtain maximum resolution and a rapid change to minimise analysis time.

Programmed temperature gas chromatography permits the separation of compounds of a very wide boiling range more rapidly than by isothermal operation of the column. The peaks on the chromatogram are also sharper and more uniform in shape so the peak heights may be used to obtain accurate quantitative analysis.[11]

9.4 Quantitative analysis

The quantitative determination of a component in gas chromatography using differential detectors of the type previously described is based on measuring the recorded peak area or peak height; the peak height is more suitable in the case of small peaks, or peaks with narrow bandwidth. In order that these quantities may be related to the amount of solute in the sample, two conditions must prevail:

(a) The response of the detector and recorder must be linear with respect to the concentration of the solute.
(b) Factors such as the rate of carrier gas flow, column temperature, etc., must be kept constant or the effect of variation must be eliminated, e.g. by using the internal standard method.

Peak area is commonly used as a quantitative measure of a particular component in the sample and can be measured by geometrical methods or automatic integration.

Geometrical methods

In the so-called triangulation methods, tangents are drawn to the inflexion points of the elution peak and these two lines together with the baseline form a triangle (Figure 9.5). The area of the triangle is calculated as one-half the product of the base length times the

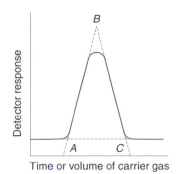

Figure 9.5   Peak area by triangulation

peak height, and the value obtained is about 97% of the actual area under the chromato-graphic peak when the peak has a Gaussian shape. The area may also be computed as the product of the peak height times the width at half the peak height, i.e. by the height × width at half-height method. Since the exact location of the tangents (required for the triangula-tion method) to the curve is not easily determined, it is in general more accurate to use the method based on width at half-height.

Automatic integration

The older methods for computing peak area are time-consuming and often unsatisfactory in terms of accuracy and reproducibility of results. The greater use of capillary column chromatography, with its resulting sharp, closely spaced peaks has accentuated the need for a rapid, automatic instrumental method for data processing. Computers are widely used now in quantitative gas chromatography, processing the analytical signal as the analyses are being run. These systems automatically identify peaks, compute peak areas and/or peak heights, and provide the results either in printed form or in one of the various computer-compatible formats.

The measurement of individual peak areas can be difficult when the chromatogram con-tains overlapping peaks. However, this problem can often be overcome by using derivative facilities which give first- or second-derivative chromatograms; Section 17.13 explains the analogous derivative procedures for spectroanalytical methods. Many computer programs will also carry out this operation, indicating the degree of peak overlap.

9.5   Quantitative procedures

Quantitative measurements depend upon correlating peak areas with the amount or con-centration of solutes in various samples. But equal areas for different solutes do not neces-sarily indicate that equal quantities of those substances are present, as the magnitude of the recorded peaks (or integration values on the recorder) depend upon the response charac-teristics of the detector(s) to those substances.

In its simplest form, quantitation may be carried out by using a series of standards of the pure substance to obtain a calibration plot and then relating any unknown concentration in a sample to an appropriate place on that plot, after allowing for any dilution factor that may have been necessary. However, this simple procedure is subject to errors, some human and some instrumental. Hence it is more common to apply area normalisation, internal standard or standard addition procedures.

Area normalisation

The composition of the mixture is obtained by expressing the area of each individual peak as a percentage of the total area of all the peaks in the chromatogram; correction should be made for any significant variation in sensitivity of the detector for the different components of the mixture (Section 9.9).

Internal standard

The height and area of chromatographic peaks are affected not only by the amount of sample but also by fluctuations of the carrier gas flow rate, the column and detector temperatures, i.e. by variations of those factors which influence the sensitivity and response of the detector. The effect of these variations can be eliminated by using the internal standard method, in which a known amount of a reference substance is added to the sample to be analysed before injection into the column. The requirements for an effective internal standard (Section 4.5) may be summarised as follows:

(a) It should give a completely resolved peak, but should be eluted close to the components to be measured.
(b) Its peak height or peak area should be similar in magnitude to those of the components to be measured.
(c) It should be chemically similar to the original sample, but not present in the original sample.

A constant amount of internal standard is added to a fixed volume of several synthetic mixtures which contain varying known amounts of the component to be determined. The resulting mixtures are chromatographed and a calibration curve is constructed of the percentage of component in the mixtures against the ratio of component peak area/standard peak area. The analysis of the unknown mixture is carried out by addition of the same amount of internal standard to the specified volume of the mixture; from the observed ratio of peak areas, the solute concentration is read off using the calibration curve. Provided a suitable internal standard is available, this is probably the most reliable method for quantitative GC. For example, the concentration of ethanol in blood samples has been determined using propan-2-ol as the internal standard.

Standard addition

The sample is chromatographed before and after the addition of an accurately known amount of the pure component to be determined, and its weight in the sample is then derived from the ratio of its peak areas in the two chromatograms. Standard addition is particularly useful in the analysis of complex mixtures where it may be difficult to find an internal standard which meets the necessary requirements. Standard addition procedures are used extensively in chemical analysis and more detailed explanations of this method are given in Sections 13.39, 15.15 and 18.11.

9.6 Elemental analysis

One of the most important quantitative applications of gas chromatography is to determine the percentage composition of the elements carbon, hydrogen, nitrogen, oxygen and sulphur in organic and organometallic compounds. Although dedicated elemental analysers vary from each other, they do follow well-established procedures, and the following are fairly typical.

Determination of C, H and N

The weighed samples (usually about 1 mg), held in a clean, dry tin container, are dropped at preset time intervals into a vertical quartz tube maintained at 1030 °C and through which flows a constant stream of helium gas. When the samples are introduced the helium stream is temporarily enriched with pure oxygen and flash combustion occurs. The mixture of gases so obtained is passed over Cr_2O_3 to obtain quantitative combustion, and then over copper at 650 °C to remove excess oxygen and reduce oxides of nitrogen to N_2. Finally the gas mixture passes through a chromatographic column (2 m long) of Porapak QS heated to approximately 100 °C. The individual components (N_2, CO_2, H_2O) are separated and eluted to a thermal conductivity detector; the detector signal is fed to a potentiometric recorder in parallel with an integrator and digital printout. The instrument is calibrated by combustion of standard compounds, such as cyclohexanone-2,4-dinitrophenylhydrazone.

Determination of **total organic carbon** (TOC) is important in water analysis and water quality monitoring, giving an indication of the overall amount of organic pollutants. The water is first acidified and purged to remove carbon dioxide from any carbonate or hydrogencarbonate present. After this treatment, a small measured volume of the water is injected into a gas stream, which then passes through a heated packed tube where the organic material is oxidised to carbon dioxide. The carbon dioxide is determined by infrared absorption or it is converted to methane for determination by gas chromatography using a flame ionisation detector (Section 9.2).

Determination of oxygen

The sample is weighed into a silver container which has been solvent-washed, dried at 400 °C and kept in a closed container to avoid oxidation. It is dropped into a reactor heated at 1060 °C, quantitative conversion of oxygen to carbon monoxide being achieved by a layer of nickel-coated carbon. The pyrolysis gases then flow into the chromatographic column (1 m long) of molecular sieves (5×10^{-8} cm) heated at 100 °C; the CO is separated from N_2, CH_4 and H_2, and is measured by a thermal conductivity detector. The addition of a chlorohydrocarbon vapour to the carrier gas is found to enhance the decomposition of the oxygen-containing compounds.

Determination of sulphur

The initial procedure for flash combustion of the sample is essentially as described for C, H and N. Quantitative conversion of sulphur to sulphur dioxide is then achieved by passing the combustion gases over tungsten(VI) oxide WO_3, and excess oxygen is removed by passing the gases through a heated reduction tube containing copper. Finally the gas mixture passes through a Porapak chromatographic column heated at 80 °C in which SO_2 is separated from other combustion gases and measured by a thermal conductivity detector.

Experimental section

> **!** **Safety.** Before carrying out any experiments in this section, pay full
> **■** attention to any safety warnings and make sure you adhere to national
> laboratory and safety regulations.

9.7 Internal normalisation method for the analysis of solvents

Introduction To obtain an accurate quantitative analysis of the composition of a mixture, a knowledge of the response of the detector to each component is required. If the detector response is not the same for each component, the areas under the peaks cannot be used as a direct measure of the proportion of the components in the mixture. This experiment illustrates the use of an internal normalisation method for the quantitative analysis of a mixture of ethyl acetate (ethanoate), octane and ethyl n-propyl ketone (hexan-3-one).

Reagents and apparatus *Reagents* Ethyl acetate (I), octane (II), ethyl n-propyl ketone (III) and toluene (IV), all GPR or comparable grade.

Microsyringe Used for injecting the samples.

Gas chromatograph Preferably equipped with a flame ionisation detector and a digital integrator.

Column Packed with stationary phase containing 10 wt% dinonyl phthalate.

Procedure Prepare mixture A containing compounds I, II and III in an unknown ratio. Prepare mixture B containing equal weights of compounds I, II and III. Set the chromatograph oven to 75 °C and the carrier gas (pure nitrogen) flow rate to 40–45 mL min^{-1}. When the oven temperature has stabilised, inject a 0.3 µL sample of mixture B and decide from the peak areas whether the detector response is the same for each component. If the detector response differs, make up by weight a 1 : 1 mixture of each of the separate components (I, II, and III) with compound IV. Inject a 0.1 µL sample of each mixture, measure the corresponding peak area, hence deduce the factors which will correct the peak areas of components I, II and III with respect to the internal standard IV. Prepare a mixture, by weight, of A with compound IV. Inject a 0.3 µL sample of this mixture, measure the various peak areas and, after making appropriate corrections for differences in detector sensitivity, determine the percentage composition of A.

9.8 Sucrose as its trimethylsilyl derivative

Introduction This experiment illustrates the value of derivatisation in the GC analysis of sugars and related substances. Silylation derivatisation is used widely in carbohydrate analysis[12] to avoid the use of high column temperatures that are likely to lead to decomposition of the underivatised compounds.

Reagents and apparatus *Pure and dry* The reagents and solvents should be pure and dry, and should be tested in advance in the gas chromatographic system which is to be used in the experiment.

Pyridine   Purify by refluxing over potassium hydroxide, followed by distillation. Store the purified pyridine over the same reagent.

Other reagents   Trimethylchlorosilane (TMCS) $(CH_3)_3SiCl$ and hexamethyldisilazane (HMDS) $(CH_3)_3Si—NH—Si(CH_3)_3$.

Reaction vessel   Use a small tube or vial fitted with a Teflon-lined screw cap.

Gas chromatograph   Operate the column isothermally at $210\,°C$ using a flame ionisation detector.

Procedure   Treat 10 mg of sucrose with 1 mL of anhydrous pyridine, 0.2 mL of HMDS, and 0.1 mL of TMCS in the plastic-stoppered vial (or similar container). Shake the mixture vigorously for about 30 s and allow it to stand for 10 min at room temperature; if the carbohydrate appears to remain insoluble in the reaction mixture, warm the vial for 2–3 min at $75–85\,°C$. Inject $0.3\,\mu L$ of the resulting mixture into the gas chromatograph. Anhydrous reaction conditions are generally essential since the silylated derivatives are sensitive to water in varying degrees.

9.9   Determination of aluminium as its tris(acetylacetonato) complex

Introduction   This experiment illustrates GC determination of trace amounts of metals as their chelate complexes. The procedure described for the determination of aluminium may be adapted for the separation and determination of aluminium and chromium(III) as their acetylacetonates.[13]

Sample   The solvent extraction of aluminium from aqueous solution using acetylacetone can provide a suitable sample solution for gas chromatographic analysis. Take 5 mL of a solution containing about 15 mg of aluminium and adjust the pH to between 4 and 6. Equilibrate the solution for 10 min with two successive 5 mL portions of a solution made up of equal volumes of acetylacetone (pure, redistilled) and chloroform. Combine the organic extracts. Fluoride ion causes serious interference to the extraction and must be previously removed.

Introduce a $0.30\,\mu L$ portion of the solvent extract into the gas chromatograph. Concentrations greater than $0.3\,M$ are unsuitable as they deposit solid and cause a blockage of the $1\,\mu L$ microsyringe used for the injection of the sample. The syringe is flushed several times with the sample solution, filled with the sample to the required volume, excess liquid wiped from the tip of the needle and the sample injected into the chromatograph.

Apparatus   Use a gas chromatograph equipped with a flame ionisation detector and data handling system. A digital integrator is particularly convenient for quantitative determinations, but other methods of measuring peak area may be used (Section 9.4). Pure nitrogen (oxygen-free), at a flow rate of $40\,mL\,min^{-1}$, is used as carrier gas. The dimensions of the glass column are 1.6 m length and 6 mm outside diameter; it is packed with 5 wt% SE30 on Chromosorb W as the stationary phase. The column is maintained at a temperature of $165\,°C$.

Procedure   Extract a series of aqueous aluminium solutions containing 5–25 mg aluminium in 5 mL, using the procedure described for the sample. Calibrate the apparatus by injecting 0.30 µL of each extract into the column and recording the peak area on the chromatogram. Plot a graph of peak area against concentration. Determine aluminium (present as its acetylacetonate) in the sample solution by injecting 0.30 µL into the column. Record the peak area obtained and read off the aluminium concentration from the calibration graph. The calibration procedure is of limited accuracy; a more accurate result may be obtained using the method of standard additions (Section 9.5).

9.10   Derivatisation and quantitation of sugar alcohols

Introduction   Kirk and Sawyer have described a detailed procedure for the GC separation of sugar alcohols in foods, such as preserves, jellies and fruits,[14] based on recommended international methods; a slightly modified form is given here. Its value lies in the use of internal standard and derivatisation techniques.

Reagents   *Sample and standard*   The sample should be macerated before use. The internal standard is *meso*-inositol.

Carrez 1 solution   Prepared from 21.9 g zinc ethanoate dihydrate, 3 g ethanoic acid made to 100 mL in a graduated flask.

Carrez 2 solution   Prepared from 10.6 g potassium ferricyanide made to 100 mL in a graduated flask.

Reference sugars   For example, glucose, fructose, sorbitol, mannitol.

Oximation solution   2.5% w/v hydroxylamine in pyridine.

Silylation reagent   Trimethylchlorosilane : *N,O*-bis-(trimethylsilyl)acetamide (1 : 5 by volume).

Apparatus   This separation is best carried out on a 200 cm × 2 mm glass column with 5% SE52 on an inert 80–100 mesh support packing, heated at 150 °C for 2 min, then temperature programmed to 250 °C at 2 °C min^{-1} using an FID at 260 °C with N$_2$ carrier gas at 30 mL min^{-1}.

Procedure   Prepare solution A by weighing 1 g of the macerated sample into a beaker with 0.3 g of *meso*-inositol and 30 mL water. Heat gently and stir to extract the sugars. Add 0.5 mL Carrez 1 solution and 0.5 mL of Carrez 2 solution then mix. Filter the mixture into a 100 mL graduated flask. Wash the solid material with a further 20 mL of warm water, filter and add to the initial filtrate. Make the total solution to 100 mL with methanol. Prepare reference solution B by dissolving approximately 0.2 g of each sugar together in 30 mL water along with 0.3 g *meso*-inositol, add 0.5 mL Carrez 1 solution and 0.5 mL of Carrez 2 solution. Transfer to a 100 mL graduated flask. Add a further 20 mL water and make to 100 mL with methanol.

   Place 0.5 mL of solution A in a sample vial suitable for fitting with a septum. In another vial do the same for a 0.5 mL sample of solution B. Evaporate the solutions under a stream of nitrogen until nearly dry. Add 0.5 mL of isopropanol and evaporate to dryness under a stream of nitrogen. Screw the septum caps onto both vials and inject 0.5 mL of the oximation solution into each. Mix and heat at 80 °C for 30 min. Allow to cool and inject 1 mL of silylation reagent into each vial, mix and heat at 80 °C for 30 min. Allow to cool. It should be possible to obtain acceptable chromatograms using 1 μL sample volumes of the solutions. The sugars in the original foodstuff can be identified by comparing their retention times with the standards, and quantitative determinations can be made by comparing chromatograms based upon the peak ratios with the mesoinositol internal standard.

9.11 References

1. A T James and A J P Martin 1952 *Biochem. J.*, **50**; 679–90
2. A J P Martin and R L M Synge 1941 *Biochem. J.*, **35**; 1358
3. K Grob 1986 *Classical split and splitless injection in capillary gas chromatography*, Huethig Verlag, New York
4. A Rizzolo, A Polesello and S Polesello 1992 *High Resolut. Chromatogr.*, **7** (2); 201
5. R C Denney 1983 *Speciality Chemicals*, **3**; 6–7, 12
6. R S Barrett 1973 *Proc. Soc. Anal. Chem.*, **45**; 167
7. K Blau and G S King (eds) 1977 *Handbook of derivatives for chromatography*, Heyden, London
8. D Knapp (ed) 1979 *Handbook of analytical derivatisation reactions*, John Wiley, New York
9. I G McWilliam 1983 *Chromatographia*, **17**; 241
10. E Bulska 1992 *J. Anal. At. Spectrom.*, **7** (2); 201
11. W E Harris and H W Habgood 1966 *Programmed temperature gas chromatography*, John Wiley, New York
12. C C Sweeley *et al.* 1963 *J. Amer. Chem. Soc.*, **85**; 2497
13. R D Hill and H Gesser 1963 *J. Gas Chromatogr.*, **1**; 11
14. R S Kirk and R Sawyer 1991 *Pearson's composition and analysis of foods*, 9th edn, Longman, Harlow, p. 203

9.12 Bibliography

F Bruner 1993 *Gas chromatographic environmental analysis*, VCH, New York

R C Denney 1982 *A dictionary of chromatography*, 2nd edn, Macmillan, London

I A Fowlis 1995 *Gas chromatography*, ACOL–Wiley, Chichester

D W Grant 1995 *Capillary gas chromatography*, John Wiley, Chichester

R L Grob (ed) 1995 *Modern practice of gas chromatography*, 3rd edn, John Wiley, Chichester

C Horváth and L S Ettre (eds) 1993 *Chromatography in Biotechnology*, American Chemical Society, Washington DC

R W Moshier and R E Sievers 1965 *Gas chromatography of metal chelates*, Pergamon, Oxford

K K Ungor (ed) 1990 *Packings and stationary phases in chromatographic techniques*, Marcel Dekker, New York

10

Titrimetric analysis

Theoretical considerations

10.1 Introduction

The methods of 'wet chemistry' such as titrimetric analysis and gravimetry still have an important role in modern analytical chemistry. There are many areas in which titrimetric procedures are invaluable. Their advantages may be listed as follows:

1. The precision (0.1%) is better than most instrumental methods.
2. Methods are usually superior to instrumental techniques for major component analysis.
3. When the sample throughput is small, e.g. for one-off analysis, simple titrations are often preferable.
4. Unlike instrumental methods, the equipment does not require constant recalibration.
5. Methods are relatively inexpensive with low unit costs per determination.
6. They are often used to calibrate and/or validate routine analysis using instruments.
7. The methods can be automated (Section 10.10).

There are, however, several disadvantages to classical titrimetric procedures. The most significant is that they are normally less sensitive and frequently less selective than instrumental methods. Furthermore, when a large number of similar determinations are required, then instrumental methods are usually much quicker and often cheaper than the more labour-intensive titrimetric methods. Nevertheless, despite the widespread popularity of instrumental techniques, it can be seen from the above that there is considerable scope for the use of titrimetric procedures, especially for practising laboratory skills. Besides giving a survey of the classical titrimetric methods, this chapter includes titrimetry based on electrochemical techniques, including automated methods, together with a brief account of spectrophotometric titrations.

10.2 Titrimetric analysis

The term 'titrimetric analysis' refers to quantitative chemical analysis carried out by determining the volume of a solution of accurately known concentration which is required to react quantitatively with a measured volume of a solution of the substance to be determined. The solution of accurately known strength is called the **standard solution** (Section 10.4). The weight of the substance to be determined is calculated from the volume of the standard solution used and the chemical equation and relative molecular masses of the reacting compounds.

The term 'volumetric analysis' was formerly used for this form of quantitative determination but it has now been replaced by **titrimetric analysis**. It is considered that 'titrimetric analysis' expresses the process of titration rather better, and 'volumetric analysis' is likely to be confused with measurements of volumes, such as those involving gases. In titrimetric analysis the reagent of known concentration is called the **titrant** and the substance being titrated is termed the **titrand**. The alternative name has not been extended to apparatus used in the various operations; so the terms 'volumetric glassware' and 'volumetric flasks' are still common, but it is better to employ the expressions 'graduated glassware' and 'graduated flasks'; they are used throughout this book.

The standard solution is usually added from a long graduated tube called a burette. The process of adding the standard solution until the reaction is just complete is called a titration, and the substance to be determined is titrated. The point at which this occurs is called the equivalence point or the **theoretical or stoichiometric end point**. The completion of the titration is detected by some physical change, produced by the standard solution itself (e.g. the faint pink colour formed by potassium permanganate) or, more usually, by the addition of an auxiliary reagent, known as an indicator; alternatively some other physical measurement may be used. After the reaction between the substance and the standard solution is practically complete, the indicator should give a clear visual change (either a colour change or the formation of turbidity) in the liquid being titrated. The point at which this occurs is called the **end point of the titration**. In the ideal titration the visible end point will coincide with the stoichiometric or theoretical end point. In practice, however, a very small difference usually occurs; this represents the **titration error**. The indicator and the experimental conditions should be selected so the difference between the visible end point and the theoretical end point is as small as possible.

For use in titrimetric analysis a reaction must fulfil the following conditions:

1. There must be a simple reaction which can be expressed by a chemical equation; the substance to be determined should react completely with the reagent in stoichiometric or equivalent proportions.
2. The reaction should be relatively fast. (Most ionic reactions satisfy this condition.) In some cases the addition of a catalyst may be necessary to increase the speed of a reaction.
3. There must be an alteration in some physical or chemical property of the solution at the equivalence point.
4. An indicator should be available which, by a change in physical properties (colour or formation of a precipitate), should sharply define the end point of the reaction. If no visible indicator is available, the detection of the equivalence point can often be achieved in other ways:

    (a) Measuring the potential between an indicator electrode and a reference electrode (**potentiometric titration**).
    (b) Developing the titrant by electrolysis (**coulometric titration**).
    (c) Measuring the current which passes through the titration cell between an indicator electrode and a depolarised reference electrode at a suitable applied e.m.f. (**amperometric titration**)

Titrimetric methods are normally capable of high precision (1 part in 1000 or better) and wherever applicable they possess obvious advantages over gravimetric methods. They need simpler apparatus and are generally quickly performed; tedious and difficult separations can often be avoided. The following apparatus is required for titrimetric analysis: (i) calibrated measuring vessels, including burettes, pipettes, and measuring flasks (Chapter 3); (ii) substances of known purity as standard solutions; (iii) a visual indicator or an instrumental method for detecting the completion of the reaction.

10.3 Classification of reactions in titrimetric analysis

The reactions employed in titrimetric analysis fall into four main classes. The first three involve no change in oxidation state as they depend on the combination of ions. But the fourth class, oxidation–reduction reactions, involves a change of oxidation state or, expressed another way, a transfer of electrons.

Neutralisation reactions, or acidimetry and alkalimetry

These include the titration of free bases, or those formed from salts of weak acids by hydrolysis, with a standard acid (acidimetry), and the titration of free acids, or those formed by the hydrolysis of salts of weak bases, with a standard base (alkalimetry). The reactions involve the combination of hydrogen and hydroxide ions to form water. They also include titrations in non-aqueous solvents, most of which involve organic compounds.

Complex formation reactions

Ethylenediaminetetra-acetic acid, largely as the disodium salt of EDTA, is a very important reagent for complex formation titrations and has become one of the most important reagents used in titrimetric analysis. Equivalence point detection by the use of metal ion indicators has greatly enhanced its value in titrimetry.

Precipitation reactions

These depend upon the combination of ions to form a simple precipitate as in the titration of silver ion with a solution of a chloride (Section 10.92). No change in oxidation state occurs.

Oxidation–reduction reactions

Under this heading come all reactions involving change of oxidation number or transfer of electrons among the reacting substances. The standard solutions are either oxidising or reducing agents. The principal oxidising agents are potassium permanganate, potassium dichromate, cerium(IV) sulphate, iodine, potassium iodate, and potassium bromate. Frequently used reducing agents are iron(II) and tin(II) compounds, sodium thiosulphate, arsenic(III) oxide and mercury(I) nitrate. The following reducing agents have rather limited use: vanadium(II) chloride or sulphate, chromium(II) chloride or sulphate, and titanium(III) chloride or sulphate.

10.4 Standard solutions

The word 'concentration' is frequently used as a general term referring to a quantity of substance in a defined volume of solution. But quantitative titrimetric analyses use standard solutions in which the base unit of quantity is the mole. This follows the definition given by the International Union of Pure and Applied Chemistry:

The mole is the amount of substance which contains as many elementary units as there are atoms in 0.012 kilogram of carbon-12. The elementary unit must be specified and may be an atom, a molecule, an ion, a radical, an electron or other particle or a specified group of such particles.[1]

As a result, standard solutions are now commonly expressed in terms of molar concentrations or molarity M. Such standard solutions are specified in terms of the number of moles of solute dissolved in 1 litre of solution; for any solution

$$\text{molarity } M = \frac{\text{moles of solute}}{\text{volume of solution in litres}}$$

10.5 Preparation of standard solutions

If a reagent is available in the pure state, a solution of definite molar strength is prepared simply by weighing out a mole, or a definite fraction or multiple thereof, dissolving it in an appropriate solvent, usually water, and making up the solution to a known volume. It is not essential to weigh out exactly a mole (or a multiple or submultiple thereof); in practice it is more convenient to prepare the solution a **little** more concentrated than is ultimately required, and then to dilute it with distilled water until the desired molar strength is obtained. If M_1 is the required molarity, V_1 the volume after dilution, M_2 the molarity originally obtained, and V_2 the original volume taken, $M_1 V_1 = M_2 V_2$, or $V_1 = M_2 V_2 / M_1$. The volume of water to be added to the volume V_2 is $(V_1 - V_2)\,\text{mL}$.

The following substances can be obtained in a state of high purity and are therefore suitable for the preparation of standard solutions: sodium carbonate, potassium hydrogenphthalate, sodium tetraborate, potassium hydrogeniodate, sodium oxalate, silver nitrate, sodium chloride, potassium chloride, iodine, potassium bromate, potassium iodate, potassium dichromate, lead nitrate and arsenic(III) oxide.

When the reagent is not available in the pure form, as in the cases of most alkali hydroxides, some inorganic acids and various deliquescent substances, solutions corresponding approximately to the molar strength required are first prepared. These are then standardised by titration against a solution of a pure substance of known concentration. It is generally best to standardise a solution by a reaction of the same type as that for which the solution is to be employed, and as nearly as possible under identical experimental conditions. The titration error and other errors are thus considerably reduced or are made to cancel out. This indirect method is employed for the preparation of, for instance, solutions of most acids, sodium hydroxide, potassium hydroxide and barium hydroxide, potassium permanganate, ammonium and potassium thiocyanates, and sodium thiosulphate.

10.6 Primary and secondary standards

In titrimetry certain chemicals are used frequently in defined concentrations as reference solutions. They are known as **primary standards** or **secondary standards**. A primary standard is a compound of sufficient purity from which a standard solution can be prepared by direct weighing of a quantity of it, followed by dilution to give a defined volume of solution. The solution produced is then a primary standard solution. A primary standard should satisfy the following requirements:

1. It must be easy to obtain, to purify, to dry (preferably at 110–120 °C), and to preserve in a pure state. (This requirement is not usually met by hydrated substances, since it is difficult to remove surface moisture completely without effecting partial decomposition.)

2. The substance should be unaltered in air during weighing; this condition implies that it should not be hygroscopic, oxidised by air, or affected by carbon dioxide. The standard should maintain an unchanged composition during storage.
3. The substance should be capable of being tested for impurities by qualitative and other tests of known sensitivity. (The total amount of impurities should not, in general, exceed 0.01–0.02%.)
4. It should have a high relative molecular mass so that the weighing errors may be negligible. (The precision in weighing is ordinarily 0.1–0.2 mg; for an accuracy of 1 part in 1000, it is necessary to employ samples weighing at least about 0.2 g.)
5. The substance should be readily soluble under the conditions in which it is employed.
6. The reaction with the standard solution should be stoichiometric and practically instantaneous. The titration error should be negligible, or easy to determine accurately by experiment.

In practice, an ideal primary standard is difficult to obtain, and a compromise between the ideal requirements is usually necessary. The substances commonly employed as primary standards are indicated below:

Acid–base reactions: sodium carbonate Na_2CO_3, sodium tetraborate $Na_2B_4O_7$, potassium hydrogenphthalate $KH(C_8H_4O_4)$, potassium hydrogeniodate $KH(IO_3)_2$.

Complex formation reactions: pure metals (e.g. zinc, magnesium, copper and manganese) and salts, depending upon the reaction used.

Precipitation reactions: silver, silver nitrate, sodium chloride, potassium chloride, and potassium bromide.

Oxidation–reduction reactions: potassium dichromate $K_2Cr_2O_7$, potassium bromate $KBrO_3$, potassium iodate KIO_3, potassium hydrogeniodate $KH(IO_3)_2$, sodium oxalate $Na_2C_2O_4$, arsenic(III) oxide As_2O_3, and pure iron.

Hydrated salts, as a rule, do not make good standards; this is because it is difficult to dry them efficiently. However, those salts which do not effloresce, such as sodium tetraborate $Na_2B_4O_7 \cdot 10H_2O$, and copper sulphate $CuSO_4 \cdot 5H_2O$, are found by experiment to be satisfactory secondary standards.[2]

A secondary standard is a substance which may be used for standardisations, and whose content of the active substance has been found by comparison against a primary standard. It follows that a secondary standard solution is a solution in which the concentration of dissolved solute has not been determined from the weight of the compound dissolved but by reaction (titration) of a volume of the solution against a measured volume of a primary standard solution.

As potentiometric, coulometric and amperometric methods can be widely applied in titrimetric analysis it is essential to provide an introduction to the underlying theory. A survey of the design of electrodes for these purposes is provided in Chapter 13.

10.7 Principles of potentiometric titrations

In a potentiometric titration the potential of an indicator electrode is measured as a function of the volume of titrant added. The equivalence point of the reaction will be revealed by a sudden change in potential in the plot of e.m.f. readings against the volume of the titrating solution; any method which will detect this abrupt change of potential may be used. One electrode must maintain a constant, but not necessarily known, potential; the other electrode, which indicates the changes in ion concentration, must respond rapidly. Throughout the titration, the analyte solution must be thoroughly stirred.

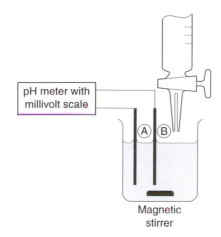

Figure 10.1   Potentiometric titration, simple apparatus: A = reference electrode, B = indicator electrode

A simple arrangement for a manual potentiometric titration is given in Figure 10.1. A is a reference electrode (e.g. a saturated calomel half-cell), B is the indicator electrode. The solution to be titrated is normally contained in a beaker fitted with a magnetic stirrer. When titrating solutions that require exclusion of air or atmospheric carbon dioxide, a three- or four-necked flask is used to enable nitrogen to be bubbled through the solution before and during the titration.

The e.m.f. of the cell containing the initial solution is determined, and relatively large increments (1–5 mL) of the titrant solution are added until the equivalence point (e.p.) is approached; the e.m.f. is determined after each addition. The approach of the e.p. is indicated by a somewhat more rapid change of the e.m.f. In the vicinity of the e.p., equal increments (e.g. 0.1 or 0.05 mL) should be added; the equal additions in the region of the e.p. are particularly important when the equivalence point is to be determined by the analytical method described below. Sufficient time should be allowed after each addition for the indicator electrode to reach a reasonably constant potential (~+1–2 mV) before the next increment is introduced. Several points should be obtained well beyond the e.p. To measure the e.m.f., the electrode system is usually connected to a pH meter that can function as a millivoltmeter so that e.m.f. values are recorded. Used as a millivoltmeter, pH meters can be employed with almost any electrode assembly to record the results of many different types of potentiometric titrations, and in many cases the instruments had provision for connection to a recorder so that a continuous record of the titration results could be obtained, usually in the form of a titration curve.

10.8   General considerations

As with classical titrimetry, potentiometric titrations involve chemical reactions which can be classified as (a) neutralisation reactions, (b) complexation reactions, (c) precipitation reactions and (d) oxidation–reduction reactions. Experimental details for potentiometric titrations will be found in the following sections: neutralisation, including non-aqueous reactions (Sections 10.55 to 10.58), complexation (Section 10.85), precipitation (Sections 10.102 to 10.103), redox (Sections 10.154 to 10.156).

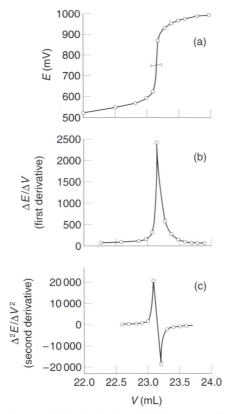

Figure 10.2   Potentiometric titration: locating the end points

10.9   Location of end points

Generally speaking, the end point of a titration can be most easily fixed by examining the titration curve, including the derivative curves to which this gives rise, or by examining a Gran's plot. When a titration curve has been obtained – i.e. a plot of e.m.f. readings obtained with the normal reference electrode–indicator electrode pair against volume of titrant added, either by manual plotting of the experimental readings, or with suitable equipment, plotted automatically during the course of the titration – it will in general be of the same form as the neutralisation curve for an acid, i.e. an S-shaped curve as shown in Figure 10.12 (Section 10.33). The central portion of this curve is shown in Figure 10.2(a), and clearly the end point will be located on the steeply rising portion of the curve; it will in fact occur at the point of inflexion. When the curve shows a very clearly marked steep portion, although one can give an approximate value of the end point as being midway along the steep part of the curve, it is usually preferred to employ **analytical (or derivat-ive) methods** of locating the end point. Analytical methods consist in plotting the first derivative curve ($\Delta E/\Delta V$ against V), or the second derivative curve ($\Delta^2 E/\Delta V^2$ against V). The first derivative curve gives a maximum at the point of inflexion of the titration curve, i.e. at the end point, whereas the second derivative curve ($\Delta^2 E/\Delta V^2$) is zero at the point where the slope of the $\Delta E/\Delta V$ curve is a maximum.

The Gran's plot procedure is a relatively simple method for fixing an end point. If a series of additions of reagent are made in a potentiometric titration, and the cell e.m.f. E is read after each addition, then if antilog ($EnF/2.303RT$) is plotted against the volume of reagent added, a straight line is obtained which, when extrapolated, cuts the volume axis at a point corresponding to the equivalence point volume of the reagent; plotting is simplified if the special semi-antilog Gran's plot paper is used. The particular advantage of this method is that the titration need not be pursued to the actual end point; it is only necessary to have the requisite number of observations before the end point to permit a straight line to be drawn, and the greatest accuracy is achieved by using results over the last 20% of the equivalence point volume.

10.10   Automatic titrators

Potentiometric titrations (Section 10.7), when performed manually, can take a considerable time. The precise location of the end point may need first or even second derivative plots (Section 10.9) which, if plotted by hand, are lengthy operations. Even classical titrations using visual indicators are sometimes time-consuming and operator-intensive. If a large number of titrations have to be performed in the course of a day, then operator fatigue can set in, which could clearly endanger the precision and accuracy of the results. Automatic titrators therefore offer great advantages, particularly when the initial financial outlay involved may be recovered in a short period of time.

A number of commercial titrators are available for potentiometric titrations. The electrical measuring unit may be coupled to a chart recorder to produce a titration curve directly. The delivery of the titrant from an automatic burette is linked to the movement of the recorder, giving an autotitrator. Instruments will also plot the first derivative curve ($\Delta E/\Delta V$) and the second derivative ($\Delta^2 E/\Delta V^2$), and will provide a Gran's plot (Section 10.9). A most important feature is the facility to stop the delivery of the titrant when the equivalence potential has been reached. This is extremely useful when a number of repetitive titrations have to be made. Commercial instruments are also available where the end point of the titration is marked by a colour change of an indicator which can be observed spectrophotometrically.

A modern automatic titrator, the Schott TitroLine Alpha, marketed by Camlab, Cambridge, has the following features. The instrument is a fully equipped titrator with an integrated burette module. Titration methods are created by means of a menu-driven system. Up to 100 preprogrammed applications are available in memory. The appropriate parameters are for type of titration, titration direction, titration end, drift and titration control. The selected application can be copied and durable-saved into one of eight memories. Fine adjustments of the titration parameters can be carried out any time and operators can create their own complete method if desired. The calculation of results can be made from eight equations which are available for each of the eight methods in the working memory, enabling blank values and calculations of back titrations. Statistical factors, mean values, standard deviations and variances from a set of measurements can also be evaluated. Additional features include a special platinum sensor which allows temperature compensation to be made for the temperature functions of eight buffers. A sample changer is also available, enabling a number of titrations to be performed in series with automatic sample changing. A number of washing positions are set automatically within the sample changer to clean the electrodes and titration tip after each titration.

The titrator can be used as a routine automatic instrument for pH, millivolt, redox, argentimetric, Karl Fischer titrations as well as titrations in non-aqueous solvents. The titration of a mixture of chloride and iodide ions can be achieved automatically for up to two equivalence points. Furthermore, up to two preselected end points can be titrated. This illustrates the versatility and scope of an automatic titration system. When given a large number of samples, it can achieve better precision than manual analyses.

The scope of potentiometric titrations has increased considerably since the introduction of a wide variety of ion selective electrodes (ISEs). Not only can ion selective electrodes be used for the **direct** determination of the analyte concentration, they can also be used as end point detectors in titrations. A range of pH and ion selective electrodes is produced by Orion Research (marketed in the UK by QuadraChem Laboratories, Forest Row, East Sussex). A series of Orion autotitrators using selected pH electrodes can perform the following neutralisation titrations in thick viscous samples, for surfactant analysis and for small-volume microtitrations. A full range of redox, chelometric and halide (argentimetric) titrations can be carried out automatically. Section 10.85 explains how to use the Orion 960 autotitrator for determining water hardness and for analysing zinc in plating baths.

10.11   Advantages of potentiometric titrations

There are many situations where potentiometric titrations have an advantage over 'classical' visual indicator methods. These include:

1. Where the end point obtained by the indicator is masked, e.g. if the analyte solution is coloured, turbid or fluorescent.
2. Where there is no suitable indicator or where the colour change is difficult to ascertain. For example, considerable skill, if not some luck, may be needed to follow the end points obtained with some indicators used in precipitation titrations.
3. In the titration of polyprotic acids, mixtures of acids, mixtures of bases, or mixtures of halides.
4. Where the process needs to be automated, with end point data stored in a computer.

Coulometry at constant current

10.12   General

Coulometry at controlled potential is applicable only to the limited number of substances which undergo quantitative reaction at an electrode during electrolysis. The reader should refer to Chapter 13 for a broader discussion on this subject. By using coulometry at controlled or constant current, the range of substances that can be determined may be extended considerably, and includes many which do not react quantitatively at an electrode. Constant-current electrolysis is employed to generate a reagent which reacts stoichiometrically with the substance to be determined. The quantity of substance reacted is calculated with the aid of Faraday's law, and the quantity of electricity passed can be evaluated simply by timing the electrolysis at constant current. Since the current can be varied from say 0.1 to 100 mA, it is possible to determine amounts of material corresponding to 1×10^{-9} to 1×10^{-6} mol s^{-1} of electrolysis time. In titrimetric analysis the reagent is added from a burette; in coulometric titrations the reagent is generated electrically and its amount is evaluated from a knowledge of the current and the generating time. The electron becomes the standard reagent. In many respects, e.g. detection of end points, the procedure differs only slightly from ordinary titrations.

The fundamental requirements of a coulometric titration are (1) that the reagent-generating electrode reaction proceeds with 100% efficiency, and (2) that the generated reagent reacts stoichiometrically and, preferably, rapidly with the substance being determined. The reagent may be generated directly within the test solution or, less frequently, it may be generated in an external solution which is allowed to run continuously into the test solution. Several methods are available for the detection of end points in coulometric titrations.

Chemical indicators   Chemical indicators must not be electroactive. Examples include starch for iodine, dichlorofluorescein for chloride, and eosin for bromide and iodide.

Potentiometric observations   Electrolytic generation is continued until the e.m.f. of a reference electrode–indicating electrode assembly placed in the test solution attains a predetermined value corresponding to the equivalence point.

Amperometric procedures   Amperometric procedures are based on establishing conditions such that either the substance being determined or, more usually, the titrant undergoes reaction at an indicator electrode to produce a current which is proportional to the concentration of the electroactive substance. With the potential of the indicator electrode maintained constant, or nearly constant, the end point can be established from the course of the current change during the titration. The voltage impressed upon the indicator electrode is well below the decomposition voltage of the pure supporting electrolyte but close to or above the decomposition voltage of the supporting electrolyte plus free titrant; consequently, as long as any of the substance being determined remains to react with the titrant, the indicator current remains very small but increases as soon as the end point is passed and free titrant is present. There is a relatively inexhaustible supply of titrant ion (e.g. bromide ion in coulometric titrations with bromine), and the indicator current beyond the equivalence point is therefore governed largely by the rate of diffusion of the free titrant (e.g. bromine) to the surface of the indicator electrode. The indicator current is consequently proportional to the concentration of the free titrant (bromine) in the bulk of the solution and to the area of the indicator electrode (cathode for bromine).

The indicator current will increase with increasing rate of stirring, since this decreases the thickness of the diffusion layer at the electrode; it is also somewhat temperature-dependent. The generation time at which the equivalence point is reached may be determined by calibrating the indicator electrode system with the supporting electrolyte alone by generating the titrant (e.g. bromine) for various times (say 10–50 s) to evaluate the constant in the relation $I_i = Kt$, where I_i is the indicator current and t is the time. The generating time to the equivalence point may then be obtained from the observed final value of the indicator current in the actual titration, calculating the excess generating time and subtracting this from the total generating time in the titration. Alternatively, and more simply, the equivalence point time may be located by measuring five values of the indicator current at five measured times beyond the equivalence point and extrapolating to zero current. The biamperometric (dead-stop) method is explained in Section 10.21.

Spectrophotometric observations   The titration cell consists of a spectrophotometer cuvette (2 cm light path). The motor-driven glass propeller stirrer and the working platinum electrode are placed in the cell in such a way as to be out of the light path: a platinum electrode in dilute sulphuric acid in an adjacent cuvette also placed in the cell holder serves as an auxiliary electrode and is connected with the titration cell by an inverted U-tube salt bridge. The appropriate wavelength is set on the instrument. Before the end point the absorbance changes only very slowly, but a rapid and linear response occurs beyond the equivalence point. Examples are the titration of Fe(II) in dilute sulphuric acid with electrogenerated Ce(IV) at 400 nm, and the titration of arsenic(III) with electrogenerated iodine at 342 nm. Further details can be found in Sections 10.24 to 10.28.

10.13 Principles

Consider the titration of iron(II) with electrogenerated cerium(IV). A large excess of Ce(III) is added to the solution containing the Fe(II) ion in the presence of say $1\,M$ sulphuric acid. Consider what happens at a platinum anode when a solution containing Fe(II) ions alone is electrolysed at constant current. Initially the reaction

$$Fe^{2+} \rightleftharpoons Fe^{3+} + e$$

will proceed with 100% current efficiency. At the anode surface the concentration of Fe(III) ions formed is relatively large, whereas the concentration of the Fe(II) ions, which is governed by the rate of transfer from the bulk of the solution, is very small; the potential of the anode gradually acquires a value which is much more positive (more oxidising) than the standard potential of the Fe(III)/Fe(II) couple (0.77 V). As electrolysis proceeds, the anode potential becomes more and more positive (oxidising) at a rate that depends on the current density, and ultimately it becomes so positive (~1.23 V) that oxygen evolution from the oxidation of water begins $(2H_2O \rightleftharpoons O_2 + 4H^+ + 4e)$, and this occurs before all the Fe(II) ions in the bulk of the solution are oxidised.

As soon as oxygen evolution commences, the current efficiency for the oxidation of Fe(II) falls below 100% and the quantity of Fe(II) initially present cannot be computed from Faraday's law. If the electrolysis is conducted in the presence of a relatively large concentration of Ce(III) ions, the following reactions will take place at the anode. At a certain potential of the anode, considerably less than required for oxygen evolution, oxidation of Ce(III) to Ce(IV) sets in, and the Ce(IV) thus produced is transferred to the bulk of the solution, where it oxidises Fe(II) to Fe(III). The potential of the working electrode is thus stabilised by the reagent-generating reaction, hence it is prevented from drifting to a value such that an interfering reaction may result.

Stoichiometrically, the total quantity of electricity passed is exactly the same as it would have been if the Fe(II) ions had been directly oxidised at the anode and the oxidation of Fe(II) proceeds with 100% efficiency. The equivalence point is marked by the first persistence of excess Ce(IV) in the solution, and may be detected by any of the methods described above. The Ce^{3+} ions added to the Fe(II) solution undergo no net change and are said to act as a **mediator**.

Side reactions are avoided at the generating electrode provided there is not complete depletion (at the electrode surface) of the substance involved in the generation of the titrant. The concentration of the titrant depends on the current through the cell, the area of the generating electrode, and the rate of stirring; the concentration of the generating substance is usually between $0.01\,M$ and $0.1\,M$.

10.14 Instrumentation

A number of coulometric titrators are available commercially and are simple to operate. Suitable apparatus can also be assembled from readily available equipment. There are two essential requirements:

A constant-current source This can be a large-capacity storage battery but is preferably an amperostat, which is an electronically controlled instrument providing a constant current; it can often provide a constant voltage (act as a potentiostat) in addition to its constant-current function.

Nitrogen

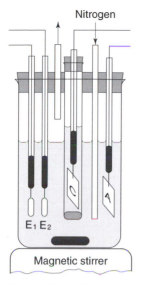

E_1 E_2

Magnetic stirrer

Figure 10.3   Coulometric titration: simple apparatus

An integrator   This is an electronic device for measuring the product (current × time), i.e. the number of coulombs. If necessary it may be replaced by an accurately calibrated milliammeter coupled with a quartz crystal clock to record the duration of the electrolysis.

The titration cell is shown in Figure 10.3. It consists of a tall-form beaker of about 200 mL capacity. Provision is made for magnetic stirring and for passing a stream of inert gas (e.g. nitrogen) through the solution. The main generator electrode (A) may consist of platinum foil (1 cm × 1 cm or 4 cm × 2.5 cm) and the auxiliary electrode (C) may consist of platinum foil (1 cm × 1 cm or 4 cm × 2.5 cm) bent into a half-cylinder so as to fit into a wide glass tube (~1 cm diameter). The isolation of electrode C within the glass cylinder (closed by a sintered-glass disc) from the bulk of the solution avoids any effects arising from undesirable reactions at this electrode.

E_1 and E_2 are the indicator electrodes. These may consist of a tungsten pair for a biamperometric end point; for an amperometric end point they may both be of platinum foil or one can be platinum and the other a saturated calomel reference electrode. The voltage impressed upon the indicator electrodes is supplied by a battery (~1.5 V) via a variable resistance. For a potentiometric end point E_1 and E_2 may consist of either platinum–tungsten bimetallic electrodes, or E_1 may be an SCE and E_2 a glass electrode. These are connected directly to a pH meter with a subsidiary scale calibrated in millivolts. The indicator electrodes should be positioned outside the electric field (current path) between the generator electrodes; otherwise, spurious indicator currents may be produced, particularly in the amperometric detection of the equivalence point.

General procedure

The electrolysis cell is set up with both generator and indicator electrodes in position and provision is made, if necessary, for passing an inert gas (e.g., nitrogen) through the solution. The titration cell is charged with the solution from which the titrant will be generated electrolytically, together with the solution to be titrated. The auxiliary electrode compartment is filled with a solution of the appropriate electrolyte at a higher level than the solution in the titration cell. The indicator electrodes are connected to a suitable apparatus for the detection of the end point, e.g. a pH meter with an additional millivolt scale. Stirring is effected with a magnetic stirrer. The reading of the digital indicator instrument is taken. The current, previously adjusted to a suitable value, is then switched on and reaction between the internally generated titrant and the test solution allowed to proceed. Readings are taken periodically (more frequently as the end point is approached) of the integrating counter and the indicating instrument (e.g. pH meter); it is usually necessary to switch off the electrolysis current while the readings of the indicating instrument are recorded. The end point of the titration is readily evaluated from the plot of the reading of the indicating instrument (e.g. millivolts) against the counter reading; the first- or second-derivative curve is drawn to locate the equivalence point accurately. It is possible to repeat the titration with a fresh volume of the test solution; if the end point is determined potentiometrically, subsequent determinations may be stopped at the potential found for the equivalence point in the initial titration.

10.15 External generation of titrant

Coulometric titration with internal generation of the titrant has the following limitations:

1. No substance may be present which undergoes reaction at the generator electrodes; for example, in acidimetric titrations the test solutions must not contain substances which are reduced at the generator cathode.
2. When applied on a macro scale, samples of 1–5 mmol, generation rates of 100–500 mA are required. Parasitic currents may be induced in the indicator electrodes at currents in excess of about 10–20 mA; consequently, precise location of the equivalence point by amperometric methods is not trustworthy.

To overcome these limitations, the reagent can be generated at constant current with 100% efficiency in an external generator cell and subsequently delivered to the titration cell. This technique is identical with an ordinary titration except the reagent is generated electrolytically. A double-arm electrolytic cell for external generation of the titrant is shown in Figure 10.4. The generator electrodes consist of two platinum spirals near the centre of the inverted U-tube. The space between the electrodes is packed with glass wool to prevent turbulent mixing; the downward legs of the generator tube are constructed of 1 mm capillary tubing to reduce the inconvenience due to hold-up. The solution of the electrolyte, which upon electrolysis will yield the desired titrant, is fed continuously into the top of the generator cell. The solution is divided at the T-joint so that about equal quantities flow through each of the arms of the cell. As these portions of the solution flow past the electrodes, electrolysis occurs; the products of electrolysis are swept along by the flow of the solution through the arms and emerge from the delivery tips. A beaker containing the substance to be titrated is placed beneath the appropriate delivery tip, and the solution from the other tip is run to waste. Thus for the titration of acids, electrode A functions as a cathode in sodium sulphate generator electrolyte and the hydroxide ion generated by the reaction

$$2H_2O + 2e \rightleftharpoons 2OH^- + H_2$$

flows into the test solution. The hydrogen ion and oxygen generated at the other electrode by the reaction

$$2H_2O \rightleftharpoons 4H^+ + O_2 + 4e$$

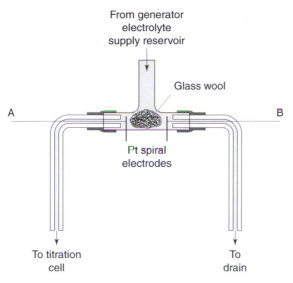

From generator
electrolyte
supply reservoir

Glass wool

A

B

Pt spiral
electrodes

To titration
cell

To
drain

Figure 10.4   External generation of titrant

are swept out of the other arm into the drain. For titration of bases, the generator electrode which delivers to the titration cell is employed as the anode. For titrations with electrically generated iodine, the generator electrolyte consists of potassium iodide solution, and the iodine solution formed at the anode flows into the titration vessel.

A minor disadvantage of external generation of titrant is the dilution of the contents of the titration cell; care is therefore necessary in suitably adjusting the rate of flow and the concentration of the generator solution. The procedure is, however, admirably suited for automatic control.

10.16   Advantages

Both time and current can be measured with a very high degree of precision and accuracy; the method has **high sensitivity**. Here are some other advantages:

1. Standard solutions are not required and in their place the coulomb becomes the primary working standard.
2. Unstable reagents, such as bromine, chlorine, silver(II) ion Ag^{2+} and titanium(III) ion, can be used, since they are generated and consumed immediately; there is no loss on storage or change in titre.
3. When necessary, very small amounts of titrants may be generated; this dispenses with the difficulties involved in the standardisation and storage of dilute solutions, and the procedure is ideally adapted for use on a micro or semimicro scale.
4. The sample solution is not diluted in the internal generation procedure.
5. By pretitration of the generating solution before the addition of the sample, more accurate results can be obtained, since the effect of impurities in the generating solution is minimised.
6. The method (which is largely electrical in nature) is readily adapted to remote control; this is significant in the titration of radioactive or dangerous materials. It may also be adapted to automatic control because of the relative ease of controlling the current automatically.

Table 10.1 *Some typical coulometric titrations*

Reagent generated	Electrolyte composition	Notes	Substances titrated	End point detection[a]
Neutralisation reactions				
H^+	Sodium sulphate (0.2 M)		OH^-, organic bases	P
OH^-	Sodium sulphate (0.2 M)		H^+, organic acids	P
Redox reactions				
Cl_2	HCl (2.0 M)		As(III), I^-	A
Br_2	KBr (0.2 M)		Sb(III), Tl(I), U(IV), I^-, SCN^-, NH_3, N_2H_4, NH_2OH	A
I_2	KI (0.1 M), phosphate buffer, pH 8		As(III), Sb(III), $S_2O_3^{2-}$, S^{2-}, ascorbic acid	A, P, I
Ce(IV)	$Ce_2(SO_4)_3$ (0.1 M)		Fe(II), Ti(III), U(IV), As(III), $Fe(CN)_6^{4-}$	P
Mn(III)	$MnSO_4$ (0.5 M), H_2SO_4 (1.8 M)		Fe(II), As(III), oxalic acid	P
Ag(II)	$AgNO_3$ (0.1 M), HNO_3 (5 M)	1	As(III), Ce(III), V(IV)	P
Cu(I)	$CuSO_4$ (0.1 M)		V(V), Cr(VI), IO_3^-, Br_2	P
Fe(II)	$Fe_2(SO_4)_3(NH_4)_2SO_4$ (0.3 M), H_2SO_4 (2 M)		V(V), Cr(VI), MnO_4^-	P
Ti(III)	Ti(IV) sulphate (0.6 M), H_2SO_4 (6 M)		Fe(II), Ce(IV), V(V), U(VI)	P
Precipitation reactions				
Ag(I)	KNO_3 (0.5 M)	2	Cl^-, Br^-, I^-, mercaptans	P
Hg(I)	$HClO_4$ (0.5 M)	3	Cl^-, Br^-	P
	$HClO_4$ (0.1 M), KNO_3 (0.4 M)		I^-	P
$Fe(CN)_6^{4-}$	$K_3Fe(CN)_6$ (0.2 M), H_2SO_4 (0.1 M)		Zn	P
Complexation reactions				
EDTA	$Hg(NH_3)Y^{2-}$ (0.1 M), NH_4NO_3 (0.1 M), pH 8.3	4[b]	Ca(II), Cu(II), Zn(II), Pb(II)	P
Miscellaneous reactions				
Br_2 (subst.)	KBr (0.2 M)	5	Aromatic amines, phenols, 'oxine'	A
Br_2 (add)	KBr (0.2 M)		Unsaturated hydrocarbons (e.g. alkenes), cyclohexene	A

Notes
(1) Gold anode. (4) Mercury cathode; for reagent see below.
(2) Silver anode. (5) Trace of mercury(II) acetate dissolved in ethanoic acid–methanol mixture added as catalyst.
(3) Mercury anode.
[a] A = amperometric, I = indicator, P = potentiometric.
[b] Mercury–EDTA reagent: prepare a stock solution containing 8.4 g mercury(II) nitrate and 9.3 g disodium-EDTA in 250 mL (each reagent 0.1 M). Mix 25 mL stock solution with 75 mL ammonium nitrate solution (0.1 M) and adjust to pH 8.3 with concentrated ammonia solution.

10.17   Applications

Coulometric titrations have been developed for all types of titration reaction. Thus, examples of neutralisation titrations (Section 10.59), complexation titrations (Section 10.86), precipitation titrations (Section 10.104) and oxidation–reduction titrations (Sections 10.157 to 10.159) are found later in this chapter. A selection of coulometric titrations are listed in Table 10.1. The important Karl Fischer method for the determination of water is described in Section 10.166.

Amperometric titrations

For a general discussion on voltammetry, the reader is referred to Chapter 13. **Amperometry** refers to measurement of current under constant applied voltage.

10.18   Principles

The limiting current is independent of the applied voltage impressed upon a dropping mercury electrode (or other indicator microelectrode). The only factor affecting the limiting current, if the migration current is almost eliminated by the addition of sufficient supporting electrolyte, is the rate of diffusion of electroactive material from the bulk of the solution to the electrode surface. Hence the diffusion current (= limiting current − residual current) is proportional to the concentration of the electroactive material in the solution. If some of the electroactive material is removed by interaction with reagent the diffusion current will decrease. This is the fundamental principle of amperometric titrations. The observed diffusion current at a suitable applied voltage is measured as a function of the volume of the titrating solution; the end point is the point of intersection of two lines giving the change of current before and after the equivalence point.

   If the current–voltage curves of the reagent and the substance being titrated are not known, the polarograms must first be determined in the supporting electrolyte in which the titration is to be carried out. The voltage applied at the beginning of the titration must be such that the total diffusion current of the substance to be titrated, or of the reagent, or of both, is obtained. Figure 10.5 shows the most common types of curve encountered in amperometric titrations together with the corresponding hypothetical polarograms of each individual substance: S refers to the solute to be titrated and R to the titrating reagent. The slight rounding off in the vicinity of the equivalence point is due to solubility of precipitate, salt hydrolysis, or dissociation of a complex; this curvature does not usually interfere, since the end point is located by extending the linear branches to the point of intersection. For each amperometric titration the applied voltage is adjusted to a value between X and Y. There are four types of end point:

A and A′ In A only the material titrated S gives a diffusion current. From A′ it can be seen that between X and Y, the titrating reagent R does not give rise to a diffusion current. Here the electroactive material is removed from the solution by precipitation with an inactive substance (e.g. lead ions titrated with sulphate ions).

B and B′ In B only the titrating reagent R gives a diffusion current. From B′ it follows that the solute S is electroinactive, between X and Y. An electroactive precipitating reagent is added to an inactive substance (e.g. sulphate ions titrated with lead or barium ions).

C and C′ In C both the solute S and the titrating reagent R give rise to diffusion currents at an applied voltage between X and Y (see C′). An example is the titration of lead ions with chromate ions where a sharp V-shaped curve is obtained.

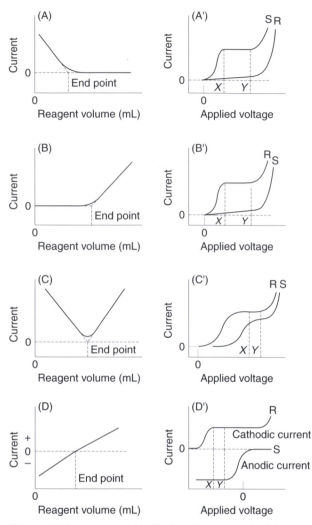

Figure 10.5   Amperometric titration: the four end points

D and D′ In (D) the solute S gives an **anodic** diffusion current (see D′); here the current changes from anodic to cathodic or vice versa, and the end point of the titration is indicated by a zero current. Examples of D include the titration of iodide ion with mercury(II) (as nitrate), of chloride ion with silver ion, and of titanium(III) in an acidified tartrate medium with iron(III). Because the diffusion coefficient of the reagent is usually slightly different from the substance being titrated, the slope of the line before the end point differs slightly from the slope after the end point; in practice it is easy to add the reagent until the current acquires a zero value or, more accurately, the value of the residual current for the supporting electrolyte.

To take into account the change in volume of the solution during the titration, the observed currents should be multiplied by the factor $(V + v)/V$, where V is the initial volume of the solution and v is the volume of the titration reagent added. Alternatively, this correction may be avoided (or considerably reduced) by adding the reagent from a semimicro burette in a concentration 10–20 times that of the solute. The use of concentrated reagents has the additional advantage that comparatively little dissolved oxygen is introduced into the system, thus rendering unnecessary prolonged bubbling with inert gas after each addition of the reagent. The migration current is eliminated by adding sufficient supporting electrolyte; if necessary, a suitable maximum suppressor is also introduced.

10.19   Amperometric titration with a dropping mercury electrode (DME)

An excellent and inexpensive titration cell consists of a 100 mL, three-necked, flat- or round-bottomed flask to which a fourth neck is sealed. The complete assembly is depicted schematically in Figure 10.6(a). The burette (preferably of the semimicro type and graduated in 0.01 mL), dropping electrode, a two-way gas-inlet tube (thus permitting nitrogen to be passed either through the solution or over its surface), and an agar–potassium salt bridge (not shown in the figure) are fitted into the four necks by means of rubber stoppers. The agar–salt bridge is connected through an intermediate vessel (a weighing bottle may be used) containing saturated potassium chloride solution to a large saturated calomel electrode. The agar–salt bridge is made from a gel which is 3% in agar and contains sufficient potassium chloride to saturate the solution at room temperature; when chloride ions interfere with the titrations, the connection is made with an agar–potassium nitrate bridge. The simple electrical circuit shown in Figure 10.6(b) is suitable for this procedure. The voltage applied to the titration cell is supplied by two 1.5 V dry cells and is controlled by the potential divider R (a 50–100 Ω variable resistance); it can be measured on the digital voltmeter V. The current flowing is read on the microammeter M.

Thermostatic control is not essential provided the cell is maintained at a fairly constant temperature during the titration. It is advantageous to store the reagent beneath an atmosphere of inert gas; this precaution is not absolutely necessary if the reagent solution has 10–20 times the concentration of the solution being titrated and is added from a semimicro burette. If the solute is electroreducible, sufficient electrolyte should be added to eliminate the migration current; if the reagent is electroreducible and the solute is not, the addition

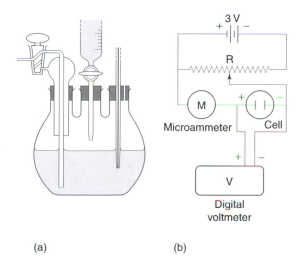

(a)   (b)

Figure 10.6   DME cell: an excellent and inexpensive method

of a supporting electrolyte is usually not required, since sufficient electrolyte is formed during the titration to eliminate the migration current beyond the end point. It may be necessary to add a suitable maximum suppressor, such as gelatin. If the polarographic characteristics of the solute and the reagent are not known, the current–voltage curve of each must be determined in the medium in which the titration is being carried out. The applied voltage is then adjusted at the beginning of the titration to such a value that the diffusion current of the unknown solute or of the reagent, or of both, is obtained. Frequently the voltage range is comparatively large and, in consequence, great accuracy is not required in adjusting the applied voltage.

The general procedure is as follows. A known volume of the solution under test is placed in the titration cell, which is then assembled as in Figure 10.6(a); the electrical connections are completed (dropping mercury electrode as cathode, saturated calomel half-cell as anode), and dissolved oxygen is removed by passing a slow stream of pure nitrogen for about 15 min. The applied voltage is then adjusted to the desired value, and the initial diffusion current is noted. A known volume of the reagent is run in from a semimicro burette, nitrogen is bubbled through the solution for about 2 min to eliminate traces of oxygen from the added liquid and to ensure complete mixing. The flow of gas through the solution is then stopped, but is allowed to pass over the surface of the solution (thus maintaining an inert, oxygen-free atmosphere). The current and burette readings are both noted. This procedure is repeated until sufficient readings have been obtained to permit the end point to be determined as the intersection of the two linear parts of the graph.

10.20   Apparatus

The dropping mercury electrode cannot be used at markedly positive potentials (say, above about 0.4 V vs SCE) because of the oxidation of the mercury. By replacing the dropping mercury electrode by an inert platinum electrode, it was hoped to extend the range of polarographic work in the positive direction to the voltage approaching that at which oxygen is evolved, namely, 1.1 V. The attainment of a steady diffusion current is slow with a stationary platinum electrode, but the difficulty may be overcome by rotating the platinum electrode at constant speed; the diffusion layer thickness is considerably reduced, thus increasing the sensitivity and the rate of attainment of equilibrium. Difficulties, however, arise in obtaining reproducible values for the diffusion currents from day to day; nevertheless, it is suitable as an indicator electrode in amperometric titrations. The larger currents (about 20 times those at the dropping mercury electrode) attained with the rotating platinum electrode allow correspondingly smaller currents to be measured without loss of accuracy and thus very dilute solutions (up to $10^{-4} M$) may be titrated. In order to obtain a linear relation between current and amount of reagent added, the speed of stirring must be kept constant during the titration; a speed of about 600 rpm is generally suitable.

Figure 10.7 shows the construction of a simple rotating platinum microelectrode. The electrode is constructed from a standard 'mercury seal'. About 5 mm of platinum wire (0.5 mm diameter) protrudes from the wall of a length of 6 mm glass tubing; the tubing is bent at an angle approaching a right angle a short distance from the lower end. Electrical connection is made to the electrode by a stout amalgamated copper wire passing through the tubing to the mercury covering the sealed-in platinum wire; the upper end of the copper wire passes through a small hole blown in the stem of the stirrer and dips into mercury contained in the mercury seal. A wire from the mercury seal is connected to the source of applied voltage. The tubing forms the stem of the electrode, which is rotated at a **constant** speed of 600 rpm.

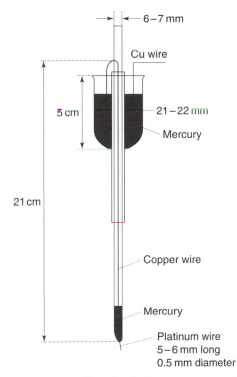

6–7 mm

Cu wire

5 cm

21–22 mm

Mercury

21 cm

Copper wire

Mercury

Platinum wire
5–6 mm long
0.5 mm diameter

Figure 10.7   Rotating platinum microelectrode

10.21   Biamperometric titrations

The titrations discussed so far have used a reference electrode (usually on SCE) in conjunction with a polarised electrode (dropping mercury electrode or rotating platinum microelectrode). Titrations may also be performed in a uniformly stirred solution by using two small but similar platinum electrodes to which a small e.m.f. (1–100 mV) is applied; the end point is usually shown by the disappearance or the appearance of a current flowing between the two electrodes. For the method to be applicable, the only requirement is that a reversible oxidation–reduction system is present either before or after the end point.

A simple apparatus suitable for this procedure is shown in Figure 10.8. B is a 3 V torch battery or 2 V accumulator, M is a microammeter, R is a 500 Ω, 0.5 W potentiometer, and E,E are platinum electrodes. The potentiometer is set so there is a potential drop of about 80–100 mV across the electrodes.

In a titration with two indicator electrodes and when the reactant involves a reversible system (e.g. $I_2 + 2e \rightleftharpoons 2I^-$), an appreciable current flows through the cell. The amount of oxidised form reduced at the cathode is equal to the amount formed by oxidation of the reduced form at the anode. Both electrodes are depolarised until the oxidised component or the reduced component of the system has been consumed by a titrant. After the end point, only one electrode remains depolarised if the titrant (e.g. thiosulphate ion, $2S_2O_3^{2-} \rightarrow S_4O_6^{2-} + 2e$) does not involve a reversible system. Current thus flows until the end point; at or after the end point the current is zero or virtually zero. In the determination of iodine by titration with thiosulphate, a rapid decrease in current is observed in the neighbourhood

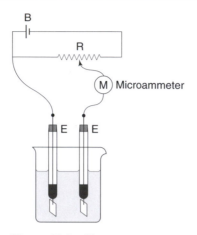

Figure 10.8 Biamperometric titration: simple apparatus

of the end point and this has led to the name **dead-stop end point**. The complementary type of end point, which resembles a reversed L-type amperometric graph, is probably more desirable in practice, and is obtained in the titration of an irreversible couple (say thiosulphate) by a reversible couple (say iodine); the current is very low before the end point, and a very rapid increase in current signals the end point. When both systems are reversible (e.g. iron(II) ions with cerium(IV) or permanganate ions; applied potential 100 mV), the current is zero or close to zero at the equivalence point and a V-shaped graph results.

10.22 Advantages

Amperometric titrations have several advantages:

1. The titration can usually be carried out rapidly, since the end point is found graphically; a few current measurements at constant applied voltage before and after the end point suffice.
2. Titrations can be carried out in cases in which the solubility relations are such that potentiometric or visual indicator methods are unsatisfactory; for example, when the reaction product is markedly soluble (precipitation titration) or appreciably hydrolysed (acid–base titration). This is because the readings near the equivalence point have no special significance in amperometric titrations. Readings are recorded in regions of excess titrant or excess reagent, where the solubility or hydrolysis is suppressed by the mass action effect; the point of intersection of these lines gives the equivalence point.
3. A number of amperometric titrations can be carried out at dilutions ($\sim 10^{-4} M$) at which many visual or potentiometric titrations no longer yield accurate results.
4. 'Foreign' salts may frequently be present without interference and, indeed, they are usually added as the supporting electrolyte in order to eliminate the migration current.

10.23 Applications

For specific examples of titrations involving complexation reactions see Sections 10.87 to 10.89; for precipitation reactions see Sections 10.105 to 10.107; and for oxidation reactions see Sections 10.161 to 10.164. The scope of the method is indicated by Table 10.2.

Table 10.2   *Examples of amperometric titrations*

Titrant	Electrode	Species determined
Complexation reactions		
EDTA	DME	Many metallic ions
Precipitation reactions		
Dimethylglyoxime	DME	Ni^{2+}
Lead nitrate	DME	SO_4^{2-}, MoO_4^{2-}, F^-
Mercury(II) nitrate	DME	I^-
Silver nitrate	Rotating Pt	Cl^-, Br^-, I^-, CN^-, thiols
Sodium tetraphenylborate	Graphite	K^+
Thorium(IV) nitrate	DME	F^-
Potassium dichromate	DME	Pb^{2+}, Ba^{2+}
Oxidation reactions		
Iodine	Rotating Pt	As(III), $Na_2S_2O_3$
KBrO$_3$/KBr	Rotating Pt	As(III), Sb(III), N_2H_4
Additions	Rotating Pt	Alkenes
Substitutions	Rotating Pt	Some phenols, aromatic amines

Spectrophotometric titrations

10.24   General

In a spectrophotometric titration the end point is evaluated from data on the absorbance of the solution. For monochromatic light passing through a solution, Beer's law may be written as

$$\text{absorbance} = \log(I_o/I_t) = \varepsilon c l$$

where I_o is the intensity of the incident light, I_t is the intensity of the transmitted light, ε is the molar absorption coefficient, c is the concentration of the absorbing species, and l is the thickness or length of the light path through the absorbing medium. Since spectro-photometric titrations are carried out in a vessel for which the light path is constant, the absorbance is proportional to the concentration. Thus in a titration where the titrant, the reactant, or a reactive product absorbs radiation, the plot of absorbance versus volume of titrant added will consist, if the reaction is complete and the volume change is small, of two straight lines intersecting at the end point.

The shape of a photometric titration curve will depend on the optical properties of the reactant, titrant and products of the reaction at the wavelength used. Some typical titration plots are given in Figure 10.9:

Figure 10.9(a) is characteristic of systems where the substance titrated is converted into a non-absorbing product.

Figure 10.9(b) is typical of the titration where the titrant alone absorbs.

Figure 10.9(c) corresponds to systems where the substance titrated and the titrant are colourless and the product alone absorbs.

Figure 10.9(d) is obtained when a coloured reactant is converted into a colourless product by a coloured titrant.

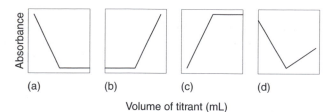

Figure 10.9 Spectrophotometric titration: typical plots

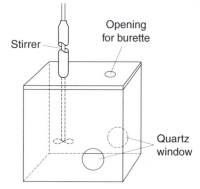

Figure 10.10 Spectrophotometric titration: Perspex cell with quartz windows

10.25 Apparatus

A special titration cell is necessary which completely fills the cell compartment of the spectrophotometer. The cell shown in Figure 10.10 can be made from 5 mm Perspex sheet, cemented together with special Perspex cement, and with dimensions suitable for the instrument to be used. Since Perspex is opaque to ultraviolet light, two openings are made in the cell to accommodate circular quartz windows* 23 mm in diameter and 1.5 mm thick; the windows are inserted in such a way that the beam of monochromatic light passes through their centres to the photoelectric cell. The Perspex cover of the cell has two small openings for the tip of a 5 mL microburette and a microstirrer, held by means of rubber bungs; the stirrer is sleeved. The whole of the cell, with the exception of the quartz windows, is covered with black paper and, as a further precaution, the top of the cell is covered with a black cloth; it is most important to exclude all extraneous light. Sometimes a probe-type photometer may be used, based on fibre optics.

10.26 Technique

The experimental technique is simple. The cell containing the solution to be titrated is placed in the light path of a spectrophotometer, a wavelength appropriate to the particular titration is selected, and the absorption is adjusted to some convenient value by means of the sensitivity and slit width controls. A measured volume of the titrant is added to the

---

* An adequate substitute are the fused silica end plates from a polarimeter sample tube.

stirred solution, and the absorbance is read again. This is repeated at several points before the end point and several more points after the end point. The end point is found graphically. The optimum concentration of the solution to be analysed depends on the molar absorption coefficient of the absorbing species involved, and is usually of the order of 10^{-4} to $10^{-5} M$. The effect of dilution can be made negligible by using a sufficiently concentrated titrant. If relatively large volumes of titrant are added, the effect of dilution may be corrected by multiplying the observed absorbances by the factor $(V + v)/V$, where V is the initial volume and v is the volume added; if the dilution is of the order of only a few percent, the lines in the titration plots appear straight. The operating wavelength is selected to avoid interference by other absorbing substances and to obtain an absorption coefficient so the change in absorbance falls within a convenient range. The range is particularly important, because serious photometric error is possible in high-absorbance regions. Light leakage must be avoided.

10.27   Advantages

Spectrophotometric titrations have several advantages:

1. The presence of other substances absorbing at the same wavelength does not necessarily cause interference, since only the **change in absorbance** is significant.
2. The precision of locating the titration line by pooling the information derived from several points is greater than the precision of any single point.
3. The method is useful for reactions which tend to be appreciably incomplete near the equivalence point.
4. A precision of 0.5% is often attainable.
5. The linear response of absorbance to concentration often produces an appreciable break in a spectrophotometric titration, even though the changes in concentration are insufficient to give a clearly defined inflexion point in a potentiometric titration.

10.28   Applications

There is usually no difficulty in using non-aqueous solvents, and this is a particularly useful feature of spectrophotometric titrations. Many organic compounds may be determined by neutralisation reactions using spectrophotometry (Section 10.60). For spectrophotometric techniques in complexation titrations see Sections 10.90 and 10.91.

Neutralisation titrations

10.29   Neutralisation indicators

The object of titrating an alkaline solution with a standard solution of an acid is the determination of the amount of acid which is exactly equivalent chemically to the amount of base present. The point at which this is reached is the equivalence point, stoichiometric point or theoretical end point; the resulting aqueous solution contains the corresponding salt. If both the acid and base are strong electrolytes, the solution at the end point will be neutral and have a pH of 7; but if either the acid or the base is a weak electrolyte, the salt will be hydrolysed to a certain degree, and the solution at the equivalence point will be either slightly alkaline or slightly acid. The exact pH of the solution at the equivalence point can be calculated from the ionisation constant of the weak acid or the weak base and the concentration of the solution. For any actual titration the correct end point will be characterised by a definite value of the hydrogen ion concentration of the solution, which depends on the nature of the acid, the nature of the base and the concentration of the solution.

A large number of substances, called **neutralisation or acid–base indicators**, change colour according to the hydrogen ion concentration of the solution. The chief characteristic of these indicators is that the change from a predominantly acid colour to a predominantly alkaline colour is not sudden and abrupt, but takes place within a small interval of pH (usually about two pH units); this is called the **colour change interval** of the indicator. The position of the colour change interval in the pH scale varies widely with different indicators. For most acid–base titrations it is possible to select an indicator which exhibits a distinct colour change at a pH close to the equivalence pH.

The first useful theory of indicator action was suggested by W. Ostwald,[3] based on the concept that indicators in general use are very weak organic acids or bases. The simple Ostwald theory has been revised, and the colour changes are believed to be due to structural changes, including the production of quinonoid and resonance forms; these may be illustrated by reference to phenolphthalein, characteristic of all phthalein indicators. In the presence of dilute alkali, the lactone ring of [10.A] opens to yield the triphenylcarbinol structure [10.B]; this undergoes loss of water to produce the resonating ion [10.C], coloured red. If phenolphthalein is treated with excess of concentrated alcoholic alkali, the red colour disappears owing to the formation of [10.D].

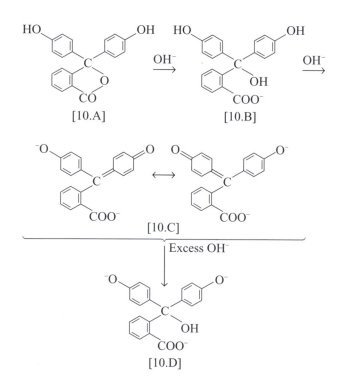

The Brønsted–Lowry concept of acids and bases[4] makes it unnecessary to distinguish between acid and base indicators; emphasis is placed on the charge types of the acid and alkaline forms of the indicator. The equilibrium between the acidic form In_A and the basic form In_B may be expressed as

$$In_A \rightleftharpoons H^+ + In_B \qquad [10.1]$$

and the equilibrium constant as

$$\frac{a_{H^+} a_{In_B}}{a_{In_A}} = K_{In} \qquad (10.1)$$

The observed colour of an indicator in solution is determined by the ratio of the concentrations of the acidic and basic forms. This is given by

$$\frac{[In_A]}{[In_B]} = \frac{a_{H^+} y_{In_B}}{K_{In} y_{In_A}} \qquad (10.2)$$

where y_{In_A} and y_{In_B} are the activity coefficients of the acidic and basic forms of the indicator. Equation (10.2) may be written in the logarithmic form

$$pH = -\log a_{H^+} = pK_{In} + \log \frac{[In_B]}{[In_A]} + \log \frac{y_{In_B}}{y_{In_A}} \qquad (10.3)$$

The pH will depend on the ionic strength of the solution (which is related to the activity coefficient). Hence, when making a colour comparison for the determination of the pH of a solution, not only must the indicator concentration be the same in the two solutions but the ionic strength must also be equal or approximately equal. The equation incidentally provides an explanation of the so-called salt and solvent effects which are observed with indicators. The colour change equilibrium at any particular ionic strength (constant activity coefficient term) can be expressed by a condensed form of equation (10.3):

$$pH = pK'_{In} + \log \frac{[In_B]}{[In_A]} \qquad (10.4)$$

where pK'_{In} is called the **apparent indicator constant**.

The value of the ratio $[In_B]/[In_A]$ (i.e. [basic form]/[acidic form]) can be determined by a visual colour comparison or, more accurately, by a spectrophotometric method. Both forms of the indicator are present at any hydrogen ion concentration. But remember that the human eye has a limited ability to detect either of two colours when one of them predominates. Experience shows that the solution will appear to have the acid colour, i.e. of In_A, when $[In_A]/[In_B]$ is above approximately 10, and the alkaline colour, i.e. of In_B, when $[In_B]/[In_A]$ is above approximately 10. Thus only the 'acid' colour will be visible when $[In_A]/[In_B] > 10$; the corresponding limit of pH given by equation (10.4) is

$$pH = pK'_{In} - 1$$

Only the alkaline colour will be visible when $[In_B]/[In_A] > 10$, and the corresponding limit of pH is

$$pH = pK'_{In} + 1$$

The colour change interval is accordingly $pH = pK'_{In} \pm 1$, i.e. over approximately two pH units. Within this range the indicator will appear to change from one colour to the other. The change will be gradual, since it depends upon the ratio of the concentrations of the

Table 10.3 Colour changes and pH range of certain indicators

Indicator	Chemical name	pH range	Colour in acid solution	Colour in alkaline solution	pK'_{In}
Brilliant cresyl blue (acid)	Aminodiethylaminomethyldiphenazonium chloride	0.0–1.0	Red orange	Blue	–
Cresol red (acid)	1-Cresolsulphonphthalein	0.2–1.8	Red	Yellow	–
m-Cresol purple	m-Cresolsulphonphthalein	0.5–2.5	Red	Yellow	–
Quinaldine red	1-(p-Dimethylaminophenylethylene)quinoline ethiodide	1.4–3.2	Colourless	Red	–
Thymol blue (acid)	Thymolsulphonphthalein	1.2–2.8	Red	Yellow	1.7
Bromophenol blue	Tetrabromophenolsulphonphthalein	2.8–4.6	Yellow	Blue	4.1
Ethyl orange	–	3.0–4.5	Red	Orange	–
Methyl orange	Dimethylaminophenylazobenzenesulphonic acid sodium salt	2.9–4.6	Red	Orange	3.7
Congo red	Diphenyldiazobis-1-naphthylaminesulphonic acid disodium salt	3.0–5.0	Blue	Red	–
Bromocresol green	Tetrabromo-m-cresolsulphonphthalein	3.6–5.2	Yellow	Blue	4.7
Methyl red	1-Carboxybenzeneazodimethylaniline	4.2–6.3	Red	Yellow	5.0
Ethyl red	–	4.5–6.5	Red	Orange	–
Chlorophenol red	Dichlorophenolsulphonphthalein	4.6–7.0	Yellow	Red	6.1
4-Nitrophenol	4-Nitrophenol	5.0–7.0	Colourless	Yellow	7.1
Bromocresol purple	Dibromo-o-cresolsulphonphthalein	5.2–6.8	Yellow	Purple	6.1
Bromophenol red	Dibromophenolsulphonphthalein	5.2–7.0	Yellow	Red	–
Azolitmin (litmus)	–	5.0–8.0	Red	Blue	–
Bromothymol blue	Dibromothymolsulphonphthalein	6.0–7.6	Yellow	Blue	7.1
Neutral red	Aminodimethylaminotoluphenazonium chloride	6.8–8.0	Red	Orange	–
Phenol red	Phenolsulphonphthalein	6.8–8.4	Yellow	Red	7.8
Cresol red (base)	1-Cresolsulphonphthalein	7.2–8.8	Yellow	Red	8.2
1-Naphtholphthalein	1-Naphtholphthalein	7.3–8.7	Yellow	Blue	8.4
m-Cresol purple	m-Cresolsulphonphthalein	7.6–9.2	Yellow	Purple	–
Thymol blue (base)	Thymolsulphonphthalein	8.0–9.6	Yellow	Blue	8.9
o-Cresolphthalein	Di-o-cresolphthalide	8.2–9.8	Colourless	Red	–
Phenolphthalein	Phenolphthalein	8.3–10.0	Colourless	Red	9.6
Thymolphthalein	Thymolphthalein	9.3–10.5	Colourless	Blue	9.3
Alizarin yellow R	p-Nitrobenzeneazosalicylic acid	10.1–12.1	Yellow	Orange red	–
Brilliant cresyl blue (base)	Aminodiethylaminomethyldiphenazonium chloride	10.8–12.0	Blue	Yellow	–
Tropaeolin O	p-Sulphobenzeneazoresorcinol	11.1–12.7	Yellow	Orange	–

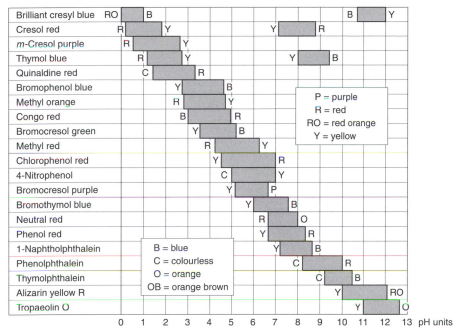

Figure 10.11   Titrimetric indicators: colour change intervals

two coloured forms (acidic form and basic form). When the pH of the solution is equal to the apparent dissociation constant of the indicator pK'_{In}, the ratio $[In_A]/[In_B] = 1$, and the indicator will have a colour due to an equal mixture of the acid and alkaline forms. This is sometimes known as the middle tint of the indicator. It applies strictly only if the two colours are of equal intensity. If one form is more intensely coloured than the other or if the eye is more sensitive to one colour than the other, then the middle tint will be slightly displaced along the pH range of the indicator.

Table 10.3 contains a list of indicators suitable for titrimetric analysis and for the colorimetric determination of pH. The colour change intervals of most indicators in the table are represented graphically in Figure 10.11. But it is a limited number of these indicators that are used routinely in neutralisation titrations.

Water may have a variable pH in quantitative analysis. Water in equilibrium with the normal atmosphere which contains 0.03 vol% of carbon dioxide has a pH of about 5.7; very carefully prepared conductivity water has a pH close to 7; water saturated with carbon dioxide under a pressure of one atmosphere has a pH of about 3.7 at 25 °C. So according to the conditions that prevail in the laboratory, the analyst may be dealing with water having a pH between the extremes pH 3.7 and pH 7. Hence for indicators which show their alkaline colours at pH values above 4.5, the effect of carbon dioxide introduced during a titration, either from the atmosphere or from the titrating solutions, must be seriously considered. This subject is discussed again in Section 10.33.

10.30   Preparation of indicator solutions

Note   Many indicator and mixed indicator solutions are available from commercial suppliers already prepared for use.

Methyl orange   Methyl orange is available either as the free acid or as the sodium salt. Dissolve 0.5 g of the free acid in 1 L of water. Filter the cold solution to remove any precipitate which separates. Dissolve 0.5 g of the sodium salt in 1 L of water, add 15.2 mL of 0.1 M hydrochloric acid, and filter if necessary when cold.

Methyl red   Dissolve 1 g of the free acid in 1 L of hot water, or dissolve in 600 mL of ethanol and dilute with 400 mL of water.

1-Naphtholphthalein   Dissolve 1 g of the indicator in 500 mL of ethanol and dilute with 500 mL of water.

Phenolphthalein   Dissolve 5 g of the reagent in 500 mL of ethanol and add 500 mL of water with constant stirring. Filter if a precipitate forms. Alternatively, dissolve 1 g of the dry indicator in 60 mL of 2-ethoxyethanol (Cellosolve), b.p. 135 °C, and dilute to 100 mL with distilled water; the loss by evaporation is less with this preparation.

Thymolphthalein   Dissolve 0.4 g of the reagent in 600 mL of ethanol and add 400 mL of water with stirring.

Sulphonphthaleins   Sulphonphthaleins are usually supplied in the acid form. They are rendered water-soluble by adding sufficient sodium hydroxide to neutralise the sulphonic acid group. One gram of the indicator is triturated in a clean glass mortar with the appropriate quantity of 0.1 M sodium hydroxide solution, and then diluted with water to 1 L. The following volumes of 0.1 M sodium hydroxide are required for 1 g of the indicators: bromophenol blue 15.0 mL, bromocresol green 14.4 mL, bromocresol purple 18.6 mL, chlorophenol red 23.6 mL, bromothymol blue 16.0 mL, phenol red 28.4 mL, thymol blue 21.5 mL, cresol red 26.2 mL, metacresol purple 26.2 mL.

Quinaldine red   Dissolve 1 g in 100 mL of 80% ethanol.

Methyl yellow, neutral red, and Congo red   Dissolve 1 g of the indicator in 1 L of 80% ethanol. Congo red may also be dissolved in water.

4-Nitrophenol   Dissolve 2 g of the solid in 1 L of water.

Alizarin yellow R   Dissolve 0.5 g of the indicator in 1 L of 80% ethanol.

Tropaeolin O   Dissolve 1 g of the solid in 1 L of water.

10.31   Mixed indicators

For some purposes it is desirable to have a sharp colour change over a narrow and selected range of pH; this is not easily seen with an ordinary acid–base indicator, since the colour change extends over two units of pH. But the required result may be achieved by using a suitable mixture of indicators; these are generally selected so that their pK'_{In} values are close together and the overlapping colours are complementary at an intermediate pH value. A few examples will be given in some detail.

The colour change of a single indicator may also be improved by the addition of a pH-sensitive dyestuff to produce the complement of one of the indicator colours. A typical example is the addition of xylene cyanol FF to methyl orange (1.0 g of methyl orange and

1.4 g of xylene cyanol FF in 500 mL of 50% ethanol); here the colour change from the alkaline to the acid side is green → grey → magenta, the middle (grey) stage being at pH = 3.8. This is an example of a **screened indicator**, and the mixed indicator solution is sometimes known as screened methyl orange. Another example is the addition of methyl green (2 parts of a 0.1% solution in ethanol) to phenolphthalein (1 part of a 0.1% solution in ethanol); the methyl green complements the red-violet basic colour of the phenolphthalein, and at a pH of 8.4–8.8 the colour change is from grey to pale blue.

Neutral red and methylene blue   A mixture of equal parts of neutral red (0.1% solution in ethanol) and methylene blue (0.1% solution in ethanol) gives a sharp colour change from violet blue to green in passing from acid to alkaline solution at pH 7. This indicator may be employed to titrate ethanoic acid with ammonia solution or vice versa. Both acid and base are approximately of the same strength, hence the equivalence point will be at a pH ≈ 7 (Section 10.36); owing to the extensive hydrolysis and the flat nature of the titration curve, the titration cannot be performed except with an indicator of very narrow range.

Phenolphthalein and 1-naphtholphthalein   A mixture of phenolphthalein (3 parts of a 0.1% solution in ethanol) and 1-naphtholphthalein (1 part of a 0.1% solution in ethanol) passes from pale rose to violet at pH = 8.9. The mixed indicator is suitable for the titration of phosphoric acid to the diprotic stage ($K_2 = 6.3 \times 10^{-8}$; the equivalence point at pH ≈ 8.7).

Thymol blue and cresol red   A mixture of thymol blue (3 parts of a 0.1% aqueous solution of the sodium salt) and cresol red (1 part of a 0.1% aqueous solution of the sodium salt) changes from yellow to violet at pH = 8.3. It has been recommended for the titration of carbonate to the hydrogencarbonate stage.

10.32   Neutralisation curves

The mechanism of neutralisation processes can be understood by studying the changes in the hydrogen ion concentration during the course of the appropriate titration. The change in pH in the neighbourhood of the equivalence point is of the greatest importance, as it enables an indicator to be selected which will give the smallest titration error. The curve obtained by plotting pH against the percentage of acid neutralised (or the number of millilitres of alkali added) is known as the neutralisation curve (or, more generally, the titration curve). This may be evaluated experimentally by determining the pH at various stages during the titration using a potentiometric method (Section 10.56), or it may be calculated from theoretical principles.

10.33   Strong acid neutralised by strong base

For this calculation it is assumed that both the acid and the base are completely dissociated and the activity coefficients of the ions are unity in order to obtain the pH values during the course of the neutralisation of the strong acid and the strong base, or vice versa, at the laboratory temperature. For simplicity of calculation consider the titration of 100 mL of 1 M hydrochloric acid with 1 M sodium hydroxide solution. The pH of 1 M hydrochloric acid is 0. When 50 mL of the 1 M base have been added, 50 mL of unneutralised 1 M acid will be present in a total volume of 150 mL:

Table 10.4   *pH during titration of 100 mL of HCl with NaOH of equal concentration*

NaOH added (mL)	1 M solution (pH)	0.1 M solution (pH)	0.01 M solution (pH)
0	0.0	1.0	2.0
50	0.5	1.5	2.5
75	0.8	1.8	2.8
90	1.3	2.3	3.3
98	2.0	3.0	4.0
99	2.3	3.3	4.3
99.5	2.6	3.6	4.6
99.8	3.0	4.0	5.0
99.9	3.3	4.3	5.3
100.0	7.0	7.0	7.0
100.1	10.7	9.7	8.7
100.2	11.0	10.0	9.0
100.5	11.4	10.4	9.4
101	11.7	10.7	9.7
102	12.0	11.0	10.0
110	12.7	11.7	10.7
125	13.0	12.0	11.0
150	13.3	12.3	11.3
200	13.5	12.5	11.5

$[H^+]$ will therefore be $50 \times 1/150 = 3.33 \times 10^{-1}$ pH = 0.48

For 75 mL of base, $[H^+] = 25 \times 1/175 = 1.43 \times 10^{-1}$ pH = 0.84

For 90 mL of base, $[H^+] = 10 \times 1/190 = 5.26 \times 10^{-2}$ pH = 1.3

For 98 mL of base, $[H^+] = 2 \times 1/198 = 1.01 \times 10^{-2}$ pH = 2.0

For 99 mL of base, $[H^+] = 1 \times 1/199 = 5.03 \times 10^{-3}$ pH = 2.3

For 99.9 mL of base, $[H^+] = 0.1 \times 1/199.9 = 5.00 \times 10^{-4}$ pH = 3.3

Upon adding 100 mL of base, the pH will change sharply to 7, i.e. the theoretical equivalence point. The resulting solution is simply one of sodium chloride. Any sodium hydroxide added beyond this will be in excess of that needed for neutralisation:

With 100.1 mL of base $[OH^-] = 0.1/200.1 = 5.00 \times 10^{-4}$
                         pOH = 3.3   and   pH = 10.7

With 101 mL of base $[OH^-] = 1/201 = 5.00 \times 10^{-3}$
                         pOH = 2.3   and   pH = 11.7

These results show that, as the titration proceeds, initially the pH rises slowly, but between the addition of 99.9 mL and 100.1 mL of alkali, the pH of the solution rises from 3.3 to 10.7, i.e. in the vicinity of the equivalence point there is a very rapid rate of pH change.

Table 10.4 shows the complete results up to the addition of 200 mL of alkali; this also includes the figures for 0.1 M and 0.01 M solutions of acid and base respectively. The

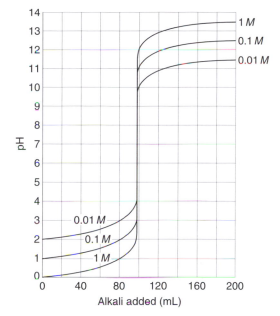

Figure 10.12   Calculated curves for 100 mL HCl with NaOH of same concentration

additions of alkali have been extended in all three cases to 200 mL; the range from 200 mL to 100 mL and beyond represents the reverse titration of 100 mL of alkali with the acid in the presence of the non-hydrolysed sodium chloride solution. The data in the table is graphed in Figure 10.12.

In quantitative analysis it is the changes of pH near the equivalence point which are of special interest, so this part of Figure 10.12 is shown on a larger scale in Figure 10.13, which also show the colour change intervals of some of the common indicators. With 1 M solutions it is possible to use any indicator having an effective pH range between 3 and 10.5. The colour change will be sharp and the titration error negligible. With 0.1 M solutions the ideal pH range is limited to 4.5–9.5. Methyl orange will exist chiefly in the alkaline form when 99.8 mL of alkali have been added, and the titration error will be 0.2%, which is negligibly small for most practical purposes; it is therefore advisable to add sodium hydroxide solution until the indicator is present completely in the alkaline form. The titration error is also negligibly small with phenolphthalein. With 0.01 M solutions the ideal pH range becomes 5.5–8.5; indicators such as methyl red, bromothymol blue or phenol red will be suitable. The titration error for methyl orange will be 1–2%.

These considerations apply to solutions which do not contain carbon dioxide. In practice, carbon dioxide is usually present (Section 10.29) arising from the small quantity of carbonate in the sodium hydroxide and/or from the atmosphere. The gas is in equilibrium with carbonic acid, and both stages of ionisation are weak. This will introduce a small error when indicators of high pH range (above pH 5) are used, e.g. phenolphthalein or thymolphthalein. More acid indicators, such as methyl orange and methyl yellow, are unaffected by carbonic acid. The difference between the amounts of sodium hydroxide solution used with methyl orange and phenolphthalein is not greater than 0.15–0.20 mL of 0.1 M sodium hydroxide when 100 mL of 0.1 M hydrochloric acid is titrated. A method of eliminating this error, besides selecting an indicator with a pH range below pH 5, is to boil the solution

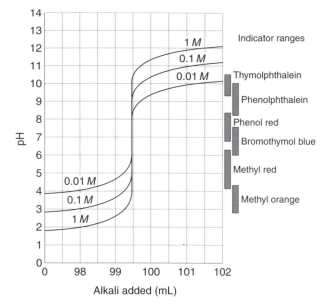

Figure 10.13   Equivalence point of Figure 10.12 shown in greater detail

while still acid to expel carbon dioxide and then to continue the titration with the cold solution. Boiling the solution is particularly efficacious when titrating dilute (e.g. $0.01\,M$) solutions.

10.34   Weak acid neutralised by strong base

Consider the neutralisation of $100\,\text{mL}$ of $0.1\,M$ ethanoic (acetic) acid with $0.1\,M$ sodium hydroxide solution; other concentrations can be treated in a similar way. The pH of the solution at the equivalence point is given by (Section 2.13)

$$\text{pH} = \tfrac{1}{2}pK_w + \tfrac{1}{2}pK_a - \tfrac{1}{2}pc = 7 + 2.37 - \tfrac{1}{2}(1.3) = 8.72$$

For other concentrations, we may employ the approximate mass action expression:

$$[H^+][CH_3COO^-]/[CH_3COOH] = K_a \tag{10.5}$$

$$\text{or}\quad [H^+] = K_a[CH_3COOH]/[CH_3COO^-]$$

$$\text{or}\quad \text{pH} = \log\,[\text{salt}]/[\text{acid}] + pK_a \tag{10.6}$$

The salt concentration (and the acid concentration) at any point is calculated from the volume of alkali added, allowing for the total volume of the solution.

   The initial pH of $0.1\,M$ ethanoic acid is computed from equation (10.5); the dissociation of the acid is relatively small, so it may be neglected in expressing the concentration of ethanoic acid. Hence from equation (10.5) we have

$$[H^+][CH_3COO^-]/[CH_3COOH] = 1.82 \times 10^{-5}$$

$$\text{or}\quad [H^+]^2/0.1 = 1.82 \times 10^{-5}$$

$$\text{or}\quad [H^+] = \sqrt{1.82 \times 10^{-6}} = 1.35 \times 10^{-3}$$

$$\text{or}\quad \text{pH} = 2.87$$

Table 10.5 *Neutralisation of 100 mL of 0.1 M CH_3COOH ($K_a = 1.82 \times 10^{-5}$) and 100 mL of 0.1 M HA ($K_a = 1 \times 10^{-7}$) with 0.1 M NaOH*

Vol. of 0.1 M NaOH used (mL)	0.1 M CH_3COOH (pH)	0.1 M HA ($K_a = 1 \times 10^{-7}$) (pH)
0	2.9	4.0
10	3.8	6.0
25	4.3	6.5
50	4.7	7.0
90	5.7	8.0
99.0	6.7	9.0
99.5	7.0	9.3
99.8	7.4	9.7
99.9	7.7	9.8
100.0	8.7	9.9
100.2	10.0	10.0
100.5	10.4	10.4
101	10.7	10.7
110	11.7	11.7
125	12.0	12.0
150	12.3	12.3
200	12.5	12.5

When 50 mL of 0.1 M alkali have been added,

$$[\text{salt}] = 50 \times 0.1/150 = 3.33 \times 10^{-2}$$

$$[\text{acid}] = 50 \times 0.1/150 = 3.33 \times 10^{-2}$$

$$\text{pH} = \log(3.33 \times 10^{-2}/3.33 \times 10^{-2}) + 4.74 = 4.74$$

The pH values at other points on the titration curve are similarly calculated. After the equivalence point has been passed, the solution contains excess of OH^- ions, which will repress the hydrolysis of the salt; it is sufficiently accurate for our purposes to assume the pH is due to the excess of base present, so in this region the titration curve will almost coincide with the curve for 0.1 M hydrochloric acid (Figure 10.12 and Table 10.4). All the results are collected in Table 10.5 and depicted graphically in Figure 10.14. Also included are results for titration of 100 mL of 0.1 M solution of a weaker acid ($K_a = 1 \times 10^{-7}$) with 0.1 M sodium hydroxide at the laboratory temperature.

For 0.1 M ethanoic acid and 0.1 M sodium hydroxide, the curve shows that neither methyl orange nor methyl red can be used as indicator. The equivalence point is at pH 8.7, so it requires an indicator with a pH range on the slightly alkaline side, e.g. phenolphthalein, thymolphthalein or thymol blue (pH range, as base, 8.0–9.6). For the acid with $K_a = 10^{-7}$, the equivalence point is at pH = 10, but here the rate of change of pH in the neighbourhood of the stoichiometric point is very much less pronounced, owing to considerable hydrolysis.

Phenolphthalein will begin to change colour after about 92 mL of alkali have been added, and this change will occur to the equivalence point; thus the end point will not be

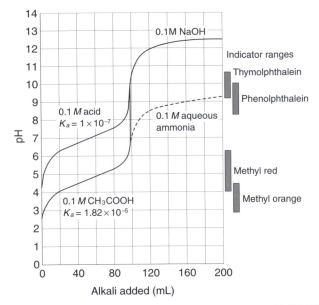

Figure 10.14 Calculated curves for 100 mL of 0.1 M CH$_3$COOH (K_a = 1.82 × 10^{-5}) and 0.1 M acid (K_a = 1 × 10^{-7}) with 0.1 M NaOH

sharp and the titration error will be appreciable. With thymolphthalein, however, the colour change covers the pH range 9.3–10.5; this indicator may be used, the end point will be sharper than for phenolphthalein, but somewhat gradual, and the titration error will be about 0.2%. Acids that have dissociation constants less than 10^{-7} cannot be satisfactorily titrated in 0.1 M solution with a simple indicator.

In general, weak acids (K_a > 5 × 10^{-6}) should be titrated with phenolphthalein, thymolphthalein or thymol blue as indicators.

10.35   Weak base neutralised by strong acid

This may be illustrated by the titration of 100 mL of 0.1 M aqueous ammonia (K_b = 1.85 × 10^{-5}) with 0.1 M hydrochloric acid at the ordinary laboratory temperature. According to Section 2.13, the pH of the solution at the equivalence point is given by

$$pH = \tfrac{1}{2}pK_w - \tfrac{1}{2}pK_b + \tfrac{1}{2}pc = 7 - 2.37 + \tfrac{1}{2}(1.3) = 5.28$$

For other concentrations the pH may be calculated using a method similar to Section 10.34:

$$[NH_4^+][OH^-]/[NH_3] = K_b \tag{10.7}$$

$$\text{or}\quad [OH^-] = K_b[NH_3]/[NH_4^+] \tag{10.8}$$

$$\text{or}\quad pOH = \log[\text{salt}]/[\text{base}] + pK_b \tag{10.9}$$

$$\text{or}\quad pH = pK_w - pK_b - \log[\text{salt}]/[\text{base}] \tag{10.10}$$

After the equivalence point has been reached, the solution contains excess of H$^+$ ions, hydrolysis of the salt is suppressed, and the subsequent pH changes may be assumed, with sufficient accuracy, to be due to the excess of acid present.

346

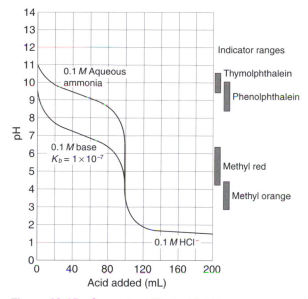

Figure 10.15   Curves for 100 mL of 0.1 M aqueous ammonia ($K_a = 1.8 \times 10^{-5}$) and 0.1 M base ($K_b = 1 \times 10^{-7}$) with 0.1 M HCl

The results are depicted graphically in Figure 10.15, along with a curve for titrating 100 mL of a 0.1 M solution of a weaker base ($K_b = 1 \times 10^{-7}$).

Neither thymolphthalein nor phenolphthalein can be used in the titration of 0.1 M aqueous ammonia. The equivalence point is at pH 5.3, and it is necessary to use an indicator with a pH range on the slightly acid side (pH 3 to 6.5), e.g. methyl orange, methyl red, bromophenol blue or bromocresol green. Bromophenol blue and bromocresol green may be used to titrate all weak bases ($K_b > 5 \times 10^{-6}$) with strong acids. Bromophenol blue or methyl orange may be used for the weak base ($K_b = 1 \times 10^{-7}$). No sharp colour change will be obtained with bromocresol green or with methyl red, and the titration error will be considerable.

10.36   Weak acid neutralised by weak base

This case is exemplified by the titration of 100 mL of 0.1 M ethanoic acid ($K_a = 1.82 \times 10^{-5}$) with 0.1 M aqueous ammonia ($K_b = 1.8 \times 10^{-5}$). The pH at the equivalence point is given by

$$pH = \tfrac{1}{2}pK_w + \tfrac{1}{2}pK_a - \tfrac{1}{2}pK_b = 7.0 + 2.38 - 2.37 = 7.1$$

The neutralisation curve up to the equivalence point is almost identical with that using 0.1 M sodium hydroxide as the base; beyond this point the titration is virtually the addition of 0.1 M aqueous ammonia solution to 0.1 M ammonium acetate solution, so equation (10.10) applies. The dashed line in Figure 10.14 shows the titration curve for the neutralisation of 100 mL of 0.1 M ethanoic acid with 0.1 M aqueous ammonia at the laboratory temperature. The chief feature is the very gradual change of pH near the equivalence point, indeed during the whole of the neutralisation. There is no sudden change in pH, hence no sharp end point can be found with any simple indicator. Sometimes it is possible to find a suitable mixed indicator which exhibits a sharp colour change over a very limited pH range. For titrations of ethanoic acid with ammonia solutions, neutral red–methylene blue mixed indicator may be used (Section 10.31), but on the whole, it is best to avoid the use of indicators in titrations involving **both** a weak acid and a weak base.

10.37 Polyprotic acid neutralised by strong base

The shape of the titration curve will depend on the relative magnitudes of the various dissociation constants. It is assumed that titrations take place at the ordinary laboratory temperature in solutions of concentration of 0.1 M or stronger. For a diprotic acid, if the difference between the primary and secondary dissociation constants is very large ($K_1/K_2 > 10\,000$), the solution behaves like a mixture of two acids with constants K_1 and K_2; the considerations given previously may be applied. For sulphurous acid $K_1 = 1.7 \times 10^{-2}$ and $K_2 = 1.0 \times 10^{-7}$, there will be a sharp change of pH near the first equivalence point followed by a less pronounced change just sufficient for an indicator such as thymolphthalein. For carbonic acid, however, $K_1 = 4.3 \times 10^{-7}$ and $K_2 = 5.6 \times 10^{-11}$, only the first stage will be just discernible in the neutralisation curve (Figure 10.14); the second stage is far too weak to exhibit any point of inflexion and there is no suitable indicator available for direct titration. Thymol blue may be used as indicator for the primary stage (Section 10.34), although a mixed indicator of thymol blue (3 parts) and cresol red (1 part) is more satisfactory (Section 10.31); with phenolphthalein the colour change will be somewhat gradual and the titration error may be several percent.

The pH at the first equivalence point for a diprotic acid is given by

$$[\text{H}^+] = \sqrt{\frac{K_1 K_2 c}{K_1 + c}}$$

Provided the first stage of the acid is weak and provided K_1 is negligible compared with the salt concentration c, this expression reduces to

$$[\text{H}^+] = \sqrt{K_1 K_2} \quad \text{or} \quad \text{pH} = \tfrac{1}{2}\text{p}K_1 + \tfrac{1}{2}\text{p}K_2$$

Knowing the pH at the stoichiometric point and knowing the course of the neutralisation curve, it should be an easy matter to select the appropriate indicator for the titration of any diprotic acid for which K_1/K_2 is at least 10^4. For many diprotic acids, however, the two dissociation constants are too close together and it is not possible to differentiate between the two stages. If $K_2 \geq \sim 10^{-7}$, all the replaceable hydrogen may be titrated, e.g. sulphuric acid (primary stage hence strong acid), oxalic acid, malonic, succinic and tartaric acids.

Similar remarks apply to triprotic acids. Consider phosphoric(V) acid (orthophosphoric acid) with $K_1 = 7.5 \times 10^{-3}$, $K_2 = 6.2 \times 10^{-8}$ and $K_3 = 5 \times 10^{-13}$. Here $K_1/K_2 = 1.2 \times 10^5$ and $K_2/K_3 = 1.2 \times 10^5$, so the acid will behave as a mixture of three monoprotic acids with the dissociation constants given above. Neutralisation proceeds almost completely to the end of the primary stage before the secondary stage is appreciably affected, and the secondary stage proceeds almost to completion before the tertiary stage is apparent. The pH at the first equivalence point is given approximately by ($\tfrac{1}{2}\text{p}K_1 + \tfrac{1}{2}\text{p}K_2$) = 4.6, and at the second equivalence point by ($\tfrac{1}{2}\text{p}K_2 + \tfrac{1}{2}\text{p}K_3$) = 9.7; in the very weak third stage, the curve is very flat and no indicator is available for direct titration. According to Section 10.34, the third equivalence point may be calculated approximately from the equation

$$\text{pH} = \tfrac{1}{2}\text{p}K_w + \tfrac{1}{2}\text{p}K_a - \tfrac{1}{2}\text{p}c = 7.0 + 6.15 - \tfrac{1}{2}(1.6) = 12.35 \quad \text{for } 0.1\,M\ \text{H}_3\text{PO}_4$$

For the primary stage (phosphoric(V) acid as a monoprotic acid), methyl orange, bromocresol green or Congo red may be used as indicators. The secondary stage of phosphoric(V) acid is very weak (acid $K_a = 1 \times 10^{-7}$ in Figure 10.16) and the only suitable simple indicator

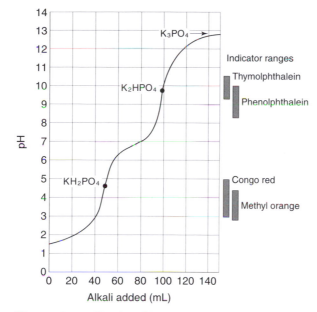

Figure 10.16   Titration of 50 mL of 0.1 M H_3PO_4 with 0.1 M KOH

is thymolphthalein (Section 10.35); with phenolphthalein the error may be several percent. A mixed indicator composed of phenolphthalein (3 parts) and 1-naphtholphthalein (1 part) is very satisfactory for determining the end point of phosphoric(V) acid as a diprotic acid (Section 10.31). Figure 10.16 shows the experimental neutralisation curve for 50 mL of 0.1 M phosphoric(V) acid with 0.1 M potassium hydroxide, determined by potentiometric titration.

There are several triprotic acids, e.g. citric acid ($K_1 = 9.2 \times 10^{-4}$, $K_2 = 2.7 \times 10^{-5}$, $K_3 = 1.3 \times 10^{-6}$), where the dissociation constants are too close together for the three stages to be differentiated easily. If $K_3 > \sim 10^{-7}$, all the replaceable hydrogen may be titrated; the indicator will be determined by the value of K_3.

10.38   Anions of weak acids titrated with strong acids

The titrations considered up to now have involved a strong base, the hydroxide ion, but titrations are also possible with weaker bases, such as the carbonate ion, the borate ion and the acetate (ethanoate) ion. Titrations involving these ions were once regarded as titrations of solutions of hydrolysed salts, and the net result was that the weak acid was displaced by the stronger acid. Thus in the titration of sodium ethanoate solution with hydrochloric acid, the following equilibria were considered:

$$CH_3COO^- + H_2O \rightleftharpoons CH_3COOH + OH^-   \text{(hydrolysis)}$$

$$H^+ + OH^- = H_2O   \text{(strong acid reacts with } OH^- \text{ from hydrolysis)}$$

The net result appeared to be

$$H^+ + CH_3COO^- = CH_3COOH$$

$$\text{or}   CH_3COONa + HCl = CH_3COOH + NaCl$$

i.e. the weak ethanoic acid was apparently displaced by the strong hydrochloric acid, and the process was known as a **displacement titration**. According to the Brønsted–Lowry theory, the so-called titration of solutions of hydrolysed salts is merely the titration of a weak base with a strong (highly ionised) acid. When the anion of a weak acid is titrated with a strong acid, the titration curve is identical with the curve from a reverse titration of a weak acid itself with a strong base (Section 10.34). Here are some examples encountered in practice.

Titration of borate ion with a strong acid

The titration of the tetraborate ion with hydrochloric acid is similar to that described above. The net result of the displacement titration is given by

$$B_4O_7^{2-} + 2H^+ + 5H_2O = 4H_3BO_3$$

Boric acid behaves as a weak monoprotic acid with a dissociation constant of 6.4×10^{-10}. The pH at the equivalence point in the titration of $0.2\,M$ sodium tetraborate with $0.2\,M$ hydrochloric acid is due to $0.1\,M$ boric acid, i.e. pH 5.6. Further addition of hydrochloric acid will cause a sharp decrease of pH and any indicator covering the pH range 3.7–5.1 (and slightly beyond this) may be used; suitable indicators are bromocresol green, methyl orange, bromophenol blue and methyl red.

Titration of carbonate ion with a strong acid

A solution of sodium carbonate may be titrated to the hydrogencarbonate stage (i.e. with 1 mol of hydrogen ions), when the net reaction is

$$CO_3^{2-} + H^+ = HCO_3^-$$

The equivalence point for the primary stage of ionisation of carbonic acid is at pH = $(\frac{1}{2}pK_1 + \frac{1}{2}pK_2) = 8.3$, and we have seen (Section 10.35) that thymol blue and, less satisfactorily, phenolphthalein or a mixed indicator (Section 10.31) may be used to detect the end point.

Sodium carbonate solution may also be titrated until all the carbonic acid is displaced. The net reaction is then:

$$CO_3^{2-} + 2H^+ = H_2CO_3$$

The same end point is reached by titrating sodium hydrogencarbonate solution with hydrochloric acid:

$$HCO_3^- + H^+ = H_2CO_3$$

The end point with 100 mL of $0.2\,M$ sodium hydrogencarbonate and $0.2\,M$ hydrochloric acid may be deduced as follows from the known dissociation constant and concentration of the weak acid. The end point will obviously occur when 100 mL of hydrochloric acid has been added, i.e. the solution now has a total volume of 200 mL. Consequently, since the carbonic acid liberated from the sodium hydrogencarbonate (0.02 mol) is now contained in a volume of 200 mL, its concentration is $0.1\,M$. K_1 for carbonic acid has a value of 4.3×10^{-7}, so we can say

$$[H^+][HCO_3^-]/[H_2CO_3] = K_1 = 4.3 \times 10^{-7}\,mol\,L^{-1}$$

and since

$$[H^+] = [HCO_3^-]$$

$$[H^+] = \sqrt{4.3 \times 10^{-7} \times 0.1} = 2.07 \times 10^{-4}$$

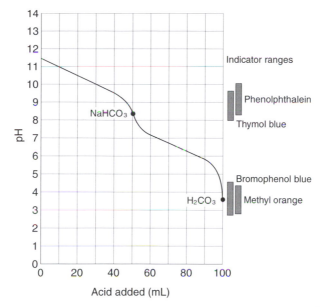

Figure 10.17   Titration of 100 mL of 0.05 M Na$_2$CO$_3$ with 0.1 M HCl

The pH at the equivalence point is thus approximately 3.7; the secondary ionisation and the loss of carbonic acid, due to any escape of carbon dioxide, have been neglected. Suitable indicators are therefore methyl yellow, methyl orange, Congo red and bromophenol blue. Figure 10.17 shows the experimental titration curve, determined potentiometrically, for 100 mL of 0.05 M sodium carbonate and 0.1 M hydrochloric acid.

Cations of weak bases (i.e. Brønsted acids such as the phenylammonium ion C$_6$H$_5$NH$_3^+$) may be titrated with strong bases, and the treatment is similar. These were formerly regarded as salts of weak bases, e.g. aniline (phenylamine) with $K_b = 4.0 \times 10^{-10}$, and strong acid; an example is aniline hydrochloride (phenylammonium chloride).

10.39   Choice of indicators in neutralisation reactions

Feasibility of titration   As a general rule, for a titration to be feasible there should be a change of approximately two units of pH at or near the stoichiometric point produced by the addition of a small volume of the reagent. The pH at the equivalence point may be calculated by using the equations of Section 2.13 (and see below); the pH at either side of the equivalence point (0.1–1 mL) may be calculated as described in the previous sections, and the difference will indicate whether the change is large enough to produce a sharp end point. Alternatively, the pH change on both sides of the equivalence point may be obtained from the neutralisation curve determined by potentiometric titration (Section 10.56). If the pH change is satisfactory, an indicator should be selected that changes at or near the equivalence point. The conclusions of the previous sections are summarised here.

Strong acid and strong base   For 0.1 M or more concentrated solutions, any indicator may be used which has a range between the limits pH 4.5 and pH 9.5. With 0.01 M solutions, the pH range is somewhat smaller (5.5–8.5). If carbon dioxide is present, the solution should be boiled while still acid and the solution titrated when cold, or an indicator with a range below pH 5 should be employed.

Weak acid and a strong base   The pH at the equivalence point is calculated from the equation

$$pH = \tfrac{1}{2}pK_w + \tfrac{1}{2}pK_a - \tfrac{1}{2}pc$$

The pH range for acids with $K_a > 10^{-5}$ is 7 to 10.5; for weaker acids ($K_a > 10^{-6}$) the range is reduced (8 to 10). The pH range 8 to 10.5 will cover most of the examples likely to be encountered; this permits the use of thymol blue, thymolphthalein or phenolphthalein.

Weak base and strong acid   The pH at the equivalence point is calculated from the equation

$$pH = \tfrac{1}{2}pK_w - \tfrac{1}{2}pK_b + \tfrac{1}{2}pc$$

The pH range for bases with $K_b > 10^{-5}$ is 3–7, and for weaker bases ($K_a > 10^{-6}$) it is 3–5. Suitable indicators are methyl red, methyl orange, methyl yellow, bromocresol green and bromophenol blue.

Weak acid and weak base   There is no sharp rise in the neutralisation curve, and generally no simple indicator can be used. The titration should therefore be avoided if possible. The approximate pH at the equivalence point can be calculated from the equation

$$pH = \tfrac{1}{2}pK_w + \tfrac{1}{2}pK_a - \tfrac{1}{2}pK_b$$

It is sometimes possible to use a mixed indicator (Section 10.31) which exhibits a colour change over a very limited pH range, e.g. neutral red–methylene blue for dilute ammonia solution and ethanoic acid.

Polyprotic acids and strong bases   For polyprotic acids with strong bases, and also for mixtures of acids with dissociation constants K_1 and K_2, the first stoichiometric end point is approximately at

$$pH = \tfrac{1}{2}(pK_1 + pK_2)$$

The second stoichiometric end point is at

$$pH = \tfrac{1}{2}(pK_2 + pK_3)$$

Anion of a weak acid titrated with a strong acid   The pH at the equivalence point is given by

$$pH = \tfrac{1}{2}pK_w - \tfrac{1}{2}pK_a - \tfrac{1}{2}pc$$

Cation of a weak base titrated with a strong base   The pH at the stoichiometric end point is given by

$$pH = \tfrac{1}{2}pK_w - \tfrac{1}{2}pK_b - \tfrac{1}{2}pc$$

No sharp end point   As a general rule, wherever an indicator does not give a sharp end point, it is advisable to prepare an equal volume of a comparison solution containing the same quantity of indicator, final products and any other components as in the solution under test. Then perform the titration until it matches the colour of this comparison solution.

No indicator can be found   Whenever it proves impossible to find a suitable indicator, e.g. with strongly coloured solutions, consider an electrometric method such as potentiometry or coulometry. Examples are given in Sections 10.56 to 10.59. Spectrophotometric titration may sometimes be preferable, especially when an indicator's colour change is difficult for the analyst to detect.

Using non-aqueous solvents   Acidic and basic properties may not be the same in a non-aqueous solvent as they are in water. Titrations which are difficult in aqueous solution may become easy in another solvent. This procedure is widely used for analysing organic materials, but it has very limited application with inorganic substances; see Sections 10.40 to 10.42.

10.40   Titrations in non-aqueous solvents

Introduction   The Brønsted–Lowry theory of acids and bases (Section 10.29) can be applied equally well to reactions occurring during acid–base titrations in non-aqueous solvents. This is because their approach considers an acid as any substance which will tend to donate a proton, and a base as any substance which will accept a proton. Substances which give poor end points due to being weak acids or bases in aqueous solution will frequently give far more satisfactory end points when titrations are carried out in non-aqueous media. An additional advantage is that many substances which are insoluble in water are sufficiently soluble in organic solvents to permit their titration in these non-aqueous media.

In the Brønsted–Lowry theory, any acid (HB) is considered to dissociate in solution to give a proton (H^+) and a conjugate base (B^-); whereas any base (B) will combine with a proton to produce a conjugate acid (HB^+):

$$HB \rightleftharpoons H^+ + B^- \qquad [10.2]$$

$$B + H^+ \rightleftharpoons HB^+ \qquad [10.3]$$

The ability of substances to act as acids or bases will very much depend on the choice of solvent system. Non-aqueous solvents are classified into the four groups: aprotic, protophilic, protogenic and amphiprotic.

Aprotic solvents   Aprotic solvents include those substances which may be considered chemically neutral and virtually unreactive under the conditions employed. Carbon tetrachloride and toluene come in this group; they possess low dielectric constants, do not cause ionisation in solutes and do not undergo reactions with acids and bases. Aprotic solvents are frequently used to dilute reaction mixtures while taking no part in the overall process.

Protophilic solvents   Protophilic solvents are substances such as liquid ammonia, amines and ketones which possess a high affinity for protons. The overall reaction can be represented as

$$HB + S \rightleftharpoons SH^+ + B^- \qquad [10.4]$$

The equilibrium in this reversible reaction will be greatly influenced by the nature of the acid and the solvent. Weak acids are normally used in the presence of strongly protophilic solvents as their acidic strengths are then enhanced and then become comparable to those of strong acids; this is known as the **levelling effect**.

353

Protogenic solvents   Protogenic solvents are acidic in nature and readily donate protons. Anhydrous acids such as hydrogen fluoride and sulphuric acid fall in this category; because of their strength and ability to donate protons, they enhance the strength of weak bases.

Amphiprotic solvents   Amphiprotic solvents consist of liquids, such as water, alcohols and weak organic acids, which are slightly ionised and combine both protogenic and protophilic properties in being able to donate protons and accept protons. Ethanoic acid displays acidic properties in dissociating to produce protons:

$$CH_3COOH \rightleftharpoons CH_3COO^- + H^+$$

But in the presence of perchloric acid, a far stronger acid, it will accept a proton:

$$CH_3COOH + HClO_4 \rightleftharpoons CH_3COOH_2^+ + ClO_4^-$$

The $CH_3COOH_2^+$ ion can very readily give up its proton to react with a base. A weak base will therefore have its basic properties enhanced, so titrations between weak bases and perchloric acid can often be readily carried out using ethanoic acid as solvent.

Levelling solvents   In general, strongly protophilic solvents force equilibrium equation [10.4] to the right. This effect is so powerful that, in strongly protophilic solvents, all acids act as if they were of similar strength. The converse occurs with strongly protogenic solvents, which cause all bases to act as if they were of similar strength. Solvents which act in this way are known as levelling solvents.

Differential titrations   Determinations in non-aqueous solvents are important for substances which may give poor end points in normal aqueous titrations and for substances which are not soluble in water. They are particularly valuable for determining the proportions of individual components in mixtures of acids or mixtures of bases. These differential titrations are carried out in solvents which do not exert a levelling effect.

Applications   Although indicators may be used to establish individual end points, as in traditional acid–base titrations, potentiometric methods of end point detection are also used extensively, especially for highly coloured solutions. Non-aqueous titrations have been used to quantify mixtures of primary, secondary and tertiary amines,[5] for studying sulphonamides, mixtures of purines and for many other organic amino compounds and salts of organic acids.

10.41   Solvents for non-aqueous titrations

Do not use a solvent until you are fully acquainted with its hazards and how to use it safely.

Introduction   A very large number of inorganic and organic solvents have been used for non-aqueous determinations, but a few have been used more frequently than most. Some of the most widely applied solvent systems are discussed below. In all instances pure, dry analytical reagent quality solvents should be used to assist in obtaining sharp end points.

Glacial ethanoic acid   Glacial ethanoic acid is by far the most frequently used non-aqueous solvent. Before it is used it is advisable to check the water content, which may be between 0.1% and 1.0%, and to add just sufficient ethanoic (acetic) anhydride to convert any water to the acid. The acid may be used by itself or in conjunction with other solvents, e.g. ethanoic anhydride, acetonitrile and nitromethane.

Acetonitrile   Acetonitrile (methyl cyanide, cyanomethane) is frequently used with other solvents such as chloroform and phenol, and especially with ethanoic acid. It enables very sharp end points to be obtained in the titration of metal ethanoates when titrated with perchloric acid.[6]

Alcohols   Salts of organic acids, especially of soaps, are best determined in mixtures of glycols and alcohols or mixtures of glycols and hydrocarbons. The most common combinations are ethylene glycol (dihydroxyethane) with propan-2-ol or butan-1-ol. The combinations provide admirable solvent power for both the polar and non-polar ends of the molecule.

Dioxan   Dioxan is another popular solvent which is often used in place of glacial ethanoic acid when mixtures of substances are to be quantified. Unlike ethanoic acid, dioxan is not a levelling solvent and separate end points are normally possible, corresponding to the individual components in the mixtures.

Dimethylformamide   Dimethylformamide (DMF) is a protophilic solvent which is frequently employed for titrations between, for instance, benzoic acid and amides, although end points may sometimes be difficult to obtain.

10.42   Indicators for non-aqueous titrations

The interconversion relationships (Section 10.29) between ionised and unionised indicator, or the different resonant forms, apply equally well to indicators used for non-aqueous titrations. But their colour changes at the end point vary from titration to titration, as they depend on the nature of the titrand. The colour corresponding to the correct end point may be established by carrying out a potentiometric titration while simultaneously observing the colour change of the indicator. The appropriate colour corresponds to the inflexion point of the titration curve (Section 10.9). The majority of non-aqueous titrations are carried out using a fairly limited range of indicators. Here are some typical examples.

Crystal violet   Used as a 0.5% w/v solution in glacial ethanoic acid. Its colour change is from violet through blue, followed by green, then to greenish yellow, in reactions in which bases such as pyridine are titrated with perchloric acid.

Methyl red   Used as a 0.2% w/v solution in dioxan with a yellow to red colour change.

1-Naphthol benzein   Gives a yellow to green colour change when employed as a 0.2% w/v solution in ethanoic acid. It gives sharp end points in nitromethane containing ethanoic anhydride for titrations of weak bases against perchloric acid.

Quinaldine red   Used as an indicator for drug determinations in dimethylformamide solution. A 0.1% w/v solution in ethanol gives a colour change from purple red to pale green.

Thymol blue   Used extensively as an indicator for titrations of substances acting as acids in dimethylformamide solution. A 0.2% w/v solution in methanol gives a sharp colour change from yellow to blue at the end point.

10.43   Standard solutions of acids and alkalis

Standard solutions of acids and alkalis are obtainable from commercial suppliers. A wide range of concentrated volumetric solutions (BDH ConvoL) can be obtained from Merck/BDH Ltd, Poole, Dorset. Supplied in sealed ampoules, they can be diluted to produce accurately standardised solutions of specific strength.

Full details on the preparation and standardisation of acids and alkalis may be found in previous editions of this book.[7]

Neutralisation: titrimetric determinations

! ■ **Safety.** Before carrying out any experiments in this section, pay full attention to any safety warnings and make sure you adhere to national laboratory and safety regulations.

The following experiments are examples of manual titrations using visual indicators. The methods described are useful for validation and should be readily adaptable to automatic titrations.

10.44   A mixture of carbonate and hydrogencarbonate

This method is particularly valuable when the sample contains relatively large amounts of carbonate and small amounts of hydrogencarbonate; the total alkali is first determined in one portion of the solution by titration with standard $0.1\,M$ hydrochloric acid using methyl orange, methyl orange–indigo carmine, or bromophenol blue as indicator:

$$CO_3^{2-} + 2H^+ = H_2CO_3$$
$$HCO_3^- + H^+ = H_2CO_3$$
$$H_2CO_3 \rightleftharpoons H_2O + CO_2$$

Let this volume correspond to V mL $1\,M$ HCl. Take another sample and add a measured excess of standard $0.1\,M$ sodium hydroxide (free from carbonate) over that required to transform the hydrogencarbonate to carbonate:

$$HCO_3^- + OH^- = CO_3^{2-} + H_2O$$

Add a slight excess of 10% barium chloride solution to the hot solution; this precipitates the carbonate as barium carbonate. Without filtering off the precipitate, immediately determine the excess sodium hydroxide solution by titration with the same standard acid; phenolphthalein or thymol blue is used as indicator. Suppose that v mL of $0.1\,M$ sodium hydroxide were added and suppose that v' mL were in excess, then the amount of hydrogencarbonate is given by $v - v'$ and the amount of carbonate by $V - (v - v')$.

10.45   Boric acid

Boric acid acts as a weak monoprotic acid ($K_a = 6.4 \times 10^{-10}$), so it cannot be titrated accurately with $0.1\,M$ standard alkali (Section 10.34). However, by the addition of certain organic polyhydroxy compounds, such as mannitol, glucose, sorbitol or glycerol, it acts as a much stronger acid (for mannitol $K_a \simeq 1.5 \times 10^{-4}$) and can be titrated to a phenolphthalein end point.

The effect of polyhydroxy compounds has been explained on the basis of the formation of 1 : 1 and 1 : 2 mole ratio complexes between the hydrated borate ion and 1,2- or 1,3-diols:

$$2 \begin{vmatrix} > C(OH) \\ \\ > C(OH) \end{vmatrix} + H_3BO_3 = \begin{bmatrix} > C-O & O-C < \\ | & \diagdown \diagup & | \\ | & B & | \\ | & \diagup \diagdown & | \\ > C-O & O-C < \end{bmatrix}^- H^+ + 3H_2O$$

Glycerol has been widely employed for this purpose but mannitol and sorbitol are more effective, and because they are solids, they do not materially increase the volume of the solution being titrated; 0.5–0.7 g of mannitol or sorbitol in 10 mL of solution is a convenient quantity.

The method may be applied to commercial boric acid, but as it may contain ammonium salts, a slight excess of sodium carbonate solution is added before boiling down to half-bulk to expel ammonia. Any precipitate which separates is filtered off and washed thoroughly, then the filtrate is neutralised to methyl red, and after boiling, mannitol is added, and the solution titrated with standard 0.1 M sodium hydroxide solution:

$$H[\text{boric acid complex}] + NaOH = Na[\text{boric acid complex}] + H_2O$$

$$1 \text{ mL } 1 M \text{ NaOH} \equiv 0.061\,84 \text{ g } H_3BO_3$$

A mixture of boric acid and a strong acid can be analysed by first titrating the strong acid using methyl red indicator, and then after adding mannitol or sorbitol, the titration is continued using phenolphthalein as indicator. Mixtures of sodium tetraborate and boric acid can be similarly analysed by titrating the salt with standard hydrochloric acid, then adding mannitol and continuing the titration with standard sodium hydroxide solution. Remember that the boric acid liberated in the first titration will react in this second titration.

10.46   Ammonia in an ammonium salt

Discussion   Two methods may be used to determine ammonia in an ammonium salt. In the direct method a solution of the ammonium salt is treated with a solution of a strong base (e.g. sodium hydroxide); the mixture is then distilled. Ammonia is quantitatively expelled and absorbed in an excess of standard acid. The excess acid is back-titrated in the presence of methyl red (or methyl orange, methyl orange–indigo carmine, bromophenol blue or bromocresol green). Each millilitre of 1 M monoprotic acid consumed in the reaction is equivalent to 0.017 032 g NH_3:

$$NH_4^+ + OH^- \rightarrow NH_3 + H_2O$$

For the indirect method the ammonium salt (other than the carbonate or hydrogen-carbonate) is boiled with a known excess of standard sodium hydroxide solution. The boiling is continued until no more ammonia escapes with the steam. The excess of sodium hydroxide is titrated with standard acid, using methyl red (or methyl orange–indigo carmine) as indicator.

Direct method   Assemble the apparatus shown in Figure 10.18; in order to provide some flexibility, the spray trap is joined to the condenser by a hemispherical ground joint, and this makes it easier to clamp the digestion flask and the condenser without introducing any strain into the assembly. The digestion flask may be a round-bottomed flask (capacity 500–1000 mL) or a Kjeldahl flask (as shown in the diagram). The Kjeldahl flask is particularly

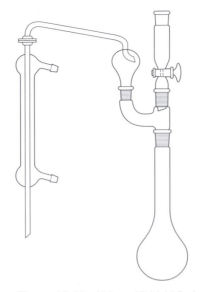

Figure 10.18 Using a Kjeldahl flask

suitable when nitrogen in organic compounds is determined by the Kjeldahl method. On completing the digestion with concentrated sulphuric acid, cooling, and dilution of the contents, the digestion flask is attached to the apparatus as shown in Figure 10.18. The purpose of the spray trap is to prevent droplets of sodium hydroxide solution being driven over during the distillation. The lower end of the condenser is allowed to dip into a known volume of standard acid contained in a suitable receiver, e.g. a conical flask. A commercial distillation assembly is available in which the tap funnel shown is replaced by a special liquid addition unit; this is similar in form to the tap funnel, but the tap and barrel are replaced by a small vertical ground-glass joint which can be closed with a tapered glass rod. This modification is especially useful when numerous determinations have to be made as it obviates the tendency of glass taps to stick after prolonged contact with concentrated solutions of sodium hydroxide.

For practice, weigh out accurately about 1.5 g of ammonium chloride, dissolve it in water, and make up to 250 mL in a graduated flask. Shake thoroughly, Transfer 50.0 mL of the solution into the distillation flask and dilute with 200 mL of water; add a few antibumping granules (fused alumina) to promote regular ebullition in the subsequent distillation. Place 100.0 mL of standard 0.1 M hydrochloric acid in the receiver and adjust the flask so the end of the condenser just dips into the acid. Make sure that all the joints are fitting tightly. Place 100 mL of 10% sodium hydroxide solution in the funnel. Run the sodium hydroxide solution into the flask by opening the tap, close the tap as soon as the alkali has entered. Heat the flask so the contents boil gently. Continue the distillation for 30–40 min, by which time all the ammonia should have passed over into the receiver; open the tap before removing the flame. Disconnect the trap from the top of the condenser. Lower the receiver and rinse the condenser with a little water. Add a few drops of methyl red* and titrate the excess of acid in the solution with standard 0.1 M sodium hydroxide. Repeat the determination.

* A sharper colour change is obtained with the mixed indicator methyl red–bromocresol green (prepared from 1 part of 0.2% methyl red in ethanol and 3 parts of 0.1% bromocresol green in ethanol).

Calculate the percentage of NH_3 in the solid ammonium salt employed:

$$1 \text{ mL } 0.1 \, M \text{ HCl} \equiv 1.703 \text{ mg } NH_3$$

Indirect method   Weigh out accurately $0.1-0.2 \text{ g}$ of the ammonium salt into a 500 mL Pyrex conical flask, and add 100 mL of standard $0.1 \, M$ sodium hydroxide. Place a small funnel in the neck of the flask in order to prevent mechanical loss, and boil the mixture until a piece of filter paper moistened with mercury(I) nitrate solution and held in the escaping steam is no longer turned black. Cool the solution, add a few drops of methyl red, and titrate with standard $0.1 \, M$ hydrochloric acid. Repeat the determination.

10.47 Organic nitrogen: the Kjeldahl procedure

Discussion   Although other chemical and physical methods now exist for the determination of organic nitrogen, the Kjeldahl procedure is still used very extensively as it remains a highly reliable technique with well-established routines. The basic concept of the method is the digestion of organic material, e.g. proteins, using sulphuric acid and a catalyst to convert any organic nitrogen to ammonium sulphate in solution. By making the mixture alkaline any ammonia can be steam-distilled off and the resulting alkaline distillate titrated with standard acid (Section 10.46).

Procedure   Weigh out accurately part of the organic sample, sufficient to contain about 0.04 g of nitrogen, and place it in the long-necked Kjeldahl digestion flask. Add 0.7 g of mercury(II) oxide, 15 g of potassium sulphate and 40 mL of concentrated sulphuric acid. Heat the flask gently in a slightly inclined position. Some frothing is likely to occur and may be controlled by the use of an antifoaming agent. When foaming ceases, boil the reactants for 2 h. After cooling, add 200 mL of water and 25 mL of $0.5 \, M$ sodium thiosulphate solution and mix well. To the mixture add a few antibumping granules, then carefully pour sufficient $11 \, M$ sodium hydroxide solution down the inside of the flask to make the mixture strongly alkaline (approximately 115 mL). Before mixing the reagents, connect the flask to a distillation apparatus (Figure 10.18) in which the tip of the delivery tube is submerged just below the surface of a measured volume of $0.1 \, M$ hydrochloric acid. Ensure the contents of the distillation flask are well mixed, then boil until at least 150 mL of liquid have been distilled into the receiver. Add methyl red indicator to the hydrochloric acid solution and titrate with $0.1 \, M$ sodium hydroxide (titration a mL). Carry out a blank titration on an equal measured volume of the $0.1 \, M$ hydrochloric acid (titration b mL).

Using the quantities and concentrations given above, the percentage of nitrogen in the sample is given by

$$\frac{(b-a) \times 0.1 \times 14 \times 100}{\text{weight of sample (g)}}$$

10.48 Nitrates

Discussion   Nitrates are reduced to ammonia by means of aluminium, zinc or, most conveniently, by Devarda's alloy (50% Cu, 45% Al, 5% Zn) in strongly alkaline solution:

$$3NO_3^- + 8Al + 5OH^- + 2H_2O = 8AlO_2^- + 3NH_3$$

The ammonia is distilled into excess of standard acid as in Section 10.35. Nitrites are similarly reduced, and must be allowed for if nitrate alone is to be determined.

Procedure Weigh out accurately about 1.0 g of the nitrate. Dissolve it in water and transfer the solution quantitatively to the distillation flask of Figure 10.18. Dilute to about 240 mL. Add 3 g of pure, finely divided Devarda's alloy (it should all pass a 20 mesh sieve). Fit up the apparatus completely and place 75–100 mL standard 0.2 M hydrochloric acid in the receiver (500 mL Pyrex conical flask). Introduce 10 mL of 20% (0.5 M) sodium hydroxide solution through the funnel, and immediately close the trap. Warm **gently** to start the reaction, and allow the apparatus to stand for an hour, by which time the evolution of hydrogen should have practically ceased and the reduction of nitrate to ammonia should be complete. Then boil the liquid gently and continue the distillation until 40–50 mL of liquid remain in the distillation flask. Open the tap before removing the flame. Wash the condenser with a little distilled water, and titrate the contents of the receiver plus the washings with standard 0.2 M sodium hydroxide, using methyl red as indicator. Repeat the deter-mination. For very accurate work, it is recommended that a blank test is carried out with distilled water:

$$1 \text{ mL } 1 M \text{ HCl} \equiv 0.062\,01 \text{ g NO}_3^-$$

10.49 Phosphate: precipitation as quinoline molybdophosphate

Discussion When a solution of an orthophosphate is treated with a large excess of ammonium molybdate solution in the presence of nitric acid at a temperature of 20–45 °C, a precipitate is obtained, which after washing is converted into ammonium molyb-dophosphate with the composition $(NH_4)_3[PO_4 \cdot 12MoO_3]$. This may be titrated with standard sodium hydroxide solution using phenolphthalein as indicator, but the end point is rather poor due to the liberation of ammonia. If, however, the ammonium molybdate is replaced by a reagent containing sodium molybdate and quinoline, then quinoline molybdophosphate is precipitated which can be isolated and titrated with standard sodium hydroxide:

$$(C_9H_7N)_3[PO_4 \cdot 12MoO_3] + 26NaOH = Na_2HPO_4 + 12Na_2MoO_4 + 3C_9H_7N + 14H_2O$$

The main advantages over the ammonium molybdophosphate method are (1) quinoline molybdophosphate is less soluble and has a constant composition, and (2) quinoline is a sufficiently weak base not to interfere in the titration.

Calcium, iron, magnesium, alkali metals and citrates do not affect the analysis. Ammo-nium salts interfere and must be eliminated by means of sodium nitrite or sodium hypobro-mite. The hydrochloric acid normally used in the analysis may be replaced by an equivalent amount of nitric acid without any influence on the course of the reaction. Sulphuric acid leads to high and erratic results and its use should be avoided.

The method may be standardised, if desired, with pure potassium dihydrogenphosphate (see below); sufficient 1 : 1 hydrochloric acid must be present to prevent precipitation of quinoline molybdate; the molybdophosphate complex is readily formed at a concentration of 20 mL of concentrated hydrochloric acid per 100 mL of solution especially when warm, and precipitation of the quinoline salt should take place **slowly** from boiling solution. A 'blank' determination should always be made; it is mostly due to silica.

Solutions required   *Sodium molybdate solution*   Prepare a 15% solution of sodium molybdate, $Na_2MoO_4 \cdot 2H_2O$. Store in a polythene bottle.

Quinoline hydrochloride solution   Add 20 mL of redistilled quinoline to 800 mL of hot water containing hydrochloric acid, and stir well. Cool to room temperature, add a little filter paper pulp (accelerator), and again stir well. Filter with suction through a paper-pulp pad, but do not wash. Dilute to 1 L with water.

Mixed indicator solution   Mix two volumes of 0.1% phenolphthalein solution and three volumes of 0.1% thymol blue solution (both in ethanol). A suitable standardising reagent is potassium dihydrogenphosphate(V), previously dried to 105 °C.

Procedure   Weigh accurately 0.20–0.25 g of dry potassium dihydrogenphosphate(V), dissolve it in water and dilute to 250 mL in a graduated flask. Transfer 25.0 mL of the solution into a 250 mL conical flask. Add 20 mL of concentrated hydrochloric acid, then 30 mL of the sodium molybdate solution. Heat to boiling, and add a few drops of the quinoline reagent from a burette while swirling the solution in the flask. Again heat to boiling and add the quinoline reagent drop by drop with constant swirling until 1 or 2 mL have been added. Boil again, and to the gently boiling solution add the reagent a few millilitres at a time, with swirling, until 60 mL in all have been introduced. A coarsely crystalline precipitate is thus produced. Allow the suspension to stand in a boiling water bath for 15 min, then cool to room temperature. Prepare a paper-pulp filter in a funnel fitted with a porcelain cone, and tamp well down. Decant the clear solution through the filter and wash the precipitate twice by decantation with about 20 mL of hydrochloric acid (1 : 9); this removes most of the excess of quinoline and of molybdate.

Transfer the precipitate to the pad with cold water, washing the flask well; wash the filter and precipitate with 30 mL portions of water, letting each washing run through before applying the next, until the washings are acid-free (test for acidity with pH test paper; about six washings are usually required). Transfer the filter pad and precipitate back to the original flask: insert the funnel into the flask and wash with about 50 mL of water to ensure the transfer of all traces of precipitate. Shake the flask well so that filter paper and precipitate are completely broken up. Run in 50.0 mL of standard (carbonate-free) 0.5 *M* sodium hydroxide, swirling during the addition. Shake until the precipitate is **completely** dissolved. Add a few drops of the mixed indicator solution and titrate with standard 0.5 *M* hydrochloric acid to an end point which changes sharply from pale green to pale yellow.

Run a blank on the reagents, but use 0.1 *M* acid and alkali solutions for the titrations; calculate the blank to 0.5 *M* sodium hydroxide. Subtract the blank (which should not exceed 0.5 mL) from the volume neutralised by the original precipitate:

$$\text{1 mL } 0.5\,M \text{ NaOH} \equiv 1.830 \text{ mg } PO_4^{3-}$$

Wilson[8] has recommended that the hydrochloric acid added before precipitation is replaced by citric acid, and the subsequent washing of the precipitate is then carried out solely with distilled water.

The method can be applied to the determination of phosphorus in a wide variety of materials, e.g. phosphate rock, phosphatic fertilisers and metals, and is suitable for use in conjunction with the oxygen flask procedure (Section 3.32). In all cases it is essential to ensure the material is appropriately treated so the phosphorus is converted to orthophosphate; this may usually be done by dissolution in an oxidising medium such as concentrated nitric acid or in 60% perchloric acid.

10.50 Relative molecular mass of an organic acid

Discussion Many of the common carboxylic acids are readily soluble in water and can be titrated with sodium hydroxide or potassium hydroxide solutions. For sparingly soluble organic acids the necessary solution can be achieved by using a mixture of ethanol and water as solvent. The theory of titrations between weak acids and strong bases is covered in Section 10.34 and is usually applicable to both monoprotic and polyprotic acids (Section 10.37). But for determinations carried out in aqueous solutions it is not normally possible to differentiate easily between the end points for the individual carboxylic acid groups in diprotic acids, such as succinic acid, as the dissociation constants are too close together. In these cases the end points for titrations with sodium hydroxide correspond to neutralisation of all the acidic groups. As some organic acids can be obtained in very high states of purity, sufficiently sharp end points can be obtained to justify their use as standards, e.g. benzoic acid and succinic acid. The titration procedure described in this section can be used to determine the relative molecular mass (RMM) of a pure carboxylic acid (if the number of acidic groups is known) or the purity of an acid of known RMM.

Procedure Weigh out accurately about 4 g of the pure organic acid, dissolve it in the minimum volume of water (note 1) or 50 vol% ethanol in water, and transfer the solution to a 250 mL graduated flask. Ensure the solution is homogeneous and make up to the required volume. Use a pipette to measure out accurately a 25 mL aliquot and transfer to a 250 mL conical flask. Using two drops of phenolphthalein solution as indicator, titrate with standard ~0.2 M sodium hydroxide solution (note 2) until the colourless solution becomes faintly pink. Repeat with further 25 mL volumes of the acid solution until two results in agreement are obtained.

The relative molecular mass is given by

$$\text{RMM} = 100\left(\frac{WP}{VM}\right)$$

where
W = the weight of the acid taken
P = the number of carboxylic acid groups
V = the volume of sodium hydroxide used
M = the molarity of the sodium hydroxide.

Notes

1. In order to obtain sharp end points, all deionised water used should be CO_2-free as far as is possible.
2. Volumes of 0.2 M sodium hydroxide required will normally be in the range from about 15 mL to 30 mL depending on the nature of the organic acid being determined.

10.51 Hydroxyl groups in carbohydrates

Discussion Hydroxyl groups present in carbohydrates can be readily acetylated by ethanoic (acetic) anhydride in ethyl ethanoate containing some perchloric acid. This reaction can be used as a basis for determining the number of hydroxyl groups in the carbohydrate molecule by carrying out the reaction with excess ethanoic anhydride followed by titration of the excess using sodium hydroxide in methyl cellosolve.

Solutions required   *Ethanoic (acetic) anhydride*   Prepare 250 mL of a 2.0 *M* solution in ethyl ethanoate containing 4.0 g of 72% perchloric acid. The solution is made by adding 4.0 g (2.35 mL) of 72% perchloric acid to 150 mL of ethyl ethanoate in a 250 mL graduated flask. Pipette 8.0 mL of ethanoic anhydride into the flask, allow to stand for half an hour. Cool the flask to 5 °C, add 42 mL of cold ethanoic anhydride. Keep the mixture at 5 °C for 1 h and then allow it to attain room temperature (note 1).

Sodium hydroxide   Prepare a solution of approximately 0.5 *M* sodium hydroxide in methyl cellosolve. This should be standardised by titration with potassium hydrogenphthalate using the mixed indicator given below.

Pyridine/water   Make up 100 mL of a mixture formed from pyridine and water in the ratio of 3 parts to 1 part by volume.

Mixed indicator   This should be prepared from 1 part of 0.1% neutralised aqueous cresol red and 3 parts of 0.1% neutralised thymol blue.

Procedure   Weigh out accurately 0.15–0.20 g of the carbohydrate into a 100 mL stoppered conical flask. Pipette into the flask exactly 5.0 mL of the ethanoic anhydride–ethyl ethanoate solution. Carefully swirl the contents until the solid has fully dissolved, or mix using a magnetic stirrer. **Do not heat the solution**. Add 1.5 mL of water and again swirl to mix the contents, then add 10 mL of the pyridine/water solution mix by swirling and allow the mixture to stand for 5 min. Titrate the excess ethanoic anhydride with the standardised 0.5 *M* sodium hydroxide using the mixed indicator to give a colour change from yellow to violet at the end point.

Carry out a blank determination on the ethanoic anhydride–ethyl ethanoate solution following the above procedure without adding the carbohydrate. Use the difference between the blank titration V_b and the sample titration V_s to calculate the number of hydroxyl groups in the sugar (note 2).

Calculation   The volume of 0.05 *M* NaOH used is given by $V_b - V_s$, so the number of moles of ethanoic anhydride used in reacting with hydroxyl group is

$$\frac{0.5(V_b - V_s)}{2 \times 1000}$$

But each ethanoic anhydride molecule reacts with two hydroxyl groups, so the number of moles of hydroxyl groups is

$$N = \frac{0.5(V_b - V_s) \times 2}{2 \times 1000} = \frac{(V_b - V_s)}{2000}$$

If the relative molecular mass (RMM) of the carbohydrate is known, then the number of hydroxyl groups per molecule is given by

$$\frac{N \times \text{RMM}}{G}$$

where G is the mass of carbohydrate taken.

Notes

1. All solutions should be freshly prepared before use. **Perchloric acid solutions must not be exposed to sunlight or elevated temperatures as they can be explosive.**
2. The solutions from the titrations should be disposed of promptly after the determination has been carried out.

10.52 Saponification value of oils and fats

Discussion For oils and fats, which are esters of long-chain fatty acids, the saponification value (or number) is defined as the number of milligrams of potassium hydroxide which will neutralise the free fatty acids obtained from the hydrolysis of 1 g of the oil or fat. This means that the saponification number is inversely proportional to the relative molecular masses of the fatty acids obtained from the esters. A typical reaction from the hydrolysis of a glyceride is

$$
\begin{array}{ll}
CH_2-O-CO-C_{17}H_{35} & CH_2OH \\
| & | \\
CH-O-CO-C_{17}H_{35} + 3KOH \longrightarrow CHOH + 3C_{17}H_{35}COOK \\
| & | \\
CH_2-O-CO-C_{17}H_{35} & CH_2OH
\end{array}
$$

$\qquad\qquad$ stearin $\qquad\qquad\qquad\qquad$ glycerol $\quad$ potassium stearate

Procedure Prepare an approximately 0.5 M solution of potassium hydroxide by dissolving 30 g potassium hydroxide in 20 mL of water and make the final volume to 1 L using 95% ethanol. Leave the solution to stand for 24 h before decanting and filtering the solution. Using 25 mL aliquots, titrate the potassium hydroxide solution with 0.5 M hydrochloric acid using phenolphthalein indicator (record as titration a mL).

For the hydrolysis, accurately weigh approximately 2 g of the fat or oil into a 250 mL conical flask with a ground-glass joint and add 25 mL of the potassium hydroxide solution. Attach a reflux condenser and heat the flask contents on a steam bath for 1 h with occasional shaking. While the solution is still hot, add phenolphthalein indicator and titrate the excess potassium hydroxide with the 0.5 M hydrochloric acid (record as titration b mL).

$$
\text{saponification value} = \frac{(a-b) \times 0.5 \times 56.1}{\text{weight of sample (mg)}}
$$

10.53 Purity of acetylsalicylic acid (aspirin)

Discussion Acetylsalicylic acid undergoes hydrolysis when treated with a warm solution of sodium hydroxide, producing sodium ethanoate and sodium salicylate according to the following reaction:

$$
CH_3COOC_6H_4COOH + 2NaOH \xrightarrow{\text{heat}} CH_3COONa + C_6H_4(OH)COONa
$$

If an excess of sodium hydroxide is used, the unreacted alkali may be determined by back titration with standard hydrochloric acid.

Procedure Weigh out accurately about 1.2 g acetylsalicylic acid and transfer to a 250 mL conical flask. Add 50 mL of 0.5 M sodium hydroxide from a pipette and heat the resulting solution on a boiling water bath for 10 min. Allow the solution to cool and back-titrate the excess alkali with standardised $0.5\,M$ hydrochloric acid using 3 drops of 0.1% phenol red as indicator (Section 10.36). Carry out a blank determination using $50\,mL$ of the $0.5\,M$ sodium hydroxide solution:

$$1\,mL\ 0.5\,M\ NaOH \equiv 0.045\,04\,g\ \text{acetylsalicylic acid.}$$

10.54   Amines using a non-aqueous titration

Discussion   Conventional acid–base titrations are seldom suitable for determining weak bases, but various non-aqueous titrations have proven admirable. If a weak base is to be determined, use an acid as strong as possible in a non-basic solvent. Perchloric acid dissolved in either glacial ethanoic (acetic) acid or dioxan are suitable for this purpose, though the glacial ethanoic acid is preferable for the titration of very weak bases.

Solutions required   *Perchloric acid*   Prepare an approximately $0.1\,M$ solution by adding slowly $2.1\,mL$ of 72% perchloric acid to dioxan and making up to $250\,mL$ in a graduated flask. Alternatively, add $2.1\,mL$ of 72% perchloric acid **cautiously** and with **continuous mixing** to $100\,mL$ of glacial ethanoic acid and **then** add $5\,mL$ of ethanoic (acetic) anyhdride. Allow the solution to cool and make up to $250\,mL$ in a graduated flask.

Warning   **Do not add the ethanoic (acetic) anhydride until the perchloric acid is well diluted with the ethanoic acid.**

Standardisation   The perchloric acid may be standardised by titrating against $25\,mL$ aliquots of a standard $0.1\,M$ solution of pure potassium hydrogenphthalate in glacial ethanoic acid (made by dissolving about $2.0\,g$, accurately weighed, of potassium hydrogenphthalate in glacial ethanoic acid and making up to $100\,mL$ with the solvent). A 0.5% solution in ethanoic acid or either crystal violet or oracet blue B may be used as indicator. The colour change using 3 drops of either indicator is blue to green (crystal violet) or blue to pink (oracet blue B).

Procedure   To determine the purity of a sample of adrenaline, accurately weigh about $0.4\,g$ of the sample of adrenaline and transfer into a $250\,mL$ conical flask. Dissolve the amine in glacial ethanoic acid, about $70\,mL$ should suffice, although gentle warming may be needed. Cool the resulting solution and titrate with $0.1\,M$ perchloric acid in glacial ethanoic acid. Use either 3 drops of oracet blue B or crystal violet as indicator. Then

$$1\,mL\ 0.1\,M\ HClO_4 \equiv 0.018\,32\,g\ \text{adrenaline}\ (C_9H_{13}O_3N)$$

Neutralisation: determinations using instruments

! **Safety.** Before carrying out any experiments in this section, pay full attention to any safety warnings and make sure you adhere to national laboratory and safety regulations.

10.55   Potentiometry: general considerations

The electrode system normally used is a glass indicator electrode and a calomel reference electrode – a combined electrode system is preferable nowadays. The accuracy with which the end point can be found potentiometrically depends on the magnitude of the change in e.m.f. in the neighbourhood of the equivalence point. This depends on the concentration and strength of the acid and alkali used. (Sections 10.33 to 10.37.)

Satisfactory results are obtained in all cases except (a) where either the acid or the base is very weak ($K < 10^{-8}$) and the solutions are dilute, and (b) those in which both the acid and the base are weak. In (b) an accuracy of about 1% may be obtained in $0.1\,M$ solution.

The method may be used to titrate a mixture of acids which differ greatly in their strengths, e.g. acetic (ethanoic) and hydrochloric acids; the first break in the titration curve occurs when the stronger of the two acids is neutralised, and the second when neutralisation is complete. For this method to be successful, the two acids or bases should differ in strength by at least 10^5 to 1. In the following experiments a pH meter in the millivolt mode or an autotitrator may be used. The first experiment describes an acid–base titration using basic equipment.

10.56   Potentiometric titration of ethanoic (acetic) acid with sodium hydroxide

Prepare solutions of ethanoic acid and sodium hydroxide, each approximately $0.1\,M$, and set up a pH meter as described in Section 13.22. The following general instructions are applicable to most potentiometric titrations and are given in detail here to avoid subsequent repetition. Other suggested experiments include titration of $0.05\,M$ Na_2CO_3 with $0.1\,M$ HCl, and titration of $0.1\,M$ boric acid in the presence of $4\,g$ of mannitol with $0.1\,M$ NaOH.

(a)   Fit up the apparatus shown in Figure 10.1 with the electrode assembly (or combination electrode) supplied with the pH meter supported inside the beaker. The beaker has a capacity of about $400\,mL$ and contains $50\,mL$ of the solution to be titrated (the ethanoic acid).

(b)   Select a burette, and using a piece of polythene tubing, attach to the jet a piece of glass capillary tubing about 8–10 cm in length. Fill the burette with the sodium hydroxide solution, taking care to remove all air bubbles from the capillary extension, and then clamp the burette so the end of the capillary is immersed in the solutions to be titrated. This procedure ensures that all additions recorded on the burette have in fact been added to the solution, and no drops have been left adhering to the tip of the burette, a factor which can be of some significance for e.m.f. readings made near the end point of the titration.

(c)   Stir the solution in the beaker gently. Read the potential difference between the electrodes with the aid of the meter. Record the reading and also the volume of alkali in the burette.

(d)   Add 2–3 mL of solution from the burette, stir for about 30 s, and after waiting for a further 30 s, measure the e.m.f. of the cell.

(e)   Repeat the addition of 1 mL portions of the base, stirring and measuring the e.m.f. after each addition until a point is reached within about 1 mL of the expected end point. Henceforth, add the solution in portions of 0.1 mL or less, and record the potentiometer readings after each addition. Continue the additions until the equivalence point has been passed by 0.5–1.0 mL.

(f)   Plot potentials against volumes of reagent added; draw a smooth curve through the points. The equivalence point is the volume corresponding to the steepest portion of the curve. In some cases the curve is practically vertical, one drop of solution causing a change of 100–200 mV in the e.m.f. of the cell; in other cases the slope is more gradual.

(g)   Locate the end point of the titration by plotting $\Delta E/\Delta V$ for small increments of the titrant in the vicinity of the equivalence point ($\Delta V = 0.1\,\text{mL}$ or $0.05\,\text{mL}$) against V. There is a maximum in the plot at the end point (Figure 10.2(b)).

(h)   Plot the second-derivative curve $\Delta^2 E/\Delta V^2$ against V; the second derivative becomes zero at the end point (Figure 10.2(c)). Although laborious, this method gives the most exact evaluation of the end point; a Gran's plot may also be made.

10.57   Potentiometric titrations in non-aqueous solvents

As indicated in Section 2.4, the strength of an acid (and of a base) depends on the solvent in which it has been dissolved, and in Sections 10.40 to 10.42 it is shown how this modification of strength can be used to carry out titrations in non-aqueous solvents which are impossible to perform in aqueous solution. Potentiometric methods can be used to determine the end point of such non-aqueous titrations, which are mainly of the acid–base type and offer very valuable methods for the determination of many organic compounds.

The general procedure for performing a potentiometric titration with non-aqueous solutions is in essence the same as that employed for aqueous solutions, but there are some important points of difference. Table 10.6 indicates common reagents and solvents and the appropriate electrode combination for a variety of acid–base titrations.

1.   An electronic millivoltmeter or a pH meter can be used to measure the e.m.f. of the titration cell, but the pH meter must be used in the millivolt mode, since pH has no significance in non-aqueous solutions.

2.   Many of the non-aqueous solvents used must be protected from exposure to the air, and titrations with such materials must be conducted in a closed vessel such as a three- or four-necked flask. Organic solvents have much greater coefficients of thermal expansion than water, and every effort must therefore be made to ensure all solutions are kept as nearly as possible at constant temperature.

Table 10.6   *Some common potentiometric non-aqueous titrations*

Reagent	Solvent	Electrodes[a]	
		Ref.	Ind.
Substances determined: acids, enols, imides, phenols, sulphonamides			
CH_3OK/toluene–methanol	Toluene–methanol	Gl.	Sb
		Cal.	Sb
	Dimethylformamide	Gl.	Sb
	Ethane-1,2-diamine	Gl.	Sb
	1-Aminobutane	Gl.	Sb
R_4NOH/toluene–methanol	Acetonitrile	Cal.	Gl.
	Pyridine	Cal.	Sb
Substances determined: amines, amine salts, amino acids, salts of acids			
$HClO_4/CH_3COOH$	Glacial ethanoic acid	Cal.	Gl.
		Ag, AgCl	Gl.

[a] Gl. = glass, Cal. = calomel.

3. Three reagents are commonly employed:

   Potassium methoxide dissolved in a mixture of toluene and methanol; in some cases potassium may be replaced by lithium or by sodium, and the methanol by ethanol or propan-1-ol. Solvents for the titrand are commonly toluene–methanol, dimethyl-formamide, ethane-1,2-diamine or 1-aminobutane.

   Quaternary ammonium hydroxides dissolved in methanol–toluene or in propan-2-ol, and with acetonitrile or pyridine as solvent for the titrand. Potassium methoxide and quaternary ammonium hydroxides are used for the titration of acids.

   Perchloric acid dissolved in glacial ethanoic (acetic) acid is used for the titration of basic substances which are themselves dissolved in glacial ethanoic acid.

4. Glass and antimony electrodes are commonly used as indicator electrodes, but in toluene–methanol solutions a glass–antimony electrode pair may be used in which the glass electrode functions as reference electrode. Glass electrodes should not be maintained in non-aqueous solvents for long periods, as the hydration layer of the glass bulb may be impaired and the electrode will then cease to function satisfactorily.

5. Reference electrodes are usually a calomel or a silver–silver chloride electrode. It is advisable to use a double-junction pattern so the potassium chloride solution from the electrode does not contaminate the test solution. In titrations involving glacial ethanoic acid as solvent, the outer vessel of the double-junction calomel electrode may be filled with glacial ethanoic acid containing a little lithium perchlorate to improve the conductance.

10.58   Non-aqueous titration of a mixture of aniline and ethanolamine

Discussion   The potentiometric procedure is most useful in determining a mixture of amines. Two individual end points are obtained by employing a glass combination electrode (Section 13.18).

Solutions required   *Perchloric acid*   Prepare an approximately $0.1\,M$ solution by adding 2.13 mL of 72% perchloric acid to dioxan and making up to 250 mL in a graduated flask. This solution should be standardised by titrating against 25 mL aliquots of a standard $0.1\,M$ solution of pure potassium hydrogenphthalate in glacial ethanoic (acetic) acid (made by dissolving 2 g of potassium hydrogenphthalate in glacial ethanoic acid and making up to 100 mL), using crystal violet indicator (Section 10.41).

Mixture of amines   A suitable mixture for analysis can be prepared by accurately weighing roughly equal amounts of aniline and ethanolamine. The determination is best carried out using a solution made from about 4 g of each amine diluted to 100 mL with acetonitrile (methyl cyanide) in a graduated flask.

Procedure   Use 5 mL aliquots of the amine mixture diluted with an additional 20 mL of acetonitrile contained in a 100 mL beaker. Carry out the titration using the $0.1\,M$ perchloric acid, following its progress by means of meter readings (in millivolts) obtained from the two electrodes dipping into the amine solution. Constant stirring is required throughout. The ethanolamine gives rise to the first end point, and the second end point corresponds to the aniline.

As the acetonitrile may contain basic impurities which also react with the perchloric acid, it is desirable to carry out a blank determination on this solvent. Subtract any value for this blank from the titration values of the amines before calculating the percentages of the two amines in the mixture.

10.59   Acids and bases by coulometry

Introduction

The limiting reactions in aqueous solution at platinum electrodes are

$$2H_2O \rightleftharpoons O_2 + 4H^+ + 4e \quad \text{(anode)}$$

$$2H_2O + 2e \rightleftharpoons H_2 + 2OH^- \quad \text{(cathode)}$$

Consequently, anodic electrogeneration of hydrogen ion for the titration of bases and cathodic electrogeneration of hydroxide ion for the titration of acids is readily accomplished. One of the many advantages of coulometric titration of acids is that difficulties associated with the presence of carbon dioxide in the test solution or carbonate in the standard titrant base are easily avoided. Carbon dioxide can be removed completely by passing nitrogen through the original acid solution before the titration is commenced. The presence of any substance that is reduced more easily than hydrogen ion or water at a platinum cathode, or which is oxidised more easily than water at a platinum anode, will interfere.

When internal generation is used in association with a platinum auxiliary electrode, the platinum auxiliary must be placed in a separate compartment; contact between the auxiliary electrode compartment and the sample solution is made through some sort of diaphragm, e.g. a tube with a sintered-glass disc or an agar–salt bridge. For titration of acids, a silver anode may be used in combination with a platinum cathode in the presence of bromide ions; the silver electrode is placed inside a straight tube closed by a sintered disc at its lower end, and this can be inserted directly into the test solution. A bromide ion concentration of about $0.05\,M$ is satisfactory.

Platinum auxiliary generating electrode

Apparatus   Use the cell (~150 mL capacity) shown in Figure 10.3 but modified by placing the auxiliary electrode in a small beaker which is connected to the cell by means of an inverted U-tube salt bridge containing a gel consisting of saturated potassium chloride solution with 3% agar. The end point detection system consists of a pH meter in conjunction with the glass electrode–saturated calomel electrode assembly supplied with the instrument.

Reagents   *Supporting electrolyte*   Use $0.1\,M$ sodium chloride solution.

Catholyte   Use $0.1\,M$ sodium chloride solution to which a little dilute sodium hydroxide solution is added.

Hydrochloric acid   Prepare $0.01\,M$ and $0.001\,M$ hydrochloric acid with boiled-out water and concentrated hydrochloric acid, then standardise.

Procedure   Place 50 mL of the supporting electrolyte in the coulometric cell, and pass nitrogen through the solution until a pH of 7.0 is attained; then pass nitrogen over the surface of the solution. Pipette 10.00 mL of the acid into the cell. Adjust the current to a suitable value (40 or 20 mA) then start the electrolysis; stop the titration when the equivalence point pH (7.00) is reached.

Silver auxiliary electrode

Apparatus   The titration cell is shown in Figure 10.3, but note that the silver anode is placed inside the glass tube with a sintered disc at the lower end. The platinum cathode and the silver anode consist of stout wires coiled into helices. The silver anode may be used repeatedly before the silver bromide coating becomes so thick that it must be removed – about 30 successive titrations of 0.1 mmol samples at 20 mA. When finally necessary, the silver bromide coating may be **carefully** removed by dissolution in potassium cyanide solution.

Supporting electrolyte   Prepare a 0.05 M sodium bromide solution.

Procedure   Place 50 mL of the supporting electrolyte in the beaker and add some of the same solution to the tube carrying the silver electrode so the liquid level in this tube is just above the beaker. Pass nitrogen into the solution until the pH is 7.0. Pipette 10.00 mL of either 0.01 M or 0.001 M hydrochloric acid into the cell. Continue the passage of nitrogen. Proceed with the titration as described for the platinum auxiliary. Several successive samples may be titrated without renewing the supporting electrolyte.

Note   These techniques are generally applicable to many other acids, both strong and weak. The only limitation is that the anion must not be reducible at the platinum cathode and must not react in any way with the silver anode or with silver bromide (e.g. by complexation).

Titration of bases

The titration of a base with electrogenerated hydrogen ions at a platinum cathode

$$2H_2O = O_2 + 4H^+ + 4e$$

may be carried out as described above for the titration of acids using an isolated auxiliary electrode; the connections to the electrodes must be reversed because it is now required to generate hydrogen ions in the coulometric cell.

Supporting electrolyte   Prepare a 0.2 M sodium sulphate solution.

Procedure   Experience in this titration may be acquired by titration of say 5.00 mL of accurately standardised 0.01 M sodium hydroxide solution. Use 50 mL of supporting electrolyte and a current of 30 mA.

10.60   Organic compounds by spectrophotometric titrations

As an example, consider the titration of phenols. This can be carried out by working at the λ_{max} value (in the ultraviolet) for the phenol being determined. By titrating with tetra-n-butylammonium hydroxide and using propan-2-ol as solvent, it is possible to differentiate between substituted phenols.[9] Similarly, aromatic amines have been titrated with perchloric acid using butanol as solvent.

Complexation titrations

! ■   **Safety.** Before carrying out any experiments in this section, pay full attention to any safety warnings and make sure you adhere to national laboratory and safety regulations.

10.61   Introduction

The nature of complexes, their stabilities and the chemical characteristics of complexones are dealt with in some detail in Chapter 2. This section looks at how complexation reactions can be employed in titrimetry, especially for determining the proportions of individual cations in mixtures. The vast majority of complexation titrations are carried out using multidentate ligands such as EDTA or similar substances as the complexone.

10.62   Titration curves

In the titration of a strong acid, if pH is plotted against the volume of the solution of the strong base added, a point of inflexion occurs at the equivalence point (Section 10.33). Similarly, in the EDTA titration, if pM (negative logarithm of the 'free' metal ion concentration, $pM = -\log[M^{n+}]$) is plotted against the volume of EDTA solution added, a point of inflexion occurs at the equivalence point; in some instances this sudden increase may exceed 10 pM units. Figure 10.19 shows the general shape of titration curves obtained by titrating 10.0 mL of a 0.01 M solution of a metal ion M with a 0.01 M EDTA solution. The apparent stability constants of various metal–EDTA complexes are indicated at the extreme right of the curves. The greater the stability constant, the sharper the end point, provided the pH is maintained constant.

In acid–base titrations the end point is generally detected by a pH-sensitive indicator. In the EDTA titration a metal-ion-sensitive indicator, shortened to **metal indicator** or **metal ion indicator**, is often employed to detect changes of pM. Such indicators (which contain types of chelate groupings and generally possess resonance systems typical of dyestuffs) form complexes with specific metal ions, which differ in colour from the free indicator and

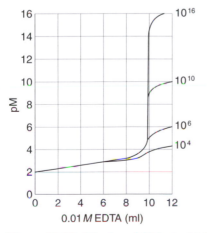

Figure 10.19   Titration of 10.0 mL of 0.01 M solution of metal ion M with a 0.01 M EDTA solution: figures on the right are apparent stability constants

produce a sudden colour change at the equivalence point. The end point of the titration can also be evaluated by other methods, including potentiometry (Section 10.85), coulometry (Section 10.86) and amperometry (Sections 10.87 to 10.89). Spectrophotometric techniques may also be used for estimating end points where the indicator colour change is difficult to visualise (Sections 10.90 and 10.91).

10.63   Types of EDTA titration

Direct titration

The solution containing the metal ion to be determined is buffered to the desired pH, e.g. to pH = 10 with NH_4^+–NH_3 (aq), and titrated directly with the standard EDTA solution. It may be necessary to prevent precipitation of the hydroxide of the metal (or a basic salt) by the addition of some auxiliary complexing agent, such as tartrate or citrate or triethanolamine. At the equivalence point, the magnitude of the concentration of the metal ion being determined decreases abruptly. This is generally determined by the change in colour of a metal indicator or by amperometric, spectrophotometric or potentiometric methods.

Back titration

Many metals cannot be titrated directly; they may precipitate from the solution in the pH range necessary for the titration, or they may form inert complexes, or a suitable metal indicator is not available. Then an excess of standard EDTA solution is added, the resulting solution is buffered to the desired pH, and the excess of the EDTA is back-titrated with a standard metal ion solution; a solution of zinc chloride or sulphate, or magnesium chloride or sulphate, is often used for this purpose. The end point is detected with the aid of the metal indicator which responds to the zinc or magnesium ions introduced in the back titration.

Replacement or substitution titration

Substitution titrations may be used for metal ions that do not react (or react unsatisfactorily) with a metal indicator, or for metal ions which form EDTA complexes that are more stable than those of other metals such as magnesium and calcium. The metal cation M^{n+} to be determined may be treated with the magnesium complex of EDTA, when the following reaction occurs:

$$M^{n+} + MgY^{2-} \rightleftharpoons (MY)^{(n-4)+} + Mg^{2+}$$

The amount of magnesium ion set free is equivalent to the cation present and can be titrated with a standard solution of EDTA and a suitable metal indicator.

   An interesting application is the titration of calcium. In the direct titration of calcium ions, solochrome black gives a poor end point; if magnesium is present, it is displaced from its EDTA complex by calcium and an improved end point results (Section 10.71).

Miscellaneous methods

Exchange reactions between the tetracyanonickelate(II) ion $[Ni(CN)_4]^{2-}$ (the potassium salt is readily prepared) and the element to be determined, whereby nickel ions are set free, have a limited application. Thus silver and gold, which themselves cannot be titrated complexometrically, can be determined in this way:

$$[Ni(CN)_4]^{2-} + 2Ag^+ \rightleftharpoons 2[Ag(CN)_2]^- + Ni^{2+}$$

These reactions take place with sparingly soluble silver salts, hence they provide a method for the determination of the halide ions Cl^-, Br^-, I^-, and the thiocyanate ion SCN^-. The anion is first precipitated as the silver salt, the silver salt is dissolved in a solution of $[Ni(CN)_4]^{2-}$, and the equivalent amount of nickel thereby set free is determined by rapid titration with EDTA using an appropriate indicator (murexide, bromopyrogallol red).

Fluoride may be determined by precipitation as lead chlorofluoride, the precipitate being dissolved in dilute nitric acid and, after adjusting the pH to 5–6, the lead is titrated with EDTA using xylenol orange indicator.[10]

Sulphate may be determined by precipitation as barium sulphate or as lead sulphate. The precipitate is dissolved in an excess of standard EDTA solution, and the excess of EDTA is back-titrated with a standard magnesium or zinc solution using solochrome black as indicator.

Phosphate may be determined by precipitating as $Mg(NH_4)PO_4 \cdot 6H_2O$, dissolving the precipitate in dilute hydrochloric acid, adding an excess of standard EDTA solution, buffering at pH = 10, and back-titrating with standard magnesium ion solution in the presence of solochrome black.

10.64   Titration of mixtures

EDTA is a very unselective reagent because it complexes with numerous doubly, triply and quadruply charged cations. When a solution containing two cations which complex with EDTA is titrated without the addition of a complex-forming indicator, and if a titration error of 0.1% is permissible, then the ratio of the stability constants of the EDTA complexes of the two metals M and N must be such that $K_M/K_N \geq 10^6$ if N is not to interfere with the titration of M. Strictly, the constants K_M and K_N considered in this expression should be the apparent stability constants of the complexes. If complex-forming indicators are used, then for a similar titration error $K_m/K_N \geq 10^8$. The following procedures will help to increase the selectivity.

Suitable control of the pH of the solution

Control of pH makes use of the different stabilities of metal–EDTA complexes. Thus bismuth and thorium can be titrated in an acidic solution (pH = 2) with xylenol orange or methylthymol blue as indicator, and most divalent cations do not interfere. A mixture of bismuth and lead ions can be successfully titrated by first titrating the bismuth at pH 2 with xylenol orange as indicator, and then adding hexamine to raise the pH to about 5, and titrating the lead.

Use of masking agents

Masking may be defined as the process in which a substance, without physical separation of it or its reaction products, is so transformed that it does not enter into a particular reaction. Demasking is the process in which the masked substance regains its ability to enter into a particular reaction. By using masking agents, some of the cations in a mixture can often be 'masked' so they can no longer react with EDTA or with the indicator. An effective masking agent is the cyanide ion; this forms stable cyanide complexes with the cations of Cd, Zn, Hg(II), Cu, Co, Ni, Ag and the platinum metals, but not with the alkaline earths, manganese and lead:

$$M^{2+} + 4CN^- \rightarrow [M(CN)_4]^{2-}$$

It is therefore possible to determine cations such as Ca^{2+}, Mg^{2+}, Pb^{2+} and Mn^{2+} in the presence of these metals by masking with an excess of potassium or sodium cyanide. A small amount of iron may be masked by cyanide if it is first reduced to the iron(II) state by adding ascorbic acid. Titanium(IV), iron(III) and aluminium can be masked with triethanolamine; mercury with iodide ions; and aluminium, iron(III), titanium(IV) and tin(II) with ammonium fluoride (the cations of the alkaline earth metals yield slightly soluble fluorides).

Sometimes the metal may be transformed into a different oxidation state: thus copper(II) may be reduced in acid solution by hydroxylamine or ascorbic acid. After rendering ammoniacal, nickel or cobalt can be titrated using, for example, murexide as indicator without interference from the copper, which is now present as Cu(I). Iron(III) can often be similarly masked by reduction with ascorbic acid.

Selective demasking

The cyanide complexes of zinc and cadmium may be demasked with methanol–ethanoic acid solution or, better, with chloral hydrate:

$$[Zn(CN)_4]^{2-} + 4H^+ + 4HCHO \rightarrow Zn^{2+} + 4HO \cdot CH_2 \cdot CN$$

The use of masking and selective demasking agents permits the successive titration of many metals. Thus a solution containing Mg, Zn and Cu can be titrated as follows:

1. Add excess of standard EDTA and back-titrate with standard Mg solution using solochrome black as indicator. This gives the sum of all the metals present.
2. Treat an aliquot portion with excess of KCN and titrate as before. This gives Mg only. **Avoid any physical contact with KCN. Keep antidotes within easy reach.**
3. Add excess of chloral hydrate (or 3 : 1 methanal : ethanoic acid) to the titrated solution in order to liberate the Zn from the cyanide complex, and titrate until the indicator turns blue. This gives the Zn only. The Cu content may then be found by difference.

Classical separation

Classical separations may be applied if they are not tedious; the following precipitates may be used for separations in which, after being redissolved, the cations can be determined complexometrically: CaC_2O_4, nickel dimethylglyoximate, $Mg(NH_4)PO_4 \cdot 6H_2O$ and CuSCN.

Solvent extraction

Solvent extraction is occasionally of value. Zinc can be separated from copper and lead by adding excess of ammonium thiocyanate solution and extracting the resulting zinc thiocyanate with 4-methylpentan-2-one (isobutyl methyl ketone); the extract is diluted with water and the zinc content determined with EDTA solution.

Choice of indicators

The indicator chosen should be one for which the formation of the metal–indicator complex is sufficiently rapid to permit establishment of the end point without undue waiting, and should preferably be reversible.

Removal of anions

Anions, such as orthophosphate, which can interfere in complexometric titrations may be removed using ion exchange resins. For the use of ion exchange resins in the separation of cations and their subsequent EDTA titration, see Section 6.5.3.

Kinetic masking

Kinetic masking is a special case in which a metal ion does not effectively enter into the complexation reaction because of its kinetic inertness (Section 2.25). Thus the slow reaction of chromium(III) with EDTA makes it possible to titrate other metal ions which react rapidly, without interference from Cr(III); this is illustrated by the determination of iron(III) and chromium(III) in a mixture (Section 10.79).

10.65   Metal ion indicators

General properties

The success of an EDTA titration depends on precise determination of the end point. The most common procedure uses metal ion indicators. For visual detection of end points, a metal ion indicator should satisfy the following criteria:

(a)   The colour reaction must be such that before the end point, when nearly all the metal ion is complexed with EDTA, the solution is strongly coloured.

(b)   The colour reaction should be specific or at least selective.

(c)   The metal–indicator complex must possess sufficient stability else, because of dissociation, a sharp colour change is not obtained. But the metal–indicator complex must be less stable than the metal–EDTA complex to ensure that, at the end point, EDTA removes metal ions from the metal–indicator complex. The change in equilibrium from the metal–indicator complex to the metal–EDTA complex should be sharp and rapid.

(d)   The colour contrast between the free indicator and the metal–indicator complex should be readily observable.

(e)   The indicator must be very sensitive to metal ions (i.e. to pM) so that the colour change occurs as near to the equivalence point as possible.

(f)   Requirements (a) to (e) must be fulfilled within the pH range at which the titration is performed.

Dyestuffs which form complexes with specific metal cations can serve as indicators of pM values; 1 : 1-complexes (metal : dyestuff = 1 : 1) are common, but 1 : 2-complexes and 2 : 1-complexes also occur. The metal ion indicators, like EDTA itself, are chelating agents; this implies that the dyestuff molecule possesses several ligand atoms suitably disposed for coordination with a metal atom. They can equally take up protons, which also produces a colour change; metal ion indicators are therefore not only pM indicators but also pH indicators.

Visual use

Discussion will be confined to the more common 1 : 1-complexes. The use of a metal ion indicator in an EDTA titration may be written as

$$M\text{–}In + EDTA \rightarrow M\text{–}EDTA + In$$

This reaction will proceed if the metal–indicator complex M–In is less stable than the metal–EDTA complex M–EDTA. M–In dissociates to a limited extent, and during the titration the free metal ions are progressively complexed by the EDTA until ultimately the metal is displaced from the complex M–In to leave the free indicator In. The stability of the metal–indicator complex may be expressed in terms of the formation constant (or indicator constant) K_{In}:

$$K_{In} = [\text{M–In}]/[\text{M}][\text{In}]$$

The indicator colour change is affected by the hydrogen ion concentration of the solution, and no account of this has been taken in the expression for the formation constant. Thus solochrome black, which may be written as H_2In^-, exhibits the following acid–base behaviour:

$$H_2In^- \underset{5.3-7.3}{\overset{pH}{\rightleftharpoons}} HIn^{2-} \underset{10.5-12.5}{\overset{pH}{\rightleftharpoons}} In^{3-}$$

$$\text{red} \qquad\qquad \text{blue} \qquad\qquad \text{yellow orange}$$

In the pH range 7–11, in which the dye itself exhibits a blue colour, many metal ions form red complexes; these colours are extremely sensitive, as shown by the fact that 10^{-6} to 10^{-7} M solutions of magnesium ion give a distinct red colour with the indicator. From the practical viewpoint, it is more convenient to define the apparent indicator constant K'_{In}, which varies with pH:

$$K'_{In} = [\text{MIn}^-]/[\text{M}^{n+}][\text{In}]$$

where
[MIn$^-$] = concentration of metal–indicator complex
[M^{n+}] = concentration of metallic ion
[In] = concentration of indicator not complexed with metallic ion

For the above indicator, this is equal to $[H_2In^-] + [HIn^{2-}] + [In^{3-}]$. The equation may be expressed as

$$\log K'_{In} = pM + \log [\text{MIn}^-]/[\text{In}]$$

and $\log K'_{In}$ gives the value of pM when half the total indicator is present as the metal ion complex. Some values for $\log K'_{In}$ for CaIn$^-$ and MgIn$^-$ respectively (where H_2In^- is the anion of solochrome black) are 0.8 and 2.4 at pH = 7; 1.9 and 3.4 at pH = 8; 2.8 and 4.4 at pH = 9; 3.8 and 5.4 at pH = 10; 4.7 and 6.3 at pH = 11; 5.3 and 6.8 at pH = 12. For a small titration error K'_{In} should be large ($> 10^4$), the ratio of the apparent stability constant of the metal–EDTA complex K'_{MY} to that of the metal–indicator complex K'_{In} should be large ($> 10^4$), and the ratio of the indicator concentration to the metal ion concentration should be small ($< 10^{-2}$).

The visual metallochromic indicators discussed above form by far the most important group of indicators for EDTA titrations and the operations subsequently described will be confined to the use of indicators of this type; nevertheless, there are certain other substances which can be used as indicators.[11]

Some examples

Numerous compounds have been proposed for use as pM indicators; a few are listed in Table 10.7. Where applicable, colour index (CI) references are given.[12] It has been pointed

Table 10.7 *Metal ion indicators*

Name and colour index (C1)	Chemical name	Titratable metal ions	pH range(s) for titration	Colour change at end point (direct titration)
Murexide C1 56085	Purpuric acid ammonium salt	Cu, Ni, Co, Ca Lanthanons	10–11	Yellow → blue violet (Ni, Co) Orange → blue violet (Cu) Red → blue violet (Ca)
Solochrome black (eriochrome black T) C1 14645	1-(1-hydroxy-2-naphthylazo)-6-nitro-2-naphthol-4-sulphonate	Mg, Mn, Zn, Cd, Hg, Pb, Ca	10	Red → blue
Patton and Reeder's indicator (HHSNNA)	2-hydroxy-1-(2-hydroxy-4-sulpho-1-naphthylazo)-3-naphthoic acid	Ca	12–14	Wine red → pure blue
Calcichrome	Cyclotris-1-(1-azo-8-hydroxynaphthalene-3-6-disulphonic acid)	Ca using CDTA as titrant	11–12	Pink → blue
Bromopyrogallol red	Dibromopyrogallolsulphon-phthalein	Many cations Useful for Bi	2–3	Blue → claret red
Xylenol orange	3–3′-[N,N-di(carboxymethyl)-aminomethyl] o-cresol sulphonphthalein	Bi, Th, Zn, Co, Cd, Pb, Sn, Ni, Mn	1–2 4–6	Red → lemon yellow Red → lemon yellow
Thymolphthalexone	Thymolphthalein di(methylimine diacetic acid)	Mn, Ca, Sr	10	Blue → slight pink
Methylthymol blue	Thymolsulphonphthalein di(methylimine diacetic acid)	Bc, Th, Zi, Hf, Hg, Zn, Co, Cd, Al, Ni, Mn, Ca, Sr, Ba, Mg	0–2 4–6 12	Blue → yellow Blue → yellow Blue → colourless or smoky grey
Zincon	1-(2-hydroxy-5-sulphophenyl)-3-phenyl-5-(2-carboxyphenyl)formazan	Ca in presence of Mg EGTA as titrant	10	Blue → orange red
Variamine blue C1 37255	4-methoxy-4′-aminodiphenylamine	Fe(III)	3	Blue → yellow

out by West[11] that, apart from a few miscellaneous compounds, the important visual metallochromic indicators fall into three main groups: (a) hydroxyazo compounds; (b) phenolic compounds and hydroxy-substituted triphenylmethane compounds; (c) compounds containing an aminomethyldicarboxymethyl group, many are also triphenylmethane compounds.

Note   In view of the varying stability of solutions of these indicators, and the possible variation in sharpness of the end point with the age of the solution, it is generally advisable (if the stability of the indicator solution is suspect) to dilute the solid indicator with 100–200 parts of potassium (or sodium) chloride, nitrate or sulphate (potassium nitrate is usually preferred) and grind the mixture well in a glass mortar. The resultant mixture is usually stable indefinitely if kept dry and in a tightly stoppered bottle.

10.66   Standard EDTA solutions

Disodium dihydrogenethylenediaminetetra-acetate of analytical reagent quality is available commercially but this may contain a trace of moisture. After drying the reagent at 80 °C its composition agrees with the formula $Na_2H_2C_{10}H_{12}O_8N_2\cdot2H_2O$ (relative molar mass 372.24), but it should not be used as a primary standard. If necessary, the commercial material may be purified by preparing a saturated solution at room temperature; this requires about 20 g of the salt per 200 mL of water. Add ethanol slowly until a permanent precipitate appears; filter. Dilute the filtrate with an equal volume of ethanol, filter the resulting precipitate through a sintered glass funnel, wash with acetone and then with diethyl ether. Air-dry at room temperature overnight and then dry in an oven at 80 °C for at least 24 h.

Solutions of EDTA of the following concentrations are suitable for most experimental work: $0.1\,M$, $0.05\,M$ and $0.01\,M$. These contain respectively 37.224 g, 18.612 g and 3.7224 g of the dihydrate per litre of solution. The dry analytical grade salt cannot be regarded as a primary standard and the solution must be standardised; this can be done by titration of nearly neutralised zinc chloride or zinc sulphate solution prepared from a known weight of zinc pellets, or by titration with a solution made from specially dried lead nitrate.

The water employed in making up solutions, particularly dilute solutions, of EDTA should contain no traces of multicharged ions. The distilled water normally used in the laboratory may require distillation in an all-Pyrex glass apparatus or, better, passage through a column of cation exchange resin in the sodium form; ion exchange will remove all traces of heavy metals. Deionised water is also satisfactory; it should be prepared from distilled water since tap water sometimes contains non-ionic impurities not removed by an ion exchange column. The solution may be kept in Pyrex (or similar borosilicate glass) vessels, which have been thoroughly steamed out before use. For prolonged storage in borosilicate vessels, the vessels should be boiled with a strongly alkaline, 2% EDTA solution for several hours and then repeatedly rinsed with deionised water. Polythene bottles are the most satisfactory, and should always be employed for the storage of very dilute (e.g. $0.001\,M$) solutions of EDTA. Vessels of ordinary (soda) glass should not be used; in the course of time such soft glass containers will yield appreciable amounts of cations (including calcium and magnesium) and anions to solutions of EDTA.

Water purified or prepared as described above should be used for the preparation of **all** solutions required for EDTA or similar titrations.

10.67   Some practical considerations

Adjustment of pH

For many EDTA titrations the pH of the solution is extremely critical; often limits of ± 1 unit of pH, and frequently limits of ± 0.5 unit of pH must be achieved for a successful titration to be carried out. To achieve such narrow limits of control, it is necessary to use a pH meter while adjusting the pH value of the solution, and even for those cases where the latitude is such that a pH test paper can be used to control the adjustment of pH, only a paper of the narrow range variety should be used.

Some of the following items refer to a buffer solution, and to ensure it works properly, make certain the original solution has first been made almost neutral by the cautious addition of sodium hydroxide or ammonium hydroxide, or dilute acid, before adding the buffer solution. When an acid solution containing a metallic ion is neutralised by the addition of alkali, take care to ensure the metal hydroxide is not precipitated.

Concentration of the metal ion to be titrated

Most titrations are successful with 0.25 mmol of the metal ion concerned in a volume of 50–150 mL of solution. If the metal ion concentration is too high, the end point may be very difficult to discern; and if difficulty is experienced with an end point, it is advisable to start with a smaller portion of the test solution, and to dilute this to 100–150 mL before adding the buffering medium and the indicator, and then repeating the titration.

Amount of indicator

The addition of too much indicator is a fault which must be guarded against; in many cases the colour due to the indicator intensifies considerably during the course of the titration, and many indicators exhibit dichroism, i.e. an intermediate colour change occurs one to two drops before the real end point. Thus, in the titration of lead using xylenol orange as indicator at pH = 6, the initial reddish purple colour becomes orange red, and then with the addition of one or two further drops of reagent, the solution acquires the final lemon yellow colour. This **end point anticipation**, which is of great practical value, may be virtually lost if too much of the indicator is added so that the colour is too intense. In general, a satisfactory colour is achieved by the use of 30–50 mg of a 1% solid mixture of the indicator in potassium nitrate.

Attainment of the end point

In many EDTA titrations the colour change in the neighbourhood of the end point may be slow. Then cautious addition of the titrant coupled with continuous stirring of the solution is advisable; the use of a magnetic stirrer is recommended. Frequently, a sharper end point may be achieved if the solution is warmed to about 40 °C. Titrations with CDTA (Section 2.26) are always slower in the region of the end point than the corresponding EDTA titrations.

Detection of the colour change

With all the metal ion indicators used in complexometric titrations, detection of the end point depends on recognising a specified change in colour; for many observers this can be a difficult task, and for those affected by colour blindness it may be virtually impossible.

These difficulties may be overcome by replacing the eye with a photocell; a photocell is much more sensitive and eliminates the human element. Any colorimeter or spectrometer should have a cell compartment large enough to accommodate the titration vessel (a conical flask or a tall-form beaker). A simple apparatus may be readily constructed in which light passing through the solution is first allowed to strike a suitable filter and then a photocell; the current generated in the photocell is measured with a galvanometer. Whatever form of instrument is used, the wavelength of the incident light is selected (either by an optical filter or by the controls on the instrument) so the titration solution (including the indicator) shows a maximum transmittance. The titration is then carried out stepwise, taking readings of the transmittance after each addition of EDTA; these readings are then plotted against volume of EDTA solution added, and at the end point (where the indicator changes colour) there will be an abrupt alteration in the transmittance, i.e. a break in the curve, from which the end point may be assessed accurately.

Alternative methods of detecting the end point

In addition to the visual and spectrophotometric detection of end points, the following methods are also available:

1. Potentiometry using a mercury electrode (Section 10.85)
2. Amperometry (Section 10.87)
3. Coulometry (Section 10.86)

The following sections are devoted to applications of ethylenediaminetetra-acetic (EDTA) and its congeners. These reagents possess great versatility arising from their inherent potency as complexing agents and from the availability of numerous metal ion indicators (Section 10.65), each effective over a limited range of pH, but together covering a wide range of pH values; to these factors must be added the additional refinements offered by masking and demasking techniques (Section 10.64).

It is impossible here to give details for all the cations (and anions) which can be determined by EDTA or similar types of titration. Accordingly, details of a few typical determinations are given which serve to illustrate the general procedures to be followed and the use of various buffering agents and some different indicators. A conspectus of some selected procedures for the commoner cations is then given, followed by some examples of using EDTA to determine the components of mixtures. The final set of examples are anion determinations. Consult Sections 2.22 to 2.27 before doing the experiments.

Complexation: determining individual cations

!   **Safety.** Before carrying out any experiments in this section, pay full
■   attention to any safety warnings and make sure you adhere to national
    laboratory and safety regulations.

10.68   Aluminium: back titration

Procedure   Pipette 25 mL of an aluminium ion solution (approximately $0.01\,M$) into a conical flask and from a burette add a slight excess of $0.01\,M$ EDTA solution; adjust the pH to between 7 and 8 by the addition of ammonia solution (test drops on phenol red paper or use a pH meter). Boil the solution for a few minutes to ensure complete complexation of the aluminium; cool to room temperature and adjust the pH to 7–8. Add 50 mg of solo-chrome black–potassium nitrate mixture (Section 10.67) and titrate rapidly with standard $0.01 M$ zinc sulphate solution until the colour changes from blue to wine red. After stand-ing for a few minutes the fully titrated solution acquires a reddish violet colour due to the transformation of the zinc dye complex into the aluminium–solochrome black complex; this change is irreversible, so overtitrated solutions are lost.

Every millilitre difference between the volume of $0.01 M$ EDTA added and the $0.01 M$ zinc sulphate solution used in the back titration corresponds to 0.2698 mg of aluminium. The standard zinc sulphate solution required is best prepared by dissolving about 1.63 g (accurately weighed) of granulated zinc in dilute sulphuric acid, nearly neutralising with sodium hydroxide solution, and then making up to 250 mL in a graduated flask; alterna-tively, the requisite quantity of zinc sulphate may be used. In either case, deionised water must be used.

10.69   Barium: direct titration

Procedure   Pipette 25 mL barium ion solution ($\sim0.01 M$) into a 250 mL conical flask and dilute to about 100 mL with deionised water. Adjust the pH of the solution to 12 by the addition of 3–6 mL of 1 M sodium hydroxide solution; the pH **must** be checked with a pH meter as it must lie between 11.5 and 12.7. Add 50 mg of methyl thymol blue–potassium nitrate mixture (Section 10.67) and titrate with standard ($0.01 M$) EDTA solution until the colour changes from blue to grey:

$$1 \text{ mol EDTA} \equiv 1 \text{ mol Ba}^{2+}$$

10.70   Bismuth: direct titration

Procedure   Pipette 25 mL of the bismuth ion solution ($\sim0.01 M$) into a 500 mL conical flask and dilute with deionised water to about 150 mL. If necessary, adjust the pH to about 1 by the cautious addition of dilute aqueous ammonia or dilute nitric acid; use a pH meter. Add 30 mg of the xylenol orange–potassium nitrate mixture (Section 10.67) and then titrate with standard $0.01 M$ EDTA solution until the red colour starts to fade. From this point add the titrant slowly until the end point is reached and the indicator changes to yellow:

$$1 \text{ mol EDTA} \equiv 1 \text{ mol Bi}^{3+}$$

10.71   Calcium: substitution titration

Discussion   When calcium ions are titrated with EDTA, a relatively stable calcium complex is formed:

$$Ca^{2+} + H_2Y^{2-} \rightleftharpoons CaY^{2-} + 2H^+$$

With calcium ions alone, no sharp end point can be obtained with solochrome black indic-ator and the transition from red to pure blue is not observed. With magnesium ions, a somewhat less stable complex is formed:

$$Mg^{2+} + H_2Y^{2-} \rightleftharpoons MgY^{2-} + 2H^+$$

and the magnesium indicator complex is more stable than the calcium–indicator complex but less stable than the magnesium–EDTA complex. Consequently, during the titration of a solution containing magnesium and calcium ions with EDTA in the presence of solochrome black, the EDTA reacts first with the free calcium ions, then with the free magnesium ions, and finally with the magnesium–indicator complex. Since the magnesium–indicator complex is wine red in colour and the free indicator is blue between pH 7 and 11, the colour of the solution changes from wine red to blue at the end point:

$$MgD^- \text{ (red)} + H_2Y^{2-} = MgY^{2-} + HD^{2-} \text{ (blue)} + H^+$$

If magnesium ions are not present in the solution containing calcium ions they must be added, since they are required for the colour change of the indicator. A common procedure is to add a small amount of magnesium chloride to the EDTA solution before it is standardised. Another procedure, which permits the EDTA solution to be used for other titrations, is to incorporate a little magnesium–EDTA (MgY^{2-}) (1–10%) in the buffer solution or to add a little $0.1\,M$ magnesium–EDTA (Na_2MgY) to the calcium ion solution:

$$MgY^{2-} + Ca^{2+} = CaY^{2-} + Mg^{2+}$$

Traces of many metals interfere in the determination of calcium and magnesium using solochrome black indicator, e.g. Co, Ni, Cu, Zn, Hg and Mn. Their interference can be overcome by the addition of a little hydroxylammonium chloride (which reduces some of the metals to their lower oxidation states), or by addition of sodium cyanide or potassium cyanide (which form very stable cyanide complexes). Iron may be rendered harmless by the addition of a little sodium sulphide.

The titration with EDTA, using solochrome black as indicator, will yield the calcium content of the sample (if no magnesium is present) or the total calcium and magnesium content if both metals are present. To determine the individual elements, calcium may be evaluated by titration using a suitable indicator, e.g., Patton and Reeder's indicator or calcon (Section 10.65) or by titration with EGTA using zincon as indicator (Section 10.75). The difference between the two titrations is a measure of the magnesium content.

Procedure   Prepare an ammonia–ammonium chloride buffer solution (pH 10) by adding 142 mL concentrated ammonia solution (specific gravity 0.88–0.90) to 17.5 g ammonium chloride and diluting to 250 mL with deionised water. Prepare the magnesium complex of EDTA, Na_2MgY, by mixing equal volumes of $0.2\,M$ solutions of EDTA and magnesium sulphate. Neutralise with sodium hydroxide solution to a pH between 8 and 9 (phenolphthalein just reddened). Take a portion of the solution, add a few drops of the buffer solution (pH 10), and a few milligrams of the solochrome black–potassium nitrate indicator mixture (Section 10.67). A violet colour should be produced which turns blue on the addition of a drop of $0.01\,M$ EDTA solution and red on the addition of a single drop of $0.01\,M$ magnesium sulphate solution; this confirms the equimolarity of magnesium and EDTA. If the solution does not pass this test, it may be treated with more EDTA or with more magnesium sulphate solution until the required condition of equimolarity is attained; this gives an approximately $0.1\,M$ solution.

Pipette 25.0 mL of the $0.01\,M$ calcium ion solution into a 250 mL conical flask, dilute it with about 25 mL of distilled water, add 2 mL buffer solution, 1 mL of $0.1\,M$ Mg–EDTA, and 30–40 mg solochrome black–potassium nitrate mixture. Titrate with the EDTA solution until the colour changes from wine red to clear blue. No tinge of reddish hue should remain at the equivalence point. Titrate slowly near the end point:

$$1 \text{ mol EDTA} \equiv 1 \text{ mol } Ca^{2+}$$

10.72   Iron(III): direct titration

Procedure   Prepare the indicator solution by dissolving 1 g variamine blue in 100 mL deionised water; variamine blue acts as a redox indicator. Pipette 25 mL iron(III) solution $(0.05\,M)$ into a conical flask and dilute to 100 mL with deionised water. Adjust the pH to 2–3; Congo red paper may be used to the first perceptible colour change. Add 5 drops of the indicator solution, warm the contents of the flask to 40 °C, and titrate with standard $(0.05\,M)$ EDTA solution until the initial blue colour of the solution turns grey just before the end point, and with the final drop of reagent changes to yellow:

$$1\text{ mol EDTA} \equiv 1\text{ mol Fe}^{3+}$$

10.73   Nickel: direct titrations

Procedure A   Prepare the indicator by grinding 0.1 g murexide with 10 g of potassium nitrate; use about 50 mg of the mixture for each titration. Also prepare a $1\,M$ solution of ammonium chloride by dissolving 26.75 g of the analytical grade solid in deionised water and making up to 500 mL in a graduated flask. Pipette 25 mL nickel solution $(0.01\,M)$ into a conical flask and dilute to 100 mL with deionised water. Add the solid indicator mixture (50 mg) and 10 mL of the $1\,M$ ammonium chloride solution, and then add concentrated ammonia solution dropwise until the pH is about 7 as shown by the yellow colour of the solution. Titrate with standard $(0.01\,M)$ EDTA solution until the end point is approached, then render the solution strongly alkaline by the addition of 10 mL of concentrated ammonia solution, and continue the titration until the colour changes from yellow to violet. The pH of the final solution must be 10; at lower pH values an orange-yellow colour develops and more ammonia solution must be added until the colour is clear yellow. Nickel complexes rather slowly with EDTA, so the EDTA solution must be added dropwise near the end point.

Procedure B   Prepare the indicator by dissolving 0.05 g bromopyrogallol red in 100 mL of 50% ethanol, and a buffer solution by mixing 100 mL of $1\,M$ ammonium chloride solution with 100 mL of $1\,M$ aqueous ammonia solution. Pipette 25 mL nickel solution $(0.01\,M)$ into a conical flask and dilute to 150 mL with deionised water. Add about 15 drops of the indicator solution, 10 mL of the buffer solution and titrate with standard EDTA solution $(0.01\,M)$ until the colour changes from blue to claret red:

$$1\text{ mol EDTA} \equiv 1\text{ mol Ni}^{2+}$$

10.74   A selection of metals by EDTA

With the detailed instructions given in Sections 10.68 to 10.73 it should be possible to carry out any of the following determinations in Table 10.8 without serious problems arising. In all cases it is recommended that the requisite pH value for the titration should be established by using a pH meter, but in the light of experience the colour of the indicator at the required pH may sometimes be a satisfactory guide. Where no actual buffering agent is specified, the solution should be brought to the required pH value by the cautious addition of dilute acid, dilute sodium hydroxide solution or aqueous ammonia solution as required.

Table 10.8 *Summarised procedures for EDTA titrations of some selected cations*

Metal[a]	Titration type[a]	pH	Buffer	Indicator[b]	Colour change[c]	
Aluminium	*back*	7–8	NH_3 (aq)	SB	B	R
Barium[d]	*direct*	12		MTB	B	Gr
Bismuth	*direct*	1		XO	R	Y
	direct	0–1		MTB	B	Y
Cadmium	direct	5	Hexamine	XO	R	Y
Calcium	direct	12		MTB	B	Gr
	substn	7–11	NH_3 (aq)/NH_4Cl	SB	R	B
Cobalt[e]	*direct*	6	Hexamine	XO	R	Y
Iron(III)[e]	*direct*	2–3		VB	B	Y
Lead	direct	6	Hexamine	XO	R	Y
Magnesium[f]	direct	10	NH_3 (aq)/NH_4Cl	SB	R	B
Manganese[g]	direct	10	NH_3 (aq)/NH_4Cl	SB	R	B
	direct	10	NH_3 (aq)	TPX	B	PP
Mercury	direct	6	Hexamine	XO	R	Y
	direct	6	Hexamine	MTB	B	Y
Nickel	*direct*	7–10	NH_3 (aq)/NH_4Cl	M	Y	V
	direct	7–10	NH_3 (aq)/NH_4Cl	BPR	B	R
	back	10	NH_3 (aq)/NH_4Cl	SB	B	R
Strontium	direct	12		MTB	B	Gr
	direct	10–11		TPX	B	PP
Thorium	direct	2–3		XO	R	Y
	direct	2–3		MTB	B	Y
Tin(II)	direct	6	Hexamine	XO	R	Y
Zinc	direct	10	NH_3 (aq)/NH_4Cl	SB	R	B
	direct	6	Hexamine	XO	R	Y
	direct	6	Hexamine	MTB	B	Y

[a] Items in italics are covered in Sections 10.68 to 10.73.
[b] BPR = bromopyrogallol red; M = murexide; MTB = methylthymol blue; SB = solochrome black; TPX = thymolphthalexone; VB = variamine blue; XO = xylenol orange.
[c] B = blue; Gr = grey; PP = pale pink; R = red; V = violet; Y = yellow.
[d] Can also be determined by precipitation as $BaSO_4$ and dissolution in excess EDTA.
[e] Temperature 40 °C.
[f] Warming optional.
[g] Add 0.5 g hydroxylammonium chloride (to prevent oxidation) and 3 mL triethanolamine (to prevent precipitation in alkaline solution); use boiled-out (air-free) water.

10.75 Calcium in the presence of magnesium using EGTA

Discussion   Calcium may be determined in the presence of magnesium by using EGTA as titrant, because whereas the stability constant for the calcium–EGTA complex is about 1×10^{11}, the stability constant of the magnesium–EGTA complex is only about 1×10^5, so magnesium does not interfere with the reagent. The method, which involves precipitation of magnesium hydroxide, is not satisfactory if the magnesium content of the mixture is much greater than about 10% of the calcium content, since coprecipitation of calcium hydroxide may occur. Hence titration with EGTA is recommended for determining small amounts of calcium in the presence of larger amounts of magnesium.

The indicator used in the titration is zincon (Section 10.65) which gives rise to an indirect end point with calcium. Detection of the end point depends on the reaction

$$ZnEGTA^{2-} + Ca^{2+} = Zn^{2+} + CaEGTA^{2-}$$

and the zinc ions liberated form a blue complex with the indicator. At the end point, the zinc–indicator complex is decomposed:

$$ZnIn^- + H_2EGTA^{2-} \rightleftharpoons ZnEGTA^{2-} + HIn^-$$

and the solution acquires the orange-red colour of the indicator.

Procedure   Prepare an EGTA solution (0.05 M) by dissolving 19.01 g in 100 mL sodium hydroxide solution (1 M) and diluting to 1 L in a graduated flask with deionised water. Prepare the indicator by dissolving 0.065 g zincon in 2 mL sodium hydroxide solution (0.1 M) and diluting to 100 mL with deionised water, and a buffer solution (pH 10) by dissolving 25 g sodium tetraborate, 3.5 g ammonium chloride, and 5.7 g sodium hydroxide in 1 L of deionised water.

Prepare 100 mL of Zn–EGTA complex solution by taking 50 mL of 0.05 M zinc sulphate solution and adding an equivalent volume of 0.05 M EGTA solution; exact equality of zinc and EGTA is best achieved by titrating a 10 mL portion of the zinc sulphate solution with the EGTA solution using zincon indicator, and from this result the exact volume of EGTA solution required for the 50 mL portion of zinc sulphate solution may be calculated.

The EGTA solution may be standardised by titration of a standard (0.05 M) calcium solution, prepared by dissolving 5.00 g calcium carbonate in dilute hydrochloric acid contained in a 1 L graduated flask, and then after neutralising with sodium hydroxide solution diluting to the mark with deionised water: use zincon indicator in the presence of Zn–EGTA solution (see below).

To determine the calcium in the calcium–magnesium mixture, pipette 25 mL of the solution into a 250 mL conical flask, add 25 mL of the buffer solution and check that the resulting solution has a pH of 9.5–10.0. Add 2 mL of the Zn–EGTA solution and 2–3 drops of the indicator solution. Titrate slowly with the standard EGTA solution until the blue colour changes to orange-red.

10.76   Total hardness of water: permanent and temporary

Discussion   The hardness of water is generally due to dissolved calcium and magnesium salts and may be determined by complexometric titration.

Procedure   To a 50 mL sample of the water to be tested add 1 mL buffer solution (ammonium hydroxide–ammonium chloride, pH 10, Section 10.71) and 30–40 mg solochrome black indicator mixture. Titrate with standard EDTA solution (0.01 M) until the colour changes from red to pure blue. Should there be no magnesium present in the sample of water, it is necessary to add 0.1 mL magnesium–EDTA solution (0.1 M) before adding the indicator (Section 10.71). The total hardness is expressed in parts of $CaCO_3$ per million of water.

If the water contains traces of interfering ions, then 4 mL of buffer solution should be added, followed by 30 mg of hydroxylammonium chloride and then 50 mg analytical grade potassium cyanide before adding the indicator. **Avoid any physical contact with potassium cyanide. Keep antidotes within easy reach.**

Notes

1.   Somewhat sharper end points may be obtained if the sample of water is first acidified with dilute hydrochloric acid, boiled for about a minute to drive off carbon dioxide, cooled, neutralised with sodium hydroxide solution, buffer and indicator solution added, and then titrated with EDTA as above.

2.   The permanent hardness of a sample of water may be determined as follows. Place 250 mL of the sample of water in a 600 mL beaker and boil gently for 20–30 min. Cool and filter it directly into a 250 mL graduated flask; do not wash the filter paper, but dilute the filtrate to volume with deionised water and mix well. Titrate 50.0 mL of the filtrate by the same procedure as was used for the total hardness. This titration measures the permanent hardness of the water. Calculate this hardness as parts per million of $CaCo_3$. Calculate the temporary hardness of the water by subtracting the permanent hardness from the total hardness.

3.   If it is desired to determine both the calcium and the magnesium in a sample of water, determine first the total calcium and magnesium content as above, and calculate the result as parts per million of $CaCO_3$. The calcium content may then be determined by titration wth EDTA using either Patton and Reeder's indicator or by titration with EGTA (Section 10.75).

10.77   Calcium in the presence of barium using CDTA

Discussion   There is an appreciable difference between the stability constants of the CDTA complexes of barium ($\log K = 7.99$) and calcium ($\log K = 12.50$), with the result that calcium may be titrated with CDTA in the presence of barium; the stability constants of the EDTA complexes of these two metals are too close together to permit independent titration of calcium in the presence of barium. The indicator calcichrome (Section 10.65) is specific for calcium at pH 11–12 in the presence of barium.

Procedure   Prepare the CDTA solution ($0.02 M$) by dissolving 6.880 g of the solid reagent in 50 mL of sodium hydroxide solution ($1 M$) and making up to 1L with deionised water; the solution may be standardised against a standard calcium solution prepared from 2.00 g of calcium carbonate (Section 10.76). The indicator is prepared by dissolving 0.5 g of the solid in 100 mL of water.

Pipette 25 mL of the solution to be analysed into a 250 mL conical flask and dilute to 100 mL with deionised water; the original solution should be about $0.02 M$ with respect to calcium and may contain barium to a concentration of up to $0.2 M$. Add 10 mL sodium hydroxide solution ($1 M$) and check that the pH of the solution lies between 11 and 12; then add three drops of the indicator solution. Titrate with the standard CDTA solution until the pink colour changes to blue.

10.78   Calcium and lead in a mixture

Discussion   With methylthymol blue, lead may be titrated at a pH of 6 without interference by calcium; the calcium is subsequently titrated at pH 12.

Procedure   Pipette 25 mL of the test solution (which may contain both calcium and lead at concentrations of up to 0.01 M) into a 250 mL conical flask and dilute to 100 mL with deionised water. Add about 50 mg of methylthymol blue–potassium nitrate mixture followed by dilute nitric acid until the solution is yellow, and then add powdered hexamine until the solution has an intense blue colour (pH ~6). Titrate with standard (0.01 M) EDTA solution until the colour turns to yellow; this gives the titration value for lead. Now carefully add sodium hydroxide solution (1 M) until the pH of the solution has risen to 12 (pH meter); 3–6 mL of the sodium hydroxide solution will be required. Continue the titration of the bright blue solution with the EDTA solution until the colour changes to smoky grey; this gives the titration value for calcium.

10.79   Chromium(III) and iron(III) in a mixture: kinetic masking

Discussion   Iron (and nickel, if present) can be determined by adding an excess of standard EDTA to the cold solution, and then back-titrating the solution with lead nitrate solution using xylenol orange as indicator; provided the solution is kept cold, chromium does not react. The solution from the back titration is then acidified, excess of standard EDTA solution added and the solution boiled for 15 min when the red-violet Cr(III)–EDTA complex is produced. After cooling and buffering to pH 6, the excess EDTA is then titrated with the lead nitrate solution.

Procedure   Place 10 mL of the solution containing the two metals (the metal concentrations should not exceed 0.01 M) in a 600 mL beaker fitted with a magnetic stirrer, and dilute to 100 mL with deionised water. Add 20 mL of standard (~0.01 M) EDTA solution and add hexamine to adjust the pH to 5–6. Then add a few drops of the indicator solution (0.5 g xylenol orange dissolved in 100 mL of water) and titrate the excess EDTA with a standard lead nitrate solution (0.01 M), i.e. to the formation of a red-violet colour.

To the resulting solution now add a further 20 mL portion of the standard EDTA solution, add nitric acid (1 M) to adjust the pH to 1–2, and then boil the solution for 15 min. Cool, dilute to 400 mL by the addition of deionised water, add hexamine to bring the pH to 5–6, add more of the indicator solution, and titrate the excess EDTA with the standard lead nitrate solution.

The first titration determines the amount of EDTA used by the iron, and the second, the amount of EDTA used by the chromium.

10.80   Manganese in the presence of iron: ferromanganese

Discussion   After dissolution of the alloy in a mixture of concentrated nitric and hydrochloric acids the iron is masked with triethanolamine in an alkaline medium, and the manganese titrated with standard EDTA solution using thymolphthalexone as indicator. The amount of iron(III) present must not exceed 25 mg per 100 mL of solution, otherwise the colour of the iron(III)–triethanolamine complex is so intense that the colour change of the indicator is obscured. Consequently, the procedure can only be used for samples of ferromanganese containing more than about 40% manganese.

Procedure   Dissolve a weighed amount of ferromanganese (about 0.40 g) in concentrated nitric acid and then add concentrated hydrochloric acid (or use a mixture of the two concentrated acids); prolonged boiling may be necessary. Evaporate to a small volume on a water bath. Dilute with water and filter directly into a 100 mL graduated flask, wash with distilled water and finally dilute to the mark. Pipette 25.0 mL of the solution into a 500 mL conical flask, add 5 mL of 10% aqueous hydroxylammonium chloride solution, 10 mL of 20% aqueous triethanolamine solution, 10–35 mL of concentrated ammonia solution, about 100 mL of water, and 6 drops of thymolphthalexone indicator solution. Titrate with standard 0.05 M EDTA until the colour changes from blue to colourless (or a very pale pink).

10.81 Nickel in the presence of iron: nickel steel

Discussion Nickel may be determined in the presence of a large excess of iron(III) in weakly acidic solution by adding EDTA and triethanolamine; the intense brown precipitate dissolves upon the addition of aqueous sodium hydroxide to yield a colourless solution. The iron(III) is present as the triethanolamine complex and only the nickel is complexed by the EDTA. The excess of EDTA is back-titrated with standard calcium chloride solution in the presence of thymolphthalexone indicator. The colour change is from colourless or very pale blue to an intense blue. The nickel–EDTA complex has a faint blue colour; the solution should contain less than 35 mg of nickel per 100 mL.

In the back titration small amounts of copper and zinc and trace amounts of manganese are quantitatively displaced from the EDTA and are complexed by the triethanolamine; small quantities of cobalt are converted into a triethanolamine complex during the titration. Relatively high concentrations of copper can be masked in the alkaline medium by the addition of thioglycollic acid until colourless. If present in quantities of more than 1 mg, manganese may be oxidised by air to manganese(III); this forms a manganese(III)–triethanolamine complex which is intensely green in colour. The problem can be avoided by adding a little hydroxylammonium chloride solution.

Procedure Prepare a standard calcium chloride solution (0.01 M) by dissolving 1.000 g of calcium carbonate in the minimum volume of dilute hydrochloric acid and diluting to 1 L with deionised water in a graduated flask. Also prepare a 20% aqueous solution of triethanolamine. Weigh out accurately a 1.0 g sample of the nickel steel and dissolve it in the minimum volume of concentrated hydrochloric acid (about 15 mL) to which a little concentrated nitric acid (~1 mL) has been added. Dilute to 250 mL in a graduated flask. Pipette 25.0 mL of this solution into a conical flask, add 25.0 mL of 0.01 M EDTA and 10 mL of triethanolamine solution. Introduce 1 M sodium hydroxide solution, with stirring, until the pH of the solution is 11.6 (use a pH meter). Dilute to about 250 mL. Add about 0.05 g of the thymolphthalexone–potassium nitrate mixture; the solution acquires a very pale blue colour. Titrate with 0.01 M calcium chloride solution until the colour changes to an intense blue. If it is felt that the end point colour change is not sufficiently distinct, add a further small amount of the indicator, a known volume of 0.01 M EDTA and titrate again with 0.01 M calcium chloride.

10.82 Lead and tin in a mixture: solder

Discussion A mixture of tin(IV) and lead(II) ions may be complexed by adding an excess of standard EDTA solution, the excess EDTA being determined by titration with a standard solution of lead nitrate; the total lead-plus-tin content of the solution is thus determined. Sodium fluoride is then added and this displaces the EDTA from the tin(IV)–EDTA complex; the liberated EDTA is determined by titration with a standard lead solution.

Procedure Prepare a standard EDTA solution (0.2 M), a standard lead solution (0.01 M), a 30% aqueous solution of hexamine, and a 0.2% aqueous solution of xylenol orange. Dissolve a weighed amount (about 0.4 g) of solder in 10 mL of concentrated hydrochloric acid and 2 mL of concentrated nitric acid; gentle warming is necessary. Boil the solution gently for about 5 min to expel nitrogen oxides and chlorine, and allow to cool slightly, whereupon some lead chloride may separate. Add 25.0 mL of standard 0.2 M EDTA and boil for 1 min; the lead chloride dissolves and a clear solution is obtained. Dilute with 100 mL of deionised water, cool and dilute to 250 mL in a graduated flask. Without delay, pipette two or three 25.0 mL portions into separate conical flasks. To each flask add 15 mL hexamine solution, 110 mL deionised water, and a few drops of xylenol orange indicator. Titrate with the standard lead nitrate solution until the colour changes from yellow to red. Now add 2.0 g sodium fluoride; the solution acquires a yellow colour owing to the liberation of EDTA from its tin complex. Titrate again with the standard lead nitrate solution until a permanent (i.e. stable for 1 min) red colour is obtained. Add the titrant dropwise near the end point; a temporary pink or red colour gradually reverting to yellow signals the approach of the end point.

Complexation: determining individual anions

> **Safety.** Before carrying out any experiments in this section, pay full attention to any safety warnings and make sure you adhere to national laboratory and safety regulations.

Anions do not complex directly with EDTA, but methods can be devised for the determinatioin of appropriate anions which involve either (i) adding an excess of a solution containing a cation which reacts with the anion to be determined, and then using EDTA to measure the excess of cation added; or (ii) the anion is precipitated with a suitable cation, the precipitate is collected, dissolved in excess EDTA solution and then the excess EDTA is titrated with a standard solution of an appropriate cation. The procedure involved in the first method is self-evident but some details are given for determinations carried out by the second method.

10.83   Phosphates

Discussion   The phosphate is precipitated as $Mg(NH_4)PO_4 \cdot 6H_2O$, the precipitate is filtered off, washed, dissolved in dilute hydrochloric acid, an excess of standard EDTA solution added, the pH adjusted to 10, and the excess of EDTA titrated with standard magnesium chloride or magnesium sulphate solution using solochrome black as indicator. The initial precipitation may be carried out in the presence of a variety of metals by first adding sufficient EDTA solution (1 M) to form complexes with all the multicharged metal cations, then adding excess of magnesium sulphate solution, followed by ammonia solution. Alternatively, the cations may be removed by passing the solution through a cation exchange resin in the hydrogen form.

Procedure   Prepare a standard (0.05 M) solution of magnesium sulphate or chloride from pure magnesium, an ammonia–ammonium chloride buffer solution (pH 10) (Section 10.71), and a standard (0.05 M) solution of EDTA. Pipette 25.0 mL of the phosphate solution (~0.05 M) into a 250 mL beaker and dilute to 50 mL with deionised water; add 1 mL of concentrated hydrochloric and a few drops of methyl red indicator. Treat with an excess of 1 M magnesium sulphate solution (~2 mL), heat the solution to boiling, and add concentrated ammonia solution dropwise and with vigorous stirring until the indicator turns yellow, followed by a further 2 mL. Allow to stand for several hours or overnight. Filter the precipitate through a sintered-glass crucible (porosity G4) and wash thoroughly with $1M$ ammonia solution (about 100 mL). Rinse the beaker (in which the precipitation was made) with 25 mL of hot $1M$ hydrochloric acid and allow the liquid to percolate through the filter crucible, thus dissolving the precipitate. Wash the beaker and crucible with a further 10 mL of $1M$ hydrochloric acid and then with about 75 mL of water. To the filtrate and washings in the filter flask add 35.0 mL of $0.05M$ EDTA, neutralise the solution with $1M$ sodium hydroxide, add 4 mL of buffer solution and a few drops of solochrome black indicator. Back-titrate with standard $0.05M$ magnesium chloride until the colour changes from blue to wine red.

10.84 Sulphates

Discussion   The sulphate is precipitated as barium sulphate from acid solution, the precipitate is filtered off and dissolved in a measured excess of standard EDTA solution in the presence of aqueous ammonia. The excess of EDTA is then titrated with standard magnesium chloride solution using solochrome black as indicator.

Procedure   Prepare a standard magnesium chloride solution $(0.05M)$ and a buffer solution (pH 10); see Section 10.83. Standard EDTA $(0.05M)$ will also be required. Pipette 25.0 mL of the sulphate solution $(0.02-0.03M)$ into a 250 mL beaker, dilute to 50 mL, and adjust the pH to 1 with $2M$ hydrochloric acid; heat nearly to boiling. Add 15 mL of a nearly boiling barium chloride solution $(\sim 0.05M)$ fairly rapidly and with vigorous stirring; heat on a steam bath for 1 h. Filter with suction through a disc of filter paper (Whatman filter paper no. 42) supported upon a porcelain filter disc or a Gooch crucible, wash the precipitate thoroughly with cold water and drain. Transfer the filter-paper disc and precipitate quantitatively to the original beaker, add 35.0 mL standard $0.05M$ EDTA solution and 5 mL concentrated ammonia solution and boil gently for 15–20 min; add a further 2 mL concentrated ammonia solution after 10–15 min to facilitate the dissolution of the precipitate. Cool the resulting clear solution, add 10 mL of the buffer solution (pH = 10), a few drops of solochrome black indicator, and titrate the excess of EDTA with the standard magnesium chloride solution to a clear red colour.

Sulphate can also be determined by an exactly similar procedure. Precipitate lead sulphate from a solution containing 50% (by volume) of propan-2-ol (to reduce the solubility of the lead sulphate), separate the precipitate, dissolve in excess of standard EDTA solution, and back-titrate the excess EDTA with a standard zinc solution using solochrome black as indicator.

Other EDTA titrations

> ! **Safety.** Before carrying out any experiments in this section, pay full attention to any safety warnings and make sure you adhere to national laboratory and safety regulations.

10.85 Potentiometric titrations*

A number of metal ions were determined potentiometrically by EDTA titration using a mercury indicator electrode.[7] A major disadvantage was the interference of halide ions even in trace amounts. More recently a variety of electrodes, including copper, cadmium and

* The experimental details in this section are courtesy of Orion Research Ltd.

calcium ion selective electrodes (ISEs), have replaced mercury as the indicator electrode. The use of ion selective electrodes as indicators is governed by the stability constant of the analyte metal ion with the complexometric reagent. There are three areas of application:

(a) The determination of a chelating agent by direct titration using a metal ion solution, and an appropriate ion selective electrode as indicator. Examples include the determination of EDTA ($\sim 10^{-4} M$) using a copper ISE and titration with a copper ion solution at pH 4.75 (acetate buffer).

(b) A back-titration procedure (Section 10.63) where the analyte concentration is determined from the difference between the concentrations of the total and unreacted complexone. For example, the determination of iron(III) ($\sim$25 ppm) by adding a known excess of EDTA, followed by back titration of the unreacted complexone with a copper solution buffered at pH 4.7 (acetate) and copper as the indicator electrode.

(c) For metals that form weaker complexes than the electroactive ion, a direct titration is performed with a complexone in the presence of a solution containing equal concentrations of the electroactive ion and the complexone.

Determining water hardness

Discussion   Calcium and magnesium hardness in water are measured by potentiometric titration with EDTA. The two end points are sensed using a calcium ion selective electrode. The ratio between EDTA's conditional stability constants (Section 2.27) for calcium and magnesium is increased by the addition of 2,4-pentanedione (acetylacetone) to the water sample so that two independent end points are produced. The pH is adjusted with TRIS (tris(hydroxymethyl)aminomethane) buffer. A first-derivative method seeking two end points may be defined and run consecutively using the 'Sequences' program, a feature of the Orion 960 software.

Hard water reagent   Prepared by adding $0.2 M$ 2,4-pentanedione (2 g) and $0.4 M$ TRIS (tris(hydroxymethyl)aminomethane) (4.4 g) to 100 mL of deionised water.

Procedure   This procedure is for the Orion 960 autotitrator. Pipette 50.00 mL of the water into a beaker (200 mL). Add 5 mL of the hard water reagent and titrate with a standard EDTA ($\sim$0.05 M) using a calcium electrode together with a single-junction silver–silver chloride reference electrode. Wash the electrodes, stirrer and dispenser probe thoroughly after use.

Determining zinc in phosphate baths

Discussion   The determination of zinc may be achieved indirectly through the titration with EDTA, using a copper electrode as end point indicator in the presence of a copper–EDTA solution. During the titration with EDTA, zinc is complexed first. When there are no more free zinc ions present, the equilibrium between free copper ions and the copper–EDTA is shifted in the direction of the complexone. The copper electrode detects the copper ions liberated by this process.

Copper–EDTA solution   Mix equal volumes of $0.05 M$ copper sulphate solution and $0.05 M$ EDTA.

Procedure The procedure is for the Orion 960 autotitrator. Use either a copper ion sensitive electrode and a reference electrode, or a solid-state combination electrode. Add 0.5 mL of the zinc solution (~1 M) to 100 mL deionised water in a 200 mL beaker. Then pipette 0.05 M copper–EDTA solution (1 mL) and **with care** add 1 mL of concentrated ammonia solution. Titrate this solution with a standard EDTA solution (~0.1 M). Use the first derivative for end point measurement.

10.86 Coulometric titrations

The generation of EDTA by an electrode process has been used in the coulometric titration of metal ions. The reagent is produced by the reduction of the amine mercury(II)–EDTA chelate ($HgNH_3Y^{2-}$) at a mercury cathode, according to the equation

$$HgNH_3Y^{2-} + NH_4^+ + 2e \rightarrow Hg^0 + 2NH_3 + HY^{3-}$$

where Y is the normal abbreviation for EDTA.

Since the mercury chelate has a higher stability (formation) constant than the corresponding complexes of Zn^{2+}, Pb^{2+}, Cu^{2+} and Ca^{2+}, these ions will only complex with the EDTA when it has been released by the electrode process. The mercury–EDTA reagent is made from a stock solution containing 8.4 g mercury(II) nitrate and 9.3 g disodium EDTA dissolved in 250 mL deionised water. The stock solution (25 mL) is mixed with 0.1 M ammonium nitrate solution (75 mL). The pH is adjusted to 8.3 (pH meter) by the careful addition of concentrated ammonia solution. The titration end point is detected potentiometrically using the mercury indicator electrode.[7]

10.87 Amperometric titrations

In order to achieve a successful amperometric titration for a metal ion with EDTA, the applied potential must be selected so that the metal ion is reduced but both the metal–EDTA complex and the free ligand are not. Hence there is a decrease in current as the titrant (EDTA) is added, producing a corresponding reduction in the concentration of the uncomplexed metal ion. Beyond the equivalence point, virtually all the metal ion has complexed with the EDTA, so there is no change in current. Thus an L-shaped curve similar to Figure 10.6(a) is obtained. The following examples illustrate the use of amperometric titrations in complexation reactions.

10.88 Zinc

Zinc ions may be titrated with EDTA solution in a strongly alkaline medium (produced with cyclohexylamine) at an applied potential of −1.4 V vs SCE.

Reagents *Zinc solution* An 'unknown' solution of zinc ions with concentration ~0.001 M (about 60 mg L^{-1} Zn^{2+}).

EDTA solution A standard 0.01 M EDTA solution.

Procedure Place 20.00 mL of the zinc ion solution in the titration flask (Figure 10.7(a)) and add 1.0 mL of pure cyclohexylamine. Set the applied potential at −1.4 V vs SCE. Deaerate the solution and titrate with the standard EDTA using a semimicro burette. Plot the titration graph and evaluate the concentration of zinc in the solution:

$$1 \text{ mL } 0.01 \text{ } M \text{ EDTA} \equiv 0.6538 \text{ mg Zn}$$

10.89   Bismuth

Bismuth ion solutions adjusted to a pH between 1 and 2 may be titrated with standard EDTA in a $0.4\,M$ sodium citrate medium at an applied potential of $-0.2\,V$ vs SCE. At this low pH, nearly all divalent ions do not interfere.

Reagents   *Bismuth solution*   Prepare an approximately $0.01\,M$ standard bismuth solution by dissolving about $2.3\,g$ of accurately weighed pure bismuth oxide in a little $1:1$ nitric acid and diluting to $1\,L$ with deionised water in a graduated flask. Dilute $25.0\,mL$ of this solution to $250\,mL$, adding sufficient sodium citrate $(20\,g)$ to render the solution $0.4\,M$ with respect to this compound.

EDTA solution   Prepare a standard $0.1\,M$ EDTA solution.

Procedure   Pipette $25.0\,mL$ of the bismuth ion solution into the titration flask (Figure 10.7(a)), adjust the pH to 2 (pH meter) with aqueous ammonia solution, add 5 drops of 1% gelatin solution. Set the applied potential at $-0.20\,V$ vs SCE and titrate with the standard EDTA using a semimicro burette. Magnetic stirring is desirable. Plot the titration graph. Compare the result obtained for the concentration of the bismuth solution with the value calculated from the weight of bismuth oxide used in the preparation of the standard solution:

$$1\,mL\ \ 0.01\,M\ \text{EDTA} \equiv 0.002\,090\,g\ \text{Bi}$$

Spectrophotometric titrations

Examples of spectrophotometric titrations with EDTA include the following.

10.90   Copper(II)

Discussion   The titration of a copper ion solution with EDTA may be carried out photometrically at a wavelength of 745 nm. At this wavelength the copper–EDTA complex has a considerably greater molar absorption coefficient than the copper solution alone. The pH of the solution should be about 2.4.

The effect of different ions upon the titration is similar to that given under iron(III) (Section 10.91). Iron(III) interferes (small amounts may be precipitated with sodium fluoride solution): tin(IV) should be masked with 20% aqueous tartaric acid solution. The procedure may be employed for the determination of copper in brass, bronze and bell metal without any previous separations except the removal of insoluble lead sulphate when present.

Reagents   *Copper ion solution 0.04 M*   Wash analytical grade copper with acetone or diethyl ether to remove any surface grease and dry at $100\,°C$. Weigh accurately about $1.25\,g$ of the pure copper, dissolve it in $5\,mL$ of concentrated nitric acid, and dilute to $1\,L$ in a graduated flask.

Other reagents $0.10\,M$ EDTA solution and pH 2.2 buffer solution (Section 10.91).

Procedure   Charge the titration cell (Figure 10.10) with $10.00\,mL$ of the copper ion solution, $20\,mL$ of the acetate buffer (pH = 2.2), and about $120\,mL$ of water. Position the cell in the spectrophotometer and set the wavelength scale at 745 nm. Adjust the slit width so the reading on the absorbance scale is zero. Stir the solution and titrate with the standard EDTA; record the absorbance every $0.50\,mL$ until the value is about 0.20 and subsequently every 0.20 mL. Continue the titration until about 1.0 mL after the end point; the end point occurs when the absorbance readings become fairly constant. Plot absorbance against the volume of titrant added; the intersection of the two straight lines is the end point (Figure 10.9(c)). Calculate the concentration of copper ion $(mg\,mL^{-1})$ in the solution and compare this with the true value.

10.91   Iron(III)

Discussion   Salicylic acid and iron(III) ions form a deep-coloured complex with a maximum absorption at about 525 nm; this complex is used as the basis for the photometric titration of iron(III) ion with standard EDTA solution. At a pH of ~2.4 the EDTA–iron complex is much more stable (higher stability constant) than the iron–salicylic acid complex. In the titration of an iron–salicylic acid solution with EDTA the iron–salicylic acid colour will therefore gradually disappear as the end point is approached. The spectrophotometric end point at 525 nm is very sharp. Considerable amounts of zinc, cadmium, tin(IV), manganese(II), chromium(III) and smaller amounts of aluminium cause little or no interference at pH 2.4; the main interferences are lead(II), bismuth, cobalt(II), nickel and copper(II).

Reagents   *EDTA solution, 0.10 M*   Standardise accurately. See Section 10.66.

Iron(III) solution, 0.05 M   Dissolve about 12.0 g, accurately weighed, of ammonium iron(III) sulphate in water to which a little dilute sulphuric acid is added, and dilute the resulting solution to 500 mL in a graduated flask. Standardise the solution with standard EDTA using variamine blue B as indicator.

Sodium ethanoate–ethanoic acid buffer   Prepare a solution which is 0.2 M in sodium acetate and 0.8 M in ethanoic acid. The pH is 4.0.

Sodium ethanoate–hydrochloric acid buffer   Add 1 M hydrochloric acid to 350 mL of 1 M sodium ethanoate until the pH of the mixture is 2.2 (pH meter).

Salicylic acid solution   Prepare a 6% solution of salicylic acid in acetone.

Procedure   Transfer 10.00 mL of the iron(III) solution to the titration cell (Figure 10.10), add about 10 mL of the buffer solution of pH = 4.0 and about 120 mL of water; the pH of the resulting solution should be 1.7–2.3. Insert the titration cell into the spectrophotometer; immerse the stirrer and the tip of the 5 mL microburette (graduated in 0.02 mL) in the solution. Switch on the tungsten lamp and allow the spectrophotometer to warm up for about 20 min. Stir the solution. Add about 4.0 mL of the standard EDTA (note the volume accurately). Set the wavelength at 525 nm, and adjust the slit width of the instrument so the reading on the absorbance scale is 0.2–0.3. Now add 1.0 mL of the salicylic acid solution; the absorbance immediately increases to a very large value (> 2). Continue the stirring. Add the EDTA solution slowly from the microburette until the absorbance approaches 1.8; record the volume of titrant. Introduce the EDTA solution in 0.05 mL aliquots and record the absorbance after each addition. Continue the titration until at least four readings are taken beyond the end point (fairly constant absorbance). Plot absorbance against volume of titrant added; the intersection of the two straight lines (Figure 10.9(a)) gives the true end point. Calculate the concentration of iron(III) $(mg\,mL^{-1})$ in the solution and compare this with the true value.

Iron(III) in the presence of aluminium   Iron(III) (concentration ~50 mg per 100 mL) can be determined in the presence of up to twice the amount of aluminium by photometric titration with EDTA in the presence of 5-sulphosalicylic acid (2% aqueous solution) as indicator at pH 1.0 at a wavelength of 510 nm. The pH of a strongly acidic solution may be adjusted to the desired value with a concentrated solution of sodium ethanoate, about 8–10 drops of the indicator solution are required. The spectrophotometric titration curve is of the form shown in Figure 10.9(a).

Precipitation titrations

> ! **Safety.** Before carrying out any experiments in this section, pay full attention to any safety warnings and make sure you adhere to national laboratory and safety regulations.

10.92   Precipitation reactions

The most important precipitation processes in titrimetric analysis use silver nitrate as the reagent (argentimetric processes), so this section concentrates on silver nitrate. Consider the changes in ionic concentration which occur during the titration of 100 mL of 0.1 M sodium chloride with 0.1 M silver nitrate. The solubility product of silver chloride at laboratory temperature is 1.2×10^{-10}. The initial concentration of chloride ions $[Cl^-]$ is 0.1 $mol\,L^{-1}$, or $pCl^- = 1$ (Section 2.17). When 50 mL of 0.1 M silver nitrate have been added, 50 mL of 0.1 M sodium chloride remain in a total volume of 150 mL; thus $[Cl^-] = 50 \times 0.1/150 = 3.33 \times 10^{-2}$, or $pCl^- = 1.48$. With 90 mL of silver nitrate solution $[Cl^-] = 10 \times 0.1/190 = 5.3 \times 10^{-3}$, or $pCl^- = 2.28$.

Now

$$a_{Ag^+}\,a_{Cl^-} \approx [Ag^+][Cl^-] = 1.2 \times 10^{-10} = K_{sol(AgCl)}$$

or

$$pAg^+ + pCl^- = 9.92 = pAgCl$$

In the last calculation, $pCl^- = 1.48$, hence $pAg^+ = 9.92 - 1.48 = 8.44$. In this manner, the various concentrations of chloride and silver ions can be computed up to the equivalence point. At the equivalence point

$$Ag^+ = Cl^- = K_{sol(AgCl)}^{1/2}$$

$$pAg^+ = pCl^- = \tfrac{1}{2}pAgCl = 9.92/2 = 4.96$$

and a saturated solution of silver chloride with no excess of silver or chloride ions is present.

Table 10.9 *Titration of 100 mL of 0.1 M NaCl and 100 mL of 0.1 M KI respectively with 0.1 M AgNO$_3$ (K$_{sol(AgCl)}$ = 1.2 × 10^{-10}, K$_{sol(AgI)}$ = 1.7 × 10^{-16})*

Vol. of 0.1 M AgNO$_3$ (mL)	Titration of chloride		Titration of iodide	
	pCl$^-$	pAg$^+$	pI$^-$	pAg$^+$
0	1.0	–	1.0	–
50	1.5	8.4	1.5	14.3
90	2.3	7.6	2.3	13.5
95	2.6	7.3	2.6	13.2
98	3.0	6.9	3.0	12.8
99	3.3	6.6	3.3	12.5
99.5	3.7	6.2	3.7	12.1
99.8	4.0	5.9	4.0	11.8
99.9	4.3	5.6	4.3	11.5
100.0	5.0	5.0	7.9	7.9
100.1	5.6	4.3	11.5	4.3
100.2	5.9	4.0	11.8	4.0
100.5	6.3	3.6	12.2	3.6
101	6.6	3.3	12.5	3.3
102	6.9	3.0	12.8	3.0
105	7.3	2.6	13.2	2.6
110	7.6	2.3	13.5	2.4

With 100.1 mL of silver nitrate solution, [Ag$^+$] = 0.1 × 0.1/200.1 = 5 × 10^{-5}, or pAg$^+$ = 4.30; pCl$^-$ = pAgCl – pAg$^+$ = 9.92 – 4.30 = 5.62.*

The values calculated in this way up to the addition of 110 mL of 0.1 M silver nitrate are collected in Table 10.9. Similar values for the titration of 100 mL of 0.1 M potassium iodide with 0.1 M silver nitrate are included in the same table (K$_{sol(AgI)}$ = 1.7 × 10^{-16}).

The silver ion exponents in the neighbourhood of the equivalence point (say between 99.8 and 100.2 mL) reveal there is a marked change in the silver ion concentration, and the change is more pronounced for silver iodide than for silver chloride, since the solubility product of the silver chloride is about 10^6 larger than the solubility of the silver iodide. This is shown more clearly by the titration curve in Figure 10.20 which represents the change of pAg$^+$ in the range between 10% before and 10% after the stoichiometric point in the titration of 0.1 M chloride and 0.1 M iodide with 0.1 M silver nitrate. An almost identical curve is obtained by potentiometric titration using a silver electrode; the pAg$^+$ values may be calculated from the e.m.f. figures as in the calculation of pH.

* This is not strictly true, since the dissolved silver chloride will contribute silver and chloride ions to the solution; the actual concentration of ions is ~1 × 10^{-5} g L^{-1}. If the excess of silver ions added is greater than 10 times this value, i.e. > 10K$_{sol(AgCl)}^{1/2}$ the error introduced by neglecting the ionic concentration produced by the dissolved salt may be taken as negligible for the purpose of the ensuing discussion.

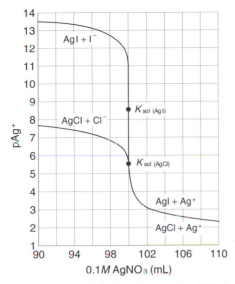

Figure 10.20   Calculated curves for 100 mL of 0.1 M NaCl and 100 mL of 0.1 M KI titrated with 0.1 M AgNO$_3$

10.93   Determining end points in precipitation reactions

Formation of a coloured precipitate

This may be illustrated by the Mohr procedure for the determination of chloride and bromide. In the titration of a neutral solution of say chloride ions with silver nitrate solution, a small quantity of potassium chromate solution is added to serve as indicator. At the end point, the chromate ions combine with silver ions to form silver chromate, red in colour and sparingly soluble.

The theory of the process is as follows. This is a case of fractional precipitation (Section 2.17), the two sparingly soluble salts being silver chloride ($K_{sol} = 1.2 \times 10^{-10}$) and silver chromate ($K_{sol} = 1.7 \times 10^{-12}$). It is best studied by considering an example, here the titration of 0.1 M sodium chloride with 0.1 M silver nitrate in the presence of a few millilitres of dilute potassium chromate solution. Silver chloride is the less soluble salt and the initial chloride concentration is high, hence silver chloride will be precipitated. At the first point where red silver chromate is just precipitated, both salts will be in equilibrium with the solution. Hence

$$[Ag^+][Cl^-] = K_{sol(AgCl)} = 1.2 \times 10^{-10}$$

$$[Ag^+]^2[CrO_4^{2-}] = K_{sol(Ag_2CrO_4)} = 1.7 \times 10^{-12}$$

$$[Ag^+] = \frac{K_{sol(AgCl)}}{[Cl^-]} = \left(\frac{K_{sol(Ag_2CrO_4)}}{[CrO_4^{2-}]}\right)^{1/2}$$

$$\frac{[Cl^-]}{[CrO_4^{2-}]^{1/2}} = \frac{K_{sol(AgCl)}}{K_{sol(Ag_2CrO_4)}^{1/2}} = \frac{1.2 \times 10^{-10}}{(1.7 \times 10^{-12})^{1/2}} = 9.2 \times 10^{-5}$$

At the equivalence point $[Cl^-] = K_{sol(AgCl)}^{1/2} = 1.1 \times 10^{-5}$. If silver chromate is to precipitate at this chloride ion concentration, then

$$[CrO_4^{2-}] = \left(\frac{[Cl^-]}{9.2 \times 10^{-5}}\right)^2 = \left(\frac{1.1 \times 10^{-5}}{9.2 \times 10^{-5}}\right)^2 = 1.4 \times 10^{-2}$$

or the potassium chromate solution should be $0.014\,M$. Note that a slight excess of silver nitrate solution must be added before the red colour of silver chromate is visible. In practice, a more dilute solution ($0.003–0.005\,M$) of potassium chromate is generally used, since a chromate solution of concentration $0.01–0.02\,M$ imparts a distinct deep orange colour to the solution, which renders the detection of the first appearance of silver chromate somewhat difficult. The error introduced can be readily calculated, for if $[CrO_4^{2-}] = 0.003$ say, silver chromate will be precipitated when

$$[Ag^+] = \left(\frac{K_{sol(Ag_2CrO_4)}}{[CrO_4^{2-}]}\right)^{1/2} = \left(\frac{1.7 \times 10^{-12}}{3 \times 10^{-3}}\right)^{1/2} = 2.4 \times 10^{-5}$$

If the theoretical concentration of indicator is used:

$$[Ag^+] = \left(\frac{1.7 \times 10^{-12}}{1.4 \times 10^{-2}}\right)^{1/2} = 1.1 \times 10^{-5}$$

The difference is $1.3 \times 10^{-5}\,mol\,L^{-1}$. If the volume of the solution at the equivalence point is $150\,mL$, then this corresponds to $1.3 \times 10^{-5} \times 150 \times 10^4/1000 = 0.02\,mL$ of $0.1\,M$ silver nitrate. This is the theoretical titration error, and is therefore negligible. In practice another factor must be considered – the small excess of silver nitrate solution which must be added before the eye can detect the colour change in the solution; this is of the order of one drop or $\sim 0.05\,mL$ of $0.1\,M$ silver nitrate.

The titration error will increase with increasing dilution of the solution being titrated and is quite appreciable ($\sim 0.4\%$) in dilute, say $0.01\,M$, solutions when the chromate concentration is of the order $0.003–0.005\,M$. This is most simply allowed for by means of an indicator blank determination, e.g. by measuring the volume of standard silver nitrate solution required to give a perceptible coloration when added to distilled water containing the same quantity of indicator as is employed in the titration. This volume is subtracted from the volume of standard solution used.

The titration should be carried out in neutral solution or in very faintly alkaline solution, i.e. within the pH range 6.5 to 9. In acid solution, the following reaction occurs:

$$2CrO_4^{2-} + 2H^+ \rightleftharpoons 2HCrO_4^- \rightleftharpoons Cr_2O_7^{2-} + H_2O$$

$HCrO_4^-$ is a weak acid, so the chromate ion concentration is reduced and the solubility product of silver chromate may not be exceeded. In markedly alkaline solutions, silver hydroxide ($K_s = 2.3 \times 10^{-8}$) might be precipitated. A simple method of making an acid solution neutral is to add an excess of pure calcium carbonate or sodium hydrogencarbonate. An alkaline solution may be acidified with ethanoic acid and then a slight excess of calcium carbonate is added. The solubility product of silver chromate increases with rising temperature; the titration should therefore be performed at room temperature. By using a mixture of potassium chromate and potassium dichromate in proportions such as to give a neutral solution, the danger of accidentally raising the pH of an unbuffered solution beyond the acceptable limits is minimised; the mixed indicator has a buffering effect and adjusts the pH of the solution to 7.0 ± 0.1. In the presence of ammonium salts, the pH must not exceed 7.2 because of the effect of appreciable concentrations of ammonia upon the solubility of silver salts. Titration of iodide and thiocyanate is not successful because silver iodide and silver thiocyanate adsorb chromate ions so strongly that a false and somewhat indistinct end point is obtained. The errors for $0.1\ M$ and $0.01\ M$ bromide may be calculated as 0.04% and 0.4% respectively.

Formation of a soluble coloured compound

This procedure is exemplified by Volhard's method for the titration of silver in the presence of free nitric acid with standard potassium thiocyanate or ammonium thiocyanate solution. The indicator is a solution of iron(III) nitrate or iron(III) ammonium sulphate. The addition of thiocyanate solution produces first a precipitate of silver thiocyanate ($K_s = 7.1 \times 10^{-13}$):

$$Ag^+ + SCN^- \rightleftharpoons AgSCN$$

When this reaction is complete, the slightest excess of thiocyanate produces a reddish brown coloration, due to the formation of a complex ion:*

$$Fe^{3+} + SCN^- \rightleftharpoons [FeSCN]^{2+}$$

This method may be applied to the determination of chlorides, bromides and iodides in acid solution. Excess of standard silver nitrate solution is added, and the excess is back-titrated with standard thiocyanate solution. For the chloride estimation, we have the following two equilibria during the titration of excess of silver ions:

$$Ag^+ + Cl^- \rightleftharpoons AgCl$$

$$Ag^+ + SCN^- \rightleftharpoons AgSCN$$

The two sparingly soluble salts will be in equilibrium with the solution, hence

$$\frac{[Cl^-]}{[SCN^-]} = \frac{K_{sol(AgCl)}}{K_{sol(AgSCN)}} = \frac{1.2 \times 10^{-10}}{7.1 \times 10^{-13}} = 169$$

When the excess of silver has reacted, the thiocyanate may react with the silver chloride, since silver thiocyanate is the less soluble salt, until the ratio $(Cl^-]/[SCN^-]$ in the solution is 169:

$$AgCl + SCN^- \rightleftharpoons AgSCN + Cl^-$$

This will take place before reaction occurs with the iron(III) ions in the solution, so there will be a considerable titration error. It is therefore absolutely necessary to prevent the reaction between the thiocyanate and the silver chloride. This may be achieved in several ways; the first one is probably the most reliable:

1.   The silver chloride is filtered off before back-titrating. Since at this stage the precipitate will be contaminated with adsorbed silver ions, the suspension should be boiled for a few minutes to coagulate the silver chloride and thus remove most of the adsorbed silver ions from its surface before filtration. The cold filtrate is titrated.
2.   After the addition of silver nitrate, potassium nitrate is added as coagulant, the suspension is boiled for about 3 min, cooled and then titrated immediately. Desorption of silver ions occurs and, on cooling, readsorption is largely prevented by the presence of potassium nitrate.

---

* This is the complex formed when the ratio of thiocyanate ion to iron(III) ion is low; higher complexes such as $[Fe(SCN)_2]^+$ are important only at higher concentrations of thiocyanate ion.

3. An immiscible liquid is added to 'coat' the silver chloride particles and thereby protect them from interaction with the thiocyanate. The most successful liquid is nitrobenzene (~1.0 mL for each 50 mg of chloride); the suspension is well shaken to coagulate the precipitate before back titration. With bromides we have

$$\frac{[Br^-]}{[SCN^-]} = \frac{K_{sol(AgBr)}}{K_{sol(AgSCN)}} = \frac{3.5 \times 10^{-13}}{7.1 \times 10^{-13}} = 0.5$$

The titration error is small, and no difficulties arise in determining the end point. Silver iodide ($K_s = 1.7 \times 10^{-16}$) is less soluble than the bromide; the titration error is negligible, but the iron(III) indicator should not be added until excess silver is present, since the dissolved iodide reacts with Fe^{3+} ions:

$$2Fe^{3+} + 2I^- \rightleftharpoons 2Fe^{2+} + I_2$$

Use of adsorption indicators

Adsorption indicators work in the following way. At the equivalence point, they are adsorbed by the precipitate and become changed, producing a new colour. Stringent conditions have to be met for an adsorption indicator to function properly. The precipitate should separate ideally in the colloidal condition; coagulation should be avoided as far as possible. The titration solution must be at a suitable pH for the indicator to be predominantly in the ionic form. Finally, titrations should be carried out in subdued light. As a result, the application of adsorption indicators is somewhat limited. Furthermore, experience allied with skill is often vital to ensure a satisfactory end point. A small selection of adsorption indicators together with their applications and end point colour changes are given in Table 10.10.

Table 10.10 *Selected adsorption indicators*

Indicator	Use	Colour change at end point[a]	Experimental conditions
Fluorescein	Cl^-, Br^-, I^- with Ag^+	Yellow green → pink	Neutral or weakly basic solution
Dichlorofluorescein	Cl^-, Br^-, with Ag^+	Yellow green → red	pH range 4.4 to 7
Tetrabromofluorescein (eosin)	Br^-, I^- with Ag^+	Pink → red violet	Best in ethanoic acid solution
Tartrazine	Ag^+ with I^- or SCN^-; $I^- + Cl^-$; excess Ag^+, back titration with I^-	Colourless solution → green solution	Sharp colour change in $I^- + Cl^-$, back titration

[a] Unless otherwise stated, the colour change occurs as the indicator passes from solution to precipitate.

10.94 Standardising silver nitrate solution

Sodium chloride has a relative molecular mass of 58.44. A 0.1000 M solution is prepared by weighing out 2.922 g of the pure dry salt (Section 10.92) and dissolving it in 500 mL of water in a graduated flask. Alternatively about 2.9 g of the pure salt is accurately weighed out, dissolved in 500 mL of water in a graduated flask and the molar concentration calculated from the weight of sodium chloride employed.

With potassium chromate as indicator

The Mohr titration See Section 10.93 for the detailed theory of the titration. Prepare the indicator solution by dissolving 5 g potassium chromate in 100 mL of water. The final volume of the solution in the titration is 50–100 mL, and 1 mL of the indicator solution is used, so the indicator concentration in the actual titration is 0.005–0.0025 M. Alternatively, and preferably, dissolve 4.2 g potassium chromate and 0.7 g potassium dichromate in 100 mL of water; use 1 mL of indicator solution for each 50 mL of the final volume of the test solution.

Pipette 25 mL of the standard 0.1 M sodium chloride into a 250 mL conical flask resting on a white tile, and add 1 mL of the indicator solution (preferably with a 1 mL pipette). Add the silver nitrate solution slowly from a burette, swirling the liquid constantly, until the red colour formed by the addition of each drop begins to disappear more slowly; this is an indication that most of the chloride has been precipitated. Continue the addition dropwise until a faint but distinct change in colour occurs. This **faint** reddish brown colour should persist after brisk shaking. If the end point is overstepped (production of a deep reddish brown colour), add more of the chloride solution and titrate again. Determine the indicator blank correction by adding 1 mL of the indicator to a volume of water equal to the final volume in the titration, and then 0.01 M silver nitrate solution until the colour of the blank matches the colour of the titrated solution. The indicator blank correction, which should not amount to more than 0.03–0.10 mL of silver nitrate, is deducted from the volume of silver nitrate used in the titration. Repeat the titration with two further 25 mL portions of the sodium chloride solution. The various titrations should agree within 0.1 mL.

With an adsorption indicator

Discussion Both fluorescein and dichlorofluorescein are suitable for the titration of chlorides. In both cases the end point is reached when the white precipitate in the greenish yellow solution suddenly assumes a pronounced reddish tint. The change may be reversed by adding chloride. With fluorescein the solution must be neutral or only faintly acidic with ethanoic acid; acid solutions should be treated with a slight excess of sodium ethanoate. The chloride solution should be diluted to about 0.01–0.05 M, for if it is more concentrated the precipitate coagulates too soon and interferes. Fluorescein cannot be used in solutions more dilute than 0.005 M. With more dilute solutions, use dichlorofluorescein, which possesses other advantages over fluorescein. Dichlorofluorescein gives good results in very dilute solutions (e.g. for drinking water) and is applicable in the presence of ethanoic acid and in weakly acid solutions. For this reason the chlorides of copper, nickel, manganese, zinc, aluminium and magnesium, which cannot be titrated according to Mohr's method, can be determined by a direct titration when dichlorofluorescein is used as indicator. For the reverse titration (chloride into silver nitrate), tartrazine (four drops of a 0.2% solution per 100 mL) is a good indicator. At the end point, the almost colourless liquid assumes a bluey-green colour.

Indicator solutions   *Fluorescein*   Dissolve 0.2 g fluorescein in 100 mL of 70% ethanol, or dissolve 0.2 g sodium fluoresceinate in 100 mL of water.

Dichlorofluorescein   Dissolve 0.1 g dichlorofluorescein in 100 mL of 60–70% ethanol, or dissolve 0.1 g sodium dichlorofluoresceinate in 100 mL of water.

Procedure   Pipette 25 mL of the standard 0.1 M sodium chloride into a 250 mL conical flask. Add 10 drops of either fluorescein or dichlorofluorescein indicator, and titrate with the silver nitrate solution in a diffuse light, while rotating the flask constantly. As the end point is approached, the silver chloride coagulates appreciably, and the local development of a pink colour upon the addition of a drop of the silver nitrate solution becomes more and more pronounced. Continue the addition of the silver nitrate solution until the precipitate suddenly assumes a pronounced pink or red colour. Repeat the titration with two other 25 mL portions of the chloride solution. Individual titrations should agree within 0.1 mL. Calculate the molar concentration of the silver nitrate solution.

10.95   Chlorides and bromides

Chlorides   Either the Mohr titration or the adsorption indicator method may be used for the determination of chlorides in neutral solution by titration with standard 0.1 M silver nitrate. If the solution is acid, neutralisation may be effected with chloride-free calcium carbonate, sodium tetraborate or sodium hydrogencarbonate. Mineral acid may also be removed by neutralising most of the acid with ammonia solution and then adding an excess of ammonium ethanoate. Titration of the neutral solution, prepared with calcium carbonate, by the adsorption indicator method is rendered easier by the addition of 5 mL of 2% dextrin solution; this offsets the coagulating effect of the calcium ion. If the solution is basic, it may be neutralised with chloride-free nitric acid, using phenolphthalein as indicator.

Bromides   Bromides are determined rather like chlorides. The Mohr titration can be used, and the most suitable adsorption indicator is eosin, which can be used in dilute solutions and even in the presence of 0.1 M nitric acid, but ethanoic acid solutions are generally preferred. Fluorescein may be used but is subject to the same limitations as experienced with chlorides (Section 10.94). With eosin indicator, the silver bromide flocculates approximately 1% before the equivalence point and the local development of a red colour becomes more and more pronounced with the addition of silver nitrate solution; at the end point the precipitate assumes a magenta colour.

   The indicator is prepared by dissolving 0.1 g eosin in 100 mL of 70% ethanol, or by dissolving 0.1 g of the sodium salt in 100 mL of water. For the reverse titration (bromide into silver nitrate), rhodamine 6G (10 drops of a 0.05% aqueous solution) is an excellent indicator. The solution is best adjusted to 0.05 M with respect to silver ion. The precipitate acquires a violet colour at the end point.

Thiocyanates   Thiocyanates may also be determined using adsorption indicators in exactly similar manner to chlorides and bromides, but an iron(III) salt indicator is usually preferred (Section 10.97).

10.96   Iodides

Discussion   The Mohr method cannot be applied to the titration of iodides (or thiocyanates), because of adsorption phenomena and the difficulty of distinguishing the colour change of the potassium chromate. Eosin is a suitable adsorption indicator.

10.97   Preparing thiocyanate solutions: Volhard's method

Discussion   Volhard's original method for the determination of silver in dilute nitric acid solution by titration with standard thiocyanate solution in the presence of an iron(III) salt as indicator has proved of great value not only for silver determinations, but also in numerous indirect analyses. The theory of the Volhard process has been given in Section 10.93. Note that the concentration of the nitric acid should be 0.5–1.5 M (strong nitric acid retards the formation of the thiocyanato–iron(III) complex ([FeSCN]$^{2+}$) and at a temperature not exceeding 25 °C (higher temperatures tend to bleach the colour of the indicator). The solutions must be free from nitrous acid, which gives a red colour with thiocyanic acid, and may be mistaken for iron(III) thiocyanate. Pure nitric acid is prepared by diluting the usual pure acid with about one-quarter of its volume of water and boiling until perfectly colourless; this eliminates any lower oxides of nitrogen which may be present.

The method may be applied to those anions (e.g. chloride, bromide and iodide) which are completely precipitated by silver and are sparingly soluble in dilute nitric acid. Excess of standard silver nitrate solution is added to the solution containing free nitric acid, and the residual silver nitrate solution is titrated with standard thiocyanate solution. This is sometimes known as **the residual process**. Anions whose silver salts are slightly soluble in water, but which are soluble in nitric acid, such as phosphate, arsenate, chromate, sulphide and oxalate, may be precipitated in neutral solution with an excess of standard silver nitrate solution. The precipitate is filtered off, thoroughly washed, dissolved in dilute nitric acid, and the silver titrated with thiocyanate solution. Alternatively, the residual silver nitrate in the filtrate from the precipitation may be determined with thiocyanate solution after acidification with dilute nitric acid.

Both ammonium and potassium thiocyanates are usually available as deliquescent solids; the analytical grade products are, however, free from chlorides and other interfering substances. An approximately 0.1 M solution is therefore first prepared, and this is standardised by titration against standard 0.1 M silver nitrate.

Procedure   *Preparation*   Weigh out about 8.5 g ammonium thiocyanate, or 10.5 g potassium thiocyanate, and dissolve it in 1 L of water in a graduated flask. Shake well.

Standardisation   Use 0.1 M silver nitrate, which has been prepared and standardised as described in Section 10.94. The iron(III) indicator solution consists of a cold, saturated solution of ammonium iron(III) sulphate in water (about 40%) to which a few drops of 6 M nitric acid have been added. One millilitre of this solution is employed for each titration. Pipette 25 mL of the standard 0.1 M silver nitrate into a 250 mL conical flask, add 5 mL of 6 M nitric acid and 1 mL of the iron(III) indicator solution. Run in the potassium or ammonium thiocyanate solution from a burette. At first a white precipitate is produced, giving the liquid a milky appearance, and as each drop of thiocyanate falls in, it produces a reddish brown cloud, which quickly disappears on shaking. As the end point approaches, the precipitate becomes flocculent and settles easily; finally one drop of the thiocyanate solution produces a faint brown colour, which no longer disappears upon shaking. This is the end point. The indicator blank amounts to 0.01 mL of 0.1 M silver nitrate. It is essential to shake vigorously during the titration in order to obtain correct results.* The standard solution thus prepared is stable for a very long period if evaporation is prevented.

Tartrazine as indicator   Satisfactory results may be obtained by the use of tartrazine as indicator. Proceed as above, but add 4 drops of tartrazine (0.5% aqueous solution) instead of the iron(III) indicator. The precipitate will appear pale yellow during the titration, but the supernatant liquid (best viewed by placing the eye at the level of the liquid and looking through it) is colourless. At the end point, the supernatant liquid assumes a bright lemon-yellow colour. The titration is sharp to one drop of 0.1 M thiocyanate solution.

10.98   Silver in a silver alloy

Procedure   A commercial silver alloy in the form of wire or foil is suitable for this determination. Clean the alloy with emery cloth and weigh it accurately. Place it in a 250 mL conical flask, add 5 mL water and 10 mL concentrated nitric acid; place a funnel in the mouth of the flask to avoid mechanical loss. Warm the flask gently until the alloy has dissolved. Add a little water and boil for 5 min in order to expel oxides of nitrogen. Transfer the cold solution quantitatively to a 100 mL graduated flask and make up to the mark with distilled water. Titrate 25 mL portions of the solution with standard 0.1 M thiocyanate.

$$1 \text{ mol KSCN} \equiv 1 \text{ mol Ag}^+$$

Note   The presence of metals whose salts are colourless does not influence the accuracy of the determination, except that mercury and palladium must be absent since their thiocyanates are insoluble. Salts of metals (e.g. nickel and cobalt) which are coloured must not be present to any considerable extent. Copper does not interfere, provided it does not form more than about 40% of the alloy.

10.99   Chlorides by Volhard's method

Discussion   The chloride solution is treated with excess of standard silver nitrate solution, and the residual silver nitrate determined by titration with standard thiocyanate solution. Now silver chloride is more soluble than silver thiocyanate, and would react with the thiocyanate:

$$\text{AgCl (solid)} + \text{SCN}^- \rightleftharpoons \text{AgSCN (solid)} + \text{Cl}^-$$

It is therefore necessary to remove the silver chloride by filtration. The filtration may be avoided by the addition of a little nitrobenzene (about 1 mL for each 0.05 g of chloride); the silver chloride particles are probably surrounded by a film of nitrobenzene. Another method, applicable to chlorides, in which filtration of the silver chloride is unnecessary, is to employ tartrazine as indicator (Section 10.97).

---

* The freshly precipitated silver thiocyanate adsorbs silver ions, thereby causing a false end point, but this disappears with vigorous shaking.

Procedure A   This determines the HCl content of concentrated hydrochloric acid. Ordinary concentrated hydrochloric acid is usually 10–11 M, and must be diluted first. Measure out accurately 10 mL of the concentrated acid from a burette into a 1 L graduated flask and make up to the mark with distilled water. Shake well. Pipette 25 mL into a 250 mL conical flask, add 5 mL 6 M nitric acid and then add 30 mL standard 0.1 M silver nitrate (or sufficient to give 2–5 mL excess). Shake to coagulate the precipitate,* filter through a quantitative filter paper (or through a porous porcelain or sintered-glass crucible), and wash thoroughly with very dilute nitric acid (1 : 100). Add 1 mL of the iron(III) indicator solution to the combined filtrate and washings, and titrate the residual silver nitrate with standard 0.1 M thiocyanate. Calculate the volume of standard 0.1 M silver nitrate that has reacted with the hydrochloric acid, and use it to obtain the percentage of HCl in the sample.

Procedure B   Pipette 25 mL of the diluted solution into a 250 mL conical flask containing 5 mL 6 M nitric acid. Add a slight excess of standard 0.1 M silver nitrate (about 30 mL in all) from a burette. Then add 2–3 mL pure nitrobenzene and 1 mL of the iron(III) indicator, and shake vigorously to coagulate the precipitate. Titrate the residual silver nitrate with standard 0.1 M thiocyanate until a permanent faint reddish brown coloration appears. From the volume of silver nitrate solution added, subtract the volume of silver nitrate solution that is equivalent to the volume of standard thiocyanate required. Then calculate the percentage of HCl in the sample.

Procedure C   Pipette 25 mL of the diluted solution into a 250 mL conical flask containing 5 mL of 6 M nitric acid, add a slight excess of 0.1 M silver nitrate (30–35 mL) from a burette, and four drops of tartrazine indicator (0.5% aqueous solution). Shake the suspension for about a minute in order to ensure the indicator is adsorbed on the precipitate as far as possible. Titrate the residual silver nitrate with standard 0.1 M ammonium or potassium thiocyanate with swirling of the suspension until the very pale yellow supernatant liquid (viewed with the eye at the level of the liquid) assumes a rich lemon-yellow colour.

Bromides   Bromides can also be determined by the Volhard method, but as silver bromide is less soluble than silver thiocyanate, it is not necessary to filter off the silver bromide (compare chloride). The bromide solution is acidified with dilute nitric acid, an excess of standard 0.1 M silver nitrate added, the mixture thoroughly shaken, and the residual silver nitrate determined with standard 0.1 M ammonium or potassium thiocyanate, using ammonium iron(III) sulphate as indicator.

Iodides   Iodides can also be determined by this method, and in this case too there is no need to filter off the silver halide, since silver iodide is very much less soluble than silver thiocyanate. In this determination the iodide solution must be very dilute in order to reduce adsorption effects. The dilute iodide solution (~300 mL), acidified with dilute nitric acid, is treated very slowly and with vigorous stirring or shaking with standard 0.1 M silver nitrate until the yellow precipitate coagulates and the supernatant liquid appears colourless. Silver nitrate is then present in excess. One millilitre of iron(III) indicator solution is added, and the residual silver nitrate is titrated with standard 0.1 M ammonium or potassium thiocyanate.

---

* It is better to boil the suspension for a few minutes to coagulate the silver chloride and thus remove most of the adsorbed silver ions from its surface before filtration.

10.100   Fluoride: Volhard titration of lead chlorofluoride

Discussion   This method is based on the precipitation of lead chlorofluoride, in which the chlorine is determined by Volhard's method, and from this result the fluorine content can be calculated. The advantages of the method are the precipitate is granular, settles readily and is easily filtered; the factor for conversion to fluorine is low; the procedure is carried out at pH 3.6–5.6, so substances which might be co-precipitated, such as phosphates, sulphates, chromates and carbonates, do not interfere. Aluminium must be entirely absent, since even very small quantities cause low results; a similar effect is produced by boron (> 0.05 g), ammonium (> 0.5 g) and sodium or potassium (> 10 g) in the presence of about 0.1 g of fluoride. Iron must be removed, but zinc is without effect. Silica does not vitiate the method, but causes difficulties in filtration.

Procedure   Pipette 25.0 mL of the solution containing between 0.01 and 0.1 g fluoride into a 400 mL beaker, add two drops of bromophenol blue indicator, 3 mL of 10% sodium chloride, and dilute the mixture to 250 mL. Add dilute nitric acid until the colour just changes to yellow, and then add dilute sodium hydroxide solution until the colour just changes to blue. Treat with 1 mL of concentrated hydrochloric acid, then with 5.0 g of lead nitrate, and heat on a water bath. Stir gently until the lead nitrate has dissolved, and then immediately add 5.0 g of crystallised sodium ethanoate and stir vigorously. Digest on the water bath for 30 min, with occasional stirring, and allow to stand overnight.

Meanwhile a washing solution of lead chlorofluoride is prepared as follows. Add a solution of 10 g of lead nitrate in 200 mL of water to 100 mL of a solution containing 1.0 g of sodium fluoride and 2 mL of concentrated hydrochloric acid, mix it thoroughly, and allow the precipitate to settle. Decant the supernatant liquid, wash the precipitate by decantation with five portions of water, each of about 200 mL. Finally add 1 L of water to the precipitate, shake the mixture at intervals during an hour, allow the precipitate to settle, and filter the liquid. Further quantities of wash liquid may be prepared as needed by treating the precipitate with fresh portions of water. The solubility of lead chlorofluoride in water is $0.325 \, g \, L^{-1}$ at 25 °C.

Separate the original precipitate by decantation through a Whatman no. 542 or no. 42 paper. Transfer the precipitate to the filter, wash once with cold water, four or five times with the saturated solution of lead chlorofluoride, and finally once more with cold water. Transfer the precipitate and paper to the beaker in which precipitation was made, stir the paper to a pulp in 100 mL of 5% nitric acid, and heat on the water bath until the precipitate has dissolved (5 min). Add a slight excess of standard 0.1 M silver nitrate, digest on the bath for a further 30 min, and allow to cool to room temperature while protected from the light. Filter the precipitate of silver chloride through a sintered-glass crucible, wash with a little cold water, and titrate the residual silver nitrate in the filtrate and washings with standard 0.1 M thiocyanate. Subtract the amount of silver found in the filtrate from the amount originally added. The difference represents the amount of silver that was required to combine with the chlorine in the lead chlorofluoride precipitate:

$$1 \text{ mol AgNO}_3 \equiv 1 \text{ mol F}^-$$

10.101   Potassium

Discussion   Potassium may be precipitated with excess of sodium tetraphenylborate solution as potassium tetraphenylborate. The excess of reagent is determined by titration with mercury(II) nitrate solution. The indicator consists of a mixture of iron(III) nitrate and dilute sodium thiocyanate solution. The end point is revealed by the decolorisation of the iron(III)–thiocyanate complex due to the formation of the colourless mercury(II) thiocyanate. The reaction between mercury(II) nitrate and sodium tetraphenylborate under the experimental conditions used is not quite stoichiometric, hence it is necessary to determine the volume in millilitres of $Hg(NO_3)_2$ solution equivalent to 1 mL of a $NaB(C_6H_5)_4$ solution. Halides must be absent.

Procedure   Prepare the sodium tetraphenylborate solution by dissolving 6.0 g of the solid in about 200 mL of distilled water in a glass-stoppered bottle. Add about 1 g of moist aluminium hydroxide gel, and shake well at 5 min intervals for about 20 min. Filter through a Whatman no. 40 filter paper, pouring the first runnings back through the filter if necessary, to ensure a clear filtrate. Add 15 mL of 0.1 M sodium hydroxide to the solution to give a pH of about 9, then make up to 1 L and store the solution in a polythene bottle.

Prepare a mercury(II) nitrate solution (0.03 M) by **carefully** dissolving 10.3 g recrystallised mercury(II) nitrate $Hg(NO_3)_2 \cdot H_2O$ in 800 mL distilled water containing 20 mL of 2 M nitric acid. Dilute to 1 L in a graduated flask and then standardise by titrating with a standard thiocyanate solution using iron(III) indicator solution. Prepare the indicator solutions for the main titration by dissolving separately 5 g hydrated iron(III) nitrate in 100 mL of distilled water and filtering, and 0.08 g sodium thiocyanate in 100 mL of distilled water.

Standardisation   Pipette 10.0 mL of the sodium tetraphenylborate solution into a 250 mL beaker and add 90 mL of water, 2.5 mL of 0.1 M nitric acid, 1.0 mL of iron(III) nitrate solution, and 10.0 mL of sodium thiocyanate solution. Without delay stir the solution mechanically, then slowly add from a burette 10 drops of mercury(II) nitrate solution. Continue the titration by adding the mercury(II) nitrate solution at a rate of 1–2 drops per second until the colour of the indicator is temporarily discharged. Continue the titration more slowly, but maintain the rapid rate of stirring. The end point is arbitrarily defined as the point when the indicator colour is discharged and fails to reappear for 1 min. Perform at least three titrations, and calculate the mean volume of mercury(II) nitrate solution equivalent to 10.0 mL of the sodium tetraphenylborate solution.

Pipette 25.0 mL of the potassium ion solution (about 10 mg K^+) into a 50 mL graduated flask, add 0.5 mL 1 M nitric acid and mix. Introduce 20.0 mL of the sodium tetraphenylborate solution, dilute to the mark, mix, then pour the mixture into a 150 mL flask provided with a ground stopper. Shake the stoppered flask for 5 min on a mechanical shaker to coagulate the precipitate, then filter most of the solution through a dry Whatman no. 40 filter paper into a dry beaker. Transfer 25.0 mL of the filtrate into a 250 mL conical flask and add 75 mL of water, 1.0 mL of iron(III) nitrate solution, and 1.0 mL of sodium thiocyanate solution. Titrate with the mercury(II) nitrate solution as described above.

Precipitation: determinations using instruments

!   **Safety.** Before carrying out any experiments in this section, pay full attention to any safety warnings and make sure you adhere to national laboratory and safety regulations.

10.102   Potentiometry: general considerations

The theory of precipitation reactions is given in Sections 10.92 and 10.93. The ion concentration at the equivalence point is determined by the solubility product of the sparingly soluble material formed during the titration. In the precipitation of an ion I from solution by the addition of a suitable reagent, the concentration of I in the solution will clearly change most rapidly in the region of the end point. The potential of an indicator electrode responsive to the concentration of I will undergo a like change, hence the change can be followed potentiometrically. Here one electrode may be a saturated calomel or silver–silver chloride electrode, and the other must be an electrode which will readily come into equilibrium with one of the ions of the precipitate. For example, in the titration of silver ions with a halide (chloride, bromide or iodide) this must be a silver electrode. It may consist of a silver wire, or a platinum wire or gauze plated with silver and sealed into a glass tube. Since a halide is to be determined, the salt bridge must be a saturated solution of potassium nitrate. Excellent results are obtained by titrating silver nitrate solutions with thiocyanate ions. It is often possible to use appropriate ion selective electrodes.

10.103   Mixtures of halides by potentiometry

Discussion   Potentiometric methods are particularly useful for titrating mixtures of halides with silver nitrate. The calculated titration curves are shown in Figure 10.20.

Procedure   Prepare a solution containing both potassium chloride and potassium iodide; weigh each substance accurately and arrange for the solution to be about $0.025\,M$ with respect to each salt. A silver nitrate solution of known concentration (about $0.05\,M$) will also be required. Pipette $10\,\text{mL}$ of the halide solution into the titration vessel and dilute to about $100\,\text{mL}$ with distilled water. Insert a combination silver electrode. The solution is ready to be titrated automatically using the silver nitrate solution. Two end points are obtained, the first for iodide I^- and the second for chloride Cl^-.

10.104   Chloride, bromide and iodide by coulometry

With Hg(I) ions

Discussion   Mercury(I) ions can be generated at 100% efficiency from mercury-coated gold or from mercury pool anodes, and used for the coulometric titration of halides. The end point is conveniently determined potentiometrically. In titrations of chloride ion, the addition of methanol (up to 70–80%) is desirable in order to reduce the solubility of the mercury(I) chloride.

The standard potentials (versus the standard hydrogen electrode) of the fundamental couples involving uncomplexed mercury(I) and mercury(II) ions are as follows:

$$Hg_2^{2+} + 2e = 2Hg \qquad E^{\ominus} = +0.80\,\text{V}$$

$$Hg^{2+} + 2e = Hg \qquad E^{\ominus} = +0.88\,\text{V}$$

$$2Hg^{2+} + 2e = Hg_2^{2+} \qquad E^{\ominus} = +0.91\,\text{V}$$

Oxidation of Hg to Hg_2^{2+} requires a smaller (less oxidising) potential than oxidation of Hg to Hg^{2+}; mercury(I) ions are the main product when a mercury electrode is subjected to anodic polarisation in a non-complexing medium. From a stoichiometric standpoint it does not matter whether oxidation of a mercury anode produces the mercury(I) or mercury(II) salt of a given anion, because the same quantity of electricity per mole of the anion is involved in either case, thus the same number of coulombs per mole of the anion are required to form either Hg_2Cl_2 or $HgCl_2$.

Apparatus   The apparatus is similar to that described in Section 10.14. The generator anode (A) now consists of a mercury pool, 0.5–1.0 cm deep, at the bottom of the cell; electrical connection is made by means of a platinum wire sealed through glass tubing and dipping into the mercury. For titrations of chloride and bromide, the mercury pool generator anode also serves as the indicator electrode and is used in conjunction with a saturated calomel reference electrode; the reference electrode is connected to the cell through a saturated potassium nitrate salt bridge. For titrations of iodide, the indicator electrode consists of a silver rod fitted through glass tubing and held by the cover of the cell. During the titration, the contents of the electrolysis cell are stirred vigorously with a magnetic stirrer; the stirrer bar floats on the surface of the mercury pool anode.

Reagents   *Supporting electrolyte*   For chloride and bromide, use 0.5 M perchloric acid. For iodide, use 0.1 M perchloric acid plus 0.4 M potassium nitrate. It is recommended to prepare a stock solution of about five times these concentrations (2.5 M perchloric acid for chloride and bromide; 0.5 M perchloric acid + 2.0 M potassium nitrate for iodide), and to dilute in the cell according to the volume of test solution used. The reagents must be chloride-free.

Catholyte   The electrolyte in the isolated cathode compartment may be either the same supporting electrolyte as in the cell or 0.1 M sulphuric acid; the formation of mercury(I) sulphate causes no difficulty.

Chloride   Experience in this determination may be obtained by the titration of carefully standardised ~0.005 M hydrochloric acid. Pipette 5.00 or 10.00 mL of the hydrochloric acid into the cell, add 35–40 mL of methanol and 10 mL of the stock solution of the supporting electrolyte. Fill the isolated cathode compartment with supporting electrolyte of the same concentration as in the main body of the solution or with 0.1 M sulphuric acid; the level of the liquid must be kept above the level in the titration cell. Note the counter reading, stir magnetically, and commence the electrolysis at about 50 mA. Stop the generating current periodically, record the counter reading, and observe the potential between the mercury pool and the SCE. Plot a potential–counter reading curve and evaluate the equivalence point by graphing its first or second derivative. The approach of the equivalence point is readily detected in practice; successive small increments of 0.05 or 0.1 counter unit result in a relatively large change of potential (~30 mV per 0.1 counter unit).

Bromide   A 0.01 M solution of potassium bromide, prepared from the pure salt previously dried at 110 °C, is suitable for practice in this determination. The experimental details are similar to those given for chloride except that no methanol is required. The titration cell may contain 10.00 mL of the bromide solution, 30 mL of water, and 10 mL of the stock solution of supporting electrolyte.

Iodide   A 0.01 M solution of potassium iodide, prepared from the dry salt with boiled-out water, is suitable for practice in this determination. The experimental details are similar to those given for bromide, except the indicator electrode consists of a silver rod immersed in the solution. The titration cell may be charged with 10.00 mL of the iodide solution, 30 mL of water, and 10 mL of the stock solution of perchloric acid + potassium nitrate. In the neighbourhood of the equivalence point it is necessary to wait at least 30–60 s before steady potentials are established.

With Ag(I) ions

Discussion   Silver ions can be electrogenerated with 100% efficiency at a silver anode and can be applied to precipitation titrations. The end points can be determined potentiometrically. The supporting electrolyte may be $0.5 M$ potassium nitrate for bromide and iodide. For chloride use $0.5 M$ potassium nitrate in 25–50% ethanol; ethanol has to be used because of the appreciable solubility of silver chloride in water.

Apparatus   Use the apparatus of Section 10.14. The generator anode is of **pure** silver foil (3 cm × 3 cm); the cathode in the isolated compartment is a platinum foil (3 cm × 3 cm) bent into a half-cylinder. For the potentiometric end point detection, use a short length of silver wire as the indicator electrode; the electrical connection to the saturated calomel reference electrode is made by means of an agar–potassium nitrate bridge.

Supporting electrolyte   Prepare a $0.5 M$ potassium nitrate solution from the pure salt; for chloride determinations the solution must be prepared with a mixture of equal volumes of distilled water and ethanol.

Procedure   The determinations are carried out as described above.

10.105   Lead by amperometry using potassium dichromate

Discussion   Both lead ion and dichromate ion yield a diffusion current at an applied potential to a dropping mercury electrode of -1.0 V vs SCE. Amperometric titration gives a V-shaped curve (Figure 10.5(C)). This experiment describes how to determine lead in lead nitrate; it can easily be modified to determine lead in other dilute aqueous solutions (10^{-3} to $10^{-4} M$).

Reagents . *Lead nitrate solution*   Dissolve an accurately weighed amount of lead nitrate in 250 mL water in a graduated flask to give an approximately $0.01 M$ solution. For the titration, dilute 10 mL of this solution (use a pipette) to 100 mL in a graduated flask, thus yielding a $\sim 0.001 M$ solution of known strength.

Potassium dichromate solution ~0.05 M   Use the appropriate quantity of the dry solid, accurately weighted.

Potassium nitrate solution ~0.01 M   Use as the supporting electrolyte.

Procedure   Use the electrical equipment of Figure 10.6. Set up the dropping mercury electrode assembly and allow the mercury to drop into distilled water for at least 5 min. Meanwhile, place 25.0 mL of the $\sim 0.001 M$ lead nitrate solution in the titration cell, add 25 mL of $0.01 M$ potassium nitrate solution, complete the cell assembly, and bubble nitrogen slowly through the solution for 15 min. Make the necessary electrical connections. Apply a potential of -1.0 V vs SCE; at this potential both the lead and the dichromate ions yield diffusion currents. Turn the three-way tap so the nitrogen now passes over the surface of the solution. Adjust the microammeter range so the reading is at the high end of the scale. Do not alter the applied voltage during the determination. Add the $\sim 0.05\,M$ dichromate solution in 0.05 mL portions until within 1 mL of the end point, and henceforth in 0.01 mL portions until about 1 mL beyond the end point, and continue with additions of 0.05 mL. After each addition pass nitrogen through the solution for 1 min to ensure thorough mixing and also deoxygenation, turn the tap so the nitrogen passes over the surface of the solution, and observe the current. Notice that a large initial current will decrease as the titration proceeds to a small value at the equivalence point, and then the current will increase again beyond the equivalence point. Plot the values of the current against the volume of reagent added; draw two straight lines through the branches of the curve. The point of intersection is the equivalence point. Calculate the percentage of lead in the sample of lead nitrate.

10.106   Sulphate by amperometry using lead nitrate

Discussion   Solutions as dilute as $0.001\,M$ with respect to sulphate may be titrated with $0.01\,M$ lead nitrate solution in a medium containing 30% ethanol with reasonable accuracy. For solutions $0.01\,M$ or higher in sulphate, the best results are obtained in a medium containing about 20% ethanol. The object of the alcohol is to reduce the solubility of the lead sulphate and thus minimise the magnitude of the rounded portion of the titration curve in the vicinity of the equivalence point. The titration is performed in the absence of oxygen at a potential of -1.2 V vs SCE, where lead ions yield a diffusion current. A reversed L-graph (Figure 10.5(B)) is obtained; the intersection of the two branches gives the end point. A supporting electrolyte need not be added, since the current does not increase appreciably until an excess of lead is present in the solution, and the amount of salt formed during the titration completely suppresses the migration current of lead ions.

Reagents   *Potassium sulphate*   Prepare an approximately $0.01\,M$ solution in a 100 mL graduated flask using an accurately weighed quantity of the solid.

Lead nitrate solution ~0.1 M   Prepare like the potassium sulphate solution in a 100 mL graduated flask from a known weight of the dry solid.

Procedure   Following the technique of Section 10.105, introduce 25.0 mL of the potassium sulphate solution into the cell, add 2–3 drops of thymol blue followed by a few drops of concentrated nitric acid until the colour is just red (pH 1.2); finally, add 25 mL of 95% ethanol. Connect the saturated calomel electrode through an agar–potassium nitrate bridge to the cell. Fill the semimicro burette with the standard lead nitrate solution. Pass nitrogen through the solution in the cell for 15 min and then over the surface of the solution. Meanwhile adjust the applied voltage to -1.2 V. Add the lead nitrate solution from the burette using the same sequence of additions as in Section 10.105. After each addition of lead nitrate pass nitrogen through the solution for about 1 min to allow time for the lead sulphate to precipitate; for more dilute solutions the nitrogen should be passed for 3 min before reading the current. Plot the titration curve, deduce the end point and calculate the percentage of SO_4 in the sample of potassium sulphate.

411

10.107 Iodide by amperometry using mercury(II) nitrate

This experiment illustrates the titration of a substance yielding an anodic step (iodide ion) with a solution of an oxidant (mercury(II) nitrate) giving a cathodic diffusion current at the same applied voltage. The magnitude of the anodic diffusion current decreases up to the end point; on adding an excess of titrant, the diffusion current increases, but in the opposite direction. The type of graph obtained is similar to Figure 10.5(D). The end point of the titration is given by the intersection of the two linear portions of the graph with the volume axis; the diffusion current is then approximately zero. The two linear parts do not usually have the same slope, because the titrant and the substance being titrated have different diffusion currents for equivalent concentrations.

Reagents *Potassium iodide solution, ~0.004* M   Dissolve 0.68 g potassium iodide, accurately weighed, in 1 L water.

Mercury(II) nitrate solution   Dissolve 17.13 g pure mercury(II) nitrate monohydrate in 500 mL of 0.05 M nitric acid. **Be careful with mercury nitrate; it is poisonous.**

Nitric acid 0.1 M.

Procedure   Equip the titration vessel with a dropping mercury electrode, an agar–potassium nitrate bridge connected to an SCE through saturated potassium chloride solution contained in a 10 mL beaker, a nitrogen gas inlet, and a magnetic stirrer. Charge the flask with 25.0 mL of the iodide solution, add 25 mL 0.1 M nitric acid, and 2.5 mL warm 1% gelatin solution. Connect the dropping mercury electrode to the negative terminal of the polarising unit and the positive terminal to the SCE. Set the applied potential at zero and adjust the microammeter reading at the centre of the scale. Pass nitrogen through the solution for at least 5 min while stirring magnetically. Run in the mercury(II) nitrate solution from a semimicro burette and take readings of the current at 0.10 mL intervals. The end point corresponds to zero current, but continue the titration beyond this point to obtain the cathodic current due to excess of mercury(II) nitrate. Plot current against volume of mercury(II) nitrate solution, and evaluate the exact end point from the graph. Calculate the concentration of the mercury(II) nitrate solution from the known concentration of the potassium iodide solution.

Oxidation–reduction titrations

! ■ **Safety.** Before carrying out any experiments in this section, pay full attention to any safety warnings and make sure you adhere to national laboratory and safety regulations.

10.108 Change of electrode potential

Sections 10.32 to 10.37 show how the change in pH during acid–base titrations may be calculated, and how the titration curves thus obtained can be used (a) to ascertain the most suitable indicator to be used in a given titration, and (b) to determine the titration error. Similar procedures may be carried out for oxidation–reduction titrations. Consider first a simple case which involves only change in ionic charge, and is theoretically independent of the hydrogen ion concentration. A suitable example is the titration of $100\,mL$ of $0.1\,M$ iron(II) with $0.1\,M$ cerium(IV) in the presence of dilute sulphuric acid:

$$Ce^{4+} + Fe^{2+} \rightleftharpoons Ce^{3+} + Fe^{3+}$$

The quantity corresponding to $[H^+]$ in acid–base titrations is the ratio $[ox]/[red]$. Two systems are involved here, the Fe^{3+}/Fe^{2+} ion electrode (1), and the Ce^{4+}/Ce^{3+} ion electrode (2).
    For (1) at $25\,°C$

$$E_1 = E_1^\ominus + \frac{0.0591}{1} \log \frac{[Fe^{3+}]}{[Fe^{2+}]} = +0.75 + 0.0591 \log \frac{[Fe^{3+}]}{[Fe^{2+}]}$$

For (2), at $25\,°C$

$$E_2 = E_2^\ominus + \frac{0.0591}{1} \log \frac{[Ce^{4+}]}{[Ce^{3+}]} = +1.45 + 0.0591 \log \frac{[Ce^{4+}]}{[Ce^{3+}]}$$

According to Section 2.33, the equilibrium constant of the reaction is given by

$$\log K = \log \frac{[Ce^{3+}][Fe^{3+}]}{[Ce^{4+}][Fe^{2+}]} = \frac{1}{0.0591}(1.45 - 0.75) = 11.84$$

or

$$K \doteq 7 \times 10^{11}$$

The reaction is therefore virtually complete.
    During the addition of the cerium(IV) solution up to the equivalence point, its only effect will be to oxidise the iron(II) (since K is large) and consequently change the ratio $[Fe^{3+}]/[Fe^{2+}]$. When $10\,mL$ of the oxidising agent have been added, $[Fe^{3+}]/[Fe^{2+}] = \sim 10/90$ and

$$E_1 = 0.75 + 0.0591 \log 10/90 = 0.75 - 0.056 = 0.69\,V$$

With $50\,mL$ of the oxidising agent, $E_1 = E_1^\ominus = 0.75\,V$

With $90\,mL$ $E_1 = 0.75 + 0.0591 \log 90/10$ $= 0.81\,V$

With $99\,mL$ $E_1 = 0.75 + 0.0591 \log 99/1$ $= 0.87\,V$

With $99.9\,mL$ $E_1 = 0.75 + 0.0591 \log 99.9./0.1$ $= 0.93\,V$

At the equivalence point $(100.0\,mL)$ $[Fe^{3+}] = [Ce^{3+}]$ and $[Ce^{4+}] = [Fe^{2+}]$, and the electrode potential is given by*

$$\frac{E_1^\ominus + E_2^\ominus}{2} = \frac{0.75 + 1.45}{2} = 1.10\,V$$

The subsequent addition of cerium(IV) solution will merely increase the ratio $[Ce^{4+}]/[Ce^{3+}]$. Thus

---

* See textbooks for a derivation and the approximations involved. For the reaction

$$a\,ox_1 + b\,red_2 \rightleftharpoons b\,ox_2 + a\,red_1$$

the potential at the equivalence point is given by

$$E_0 = \frac{bE_1^\ominus + aE_2^\ominus}{a + b}$$

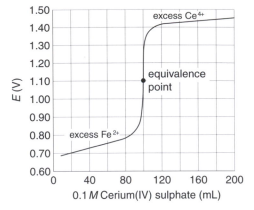

Figure 10.21　Calculated curve for 100 mL of 0.1 M iron(II) titrated with 0.1 M cerium sulphate

With 100.1 mL　$E_2 = 1.45 + 0.0591 \log 0.1/100 = 1.27$ V

With 101 mL　$E_2 = 1.45 + 0.0591 \log 1/100 = 1.33$ V

With 110 mL　$E_2 = 1.45 + 0.0591 \log 10/100 = 1.39$ V

With 190 mL　$E_2 = 1.45 + 0.0591 \log 90/100 = 1.45$ V

These results are shown in Figure 10.21.

It is of interest to calculate the iron(II) concentration in the neighbourhood of the equivalence point. When 99.9 mL of the cerium(IV) solution have been added, $[Fe^{2+}] = 0.1 \times 0.1/199.9 = 5 \times 10^{-5}$, or $pFe^{2+} = 4.3$. According to Section 2.33, the concentration at the equivalence point is given by

$$[Fe^{3+}]/[Fe^{2+}] = K^{1/2} = (7 \times 10^{11})^{1/2} = 8.4 \times 10^5$$

Now $[Fe^{3+}] = 0.05\,M$, hence $[Fe^{2+}] = 5 \times 10^{-2}/8.4 \times 10^5 = 6 \times 10^{-8}\,M$, or $pFe^{2+} = 7.2$. Upon addition of 100.1 mL of cerium(IV) solution, the reduction potential (see above) is 1.27 V. The $[Fe^{3+}]$ is practically unchanged at $5 \times 10^{-2}\,M$, and we may calculate $[Fe^{2+}]$ with sufficient accuracy for our purpose from the equations

$$E = E_1^{\ominus} + 0.0591 \log \frac{[Fe^{3+}]}{[Fe^{2+}]}$$

$$1.27 = 0.75 + 0.0591 \log \frac{5 \times 10^{-2}}{[Fe^{2+}]}$$

$$[Fe^{2+}] = 1 \times 10^{-10}$$

or

$$pFe^{2+} = 10$$

Thus pFe^{2+} changes from 4.3 to 10 between 0.1% before and 0.1% after the stoichiometric end point. These quantities are important when using indicators to detect the equivalence point.

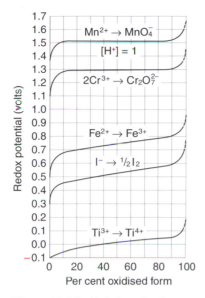

Figure 10.22 Variation of redox potentials with oxidant/reductant ratio

The abrupt change of the potential in the neighbourhood of the equivalence point depends on the standard potentials of the two oxidation–reduction systems that are involved, and therefore on the equilibrium constant of the reaction; it is independent of the concentrations unless they are extremely small. The changes in redox potential for a number of typical oxidation–reduction systems are shown graphically in Figure 10.22. For MnO_4^-/Mn^{2+} and other systems which depend on the solution pH, the hydrogen ion concentration is assumed to be molar; lower acidities give lower potentials. The value at 50% oxidised form will correspond to the standard redox potential. Consider the titration of iron(II) with potassium dichromate. The titration curve would follow the curve of the Fe(II)/Fe(III) system until it reached the end point, then it would rise steeply and continue along the curve for the $Cr_2O_7^{2-}/Cr^{3+}$ system; the potential at the equivalence point can be determined as already described.

It is possible to titrate two substances by the same titrant, provided the standard potentials of the substances being titrated, and their oxidation or reduction products, differ by about 0.2 V. Stepwise titration curves are obtained in the titration of mixtures or substances having several oxidation states. The titration of a solution containing Cr(VI), Fe(III) and V(V) by an acid titanium(III) chloride solution is an example. In the first step Cr(VI) is reduced to Cr(III) and V(V) to V(IV); in the second step Fe(III) is reduced to Fe(II); in the third step V(IV) is reduced to V(III); chromium is evaluated by difference of the volumes of titrant used in the first and third steps. Another example is the titration of a mixture of Fe(II) and V(IV) sulphates with Ce(IV) sulphate in dilute sulphuric acid. In the first step Fe(II) is oxidised to Fe (III) and in the second 'jump' V(IV) is oxidised to V(V); the vanadium oxidation is accelerated by heating the solution after oxidation of the Fe(II) ion is complete. For titration of a substance having several oxidation states, an example is the stepwise reduction by acid chromium(II) chloride of Cu(II) ion to the Cu(I) state and then to the metal.

10.109 Formal potentials

Standard potentials $E^{\ominus}$ are evaluated with full regard to activity effects and with all ions present in simple form; they are really limiting or ideal values and are rarely observed in a potentiometric measurement. In practice the solutions may be quite concentrated and frequently contain other electrolytes; under these conditions the activities of the pertinent species are much smaller than the concentrations, so standard potentials may lead to unreliable conclusions. Also, the actual active species present (see example below) may differ from those to which the ideal standard potentials apply. That is why formal potentials have been proposed to supplement standard potentials. The formal potential is the potential observed experimentally in a solution containing one mole each of the oxidised and reduced substances together with other specified substances at specified concentrations. Formal potentials vary appreciably with the nature and concentration of the acid that is present. The formal potential incorporates in one value the effects resulting from variation of activity coefficients with ionic strength, acid–base dissociation, complexation, liquid junction potentials, etc., and thus has a real practical value. Formal potentials do not have the theoretical significance of standard potentials, but they are observed values in actual potentiometric measurements. In dilute solutions they usually obey the Nernst equation fairly closely in the form:

$$E = E^{\ominus\prime} + \frac{0.0591}{n} \log \frac{[\text{ox}]}{[\text{red}]} \qquad (\text{at } 25\,^{\circ}\text{C})$$

where $E^{\ominus\prime}$ is the formal potential and corresponds to the value of E at **unit** concentrations of oxidant and reductant, and the quantities in square brackets refer to molar concentrations. It is useful to determine and to tabulate $E^{\ominus\prime}$ with equivalent amounts of various oxidants and their conjugate reductants at various concentrations of different acids. If one is dealing with solutions whose composition is identical with or similar to the composition for the formal potential, more trustworthy conclusions can be derived from formal potentials than from standard potentials.

To illustrate how the use of standard potentials may occasionally lead to erroneous conclusions, consider the hexacyanoferrate(II)–hexacyanoferrate(III) and the iodide–iodine systems. The standard potentials are

$$[\text{Fe(CN)}_6]^{3-} + \text{e} \rightleftharpoons [\text{Fe(CN)}_6]^{4-} \qquad E^{\ominus} = +0.36\text{ V}$$

$$I_2 + 2\text{e} \rightleftharpoons 2I^{-} \qquad E^{\ominus} = +0.54\text{ V}$$

It might be expected that iodine would quantitatively oxidise hexacyanoferrate(II) ions:

$$2[\text{Fe(CN)}_6]^{4-} + I_2 = 2[\text{Fe(CN)}_6]^{3-} + 2I^{-}$$

In fact $[\text{Fe(CN)}_6]^{4-}$ ion oxidises iodide ion quantitatively in media containing $1\,M$ hydrochloric, sulphuric or perchloric acid. This is because in solutions of low pH, protonation occurs and the species derived from $H_4\text{Fe(CN)}_6$ are weaker than those derived from $H_3\text{Fe(CN)}_6$; the activity of the $[\text{Fe(CN)}_6]^{4-}$ ion is decreased to a greater extent than the activity of the $[\text{Fe(CN)}_6]^{3-}$ ion, so the reduction potential is increased. The actual redox potential of a solution containing equal concentrations of both cyanoferrates in $1\,M$ HCl, H_2SO_4 or $HClO_4$ is $+0.71$ V, greater than the potential of the iodine–iodide couple.

If there is no great difference in complexation of either the oxidant or its conjugate reductant in various acids, the formal potentials lie close together in these acids. Thus for the Fe(II)/Fe(III) system $E^{\ominus} = +0.77$ V, $E^{\ominus\prime} = +0.73$ V in $1\,M$ $HClO_4$, $+0.70$ V in $1\,M$ HCl,

+0.68 V in 1 M H_2SO_4, and +0.61 V in 0.5 M H_3PO_4 + 1 M H_2SO_4. It would seem that complexation is least in perchloric acid and greatest in phosphoric(V) acid.

For the Ce(III)/Ce(IV) system $E^{\ominus\prime} = +1.44$ V in 1 M H_2SO_4, +1.61 V in 1 M HNO_3, and +1.70 V in 1 M $HClO_4$. Perchloric acid solutions of cerium(IV) perchlorate, although unstable on standing, react rapidly and quantitatively with many inorganic compounds and have greater oxidising power than cerium(IV) sulphate–sulphuric acid or cerium(IV) nitrate–nitric acid solutions.

10.110   Detecting the end point in oxidation-reduction titrations*

Internal oxidation–reduction indicators

As discussed in Sections 10.31 to 10.37, acid–base indicators are employed to mark the sudden change in pH during acid–base titrations. Similarly an oxidation–reduction indicator should mark the sudden change in the oxidation potential in the neighbourhood of the equivalence point in an oxidation–reduction titration. The ideal oxidation–reduction indicator will have an oxidation potential intermediate between the values for the solution titrated and the titrant, and it should exhibit a sharp, readily detectable colour change.

An oxidation–reduction indicator (redox indicator) is a compound which exhibits different colours in the oxidised and reduced forms:

$$In_{ox} + ne \rightleftharpoons In_{red}$$

The oxidation and reduction should be reversible. At a potential E the ratio of the concentrations of the two forms is given by the Nernst equation:

$$E = E_{In}^{\ominus} + \frac{RT}{nF} \ln a_{In.ox}/a_{In.red}$$

$$E \approx E_{In}^{\ominus} + \frac{RT}{nF} \ln \frac{[In_{ox}]}{[In_{red}]}$$

where $E_{In}^{\ominus}$ is the standard (strictly the formal) potential of the indicator. If the colour intensities of the two forms are comparable, a practical estimate of the colour change interval corresponds to the change in the ratio $[In_{ox}]/[In_{red}]$ from 10 to 1/10; this leads to the following interval of potential:

$$E = E_{In}^{\ominus} \pm \frac{0.0591}{1} \qquad \text{(at 25 °C)}$$

If the colour intensities of the two forms differ considerably, the intermediate colour is attained at a potential somewhat removed from $E_{In}^{\ominus}$, but the error is unlikely to exceed 0.06 V. For a sharp colour change at the end point, $E_{In}^{\ominus}$ should differ by at least 0.15 V from the standard (formal) potentials of the other systems involved in the reaction.

One of the best oxidation–reduction indicators is the 1,10-phenanthroline–iron(II) complex. The base 1,10-phenanthroline combines readily in solution with iron(II) salts in the ratio 3 molecules base to 1 molecule iron(II) ion, forming the intensely red 1,10-phenanthroline–iron(II) complex ion; with strong oxidising agents the iron(III) complex ion is formed, which has a pale blue colour. The colour change is very striking:

$$[Fe(C_{12}H_8N_2)_3]^{3+} + e \rightleftharpoons [Fe(C_{12}H_8N_2)_3]^{2+}$$
$$\text{pale blue} \qquad\qquad\qquad \text{deep red}$$

* See Sections 10.7 to 10.11 for potentiometric methods.

Table 10.11   *Some oxidation–reduction indicators*

Indicator	Colour change		Formal potential (V) at pH = 0
	Oxidised form	Reduced form	
5-Nitro-1,10-phenanthroline iron(II) sulphate (nitroferroin)	Pale blue	Red	1.25
1,10-Phenanthroline iron(II) sulphate (ferroin)	Pale blue	Red	1.06
N-Phenylanthranilic acid	Purple red	Colourless	0.89
Diphenylaminesulphonic acid	Red violet	Colourless	0.85
Diphenylamine	Violet	Colourless	0.76
Starch–I_3^-, KI	Blue	Colourless	0.53
Methylene blue	Blue	Colourless	0.52

The standard redox potential is 1.14 V; the formal potential is 1.06 V in 1 M hydrochloric acid solution. The colour change, however, occurs at about 1.12 V, because the colour of the reduced form (deep red) is so much more intense than the colour of the oxidised form (pale blue). The indicator is of great value in the titration of iron(II) salts and other substances with cerium(IV) sulphate solutions. It is prepared by dissolving 1,10-phenanthroline hydrate (relative molecular mass = 198.1) in the calculated quantity of 0.02 M acid-free iron(II) sulphate, and is therefore 1,10-phenanthroline–iron(II) complex sulphate (known as ferroin). One drop is usually sufficient in a titration; this is equivalent to less than 0.01 mL of 0.05 M oxidising agent, hence the indicator blank is negligible at this or higher concentrations.

The potential at the equivalence point is the mean of the two standard redox potentials (Section 10.108). Figure 10.21 shows the variation of the potential during the titration of 0.1 M iron(II) ion with 0.1 M cerium(IV) solution, and the equivalence point is at 1.10 V. Ferroin changes from deep red to pale blue at a redox potential of 1.12 V; the indicator will therefore be present in the red form. After the addition of say a 0.1% excess of cerium(IV) sulphate solution, the potential rises to 1.27 V and the indicator is oxidised to the pale blue form. The titration error is negligibly small. The disadvantage of diphenylamine is its slight solubility in water. This has been overcome by the use of the soluble barium or sodium diphenylaminesulphonate, which is employed in 0.2% aqueous solution. The redox potential $E_{In}^{\ominus}$ is slightly higher (0.85 V in 0.5 M sulphuric acid), and the oxidised form has a reddish violet colour, resembling potassium permanganate, but the colour slowly disappears on standing; the presence of phosphoric(V) acid is desirable in order to lower the redox potential of the system. Selected redox indicators, together with their colour changes and reduction potentials in an acidic medium, are listed in Table 10.11.

Self-indicating reagents

Potassium permanganate is a good example of a self-indicating reagent; one drop will impart a visible pink coloration to several hundred millilitres of solution, even in the presence of slightly coloured ions, such as iron(III). The colours of cerium(IV) sulphate and iodine solutions have also been employed in the detection of end points, but the colour change is not so marked as for potassium permanganate; here, however, sensitive internal indicators are available (1,10-phenanthroline–iron(II) ion or *N*-phenylanthranilic acid). Self-indicating reagents have the drawback that an excess of oxidising agent is always present at the end point. For work of the highest accuracy, the indicator blank may be determined and allowed for, or the error may be considerably reduced by performing the standardisation and determination under similar experimental conditions.

Oxidations with potassium permanganate

! **Safety.** Before carrying out any experiments in this section, pay full attention to any safety warnings and make sure you adhere to national laboratory and safety regulations.

10.111 Discussion

Potassium permanganate is not a primary standard. It is difficult to obtain the substance perfectly pure and completely free from manganese dioxide. Moreover, ordinary distilled water is likely to contain reducing substances (traces of organic matter, etc.) which will react with the potassium permanganate to form manganese dioxide. The presence of manganese dioxide is very objectionable because it catalyses the auto decomposition of the permanganate solution on standing:

$$4MnO_4^- + 2H_2O = 4MnO_2 + 3O_2 + 4OH^-$$

Permanganate is inherently unstable in the presence of manganese(II) ions:

$$2MnO_4^- + 3Mn^{2+} + 2H_2O = 5MnO_2 + 4H^+$$

This reaction is slow in acid solution, but it is very rapid in neutral solution. For these reasons, potassium permanganate solution is rarely made up by dissolving weighed amounts of the purified solid in water; it is more usual to heat a freshly prepared solution to boiling for 15–30 min, allow it to cool to room temperature, and then filter the solution through a sintered-glass crucible (porosity no. 4). Alternatively the solution is allowed to stand for 2–3 days at room temperature before filtration. Solutions of potassium permanganate should be stored in a clean, glass-stoppered, dark-coloured bottle that has been cleaned with a cleaning solution and then thoroughly rinsed with deionised water. Acidic and alkaline solutions are less stable than neutral solutions. Solutions of permanganate should be protected from unnecessary exposure to light; a dark-coloured bottle is recommended. Diffuse daylight causes no appreciable decomposition, but bright sunlight slowly decomposes even pure solutions.

10.112 Preparing 0.02 *M* potassium permanganate

Weigh out about 3.2–3.5 g potassium permanganate, transfer it to a 1500 mL beaker, add 1 L water, cover the beaker with a clockglass, heat the solution to boiling, boil gently for 15–30 min and allow the solution to cool to laboratory temperature. Filter the solution through a sintered-glass crucible or funnel. Collect the filtrate in a clean glass vessel. The filtered solution should be stored in a clean, glass-stoppered bottle, and kept in the dark or diffuse light except when in use. Alternatively, it may be kept in a dark brown glass bottle. Standard soluions of potassium permanganate are available from many commercial suppliers, who recommend periodic restandardisation of the solution.

10.113 Standardising permanganate solutions

Discussion   Sodium oxalate is readily obtained pure and anhydrous, and the ordinary material has a purity of at least 99.9%. The original procedure used a solution of the oxalate, acidified with dilute sulphuric acid and warmed to 80–90 °C; it was slowly titrated with the permanganate solution (10–15 mL min^{-1}) and with constant stirring until the first permanent faint pink colour was obtained; the temperature near the end point was not allowed to fall below 60 °C. However, with this procedure the results may be 0.1–0.45% high; the titre depends upon the acidity, the temperature, the rate of addition of the permanganate solution, and the speed of stirring. Because of this it is best to make a **more rapid** addition of 90–95% of the permanganate solution (about 25–35 mL min^{-1}) to a solution of sodium oxalate in 1 M sulphuric acid at 25–30 °C, the solution is then warmed to 55–60 °C and the titration completed, the last 0.5–1.0 mL portion being added dropwise. The method is accurate to 0.06%. Full experimental details are given below.

$$2Na^+ + C_2O_4^{2-} + 2H^+ \rightleftharpoons H_2C_2O_4 + 2Na^+$$

$$2MnO_4^- + 5H_2C_2O_4 + 6H^+ = 2Mn^{2+} + 10CO_2 + 8H_2O$$

If oxalate is to be determined, it is often not convenient to use the room temperature technique for unknown amounts of oxalate. The permanganate solution may then be standardised against sodium oxalate at about 80 °C using the same procedure in the standardisation as in the analysis.

Procedure   Dry some analytical grade sodium oxalate at 105–110 °C for 2 h, and allow it to cool in a covered vessel in a desiccator. Weigh out accurately from a weighing bottle about 0.3 g of the dry sodium oxalate into a 600 mL beaker, add 240 mL of recently prepared distilled water, and **carefully add** 12.5 mL of concentrated sulphuric acid or 250 mL of 1 M sulphuric acid. Cool to 25–30 °C and stir until the oxalate has dissolved. Add 90–95% of the required quantity of permanganate solution from a burette at a rate of 25–35 mL min^{-1} while stirring slowly. Heat to 55–60 °C (use a thermometer as stirring rod), and complete the titration by adding permanganate solution until a faint pink colour persists for 30 s. Add the last 0.5–1.0 mL dropwise, with particular care to allow each drop to become decolorised before the next is introduced. For the most exact work, it is necessary to determine the excess of permanganate solution required to impart a pink colour to the solution. This is done by matching the colour produced by adding permanganate solution to the same volume of boiled and cooled dilute sulphuric acid at 55–60 °C. This correction usually amounts to 0.03–0.05 mL. Repeat the determination with two other similar quantities of sodium oxalate. Provided it is stored with due regard to the precautions in Section 10.111, the standardised permanganate solution will keep for a long time, but it is advisable to restandardise the solution frequently to confirm that no decomposition has set in.

10.114 Hydrogen peroxide

Discussion   Hydrogen peroxide is usually encountered in the form of an aqueous solution containing about 6%, 12% or 30% hydrogen peroxide, frequently known as 20-volume, 40-volume and 100-volume hydrogen peroxide respectively; this terminology is based on the volume of oxygen liberated when the solution is decomposed by boiling. Thus 1 mL of 100-volume hydrogen peroxide will yield 100 mL of oxygen measured at standard temperature and pressure.

The following reaction occurs when potassium permanganate solution is added to hydrogen peroxide solution acidified with dilute sulphuric acid:

$$2MnO_4^- + 5H_2O_2 + 6H^+ = 2Mn^{2+} + 5O_2 + 8H_2O$$

This forms the basis of the method of analysis given below.

It is good practice to use a fairly high concentration of acid and a reasonably low rate of addition in order to reduce the danger of forming manganese dioxide, which is an active catalyst for the decomposition of hydrogen peroxide. For slightly coloured solutions or for titrations with dilute permanganate, the use of ferroin as indicator is recommended. Organic substances may interfere. A fading end point indicates the presence of organic matter or other reducing agents, in which case the iodometric method is better (Section 10.135).

Procedure   Transfer 25.0 mL of the 20-volume solution by means of a burette to a 500 mL graduated flask, and dilute with water to the mark. Shake thoroughly. Pipette 25.0 mL of this solution to a conical flask, dilute with 200 mL water, add 20 mL dilute sulphuric acid (1 : 5), and titrate with standard 0.02 M potassium permanganate to the first permanent, faint pink colour. Repeat the titration; two consecutive determinations should agree within 0.1 mL.

10.115   Nitrites

Discussion   Nitrites react in warm acid solution (~40 °C) with permanganate solution in accordance with the equation

$$2MnO_4^- + 5NO_2^- + 6H^+ = 2Mn^{2+} + 5NO_3^- + 3H_2O$$

If a solution of a nitrite is titrated in the ordinary way with potassium permanganate, poor results are obtained, because the nitrite solution has first to be acidified with dilute sulphuric acid. Nitrous acid is liberated, which being volatile and unstable, is partially lost. If, however, a measured volume of standard potassium permanganate solution, acidified with dilute sulphuric acid, is treated with the nitrite solution, added from a burette, until the permanganate is just decolorised, results accurate to 0.5–1.0% may be obtained. This is due to the fact that nitrous acid does not react instantaneously with the permanganate. This method may be used to determine the purity of commercial potassium nitrite.

Procedure   Weigh out accurately about 1.1 g of commercial potassium nitrite, dissolve it in cold water, and dilute to 250 mL in a graduated flask. Shake well. Measure out 25.0 mL of standard 0.02 M potassium permanganate into a 500 mL flask, add 225 mL of 0.5 M sulphuric acid, and heat to 40 °C. Place the nitrite solution in the burette, and add it slowly and with constant stirring until the permanganate solution is just decolorised. Better results are obtained by allowing the tip of the burette to dip under the surface of the diluted permanganate solution. The reaction is sluggish towards the end, so the nitrite solution must be added very slowly.

More accurate results may be secured by adding the nitrite to an acidified solution in which permanganate is present in excess (the tip of the pipette containing the nitrite solution should be below the surface of the liquid during the addition), and back-titrating the excess potassium permanganate with a solution of ammonium iron(II) sulphate which has recently been compared with the permanganate solution.

10.116   Persulphates

Discussion   Alkali persulphates (peroxydisulphates) can readily be evaluated by adding to their solutions a known excess of an acidified iron(II) salt solution, and determining the excess of iron(II) by titration with standard potassium permanganate solution:

$$S_2O_8^{2-} + 2Fe^{2+} + 2H^+ = 2Fe^{3+} + 2HSO_4^-$$

By adding phosphoric(V) acid reduction is complete in a few minutes at room temperature. Many organic compounds interfere.

Another procedure uses standard oxalic acid solution. When a sulphuric acid solution of a persulphate is treated with excess of standard oxalic acid solution in the presence of a little silver sulphate as catalyst, the following reaction occurs:

$$H_2S_2O_8 + H_2C_2O_4 = 2H_2SO_4 + 2CO_2$$

The excess of oxalic acid is titrated with standard potassium permanganate solution.

Procedure A   Prepare an approximately $0.1 M$ solution of ammonium iron(II) sulphate by dissolving about 9.8 g of the solid in 200 mL of sulphuric acid ($0.5 M$) in a 250 mL graduated flask, and then making up to the mark with freshly boiled and cooled distilled water. Standardise the solution by titrating 25 mL portions with standard potassium permanganate solution ($0.02 M$) after the addition of 25 mL sulphuric acid ($0.5 M$).

Weigh out accurately about 0.3 g potassium persulphate into a conical flask and dissolve it in 50 mL of water. **Carefully add** 5 mL syrupy phosphoric(V) acid, 10 mL $2.5 M$ sulphuric acid, and 50.0 mL of the $\sim 0.1 M$ iron(II) solution. After 5 min titrate the excess of Fe^{2+} ion with standard $0.02 M$ potassium permanganate.

Calculate the percentage purity of the sample from the difference between the volume of $0.02 M$ permanganate required to oxidise 50 mL of the iron(II) solution and the volume required to oxidise the iron(II) salt remaining after the addition of the persulphate.

Procedure B   Prepare an approximately $0.05 M$ solution of oxalic acid by dissolving about 1.6 g of the compound and making up to 250 mL in a graduated flask. Standardise the solution with standard ($0.02 M$) potassium permanganate solution using the procedure described in Section 10.113.

Weigh out accurately 0.3–0.4 g potassium persulphate into a 500 mL conical flask, add 50 mL of $0.05 M$ oxalic acid, followed by 0.2 g of silver sulphate dissolved in 20 mL of 10% sulphuric acid. Heat the mixture in a water bath until no more carbon dioxide is evolved (15–20 min), dilute the solution to about 100 mL with water at about 40 °C, and titrate the excess of oxalic acid with standard $0.02 M$ potassium permanganate.

Oxidations with potassium dichromate

> **!**   **Safety.** Before carrying out any experiments in this section, pay full attention to any safety warnings and make sure you adhere to national laboratory and safety regulations.

10.117   Discussion

Potassium dichromate is not such a powerful oxidising agent as potassium permanganate (compare reduction potentials in Table 2.6 in Section 2.31), but it has several advantages over it. It can be obtained pure, it is stable up to its fusion point, and it is therefore an excellent primary standard. Standard solutions of exactly known concentration can be prepared by weighing out the pure dry salt and dissolving it in the proper volume of water. Furthermore, the aqueous solutions are stable indefinitely if adequately protected from evaporation. Potassium dichromate is used only in acid solution, and is reduced rapidly at the ordinary temperature to a green chromium(III) salt. It is not reduced by cold hydrochloric acid, provided the acid concentration does not exceed 1 or $2\,M$. Compared with permanganate solutions, dichromate solutions are less easily reduced by organic matter and they are also stable towards light. Potassium dichromate is therefore of particular value in the determination of iron in iron ores; the ore is usually dissolved in hydrochloric acid, the iron(III) reduced to iron(II), and the solution then titrated with standard dichromate solution:

$$Cr_2O_7^{2-} + 6Fe^{2+} + 14H^+ = 2Cr^{3+} + 6Fe^{3+} + 7H_2O$$

In acid solution, the reduction of potassium dichromate may be represented as

$$Cr_2O_7^{2-} + 14H^+ + 6e \rightleftharpoons 2Cr^{3+} + 7H_2O$$

The green colour due to the Cr^{3+} ions formed by the reduction of potassium dichromate makes it impossible to ascertain the end point of a dichromate titration by simple visual inspection of the solution, so a redox indicator must be employed which gives a strong and unmistakable colour change; this procedure has rendered obsolete the external indicator method which was formerly widely used. Suitable indicators for use with dichromate titrations include _N_-phenylanthranilic acid (0.1% solution in $0.005\,M$ NaOH) and sodium diphenylamine sulphonate (0.2% aqueous solution); the sodium diphenylamine sulphonate must be used in the presence of phosphoric(V) acid.

10.118 Preparing 0.02 _M_ potassium dichromate

Analytical grade potassium dichromate has a purity of not less than 99.9%; it is a satisfactory primary standard. Powder finely about 6 g of the analytical grade material in a glass or agate mortar, and heat for 30–60 min in an oven at 140–150 °C. Allow to cool in a closed vessel in a desiccator. Weigh out accurately about 5.88 g of the dry potassium dichromate into a weighing bottle and transfer the salt quantitatively to a 1 L graduated flask, using a small funnel to avoid loss. Dissolve the salt in the flask in water and make up to the mark; shake well. Alternatively, place a little over 5.88 g of potassium dichromate in a weighing bottle, and weigh accurately. Empty the salt into a 1 L graduated flask, and weigh the bottle again. Dissolve the salt in water, and make up to the mark. The molarity of the solution can be calculated directly from the weight of salt taken.

10.119 Iron in an ore

Procedure   The iron ore is brought into solution with concentrated hydrochloric acid and the resulting iron(III) solution is reduced to iron(II) by the use of tin(II) chloride. See Section 10.153 for full experimental details. After standing for 5 min the iron(II) solution, is further diluted with 100 mL of water, 100 mL of sulphuric acid ($1.5\,M$), and 5 mL of 85% phosphoric(V) acid. Finally, add 6–8 drops of sodium diphenylamine sulphonate indicator and titrate slowly with standard dichromate until the colour changes from green to violet red.

10.120 Chromium in a chromium(III) salt

Discussion Chromium(III) salts are oxidised to dichromate by boiling with excess of a persulphate solution in the presence of a little silver nitrate (catalyst). The excess of persulphate remaining after the oxidation is complete is destroyed by boiling the solution for a short time. The dichromate content of the resultant solution is determined by adding excess of a standard iron(II) solution and titrating the excess with standard $0.02\,M$ potassium dichromate:

$$2Cr^{3+} + 3S_2O_8^{2-} + 7H_2O \xrightarrow{(AgNO_3)} Cr_2O_7^{2-} + 6HSO_4^- + 8H^+$$

$$2S_2O_8^{2-} + 2H_2O = O_2(g) + 4HSO_4^-$$

Procedure Weigh out accurately an amount of the salt which will contain about $0.25\,g$ of chromium, and dissolve it in $50\,mL$ distilled water. Add $20\,mL$ of $\sim0.1\,M$ silver nitrate solution, followed by $50\,mL$ of a 10% solution of ammonium or potassium persulphate. Boil the liquid gently for $20\,min$. Cool and dilute to $250\,mL$ in a graduated flask. Remove $50\,mL$ of the solution with a pipette, add $50\,mL$ of a $0.1\,M$ ammonium iron(II) sulphate solution, $200\,mL$ of $1\,M$ sulphuric acid, and $0.5\,mL$ of N-phenylanthranilic acid indicator. Titrate the excess of the iron(II) salt with standard $0.02\,M$ potassium dichromate until the colour changes from green to violet red.

Standardise the ammonium iron(II) sulphate solution against the $0.02\,M$ potassium dichromate, using N-phenylanthranilic acid as indicator. Calculate the volume of the iron(II) solution which was oxidised by the dichromate originating from the chromium salt, and from this the percentage of chromium in the sample.

Note Lead or barium can be determined by precipitating the sparingly soluble chromate, dissolving the washed precipitate in dilute sulphuric acid, adding a known excess of ammonium iron(II) sulphate solution, and titrating the excess of Fe^{2+} ion with $0.02\,M$ potassium dichromate in the usual way:

$$2PbCrO_4 + 2H^+ = 2Pb^{2+} + Cr_2O_7^{2-} + H_2O$$

10.121 Chemical oxygen demand

Discussion One very important application of potassium dichromate is in a back titration for the environmental determination[13] of the amount of oxygen required to oxidise all the organic material in a sample of impure water, such as sewage effluent. This is known as the chemical oxygen demand (COD) and is expressed in terms of milligrams of oxygen required per litre of water, $mg\,L^{-1}$. The analysis of the impure water sample is carried out in parallel with a blank determination on pure, double-distilled water.

Procedure Place a $50\,mL$ volume of the water sample in a $250\,mL$ conical flask with a ground-glass neck which can be fitted with a water condenser for refluxing. Add $1\,g$ of mercury(II) sulphate, followed by $80\,mL$ of a silver sulphate–sulphuric acid solution (note 1). Then add $10\,mL$ of approximately $0.008\,33\,M$ standard potassium dichromate solution (note 2), fit the flask with the reflux condenser and boil the mixture for $15\,min$. On cooling, rinse the inside of the condenser with $50\,mL$ of water into the flask contents. Add either diphenylamine indicator ($1\,mL$) or ferroin indicator and titrate with $0.025\,M$ ammonium iron(II) sulphate solution (note 3). Diphenylamine gives a colour change from blue to green at the end point; the ferroin colour change is blue green to red brown. Call this titration A mL. Repeat the back titration for the blank (titration B mL). The difference between the two values is the amount of potassium dichromate used up in the oxidation. The COD is calculated from the relationship

$$COD = (B - A) \times 0.2 \times 20 \, \text{mg L}^{-1}$$

as a 1 mL difference between the titrations corresponds to 0.2 mg of oxygen required by the 50 mL sample; a correction must be made with solutions of slightly different molarities (note 4).

Notes

1. This solution is prepared by dissolving 5 g of silver sulphate in 500 mL of concentrated sulphuric acid.
2. The required concentration is obtained by weighing out 1.225 g of potassium dichromate and diluting to 500 mL with deionised water in a graduated flask.
3. Dissolve 4.9 g of ammonium iron(II) sulphate heptahydrate in 150 mL of water and add 2.5 mL of concentrated sulphuric acid. Dilute the solution to 500 mL in a graduated flask.
4. This method gives high results with samples possessing a high chloride content due to reaction between the mercury(II) sulphate and the chloride ions. In these cases the problem can be overcome by following a procedure using chromium(III) potassium sulphate, $Cr(III)K(SO_4)_2 \cdot 12H_2O$.[14]

Oxidations with cerium(IV) sulphate solution

! **Safety.** Before carrying out any experiments in this section, pay full
■ attention to any safety warnings and make sure you adhere to national
laboratory and safety regulations.

10.122   General discussion

Cerium(IV) sulphate is a powerful oxidising agent; its reduction potential in $0.5-4.0\,M$ sulphuric acid at 25 °C is 1.43 ± 0.05 V. It can be used only in acid solution, best in $0.5\,M$ or higher concentrations; as the solution is neutralised, cerium(IV) hydroxide (hydrated cerium(IV) oxide) or basic salts precipitate. The solution has an intense yellow colour, and in hot solutions which are not too dilute the end point may be detected without an indicator; but this procedure requires a blank correction, so it is preferable to add a suitable indicator.
    As a standard oxidising agent, cerium(IV) sulphate has four advantages:

1. Cerium(IV) sulphate solutions are remarkably stable over prolonged periods. They need not be protected from light, and may even be boiled for a short time without appreciable change in concentration. The stability of sulphuric acid solutions covers the wide range of 10–40 mL of concentrated sulphuric acid per litre. Therefore, an acid solution of cerium(IV) sulphate surpasses a permanganate solution in stability.
2. Cerium(IV) sulphate may be used to determine reducing agents in the presence of a high concentration of hydrochloric acid (contrast this with potassium permanganate).
3. Cerium(IV) solutions in $0.1\,M$ solution are not too highly coloured to obstruct vision when reading the meniscus in burettes and other titrimetric apparatus.

4. In the reaction of cerium(IV) salts in acid solution with reducing agents, the simple change

$$Ce^{4+} + e \rightleftharpoons Ce^{3+}$$

is assumed to take place. Permanganate leads to several reduction products, according to the experimental conditions.

Solutions of cerium(IV) sulphate in dilute sulphuric acid are stable even at boiling temperatures. Hydrochloric acid solutions of the salt are unstable because of reduction to cerium(III) by the acid with the simultaneous liberation of chlorine:

$$2Ce^{4+} + 2Cl^- = 2Ce^{3+} + Cl_2$$

This reaction takes place quite rapidly on boiling, hence hydrochloric acid cannot be used in oxidations which require boiling with excess of cerium(IV) sulphate in acid solution; sulphuric acid must be used in those cases. However, direct titration with cerium(IV) sulphate in a dilute hydrochloric acid medium, e.g. for iron(II) may be accurately performed at room temperature, and in this respect cerium(IV) sulphate is superior to potassium permanganate; see item 2 in the list above. The presence of hydrofluoric acid is harmful, since fluoride ion forms a stable complex with Ce(IV) and decolorises the yellow solution.

Solutions of cerium(IV) sulphate may be prepared by dissolving cerium(IV) sulphate or the more soluble ammonium cerium(IV) sulphate in dilute (0.5–1.0 M) sulphuric acid. Internal indicators suitable for use with cerium(IV) sulphate solutions include N-phenylanthranilic and ferroin. Ferroin is 1,10-phenanthroline iron(II).

10.123 Preparing 0.1 M cerium(IV) sulphate

Weigh out 35–36 g of pure cerium(IV) sulphate into a 500 mL beaker, add 56 mL of a 1:1 mixture of sulphuric acid and water, then stir with frequent additions of water and gentle warming, until the salt is dissolved. Transfer to a 1 L graduated flask, and when cold, dilute to the mark with distilled water. Shake well. Alternatively, weigh out 64–66 g of ammonium cerium(IV) sulphate into a solution prepared by adding 28 mL of concentrated sulphuric acid to 500 mL of water; stir the mixture until the solid has dissolved. Transfer to a 1 L graduated flask, and make up to the mark with distilled water.

The relative molecular masses of cerium(IV) sulphate $Ce(SO_4)_2$ and ammonium cerium(IV) sulphate $(NH_4)_4[Ce(SO_4)_4] \cdot 2H_2O$ are 333.25 and 632.56 respectively.

10.124 Standardising cerium(IV) sulphate solutions

Standard solutions of cerium(IV) sulphate are available from many commercial suppliers. Solutions of cerium(IV) in sulphuric acid are extremely stable, so they can be stored for many months without any significant change in molarity. Details of how to standardise cerium(IV) sulphate solutions can be found in the previous edition of this book.[7]

10.125 Copper

Discussion Copper(II) ions are quantitatively reduced in 2 M hydrochloric acid solution by means of the silver reductor (Section 10.152) to the copper(I) state. The solution, after reduction, is collected in a solution of ammonium iron(III) sulphate, and the Fe^{2+} ion formed is titrated with standard cerium(IV) sulphate solution using ferroin or N-phenylanthranilic acid as indicator. Comparatively large amounts of nitric acid, and also zinc, cadmium, bismuth, tin and arsenate have no effect on the determination; the method may therefore be applied to determine copper in brass.

Procedure   *Copper in crystallised copper sulphate*   Weigh out accurately about 3.1 g of copper sulphate crystals, dissolve in water, and make up to 250 mL in a graduated flask. Shake well. Pipette 50 mL of this solution into a small beaker, add an equal volume of ~4 M hydrochloric acid. Pass this solution through a silver reductor at the rate of 25 mL min^{-1}, and collect the filtrate in a 500 mL conical flask charged with 20 mL of 0.5 M iron(III) ammonium sulphate solution (prepared by dissolving the appropriate quantity of the analytical grade iron(III) salt in 0.5 M sulphuric acid). Wash the reductor column with six 25 mL portions of 2 M hydrochloric acid. Add 1 drop of ferroin indicator or 0.5 mL N-phenylanthranilic acid, and titrate with 0.1 M cerium(IV) sulphate solution. The end point is sharp, and the colour imparted by the Cu^{2+} ions does not interfere with the detection of the equivalence point.

Copper in copper(I) chloride   Prepare an ammonium iron(III) sulphate solution by dissolving 10.0 g of the salt in about 80 mL of 3 M sulphuric acid and dilute to 100 mL with acid of the same strength. Weigh out accurately about 0.3 g of the sample of copper(I) chloride into a dry 250 mL conical flask and add 25.0 mL of the iron(III) solution. Swirl the contents of the flask until the copper(I) chloride dissolves, add a drop or two of ferroin indicator, and titrate with standard 0.1 M cerium(IV) sulphate. Repeat the titration with 25.0 mL of the iron solution, omitting the addition of the copper(I) chloride. The difference in the two titrations gives the volume of 0.1 M cerium(IV) sulphate which has reacted with the known weight of copper(I) chloride.

10.126   Molybdate

Discussion   Mo(VI) is quantitatively reduced in 2 M hydrochloric acid solution at 60–80 °C by the silver reductor to Mo(V). The reduced molybdenum solution is sufficiently stable over short periods of time in air to be titrated with standard cerium(IV) sulphate solution using ferroin or N-phenylanthranilic acid as indicator. Nitric acid must be completely absent; the presence of a little phosphoric(V) acid during the reduction of the molybdenum(VI) is not harmful; indeed, it appears to increase the rapidity of the subsequent oxidation with cerium(IV) sulphate. Elements such as iron, copper and vanadium interfere; nitrate interferes, since its reduction is catalysed by the presence of molybdates.

Procedure   Weigh out accurately about 2.5 g ammonium molybdate $(NH_4)_6Mo_7O_{24} \cdot 4H_2O$, dissolve in water and make up to 250 mL in a graduated flask. Pipette 50 mL of this solution into a small beaker, add an equal volume of 4 M hydrochloric acid, then 3 mL of 85% phosphoric(V) acid, and heat the solution to 60–80 °C. Pour hot 2 M hydrochloric acid through a silver reductor, and then pass the molybdate solution through the hot reductor at the rate of about 10 mL min^{-1}. Collect the reduced solution in a 500 mL beaker or 500 mL conical flask, and wash the reductor with six 25 mL portions of 2 M hydrochloric acid; the first two washings should be made with the hot acid (rate 10 mL min^{-1}) and the last four washings with the cold acid (rate 20–25 mL min^{-1}). Cool the solution, add one drop of ferroin or 0.5 mL N-phenylanthranilic acid, and titrate with standard 0.1 M cerium(IV) sulphate. The precipitate of cerium(IV) phosphate, which is initially formed, dissolves on shaking. Add the last 0.5 mL of the reagent dropwise and with vigorous stirring or shaking.

Iodometric titrations

> **!** **Safety.** Before carrying out any experiments in this section, pay full
> ■ attention to any safety warnings and make sure you adhere to national
> laboratory and safety regulations.

10.127 General discussion

The direct iodometric titration method, sometimes called **iodimetry**, refers to titrations with a standard solution of iodine. The indirect iodometric titration method, sometimes termed **iodometry**, deals with the titration of iodine liberated in chemical reactions. The normal reduction potential of the reversible system.

$$I_2 \text{ (s)} + 2e \rightleftharpoons 2I^-$$

is 0.5345 V. This equation refers to a saturated aqueous solution in the presence of solid iodine; this half-cell reaction will occur towards the end of a titration of iodide with an oxidising agent such as potassium permanganate, when the iodide ion concentration becomes relatively low. Near the beginning, or in most iodometric titrations, when an excess of iodide ion is present, the tri-iodide ion is formed:

$$I_2 \text{ (aq)} + I^- \rightleftharpoons I_3^-$$

since iodine is readily soluble in a solution of iodide. The half-cell reaction is better written

$$I_3^- + 2e \rightleftharpoons 3I^-$$

and the standard reduction potential is 0.5355 V. Iodine or the tri-iodide ion is therefore a much weaker oxidising agent than potassium permanganate, potassium dichromate and cerium(IV) sulphate.

In most direct titrations with iodine (iodimetry) a solution of iodine in potassium iodide is employed, and the reactive species is therefore the tri-iodide ion I_3^-. Strictly speaking, all equations involving reactions of iodine should be written with I_3^- rather than with I_2, e.g.

$$I_3^- + 2S_2O_3^{2-} = 3I^- + S_4O_6^{2-}$$

is more accurate than

$$I_2 + 2S_2O_3^{2-} = 2I^- + S_4O_6^{2-}$$

For the sake of simplicity, however, the equations in this book will usually be written in terms of molecular iodine rather than the tri-iodide ion.

Strong reducing agents (substances with a much lower reduction potential), such as tin(II) chloride, sulphurous acid, hydrogen sulphide and sodium thiosulphate, react completely and rapidly with iodine even in acid solution. With somewhat weaker reducing agents, e.g. arsenic(III) or antimony(III), complete reaction occurs only when the solution is kept neutral or very faintly acid; under these conditions the reduction potential of the reducing agent is a minimum, or its reducing power is a maximum.

If a strong oxidising agent is treated in neutral or (more usually) acid solution with a large excess of iodide ion, the iodide ion reacts as a reducing agent and the oxidant will be quantitatively reduced. In these cases an equivalent amount of iodine is liberated, and is then titrated with a standard solution of a reducing agent, usually sodium thiosulphate.

The normal reduction potential of the iodine–iodide system is independent of the pH of the solution, so long as the pH is less than about 8; at higher pH iodine reacts with hydroxide ions to form iodide and the extremely unstable hypoiodite, which is rapidly transformed into iodate and iodide by self-oxidation and reduction:

$$I_2 + 2OH^- = I^- + IO^- + H_2O$$

$$3IO^- = 2I^- + IO_3^-$$

The reduction potentials of certain substances increase considerably with increasing hydrogen ion concentration of the solution. This is the case with systems containing permanganate, dichromate, arsenate, antimonate, bromate, etc., i.e. systems with anions containing oxygen; hydrogen is then required for complete reduction. Many weak oxidising anions are completely reduced by iodide ions if their reduction potentials are raised considerably by the presence in solution of a large amount of acid.

By suitable control of the solution pH, it is sometimes possible to titrate the reduced form of a substance with iodine, and the oxidised form, after the addition of iodide, with sodium thiosulphate. For the arsenic(III)–arsenic(V) system, the reaction is completely reversible:

$$H_3AsO_3 + I_2 + H_2O \rightleftharpoons H_3AsO_4 + 2H^+ + 2I^-$$

At pH values between 4 and 9, arsenite can be titrated with iodine solution. But in strongly acid solutions, arsenate is reduced to arsenite, and iodine is liberated. Upon titration with sodium thiosulphate solution, the iodine is removed and the reaction proceeds from right to left.

Two important **sources of error** in titrations involving iodine are loss of iodine owing to its appreciable volatility; and oxidation of iodide in acid solutions by oxygen from the air:

$$4I^- + O_2 + 4H^+ = 2I_2 + 2H_2O$$

In the presence of excess iodide, the volatility is decreased markedly through formation of the tri-iodide ion; at room temperature the loss of iodine by volatilisation from a solution containing at least 4% of potassium iodide is negligible provided the titration is not prolonged unduly. Titrations should be performed in cold solutions in conical flasks, not in open beakers. If a solution is to stand, it should be kept in a glass-stoppered vessel. The atmospheric oxidation of iodide is negligible in neutral solution in the absence of catalysts, but the rate of oxidation increases rapidly with decreasing pH. The reaction is catalysed by certain metal ions of variable charge (particularly copper), by nitrite ion, and also by strong light. For this reason titrations should not be performed in direct sunlight, and solutions containing iodide should be stored in amber glass bottles.

Furthermore, the air oxidation of iodide ion may be induced by the reaction between iodide and the oxidising agent, especially when the main reaction is slow. Solutions containing an excess of iodide and acid must therefore not be allowed to stand longer than necessary before titration of the iodine. If prolonged standing is necessary (as in the titration of vanadate or Fe^{3+} ions) the solution should be free from air before the addition of iodide and the air displaced from the titration vessel by carbon dioxide, e.g. by adding small portions (0.2–0.5 g) of pure sodium hydrogencarbonate to the acid solution, or a little solid carbon dioxide, dry ice. Potassium iodide is then introduced and the glass stopper replaced immediately.

429

A standard solution containing **potassium iodide and potassium iodate** is quite stable and yields iodine when treated with acid:

$$IO_3^- + 5I^- + 6H^+ = 3I_2 + 3H_2O$$

The standard solution is prepared by dissolving a weighed amount of pure potassium iodate in a solution containing a slight excess of pure potassium iodide, and diluting to a definite volume. This solution has two important uses. The first is as a source of a known quantity of iodine in titrations (Section 10.132). It must be added to a solution containing strong acid; it cannot be employed in a medium which is neutral or possesses a low acidity.

The second use is to determine the acid content of solutions iodometrically or in the **standardisation of solutions of strong acids**. The amount of iodine liberated is equivalent to the acid content of the solution. Suppose 25 mL of an approximately $0.1\,M$ solution of a strong acid are treated with a slight excess of potassium iodate (25 mL of $0.02\,M$ potassium iodate solution, Section 10.141) and a slight excess of potassium iodide solution (10 mL of a 10% solution), and the liberated iodine titrated with standard $0.1\,M$ sodium thiosulphate with the aid of starch as an indicator, the concentration of the acid may be readily evaluated.

10.128   Detecting the end point

Discussion   A solution of iodine in aqueous iodide has an intense yellow to brown colour. One drop of $0.05\,M$ iodine solution imparts a perceptible pale yellow colour to 100 mL of water, so in otherwise colourless solutions iodine can serve as its own indicator. The test is made much more sensitive by using a solution of starch as indicator. Starch reacts with iodine in the presence of iodide to form an intensely blue-coloured complex, which is visible at very low concentrations of iodine. The sensitivity of the colour reaction is such that a blue colour is visible when the iodine concentration is $2 \times 10^{-5}\,M$ and the iodide concentration is greater than $4 \times 10^{-4}\,M$ at 20 °C. The colour sensitivity decreases with increasing temperature of the solution; thus at 50 °C it is about 10 times less sensitive than at 25 °C. The sensitivity decreases on addition of solvents such as ethanol; no colour is obtained in solutions containing 50% ethanol or more. It cannot be used in a strongly acid medium because hydrolysis of the starch occurs.

Starches can be separated into two major components, amylose and amylopectin, which exist in different proportions in various plants. Amylose, which is a straight-chain compound and is abundant in potato starch, gives a blue colour with iodine and the chain assumes a spiral form. Amylopectin, which has a branched-chain structure, forms a red-purple product, probably by adsorption.

The great merit of starch is that it is inexpensive. It possesses the following disadvantages: insolubility in cold water; instability of suspensions in water; it gives a water-insoluble complex with iodine, which means the indicator has to be added comparatively late in the titration, sometimes just before the end point; and there is sometimes a 'drift' end point, which is marked when the solutions are dilute.

Most of the shortcomings of starch as an indicator are absent in **sodium starch glycollate**. This is a white, non-hygroscopic powder, readily soluble in hot water to give a faintly opalescent solution, which is stable for many months; it does not form a water-insoluble complex with iodine, hence the indicator may be added at any stage of the reaction. With excess of iodine (e.g. at the beginning of a titration with sodium thiosulphate) the colour of the solution containing 1 mL of the indicator (0.1% aqueous solution) is green; as the iodine concentration diminishes the colour changes to blue, which becomes intense just before the end point is reached. The end point is very sharp and reproducible and there is no drift in dilute solution.

Preparation and use   *Starch solution*   Make a paste of 0.1 g of soluble starch with a little water, and pour the paste, with constant stirring, into 100 mL of boiling water, and boil for 1 min. Allow the solution to cool and add 2–3 g of potassium iodide. Keep the solution in a stoppered bottle.

Only freshly prepared starch solution should be used. Two millilitres of a 0.1% solution per 100 mL of the solution to be titrated is a satisfactory amount; the same volume of starch solution should always be added in a titration. In the titration of iodine, starch must not be added until just before the end point is reached. Apart from the fact that the fading of the iodine colour is a good indication of the approaching end point, if the starch solution is added when the iodine concentration is high, some iodine may remain adsorbed even at the end point. The indicator blank is negligibly small in iodimetric and iodometric titrations of 0.05 *M* solutions; with more dilute solutions the blank must be determined in a liquid having the same composition as the titrated solution at its end point.

A solid solution of starch in urea may also be employed. Reflux 1 g of soluble starch and 19 g of urea with xylene. At the boiling point of the organic solvent, the urea melts with little decomposition, and the starch dissolves in the molten urea. Allow to cool, then remove the solid mass and powder it; store the product in a stoppered bottle. A few milligrams of this solid added to an aqueous solution containing iodine then behaves like the usual starch indicator.

Sodium starch glycollate indicator   Sodium starch glycollate, prepared as described below, dissolves slowly in cold water but rapidly in hot water. It is best dissolved by mixing 5.0 g of the finely powdered solid with 1–2 mL ethanol, adding 100 mL cold water, and boiling for a few minutes with vigorous stirring; a faintly opalescent solution results. This 5% stock solution is diluted to 1% concentration as required. The most convenient concentration for use as an indicator is 0.1 mg mL^{-1}, i.e. 1 mL of the 1% aqueous solution is added to 100 mL of the solution being titrated.

10.129   Preparing 0.05 *M* iodine solution

Discussion   In addition to a small solubility (0.335 g of iodine dissolves in 1 L of water at 25 °C), aqueous solutions of iodine have an appreciable vapour pressure of iodine, and therefore decrease slightly in concentration on account of volatilisation when handled. Both difficulties are overcome by dissolving the iodine in an aqueous solution of potassium iodide. Iodine dissolves readily in aqueous potassium iodide; the more concentrated the solution, the greater the solubility of the iodine. The increased solubility is due to the formation of a tri-iodide ion:

$$I_2 + I^- \rightleftharpoons I_3^-$$

The resulting solution has a much lower vapour pressure than a solution of iodine in pure water, so the loss by volatilisation is considerably diminished. Nevertheless, the vapour pressure is still appreciable so with vessels containing iodine, **always keep them closed during the actual titrations**. When an iodide solution of iodine is titrated with a reductant, the free iodine reacts with the reducing agent, this displaces the equilibrium to the left, and eventually all the tri-iodide is decomposed; the solution therefore behaves as though it were a solution of free iodine.

For preparing standard iodine solutions, use resublimed iodine and iodate-free potassium iodide. The solution may be standardised against pure arsenic(III) oxide or with a sodium thiosulphate solution which has been recently standardised against potassium iodate.

The equation for the ionic reaction is

$$I_2 + 2e \rightleftharpoons 2I^-$$

Procedure Dissolve 20 g of iodate-free potassium iodide in 30–40 mL of water in a glass-stoppered 1 L graduated flask. Weigh out about 12.7 g of resublimed iodine on a watchglass on a rough balance (never on an analytical balance on account of the iodine vapour), and transfer it by means of a small dry funnel into the concentrated potassium iodide solution. Insert the glass stopper into the flask, and shake in the cold until all the iodine has dissolved. Allow the solution to acquire room temperature, and make up to the mark with distilled water. The iodine solution is best preserved in small glass-stoppered bottles; they should be filled completely and kept in a cool, dark place.

10.130 Standardising iodine solutions

Use sodium thiosulphate solution which has been recently standardised, preferably against pure potassium iodate. Transfer 25 mL of the iodine solution to a 250 mL conical flask, dilute to 100 mL and add the standard thiosulphate solution from a burette until the solution has a pale yellow colour. Add 2 mL of starch solution, and continue the addition of the thiosulphate solution slowly until the solution is just colourless.

10.131 Preparing 0.1 *M* sodium thiosulphate

Discussion Sodium thiosulphate ($Na_2S_2O_3 \cdot 5H_2O$) is readily obtainable in a state of high purity, but there is always some uncertainty as to the exact water content because of the efflorescent nature of the salt and for other reasons. The substance is therefore unsuitable as a primary standard. It is a reducing agent by virtue of the half-cell reaction

$$2S_2O_3^{2-} \rightleftharpoons S_4O_6^{2-} + 2e$$

An approximately 0.1 *M* solution is prepared by dissolving about 25 g crystallised sodium thiosulphate in 1 L of water in a graduated flask. The solution is standardised by any of the methods described below.

Before dealing with these methods, consider the stability of thiosulphate solutions. Solutions prepared with conductivity (equilibrium) water are perfectly stable. However, ordinary distilled water usually contains an excess of carbon dioxide; this may cause a slow decomposition to take place with the formation of sulphur:

$$S_2O_3^{2-} + H^+ = HSO_3^- + S$$

Moreover, decomposition may also be caused by bacterial action, e.g. *Thiobacillus thioparus*, particularly if the solution has been standing for some time. That is why the following recommendations are made:

1. Prepare the solution with recently boiled distilled water.
2. Add 3 drops of chloroform or $10\,mg\,L^{-1}$ of mercury(II) iodide; these compounds improve the keeping qualities of the solution. Bacterial activity is least when the pH lies between 9 and 10. Adding a **small** amount of sodium carbonate ($0.1\,g\,L^{-1}$) is advantageous to ensure the correct pH. In general, alkali hydroxides, sodium carbonate ($> 0.1\,g\,L^{-1}$) and sodium tetraborate should not be added, since they tend to accelerate the decomposition:

$$S_2O_3^{2-} + 2O_2 + H_2O \rightleftharpoons 2SO_4^{2-} + 2H^+$$

3. Avoid exposure to light, as this tends to hasten the decomposition.

The standardisation of thiosulphate solutions may be effected with potassium iodate, potassium dichromate, copper and iodine as primary standards, or with potassium permanganate as a secondary standard. Owing to the volatility of iodine and the difficulty of preparation of perfectly pure iodine, this method is not suitable for beginners. But if a standard solution of iodine (Sections 10.129 and 10.130) is available, this may be used for standardising thiosulphate solutions.

Procedure   Weigh out 25 g of sodium thiosulphate crystals ($Na_2S_2O_3 \cdot 5H_2O$), dissolve in boiled-out distilled water, and make up to 1 L in a graduated flask with boiled-out water. If the solution is to be kept for more than a few days, add 0.1 g sodium carbonate.

10.132   Standardising sodium thiosulphate solutions

With potassium iodate   Potassium iodate has a purity of at least 99.9%; it can be dried at $120\,^{\circ}C$. This reacts with potassium iodide in acid solution to liberate iodine:

$$IO_3^- + 5I^- + 6H^+ = 3I_2 + 3H_2O$$

Its relative molecular mass is 214.00; a $0.02\,M$ solution therefore contains 4.28 g of potassium iodate per litre. Weigh out accurately 0.14–0.15 g of pure dry potassium iodate, dissolve it in 25 mL of cold, boiled-out distilled water, add 2 g of iodate-free potassium iodide (note 1) and 5 mL of $1\,M$ sulphuric acid (note 2). Titrate the liberated iodine with the thiosulphate solution with constant shaking. When the colour of the liquid has become a pale yellow, dilute to ~200 mL with distilled water, add 2 mL of starch solution, and continue the titration until the colour changes from blue to colourless. Repeat with two other similar portions of potassium iodate.

Notes

1. The absence of iodate is indicated by adding dilute sulphuric acid when no immediate yellow coloration should be obtained. If starch is added, no immediate blue coloration should be produced.
2. Only a small amount of potassium iodate is needed, so the error in weighing 0.14–0.15 g may be appreciable. In this case it is better to weigh out accurately 4.28 g of the salt (if a slightly different weight is used, the exact molarity is calculated), dissolve it in water, and make up to 1 L in a graduated flask. Take 25 mL of this solution and treat with excess pure potassium iodide (1 g of the solid or 10 mL of 10% solution), followed by 3 mL of $1\,M$ sulphuric acid, and titrate the liberated iodine as detailed above.

With a standard solution of iodine   If a standard solution of iodine is available (Section 10.129), this may be used to standardise the thiosulphate solution. Measure a 25.0 mL portion of the standard iodine solution into a 250 mL conical flask, add about 150 mL distilled water and titrate with the thiosulphate solution, adding 2 mL of starch solution when the liquid is pale yellow in colour.

When thiosulphate solution is added to a solution containing iodine, the overall reaction occurs rapidly and stoichiometrically under the usual experimental conditions (pH < 5):

$$2S_2O_3^{2-} + I_2 = S_4O_6^{2-} + 2I^-$$

or

$$2S_2O_3^{2-} + I_3^- = S_4O_6^{2-} + 3I^-$$

The colourless intermediate $S_2O_3I^-$ is formed by a rapid reversible reaction

$$S_2O_3^{2-} + I_2 \rightleftharpoons S_2O_3I^- + I^-$$

The intermediate reacts with thiosulphate ion to provide the main course of the overall reaction

$$S_2O_3I^- + S_2O_3^{2-} = S_4O_6^{2-} + I^-$$

The intermediate also reacts with iodide ion:

$$2S_2O_3I^- + I^- = S_4O_6^{2-} + I_3^- ;$$

This explains the reappearance of iodine after the end point in the titration of very dilute iodine solutions by thiosulphate.

10.133   Copper in crystallised copper sulphate

Procedure   Weigh out accurately about 3.0 g of the salt, dissolve it in water, and make up to 250 mL in a graduated flask. Shake well. Pipette 50.0 mL of this solution into a 250 mL conical flask, add 1 g potassium iodide or 10 mL of a 10% solution (note 1), and titrate the liberated iodine with standard $0.1 M$ sodium thiosulphate (note 2). Repeat the titration with two other 50 mL portions of the copper sulphate solution.

Written in molecular form, the reaction is

$$2CuSO_4 + 4KI = 2CuI + I_2 + 2K_2SO_4$$

from which it follows that

$$2CuSO_4 \equiv I_2 \equiv 2Na_2S_2O_3$$

Notes

1. If in a similar determination, free mineral acid is present, a few drops of dilute sodium carbonate solution must be added until a **faint** permanent precipitate remains, and this is removed by means of a drop or two of ethanoic acid. The potassium iodide is then added and the titration continued. For accurate results, the solution should have a pH of 4 to 5.5.
2. After adding the potassium iodide solution, run in standard $0.1 M$ sodium thiosulphate until the brown colour of the iodine fades, then add 2 mL of starch solution, and continue adding the thiosulphate solution until the blue colour begins to fade. Then

add about 1 g of potassium thiocyanate or ammonium thiocyanate, preferably as a 10% aqueous solution; the blue colour will instantly become more intense. Complete the titration as quickly as possible. The precipitate possesses a pale pink colour, and a distinct permanent end point is readily obtained.

10.134   Chlorates

Discussion   One procedure is based on the reaction between chlorate and iodide in the presence of concentrated hydrochloric acid:

$$ClO_3^- + 6I^- + 6H^+ = Cl^- + 3I_2 + 3H_2O$$

The liberated iodine is titrated with standard sodium thiosulphate solution.

In another method the chlorate is reduced with bromide in the presence of ~8 M hydrochloric acid, and the bromine liberated is determined iodimetrically:

$$ClO_3^- + 6Br^- + 6H^+ = Cl^- + 3Br_2 + 3H_2O$$

Procedure 1   Place 25 mL of the chlorate solution (~0.02 M) in a glass-stoppered conical flask and add 3 mL of concentrated hydrochloric acid followed by two portions of about 0.3 g each of pure sodium hydrogencarbonate to remove air. Add immediately about 1.0 g of iodate-free potassium iodide and 22 mL of concentrated hydrochloric acid. Stopper the flask, shake the contents, and allow it to stand for 5–10 min. Titrate the solution with standard 0.1 M sodium thiosulphate in the usual manner.

Procedure 2   Place 10.0 mL of the chlorate solution in a glass-stoppered flask, add ~1.0g potassium bromide and 20 mL concentrated hydrochloric acid (the final concentration of acid should be about 8 M). Stopper the flask, shake well, and allow to stand for 5–10 minutes. Add 100 mL of 1% potassium iodide solution, and titrate the liberated iodine with standard 0.1 M sodium thiosulphate.

10.135   Hydrogen peroxide

Discussion   Hydrogen peroxide reacts with iodide in acid solution in accordance with the equation

$$H_2O_2 + 2H^+ + 2I^- = I_2 + 2H_2O$$

The reaction velocity is comparatively slow, but increases with increasing concentration of acid. Addition of 3 drops of a neutral 20% ammonium molybdate solution renders the reaction almost instantaneous, but as it also accelerates the atmospheric oxidation of the hydriodic acid, the titration is best conducted in an inert atmosphere (nitrogen or carbon dioxide).

The iodometric method has the advantage over the permanganate method (Section 10.114) that it is less affected by stabilisers which are sometimes added to commercial hydrogen peroxide solutions. These preservatives are often boric acid, salicylic acid and glycerol, and they render the permanganate results less accurate.

Procedure   Dilute the hydrogen peroxide solution to ~0.3% H_2O_2. Thus, if 20-volume hydrogen peroxide is used, transfer 10.0 mL by means of a burette or pipette to a 250 mL graduated flask, and make up to the mark. Shake well. Remove 25.0 mL of this diluted solution, and add it gradually and with constant stirring to a solution of 1 g of pure potassium iodide in 100 mL of 1 M sulphuric acid (1 : 20) contained in a stoppered bottle. Allow the mixture to stand for 15 min, and titrate the liberated iodine with standard 0.1 M sodium thiosulphate, adding 2 mL starch solution when the colour of the iodine has been nearly discharged. Run a blank determination at the same time.

Better results are obtained by transferring 25.0 mL of the diluted hydrogen peroxide solution to a conical flask, and adding 100 mL 1 M sulphuric acid (1 : 20). Pass a slow stream of carbon dioxide or nitrogen through the flask, add 10 mL of 10% potassium iodide solution, followed by 3 drops of 3% ammonium molybdate solution. Titrate the liberated iodine immediately with standard 0.1 M sodium thiosulphate in the usual way. This method may also be used for all per-salts.

10.136 Dissolved oxygen

Discussion   One of the most useful titrations involving iodine was originally developed by Winkler[15] to determine the amount of oxygen in samples of water. The dissolved oxygen content is not only important with respect to the species of aquatic life which can survive in the water, but is also a measure of its ability to oxidise organic impurities in the water (Section 10.121). Despite the advent of the oxygen selective electrode, direct titrations on water samples are still used extensively.[16]

In order to avoid loss of oxygen from the water sample, it is 'fixed' by its reaction with manganese(II) hydroxide which is converted rapidly and quantitatively to manganese(III) hydroxide:

$$4Mn(OH)_2 + O_2 + 2H_2O \rightarrow 4Mn(OH)_3$$

The brown precipitate obtained dissolves on acidification and oxidises iodide ions to iodine:

$$Mn(OH)_3 + I^- + 3H^+ \rightarrow Mn^{2+} + \tfrac{1}{2}I_2 + 3H_2O$$

The free iodine may then be determined by titration with sodium thiosulphate (Section 10.113).

$$2S_2O_3^{2-} + I_2 \rightarrow S_4O_6^{2-} + 2I^-$$

This means that 4 mol thiosulphate correspond to 1 mol dissolved oxygen. The main interference in this process is due to the presence of nitrites (especially in waters from sewage treatment). This is overcome by treating the original water sample with sodium azide, which destroys any nitrite when the sample is acidified:

$$HNO_2 + HN_3 \rightarrow N_2 + N_2O + H_2O$$

Procedure   The water sample should be collected by carefully filling a 200–250 mL bottle to the very top and stoppering it while it is below the water surface. This should eliminate any further dissolution of atmospheric oxygen. By using a dropping pipette placed below the surface of the water sample, add 1 mL of a 50% manganese(II) solution (note 1) and in a similar way add 1 mL of alkaline iodide–azide solution (note 2). Restopper the water sample and shake the mixture well. The manganese(III) hydroxide forms as a brown precipitate. Allow the precipitate to settle completely for 15 min and add 2 mL of concentrated phosphoric(V) acid (85%). Replace the stopper and turn the bottle upside down two or three times in order to mix the contents. The brown precipitate will dissolve and release iodine in the solution (note 3).

Measure out a 100 mL portion of the solution with a pipette and titrate the iodine with approximately 0.0125 M standard sodium thiosulphate solution adding 2 mL of starch solution indicator as the titration proceeds and **after** the titration liquid has become pale yellow in colour. Calculate the dissolved oxygen content and express it as mg L^{-1}; 1 mL of 0.0125 M thiosulphate $\equiv$ 1 mg dissolved oxygen.

Notes

1. Prepared by dissolving 50 g of manganese(II) sulphate pentahydrate in water and making up to 100 mL.
2. Prepared from 49 g of sodium hydroxide, 20 g of potassium iodide and 0.5 g of sodium azide made up to 100 mL with water.
3. If the brown precipitate has not completely dissolved, add a little more phosphoric(V) acid (a few drops).

10.137   Available chlorine in hypochlorites

Discussion   Most hypochlorites are normally obtained only in solution, but calcium hypochlorite exists in the solid form in commercial bleaching powder which consists essentially of a mixture of calcium hypochlorite $Ca(OCl)_2$ and the basic chloride $CaCl_2 \cdot Ca(OH)_2 \cdot H_2O$; some free slaked lime is usually present. The active constituent is the hypochlorite, which is responsible for the bleaching action. Upon treating bleaching powder with hydrochloric acid, chlorine is liberated:

$$OCl^- + Cl^- + 2H^+ = Cl_2 + H_2O$$

The **available chlorine** refers to the chlorine liberated by the action of dilute acids on the hypochlorite; it is expressed as a percentage by weight in the case of bleaching powder. Commercial bleaching powder contains 36–38% of available chlorine. The hypochlorite solution or suspension is treated with excess of a potassium iodide solution and strongly acidified with ethanoic acid:

$$OCl^- + 2I^- + 2H^+ \rightleftharpoons Cl^- + I_2 + H_2O$$

The liberated iodine is titrated with standard sodium thiosulphate solution. The solution should not be strongly acidified with hydrochloric acid, for the little calcium chlorate which is usually present, by virtue of the decomposition of the hypochlorite, will react slowly with the potassium iodide and liberate iodine:

$$ClO_3^- + 6I^- + 6H^+ = Cl^- + 3I_2 + 3H_2O$$

Procedure (iodometric method)   Weigh out accurately about 5.0 g of the bleaching powder into a clean glass mortar. Add a little water, and rub the mixture to a smooth paste. Add a little more water, triturate with the pestle, allow the mixture to settle, and pour off the milky liquid into a 500 mL graduated flask. Grind the residue with a little more water, and repeat the operation until the whole of the sample has been transferred to the flask, either in solution or in a state of very fine suspension, and the mortar washed quite clean. The flask is them filled to the mark with distilled water, well shaken, and 50.0 mL of the turbid liquid immediately withdrawn with a pipette. This is transferred to a 250 mL conical flask, 25 mL of water added, followed by 2 g of iodate-free potassium iodide (or 20 mL of a 10% solution) and 10 mL of glacial ethanoic acid. Titrate the liberated iodine with standard 0.1 M sodium thiosulphate.

10.138 Hexacyanoferrates(III)

Discussion The reaction between hexacyanoferrates(III) (ferricyanides) and soluble iodides is a reversible one:

$$2[Fe(CN)_6]^{3-} + 2I^- \rightleftharpoons 2[Fe(CN)_6]^{4-} + I_2$$

In strongly acid solution the reaction proceeds from left to right, but is reversed in almost neutral solution. Oxidation also proceeds quantitatively in a slightly acid medium in the presence of a zinc salt. The very sparingly soluble potassium zinc hexacyanoferrate(II) is formed, and the hexacyanoferrate(II) ions are removed from the sphere of action:

$$2[Fe(CN)_6]^{4-} + 2K^+ + 3Zn^{2+} = K_2Zn_3[Fe(CN)_6]_2$$

The procedure may be used to determine the purity of potassium hexacyanoferrate(III).

Procedure Weigh out accurately about 10 g of the salt and dissolve it in 250 mL of water in a graduated flask. Pipette 25 mL of this solution into a 250 mL conical flask, add about 20 mL of 10% potassium iodide solution, 2 mL of 1 M sulphuric acid, and 15 mL of a solution containing 2.0 g crystallised zinc sulphate. Titrate the liberated iodine immediately with standard 0.1 M sodium thiosulphate and starch; and the starch solution (2 mL) after the colour has faded to a pale yellow. The titration is complete when the blue colour has just disappeared.

10.139 Vitamin C tablets

Discussion Ascorbic acid (vitamin C) rapidly reduces iodine to iodide. This reaction can form the basis of a direct titration of vitamin C with a standard iodine solution using starch as the indicator. An alternative procedure is to generate excess iodine (by reaction of iodate with iodide), which then reacts with the ascorbic acid.[17] The excess iodine is then titrated with standard thiosulphate solution.

Procedure Prepare an approximately 0.01 M solution of potassium iodate by weighing out accurately about 2.0 g of the solid and making up to the mark in a 1 L graduated flask with deionised water. Use a previously standardised thiosulphate solution (~0.08 M); see Sections 10.131 and 10.132. Use starch indicator solution; see Section 10.128. And use commercial tablets, each containing about 100 mg of vitamin C. Dissolve two tablets in 70 mL of sulphuric acid (0.25 M); some binders may not dissolve. Add to this solution (suspension) potassium iodide (2 g) and 50.00 mL of the standard potassium iodate solution. Finally, back-titrate with the standardised thiosulphate solution, adding 2 mL starch indicator just before the end point (when the solution is a pale yellow colour). This procedure can be adapted to determine glucose and other reducing sugars. In this case the reaction takes place in sodium hydroxide solution, which is then acidified (after 5 min) with hydrochloric acid. The solution is then back-titrated with standard thiosulphate.

Oxidations with potassium iodate

! **Safety.** Before carrying out any experiments in this section, pay full attention to any safety warnings and make sure you adhere to national laboratory and safety regulations.

10.140   General discussion

Potassium iodate is a powerful oxidising agent, but the course of the reaction is governed by the conditions under which it is employed. The reaction between potassium iodate and reducing agents such as iodide ion or arsenic(III) oxide in solutions of moderate acidity (0.1–2.0 M hydrochloric acid) stops at the stage when the iodate is reduced to iodine:

$$IO_3^- + 5I^- + 6H^+ = 3I_2 + 3H_2O$$

$$2IO_3^- + 5H_3AsO_3 + 2H^+ = I_2 + 5H_3AsO_4 + H_2O$$

The first of these reactions if very useful for the generation of known amounts of iodine, and it also serves as the basis of a method for standardising solutions of acids (Section 10.127).

With a more powerful reductant, e.g. titanium(III) chloride, the iodate is reduced to iodide:

$$IO_3^- + 6Ti^{3+} + 6H^+ = I^- + 6Ti^{4+} + 3H_2O$$

In more strongly acid solutions (3–6 M hydrochloric acid) reduction occurs to iodine monochloride, and it is under these conditions that it is most widely used:[18,19]

$$IO_3^- + 6H^+ + Cl^- + 4e \rightleftharpoons ICl + 3H_2O$$

In hydrochloric acid solution, iodine monochloride forms a stable complex ion with chloride ion:

$$ICl + Cl^- \rightleftharpoons ICl_2^-$$

The overall half-cell reaction may therefore be written as

$$IO_3^- + 6H^+ + 2Cl^- + 4e \rightleftharpoons ICl_2^- + 3H_2O$$

The reduction potential is 1.23 V; hence under these conditions potassium iodate acts as a very powerful oxidising agent.

Oxidation by iodate ion in a strong hydrochloric acid medium proceeds through several stages:

$$IO_3^- + 6H^+ + 6e \rightleftharpoons I^- + 3H_2O$$

$$IO_3^- + 5I^- + 6H^+ = 3I_2 + 3H_2O$$

$$IO_3^- + 2I_2 + 6H^- = 5I^+ + 3H_2O$$

In the initial stages of the reaction free iodine is liberated[20] as more titrant is added, oxidation proceeds to iodine monochloride, and the dark colour of the solution gradually disappears. The overall reaction may be written as

$$IO_3^- + 6H^+ + 4e \rightleftharpoons I^+ + 3H_2O$$

The reaction has been used for the determination of many reducing agents; the optimum acidity for reasonably rapid reaction varies from one reductant to another within the range 2.5 to 9 M hydrochloric acid. In many cases the concentration of acid is not critical, but for Sb(III) it is 2.5–3.5 M.

Under these conditions starch cannot be used as indicator because the characteristic blue colour of the starch–iodine complex is not formed at high concentrations of acid. In the original procedure, a few millilitres of an immiscible solvent (chloroform) were added to the solution being titrated contained in a glass-stoppered bottle or conical flask. The end point is marked by the disappearance of the last trace of violet colour, due to iodine, from the solvent; iodine monochloride is not extracted and imparts a pale yellowish colour to the aqueous phase. The extraction end point is very sharp. The main disadvantage is the inconvenience of vigorous shaking with the extraction solvent in a stoppered vessel after each addition of the reagent near the end point.

The immiscible solvent may be replaced by certain dyes, e.g. amaranth (colour index 16185), colour change red to colourless; xylidine ponceau (colour index 16150) colour change orange to colourless. The indicators are used as 0.2% aqueous solutions and about 0.5 mL per titration is added near the end point. The dyes are destroyed by the first excess of iodate, hence the indicator action is irreversible. The indicator blank is equivalent to 0.05 mL of 0.025 M potassium iodate per 1.0 mL of indicator solution, and is therefore virtually negligible.

10.141   Preparing 0.025 M potassium iodate

Dry some potassium iodate at 120 °C for 1 h and allow it to cool in a covered vessel in a desiccator. Weigh out exactly 5.350 g of the finely powdered potassium iodate on a watch-glass, and transfer it by means of a clean camel-hair brush directly into a dry 1 L graduated flask. Add about 400–500 mL of water, and gently rotate the flask until the salt is completely dissolved. Make up to the mark with distilled water. Shake well. The solution will keep indefinitely. This 0.025 M solution is intended for the reaction

$$IO_3^- + 6H^+ + Cl^- + 4e \rightleftharpoons ICl + 3H_2O$$

but when used in solutions of moderate acidity leading to the liberation of free iodine, ideally the solution then requires $4.28 \, g \, L^{-1}$ potassium iodate; the method of preparation will be as described above with suitable adjustment of the weight of salt taken.

10.142   Arsenic or antimony

Discussion   The determination of arsenic in arsenic(III) compounds is based on the following reaction:

$$IO_3^- + 2H_3AsO_3 + 2H^+ + Cl^- = ICl + 2H_3AsO_4 + H_2O$$

A similar reaction occurs with antimony compounds in which the end point is best obtained by using amaranth as the indicator:

$$IO_3^- + 2[SbCl_4]^- + 6H^+ + 5Cl^- = ICl + 2[SbCl_6]^- + 3H_2O$$

Procedure   Weigh out accurately about 1.1 g of the oxide sample, dissolve in a small quantity of warm 10% sodium hydroxide solution, cool and then make up to 250 mL in a graduated flask. **Carefully** pipette 25.0 mL into a 250 mL conical flask; this solution is **poisonous**. Add 25 mL water, 60 mL of concentrated hydrochloric acid. Cool to room temperature. Titrate with the standard 0.025 M potassium iodate until the deeply coloured solution becomes pale brown. Then add 0.5 mL of 0.2% aqueous solution of either amaranth (A) or xylidine ponceau (XP) indicator, when the colour change at the end point is red to colourless (A) or orange to colourless (XP). Continue the addition dropwise until the end point is reached. Remember, the indicator reaction is non-reversible – the dyestuff is colour-destroyed by the first excess iodate solution.

10.143   Hydrazine

Discussion   Hydrazine reacts with potassium iodate under the usual Andrews[23] conditions, thus

$$IO_3^- + N_2H_4 + 2H^+ + Cl^- = ICl + N_2 + 3H_2O$$

And

$$KIO_3 \equiv N_2H_4$$

To determine the $N_2H_4 \cdot H_2SO_4$ content of hydrazinium sulphate, use the following method.

Procedure   Weigh out accurately 0.08–0.10 g of hydrazinium sulphate into a 250 mL conical flask, add 30 mL concentrated hydrochloric acid then add 20 mL water. Run in the standard 0.025 M potassium iodate slowly from a burette until the solution is pale brown. Then add 0.5 mL of 0.2% aqueous solution of amaranth indicator. Continue to add the iodate dropwise until the colour changes from red to colourless.

10.144   Other ions

Copper(II) compounds   Many other metallic ions which are capable of undergoing oxidation by potassium iodate can also be determined. Thus, copper(II) compounds can be analysed by precipitation of copper(I) thiocyanate which is titrated with potassium iodate:

$$7IO_3^- + 4CuSCN + 18H^+ + 7Cl^- = 7ICl + 4Cu^{2+} + 4HSO_4^- + 4HCN + 5H_2O$$

As a typical example, 0.8 g of copper(II) sulphate ($CuSO_4 \cdot 5H_2O$) is dissolved in water, 5 mL of 0.5 M sulphuric acid added, and the solution made up to 250 mL in a graduated flask; 25.0 mL of the resulting solution are pipetted into a 250 mL conical flask, 10–15 mL of freshly prepared sulphurous acid solution added, and then after heating to boiling, 10% ammonium thiocyanate solution is added slowly from a burette with constant stirring until there is no further change in colour, and then 4 mL of reagent is added in excess. After allowing the precipitate to settle for 10–15 min, it is filtered through a fine filter paper then washed with cold 1% ammonium sulphate solution until free from thiocyanate. It is then transferred quantitatively into the vessel in which the titration is to be performed, and after adding 30 mL of concentrated hydrochloric acid, followed by 20 mL of water, the titration is carried out in the usual manner either with an organic solvent present, or with an internal indicator being added as the end point is approached. For the internal indicator use either amaranth or xylidine ponceau (Section 10.142).

Thallium(I) salts   These are oxidised in accordance with the equation

$$IO_3^- + 2Tl^+ + 6H^+ + Cl^- = ICl + 2Tl^{3+} + 3H_2O$$

so that

$$KIO_3 \equiv 2Tl$$

The solution should contain 0.25–0.30 g Tl^+ in 20 mL plus 60 mL of concentrated hydrochloric acid; it is titrated as usual with 0.025 M KIO_3 solution.

Warning   **Thallium salts are extremely poisonous and should be handled with care.**

441

Oxidations with potassium bromate

!
■
Safety. Before carrying out any experiments in this section, pay full attention to any safety warnings and make sure you adhere to national laboratory and safety regulations.

10.145   General discussion

Potassium bromate is a powerful oxidising agent which is reduced smoothly to bromide:

$$BrO_3^- + 6H^+ + 6e \rightleftharpoons Br^- + 3H_2O$$

The relative molecular mass is 167.00, and a $0.02\,M$ solution contains $3.34\,g\,L^{-1}$ potassium bromate. At the end of the titration free bromine appears:

$$BrO_3^- + 5Br^- + 6H^+ = 3Br_2 + 3H_2O$$

The presence of free bromine, and consequently the end point, can be detected by its yellow colour, but it is better to use indicators such as methyl orange, methyl red, naphthalene black 12B, xylidine ponceau and fuchsine. These indicators have their usual colour in acid solution, but are destroyed by the first excess of bromine. With all irreversible oxidation indicators the destruction of the indicator is often premature to a slight extent; a little additional indicator is usually required near the end point. The quantity of bromate solution consumed by the indicator is exceedingly small, and the blank can be neglected for $0.02\,M$ solutions. Direct titrations with bromate solution in the presence of irreversible dyestuff indicators are usually made in hydrochloric acid solution, concentration at least 1.5 to $2\,M$. At the end of the titration some chlorine may appear by virtue of the reaction

$$10Cl^- + 2BrO_3^- + 12H^+ = 5Cl_2 + Br_2 + 6H_2O$$

This immediately bleaches the indicator.

The titrations should be carried out slowly so the indicator change, which is a time reaction, may be readily detected. If the determinations are to be executed rapidly, the volume of the bromate solution to be used must be known approximately, since ordinarily with irreversible dyestuff indicators there is no simple way of ascertaining when the end point is close at hand. With the highly coloured indicators (xylidine ponceau, fuchsine and naphthalene black 12B), the colour fades as the end point is approached (owing to local excess of bromate) and another drop of indicator can be added. At the end point the indicator is irreversibly destroyed and the solution becomes colourless or almost colourless. If the fading of the indicator is confused with the equivalence point, another drop of the indicator may be added. If the indicator has faded, the additional drop will colour the solution; if the end point has been reached, the additional drop of indicator will be destroyed by the slight excess of bromate present in the solution.

The introduction of reversible redox indicators for the determination of arsenic(III) and antimony(III) has considerably simplified the procedure; those at present available include 1-naphthoflavone, and p-ethoxychrysoidine. The addition of a little tartaric acid or potassium sodium tartrate is recommended when antimony(III) is titrated with bromate in the presence of the reversible indicators; this will prevent hydrolysis at the lower acid concentrations. The end point may be determined with high precision by potentiometric titration.

Examples of some direct titrations with bromate solutions are given by the following equations:

$$BrO_3^- + 3H_3AsO_3 \xrightarrow{(HCl)} Br^- + 3H_3AsO_4$$

$$2BrO_3^- + 3N_2H_4 \xrightarrow{(HCl)} 2Br^- + 3N_2 + 6H_2O$$

$$BrO_3^- + NH_2OH \xrightarrow{(HCl)} Br^- + NO_3^- + H^+ + H_2O$$

$$BrO_3^- + 6[Fe(CN)_6]^{4-} + 6H^+ \rightarrow Br^- + 6[Fe(CN)_6]^{3-} + 3H_2O$$

Various substances cannot be oxidised directly with potassium bromate, but react quantitatively with an excess of bromine. Acid solutions of bromine of exactly known concentration are readily obtainable from a standard potassium bromate solution by adding acid and an excess of bromide:

$$BrO_3^- + 5Br^- + 6H^+ = 3Br_2 + 3H_2O$$

In this reaction 1 mol of bromate yields six atoms of bromine. Bromine is very volatile, hence operations should be conducted at temperatures as low as possible and in conical flasks fitted with ground-glass stoppers. The excess of bromine may be determined iodometrically by the addition of excess potassium iodide and titration of the liberated iodine with standard thiosulphate solution:

$$2I^- + Br_2 = I_2 + 2Br^-$$

Potassium bromate is readily available in a high state of purity; it has an assay value of at least 99.9%. It can be dried at 120–150 °C, it is anhydrous, and the aqueous solution keeps indefinitely. Potassium bromate can therefore be used as a primary standard. Its only disadvantage is that one-sixth of the relative molecular mass is a comparatively small quantity.

10.146   Preparing 0.02 M potassium bromate

Dry some finely powdered potassium bromate for 1–2 h at 120 °C, and allow to cool in a closed vessel in a desiccator. Weigh out accurately 3.34 g of the pure potassium bromate, and dissolve in 1 L of water in a graduated flask.

10.147   Metals: using 8-hydroxyquinoline (oxine)

Discussion   Various metals (e.g. aluminium, iron, copper, zinc, cadmium, nickel, cobalt, manganese and magnesium) under specified conditions of pH yield well-defined crystalline precipitates with 8-hydroxyquinoline. These precipitates have the general formula $M(C_9H_6ON)_n$, where n is the charge on the M ion, see Section 11.3. Upon treatment of the oxinates with dilute hydrochloric acid, the oxine is liberated. One molecule of oxine reacts with two molecules of bromine to give 5,7-dibromo-8-hydroxyquinoline:

$$C_9H_7ON + 2Br_2 = C_9H_5ONBr_2 + 2H^+ + 2Br^-$$

Hence 1 mol of the oxinate of a doubly charged metal corresponds to 4 mol of bromine, whereas 1 mol of a triply charged metal corresponds to 6 mol. The bromine is derived by adding standard 0.02 M potassium bromate and excess potassium bromide to the acid solution:

$$BrO_3^- + 5Br^- + 6H^+ = 3Br_2 + 3H_2O$$

Full details are given for the determination of aluminium by this method. Many other metals may be determined by this same procedure, but in many cases complexometric titration offers a simpler method of determination. In cases where the oxine method offers advantages, the experimental procedure may be readily adapted from the details given for aluminium.

Determination of aluminium   Prepare a 2% solution of 8-hydroxyquinoline (Section 11.3) in $2\,M$ ethanoic (acetic) acid; add ammonia solution until a **slight** precipitate persists, then redissolve it by warming the solution. Transfer 25 mL of the solution to be analysed, containing about 0.02 g of aluminium, to a conical flask, add 125 mL of water and warm to 50–60 °C. Then add a 20% excess of the oxine solution (1 mL will precipitate 0.001 g of Al), when the complex $Al(C_9H_6ON)_3$ will be formed. Complete the precipitation by adding a solution of 4.0 g of ammonium ethanoate in the minimum quantity of water, stir the mixture, and allow to cool. Filter the granular precipitate through a sintered-glass crucible of porosity no. 4 and wash with warm water (see note).

Dissolve the complex in warm concentrated hydrochloric acid, collect the solution in a 250 mL reagent bottle, add a few drops of indicator (0.1% solution of the sodium salt of methyl red or 0.1% methyl orange solution), and 0.5 to 1 g of pure potassium bromide. Titrate slowly with standard $0.02\,M$ potassium bromate until the colour becomes pure yellow (with either indicator). The exact end point is not easy to detect, and the best procedure is to add an excess of potassium bromate solution, i.e. a further 2 mL beyond the estimated end point, so the solution now contains free bromine. Dilute the solution considerably with $2\,M$ hydrochloric acid (to prevent the precipitation of 5,7-dibromo-8-hydroxyquinoline during the titration), then after 5 min add 10 mL of 10% potassium iodide solution, and titrate the liberated iodine with standard $0.1\,M$ sodium thiosulphate, using starch as indicator to determine the excess bromate (Section 10.145). It is evident that $Al \equiv 12Br$.

Note   This will remove the excess of oxine. Complications due to adsorption of iodine will thus be avoided.

10.148   Hydroxylamine

Procedure   The method based on the reduction of iron(III) solutions in the presence of sulphuric acid, boiling, and subsequent titration in the cold with standard $0.02\,M$ potassium permanganate frequently yields high results unless the experimental conditions are closely controlled:

$$2NH_2OH + 4Fe^{3+} = N_2O + 4Fe^{2+} + 4H^+ + H_2O$$

Better results are obtained by oxidation with potassium bromate in the presence of hydrochloric acid:

$$NH_2OH + BrO_3^- = NO_3^- + Br^- + H^+ + H_2O$$

The hydroxylamine solution is treated with a measured volume of $0.02\,M$ potassium bromate so as to give 10–30 mL excess, followed by 40 mL of $5\,M$ hydrochloric acid. After 15 min the excess bromate is determined by adding potassium iodide solution and titrating with standard $0.1\,M$ sodium thiosulphate (Section 10.147).

10.149   Phenol

Discussion   A number of phenols can be substituted rapidly and quantitatively with bromine produced from bromate and bromide[21] in acid solution (Section 10.145). The determination involves treating phenol (note 1) with an excess of potassium bromate and potassium bromide; when bromination of the phenol is complete the unreacted bromine is then determined by adding excess potassium iodide and back-titrating the liberated iodine with standard sodium thiosulphate.

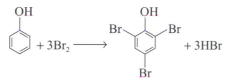

Procedure   Prepare a ~0.02 M standard solution of potassium bromate by weighing accurately about 1.65 g of the analytical grade reagent, dissolving it in water and making it up to 500 mL in a graduated flask (note 2).

To determine the purity of a sample of phenol, weigh out accurately approximately 0.3 g of phenol, dissolve it in water and make the volume to 250 mL in a graduated flask. Pipette 25 mL volumes of this solution into 250 mL stoppered (ground-glass) conical flasks. To each flask pipette 25 mL of the standard potassium bromate solution and add 0.5 g of potassium bromide and 5 mL of 3 M sulphuric acid (note 3). Mix the reagents and let them stand for 15 min, then rapidly add about 2.5 g of potassium iodide to each flask, immediately restoppering and swirling the contents to dissolve the solid. Titrate the liberated iodine with the standard 0.1 M sodium thiosulphate until the solution is only slightly yellow, then add 5 mL of starch indicator solution and continue the titration until the blue colour disappears.

Calculate the amount of excess bromate from the amount of thiosulphate needed for the back titration of the free iodine, hence the quantity of bromate which reacted with the phenol.

Notes

1.   Other phenols which undergo this type of reaction include 4-chlorophenol, *m*-cresol (3-methylphenol) and 2-naphthol.
2.   The concentration of the potassium bromate can be checked by the following method. Pipette 25 mL of the solution into a 250 mL conical flask, add 2.5 g of potassium iodide and 5 mL of 3 M sulphuric acid. Titrate the liberated iodine with standard 0.1 M sodium thiosulphate (Section 10.131) until the solution is faintly yellow. Add 5 mL of starch indicator solution and continue the titration until the blue colour disappears.
3.   The flasks must be stoppered at all times after the addition of reagents to prevent the loss of bromine.

Reduction of higher oxidation states

!   **Safety.** Before carrying out any experiments in this section, pay full attention to any safety warnings and make sure you adhere to national laboratory and safety regulations.

10.150   General discussion

Before titration with an oxidising agent can be carried out, it may sometimes be necessary to reduce the compound supplied to a lower state of oxidation. Such a situation is frequently encountered with the determination of iron; iron(III) compounds must be reduced to iron(II) before titration with potassium permanganate or potassium dichromate can be performed. It is possible to carry out such determinations directly as a **reductimetric titration** by the use of solutions of powerful reducing agents such as chromium(II) chloride, titanium(III) chloride or vanadium(II) sulphate, but the problems associated with the preparation, storage and handling of these reagents have militated against their widespread use.

Titanium(III) sulphate has found application in the analysis of certain types of organic compound,[22] but is of limited application in the inorganic field. An apparatus suitable for the preparation, storage and manipulation of chromium(II) and vanadium(II) solutions is described in the literature;[23] with both these reagents it is necessary (and it is also advisable with Ti(III) solutions) to carry out titrations in an atmosphere of hydrogen, nitrogen or carbon dioxide, and in view of the instability of most indicators in the presence of these powerful reducing agents, it is frequently necessary to determine the end point potentiometrically.

The most important method for reduction of compounds to an oxidation state suitable for titration with one of the common oxidising titrants is based on the use of metal amalgams, but there are various other methods which can be used, and these will be discussed in the following sections.

10.151   Reduction with amalgamated zinc: Jones reductor

Amalgamated zinc is an excellent reducing agent for many metallic ions. Zinc reacts rather slowly with acids, but upon treatment with a dilute solution of a mercury(II) salt, the metal is covered with a thin layer of mercury; the amalgamated metal reacts quite readily. Reduction with amalgamated zinc is usually carried out in the reductor due to C. Jones. This consists of a column of amalgamated zinc contained in a long glass tube provided with a stopcock, through which the solution to be reduced may be drawn. A large surface is exposed, so this zinc column is much more efficient than pieces of zinc placed in the solution.

Figure 10.23 shows a suitable form of the Jones reductor and some approximate dimensions. A sintered-glass disc supports the zinc column. The tube below the tap passes through a tightly fitting one-holed rubber stopper into a 750 mL filter flask. It is advisable to connect another filter flask in series with the water pump, so that if any water sucks back it will not spoil the determination. The amalgamated zinc is prepared as follows. About 300 g of granulated zinc (or zinc shavings, or pure 20–30 mesh zinc) are covered with 2% mercury(II) chloride solution, **caution**, in a beaker. The mixture is stirred for 5–10 min, then the solution is decanted from the zinc, which is washed three times with water by decantation. The resulting amalgamated zinc should have a bright silvery lustre. Then the amalgamated zinc is added to the glass tube, up to its shoulder. The zinc is washed with distilled water (500 mL) using gentle suction. If the reductor is not to be used immediately, it must be left full of water in order to prevent the formation of basic salts by atmospheric oxidation, which impair the reducing surface. If the moist amalgam is exposed to the oxygen of the atmosphere, hydrogen peroxide may be generated:

$$Zn + O_2 + 2H_2O = Zn(OH)_2 + H_2O_2$$

but no hydrogen peroxide is formed if acid is present.

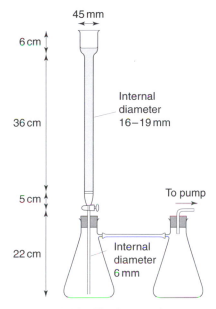

Figure 10.23   The Jones reductor

Reduction of iron(III)

Proceed as follows. The zinc is activated by filling the cup (which holds about 50 mL) with 1 M (~5%) sulphuric acid, the tap being closed. The flask is connected to a filter pump, the tap opened, and the acid **slowly** drawn through the column until it has fallen to **just above** the level of the zinc; the tap is then closed and the process repeated twice. The tap is shut, and the flask detached, cleaned, and replaced. The reductor is now ready for use. Note that during use the level of the liquid should always be just above the top of the zinc column. The solution to be reduced should have a volume of 100–150 mL, contain not more than 0.25 g of iron, and be about 1 M in sulphuric acid. The cold iron solution is passed through the reductor, using gentle suction, at a flow rate not exceeding 75–100 mL min^{-1}. As soon as the reservoir is nearly emptied of the solution, 100 mL of 2.5% (~0.5M) sulphuric acid is passed through in two portions, followed by 100–150 mL of water. The last washing is necessary in order to wash out all the reduced compound and also the acid, which would otherwise cause unnecessary consumption of the zinc. Disconnect the flask from the reductor, wash the end of the delivery tube, and titrate immediately with standard 0.02 M potassium permanganate.

Carry out a blank determination, preferably before passing the iron solution through the reductor, by running the same volumes of acid and water through the apparatus as are used in the actual determination. This should not amount to more than about 0.1 mL of 0.02 M permanganate, and should be deducted from the volume of permanganate solution used in the subsequent titration.

If hydrochloric acid has been used in the original solution of the iron-bearing material, the volume should be reduced to ~25 mL then diluted to ~150 mL with 5% sulphuric acid. The determination is carried out as detailed above, but 25 mL of Zimmermann–Reinhardt or 'preventive solution' must be added before titration with standard potassium permanganate

447

solution. For the determination of iron in hydrochloric acid solution, it is more convenient to reduce the solution in a silver reductor (Section 10.152) and to titrate the reduced solution with either standard potassium dichromate or standard cerium(IV) sulphate solution.

Applications and limitations of the Jones reductor

Solutions containing 1–10 vol% sulphuric acid or 3–15 vol% concentrated hydrochloric acid can be used in the reductor. Sulphuric acid is generally used, as hydrochloric acid may interfere in the subsequent titration, e.g. with potassium permanganate.

Nitric acid must be absent, for this is reduced to hydroxylamine and other compounds which react with permanganate. If nitric acid is present, evaporate the solution just to dryness, wash the sides of the vessel with about 3 mL of water, carefully add 3–4 mL of concentrated sulphuric acid, and evaporate until sulphuric fumes are evolved. Repeat this operation twice to ensure complete removal of the nitric acid, dilute to 100 mL with water, add 5 mL of concentrated sulphuric acid, and proceed with the reduction.

Organic matter (ethanoates, etc.) must be absent. It is removed by heating to fumes of sulphuric acid in a covered beaker, then carefully adding drops of a saturated solution of potassium permanganate until a permanent colour is obtained, and finally continuing the fuming for a few minutes.

Solutions containing compounds of copper, tin, arsenic, antimony and other reducible metals must never be used. They must be removed before the reduction by treatment with hydrogen sulphide.

Other ions which are reduced in the reductor to a definite lower oxidation state are those of titanium to Ti(III), chromium to Cr(II), molybdenum to Mo(III), niobium to Nb(III), and vanadium to V(II). Uranium is reduced to a mixture of U(III) and U(IV), but by bubbling a stream of air through the solution in the filter flask for a few minutes, the dirty dark-green colour changes to the bright apple-green colour characteristic of pure uranium(IV) salts. Tungsten is reduced, but not to any definite lower oxidation state.

With the exception of iron(II) and uranium(IV), the reduced solutions are extremely unstable and readily reoxidise upon exposure to air. They are best stabilised in a fivefold excess of a solution made from 150 g of ammonium iron(III) sulphate and 150 mL of concentrated sulphuric acid per litre (approximately 0.3 M with respect to iron) contained in the filter flask. The iron(II) formed is then titrated with a standard solution of a suitable oxidising agent. Titanium and chromium are completely oxidised and produce an equivalent amount of iron(II) sulphate; molybdenum is reoxidised to the Mo(V) stage (red), which is fairly stable in air, and complete oxidation is effected by the permanganate, but the net result is the same, i.e. Mo(III) → Mo(VI); vanadium is reoxidised to the V(IV) condition, which is stable in air, and the final oxidation is completed by slow titration with potassium permanganate solution or with cerium(IV) sulphate solution.

10.152   The silver reductor

The silver reductor has a relatively low reduction potential (the Ag/AgCl electrode potential in 1 M hydrochloric acid is 0.2245 V), so it cannot perform many of the reductions achievable with amalgamated zinc. The silver reductor is preferably used with hydrochloric acid solutions, and this is frequently an advantage. The various reductions are summarised in Table 10.12.

The silver reductor (shaped like a short, squat Jones reductor tube) may be constructed from a tube 12 cm long and 2 cm internal diameter fused to a reservoir bulb of 50–75 mL capacity. Suction is not always required. The silver is conveniently prepared as follows on a large scale; for preparations on a smaller scale the procedure must be appropriately

Table 10.12   *Reductions with the silver reductor and the Jones reductor*

Silver reductor Hydrochloric acid solution	Amalgamated zinc (Jones) reductor Sulphuric acid solution
$Fe^{3+} \rightarrow Fe^{2+}$	$Fe^{3+} \rightarrow Fe^{2+}$
Ti(IV) not reduced	Ti(IV) $\rightarrow$ Ti(III)
Mo(VI) $\rightarrow$ Mo(V) ($2\,M$ HCl, 60–80 °C)	Mo(VI) $\rightarrow$ Mo(III)
Cr(III) not reduced	Cr(III) $\rightarrow$ Cr(II)
$UO_2^{2+} \rightarrow$ U(IV) ($4\,M$ HCl, 60–90 °C)	$UO_2^{2+} \rightarrow$ U(III) + U(IV)
V(V) $\rightarrow$ V(IV)	V(V) $\rightarrow$ V(II)
$Cu^{2+} \rightarrow Cu^+$ ($2\,M$ HCl)	$Cu^{2+} \rightarrow Cu^0$

adapted. A solution of 500 g of silver nitrate in 2500 mL of water, slightly acidified with dilute nitric acid, is placed in a 4 L beaker. Cathodes consisting of two heavy-gauge platinum plates, each 10 cm square, are suspended in the electrolyte by the use of a heavy copper busbar connection to a source of current. The anode consists of a silver rod 200 mm long and 10–25 mm in diameter, or a similar weight of silver as a heavy-gauge rectangular sheet; it is suspended in the centre of the electrolyte with the platinum cathodes placed at the outer edges of the deposition cell. Silver is deposited as granular crystals with high surface-to-mass ratio by a current of 60–70 A at 5–6 V. These crystals, obtained in excellent yield, are deposited on the four outside edges of the cathodes; they should be dislodged by gentle tapping, and washed by decantation with dilute sulphuric acid. About 30 g of silver in this form occupy a volume of 40–50 mL – enough to fill one reductor tube.

The necessary quantity of silver is introduced into the reductor above the sintered disc; using a glass rod flattened at one end, it is compressed to as great an extent as necessary but without restricting the free flow of solution through the column. The reductor is rinsed with 100 mL of $1\,M$ hydrochloric acid, added in five equal portions, each consecutive portion being allowed to pass through the reductor to just above the level of the silver.

With hydrochloric acid solutions, a dark silver chloride coating covers the silver of the upper part of the reductor, which moves further down the column during use. When it extends to about three-quarters of the length of the column, the reductor must be regenerated by the following method. The reductor is rinsed with water and filled completely with 1 : 3 ammonia solution. The silver chloride dissolves; after 10 min the solution is rinsed out of the reductor tube with water, followed by $1\,M$ hydrochloric acid and is then ready for reuse. As a precautionary measure, the ammoniacal solution of silver chloride should be immediately acidified. The wastage of silver associated with this method may be avoided by filling the tube with sulphuric acid ($0.1\,M$) and then inserting a rod of zinc with its lower end well buried in the silver; when the reduction is complete (as evidenced by loss of the dark colour), the column is well washed with water and is then ready for use.

Examples of using the silver reductor are given in Sections 10.125 and 10.126.

10.153   Other methods of reduction

Although metal amalgams, particularly the Jones reductor or the related silver reductor, are the best way to reduce solutions before titration with an oxidant, sometimes a Jones reductor is not available, so a simpler procedure may be needed. This is most likely to arise when determining iron, where iron(III) may need to be reduced to iron(II).

Tin(II) chloride solution

Many iron ores are brought into solution with concentrated hydrochloric acid, and the resulting solution may be readily reduced with tin(II) chloride:

$$2Fe^{3+} + Sn^{2+} = 2Fe^{2+} + Sn^{4+}$$

The hot solution (70–90 °C) from about 0.3 g of iron ore, which should occupy a volume of 25–30 mL and should be 5–6 M with respect to hydrochloric acid, is reduced by adding concentrated tin(II) chloride solution dropwise from a separatory funnel or a burette, with stirring, until the yellow colour of the solution has **nearly** disappeared. The reduction is then completed by diluting the concentrated solution of tin(II) chloride with 2 volumes of dilute hydrochloric acid, and adding the dilute solution dropwise, with agitation after each addition, until the liquid has a faint green colour, quite free from any tinge of yellow. The solution is then rapidly cooled under the tap to about 20 °C, with protection from the air, and the slight excess of tin(II) chloride present removed by adding 10 mL of a saturated solution (~5%) of mercury(II) chloride rapidly in one portion, **caution**, and with thorough mixing; a **slight** silky white precipitate of mercury(I) chloride should be obtained.

The small amount of mercury(I) chloride in suspension has no appreciable effect on the oxidising agent used in the subsequent titration, but if a heavy precipitate forms, or a grey or black precipitate is obtained, too much tin(II) solution has been used; the results are inaccurate and the reduction must be repeated. Finely divided mercury reduces permanganate or dichromate ions and slowly reduces Fe^{3+} ions in the presence of chloride ion.

After adding the mercury(II) chloride solution, the whole is allowed to stand for 5 min then diluted to about 400 mL and titrated with standard potassium dichromate solution (Section 10.118) or with standard permanganate solution in the presence of 'preventive solution' (Section 10.111). Blank runs on the reagents should be carried through all the operations, and corrections made, if necessary. The concentrated solution of tin(II) chloride is prepared by dissolving 12 g of pure tin or 30 g of crystallised tin(II) chloride ($SnCl_2 \cdot 2H_2O$) in 100 mL of concentrated hydrochloric acid and diluting to 200 mL with water.

Redox reactions: determinations using instruments

! **Safety.** Before carrying out any experiments in this section, pay full attention to any safety warnings and make sure you adhere to national laboratory and safety regulations.

10.154 Potentiometry: general considerations

The theory of oxidation–reduction reactions is given in Section 2.31. The determining factor is the ratio of the concentrations of the oxidised and reduced forms of certain ion species. For the reaction

oxidised form + n electrons $\rightleftharpoons$ reduced form

the potential E acquired by the indicator electrode at 25 °C is given by

$$E = E^{\ominus} + \frac{0.0591}{n} \log \frac{[\text{ox}]}{[\text{red}]}$$

where $E^{\ominus}$ is the standard potential of the system. The potential of the immersed electrode is thus controlled by the **ratio** of these concentrations. During the oxidation of a reducing agent or the reduction of an oxidising agent the ratio, and therefore the potential, changes more rapidly in the vicinity of the end point of the reaction. Thus titrations involving these reactions (e.g. iron(II) with potassium permanganate or potassium dichromate or cerium(IV) sulphate) may be followed potentiometrically and produce titration curves characterised by a sudden change of potential at the equivalence point.

Orion Research Ltd produces combination redox electrodes which combine a platinum sensor with a reference electrode. Autotitrators are widely used to study redox reactions. Determinations include (a) chromium in copper–chromium alloys using iron(II) ammonium sulphate; (b) iron(II)/iron(III) using cerium(IV) sulphate; (c) manganese in an ore using potassium permanganate as titrant (Section 10.155).

10.155   Manganese by potentiometry

Discussion   The method is based on titrating manganese(II) ions with permanganate in neutral pyrophosphate solution:

$$4Mn^{2+} + MnO_4^- + 8H^+ + 15H_2P_2O_7^{2-} = 5Mn(H_2P_2O_7)_3^{3-} + 4H_2O$$

The manganese(III) pyrophosphate complex has an intense reddish violet colour, so the titration must be performed potentiometrically; a combination redox electrode would be used. With relatively pure manganese solutions, a sodium pyrophosphate concentration of $0.2–0.3\,M$, the potential at the equivalence point can easily be measured at pH 6–7. But at pH > 8 the pyrophosphate complex dissociates, hence the method cannot be used.

Large amounts of chloride, cobalt(II) and chromium(III) do not interfere; iron(III), nickel, molybdenum(VI), tungsten(VI) and uranium(VI) are innocuous; nitrate, sulphate and perchlorate ions are harmless. Large quantities of magnesium, cadmium, and aluminium yield precipitates which may coprecipitate manganese and should therefore be absent. Vanadium causes difficulties only when the amount is greater than or equal to the amount of manganese; when it is present originally in the +4 state, it is oxidised slowly in the titration to the +5 state along with the manganese. Small amounts of vanadium (up to about one-fifth the amount of manganese) cause little error. The interference of large amounts of vanadium(V) can be circumvented by performing the titration at a pH of 3 to 3.5. Oxides of nitrogen interfere because of their reaction with potassium permanganate. So when nitric acid is used to dissolve the sample, the resulting solution must be boiled thoroughly and a small mount of urea or sulphamic acid must be added to the acid solution to remove the last traces of oxides of nitrogen before introducing the sodium pyrophosphate solution.

Reagents   *Potassium permanganate*   Use a standard solution, concentration ~$0.02\,M$.

Sodium pyrophosphate   Use a saturated solution of sodium pyrophosphate $Na_4P_2O_7 \cdot 10H_2O$ (about $12\,g$ in $100–150\,mL$ water); it must be freshly made.

Manganese(II)   Use a test solution of manganese(II) ions, concentration $0.05–0.10\,M$.

Procedure   Place $150\,mL$ of the sodium pyrophosphate solution in a $250–400\,mL$ beaker, adjust the pH to 6–7 by adding concentrated sulphuric acid from a $1\,mL$ graduated pipette (use a pH meter). Add $25\,mL$ of the manganese(II) sulphate solution and adjust the pH again to 6–7 by adding $5\,M$ sodium hydroxide solution. Place the combination redox electrode into the solution. It is now ready for autotitration with the standardised permanganate solution. The end point can be obtained either directly or using derivatives. The method can be adapted for manganese in steel or in manganese ores.

Pyrolusite   Accurately weigh out 1.5 to 2 g of pyrolusite and **carefully** dissolve it in a mixture made from 25 mL of 1 : 1 hydrochloric acid and 6 mL of concentrated sulphuric acid. Dilute to 250 mL. Filtration is unnecessary. Titrate an aliquot part containing 80–100 mg manganese: add 200 mL freshly prepared, saturated sodium pyrophosphate solution, adjust the pH to a value between 6 and 7, and perform the potentiometric titration.

Steel   Accurately weigh 5 g of steel and dissolve it in 1 : 1 nitric acid using the minimum volume of hydrochloric acid in a Kjeldahl flask. Boil the solution down to a small volume with excess concentrated nitric acid to reoxidise any vanadium present reduced by the hydrochloric acid; this step is not necessary if vanadium is absent. Dilute, boil to remove gaseous oxidation products, allow to cool, add 1 g of urea and dilute to 250 mL. Titrate 50.0 mL portions as above.

10.156   Copper by potentiometry

Following the usual methods, prepare a sample solution containing about 0.1 g copper and without interfering elements; any large excess of nitric acid and all traces of nitrous acid must be removed. Boil the solution to expel most of the acid, add about 0.5 g urea (to destroy the nitrous acid) and boil again. Treat the cooled solution with concentrated ammonia solution dropwise until the deep blue cuprammonium compound is formed, and then add a further two drops. Decompose the cuprammonium complex with glacial ethanoic acid and add 0.2 mL in excess. Too great a dilution of the final solution should be avoided, otherwise the reaction between the copper(II) ethanoate and the potassium iodide may not be complete.

Place the prepared copper ethanoate solution in the beaker and add 10 mL of 20% potassium iodide solution. Using a combination redox electrode carry out the normal potentiometric titration procedure with a standard sodium thiosulphate solution as titrant.

10.157   Coulometry: general considerations

There are numerous reagents generated coulometrically that can be used in oxidation–reduction titrations. As well as the frequently used electrolytically generated cerium(IV) and iron(II) ions, the less common silver(II) and chromium(II) ions may be produced. However, perhaps the most important are electrogenerated bromine and iodine. Bromine, generated by the oxidation of the bromide ion at a platinum anode according to the equation

$$2Br^- = Br_2 + 2e$$

is a most versatile reagent. It can be used for the determination of a variety of organic compounds, including phenols, aromatic amines and olefins. The following examples use either bromine or iodine as electrogenerated reagents; the end points are all detected amperometically.

10.158   Cyclohexene by coulometry

Discussion   Cyclohexene may be titrated with bromine generated by the oxidation of bromide ion ($2Br^- = Br_2 + 2e$). The bromination of cyclohexene, catalysed by mercury(II), results in the formation of *trans*-1,2-dibromocyclohexene. The bromine reaction liberates

2 mol of electrons for 1 mol of bromine, so 2 mol of electrons are equivalent to 1 mol of cyclohexene.

Apparatus   Set up the apparatus as in Figure 10.8 with two small platinum electrodes used for the amperometric detection of the end point.

Reagents   *Warning*   **Mercury compounds are very toxic. If the solvent mixture comes into contact with the skin, wash the affected area with plenty of water.**

Supporting electrolyte   Take a mixture of glacial ethanoic (acetic) acid (300 mL), methanol (130 mL) and water (65 mL). Add potassium bromide (9.0 g) and mercury(II) ethanoate (0.5 g). Stir cautiously until the solids dissolve.

Cyclohexene in methanol   Prepare a stock solution from about 0.5 g cyclohexene accurately weighed and dissolved in methanol using a 100 mL graduated flask. Pipette a 10 mL aliquot of the stock solution into another 100 mL flask and make up to the mark with methanol. (This solution contains about $0.5 \, mg \, mL^{-1}$ cyclohexene.)

Procedure   The following experiment is adapted from a method described by D. H. Evans.[24] Add sufficient supporting electrolyte to the electrolysis vessel in order to cover the electrodes. Stir the solution magnetically and apply a potential of about 0.25 V to the indicator electrodes. Switch on the generator electrode system to produce bromine until the microammeter registers a current of $20 \, \mu A$ – the generator current should be adjusted to 5–10 mA. Pipette a 5.0 mL aliquot of the methanolic cyclohexene solution into the electrolysis vessel. This causes the indicator current to decrease almost to zero as the bromine reacts with the cyclohexene. Turn on the generator and start timing. When the original indicator current of $20 \, \mu A$ is re-established, record the time. To test the reproducibility of the determination, repeat the procedure twice.

Example calculation   In a coulometric determination of a cyclohexene solution in methanol, a generator current of 9.2 mA was used. The time recorded to complete the titration was 700 s. What is the concentration of cyclohexene in the solution?

$$\text{moles of electrons} = \frac{\text{charge}}{\text{Faraday constant}} = \frac{It}{F}$$

where I is the current (A) and t is the time in (s). So

$$\text{moles of electrons} = \frac{9.2 \times 10^{-3} \times 700}{96\,487}$$

$$= 6.674 \times 10^{-5}$$

There are 2 mol of electrons for 1 mol of bromine, hence there are 2 mol of electrons for 1 mol of cyclohexene, so 6.674×10^{-5} mol electrons imply 3.337×10^{-5} mol cyclohexene. And the RMM of cyclohexene is 82.146, therefore

$$\text{mass of cyclohexene in 5 mL} = 3.337 \times 10^{-5} \, \text{mol} \times 82.146 \, \text{g mol}^{-1}$$

$$= 2.74 \, \text{mg}$$

$$\text{mass of cyclohexene in 1 L} = 2.74 \times \left(\frac{1000}{5}\right)$$

$$= 0.548 \, \text{g}$$

$$\text{concentration of cyclohexene} = 0.548 \, \text{g L}^{-1}$$

10.159 8-Hydroxyquinoline (oxine)

Discussion Bromine may be electrogenerated with 100% current efficiency by the oxidation of bromide ion at a platinum anode. Bromination of oxine proceeds according to the equation

$$C_9H_7ON + 2Br_2 = C_9H_5ONBr_2 + 2H^+ + 2Br^-$$

and thus four Faraday constants are required per mole of oxine. The end point is detected amperometrically.

Apparatus Set up the apparatus as in Figure 10.3 with two small platinum plates connected to apparatus for the amperometric detection of the end point.

Reagents *Supporting electrolyte* Prepare $0.2\,M$ potassium bromide from the analytical grade salt.

Oxine solution $0.003\,M$ oxine (use the analytical grade material) in $0.0025\,M$ hydrochloric acid.

Procedure Place 40 mL of the supporting electrolyte in the coulometric cell and pipette 10.00 mL of the oxine solution into it. Charge the cathode compartment with the $0.2\,M$ potassium bromide. Pass a current of 30 mA while stirring the solution magnetically. Adjust the sensitivity of the indicating apparatus to a suitable value. Transient deflections occur near the end point, warning of its approach. The end point occurs at the first permanent deflection, and this is where the counter reading is taken.

10.160 Amperometry: general considerations

Amperometric titrations are used for many determinations based on oxidation–reduction reactions. The titrations described below use the rotating platinum microelectrode (Section 10.20). Other experiments (Sections 10.163 to 10.165) are biamperometric titrations using the dead-stop end point of Section 10.21. The important Karl Fischer method for determination of water is given in Section 10.166.

10.161 Thiosulphate by amperometry with iodine

Discussion Dilute solutions of sodium thiosulphate (e.g. $0.001\,M$) may be titrated with dilute iodine solutions (e.g. $0.005\,M$) at zero applied voltage. For satisfactory results, the thiosulphate solution should be present in a supporting electrolyte which is $0.1\,M$ in potassium chloride and $0.004\,M$ in potassium iodide. Under these conditions no diffusion current is detected until after the equivalence point when excess of iodine is reduced at the electrode; the titration graph is a reversed L-shape.

Dilute solutions of iodine, e.g. $0.0001\,M$, may be titrated similarly with standard thiosulphate. The supporting electrolyte consists of $1.0\,M$ hydrochloric acid and $0.004\,M$ potassium iodide. No external e.m.f. is required when an SCE is used as reference electrode.

Reagents *Sodium thiosulphate* Use ~$0.001\,M$ sodium thiosulphate solution, $0.1\,M$ with respect to potassium chloride and $0.004\,M$ with respect to potassium iodide.

Iodine solution Use standard $0.05\,M$ iodine solution in $0.004\,M$ potassium iodide.

Procedure   Place 25.0 mL of the thiosulphate solution in the titration cell. Set the applied voltage to zero with respect to the SCE after connecting the rotating platinum microelectrode (Section 10.20) to the polarising unit. Adjust the range of the microammeter. Titrate with the standard 0.005 M iodine solution in the usual manner. Plot the titration graph, evaluate the end point, and calculate the exact concentration of the thiosulphate solution. As a check, repeat the titration using freshly prepared starch indicator solution.

10.162   Antimony by amperometry with potassium bromate

Discussion   Dilute solutions of antimony(III) and arsenic(III) (~0.0005 M) may be titrated with standard 0.002 M potassium bromate in a supporting electrolyte of 1 M hydrochloric acid containing 0.05 M potassium bromide. The two electrodes are a rotating platinum microelectrode and an SCE: the former is polarised to +0.2 volt. A reversed L-type of titration graph is obtained.

Reagents   *Potassium antimonyl tartrate*   Prepare a 0.005 M potassium antimonyl tartrate solution. Dissolve 1.625 g of the solid in 1 L of distilled water. Dilute 25.0 mL of this solution to 250 mL with 1 M hydrochloric acid which is 0.05 M in potassium bromide.

Potassium bromate   Prepare a 0.002 M standard potassium bromate solution from the pure solid.

Procedure   Pipette 25.0 mL of the antimony solution into the titration cell. Set the applied voltage at 0.2 V vs SCE, and adjust the range of the microammeter. Titrate in the usual manner, and calculate the concentration of the antimony solution.

10.163   Thiosulphate with iodine: dead-stop end point

Reagents   Prepare a ~0.001 M sodium thiosulphate solution and a standard 0.005 M iodine solution.

Procedure   Pipette 25.0 mL of the thiosulphate solution into the titration cell, e.g. a 150 mL Pyrex beaker. Insert two similar platinum wire or foil electrodes into the cell and connect to the apparatus of Figure 10.8. Apply 0.10 V across the electrodes. Adjust the range of the microammeter to obtain full-scale deflection for a current of 10–25 mA. Stir the solution with a magnetic stirrer. Add the iodine solution from a 5 mL semimicro burette slowly in the usual manner and read the current (galvanometer deflection) after each addition of the titrant. When the current begins to increase, stop the addition; then add the titrant by small increments of 0.05 or 0.10 mL. Plot the titration graph, evaluate the end point, and calculate the concentration of the thiosulphate solution. The current is fairly constant until the end point is approached and increases rapidly beyond it.

10.164   Glucose by amperometry with an enzyme electrode

In the presence of the enzyme glucose oxidase, an aqueous solution of glucose undergoes oxidation to gluconic acid, forming hydrogen peroxide which can be determined by anodic oxidation at a fixed potential:

$$C_6H_{12}O_6 + O_2 \xrightarrow[\text{phosphate buffer}]{\text{enzyme}} \text{gluconic acid} + H_2O_2$$

The enzyme is used in an enzyme electrode in which a tube is sealed at its lower end with a cellulose acetate membrane. An outer membrane of collagen is also attached to the end of the electrode tube and glucose oxidase is contained in the space between the two diaphragms.

When the electrode is placed in an aqueous solution of glucose which has been suitably diluted with a phosphate buffer (pH 7.3), solution passes through the outer membrane into the enzyme where hydroxen peroxide is produced. The membrane allows hydrogen peroxide to diffuse, but it is impermeable to other components of the solution. The electrode vessel contains a phosphate buffer along with a platinum wire and a silver wire which act as electrodes. A potential of 0.7 V is applied to the electrodes with the platinum wire as anode (the apparatus in Figure 10.8 is suitable). Oxygen is produced at the anode in the half-reaction $H_2O_2 \rightarrow O_2 + 2H^+ + 2e$; it is reduced at the cathode in the half-reaction $\frac{1}{2}O_2 + 2H^+ + 2e \rightarrow H_2O$.

After a short time to allow for equilibration, the current settles down to a steady value; its magnitude is governed by the concentration of hydrogen peroxide in the electrode, and this is proportional to the glucose concentration of the test solution. The unknown concentration can be deduced by making readings with a series of standard glucose solutions (prepared in the same phosphate buffer solution), and then plotting the observed steady currents against concentration of glucose (Section 13.21).[25]

10.165 The Clark cell for oxygen determinations

A gold disc to which a gold wire is attached is placed inside a glass tube about 1.5 cm in diameter. A thin plastic film (Teflon is suitable) is stretched tightly over the end of the tube and fixed in position by an O-ring. A conducting solution is placed in the tube (0.1 M potassium chloride), and a silver–silver chloride electrode inserted; the electrode consists of a silver wire coated with silver chloride by electrolysis of a chloride solution. The lower end of the silver wire is coiled into a helix which is placed around (but does not touch) the gold disc and the lower end of the gold wire; both wires are inserted through a plastic closure which seals the top end of the tube.

When the tube is placed in a solution which contains oxygen, oxygen will pass through the membrane to the internal solution, and on applying a voltage (0.6–0.8 V) to the two electrodes, the oxygen undergoes reduction at the gold cathode:

$$\tfrac{1}{2}O_2 + 2H^+ + 2e \rightarrow H_2O$$

The amperometric current is read on a microammeter; when a steady current is obtained. It is controlled by the rate of diffusion of oxygen to the cathode. This is determined by the concentration of dissolved oxygen inside the cell, in turn related to the oxygen content of the test solution. Calibration is carried out by making readings with solutions which have been saturated with oxygen at varying partial pressures.

The apparatus is sometimes known as an oxygen electrode, but it is actually a cell. Although the Teflon membrane is impermeable to water, hence impermeable to most substances dissolved in water, dissolved gases can pass through, and gases such as chlorine, sulphur dioxide and hydrogen sulphide can affect the electrode. The apparatus can be made portable, so it may be used for monitoring the oxygen content of rivers and lakes.[26]

10.166 Water by amperometry using the Karl Fischer reagent

For determining small amounts of water, Karl Fischer (1935) proposed a reagent prepared by the action of sulphur dioxide on a solution of iodine in a mixture of anhydrous pyridine and anyhdrous methanol. Water reacts with this reagent in a two-stage process whereby one molecule of iodine disappears for each molecule of water present. The mechanism of the reaction is usually represented by the following equations:

$$3C_5H_5N + I_2 + SO_2 + H_2O = 2C_5H_5\overset{+}{N}H\overset{-}{I} + C_5H_5\overset{+}{N}\underset{\overset{|}{\,^-O}}{\overset{\diagup SO_2}{|}}$$

$$C_5H_5\overset{+}{N}\underset{\overset{|}{\,^-O}}{\overset{\diagup SO_2}{|}} + CH_3OH = C_5H_5N\overset{\diagup OSO_2OCH_3}{\diagdown H}$$

The end point of the reaction is conveniently determined electrometrically using the dead-stop procedure. If a small e.m.f. is applied across two platinum electrodes immersed in the reaction mixture, a current will flow as long as free iodine is present, to remove hydrogen and depolarise the cathode. When the last trace of iodine has reacted, the current will decrease to zero or very close to zero. Conversely, the technique may be combined with a direct titration of the sample with the Karl Fischer reagent; here the current in the electrode circuit suddenly increases at the first appearance of unused iodine in the solution.

The original Karl Fischer reagent prepared with an excess of methanol was somewhat unstable and required frequent standardisation. The stability was improved by replacing the methanol by 2-methoxyethanol, 2-chloroethanol or trifluoroethanol; the reagent was then described as 'methanol-free'. It is essential to have an alcohol in the Karl Fischer (KF) reagent to esterify the sulphur dioxide. A base is necessary to neutralise the acids produced in the reaction. The KF reagent has recently been modified because pyridine is too weak a base to completely neutralise the acids (causing sluggish end points). The titration should be in the pH range 5–7 and the base used now is imidazole. This KF reagent is described as 'pyridine-free'.

The method is confined to those cases where the test substance does not react with any component of the reagent, or with the hydrogen iodide formed during the reaction with water. The following compounds interfere in the Karl Fischer titration:

1. Oxidising agents such as chromates, dichromates, copper(II) and iron(III) salts, higher oxides and peroxides:

$$MnO_2 + 4C_5H_5NH^+ + 2I^- = Mn^{2+} + 4C_5H_5N + I_2 + 2H_2O$$

2. Reducing agents such as thiosulphates, tin(II) salts and sulphides.
3. Compounds which can be regarded as forming water with the components of the Karl Fischer reagent. Two examples are basic oxides and salts of weak oxyacids:

$$ZnO + 2C_5H_5NH^+ = Zn^{2+} + 2C_5H_5N + H_2O$$

$$NaHCO_3 + C_5H_5NH^+ = Na^+ + H_2O + CO_2 + C_5H_5N$$

4. Aldehydes form a bisulphite product:

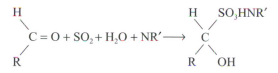

457

5. Ketones react with methanol to produce ketal and water:

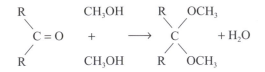

The Karl Fischer procedure has been improved by a modification to a coulometric method. In this procedure the sample under test is added to the alcohol–base solution (see above) containing sulphur dioxide and a soluble iodide. Upon electrolysis, iodine is liberated at the anode; the end point is detected by a pair of electrodes which function as a biamperometric detection system and indicate the presence of free iodine. Since 1 mol of iodine reacts with 1 mol of water, it follows that 1 mg of water is equivalent to 10.71 coulombs.

A number of dedicated Karl Fischer titrators are marketed by Orion Research Ltd. Two instruments are volumetric titrators using a platinum electrode system. As well as the Orion AF8, used for rapid moisture determination in samples ranging from 50 ppm to 100% H_2O, the Orion Turbo 2 is designed to deal with difficult solid and viscous samples. A high-speed homogeniser (speeds up to 7500 rpm) ensures complete moisture extraction from sample into solution, with minimal or no sample preparation. The microprocessor simplifies titration operating procedures and allows greater versatility of sample handling. The time from weighing to final evaluation is typically only a few minutes. Since the sample is blended, extracted and its moisture determined all in the same vessel, this instrument is particularly useful for moisture determinations in pharmaceutical and confectionery products.

Iodine generation enables fast and accurate moisture measurement over a wide range of concentrations (from 1 ppm to 100% H_2O). Orion supplies two coulometric Karl Fischer titrators. Model AF7 works over a wide concentration range, whereas model AF7LC works in the 1–10 ppm range. Model AF7 has a completely sealed chemical handling system with an integral pump controlled from a keypad; this allows safe reagent replenishment without opening the vessel to environmental humidity. Coulometric titrations have the advantage of being absolute, so there is no need to calibrate the reagents. All the above instruments are microprocessor controlled from sample introduction to the analysis of results. Furthermore, a built-in printer provides a comprehensive record of the method, the results and any statistical analysis.

10.167 Robotics

Sample preparation can often be the hardest analytical step, especially with solids, and several stages may be required, e.g. reduction of particle size (grinding), mixing for homogeneity, drying and weighing. Operations such as heating, ignition, fusion and the use of solvents are often necessary for sample dissolution. Further, there may be need for samples to be moved to various points within the laboratory. A fully automated analysis requires every operation to be automated. But laboratory robots, developed during the last decade, have enabled routine tasks to be performed sequentially without human involvement.

Owens and Eckstein have described an early robotic system that performed automatic weighing of solids and liquids, pH determination, dilution and dissolution.[27] Controlled by a microcomputer, it totally automated the pH titration of a solid sample. It calibrated the pH meter, prepared standard solutions, determined the end point, presented the data and recorded the results. An important feature is the robot arm, whose movements can often replace human hand, wrist and arm functions. Hence the robot arm can pour liquids or solids, mix liquids by shaking and transfer samples from one laboratory station to another. Some designs allow a robot hand to be interchanged automatically for a syringe that dispenses a given volume of liquid.

The first commercial laboratory robot, the Zymate Laboratory Automation System, was introduced in 1982 by the Zymark Corporation, Hopkinton MA. Strimaltis has made a survey of commercially available robot systems together with their areas of application in analytical science.[28,29] Examples include automatic Karl Fischer titrations, and determinations of dissolved oxygen and biochemical oxygen demand (BOD). The most commonly automated methods include HPLC, GC and UV/visible sample preparation and analysis. Robotic methods have several advantages over manual procedures:

1. A considerable improvement in the analytical precision.
2. A far larger sample throughput and a faster availability of analytical results.
3. An economic saving which is often achieved by an improvement in productivity, particularly if the method requires extensive and repetitive operations.

When integrated with instrumentation and computer control, laboratory robotics allows complete automation of many analytical methods. Freedom from tedious and repetitive tasks will enable the analytical scientist to develop innovatory techniques and to focus on problem solving. In the near future, developments in robot technology will undoubtedly generate new instruments, leading to major advances in laboratory practice.

10.168   References

1. Anon 1974 *Information Bulletin 36*, International Union of Pure and Applied Chemistry
2. C Woodward and H N Redman 1973 *High precision titrimetry*, Society for Analytical Chemistry, London
3. W Ostwald 1895 *Scientific foundations of analytical chemistry*, p. 118; A R Hantzch 1908 *Ber. Dtsch. Chem. Gesell.*, **61**; 1171, 1187
4. J N Brønsted 1923 *Rev. Trav. Chim.*, **42**; 718. T M Lowry 1924 *Trans. Farad. Soc.*, **20**; 13
5. S Siggia, J C Hanna and I R Kervenski 1950 *Anal. Chem.*, **22**; 1295
6. J S Fritz 1954 *Anal. Chem.*, **26**; 1701
7. G H Jeffrey, J Bassett, J Mendham and R C Denney 1989 *Vogel's quantitative chemical analysis*, 5th edn, Longman, Harlow pp. 284–94
8. H N Wilson 1951 *Analyst*, **76**; 65
9. L E Hummelstedt and D N Hume 1960 *Anal. Chem.*, **32**; 1792
10. W J Williams 1979 *Handbook of anion determinations*, Butterworth, London, p. 350
11. T S West 1969 *Complexometry*, 3rd edn, BDH Chemicals Ltd, Poole
12. Anon 1956 *Colour index*, 2nd edn, Society of Dyers and Colourists, Bradford
13. I L Marr and M S Cresser 1983 *Environmental chemical analysis*, International Textbook Company, Glasgow, p. 121
14. K C Thompson, D Mendham, D Best and K-E de Casseres 1986 *Analyst*, **111**; 483
15. L W Winkler 1888 *Ber. Dtsch. Chem. Gesell.*, **21**; 2843
16. I L Marr and M S Cresser 1983 *Environmental chemical analysis*, International Textbook Company, Glasgow, pp. 116–17
17. D N Bailey 1974 *J. Chem. Educ.*, **51**; 488
18. L W Andrews 1903 *J. Am. Chem. Soc.*, **25**; 76
19. G S Jamieson 1926 *Volumetric iodate methods*, Reinhold, New York

20.   G J Moody and J D R Thomas 1963 *J. Chem. Educ.*, **40**; 151. G J Moody and J D R Thomas 1964 *Education in Chemistry*, **1**; 214
21.   D A Skoog, D M West and F J Holler 1992 *Fundamentals of analytical chemistry*, 6th edn, Holt, Rinehart and Winston, New York
22.   A I Vogel 1958 *Elementary practical organic chemistry*, Part III, *Quantitative organic analysis*, Longman, London
23.   C M Ellis and A I Vogel 1956 *Analyst*, **81**; 693
24.   D H Evans 1968 *J. Chem. Educ.*, **45**; 88
25.   G Sittapalam and G S Wilson 1982 *J. Chem. Educ.*, **59**; 70
26.   I Fatt 1976 *Polarographic oxygen sensors*, CRC Press, Cleveland OH
27.   G D Owens and R J Eckstein 1982 *Anal. Chem.*, **54**; 2347
28.   J R Strimaltis 1989 *J. Chem. Educ.*, **66**; A8
29.   J R Strimaltis 1990 *J. Chem. Educ.*, **67**; A20

10.169   Bibliography

O Budevsky 1979 *Foundations of chemical analysis*, Ellis Horwood, Chichester

G Christian 1994 *Analytical chemistry*, 5th edn, John Wiley, Chichester

D Cooper and C Doran 1987 *Classical methods of chemical analysis*, Volume I ACOL–Wiley, Chichester

J S Fritz and G H Schenk 1987 *Quantitative analytical chemistry*, 5th edn, Allyn and Bacon, Boston MA

D C Harris 1998 *Quantitative chemical analysis*, 5th edn, W H Freeman, San Francisco

J Mendham, D Dodd and D Cooper 1987 *Classical methods of chemical analysis*, Volume II, ACOL–Wiley, Chichester

D A Skoog and D M West 1992 *Analytical chemistry: an introduction*, 6th edn, Saunders, New York

C L Wilson and D W Wilson 1962 *Comprehensive analytical chemistry*, Elsevier, Amsterdam

11

Gravimetric analysis

11.1 Introduction

Gravimetric, electrogravimetric and some forms of thermal analysis are concerned with the process of producing and weighing a compound or element in as pure a form as possible after some form of chemical treatment has been carried out on the substance to be examined. Traditional gravimetric determinations have been concerned with the transformation of the element, ion or radical to be determined into a pure stable compound which is suitable for direct weighing or for conversion into another chemical form that can be readily quantified. The mass of the element, ion or radical in the original substance can then be readily calculated from a knowledge of the formula of the compound and the relative atomic masses of the constituent elements.

This chapter deals with the methods used for the production and separation of substances containing the required element or compound, usually by precipitation, in forms which are relatively easy to handle. Electrogravimetric methods are dealt with in Chapter 13 and thermal analysis in Chapter 12. Traditional gravimetric procedures are essentially manual in character and labour-intensive, whereas electrogravimetric methods may be considered as partially instrumental and thermal methods as highly instrumental. It is appropriate to mention at this stage the reasons for the continuing use of gravimetric analysis despite the disadvantage that it is generally somewhat time-consuming. The advantages offered by gravimetric analysis are:

1. It is accurate and precise when using modern analytical balances.
2. Possible sources of error are readily checked, since filtrates can be tested for completeness of precipitation and precipitates may be examined for the presence of impurities.
3. It is an absolute method, i.e. it involves direct measurement without any form of calibration being required.
4. Determinations can be carried out with relatively inexpensive apparatus; the most expensive items are a muffle furnace and sometimes platinum crucibles.

Gravimetric analysis is a macroscopic method usually involving relatively large samples compared with many other quantitative analytical procedures. It is possible to achieve a very high level of accuracy, and even under normal laboratory conditions, it should be possible to obtain repeatability of results within 0.3–0.5%. There are two main areas of application for gravimetric methods:

1. Analysis of standards to be used for the testing and/or calibration of instrumental techniques.

2.   Analyses requiring high accuracy, although the time-consuming nature of gravimetry limits this application to small numbers of determinations.

Besides this, gravimetric and electrogravimetric procedures provide a very broad training experience in laboratory procedures. Thermal analysis is providing more and more insights into chemical, and increasingly into biochemical, structures and reactions occurring under thermal conditions.

11.2   Fundamentals

Gravimetric analysis is concerned ultimately with the weighing of a substance that has been either precipitated from solution or volatilised and absorbed.[1] Many methods exist for the precipitation of metals and compounds but there is an increasing development of procedures and instruments for gas and vapour analysis in relation to environmental monitoring. However, many of these methods are now based on chromatography rather than traditional absorption and weighing (Chapter 9).

It is essential that the chosen method precipitates the element or ion being determined in a form which is so slightly soluble that no appreciable loss occurs when the precipitate is separated by filtration and weighed. Thus, in the determination of silver, a solution of the substance is treated with an excess of sodium chloride or potassium chloride solution, the precipitate is filtered off, well washed to remove soluble salts, dried at $130-150\,°C$, and weighed as silver chloride. Frequently the constituent being determined is weighed in a form that is different from its precipitate. Thus magnesium is precipitated, as ammonium magnesium phosphate $Mg(NH_4)PO_4·6H_2O$, but is weighed after ignition as the pyrophosphate $Mg_2P_2O_7$. Three factors determine a successful analysis by precipitation:

1.   The precipitate must be so insoluble that no appreciable loss occurs when it is collected by filtration. In practice this usually means that the quantity remaining in solution does not exceed 0.1 mg, the minimum detectable by the ordinary analytical balance.
2.   The physical nature of the precipitate must be such that it can be readily separated from the solution by filtration, and can be washed free of soluble impurities. These conditions require that the particles do not pass through the filtering medium, and that the particle size is unaffected (or, at least, not diminished) by the washing process.
3.   The precipitate must be convertible into a pure substance of definite chemical composition; this may be effected either by ignition or by a simple chemical operation, such as evaporation, with a suitable liquid.

Problems which arise with certain precipitates include the coagulation or flocculation of a colloidal dispersion of a finely divided solid to permit its filtration and to prevent its repeptisation upon washing the precipitate. Colloidal properties are, in general, exhibited by substances of particle size ranging between 0.1 µm and 1 nm. Ordinary quantitative filter paper will retain particles down to a diameter of about 10^{-2} mm or 10 µm, so that colloidal solutions in this respect behave like true solutions and are not filterable (size of molecules is of the order of 0.1 nm or 10^{-8} cm).

Supersaturation is another difficulty which can occur in gravimetric analysis. A supersaturated solution contains a greater concentration of solute than expected for equilibrium solubility at the temperature under consideration. It is therefore an unstable state and the corresponding state of stable equilibrium can be established by the addition of a crystal of the pure solute (known as seeding the solution) or by creating points for crystallisation to occur, perhaps by scratching the inside of the vessel.

Many of the possible problems associated with gravimetric analysis are overcome by following well-established procedures:

1. Precipitation should be carried out in dilute solution, due regard being paid to the solubility of the precipitate, the time required for filtration, and the subsequent operations to be carried out with the filtrate. This will minimise the errors due to co-precipitation.

2. The reagents should be mixed slowly and with constant stirring. This will keep the degree of supersaturation small and will assist the growth of large crystals. A slight excess of the reagent is all that is generally required; in exceptional cases a large excess may be necessary. In some instances the order of mixing the reagents may be important. Precipitation may be effected under conditions which increase the solubility of the precipitate, further reducing the degree of supersaturation.

3. Precipitation is effected in hot solutions, provided the solubility and the stability of the precipitate permit. Either one or both of the solutions should be heated to just below the boiling point or other more favourable temperature. At the higher temperature: (a) the solubility is increased with a consequent reduction in the degree of supersaturation, (b) coagulation is assisted and sol formation decreased, and (c) the velocity of crystallisation is increased, thus leading to better-formed crystals.

4. Crystalline precipitates should be digested for as long as practical, preferably overnight, except in those cases where post-precipitation may occur. As a rule, digestion on the steam bath is desirable. This process decreases the effect of co-precipitation and gives more readily filterable precipitates. Digestion has little effect upon amorphous or gelatinous precipitates.

5. The precipitate should be washed with the appropriate dilute solution of an electrolyte. Pure water may tend to cause peptisation.

6. If the precipitate is still appreciably contaminated as a result of coprecipitation or other causes, the error may often be reduced by dissolving it in a suitable solvent and then reprecipitating it. The amount of foreign substance present in the second precipitation will be small, hence the amount of entrainment by the precipitate will also be small.

7. Precipitation from a homogeneous solution is commonly employed to prevent supersaturation.[2] The precipitating agent is generated within the solution by means of a homogeneous reaction at a rate similar to that required for precipitation of the species.

Once a precipitate has been obtained and filtered, it still requires further treatment. Besides water retained from the solution it may also contain four other kinds of water:

Adsorbed water is present on all solid surfaces in amounts that depend on the humidity of the atmosphere.

Occluded water is present in solid solution or in cavities within crystals.

Sorbed water is associated with substances having a large internal surface development, e.g. hydrous oxides.

Essential water is present as water of hydration or crystallisation, e.g. $CaC_2O_4 \cdot H_2O$ or $Mg(NH_4)PO_4 \cdot 6H_2O$, or as water of constitution; water of constitution is not present as such but is formed on heating, e.g. $Ca(OH)_2 \rightarrow CaO + H_2O$.

In addition to the evolution of water, the ignition of precipitates often results in thermal decomposition reactions involving the dissociation of salts into acidic and basic components, e.g. the decomposition of carbonates and sulphates; the decomposition temperatures will obviously be related to the thermal stabilities.

The temperatures at which precipitates may be dried, or ignited to the required chemical form, can be determined from a study of the **thermogravimetric curves** for the individual substances. They are dealt with in Section 12.2.

11.3   Precipitation reagents

The vast majority of gravimetric precipitations are carried out using a range of organic reagents. Although some very well-established determinations, such as barium (as sulphate) and lead (as chromate) do involve the use of inorganic reagents. However, organic reagents not only have the expected advantage of producing compounds which are sparingly soluble and usually coloured, but also have high relative molecular masses which will yield a correspondingly large amount of precipitate from a small amount of available ions being measured.

The ideal organic precipitant should be **specific** in character, i.e. it should give a precipitate with only one particular ion. But rarely has this ideal been attained; it is more usual to find that the organic reagent will react with a group of ions, but frequently by a rigorous control of the experimental conditions it is possible to precipitate only one of the ions of the group. Sometimes the precipitated compound may be weighed after drying at a suitable temperature; in other cases the composition is not quite definite and the substance is converted by ignition to the oxide of the metal; in a few instances, a titrimetric method is employed which utilises the quantitatively precipitated organic complex.

It is difficult to give a rigid classification of the numerous organic reagents. The most important, however, are those which form chelate complexes, which involve the formation of one or more (usually five- or six-membered) rings incorporating the metal ion; ring formation leads to a relatively great stability. One classification of these reagents is concerned with the number of hydrogen ions displaced from a neutral molecule in forming one chelate ring. A guide to the applicability of organic reagents for analytical purposes may be obtained from a study of the formation constant of the coordination compound (which is a measure of its stability), the effect of the nature of the metallic ion and of the ligand on the stability of complexes, and of the precipitation equilibria involved, particularly in the production of uncharged chelates. For further details, consult the works listed in Section 11.12. A typical chelate structure is formed by dimethylglyoxime with nickel, and this is an ideal process for gaining experience (Section 11.8). The following reagents are typical of those used for chelate formation in metal analysis.

Dimethylglyoxime

Dimethylglyoxime [11.A] gives a bright red precipitate of $Ni(C_4H_7O_2N_2)_2$ [11.B] with nickel salt solutions; precipitation is usually carried out in ammoniacal solution or in a buffer solution containing ammonium ethanoate and ethanoic acid. Solutions of palladium (II) salts give a characteristic yellow precipitate in dilute hydrochloric or sulphuric acid solution; the composition $Pd(C_4H_7O_2N_2)_2$ is similar to the nickel complex.

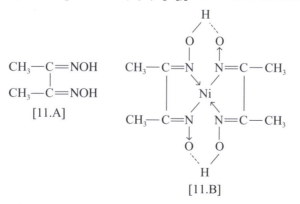

Cupferron

Cupferron [11.C] is the ammonium salt of *N*-nitroso-*N*-phenylhydroxylamine. It precipitates iron(III), vanadium(V), titanium(IV), zirconium(IV), cerium(IV), niobium(V), tantalum(V), tungsten(VI), gallium(III) and tin(IV), separating these elements from aluminium, beryllium, chromium, manganese, nickel, cobalt, zinc, uranium(VI), calcium, strontium and barium in both weakly acid and strongly acid solutions.

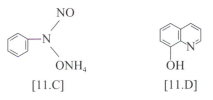

[11.C]   [11.D]

8-Hydroxyquinoline (oxine)

8-Hydroxyquinoline [11.D] has molecular formula C_9H_7ON and is also known as **oxine**. It forms sparingly soluble derivatives with metallic ions, which have the composition $M(C_9H_6ON)_2$ if the coordination number of the metal is 4 (e.g. magnesium, zinc, copper, cadmium, lead, and indium), $M(C_9H_6ON)_3$ if the coordination number is 6 (e.g. aluminium, iron, bismuth, and gallium), and $M(C_9H_6ON)_4$ if the coordination number is 8 (e.g. thorium and zirconium). But there are some exceptions, e.g. $TiO(C_9H_6ON)_2$, $MnO_2(C_9H_6ON)_2$, $WO_2(C_9H_6ON)_2$ and $UO_2(C_9H_6ON)_2$. Table 11.1 gives the pH ranges for precipitation of the various metal **oxinates** that may be formed.

Table 11.1 *pH range for precipitation of metal **oxinates***

Metal	pH		Metal	pH	
	Initial precipitation	Complete precipitation		Initial precipitation	Complete precipitation
Aluminium	2.9	4.7–9.8	Manganese	4.3	5.9–9.5
Bismuth	3.7	5.2–9.4	Molybdenum	2.0	3.6–7.3
Cadmium	4.5	5.5–13.2	Nickel	3.5	4.6–10.0
Calcium	6.8	9.2–12.7	Thorium	3.9	4.4–8.8
Cobalt	3.6	4.9–11.6	Titanium	3.6	4.8–8.6
Copper	3.0	> 3.3	Tungsten	3.5	5.0–5.7
Iron(III)	2.5	4.1–11.2	Uranium	3.7	4.9–9.3
Lead	4.8	8.4–12.3	Vanadium	1.4	2.7–6.1
Magnesium	7.0	> 8.7	Zinc	3.3	> 4.4

Benzoin-α-oxime

Benzoin-α-oxime [11.E], also known as cupron, yields a green precipitate of $CuC_{14}H_{11}O_2N$ with copper in dilute ammoniacal solution. Copper may thus be separated from cadmium, lead, nickel, cobalt, zinc, aluminium, and small amounts of iron.

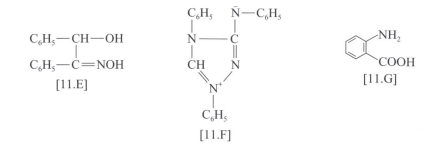

$$C_6H_5\text{---}CH\text{---}OH$$
$$C_6H_5\text{---}C\text{=}NOH$$
$$[11.E]$$

$$[11.F]$$

$$[11.G]$$

Nitron

Nitron [11.F], or 1,4-diphenyl-3-phenylamino-1H-1,2,4-triazole, yields a sparingly soluble crystalline nitrate $C_{20}H_{16}N_4,HNO_3$ in solutions acidified with ethanoic acid or sulphuric acid.

Anthranilic acid

In neutral or weakly acid solution the sodium salt of anthranilic acid [11.G] precipitates the anthranilates of cadmium, zinc, nickel, cobalt and copper, all of which are suitable for quantitative analysis. The salts have the general formula $M(C_7H_6O_2N)_2$ and may be dried at 105–110 °C.

Determination of chloride, sulphate and metal ions

! **Safety.** Before carrying out any experiments in this section, pay full attention to any safety warnings and make sure you adhere to national laboratory and safety regulations.

11.4 Gravimetric experiments

Experimental work in gravimetric analysis requires very careful use of pipettes, burettes and balances and it is essential to become familiar with these pieces of laboratory equipment before embarking on this type of experimental work. Accurate results require considerable care and the development of specialist skills. For all gravimetric determinations described in this chapter, the phrase 'allow to cool in a desiccator' should be interpreted as cooling the crucible, etc., **provided with a well-fitting cover** in a desiccator. The crucible, etc., should be weighed as soon as it has acquired the laboratory temperature (for a detailed discussion, see Section 3.22).

The following sections detail a number of determinations. They are particularly suitable for students to gain experience in the technique of gravimetric analysis and they indicate the types of conditions and processes involved in many other precipitation methods. Besides these determinations, Table 11.2 lists the most common cations and anions that can be determined using gravimetric methods, along with the appropriate reagent, the formula of the product and any special conditions that might be applied to improve its quality.

11.5   Aluminium as the 8-hydroxyquinolate (oxinate)

Discussion   This procedure separates aluminium from beryllium, the alkaline earths, magnesium and phosphate and involves precipitation from homogeneous solution. For the gravimetric determination a 2% or 5% solution of oxine (8-hydroxyquinoline) in $2M$ ethanoic acid may be used: 1 mL of the 5% solution is sufficient to precipitate 3 mg of aluminium. For practice in this determination, use about 0.40 g, accurately weighed of aluminium ammonium sulphate. Dissolve it in 100 mL of water, heat to 70–80 °C, add the appropriate volume of the oxine reagent, and (if a precipitate has not already formed) slowly introduce $2M$ ammonium ethanoate solution until a precipitate just appears, heat to boiling, and then add 25 mL of $2M$ ammonium ethanoate solution dropwise and with constant stirring (to ensure complete precipitation). If the supernatant liquid is yellow, enough oxine reagent has been added. Allow to cool, and collect the precipitated aluminium 'oxinate' on a weighed sintered glass (porosity no. 4) or porous porcelain filtering crucible, and wash well with cold water. Dry to constant weight at 130–140 °C. Weigh as $Al(C_9H_6ON)_3$.

Precipitation may also be effected from homogeneous solution. The solution containing 25–50 mg of aluminium should also contain 1.25–2.0 mL of concentrated hydrochloric acid in a total volume of 150–200 mL. After addition of excess of the oxine reagent, 5 g of urea is added for each 25 mg of aluminium present, and the solution is heated to boiling. The beaker is covered with a clockglass and heated for 2–3 hours at 95 °C. Precipitation is complete when the supernatant liquid, originally greenish yellow, acquires an orange-yellow colour. The cold solution is filtered through a sintered-glass filtering crucible (porosity no. 3 or 4), the precipitate washed well with cold water, and dried to constant weight at 130 °C.

Procedure   For practice in this determination, weigh out accurately about 0.45 g of aluminium ammonium sulphate, dissolve it in water containing about 1.0 mL of concentrated hydrochloric acid, and dilute to about 200 mL. Add 5–6 mL of oxine reagent (a 10% solution in 20% ethanoic acid) and 5 g of urea. Cover the beaker with a clockglass and heat on an electric hotplate at 95 °C for 2.5 h. Precipitation is complete when the supernatant liquid, originally greenish yellow, acquires a pale orange-yellow colour. The precipitate is compact and filters easily. Allow to cool and collect the precipitate in a sintered-glass filtering crucible (porosity no. 3 or 4), wash with a little hot water and finally with cold water. Dry at 130 °C. Weigh as $Al(C_9H_6ON)_3$.

11.6   Chloride as silver chloride

Discussion   The aqueous solution of the chloride is acidified with dilute nitric acid in order to prevent the precipitation of other silver salts, such as the phosphate and carbonate, which might form in neutral solution, and also to produce a more readily filterable precipitate. A slight excess of silver nitrate solution is added, whereupon silver chloride is precipitated:

$$Cl^- + Ag^+ = AgCl$$

The precipitate, which is initially colloidal, is coagulated into curds by heating the solution and stirring the suspension vigorously; the supernatant liquid becomes almost clear. The precipitate is collected in a filtering crucible, washed with very dilute nitric acid, in order to prevent it from becoming colloidal, dried at 130–150 °C, and finally weighed as AgCl. If silver chloride is washed with pure water, it may become colloidal and run through the filter. For this reason the wash solution should contain an electrolyte. Nitric acid is generally employed because it is without action on the precipitate and is readily volatile; its concentration need not be greater than 0.01 M. Completeness of washing of the precipitate is tested for by determining whether the excess of the precipitating agent, silver nitrate, has been removed. This may be done by adding one or two drops of 0.1 M hydrochloric acid to 3–5 mL of the washings collected after the washing process has been continued for some time; if the solution remains clear or exhibits only a very slight opalescence, all the silver nitrate has been removed.

Silver chloride is light-sensitive; decomposition occurs into silver and chlorine, and the silver remains colloidally dispersed in the silver chloride and thereby imparts a purple colour to it. The decomposition by light is only superficial, and is negligible unless the precipitate is exposed to direct sunlight and is stirred frequently. Hence the determination must be carried out in as subdued a light as possible, and when the solution containing the precipitate is set aside, it should be placed in the dark (e.g. in a locker), or the vessel containing it should be covered with thick brown paper.

Procedure Weigh out accurately about 0.2 g of the solid chloride (or an amount containing approximately 0.1 g of chlorine) into a 250 mL beaker provided with a stirring rod and covered with a clockglass. Add about 150 mL of water, stir until the solid has dissolved, and add 0.5 mL of concentrated nitric acid. To the cold solution add 0.1 M silver nitrate slowly and with constant stirring. Only a slight excess should be added; this is readily detected by allowing the precipitate to settle and adding a few drops of silver nitrate solution, when no further precipitate should be obtained. **Carry out the determination in subdued light**. Heat the suspension nearly to boiling, while stirring constantly, and maintain it at this temperature until the precipitate coagulates and the supernatant liquid is clear (2–3 min). Make certain that precipitation is complete by adding a few drops of silver nitrate solution to the supernatant liquid. If no further precipitate appears, set the beaker aside in the dark, and allow the solution to stand for about 1 h before filtration. In the meantime prepare a sintered-glass filtering crucible; the crucible must be dried at the same temperature as is employed in heating the precipitate (130–150 °C) and allowed to cool in a desiccator. Collect the precipitate in the weighed filtering crucible. Wash the precipitate two or three times by decantation with about 10 mL of cold very dilute nitric acid (0.5 mL of the concentrated acid added to 200 mL of water) before transferring the precipitate to the crucible. Remove the last small particles of silver chloride adhering to the beaker with a policeman (Section 3.13). Wash the precipitate in the crucible with very dilute nitric acid added in small portions until 3–5 mL of the washings, collected in a test tube, give no turbidity with one or two drops of 0.1 M hydrochloric acid. Place the crucible and contents in an oven at 130–150 °C for 1 h, allow to cool in a desiccator, and weigh. Repeat the heating and cooling until constant weight is attained. Calculate the percentage of chlorine in the sample.

Gravimetric standardisation of HCl The gravimetric standardisation of hydrochloric acid by precipitation as silver chloride is a convenient and accurate method, which has the additional advantage of being independent of the purity of any primary standard. Measure out from a burette 30–40 mL of say 0.1 M hydrochloric acid which is to be standardised. Dilute to 150 mL, precipitate (but omit the addition of nitric acid), and weigh the silver chloride. From the weight of the precipitate, calculate the chloride concentration of the solution, hence the concentration of the hydrochloric acid.

11.7   Lead as chromate

Discussion   Although this method is limited in its applicability because of the general insolubility of chromates, it is a useful procedure for gaining experience in gravimetric analysis. The best results are obtained by precipitating from homogeneous solution using the homogeneous generation of chromate ion produced by slow oxidation of chromium(III) by bromate at 90–95 °C in the presence of an ethanoate buffer.

Procedure   Use a sample solution containing 0.1–0.2 g lead. Neutralise the solution by adding sodium hydroxide until a precipitate just begins to form. Add 10 mL ethanoate buffer solution (6 M in ethanoic acid and 0.6 M in sodium ethanoate); 10 mL chromium nitrate solution (2.4 g per 100 mL); and 10 mL potassium bromate solution (2.0 g per 100 mL). Heat to 90–95 °C. After generation (of chromate) and precipitation are complete (about 45 min) as shown by a clear supernatant liquid, cool, filter through a weighed sintered-glass or porcelain filtering crucible, wash with a little 1% nitric acid, and dry at 120 °C. Weigh as $PbCrO_4$.

11.8   Nickel as the dimethylglyoximate

Discussion   Nickel is precipitated by the addition of an ethanolic solution of dimethylglyoxime (H_2DMG) to a hot, faintly acid solution of the nickel salt, and then adding a slight excess of aqueous ammonia solution (free from carbonate). The precipitate is washed with cold water and then weighed as nickel dimethylglyoximate after drying at 110–120 °C. With large precipitates, or in work of high accuracy, a temperature of 150 °C should be used; this volatilises any reagent that may have been carried down by the precipitate. The equation is

$$Ni^{2+} + 2H_2DMG = Ni(HDMG)_2 + 2H^+$$

For the structure of the complex and further details about the reagent, see Section 11.3. The precipitate is insoluble in dilute ammonia solution, in solutions of ammonium salts, and in dilute ethanoic acid–sodium ethanoate solutions. Large amounts of aqueous ammonia and of cobalt, zinc or copper retard the precipitation; extra reagent must be added, because these elements consume dimethylglyoxime to form various soluble compounds.

Dimethylglyoxime is almost insoluble in water, and is added in the form of a 1% solution in 90% ethanol (rectified spirit) or absolute ethanol; 1 mL of this solution is sufficient for the precipitation of 0.0025 g of nickel. The reagent is added to a hot feebly acid solution of a nickel salt, and the solution is then rendered faintly ammoniacal. This procedure gives a more easily filterable precipitate than direct precipitation from cold or from ammoniacal solutions. Only a slight excess of the reagent should be used, since dimethylglyoxime is not very soluble in water or in very dilute ethanol and may precipitate; if a very large excess is added (such that the alcohol content of the solution exceeds 50%), some of the precipitate may dissolve.

Procedure   Weigh out accurately 0.3–0.4 g of pure ammonium nickel sulphate $(NH_4)_2SO_4 \cdot NiSO_4 \cdot 6H_2O$ into a 500 mL beaker provided with a clockglass cover and stirring rod. Dissolve it in water, add 5 mL of dilute hydrochloric acid (1 : 1) and dilute to 200 mL. Heat to 70–80 °C, add a slight excess of the dimethylglyoxime reagent (at least 5 mL for every 10 mg of Ni present), and immediately add dilute ammonia solution dropwise, directly to the solution and not down the beaker wall, and with constant stirring until precipitation takes place, and then in slight excess. Allow to stand on the steam bath for 20–30 min, and test the solution for complete precipitation when the red precipitate has settled out. Allow the precipitate to stand for 1 h, cooling at the same time. Filter the cold solution through a sintered-glass or porcelain filtering crucible, previously heated to 110–120 °C and weighed after cooling in a desiccator. Wash the precipitate with cold water until free from chloride, and dry it at 110–120 °C for 45–50 min. Allow to cool in a desiccator and weigh. Repeat the drying until constant weight is attained. Weigh as $Ni(C_4H_7O_2N_2)_2$, which contains 20.32% Ni.

11.9   Sulphate as barium sulphate

Discussion   The method consists in slowly adding a dilute solution of barium chloride to a hot solution of the sulphate slightly acidified with hydrochloric acid:

$$Ba^{2+} + SO_4^{2-} \rightarrow BaSO_4$$

The precipitate is filtered off, washed with water, carefully ignited at a red heat, and weighed as barium sulphate. It is customary to carry out the precipitation in weakly acid solution in order to prevent the possible formation of the barium salts of such anions as chromate, carbonate and phosphate, which are insoluble in neutral solutions; moreover, the precipitate so obtained consists of large crystals and is therefore more easily filtered. It is also of great importance to carry out the precipitation at boiling temperature, for the relative supersaturation is less at higher temperatures. The concentration of hydrochloric acid is limited by the solubility of the barium sulphate, but a concentration of $0.05 M$ is suitable; the solubility of the precipitate in the presence of barium chloride at this acidity is negligible. The precipitate may be washed with cold water, and any losses owing to solubility influences may be neglected, except for the most accurate work.

Barium sulphate exhibits a marked tendency to carry down other salts. Whether the results will be low or high will depend upon the nature of the co-precipitated salt. Thus barium chloride and barium nitrate are readily co-precipitated. These salts will be an addition to the true weight of the barium sulphate, hence the results will be high, since the chloride is unchanged upon ignition and the nitrate will yield barium oxide. The error due to the chloride will be considerably reduced by the very slow addition of hot dilute barium chloride solution to the hot sulphate solution, which is constantly stirred; that due to the nitrate cannot be avoided, hence nitrate ion must always be removed by evaporation with a large excess of hydrochloric acid before precipitation.

Pure barium sulphate is not decomposed when heated in dry air until a temperature of about 1400 °C is reached:

$$BaSO_4 = BaO + SO_3$$

But the precipitate is easily reduced to sulphide at temperatures above 600 °C by the carbon of the filter paper:

$$BaSO_4 + 4C = BaS + 4CO$$

The reduction is avoided by first charring the paper without inflaming, and then burning off the carbon slowly at a low temperature with free access of air. If a reduced precipitate is obtained, it may be reoxidised by treatment with sulphuric acid, followed by volatilisation of the acid and reheating. The final ignition of the barium sulphate need not be made at a higher temperature than 600–800 °C (dull red heat). A Vitreosil or porcelain filtering crucible may be used, and the difficulty of reduction by carbon is entirely avoided.

Procedure   Weigh out accurately about 0.3 g of the solid (or a sufficient amount to contain 0.05–0.06 g of sulphur) into a 400 mL beaker, provided with a stirring rod and clockglass cover. Dissolve the solid in about 25 mL of water, add 0.3–0.6 mL of concentrated hydrochloric acid, and dilute to 200–255 mL. Heat the solution to boiling, add dropwise from a burette or pipette 10–12 mL of warm 5% barium chloride solution (5 g $BaCl_2 \cdot 2H_2O$ in 100 mL of water, ~0.2 M). Stir the solution constantly during the addition. Allow the precipitate to settle for a minute or two. Then test the supernatant liquid for complete precipitation by adding a few drops of barium chloride solution. If a precipitate is formed, add slowly a further 3 mL of the reagent, allow the precipitate to settle as before, and test again; repeat this operation until an excess of barium chloride is present. When an excess of the precipitating agent has been added, keep the covered solution hot, but not boiling, for an hour (steam bath, low-temperature hotplate or small flame) in order to allow time for complete precipitation. The volume of the solution should not be allowed to fall below 150 mL; if the clockglass covering the beaker is removed, the underside must be rinsed off into the beaker by means of a stream of water from a wash bottle. The precipitate should settle readily, and a clear supernatant liquid should be obtained. Test the supernatant liquid with a few drops of barium chloride solution for complete precipitation. If no precipitate is obtained, the barium sulphate is ready for filtration.[3,4]

Clean, ignite, and weigh either a porcelain filtering crucible or a Vitreosil filtering crucible (porosity no. 4). Carry out the ignition either upon a crucible ignition dish or by placing the crucible inside a nickel crucible at red heat (or, if available, in an electric muffle furnace at 600–800 °C), allow to cool in a desiccator and weigh. After digestion of the precipitate, filter the supernatant liquid through the weighed crucible using gentle suction. Reject the filtrate, after testing for complete precipitation with a little barium chloride solution. Transfer the precipitate to the crucible and wash with warm water until 3–5 mL of the filtrate give no precipitate with a few drops of silver nitrate solution. Dry the crucible and precipitate in an oven at 100–110 °C, and then ignite in a manner similar to that used for the empty crucible for periods of 15 min until constant weight is attained.

11.10   Procedures for other ions

The procedures given in Sections 11.5 to 11.9 are fairly typical of those needed to carry out many similar gravimetric determinations of other ions. In most cases precipitation has to be carried out at specified conditions of pH and temperature in order to ensure complete precipitation of the required complex and to prevent co-precipitation of other ions that might be present. Table 11.2 gives the chemical formulae for products obtained by precipitation processes in gravimetric determinations for an extensive range of cations and anions. However, the reader should consult the sources of data in Section 11.12 for full details of any individual procedure for the gravimetric determination of a substance.*

---

* Many detailed procedures for gravimetric analysis will be found in the fifth edition of *Vogel's Quantitative Chemical Analysis* (Longman, Harlow, 1989).

Table 11.2   *Products of precipitation*

Ion	Reagent	Product formula	Special conditions
Cations			
Aluminium	8-hydroxyquinoline (oxine)	$Al(C_9H_6ON)_3$	See Section 11.5
Ammonium	sodium tetraphenylborate[a]	$NH_4[B(C_6H_5)_4]$	Dry at 100 °C, K, Rb, Cs interfere
Antimony	pyrogallol	$Sb(C_6H_5O_3)$	Suitable for separation from As
Arsenic	hydrogen sulphide	As_2S_3	Dry at 105 °C
Barium	sulphuric acid	$BaSO_4$	See Section 11.9
Beryllium	ammonium hydroxide	BeO	The hydroxide is heated to 700 °C and ignited at 1000 °C
Bismuth	potassium iodide	$BiOI$	$K[BiI_4]$ gives product on dilution and boiling
Cadmium	quinaldic acid	$Cd(C_{10}H_6O_2N)_2$	Dry at 125 °C
Calcium	ammonium oxalate	$CaC_2O_4 \cdot H_2O$ or $CaCO_3$	Calcium oxalate obtained by homogeneous precipitation[b] $CaCO_3$ obtained above 475 °C
Cerium	potassium iodate	CeO_2	$Ce(IO_3)_4$ is ignited at 500 °C
Chromium	lead nitrate	$PbCrO_4$	Dry at 120 °C. See Section 11.7
Cobalt	mercury(II)chloride and ammonium thiocyanate	$Co[Hg(SCN)_4]$	Dry at 100 °C
Copper	ammonium thiocyanate	$CuSCN$	Dry at 110–120 °C
Gold	metal	SO_2, $(COOH)_2$, $FeSO_4$	Ignite
Iron	methanoic acid and urea	Fe_2O_3 via the methanoate	Ignite at 850 °C
Lead	chromium nitrate and potassium bromate	$PbCrO_4$	See Section 11.7
Lithium	sodium aluminate	$2Li_2O \cdot 5Al_2O_3$	Ignite at 500–550 °C
Magnesium	8-hydroxyquinoline	$Mg(C_9H_6ON)_2$	Dry at 155–160 °C
Manganese	diammonium hydrogen phosphate	$MnNH_4PO_4 \cdot H_2O$ or $Mn_2P_2O_7$	Dry at 100–105 °C or heat to 700–800 °C
Mercury	thionalide	$Hg(C_{12}H_{10}ONS)_2$	Dry at 105 °C
Molybdenum	8-hydroxyquinoline	$MoO_2(C_9H_6ON)_2$	Dry at 130–140 °C
Nickel	dimethylglyoxime	$Ni(C_4H_7O_2N_2)_2$	See Section 11.8
Palladium	dimethylglyoxime	$Pd(C_4H_7O_2N_2)_2$	See Section 11.8
Platinum	methanoic acid	Pt metal	Ignite
Potassium	sodium tetraphenylborate[a]	$K[B(C_6H_5)_4]$	Filter cold, wash with saturated solution of potassium tetraphenylboron, dry at 120 °C
Selenium	sulphur dioxide	Se metal	Dry at 100–110 °C
Silver	hydrochloric acid	$AgCl$	Dry at 130–150 °C. See Section 11.6
Sodium	zinc uranyl ethanoate	$NaZn(UO_2)_3(C_2H_3O_2)_9 \cdot 6H_2O$	Dry at 55–60 °C

Table 11.2   *(cont'd)*

Ion	Reagent	Product formula	Special conditions
Strontium	potassium dihydrogenphsphate	$SrHPO_4$	Dry at 120°C
Tellurium	sulphur dioxide	Te metal	Dry at 105°C
Thallium	potassium chromate	Tl_2CrO_4	Dry at 120°C
Thorium	sebacic acid	ThO_2	Ignite at 700–800°C
Tin	*N*-benzoyl-*N*-phenylhydroxylamine	$(C_{13}H_{11}O_2N)_2SnCl_2$	Dry at 110°C
Titanium	tannic acid and phenazone	TiO_2	Dry at 100°C then ignite at 700–800°C
Tungsten	tannic acid and phenazone	WO_3	Ignite at 800–900°C
Uranium	cupferron	U_3O_8	Dry at 100°C and ignite at 1000°C
Vanadium	silver nitrate and sodium ethanoate	Ag_3VO_4	Dry at 110°C
Zinc	8-hydroxy-quinaldinate	$Zn(C_{10}H_8ON)_2$	Dry at 130–140°C
Zirconium	mandelic acid	ZrO_2	Ignite at 900–1000°C

Anions

Ion	Reagent	Product formula	Special conditions
Borate	nitron	$C_{20}H_{16}N_4·HBF_4$	Dry at 105–110°C – this involves HF!
Bromate and bromide	silver nitrate in nitric acid	AgBr	Bromate is initially reduced to bromide; protect product from light
Carbonate	dilute hydrochloric acid or phosphoric acid	CO_2	Decomposition and absorption in soda-lime
Chlorate	iron(II) sulphate and silver nitrate	AgCl	Chlorate is reduced then chloride precipitated
Chloride	silver nitrate	AgCl	See Section 11.6
Cyanide	silver nitrate	AgCN	Danger! Do not heat solutions; dry at 100°C
Fluoride	sodium chloride and lead nitrate	PbClF	Dry at 140–150°C. See Section 10.100
Hypophosphite	mercury(II) chloride	Hg_2Cl_2	An indirect method
Iodate	sulphur dioxide and silver nitrate	AgI	Iodate is initially reduced to iodide
Iodide	silver nitrate	AgI	Dry below 150°C
Nitrate	nitron	$C_{20}H_{16}N_4·HNO_3$	Dry at 105°C; see Section 11.3
Nitrite	potassium permanganate	$C_{20}H_{16}N_4·HNO_3$	Indirectly by oxidising to nitrate
Oxalate	calcium chloride	$CaCO_3$ or CaO	Leave precipitate to stand 12 h, decompose to CaO
Perchlorate	ammonium chloride and silver nitrate	AgCl	Indirect method via reduction to chloride
Phosphate	ammonium molybdate	$(NH_4)_3[PMo_{12}O_{40}]$ or $P_2O_5·24MoO_3$	Dry at 200–400°C or heat to 800–825°C

Table 11.2 *(cont'd)*

Ion	Reagent	Product formula	Special conditions
Phosphite	mercury(II) chloride	Hg_2Cl_2	Indirect method; dry at 105–110 °C
Silicate	sodium hydroxide, ammonium molybdate and quinoline	$(C_9H_7)_4H_4[SiO_4 \cdot 12MoO_3]$	For ceramic materials
Sulphate	barium chloride	$BaSO_4$	See Section 11.9
Sulphide	sodium peroxide and potassium nitrate	$BaSO_4$	Initial oxidation to sulphuric acid
Sulphite	ammoniacal hydrogen peroxide	$BaSO_4$	Initial oxidation to sulphate
Thiocyanate	copper sulphate	$CuSCN$	Dry at 110–120 °C
Thiosulphite	ammoniacal hydrogen peroxide	$BaSO_4$	Initial oxidation to sulphate

[a] Flashka and Barnard have written an article on tetraphenylboron as an analytical reagent.[5]
[b] See the *Talanta* article by Grzeskowiak and Turner.[6]

11.11   References

1. C L Wilson and D W Wilson (eds) 1960 *Comprehensive analytical chemistry*, Volume 1A, *Classical analysis*, Elsevier, Amsterdam
2. L Gordon, M L Salutsky and H H Willard 1959 *Precipitation from Homogeneous solution*, John Wiley, New York
3. T B Smith 1940 *Analytical processes*, 2nd edn, Edward Arnold, London
4. H H Laitinen and W E Harris 1975 *Chemical analysis*, 2nd edn, McGraw-Hill, New York
5. H Flashka and A J Barnard 1960 Tetraphenylboron (TPB) as an analytical reagent. In C N Reilley (ed) *Advances in analytical chemistry and instrumentation*, Volume I, Interscience, New York
6. R Grzeskowiak and T A Turner 1973 *Talanta*, **20**; 351

11.12   Bibliography

D H Everett 1988 *Basic principles of colloid science*, Royal Society of Chemistry, London
J Mendham, D Dodd and D Cooper 1987 *Classical methods of chemical analysis*, Volume II, ACOL–Wiley, Chichester
A G Sharpe 1992 *Inorganic chemistry*, 3rd edn, Longman, Harlow
W F Smith 1996 *Analytical chemistry of complex matrices*, John Wiley, Chichester
J Tyson 1994 *Analysis*, Royal Society of Chemistry, London

12

Thermal analysis

12.1 General discussion

Thermal methods of analysis have been included here because thermogravimetry (TG) forms a link between classical gravimetry and it is often used in conjunction with other important methods of thermal analysis. Only a fairly limited coverage of the wide diversity of thermal methods and their applications can be included in this chapter. For more detailed information, consult the texts and review articles in Section 12.13.

Thermal methods of analysis may be defined as those techniques in which changes in physical and/or chemical properties of a substance are measured as a function of temperature. Methods that involve changes in weight or changes in energy come within this definition. Thermomechanical analysis (TMA), in which changes in dimensions of a substance are measured as a function of temperature, are outside the range of this book. Four techniques are discussed in this chapter:

Thermogravimetry (TG) a technique in which a change in the weight of a substance is recorded as a function of temperature or time.

Differential thermal analysis (DTA) a technique in which the temperature difference between a substance and a reference material is measured as a function of temperature whilst the substance and reference are subjected to a controlled temperature programme.[1]

Differential scanning calorimetry (DSC) a technique in which the difference in energy inputs into a substance and a reference material is measured as a function of temperature whilst the substance and reference material are subjected to a controlled temperature programme. (Two modes, power compensation DSC and heat flux DSC, can be distinguished, depending on the methods of measurement used.)[1]

Evolved gas analysis (EGA) where qualitative and quantitative evaluations of volatile products formed during thermal analysis are made.

12.2 Thermogravimetry

The basic instrumental requirement for thermogravimetry is a precision balance with a furnace programmed for a linear rise of temperature with time. The results may be presented as a thermogravimetric (TG) curve, in which the weight change is recorded as a function of temperature or time; or as a derivative thermogravimetric (DTG) curve, where the first derivative of the TG curve is plotted with respect to either temperature or time. Figure 12.1 shows a typical thermogravimetric curve for copper sulphate pentahydrate $CuSO_4 \cdot 5H_2O$. It can be divided into horizontal portions and curved portions:

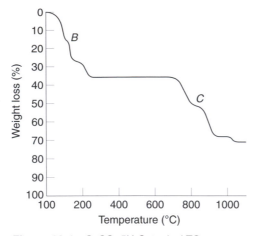

Figure 12.1 $CuSO_4 \cdot 5H_2O$: typical TG curve

Horizontal portions (plateaus) indicate regions where there is no weight loss.
Curved portions indicate regions of weight loss.

Since the TG curve is quantitative, calculations on compound stoichiometry can be made at any given temperature. Copper sulphate pentahydrate has four distinct regions of decomposition. They are listed here along with the approximate temperature ranges:

$$CuSO_4 \cdot 5H_2O \rightarrow CuSO_4 \cdot H_2O \qquad\qquad 90-150\,^{\circ}C$$
$$CuSO_4 \cdot H_2O \rightarrow CuSO_4 \qquad\qquad 200-275\,^{\circ}C$$
$$CuSO_4 \rightarrow CuO + SO_2 + \tfrac{1}{2}O_2 \qquad\qquad 700-900\,^{\circ}C$$
$$2CuO \rightarrow Cu_2O + \tfrac{1}{2}O_2 \qquad\qquad 1000-1100\,^{\circ}C$$

The precise temperature regions for each of the reactions depend upon the experimental conditions (Section 12.5). Although in Figure 12.1 the ordinate is shown as the percentage weight loss, the scale on this axis may take other forms:

1.   True weight
2.   Percentage of the total weight
3.   Relative molecular mass units

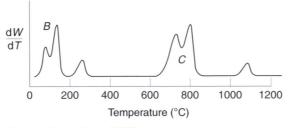

Figure 12.2   Typical DTG curve

Notice regions B and C in Figure 12.1, where there are changes in the slope of the weight loss curve. If the rate of change of weight with time dW/dT is plotted against temperature, a derivative thermogravimetric (DTG) curve is obtained (Figure 12.2). In the DTG curve when there is no weight loss then $dW/dT = 0$. The peak on the derivative curve corresponds to a maximum slope on the TG curve. When dW/dT is a minimum but not zero there is an inflexion, i.e. a change of slope on the TG curve. Inflexions B and C on Figure 12.1 may imply the formation of intermediate compounds. In fact, the inflexion at B arises from the formation of the trihydrate $CuSO_4 \cdot 3H_2O$ and at point C it is reported by Duval[2] as due to formation of a golden-yellow basic sulphate of composition $2CuO \cdot SO_3$. Derivative thermogravimetry is useful for many complicated determinations and any change in the rate of weight loss may be readily identified as a trough indicating consecutive reactions; hence weight changes occurring at close temperatures may be ascertained.

When a variety of commercial thermobalances became available in the early 1960s it was soon realised that a wide range of factors could influence the results obtained. Reviews of these factors have been made by Simons and Newkirk[3] and by Coats and Redfern[4] as a basis for establishing criteria necessary to obtain meaningful and reproducible results. The factors which may affect the results can be classified into instrumental effects – heating rate, furnace atmosphere, crucible geometry – and sample characteristics.

Heating rate

When a substance is heated at a fast rate, the temperature of decomposition will be higher than that obtained at a slower rate of heating. The effect is shown for a single-step reaction in Figure 12.3. The curve AB represents the decomposition curve at a slow heating rate, whereas the curve CD is due to the faster heating rate. If T_A and T_C are the decomposition temperatures at the start of the reaction and the final temperatures on completion of the decomposition are T_B and T_D, then

$$T_A < T_C$$
$$T_B < T_D$$
$$T_B - T_A < T_D - T_C$$

The heating rate has only a small effect when a fast reversible reaction is considered. The points of inflexion B and C obtained on the thermogravimetric curve for copper sulphate pentahydrate (Figure 12.1) may be resolved into a plateau if a slower heating rate is used. Hence the detection of intermediate compounds by thermogravimetry is very dependent upon the heating rate employed.

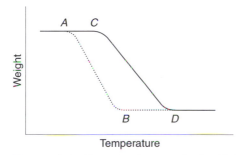

Figure 12.3 Faster heating rates lead to higher decomposition temperatures

Furnace atmosphere

The nature of the surrounding atmosphere can have a profound effect upon the temperature of a decomposition stage. For example, the decomposition of calcium carbonate occurs at a much higher temperature if carbon dioxide rather than nitrogen is employed as the surrounding atmosphere. Normally the function of the atmosphere is to remove the gaseous products evolved during thermogravimetry, in order to ensure the nature of the surrounding gas remains as constant as possible throughout the experiment. This condition is achieved in many modern thermobalances by heating the test sample in vacuo.

There are three atmospheres most commonly employed in thermogravimetry:

Static air: air from the surroundings flows through the furnace.
Dynamic air: compressed air from a cylinder is passed through the furnace at a measured flow rate.
Nitrogen gas: oxygen-free nitrogen gas provides an inert environment.

Atmospheres that take part in the reaction, e.g. humidified air, have been used to study the decomposition of hydrated metal salts and similar compounds.

Since thermogravimetry is a dynamic technique, convection currents arising in a furnace will cause a continuous change in the gas atmosphere. The exact nature of this change further depends upon the furnace characteristics, so widely differing thermogravimetric data may be obtained from different designs of thermobalance.

Crucible geometry

The geometry of the crucible can alter the slope of the thermogravimetric curve. A flat, plate-shaped crucible is generally preferred to a 'high-form' cone shape because the diffusion of any evolved gases is easier with a flat shape.

Sample characteristics

The weight, the particle size and the mode of preparation (the prehistory) of a sample all govern the thermogravimetric results. A large sample can often create a deviation from linearity in the temperature rise. This is particularly true when a fast exothermic reaction is studied, e.g. the evolution of carbon monoxide during the decomposition of calcium oxalate to calcium carbonate. A large volume of sample in a crucible can impede the diffusion of evolved gases through the bulk of the solid crystals, especially those of certain metallic nitrates which may undergo decrepitation (spitting or spattering) when heated. Other samples may swell, or foam and even bubble. In practice it is better to use a small sample weight with a particle size as small as practicable.

Diverse thermogravimetric results can be obtained from samples with different prehistories; for example, TG and DTG curves have shown that magnesium hydroxide prepared by precipitation methods has a different temperature of decomposition from that for the naturally occurring material.[5] It follows that the source and/or the method of formation of the sample should be ascertained.

12.3   Instruments for thermogravimetry

12.3.1   Good thermal balance design

Lukaszewski and Redfern's criteria[6]

(a)   The thermobalance should be capable of continuously registering the weight change of the sample studied as a function of temperature and time.
(b)   The furnace should reach the maximum desired temperature. (With some modern thermobalances a temperature range of −150 to 2400 °C can be obtained.)

(c)   The rate of heating is linear and reproducible.

(d)   The sample holder should be in the hot zone of the furnace and this zone should be of uniform temperature.

(e)   The thermobalance should have facilities for the provision of variable heating rates to permit heating in a variety of controlled atmospheres and for heating in vacuo. The instrument should also be capable of carrying out accurate isothermal studies.

(f)   The balance mechanism should be protected from the furnace and from the effect of corrosive gases.

(g)   The temperature of the sample must be measured as accurately as possible.

(h)   Use a balance sensitivity suitable for studying small sample weights.

Extra criteria for modern balances

(a)   A facility for rapid heating and cooling of the thermobalance; this permits several analyses to be carried out in a relatively short period of time.

(b)   The instrument should be capable of plotting derivative DTG curves.

(c)   Make sure there is a straightforward coupling with a gas analyser for EGA (gas chromatograph GC, mass spectrometer MS, Fourier transform infrared FTIR).

(d)   A dynamic heating rate. This allows high heating rates in regions where there are no weight changes, but it can be continuously varied in response to the decomposition rate of the sample.

12.3.2   Major components of a thermobalance

A wide range of commercial instruments is available and have many similar features. Modern thermobalances are computer-controlled, hence they can generate almost any temperature profile. Figure 12.4 is a block diagram showing the main components of a thermobalance. The major components of a thermobalance are the balance and the furnace assembly. Samples are placed in a shallow platinum crucible (sample container) that is connected to an automatic recording microbalance. The most usual type of balance system employed in thermogravimetry is the **null point** balance. In the null point system, when there is a change in weight, the balance beam will deviate from its usual position. A sensor detects this deviation and initiates a force that will restore the balance to the null position. This restoring force is proportional to the change in weight.

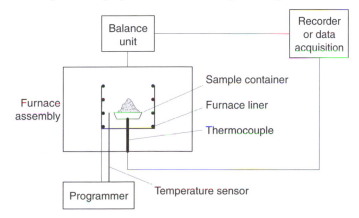

Figure 12.4   Main components of a thermal balance

479

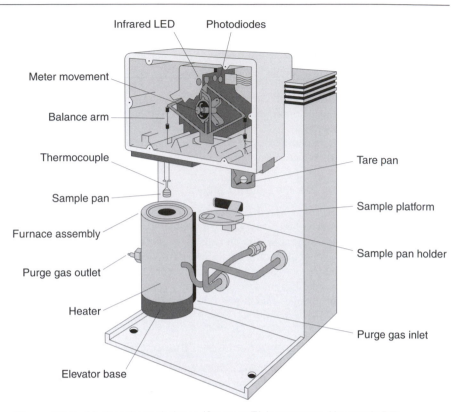

Infrared LED　　Photodiodes

Meter movement

Balance arm

Thermocouple

Sample pan

Furnace assembly

Purge gas outlet

Heater

Elevator base

Tare pan

Sample platform

Sample pan holder

Purge gas inlet

Figure 12.5　Modern thermobalance (Courtesy TA Instruments, Newcastle DE)

The sample container is placed in a quartz or Pyrex glass housing located within the furnace. A thermocouple, located immediately below the sample container, is used to monitor the furnace temperature. The resulting signal is connected directly to the x-axis of the recorder. A modern thermobalance is fully computer-controlled, allowing a wide variety of heating, cooling and isothermal modes. The computer control enables the user to change sample atmosphere conditions automatically. Another important feature is that the cooling time between experiments is a matter of a few minutes in going from 1000 °C to ambient temperature.

An example of a modern thermobalance is the TA Instruments TGA 2950; its basic components are shown in Figure 12.5. The system operates on a null balance principle using a highly sensitive transducer coupled to a taut-band suspension system allowing the detection of minute changes in the mass of the sample. An optically actuated servo loop maintains the balance arm in the horizontal reference (null) position by regulating the amount of current flowing through the transducer coil. An infrared LED light source and a pair of photosensitive diodes detect movement of the beam. A flag at the top of the balance arm controls the amount of light reaching each photosensor. As sample weight is lost or gained, the beam becomes unbalanced, causing the light to strike the photodiodes unequally. The unbalanced signal is fed into the control programme, where the amount of current supplied to the meter movement is altered, causing the balance to rotate back to its null position. The amount of current required is directly proportional to the change in the mass of the sample. Heating rate and sample temperature are measured by a thermocouple

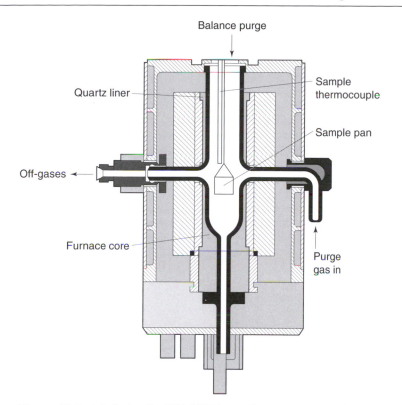

Figure 12.6   Interfacing the TGA 2950 to another system, e.g. MS or FT-IR (Courtesy TA Instruments, Newcastle DE)

located immediately adjacent to the sample. This ensures accurate measure of sample temperature as well as enabling the control unit to set up and maintain the temperature environment and heating rate selected by the operator.

The sample pan is filled and placed on the sample platform, where its orientation may be adjusted for automatic pickup. Pressing the START key begins an automatic series of events: rotation of the sample platform, pickup of the sample, raising of the furnace and initiation of the previously programmed experimental conditions and procedures. Sample unloading and rapid furnace cooling can also be programmed by the user. During an experiment, data is automatically collected and stored in the electronic memory, where it can be accessed for analysis and presentation of results. Optional features include automatic change of atmosphere using a gas switching accessory, and multiple samples can be run unattended using an autosampler accessory.

Figure 12.6 shows how the TGA 2950 may be interfaced with FT-IR (Fourier transform infrared spectroscopy), or with MS (mass spectrometry) for the analysis of evolved gases (EGA). All it requires is an interconnection from the purge gas outflow port. Interfacing TGA with FT-IR poses far less of an instrumental problem than coupling TGA with MS, where there is an appreciable pressure difference between the two components. FT-IR is invaluable in the identification of toxic substances that may be evolved during the thermal decomposition of polymeric materials. Furthermore, when molecules have identical or nearly the same relative molecular mass, e.g. water and ammonia, it is sometimes difficult

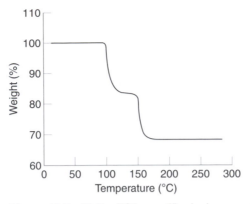

Figure 12.7 Hi-Res TGA quantifies hydrogencarbonate mixtures: constant reaction rate, hermetic pan with pinhole, 133 min (Courtesy TA Instruments, Newcastle DE)

to ascertain exactly which species has been evolved. FT-IR could remove any ambiguity that might arise in the identity of thermal decomposition products.

A recent innovation developed by TA Instruments is called Hi-Res TGA. Four variable heating rate algorithms can be used to optimise the resolution and the analysis time. Using a constant reaction rate approach – in which the control system varies the furnace temperature to maintain a constant rate of weight change – a mixture of sodium and potassium hydrogencarbonates can be resolved and readily quantified (Figure 12.7).

12.4   Applications of thermogravimetry

Four applications of thermogravimetry are of particular importance to the analyst:

1. Determining the purity and thermal stability of both primary and secondary standards.
2. Investigating the correct drying temperatures and the suitability of various weighing forms for gravimetric analysis.
3. Direct application to analytical problems (automatic thermogravimetric analysis).
4. Determining the composition of alloys and mixtures.

Thermogravimetry is a valuable technique for assessing the purity of materials. Analytical reagents, especially those used in titrimetric analysis as primary standards, e.g. sodium carbonate, sodium tetraborate and potassium hydrogenphthalate, have been examined. Many primary standards absorb appreciable amounts of water when exposed to moist atmospheres. TG data can show the extent of this absorption, hence the most suitable drying temperature for a given reagent may be determined.

The thermal stability of EDTA as the free acid and also as the more widely used disodium salt $Na_2EDTA \cdot 2H_2O$ has been reported by Wendlandt.[7] He showed that the dehydration of the disodium salt commences at between 110 and 125 °C, which confirmed the view of Blaedel and Knight[8] that $Na_2EDTA \cdot 2H_2O$ could be safely heated to constant weight at 80 °C.

Initially, the most widespread application of thermogravimetry in analytical chemistry has been in the study of the recommended drying temperatures of gravimetric precipitates. Duval studied over a thousand gravimetric precipitates by this method and gave the recommended drying temperatures. He further concluded that only a fraction of these precipitates are suitable weighing forms for the elements. The results recorded by Duval were obtained with materials prepared under specified conditions of precipitation, and this must be borne

in mind when assessing the value of a given precipitate as a weighing form, since conditions of precipitation can have a profound effect on the pyrolysis curve. It is unjustified to
reject a precipitate because it does not give a stable plateau on the pyrolysis curve at one
given rate. Furthermore, the limits of the plateau should not be taken to indicate thermal
stability within the complete temperature range. The weighing form is not necessarily isothermally stable at all temperatures that lie on the horizontal portion of a thermogravimetric
curve. A slow rate of heating is to be preferred, especially with a large sample weight, over
the temperature ranges in which chemical changes take place. Thermogravimetric curves
must be interpreted with due regard to the fact that while they are being obtained the
temperature is usually changing at a uniform rate, whereas in routine gravimetric analysis
the precipitate is often brought rapidly to a specified temperature and maintained at that
temperature for a definite time.

Thermogravimetry may be used to determine the composition of binary mixtures. If each
component possesses a unique pyrolysis curve, then a resultant curve for the mixture will afford
a basis for determining its composition. In such an automatic gravimetric determination
the initial weight of the sample need not be known. A simple example is given by the
automatic determination of a mixture of calcium and strontium as their carbonates. Both
carbonates decompose to their oxides with the evolution of carbon dioxide. The decomposition temperature for calcium carbonate is in the temperature range 650–850 °C, whereas
strontium carbonate decomposes in the range 950–1150 °C. Hence the amount of calcium
and strontium present in a mixture may be calculated from the weight losses due to the
evolution of carbon dioxide at the lower and higher temperature ranges respectively. This
method can be extended to the analysis of a three-component mixture, as barium carbonate is
reported to decompose at an even higher temperature (~1300 °C) than strontium carbonate.

A further example, cited by Duval,[9] is the automatic determination of a mixture of
calcium and magnesium as their oxalates. Calcium oxalate monohydrate has three distinct
regions of decomposition; magnesium oxalate dihydrate has only two:

(a) $CaC_2O_4 \cdot H_2O$ $\rightarrow CaC_2O_4 + H_2O$ 100–250 °C
(b) CaC_2O_4 $\rightarrow CaCO_3 + CO$ 400–500 °C
(c) $CaCO_3$ $\rightarrow CaO + CO_2$ 650–850 °C
(d) $MgC_2O_4 \cdot 2H_2O \rightarrow MgC_2O_4 + 2H_2O$ 100–250 °C
(e) MgC_2O_4 $\rightarrow MgO + CO + CO_2$ 400–500 °C

A pyrolysis curve for a mixture of these two oxalates would thus show three decomposition
steps. The final step would be due entirely to the loss of carbon dioxide from calcium
carbonate, hence the amount of calcium present in the mixture may be calculated. The
amount of magnesium in the oxalate mixture may be calculated from the second step,
where stages (b) and (e) occur; this is because the amount of carbon monoxide due to
calcium carbonate may be subtracted from the total observed weight loss, giving the loss
of carbon dioxide and carbon monoxide from anhydrous magnesium oxalate. Since stages
(a) and (d) do not feature in the calculation, the mixture of the calcium and magnesium
oxalates does not require drying before the automatic determination.

Thermogravimetric studies have been made on mineralogical, metallurgical and polymeric materials. Selected examples include the study of clays and soils by Hoffman et al.[10]
The pyrolysis curves of most soils examined showed plateaus starting at 150–180 °C and
extending to 210–240 °C, indicating that hygroscopic moisture and or easily volatile
organic compounds had been removed. When the clay content of a soil was studied, the
loss in weight at 500 °C read from a pyrolysis curve gave an estimate of the organic matter
which was in reasonable agreement with dry combustion and wet oxidation data. An

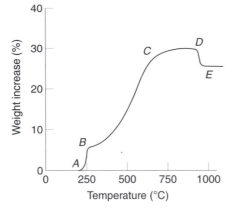

Figure 12.8 Oxidation of Co_5Sm by thermogravimetry in an air atmosphere: *AB* shows oxidation of Sm to Sm_2O_3, *BC* shows oxidation of Co to Co_3O_4 and formation of the mixed oxide $CoSmO_4$, *DE* shows the conversion of Co_3O_4 to CoO. The formation of Sm_2O_3 and the mixture of Co_3O_4 and $CoSmO_3$ were confirmed by X-ray diffraction studies

additional feature of the work suggests that lattice water may be quantitatively determined in pure clays. Because lattice water came off from different clays at different temperatures, these temperatures may possibly be used as a method of identification.

The oxidation of the cobalt–samarium alloy Co_5Sm was examined by thermogravimetry using an air atmosphere.[11] The results are shown in Figure 12.8. The weight increase *AB* resulted from the oxidation of Sm to Sm_2O_3. The increase *BC* was due to the oxidation of Co to Co_3O_4 and to the formation of the mixed oxide $CoSmO_3$. The final weight loss *DE* was obtained from the conversion of Co_3O_4 to CoO. The formation of Sm_2O_3 and the mixture of Co_3O_4 and $CoSmO_3$ were confirmed by X-ray diffraction studies.

One of the most important applications of thermogravimetry is in examining the thermal stability of polymers. Thermogravimetry is also capable of giving information concerning polymer identity. Since derivative thermogravimetry (DTG) is able to show the temperature where the maximum weight change appears to take place, it is a valuable method for distinguishing polymers. Thus, TG and/or DTG are useful in identifying the components of a polymer blend; the quantitative determination can be performed by differential scanning calorimetry (DSC); see Section 12.8.

Experimental section

!   **Safety.** Before carrying out any experiments in this section, pay full attention to any safety warnings and make sure you adhere to national laboratory and safety regulations.

12.5   Thermogravimetric experiments

A small number of thermogravimetric experiments will be outlined below. For more detailed information on these and other studies, consult the publications listed in Section 12.14. Here are some precautions to follow when using a modern thermobalance incorporating an electronic microbalance that requires small sample weights:

(a)   Select the weight of sample according to the actual weight loss anticipated.
(b)   Try not to handle the crucible; handling may transfer grease or moisture to it. A platinum crucible may be cleaned by placing it in dilute nitric acid.
(c)   Take a representative sample from the original batch. If the material is thought to be inhomogeneous, several samples should be run; different results will confirm inhomogeneity.

The method of obtaining a sample depends upon the nature of the material.

A circular disc may be cut from a film of material by the use of an appropriate cork borer or leather punch.
Fibrous material which does not pack easily, may be squeezed between metal foil before being transferred to the crucible.
Liquid samples may be transferred to the crucible by means of a hypodermic syringe.
Air-sensitive samples should be loaded on to the crucible in a glove box and transferred rapidly to the thermobalance which should be set up all ready for a dry inert gas flow.
Materials which creep or froth should not be used in a thermobalance.

It is always sound practice to heat the test material in a small crucible in an oven or muffle furnace to ascertain whether or not there is any creeping or frothing before using the sample for thermogravimetry. Considerable damage can occur to a thermobalance if samples are not checked in this way.

12.5.1   Thermal decomposition of calcium oxalate monohydrate

This determination may be carried out on any standard thermobalance. In all cases the manufacturer's handbook should be consulted for detailed instructions on operating the instrument. Zero the balance on the 10 mg range with an empty crucible in position and use an airflow of 10 mL min^{-1}. Weigh accurately about 2 mg of the calcium oxalate monohydrate directly into the crucible and record the weight on the chart. The recorder variable range may now be used to expand the sample weight to 100% of full scale. Select a suitable heating rate (20 °C min^{-1}) and record the pyrolysis curve of calcium oxalate monohydrate from ambient temperature to 1000 °C in terms of percentage sample weight loss. From the TG curve estimate the purity of the calcium oxalate (Section 12.4). This experiment may be extended for the automatic gravimetric determination of calcium and magnesium as their oxalates. Details of the thermal decomposition of both oxalates are given in Section 12.4. From the TG curve calculate the amounts of calcium and magnesium.

12.5.2   Thermal decomposition of copper sulphate pentahydrate

Follow the procedure outlined in Section 12.5.1 but weigh out accurately about 6 mg of the copper sulphate. Record the thermal decomposition of copper sulphate from ambient temperature to 1000 °C using a heating rate of 10 °C min^{-1} and an air atmosphere with a flow rate of 10 cm^3 min^{-1}. Examine the effect of varying the heating rate on the dehydration reactions by selecting rates of 2, 20 and 100 °C min^{-1} in addition to the 10 °C min^{-1} used previously. The enhanced resolution obtained by slowing the heating rates is better displayed using DTG plots.

12.5.3   Thermal decomposition of nickel oxalate dihydrate

This experiment illustrates the effect of using different furnace atmospheres in thermogravimetry. Follow the procedure outlined in the previous experiments. In this case record the decomposition of nickel oxalate from ambient temperature to $1000\,°C$ using a heating rate of $10\,°C\,min^{-1}$ with an air atmosphere at a flow rate of $10\,cm^3\,min^{-1}$. Repeat the experiment with the same conditions but using a nitrogen atmosphere with a flow rate of $10\,cm^3\,min^{-1}$. Confirm that in an air atmosphere the final decomposition product is nickel(II) oxide, whereas using a nitrogen atmosphere the final product is nickel metal.

12.5.4 Determination of calcium and magnesium in dolomite

Dolomite is strictly an equimolecular compound of calcium and magnesium carbonates ($CaCO_3$, $MgCO_3$). Follow the procedure outlined in the previous experiments using a dynamic air atmosphere and a heating rate of $30\,°C\,min^{-1}$. Magnesium carbonate is converted to magnesium oxide by about $480\,°C$ and calcium carbonate decomposes to calcium oxide between 650 and $850\,°C$. It is thus possible to calculate the percentage of CaO and MgO in the sample of dolomite. The values of calcium and magnesium obtained may be compared with the results from an EDTA titration for $Ca^{2+} + Mg^{2+}$ (Section 10.76) and Ca^{2+} alone using EGTA as titrant (Section 10.75).

12.6 Differential techniques

Differential thermal analysis

In differential thermal analysis (DTA) both the test sample and an inert reference material (usually α-alumina) undergo a controlled heating or cooling programme which is usually linear with respect to time. There is a zero temperature difference between the sample and the reference material when the sample does not undergo any chemical or physical change. But if any reaction takes place, then a temperature difference ΔT will occur between the sample and the reference material. Thus in an endothermic change, e.g. when the sample melts or is dehydrated, the sample is at a lower temperature than the reference material. This condition is only transitory because, on completion of the reaction, the sample will again show zero temperature difference compared with the reference.

In DTA a plot is made of ΔT against temperature or time, if the heating or cooling programme is linear with respect to time. An idealised DTA curve is shown in Figure 12.9: peak 1 is an exothermic peak and peak 2 is an endothermic peak. The shape and size of the peaks can give a large amount of information about the nature of the test sample. Thus, sharp endothermic peaks often signify changes in crystallinity or fusion processes, whereas broad endotherms arise from dehydration reactions. Physical changes usually result in

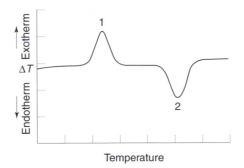

Figure 12.9 Idealised DTA curve: (1) exothermic peak, (2) endothermic peak

endothermic curves whereas chemical reactions, particularly those of an oxidative nature, are predominantly exothermic.

Differential scanning calorimetry (DSC)

In power compensation DSC – developed by Perkin-Elmer USA and the first technique to be called differential scanning calorimetry – the energy necessary to establish a zero temperature difference between the sample and a reference material is measured as a function of temperature or time. Thus, when an endothermic transition occurs, the energy absorbed by the sample is compensated by an increased energy input to the sample in order to maintain a zero temperature difference. Because this energy input is precisely equivalent in magnitude to the energy absorbed in the transition, direct calorimetric measurement of the energy of transition is obtained from this balancing energy. On the DSC chart recording, the abscissa indicates the transition temperature and the peak area measures the total energy transfer to or from the sample. Heat **flux measurement** DSC is detailed in Section 12.7.

12.7   Instruments for DTA and DSC

Modern thermal instruments usually have a modular design, hence the same furnace, programmer and recorder system may be used simultaneously or sequentially to obtain TG, DTG and DSC data from a single sample. Nowadays there is considerable scope for the operational conditions that can be used. With the aid of microprocessor control and user programmes, the operator may vary heating and/or cooling rates as well as performing isothermal studies. Furthermore, derivative plots can be displayed and, most important for DSC, peak areas may be evaluated. Thermoanalytical results can be retained by storage on floppy disks with hard copy obtained from a printer–plotter.

The exact distinction between DSC and DTA instrumentation was the subject of controversy for many years; it was eventually resolved by Mackenzie.[12] At one end of the scale there is conventional (classical) DTA, where ΔT is the difference between T_S (the sample temperature) and T_R (the reference temperature); see Figure 12.10. The junctions of the difference thermocouple are located in the centre of the sample and reference specimens. In this arrangement, ΔT cannot be directly related to an enthalpy change, thus the peak area cannot be reliably converted to energy units. Classical DTA can provide useful qualitative information, but it can never be more than semiquantitative. For quantitative measurements the thermocouple junctions should be sited immediately below and in thermal contact with two separate metal containers enclosed in an air space.

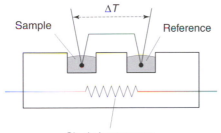

Figure 12.10   Apparatus for conventional DTA

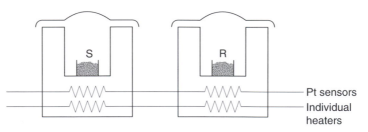

Figure 12.11 Apparatus for power compensation DTA

Quite different from DTA is power compensation DSC developed by Perkin Elmer USA. This makes a direct measurement of the enthalpy change. The sample and reference materials are supplied with separate heaters maintained at the same temperature by a servo system (Figure 12.11). Power is supplied to two heaters to maintain the samples and reference R at the same temperature. When an endothermic reaction occurs, the energy absorbed by the sample is replenished by an increased energy input to the sample, thus maintaining the temperature balance. Since the input of energy is equivalent to the energy absorbed in the transition, the method of power compensation yields a direct calorimetric measure of the energy of transition.

Heat flux DSC

Some variants of DSC are based on heat flux measurements. The system initially proposed by Boersma[13] demonstrated that the introduction of a controlled heat leak between sample and reference holders enabled a quantitative measurement of energy changes to be made. The peak area is therefore related to the enthalpy change by a calibration factor which is partially temperature dependent (Section 12.8). Heat flux can be measured directly if a sample is surrounded by a thermopile. The heat flux between separate samples and reference holders and a surrounding block are measured with two thermopiles connected in opposition. This is the underlying principle behind the DSC III (Setaram, France). It is now established that instruments based on power compensation or on heat flux measurement both come under the remit of DSC.[13]

The heat flux DSC developed by TA Instruments is shown in Figure 12.12. A metallic disc (made of constantan alloy) is the primary means of heat transfer to and from the sample and reference. The sample, contained in a metal pan, and the reference, an empty pan, sit on raised platforms in the constantan disc. As heat is transferred through the disc, the differential heat flow to sample and reference is measured by thermocouples formed by the junction of the constantan disc and chromel wafers which cover the underside of the platforms. These thermocouples are connected in series and measure the differential heat flow. Chromel and alumel wires attached to the chromel wafers form thermocouples which directly measure the sample temperature. Purge gas is admitted to the sample chamber through an orifice in the heating block. With this equipment, it is possible to achieve heating or cooling rates of $100\,^{\circ}\text{C}\,\text{min}^{-1}$ down to rates of $0\,^{\circ}\text{C}\,\text{min}^{-1}$ (isothermal).

Another technique is modulated DSC (MDSC); also developed by TA Instruments, it uses the heat flux DSC cell shown in Figure 12.11 but a different furnace heating profile. A sinusoidal modulation (oscillation) is overlaid on the traditional linear heating ramp to yield a heating profile in which the sample temperature still increases with time but not in a linear fashion. The overall effect of this heating profile on the sample is the same as if two

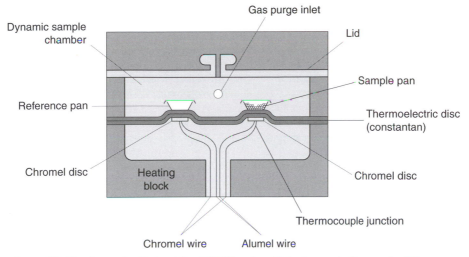

Figure 12.12   Apparatus for heat flux DSC (Courtesy TA Instruments, Newcastle DE)

simultaneous experiments were performed. This produces a slow underlying heating rate (improving resolution) as well as a faster instantaneous heating rate (improving sensitivity).

12.8   Experimental and instrumental factors

DTA and DSC peaks are also governed by the factors affecting TG curves (Section 12.2). Hence heating rates, atmosphere and geometry of sample holders can alter the position of DTA and DSC peaks. However, the most important factor in obtaining reliable results for both techniques is the preparation of the sample and reference material. Great care should be taken in the preparation of the sample and in the way the crucible or ampoule is loaded. To obtain reproducible results between successive experiments, it is essential to carry out the same packing procedure each time. The selection and handling of samples for DTA is similar to Section 12.5. It is possible, however, to use materials which creep, froth or boil if sealed sample containers are used to ensure no damage occurs to the sample holder assembly. Most modern DTA apparatus includes a device for sample encapsulation. It is usual practice to encapsulate the sample in metal pans of high thermal conductivity; this ensures the sample is in the form of a thin wafer which enables the best thermal contact between sample and temperature sensor.

It is now standard practice to use an empty pan as the reference in DSC (a similar practice is made in DTA when the sample weight is of the order of 1 mg). With higher sample weights it is necessary to use a reference material, because the total weight of the sample and its container should be approximately the same as the total weight of the reference and its container. The reference material should be selected so it possesses similar thermal characteristics to the sample. The most widely used reference material is α-alumina, which must be of analytical reagent quality. Before use, α-alumina should be re-calcined and stored over magnesium perchlorate in a desiccator. Kieselguhr is another reference material normally used when the sample has a fibrous nature. If there is an appreciable difference between the thermal characteristics of the sample and reference materials, or if values of ΔT are large, then dilution of the sample with the reference substance is sensible practice. Dilution may be accomplished by thoroughly mixing suitable proportions of sample and reference material.

12.9   Applications of DTA and DSC

The weight changes monitored by thermogravimetry invariably involve the absorption or release of energy, hence they can be measured by either DSC or DTA. But there are many changes in energy that are not accompanied by a gain or loss in weight. For example, melting, crystallisation, fusion and solid-state transitions do not involve weight changes. Thus, TG is often used in conjunction with either DTA or DSC.

Among the early applications of DTA was the qualitative analysis of materials. A rapid method for the fingerprinting of minerals, clays and polymeric materials was provided by DTA measurements. For example, using DTA with a nitrogen atmosphere, chrysotile (white asbestos) was shown to give a dehydroxylation endotherm at 650 °C and a characteristic crystallisation exotherm at 845 °C. A sample containing 1 % w/w chrysotile in talc was detected by DTA. This fast technique does not involve the extensive sample preparation required by competitive methods. Polymers can exhibit their own characteristic melting point endotherms. Thus, in a blend containing seven commercial polymers, each component was readily identified by the endotherm of melting using DTA.

DTA and particularly DSC have been used in pharmaceutical chemistry for the investigation of product purity, the identification of optical isomers, polymorphism and eutectic formation. In the food industry, edible fats and oils have been characterised by differential thermal methods. DSC is preferred for quantitative measurements because it requires only one standard for peak area calibration. In DTA and DSC the peak area A depends on the sample mass m, the heat of reaction ΔH and an empirical constant K, where

$$A = \pm \Delta H m K$$

In DTA the constant K depends on the temperature. This involves the calibration of peak areas using a standard in the same temperature range as the analyte sample. With an efficient DSC the constant K is effectively independent of temperature, hence only one standard is required for peak area calibration. Thus, for DSC, the constant K is determined from a standard of high purity with an accurately known enthalpy of fusion (ΔH_f). Pure indium metal with a melting point of 156.5 °C is widely used for this purpose. A modern instrument under microprocessor control will evaluate K using the peak area obtained from the melting endotherm of indium. So what role is there for DTA in modern thermal analysis? At the time of writing, the maximum temperature attainable by DSC is the order of 800 °C, whereas some DTA instruments can function up to 2000 °C. Thus, DTA is still used for the high-temperature studies of minerals, refractory materials and ceramics.

DSC analysis on a blend of synthetic fibres was an early application, and it shows just how versatile the technique can be. Using a Perkin-Elmer differential scanning calorimeter, a blend containing Nylon 66, Orlon and Vycron polyester was determined. The ΔH values per gram of sample were compared with the corresponding ΔH per gram of each pure component. Hence the ΔH values for the crystallisation peaks of nylon and the polyester were measured together with a cross-linking exotherm for Orlon. A quantitative analysis was then made on the fibre blend. For example, the ratio of the ΔH value for the nylon crystallisation peak in the fibre blend over the ΔH value for the pure nylon multiplied by 100 gave the percentage nylon in the fibre blend. The total time for the analysis, performed without sample treatment or any separatory procedure, was less than 30 min. The repeatability of the experiment was found to be within 5% of the amount of each component present.

Polymer blends difficult to evaluate by conventional DSC have been successfully analysed by modulated DSC (MDSC). For example, a polymer blend containing polyethylene terephthalate (PET) and acrylonitrile–butadiene–styrene (ABS) has been separated and evaluated using MDSC. Differential scanning calorimetry can be used to study the number and temperature range of polymorphs, since each polymorphic transition causes an energy change that may be detected by DSC. Various forms of digoxin have been investigated,[14] although grinding the material can affect crystallinity and may lead to the formation of an amorphous phase.[15] There is no doubt that thermal analysis is extremely versatile and able to address a wide variety of analytical problems. Modern instruments have become relatively simple to operate without sacrificing flexibility.

Experimental section

! **Safety.** Before carrying out any experiments in this section, pay full attention to any safety warnings and make sure you adhere to national laboratory and safety regulations.

12.10   DTA studies of hydrated copper sulphate and sodium tungstate

Introduction   These experiments provide initial experience with the technique and some information to interpret thermal events. Consult the manufacturer's handbook for detailed operating instructions.

Procedure   Accurately weigh the pair of empty crucibles and record each individual weight. Accurately weigh out ~50 mg of copper sulphate pentahydrate into one of the previously weighed crucibles (sample pan). Load the other crucible (reference pan) with an equal amount of α-alumina. Locate the sample holder assembly in the furnace mounting. Select the appropriate gas atmosphere (e.g. nitrogen) at a flow rate of $100\,mL\,min^{-1}$ and a typical heating rate of $20\,°C\,min^{-1}$. Record the DTA results for $CuSO_4 \cdot 5H_2O$ over the desired range of temperature (ambient to 1000 °C). With modern instruments, the whole of this procedure may be fully automated. Repeat the procedure with sodium tungstate dihydrate and the same experimental conditions. Results obtained with a simultaneous TG–DTA, the TA Instruments SDT 2960, are shown in Figure 12.13.

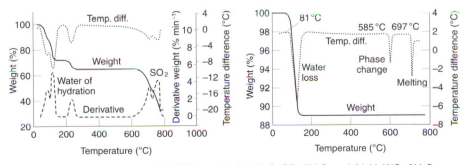

Figure 12.13   Simultaneous TG–DTA results for (a) $CuSO_4 \cdot 5H_2O$ and (b) $NaWO_4 \cdot 2H_2O$ (Courtesy TA Instruments, Newcastle DE)

In Figure 12.13(a) the TG plot of $CuSO_4 \cdot 5H_2O$ is represented by the bold line; the dashed line shows the DTG profile. The DTA plot is the chain dotted line in the top section of the diagram. Notice that all DTA endotherms have corresponding weight losses and are therefore decomposition stages. In Figure 12.13(b) the bold line shows the TG plot of $NaWO_4 \cdot 2H_2O$ with a single dehydration step at $81\,°C$. Besides the dehydration peak, the DTA results (dashed line) show two additional endotherms, a crystalline phase change at $585\,°C$ and a melting at $697\,°C$.

12.11   DSC determination of calcium sulphate hydrates in cement*

Introduction   In the manufacture of Portland cement, around 5% gypsum ($CaSO_4 \cdot 2H_2O$) is added to reduce the rate of setting. The gypsum is added to the fused clinker during processing, and the two components are subsequently milled to obtain uniform mixing and the required particle size. During milling, the thermal energy generated may cause partial dehydration of gypsum to the hemihydrate $CaSO_4 \cdot \frac{1}{2}H_2O$ which adversely affects (increases) the rate of setting of the cement. Hence it is important to monitor the amounts of each hydrate present in the final cement. In order to provide quantitation at the required levels, this problem can only be solved by DSC. The dehydration of gypsum occurs as a two-stage endothermic process:

$$CaSO_4 \cdot 2H_2O \rightarrow CaSO_4 \cdot \tfrac{1}{2}H_2O \rightarrow CaSO_4$$

Procedure   Follow the manufacturer's operating instructions. The following experimental conditions were adopted using the TA Instruments DSC2920.[16] Encapsulate the cement sample (between 5 and 8 mg) in a hermetically sealed aluminium DSC pan. Select a temperature range between 100 and $240\,°C$ with a heating rate of $15\,°C\,min^{-1}$, using an air atmosphere. Obtain the peak area of the endotherm at approximately $150\,°C$ for the conversion $CaSO_4 \cdot 2H_2O \rightarrow CaSO_4 \cdot \tfrac{1}{2}H_2O$ and the area of the endotherm at approximately $200\,°C$ for the dehydration step $CaSO_4 \cdot \tfrac{1}{2}H_2O \rightarrow CaSO_4$.

Using a calibration curve constructed from similar experiments on pure gypsum, the percentage gypsum in the cement is determined by graphical interpolation of the value obtained from the peak area of the low-temperature endotherm. By using the peak ratio for the two-stage dehydration of pure gypsum, determine the amount of heat in the second stage of the cement dehydration associated with the gypsum initially present. Subtract this value from the total area of the second dehydration stage for the cement; this yields the heat associated with the hemihydrate initially present. Using a calibration curve constructed from evaluations on the pure hemihydrate, it is possible to determine the percentage hemihydrate present in the cement.

12.12   Determining the purity of pharmaceuticals by DSC

Introduction   DSC provides a rapid yet reliable method for determining the purity of materials, particularly pharmaceuticals; the presence of minor impurities may reduce the effectiveness of a drug, or even cause adverse side effects on a patient. The DSC technique allows the melting curve to be determined for the sample as it is heated through its melting point. It is well known that the higher the concentration of impurity present in a sample, the lower its melting point and the broader its melting range. The data obtained by DSC

---

* Details courtesy TA Instruments.

includes the complete melting curve and the latent heat of fusion (ΔH_f) of the sample. The interpretation of the DSC curve is based on a modified form of the van't Hoff equation:

$$T_S = T_0 - \frac{RT_0^2 X_1}{\Delta H_f}\left(\frac{1}{F}\right)$$

where

ΔH_f = heat of fusion of pure major component ($J\,mol^{-1}$)
  R = gas constant ($8.314\ J\,mol^{-1}\,K^{-1}$)
  T_S = sample temperature (K)
  T_0 = theoretical melting point of pure compound (K)
  X_1 = mole fraction of impurity
  F = fraction of sample melted at T_S

A plot of T_S against $1/F$ should produce a straight line of intercept T_0, and slope $RT_0^2 X_1/\Delta H_f$ may be obtained from the DSC scan or from the literature, allowing X_1 to be calculated.

Procedure for phenacetin   Weigh the sample of phenacetin (between 1 and 5 mg) into an aluminium DSC pan and crimp a lid on. The reference is an empty DSC pan with lid. Use a nitrogen atmosphere ($50\,mL\,min^{-1}$) and a slow heating rate of $1\,°C\,min^{-1}$. Start the DSC run at a temperature about $10\,°C$ below the melting point (from 124 to $140\,°C$). With a modern DSC (TA Instruments DSC 2920) the calorimetric purity software will determine the purity of phenacetin expressed as mol% pure. Figure 12.14 shows results for the purity of a phenacetin sample. The solid line shows the melting endotherm of phenacetin with a melting point of $134.9\,°C$. The open triangles are the result of plotting T_S against $1/F$. This curve is non-linear, but it can be straightened to the theoretically predicted straight line using successive approximations. The slope of corrected line (open squares) is used to calculate the mol% purity.

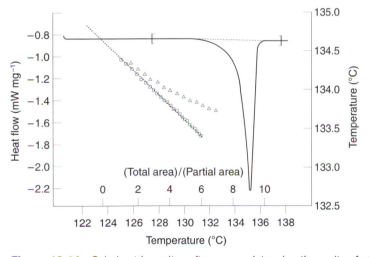

Figure 12.14 Calorimetric purity software can determine the purity of phenacetin: purity = $99.55\,mol\%$, melting point = $134.9\,°C$, depression = $0.24\,°C$, ΔH = $26.4\,kJ\,mol^{-1}$, correction = 8.11%, RMM = 179.2, cell constant = 0.977, onset slope = $10.14\,mW\,°C^{-1}$. The open triangles are the plot of T_s against $1/F$ and the open squares its theoretical straightening (Courtesy TA Instruments, Newcastle DE)

Manual calculations of purity from DSC scans are usually lengthy and laborious. Thus, the appropriate computer software is strongly recommended. It is most important to realise that not all samples are suitable for purity determination by DSC methods. Decomposition of the sample can produce a misshapen DSC endotherm. Also, the formation of solid solution must be absent; the pure major component must exist in the solid state as a single phase. This is difficult to detect from the DSC endotherm and it is often expedient to dope the sample with known amounts of impurity in order to check the results.[17]

12.13   References

1. R C Mackenzie 1979 *Thermochim. Acta*, **28**; 1
2. C Duval and M de Clercq 1951 *Anal. Chim. Acta*, **5**; 282
3. E K Simons and A E Newkirk 1964 *Talanta*, **11**; 549
4. A W Coats and J P Redfern 1963 *Analyst*, **88**; 906
5. C Turner, I Hoffman and D Chen 1963 *Can. J. Chem.*, **41**; 243
6. G M Lukaszewski and J P Redfern 1961 *Lab Practice*, **10**; 552
7. W W Wendlandt 1960 *Anal. Chem.*, **32**; 848
8. W J Blaedel and H T Knight 1954 *Anal. Chem.*, **26**; 741
9. C Duval 1963 *Inorganic thermogravimetric analysis*, 2nd edn, Elsevier, Amsterdam, p. 93
10. I Hoffman, M Schnitzer and J R Wright 1959 *Anal. Chem.*, **31**; 440
11. D M Nicholas, P Barnfield and J Mendham 1988 *Mater. Sci. Lett.*, **7**; 217
12. R C Mackenzie 1980 *Anal. Prac.*, **17**; 217
13. S L Boersma 1955 *J. Am. Ceram. Soc.*, **38**; 281
14. A T Florence, E G Salole and J B Stenlake 1974 *J. Pharm. Pharmac.*, **26**; 479
15. A T Florence and E G Salole 1976 *J. Pharm. Pharmac.*, **28**; 637
16. J G Dunn, K Oliver and I Sills 1989 *Thermochim. Acta*, **155**; 93
17. P Burroughs 1980 *Anal. Prac.*, **17**; 231

12.14   Bibliography

J W Dodd and K H Tonge 1986 *Thermal methods*, ACOL–Wiley, Chichester

T Hatakeyama and F X Quinn 1995 *Thermal analysis: fundamentals and applications to polymer science*, John Wiley, Chichester

F Paulik 1995 *Special trends in thermal analysis*, John Wiley, Chichester

W W Wendlandt 1997 *Thermal analysis*, 3rd edn, John Wiley, Chichester

J D Wineforder, D Dollimore, J Dunn and I M Kolthoff 1998 *Treatise on analytical chemistry*, 2nd edn, Volume 13, Part 1, *Thermal methods*, John Wiley, Chichester

13

Direct electroanalytical methods

13.1 Introduction

This chapter deals with the electroanalytical techniques that can determine an ion or molecule by **direct** measurements. Analytical methods involving titrimetric procedures are dealt with in Chapter 10. There are four electrochemical methods where direct measurements can be made:

Electrogravimetry In these processes the analyte element is weighed after it has been electrolytically deposited upon a suitable electrode (Sections 13.2 to 13.7).

Coulometry In controlled potential coulometry the analyte is determined by quantitative reaction at an electrode during electrolysis (Sections 13.8 to 13.12).

Potentiometry The use of single measurements of electrode potentials are the basis of determinations of ionic species in solution and these are compared with potentials developed by standard solution(s) of the analyte (Sections 13.13 to 13.25).

Voltammetry The magnitude of the limiting current is related to the concentration of the analyte. Hence by using standard solutions, or internal standard and standard addition techniques, a quantitative evaluation may be carried out (Sections 13.26 to 13.46).

Consult Chapter 2, especially Sections 2.28 to 2.33, for the basic concepts behind cell processes and electrode potentials. These concepts underpin the methods described in this chapter.

Electrogravimetric analysis

13.2 Electrogravimetry: theory

Electrogravimetric procedures require the use of carefully controlled experimental conditions to deposit an element electrolytically upon a suitable electrode. Under specified conditions it is possible to avoid the codeposition of two metals when they are present in the same solution. The use of ion selective electrodes has superseded this form of analysis, especially environmental analyses for substances in solution. However, there are still occasions when the quantitative removal of an element from solution is required. A brief outline of the theory is given here; more detailed treatments can be found in the literature.[1–4]

Electrodeposition is governed by Ohm's law and by Faraday's two laws of electrolysis. Ohm's law expresses the relation between the three fundamental quantities, current, electromotive force, and resistance. The current I is directly proportional to the electromotive force E and indirectly proportional to the resistance R:

$$I = E/R$$

Faraday's Laws state:

1. The amounts of substances liberated (or dissolved) at the electrodes of a cell are directly proportional to the quantity of electricity which passes through the solution.
2. The amounts of different substances liberated or dissolved by the same quantity of electricity are proportional to their relative atomic (or molar) masses divided by the number of electrons involved in the respective electrode processes.

It follows from Faraday's second law that when a given current is passed in series through solutions containing copper(II) sulphate and silver nitrate respectively, then the weights of copper and silver deposited in a given time will be in the ratio of 63.55/2 to 107.87.

SI units are used in electrogravimetric calculations. The fundamental unit of current is called the **ampere** (A); it is defined as the constant current which, if maintained in two parallel rectilinear conductors of negligible cross-section and of infinite length and placed one metre apart in a vacuum, would produce between these conductors a force equal to 2×10^{-7} N per metre length. The unit of electrical potential is the **volt** (V); it is the difference of potential between two points of a conducting wire which carries a constant current of 1 A, when the power dissipitated between these two points is $1 \, \mathrm{J \, s^{-1}}$. The unit of electrical resistance is the **ohm** (Ω) which is the resistance between two points of a conductor when a constant difference of potential of 1 V applied between these two points produces a current of 1 A.

The unit quantity of electricity is the **coulomb** (C), and is defined as the quantity of electricity passing when a current of 1 A flows for 1 s. To liberate one mole of electrons, or one mole of a singly charged ion, will require Le coulombs, where L is the Avogadro constant ($6.022 \times 10^{23} \, \mathrm{mol^{-1}}$) and e is the elementary charge (1.602×10^{-10} C); the product Le ($9.647 \times 10^4 \, \mathrm{C \, mol^{-1}}$) is called the Faraday constant F.

13.3   Electrogravimetry: apparatus

Electrogravimetry is carried out in an electrolytic cell. This consists of two electrodes with an external electrical energy supply. The two main parts of the system are the electrodes. The cathode is the electrode on which metal deposition occurs due to reduction of the ions; it is attached to the negative terminal of the energy source. The anode is the electrode at which oxidation occurs; it is attached to the positive terminal of the energy source. The cathode and the anode are illustrated in Figure 13.1.

The electrodes are made of platinum gauze as the open construction assists the circulation of the solution. It is possible to use one of the electrodes as stirrer for the solution, but special arrangements must then be made for connection of the electrolysis current to this electrode, and an independent glass-paddle stirrer or a magnetic stirrer offer a simple alternative. Typical electrodes are the Fischer type (Figure 13.2); a glass tube is slid into the loops on the wire of the outer electrode, and the wire to the inner electrode passes through this tube. When it is necessary to know the current density for a particular determination, the area of a gauze electrode may be regarded as approximately equal to twice the area of a foil electrode of the same dimensions.

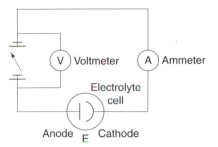

Figure 13.1   Cathode and anode: reduction occurs at the cathode, it is connected to the negative terminal of the energy source; oxidation occurs at the anode, it is connected to the positive terminal of the energy source

Figure 13.2   Fischer electrode

Besides the electrodes, the circuit also requires a d.c. current source rated at 3–15 V, a variable resistance, an ammeter, a voltmeter, a magnetic stirrer and a hotplate. All these components are provided in commercial electrolysis units and may sometimes include a spinning inner electrode (anode) to assist in stirring the solution undergoing electrolysis.

13.4   Cell processes

During electrolysis a number of things may occur in addition to the deposition of the required species. Not only may co-deposition of other metals take place, but the water itself may decompose under the applied conditions and produce hydrogen and oxygen, which may interfere with the deposition of the metals on the cathode. Because of the possible complications, the conditions for a particular determination must be very carefully controlled.

Some of the features arising during electrolysis can be observed if a small potential of, say 0.5 V, is applied to two smooth platinum electrodes immersed in a solution of $1\,M$ sulphuric acid, then an ammeter placed in the circuit will at first show that an appreciable current is flowing, but its strength decreases rapidly, and after a short time it becomes virtually

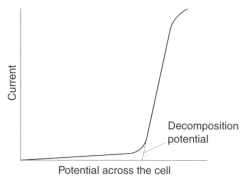

Figure 13.3   Back e.m.f.

equal to zero. If the applied potential is gradually increased, there is a slight increase in the current until, when the applied potential reaches a certain value, the current suddenly increases rapidly with increase in the e.m.f. It will be observed, in general, that when there is a sudden increase in current, bubbles of gas begin to be freely evolved at the electrodes. A plot of the current against the applied potential should produce a curve similar to Figure 13.3. The point at which the current suddenly increases is evident, and in this instance it is at about 1.7 V. The potential at this point is called the 'decomposition potential' and it is at this point that the evolution of both hydrogen and oxygen in the form of bubbles is first observed. We may define the **decomposition potential** of an electrolyte as the minimum external potential that must be applied in order to bring about continuous electrolysis.

If the circuit is broken after the e.m.f. has been applied, the reading on the voltmeter will at first be fairly steady, and then it will decrease, more or less rapidly, to zero. The cell is now clearly behaving as a source of current, and is said to exert a **back** or **counter** or **polarisation** e.m.f., since it acts in a direction opposite to the applied e.m.f. This back e.m.f. arises from the accumulation of oxygen and hydrogen at the anode and cathode respectively; two gas electrodes are consequently formed, and the potential difference between them opposes the applied e.m.f. When the primary current from the battery is shut off, the cell produces a moderately steady current until the gases at the electrodes are either used up or have diffused away; the voltage then falls to zero. This back e.m.f. is present even when the current from the battery passes through the cell and accounts for the shape of the curve in Figure 13.3.

It has been found by experiment that the decomposition voltage of an electrolyte varies with the nature of the electrodes employed for the electrolysis and is, in many instances, higher than the value calculated from the difference of the **reversible** electrode potentials. The excess voltage over the calculated back e.m.f. is called the overpotential. Overpotential may occur at the anode as well as at the cathode. The decomposition voltage E_D is therefore

$$E_D = E_{cathode} + E_{oc} - (E_{anode} + E_{oa})$$

where E_{oc} and E_{oa} are the overpotentials at the cathode and anode respectively.

The overpotential of hydrogen is of great importance in electrolytic determinations and separations. It is greatest with the relatively soft metals, such as bismuth (0.4 V), lead (0.4 V), tin (0.5 V), zinc (0.7 V), cadmium (1.1 V) and mercury (1.2 V): these overvoltage values refer to the electrolysis of $0.05 M$ sulphuric acid with a current density of $0.01 \, A \, cm^{-2}$, and can be compared with 0.09 V, the value for a bright platinum electrode under similar conditions. The existence of hydrogen overpotential renders possible the electrogravimetric determination of metals, such as cadmium and zinc, which otherwise would not be deposited before the reduction of hydrogen ion. In alkaline solution the hydrogen overpotential is slightly higher (0.05–0.03 V) than in acid solution.

Oxygen overpotential is about 0.4–0.5 V at a polished platinum anode in acid solution, and is of the order of 1 V in alkaline solution with current densities of 0.02–0.03 A cm^{-2}. As a rule the overpotential associated with the deposition of metals on the cathode is quite small (about 0.1–0.3 V) because the depositions proceed nearly reversibly.

Electrogravimetric determinations may be carried out either at **constant current** or with a **controlled-potential** procedure. Constant current is really limited to determining the quantity of an ion in a solution containing a single species, whereas controlled potential is more suited to the separation of the components of mixtures in which the decomposition potentials are not widely separated. A typical determination that may be carried out at constant current is copper from acidic solution. The metal may be deposited from either nitric or sulphuric acid solution, but a mixture of the two acids is usually employed. If this solution is electrolysed with an e.m.f. of 2–3 V, the following reactions occur:

Cathode $\quad Cu^{2+} + 2e \rightleftharpoons Cu$

$\qquad\qquad 2H^+ + 2e \rightleftharpoons H_2$

Anode $\quad\;\; 4OH^- \rightleftharpoons O_2 + 2H_2O + 4e$

The acid concentration of the solution must not be too great, otherwise the deposition of the copper may be incomplete or the deposit will not adhere satisfactorily to the cathode. The beneficial effect of nitrate ion is due to its depolarising action at the cathode:

$$NO_3^- + 10H^+ + 8e \rightleftharpoons NH_4^+ + 3H_2O$$

The reduction potential of the nitrate ion is lower than the discharge potential of hydrogen, and therefore hydrogen is not liberated.

For separations using a controlled cathode potential, it is necessary to incorporate a reference electrode into the circuit to measure the voltage between the cathode and the reference half-cell. It is then possible to control the cathode potential. Provided the cathode

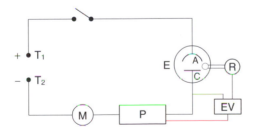

Figure 13.4 Electrogravimetry with controlled cathode potential: T_1, T_2 = d.c. supply terminals, M = ammeter, P = potentiostat, EV = electronic voltmeter, E = electrolysis cell, A = anode, C = cathode, R = reference electrode

potential is not allowed to fall below the deposition potential E_D for the ion M^{2+}, the metal M will be deposited free of contamination from any other metal that may be present. Compare the circuitry for controlled cathode potential electrogravimetry (Figure 13.4) with the much more basic circuit used for the constant-current procedures (Figure 13.1).

13.5 Deposition and separation

For the electrolysis of a solution to be maintained, the potential applied to the electrodes of the cell E_{app} must overcome the decomposition potential of the electrolyte E_D (which includes the back e.m.f. and any overpotential effects), as well as the electrical resistance of the solution. Thus, E_{app} must be greater than or equal to $(E_D + IR)$, where I is the electrolysis current and R is the cell resistance. As electrolysis proceeds, the concentration of the cation which is being deposited decreases, hence the cathode potential changes.

If the relevant ionic concentration in the solution is c_i, and the ion concerned has a charge number of 2, then at a temperature of 25 °C the cathode potential will have a value given by

$$E_1 = E_M^\ominus + \frac{0.0591}{2} \log c_i = E_M + 0.0296 \log c_i$$

If the ionic concentration is reduced by deposition to 1/10 000 of its original value, thus giving an accuracy of 0.01% in the determination, the new cathode potential will be

$$E_2 = E_M + 0.926 \log (c_i \times 10^{-4}) = E_M + 0.0296 \log c_i + 0.0296 \log 10^{-4}$$

$$= (E_M + 0.0296 \log c_i) - 4 \times 0.0296 = E_1 - 0.118 \, \text{V}$$

It follows that if the original solution contains two cations whose deposition potentials differ by about 0.25 V, then the cation of higher deposition potential should be deposited without any contamination by the ion of lower deposition potential. In practice it may be necessary to ensure the cathode potential is unable to fall to a level where deposition of the second ion may occur.

13.6 Electrolytic separation of metals

If a current is passed through a solution containing copper(II), hydrogen and cadmium(II) ions, copper will be deposited first at the cathode. As the copper deposits, the electrode potential decreases, and when it equals the potential of the hydrogen ions, hydrogen gas will form at the cathode. The potential at the cathode will remain virtually constant as long as hydrogen is evolved, which would mean as long as any water remains, and it is therefore unable to become sufficiently negative to permit the deposition of cadmium. Thus, metal ions with positive reduction potentials may be separated, without external control of the cathode potential, from metal ions having negative reduction potentials.

Silver can be readily separated from copper, even though they both have positive reduction potentials, because the difference between the two values is large (silver +0.779 V, copper +0.337 V), but when the standard potentials of the two metals differ only slightly, the electroseparation is more difficult. An obvious solution to this problem is to decrease the concentration of one of the ions being discharged by incorporating it in a complex ion of large stability constant (Section 2.23). As an illustration of the kind of result achieved, the deposition potentials for 0.1 M solutions of ions M^{2+} of the following metals have the values indicated: zinc +0.79 V, cadmium +0.44 V, copper +0.34 V. When 0.1 mol of the corresponding cyanides is dissolved in potassium cyanide to give an excess concentration of potassium cyanide of 0.4 M, the deposition potentials become zinc +1.18 V, cadmium +0.87 V, copper +0.96 V.

An interesting application of these results is to the direct quantitative separation of copper and cadmium. The copper is first deposited in acid solution; the solution is then made slightly alkaline with pure aqueous sodium hydroxide, potassium cyanide is added until the initial precipitate just redissolves, and the cadmium is deposited electrolytically.

13.7   Some metals which can be determined

Constant-current procedure

With the exception of lead, which from nitric acid solutions is deposited on the **anode** as PbO_2, the ions listed in Table 13.1 are deposited as metal on the cathode. With the ions indicated by an asterisk in Table 13.1, it is advisable to use a platinum cathode which has been plated with copper before the initial weighing; this is because the deposited metals cannot be readily distinguished on a platinum surface and it is difficult to be certain when deposition is complete. The separations in Table 13.2 can be readily accomplished electrolytically; the first one depends on a large difference in deposition potentials (Section 13.5), and the second depends on the fact that Pb^{2+} can be anodically deposited as PbO_2.

Table 13.1   *Conditions for determining metals by electrogravimetry*

Ion	Electrolyte	Electrical details
Cd^{2+a}	Potassium cyanide forming $K_2[Cd(CN)_4]$	1.5 to 2 A, 2.5 to 3 V
Co^{2+a}	Ammoniacal sulphate	4 A, 3 to 4 V
Cu^{2+}	Sulphuric acid–nitric acid	2 to 4 A, 3 to 4 V
Pb^{2+a}	Tartrate buffer or chloride solution (solubility limits the amount of lead to less than 50 mg per 100 mL)	2 A, 2 to 3 V
Pb^{2+}	Nitric acid. PbO_2 deposited on anode; use empirical conversion factor of 0.864	5 A, 2 to 3 V
Ni^{2+a}	Ammoniacal sulphate	4 A, 3 to 4 V
Ag^+	Potassium cyanide forming $K[Ag(CN)_2]$	0.5 to 1.0 A, 2.5 to 3 V
Zn^{2+a}	Potassium hydroxide solution	4 A, 3.5 to 4.5 V

[a] Advisable to use a platinum cathode plated with copper before initial weighing: see text.

Table 13.2   *Simple electrolytic separations*

Ions	Electrolyte	Electrical details
Cu/Ni	Deposit Cu from H_2SO_4 solution Neutralise with NH_3 Add 15 mL conc. NH_3 (aq) Deposit Ni	2 to 4 A, 3 to 4 V 4 A, 3 to 4 V
Cu/Pb	Nitric acid solution Deposit Cu on cathode, PbO_2 on anode	1.5 to 2 A, 2 V

Table 13.3 *Examples of controlled cathode potential determinations*

Metal	Electrolyte	$E_{cathode}$ vs SCE (V)	Separated from
Antimony	Hydrazine–HCl	−0.3	Pb, Sn
Cadmium	Ethanoate buffer	−0.8	Zn
Copper	Tartrate–hydrzine–Cl⁻	−0.3	Bi, Cd, Pb, Ni, Sn, Zn
Lead	Tartrate–hydrazine–Cl⁻	−0.6	Cd, Fe, Mn, Ni, Sn, Zn
Nickel	Tartrate–NH₄OH	−1.1	Al, Fe, Zn
Silver	Ethanoate buffer	+0.1	Cu, heavy metals

Controlled-potential procedure

The controlled-potential procedure is particularly suited to the study of alloys such as antimony, copper, lead and tin in bearing metals, and for alloys of copper, bismuth, lead and tin. In the determination described by Lingane and Jones,[5] carried out in acid solution, the metals were deposited in the order copper at −0.3 V, bismuth at −0.4 V, lead at −0.6 V and tin at −0.65 V vs the standard calomel electrode. Examples of controlled cathode potential determinations are given in Table 13.3.

Coulometry

> **Safety.** Before carrying out any experiments in this section, pay full attention to any safety warnings and make sure you adhere to national laboratory and safety regulations.

13.8 General discussion

Coulometric analysis is an application of Faraday's first law of electrolysis – the extent of chemical reaction at an electrode is directly proportional to the quantity of electricity passing through the electrode. For each mole of chemical change at an electrode $(96\,487 \times n)$ coulombs are required, i.e. the Faraday constant multiplied by the number of electrons involved in the electrode reaction. The weight of substance produced or consumed in an electrolysis involving Q coulombs is therefore given by the expression

$$W = \frac{M_r Q}{96\,487n}$$

where M_r is the relative atomic (or molecular) mass of the substance liberated or consumed. Analytical methods based upon the measurement of a quantity of electricity and the application of this equation are called **coulometric methods**, from the word 'coulomb'.

The fundamental requirement of a coulometric analysis is that the electrode reaction used for the determination proceeds with 100% efficiency, so the quantity of substance reacted can be expressed by means of Faraday's law from the measured quantity of electricity (coulombs) passed. The substance being determined may directly undergo reaction at one of the electrodes; this is **primary coulometric analysis**. Or it may react in solution with another substance generated by an electrode reaction; this is **secondary coulometric analysis**.

Coulometry may be achieved in two ways:

1. With controlled potential of the working electrode – direct methods
2. With constant current (Sections 10.12 to 10.17)

In method 1 the substance being determined reacts with 100% current efficiency at a working electrode, the potential of which is controlled. The completion of the reaction is indicated by the current decreasing to practically zero, and the quantity of the substance reacted is obtained from the reading of a coulometer in series with the cell or by means of a current–time integrating device. In method 2 a solution of the substance to be determined is electrolysed with constant current until the reaction is completed (as detected by a visual indicator in the solution or by amperometric, potentiometric or spectrophotometric methods) and the circuit is then opened. The total quantity of electricity passed is derived from the product current (amperes) × time (seconds); the present practice is to include an electronic integrator in the circuit.

13.9   Coulometry at controlled potential

In a controlled-potential coulometric analysis, the current generally decreases exponentially with time according to the equation

$$I_t = I_0 e^{-k't} \quad \text{or} \quad I_t = I_0 10^{-kt}$$

where I_0 is the initial current, I_t the current at time t, and k, k' are constants. A typical curve is shown in Figure 13.5; the current decreases more or less exponentially to almost zero. In many cases an appreciable background current is observed with the supporting electrolyte alone, and in these instances the current finally decays to the background current rather than to zero; a correction can be applied by assuming the background current is constant during the electrolysis.

In electrolysis at controlled potential, the quantity of electricity Q (coulombs) passed from the beginning of the determination to time t is given by

$$Q = \int_0^t I_t \, dt$$

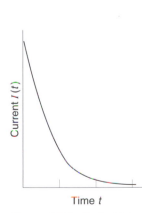

Figure 13.5   Typical *I–t* curve for controlled potential coulometry

where I_t is the current at time t. Equations relating the variation of current with time can also be expressed in terms of the concentration of electrolyte at time t, C_t, and the initial concentration C_0:

$$C_t = C_0 e^{-k't}$$

This equation represents a first-order reaction process, so the fraction of material electrolysed at any instant is independent of the initial concentration. If the limit of accuracy for the determination is set at $C_t = 0.001C_0$, the time t required to achieve this result will be independent of the initial concentration. The constant k' is equal to Am/V, where A is the area of the relevant electrode, V the volume of the solution and m the mass transfer coefficient of the electrolyte.[6] It follows that to make t small, A and m must be large and V small, and this leads to the subdivision of controlled-potential coulometry into stirred solutions, flowing streams, and thin-layer cavity cells.

13.10 Apparatus and general technique

The number of coulombs passed was originally determined by including a coulometer in the circuit, e.g. a silver, an iodine or a hydrogen–oxygen coulometer. The amount of chemical change taking place in the coulometer can be ascertained, and from this result the number of coulombs passed can be calculated, but with modern equipment an electronic integrator is used to measure the quantity of electricity passed.

Apparatus

The source of current is a **potentiostat**, which is used in conjunction with a reference electrode (commonly a saturated calomel electrode) to control the potential of the working electrode. The circuit will be essentially that shown in Figure 13.4 but with the addition of the integrator or of a coulometer.

For stirred solution coulometry a mercury cathode is commonly used, suitably equipped for coulometric determinations. The cell has a capacity of about 100 ml, and is closed with a Teflon cover (fitted on to the top of the glass cell, which is ground flat) and a gas delivery tube is provided for removing dissolved air from the solution with nitrogen or other inert gas; the excess nitrogen escapes through the loosely fitting glass sleeve through which the shaft of a glass stirrer passes. Removal of the air is necessary, because oxygen is reduced at the mercury cathode at about -0.05 Volt versus the saturated calomel electrode (SCE) and this would interfere with the determination of most substances. The area of the mercury cathode is about $20\,cm^2$. Two kinds of anode, immersed directly in the test solution, may be used: a large helical silver wire (~2.6 mm diameter, helix 5 cm long, and 3 cm diameter; area about $100\,cm^2$) or a platinum gauze cylinder (area $75\,cm^2$) mounted vertically and coaxially with the stirrer shaft. The silver anode is employed when the solution contains metals, such as bismuth, which tend to be oxidised to insoluble higher oxides at a platinum anode; chloride ion, at least equivalent to the quantity of the cathode reaction (but preferably in 50–100% excess) is added. The reaction at the silver anode is

$$Ag + Cl \rightleftharpoons AgCl + e$$

For metals which are not reduced by hydrazine, hydrazine is used as depolariser at the platinum anode:

$$N_2H_5^+ \rightleftharpoons N_2 + 5H^+ + 4e$$

Table 13.4   *Deposition of metals at controlled potential of the mercury cathode*

Element	Supporting electrolyte	Potential vs SCE (V)	
		$E_{1/2}$	$E_{cathode}$
Cu	0.5 M acid sodium tartrate, pH 4.5	−0.09	−0.16
Bi	0.5 M acid sodium tartrate, pH 4.5	−0.23	−0.40
Pb	0.5 M acid sodium tartrate, pH 4.5	−0.48	−0.56
Cd	1 M NH$_4$Cl + 1 M NH$_3$ (aq)	−0.81	−0.85
Zn	1 M NH$_4$Cl + 1 M NH$_3$ (aq)	−1.33	−1.45
Ni	1 M pyridine + HCl, pH 7.0	−0.78	−0.95
Co	1 M pyridine + HCl, pH 7.0	−1.06	−1.20

The evolution of nitrogen aids in removing dissolved air. A salt bridge (4 mm tube) attached to the saturated calomel electrode is filled with 3% agar gel saturated with potassium chloride and its tip is placed within 1 mm of the mercury cathode when the mercury is not being stirred; this ensures the tip trails in the mercury surface when the mercury is stirred. The mercury–solution interface (not merely the solution) must be vigorously stirred, and for this purpose the propeller blades of the glass stirrer are partially immersed in the mercury.

One of the outstanding advantages of the mercury cathode is that the optimum control potential for a given separation is easily determinable from polarograms recorded with the dropping mercury electrode. This potential corresponds to the beginning of the polarographic diffusion current plateau (Section 13.28); there is usually no advantage in employing a control potential more than about 0.15 V greater than the half-wave potential. Table 13.4 gives some values for the half-wave potential $E_{1/2}$ and suitable values for the cathode potential.

Details have been collected for the determination of some 50 elements by this technique,[7,8] and it is possible to effect many difficult separations, such as Cu and Bi, Cd and Zn, Ni and Co; it has been widely used in the nuclear energy industry. A number of organic compounds can also be determined by this procedure, e.g. trichloroacetic acid and 2.4.6-trinitrophenol are reduced at a mercury cathode in accordance with the equations

$$Cl_3CCOO^- + H^+ + 2e \rightleftharpoons CHCl_2COO^- + Cl^-$$

$$C_6H_2(NO_2)_3(OH) + 18H^+ + 18e \rightleftharpoons C_6H_2(NH_2)_3(OH) + 6H_2O$$

and so can be determined coulometrically.[9]

General technique

A coulometric determination at controlled potential of the mercury cathode is performed by the following general method. The supporting electrolyte (50–60 mL) is first placed in the cell and the air is removed by passing a rapid stream of nitrogen through the solution for about 5 min. The cathode mercury is then introduced through the stopcock at the bottom of the cell by raising the mercury reservoir. The stirrer is started and the tip of the bridge from the reference electrode is adjusted so it just touches, or trails slightly in, the stirred mercury cathode. The potentiostat is adjusted to maintain the desired control potential and the solution is electrolysed, with nitrogen passing continuously, until the current decreases to a very small constant value (the background current). This preliminary electrolysis removes traces of reducible impurities; the current usually decreases to 1 mA or less after about 10 min. A known volume (say 10–40 mL) of the sample solution is then pipetted into the cell, and the electrolysis is allowed to proceed until the current decreases to the same small value observed with the supporting electrolyte alone. Electrolysis is usually complete within an hour. The electronic integrator is read and the weight of metal deposited is calculated as explained in Section 13.8.

Experimental section

> **!** **Safety.** Before carrying out any experiments in this section, pay full attention to any safety warnings and make sure you adhere to national laboratory and safety regulations.

13.11 Separation of nickel and cobalt

Reagents

Standard nickel and cobalt ion solutions Prepare solutions of nickel and cobalt ion ($\sim$10 mg mL^{-1}) from pure ammonium nickel sulphate and pure ammonium cobalt sulphate respectively.

Pyridine Redistil pyridine and collect the middle fraction boiling within a 2 °C range, namely 113–115 °C.

Supporting electrolyte Prepare a supporting electrolyte composed of 1.00 M pyridine and 0.50 M chloride ion, adjusted to a pH of 7.0 ± 0.2 for use with a silver anode, or 1.00 M pyridine, 0.30 M chloride ion and 0.20 M hydrazinium sulphate, adjusted to a pH of 7.0 ± 0.2, for use with a platinum cathode. A small background current is obtained with the platinum cathode.

Procedure

Place 90 mL of the supporting electrolyte in the cell, remove dissolved air with pure nitrogen and subject the solution to a preliminary electrolysis with the potential of the mercury cathode −1.20 V vs SCE to remove traces of reducible impurities; stop the electrolysis when the background current ($\sim$2 mA) has decreased to a constant value (30–60 min). Prepare the coulometer, adjust the potentiostat to maintain the potential of the cathode at the value to be used in the determination (−0.95 V vs SCE for nickel) and add 10 ml of each of the prepared solutions to the cell. Electrolyse until the current has decreased to the value of the background current. Note the number of coulombs passed and calculate the weight of nickel deposited. Now adjust the potential to −1.20 V and continue the electrolysis until the current again falls to the background current value, and from the number of coulombs passed in this second electrolysis, calculate the weight of cobalt present. If necessary, for each determination, correction for the background current can be carried out by subtracting the quantity $I_b t$ from Q (the number of coulombs recorded), where I_b is the base current, and t is the duration of the electrolysis in seconds.

13.12 Flowing stream coulometry

The time required for a coulometric determination can be reduced by using a flowing stream technique. In this technique the electrode is a cylinder of **reticular vitreous carbon**,[10] a material of porous structure having a porosity factor of 0.95, which indicates a large open internal volume. This means that if a solution is caused to flow through the electrode, the effective cross-sectional area of the stream of liquid is comparatively large. This electrode (which is the working electrode) is surrounded by a metal cylinder (usually of stainless steel) which acts as the auxiliary electrode. For a reasonable flow rate (about $1 \, mL \, s^{-1}$) the solution must be pumped through the electrode, and the assembly is mounted inside a small chamber in which connection to the reference electrode is made. With a suitable flow rate and a small concentration of the electrolyte under determination entering the electrode, the effluent will be virtually free of this material, which means the electrolysis is completed inside the electrode.[11] The usual procedure is to take a small known volume (say 20 µL) of the solution to be analysed and to inject this into a stream of supporting electrolyte flowing in a tube of about 0.5 mm diameter. Mixing takes place, and at the entry to the working electrode the diluted test solution may occupy a volume of about 200 µL. At the specified flow rate, all the material to be determined will pass through the electrode in about 12–15 s, so a rapid rate of determination has been achieved. The electrode will be incorporated in a circuit similar to that previously described in which a potentiostat controls the potential of the working electrode and may also provide a countercurrent facility to nullify the background current; an electronic integrator will also be included.

Thin-layer cavity cell

In this technique a cell is constructed to give a thin cavity on one wall of which the metal-plate working electrode is mounted. This wall is separated by a Teflon sheet in which a central aperture has been cut out, from the opposite wall of the cavity. This wall contains entry and exit tubes for the test solution which is caused to flow past the working electrode; provision is made for connections to the other electrodes. If the Teflon sheet is thin enough (about 0.05 mm), the distance between the two walls of the cavity is less than the normal thickness of the diffusion layer of the electrolyte when undergoing electrolysis, and so electrolysis within the cavity is rapid.[12]

It is also possible to reduce the time required for conventional controlled-potential coulometry by adopting the procedure of **predictive coulometry**. A given determination will need a certain number of coulombs Q for completion, and if at time t, Q_t coulombs have been passed, then Q_R further coulombs will be required to complete the determination, and $Q_R = Q_\infty - Q_t$. By choosing a number of times t_1, t_2, t_3 separated by a common interval (say 10 s) and measuring the corresponding numbers of coulombs passed Q_1, Q_2, Q_3, it can be shown that

$$Q_\infty = Q_3 + \frac{(Q_2 - Q_3)^2}{2Q_2 - (Q_1 + Q_3)}$$

A computer is programmed to calculate Q_∞ from the value of Q_t at successive intervals of 10 s until a constant value is obtained for Q_∞ thus completing the determination.

13.13 Evaluation of direct coulometry

In controlled-potential coulometry, unlike electrogravimetry, there is no need to generate a product that requires weighing. Indeed, reactions where no solid products are formed, e.g. the oxidation of iron(II) to iron(III) can be determined by direct coulometric methods. Coulometric analysis at controlled potential is more selective than constant-current coulometry. Thus, the determination of metal ions that are normally difficult to separate, e.g. nickel and cobalt (Section 13.11), can be achieved by direct coulometric analysis. But coulometric determinations are often time-consuming, although the process can be automated.

Analysis time can be reduced by the flowing stream technique (Section 13.12). A major disadvantage is that side reactions can occur before completion of the desired electrochemical process, thus inhibiting the quantitative reaction at an electrode. The range of substances determined by direct methods is fairly limited. Many more applications are possible using constant-current coulometry. Details of coulometric titrations may be found in Chapter Sections 10.16 to 10.20, 10.74, 10.107 and 10.184 to 10.187.

Potentiometry

13.14   Fundamentals of potentiometry

When a metal M is immersed in a solution containing its own ion M^{n+}, then an electrode potential is established (Section 2.28). The value of this electrode potential is given by the **Nernst equation**

$$E = E^{\ominus} + (RT/nF)\ln a_{M^{n+}}$$

The standard electrode potential for the metal M is the constant $E^{\ominus}$. A value for E can be established by linking the solution (an electrode in its own right) with a reference electrode, commonly a saturated calomel electrode (Section 13.16), and measuring the e.m.f. of the resultant cell. As the potential E_r of the reference electrode is known, it is possible to deduce the value of the electrode potential E; and provided the standard electrode potential $E^{\ominus}$ of the given metal is known, it is then possible to calculate the metal ion activity $a_{M^{n+}}$ in the solution. For a dilute solution the measured ionic activity will be virtually the same as the ionic concentration, and for stronger solutions, given the value of the activity coefficient, we can convert the measured ionic activity into the corresponding concentration.

This procedure of using a single measurement of electrode potential to determine the concentration of an ionic species in solution is called **direct potentiometry**. The electrode whose potential depends on the concentration of the ion to be determined is called the **indicator electrode**. And when the ion to be determined is directly involved in the electrode reaction, the electrode is an **electrode of the first kind**; this is the case for metal M immersed in a solution of M^{n+} ions.

It is also possible in appropriate cases to measure by direct potentiometry the concentration of an ion which is not directly concerned in the electrode reaction. This involves the use of an **electrode of the second kind**. An example is the silver–silver chloride electrode formed by coating a silver wire with silver chloride; this electrode can be used to measure the concentration of chloride ions in solution.

The silver wire can be regarded as a silver electrode with a potential given by the Nernst equation as

$$E = E^{\ominus}_{Ag} + (RT/nF)\ln a_{Ag^+}$$

The silver ions involved are derived from the silver chloride, and by the solubility product principle (Section 2.14), the activity of these ions will be governed by the activity of the chloride ion

$$a_{Ag^+} = K_{s(AgCl)}/a_{Cl^-}$$

Hence the electrode potential can be expressed as

$$E = E^{\ominus}_{Ag} + (RT/nF)\ln K_s - (RT/nF)\ln a_{Cl^-}$$

and is clearly governed by the activity of the chloride ions, so the activity of the chloride ions can be deduced from the measured electrode potential.

In the Nernst equation the term RT/nF involves known constants, and introducing the factor for converting natural logarithms to logarithms to base 10, the term has a value at a temperature of 25 °C of 0.0591 V when n is equal to 1. Hence, for an ion M^+ a tenfold change in ionic activity will alter the electrode potential by about 60 mV, whereas for an ion M^{2+} a similar change in activity will alter the electrode potential by approximately 30 mV. So to achieve an accuracy of 1% in the value determined for the ionic concentration by direct potentiometry, the electrode potential must be capable of measurement to within 0.26 mV for the ion M^+, and to within 0.13 mV for the ion M^{2+}.

The **liquid junction potential** at the interface between the two solutions creates uncertainty in the e.m.f. measurements, some pertaining to the reference electrode and some to the indicator electrode. But this liquid junction potential can be largely eliminated if one solution contains a high concentration of potassium chloride or ammonium nitrate, electrolytes in which the ionic conductivities of the cation and the anion have very similar values.

One way to overcome the problem of the liquid junction potential is to replace the reference electrode by an electrode composed of a solution containing the same cation as in the solution under test, but at a known concentration, together with a rod of the same metal as used in the indicator electrode; in other words, we set up a concentration cell (Section 2.29). The activity of the metal ion in the solution under test is given by

$$E_{cell} = (RT/nF) \ln \left(\frac{a_{known}}{a_{unknown}} \right)$$

In view of these problems with direct potentiometry, much attention has been directed to the procedure of **potentiometric titration** as an analytical method. As the name implies, it is a titrimetric procedure in which potentiometric measurements are carried out in order to fix the end point. In this procedure we are concerned with changes in electrode potential rather than in an accurate value for the electrode potential with a given solution, and under these circumstances the effect of the liquid junction potential may be ignored. Titrations of this type, in which the most rapid change in the cell e.m.f. occurs at the end point of the titration, are described in Sections 10.8 to 10.11. All these procedures require reference electrodes and instrumentation for measuring cell e.m.f., and they are described below along with some applications.

Reference electrodes

13.15   The hydrogen electrode

Electrode potentials are all quoted with reference to the standard hydrogen electrode (SHE) (the Hildebrand bell-type hydrogen electrode is illustrated in Section 2.28), regarded as the primary reference electrode. It consists of a platinum electrode surrounded by an outer tube along which hydrogen passes, entering through a side inlet and escaping at the bottom through the test solution.

There are several small holes near the bottom of its bell; when the speed of the gas is suitably adjusted, the hydrogen escapes through the small openings only. Because of the periodic formation of bubbles, the level of the liquid inside the tube fluctuates, and a part of the foil is alternately exposed to the solution and to hydrogen. The lower end of the foil is continuously immersed in the solution to avoid interruption of the electric current. Although Figure 2.2 shows an open vessel, in practice the electrode will be used in a stoppered flask with a suitable exit for the hydrogen; this allows an oxygen-free atmosphere to be maintained in the flask.

Preparation and the use

The hydrogen ions of the solution are brought into equilibrium with the gaseous hydrogen by means of platinum black; the platinum black adsorbs hydrogen and acts catalytically. It may be supported on platinum foil of about $1\,cm^2$ total area but a platinum wire, 1 cm long and 0.3 mm in diameter, is often satisfactory. The platinum electrode is first **carefully** cleaned with hot chromic acid mixture (Section 3.8) **caution** and thoroughly washed with distilled water. Then it is plated from a solution containing 3.0 g of chloroplatinic acid and 25 mg of lead acetate per 100 mL distilled water with platinum foil as an anode. The current may be obtained from a 4 V battery connected to a suitable sliding resistance; the current is adjusted to produce a moderate evolution of hydrogen, and the process is complete in about 2 min. It is important that only a **thin**, jet-black deposit is made; thick deposits lead to unsatisfactory hydrogen electrodes. After platinising, the electrode must be freed from traces of chlorine; it is washed thoroughly with water, electrolysed in ~0.25 M sulphuric acid as cathode for about 30 min, and again well washed with water. Hydrogen electrodes should be stored in distilled water; they should never be touched with the fingers. It is advisable to have two hydrogen electrodes so that the readings obtained with one can be periodically checked against the other. In operation, hydrogen is supplied to the electrode from a cylinder of the compressed gas.

When used as a standard electrode, the hydrogen electrode operates in a solution containing hydrogen ions at constant (unit) activity based usually on hydrochloric acid, and the hydrogen gas must be at a pressure of 1 atm (100 kPa); the effect of change in gas pressure is discussed in the literature.[13] Although it is the primary reference electrode, the hydrogen electrode is rarely used; this is because the platinum black coating of the electrode is easily poisoned by substances such as mercury and hydrogen sulphide, and it cannot be used in the presence of oxidising or reducing agents. Two common alternatives are the calomel electrode and the silver–silver chloride electrode.

13.16   The calomel electrode

Ease of preparation and constancy of potential make the calomel electrode a very reliable standard. A calomel half-cell is one in which mercury and calomel (mercury(I) chloride) are covered with potassium chloride solution of definite concentration; this may be 0.1 M, 1 M or saturated. These electrodes are known as the decimolar, the molar and the saturated calomel electrode (SCE) and relative to the standard hydrogen electrode at 25 °C they have potentials of 0.3358, 0.2824 and 0.2444 V.* Of these electrodes the SCE is most commonly used, largely because of the suppressive effect of saturated potassium chloride solution on liquid junction potentials. However, it does suffer from the drawback that its potential varies rapidly with alteration in temperature owing to changes in the solubility of potassium chloride, and restoration of a stable potential may be slow owing to the disturbance of the calomel–potassium chloride equilibrium. The potentials of the decimolar and molar electrodes are less affected by change in temperature and are to be preferred in cases where accurate values of electrode potentials are required. The electrode reaction is

$$Hg_2Cl_2(s) + 2e \rightleftharpoons 2Hg\ (liq) + 2Cl^-$$

and the electrode potential is governed by the chloride ion concentration of the solution.

* These figures include the liquid junction potential.[14]

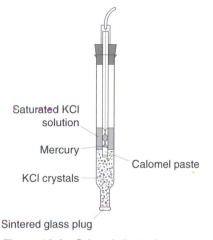

Saturated KCl
solution

Mercury

KCl crystals

Calomel paste

Sintered glass plug

Figure 13.6   Calomel electrode

Compact, ready-prepared calomel electrodes are available commercially and find wide application especially in conjunction with pH meters and ion selective meters. A typical electrode is shown in Figure 13.6. With time, the porous contact disc at the base of the electrode may become clogged, thus giving rise to a very high resistance. In some forms of the electrode the sintered disc may be removed and a new porous plate inserted, and in some modern electrodes an ion exchange membrane is incorporated in the lower part of the electrode which prevents any migration of mercury(I) ions to the sintered disc and thus to the test solution. These commercially available electrodes are normally supplied with saturated potassium chloride solution.

Some commercial electrodes are supplied with a double junction. In these arrangements the electrode depicted in Figure 13.6 is mounted in a wider vessel of similar shape which also carries a porous disc at the lower end. This outer vessel may be filled with the same solution (e.g. saturated potassium chloride solution) as is contained in the electrode vessel: in this case the main function of the double junction is to prevent the ingress of ions from the test solution which may interfere with the electrode. Alternatively, the outer vessel may contain a different solution from that involved in the electrode (e.g. 3 M potassium nitrate or 3 M ammonium nitrate solution), thus preventing chloride ions from the electrode entering the test solution. This last arrangement has the disadvantage that a second liquid junction potential is introduced into the system, and on the whole it is preferable wherever possible to choose a reference electrode which will not introduce interferences.

For some purposes, modifications of the calomel electrode may be preferred. Thus, if it is necessary to avoid the presence of potassium ions, the electrode may be prepared with sodium chloride solution replacing the potassium chloride. In some cases the presence of chloride ions may be undesirable and a mercury(I) sulphate electrode may then be used; this is prepared in similar manner to a calomel electrode using mercury(I) sulphate and potassium sulphate or sodium sulphate solution.

13.17   The silver–silver chloride electrode

Perhaps just as important as the calomel electrode is the silver–silver chloride electrode. It consists of a silver wire or a silver-plated platinum wire, coated electrolytically with a thin

Table 13.5   *Potentials of common reference electrodes*

Electrode	Potential vs SHE (V)			
	15 °C	20 °C	25 °C	30 °C
Calomel				
KCl(sat) (SCE)	0.2512	0.2477	0.2444	0.2409
1.0 *M* KCl	0.2852	0.2838	0.2824	0.2810
0.1 *M* KCl	0.3365	0.3360	0.3358	0.3356
Mercury(I) sulphate				
K₂SO₄(sat)	–	–	0.656	–
0.05 *M* H₂SO₄	–	–	0.680	–
Silver–silver chloride				
KCl(sat)	0.2091	0.2040	0.1989	0.1939
1.0 *M* KCl	–	–	0.2272	–
0.1 *M* KCl	–	–	0.2901	–

layer of silver chloride, dipping into a potassium chloride solution of known concentration which is saturated with silver chloride; this is achieved by adding two or three drops of $0.1\,M$ silver nitrate solution. Saturated potassium chloride solution is most commonly employed in the electrode, but $1\,M$ or $0.1\,M$ solutions can equally well be used; the potential of the electrode is governed by the activity of the chloride ions in the potassium chloride solution.

Commercial forms of the electrode are available and in general are similar to the calomel electrode, with replacement of the mercury by a silver electrode, and calomel by silver chloride. Just as with the calomel electrode, ion exchange membranes and double junctions are used to reduce clogging of the sintered disc in the silver–silver chloride electrode.

Values of the electrode potentials for the more common reference electrodes are given in Table 13.5 together with an indication of the effect of temperature for the most important electrodes.

Indicator and ion selective electrodes

13.18   General discussion

The indicator electrode of a cell is an electrode in which the potential depends on the activity (and therefore the concentration) of a particular ionic species which it is desired to quantify. In direct potentiometry or the potentiometric titration of a metal ion, a simple indicator electrode will usually consist of a carefully cleaned rod or wire of the appropriate metal; it is most important that the surface of the metal to be dipped into the solution is free from oxide films or any corrosion products. In some cases a more satisfactory electrode can be prepared by using a platinum wire which has been coated with a thin film of the appropriate metal by electrodeposition.

When hydrogen ions are involved, a hydrogen electrode can obviously be used as indicator electrode, but its function can also be performed by other electrodes, foremost among them the glass electrode. This is an example of a membrane electrode in which the potential developed between the surface of a glass membrane and a solution is a linear function of the pH of the solution, and so can be used to measure the hydrogen ion concentration of the solution. Since the glass membrane contains alkali metal ions, it is also possible to develop glass electrodes which can be used to determine the concentration of these ions in solution, and from this development (which is based upon an ion exchange mechanism), a whole range of membrane electrodes have evolved based upon both solid-state and liquid membrane ion exchange materials; these electrodes constitute the important series of ion selective electrodes[15] (sometimes called ion sensitive electrodes) which are now available for many different ions (Sections 13.19 to 13.21).

Indicator electrodes for anions may take the form of a gas electrode (e.g. oxygen electrode for OH^-; chlorine electrode for Cl^-), but in many instances consist of an appropriate electrode of the second kind; thus, as shown in Section 13.14, the potential of a silver–silver chloride electrode is governed by the chloride ion activity of the solution. Ion selective electrodes are also available for many anions.

The indicator electrode employed in a potentiometric titration will depend on the type of reaction under investigation. Thus, for an acid–base titration the indicator electrode is usually a glass electrode (Section 13.19), for a precipitation titration (halide with silver nitrate, or silver with chloride) a silver electrode will be used, and for a redox titration, e.g. iron(II) with dichromate, a plain platinum wire is used as the redox electrode.

13.19 The glass electrode

The glass electrode is the most widely used hydrogen ion responsive electrode, and its use depends on the fact that when a glass membrane is immersed in a solution, a potential is developed which is a linear function of the hydrogen ion concentration of the solution. The basic arrangement of a glass electrode is shown in Figure 13.7(a). The bulb (A) is immersed in the test solution, and the electrical circuit is completed by filling the bulb with a solution of hydrochloric acid (usually $0.1\,M$), and inserting a silver–silver chloride electrode. Provided the internal hydrochloric acid solution is maintained at constant concentration, the potential of the silver–silver chloride electrode inserted into it will be constant, and so too will the potential between the hydrochloric acid solution and the inner surface of the glass bulb. Hence the only potential which can vary is the potential between the outer surface of the glass bulb and the test solution in which it is immersed, so the overall potential of the electrode is governed by the hydrogen ion concentration of the test solution.

Glass electrodes are also available as **combination electrodes**[16] which contain the indicator electrode (a thin glass bulb) and a reference electrode (silver–silver chloride) combined in a single unit as depicted in Figure 13.7(b). The thin glass bulb (A) and the narrow tube (B) to which it is attached are filled with hydrochloric acid and carry a silver–silver chloride electrode (C). The wide tube (D) is fused to the lower end of tube B and contains saturated potassium chloride solution which is also saturated with silver chloride; it carries a silver–silver chloride electrode (E). The assembly is sealed with an insulating cap. This produces a more robust piece of equipment with the added convenience of only having to insert and support a single probe into the test solution instead of two separate components.

The nature of the glass used for construction of the glass electrode is very important. Hard glasses of the Pyrex type are not suitable, and for many years a soda-lime glass of the approximate composition SiO_2 72%, Na_2O 22%, CaO 6% was universally used for the

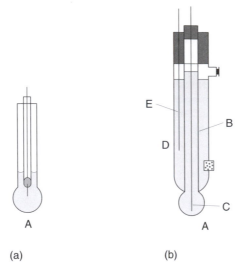

Figure 13.7 Glass electrodes in (a) basic arrangement and (b) as a combination electrode: A = glass bulb, B = narrow tube, C = silver–silver chloride electrode, D = wide tube, E = silver–silver chloride electrode

manufacture of glass electrodes. Such electrodes were extremely satisfactory over the pH range 1–9, but in solutions of higher alkalinity the electrode was subject to an 'alkaline error' and tended to give low values for the pH. Attempts were therefore made to discover glasses which would give electrodes free from this alkaline error, and it was found that the required result could be achieved by replacing most or all of the sodium content of the glass by lithium. An electrode constructed of a glass having a composition SiO_2 63%, Li_2O 28%, Cs_2O 2%, BaO 4%, La_2O_3 3% has an error of only -0.12 pH at pH 12.8 in the presence of sodium ions at a concentration of $2\,M$. Lithium-based glasses are now exclusively used for hydrogen ion responsive glass electrodes required for use at high pH values.

To measure the hydrogen ion concentration of a solution, the glass electrode must be combined with a reference electrode, and the saturated calomel electrode is most commonly used, thus giving the cell

$$\text{Ag,AgCl(s)}\,|\,\text{HCl}(0.1\,M)\,|\,\text{glass}\,|\,\text{test solution}\,\vdots\,\text{KCl(sat),Hg}_2\text{Cl}_2\text{(s)}\,|\,\text{Hg}$$

Owing to the high resistance of the glass membrane, a simple potentiometer cannot be employed for measuring the cell e.m.f. and specialised instrumentation must be used. The e.m.f. of the cell may be expressed by the equation

$$E = K + (RT/F)\ln a_{H^+}$$

or at a temperature of 25 °C by the equation

$$E = K + 0.0591\,\text{pH}$$

In these equations K is a constant that partly depends on the nature of the glass used in the construction of the membrane, and partly depends on the individual character of each electrode; its value may vary slightly with time. This variation of K with time is related to the existence of an **asymmetry potential** in a glass electrode which is determined by the differing responses of the inner and outer surfaces of the glass bulb to changes in hydrogen ion activity; this may originate as a result of differing conditions of strain in the two glass surfaces. Owing to the asymmetry potential, if a glass electrode is inserted into a test solution which is in fact identical with the internal hydrochloric acid solution, then the electrode has a small potential which is found to vary with time. On account of the existence of this asymmetry potential of time-dependent magnitude, a constant value cannot be assigned to K, and every glass electrode must be standardised frequently by placing in a solution of known hydrogen ion activity (a buffer solution).

The operation of a glass electrode is related to the situations existing at the inner and outer surfaces of the glass membrane. Glass electrodes require soaking in water for some hours before use, and it is concluded that a hydrated layer is formed on the glass surface, where an ion exchange process can take place. If the glass contains sodium, the exchange process can be represented by the equilibrium

$$H^+_{soln} + Na^+_{glass} \rightleftharpoons H^+_{glass} + Na^+_{soln}$$

The concentration of the solution within the glass bulb is fixed, hence an equilibrium condition is established on the inner side of the bulb, leading to a constant potential. On the outside of the bulb, the potential developed will depend on the hydrogen ion concentration of the solution in which the bulb is immersed. Within the layer of 'dry' glass which exists between the inner and outer hydrated layers, the conductivity is due to the interstitial migration of sodium ions within the silicate lattice. For a detailed account of the theory of the glass electrode, consult a textbook of electrochemistry.

In view of the equilibrium, it is not surprising that if the solution to be measured contains a high concentration of sodium ions, say a sodium hydroxide solution, the pH determined is too low. Under these conditions sodium ions from a solution pass into the hydrated layer in preference to hydrogen ions, so the measured e.m.f. (hence the pH) is too low. This is the reason for the 'alkaline error' encountered with the glass electrode constructed from soda-lime glass. Likewise, errors also arise in strongly acid solutions (hydrogen ion concentration in excess of $1 M$), but to a much smaller degree; this effect is related to the fact that in the relatively concentrated solutions involved, the activity of the water in the solution is reduced and this can affect the hydrated layer of the electrode which is involved in the ion exchange reaction.

The glass electrode can be used in the presence of strong oxidants and reductants, in viscous media, and in the presence of proteins and similar substances which seriously interfere with other electrodes. It can also be adapted for measurements with small volumes of solutions. It may give erroneous results when used with very poorly buffered solutions which are nearly neutral.

The glass electrode should be thoroughly washed with distilled water after each measurement and then rinsed with several portions of the next test solution before making the following measurement. The glass electrode should not be allowed to become dry, except during long periods of storage; it will return to its responsive condition when immersed in distilled water for at least 12 h prior to use.[17]

Ion selective glass electrodes

By reducing the preference of soda-lime glasses for hydrogen ion exchange it is possible for other ions to become involved in the ion exchange process. The possibility of electrode response to metallic ions like sodium and potassium was achieved by introducing aluminium oxide into the glass. Typical glass compositions for cation sensitive glass electrodes are shown in Table 13.6. In all cases some sensitivity to hydrogen ions remains; in any

Table 13.6 *Composition of glasses for cation sensitive glass electrodes*

Composition	For determination of
Na_2O 22%, CaO 6%, SiO_2 72%	H^+ (subject to alkaline error)
Li_2O 28%, Cs_2O 2%, BaO 4%, La_2O_3 3%, SiO_2 63%	H^+ (alkaline error reduced)
Li_2O 15%, Al_2O_3 25%, SiO_2 60%	Li^+
Na_2O 11%, Al_2O_3 18%, SiO_2 71%	Na^+, Ag^+
Na_2O 27%, Al_2O_3 5%, SiO_2 68%	K^+, NH_4^+

potentiometric determination with these modified glass electrodes the hydrogen ion concentration of the solution must be reduced so as to be not more than 1% of the concentration of the ion being determined, and in a solution containing more than one kind of alkali metal cation, some interference will be encountered.

The construction of these electrodes is exactly similar to the pH responsive glass electrode. They must of course be used in conjunction with a reference electrode, and a silver–silver chloride electrode is usually preferred. A double-junction reference electrode is often used. The electrode response to the activity of the appropriate cation is given by the usual Nernst equation:

$$E = k + (RT/nF)\log a_{M^{n+}}$$

and for a singly charged cation, since $-\log a_{M^+} = pM$ (cf. pH)

$$E = k - 0.0591\,pM \qquad \text{(at 25\,°C)}$$

Such an electrode may, however, also show a response to certain other singly charged cations, and when an interfering cation C^+ is present in the test solution, an equilibrium is established between ions M^+ in the glass surface in contact with the solution, and the ions C^+ in the solution:

$$M^+_{gl} + C^+_{soln} \rightleftharpoons C^+_{gl} + M^+_{soln}$$

The equilibrium constant (exchange constant) for this equilibrium is given by

$$K_{ex} = \frac{a_{M^+}a'_{C^+}}{a'_{M^+}a_{C^+}}$$

where a_{M^+} and a_{C^+} are the activities of the ions in the test solution and the corresponding a' values are the activities of those ions in the surface layer of glass.

The electrode potential under these conditions is given by

$$E = K_M + \frac{2.303\,RT}{nF}\ln(a_M + k^{pot}_{M,C}a_C)$$

where K_M is the asymmetry potential of the electrode in presence of the ion M^+, and $k^{pot}_{M,C}$ is called the **selectivity coefficient** of the electrode for M over C. Both K_M and $k^{pot}_{M,C}$ can be evaluated by making e.m.f. measurements with two solutions containing different known amounts of the two ions:

$$k_{M,C} = \frac{\text{response to } C^+}{\text{response to } M^+}$$

The selectivity coefficient is a measure of the interference of the ion C^+ in the determination of the ion M^+, but the value depends on several variables, such as the total ion concentration of the solution and the ratio of the activity of the ion being determined to the activity of the interfering ion. A small selectivity coefficient indicates an electrode which is not greatly susceptible to interference by the specified ion. For instance, a typical ion sensitive sodium electrode will have a selectivity coefficient of the order of less than 0.005 if it is not to produce misleading results due to interference from potassium ions.

Much of the early work on the development of a range of ion selective electrodes was carried out by Pungor and coworkers.[18] They showed that it was possible to use silcone rubber and plastic membranes impregnated with such compounds as silver salts for the measurement of anion concentrations. The glass membrane in the electrodes may actually be modified or replaced in many ways, such as by a single crystal, a pressed disc of crystalline material or even by a liquid ion exchanger held in place by a polymeric membrane.

An iodide ion selective electrode can be formed by incorporating finely dispersed silver iodide into a silicone rubber monomer and then carrying out polymerisation. A circular portion of the resultant silver iodide impregnated polymer used to seal the lower end of a glass tube, which was then partly filled with potassium iodide solution (0.1 M), and a silver wire was inserted to dip into the potassium iodide solution. When the membrane end of the assembly is inserted into a solution containing iodide ions, we have a situation exactly similar to using glass membrane electrodes. The silver iodide particles in the membrane set up an exchange equilibrium with the solutions on either side of the membrane. Inside the electrode, the iodide ion concentration is fixed and a stable situation results. Outside the electrode, the position of equilibrium will be governed by the iodide ion concentration of the external solution, and a potential will therefore be established across the membrane; this potential will vary according to the iodide ion concentration of the test solution.

A single-crystal electrode is exemplified by the lanthanum fluoride electrode in which a crystal of lanthanum fluoride is sealed into the bottom of a plastic container to produce a fluoride ion electrode. The container is charged with a solution containing potassium chloride and potassium fluoride and carries a silver wire coated with silver chloride at its lower end; it thus constitutes a silver–silver chloride reference electrode. The lanthanum fluoride crystal is a conductor for fluoride ions which, being small, can move through the crystal from one lattice defect to another, and equilibrium is established between the crystal face inside the electrode and the internal solution. Likewise, when the electrode is placed in a solution containing fluoride ions, equilibrium is established at the external surface of the crystal. In general, the fluoride ion activities at the two faces of the crystal are different, so a potential is established; and since the conditions at the internal face are constant, the resultant potential is proportional to the fluoride ion activity of the test solution.

The pressed disc (or pellet) type of crystalline membrane electrode is illustrated by silver sulphide, in which silver ions can migrate. The pellet is sealed into the base of a plastic container as in the case of the lanthanum fluoride electrode, and contact is made by means of a silver wire with its lower end embedded in the pellet; this wire establishes equilibrium with silver ions in the pellet and thus functions as an internal reference electrode. Placed in a solution containing silver ions, the electrode acquires a potential which is dictated by the activity of the silver ions in the test solution. Placed in a solution containing sulphide ions, the electrode acquires a potential which is governed by the silver ion activity in the solution, and this is itself dictated by the activity of the sulphide ions in the test solution and the solubility product of silver sulphide, i.e. it is an electrode of the second kind (Section 13.14). If the pellet contains a mixture of silver sulphide and silver chloride (or bromide or iodide), the electrode acquires a potential which is determined by the activity

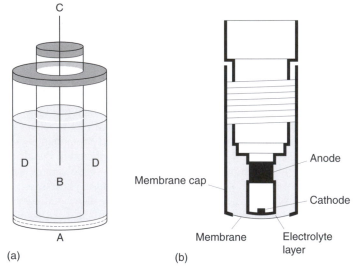

C

D   D
      B

        A

(a)

Membrane cap

Anode

Cathode

Membrane   Electrolyte
                     layer

(b)

Figure 13.8   (a) Ion exchange electrode where A = membrane, B = vessel, C = silver electrode and D = outer compartment; (b) dissolved oxygen electrode (courtesy QuadraChem Laboratories Ltd, Forest Row, E. Sussex)

of the appropriate halide ion in the test solution. Likewise, if the pellet contains silver sulphide together with the insoluble sulphide of copper(II), cadmium(II) or lead(II), electrodes are produced which respond to the activity of the appropriate metal ion in a test solution.

Ion exchange electrodes can be prepared using an organic liquid ion exchanger which is immiscible with water, or an ion-sensing material is dissolved in an organic solvent which is immiscible with water, and placed in a tube sealed at the lower end by a thin hydrophobic membrane such as 'millipore' cellulose acetate filter; aqueous solutions will not penetrate this film. The construction of such an electrode is indicated in Figure 13.8(a). The membrane (A) seals the bottom of the electrode vessel, which is divided by the central tube into an inner compartment (B) and an outer compartment (D). Compartment B contains an aqueous solution of known concentration of the chloride of the metal ion to be determined; this solution is also saturated with silver chloride and carries a silver electrode (C), which thus forms a reference electrode. The liquid ion exchange material is placed in reservoir D and the pores of the membrane become impregnated with the organic liquid which thus makes contact with the aqueous test solution in which the electrode is placed; this solution also carries a suitable reference electrode, e.g. a calomel electrode. This kind of electrode is known as a liquid membrane electrode.

Following a design by Thomas and coworkers,[19] it is now usual to prepare solid ion exchange membranes by dissolving the liquid ion exchange material together with polyvinyl chloride (PVC) in a suitable organic solvent such as tetrahydrofuran and then allowing the solvent to evaporate. A disc is cut from the flexible residue and cemented to a PVC tube to produce an electrode vessel, in which the PVC membrane replaces the cellulose acetate and reservoir material previously used, so that only a single compartment is needed. Clearly it is no longer possible to refer to a liquid membrane electrode; most ion exchange electrodes are now of this type.

A typical example of a PVC matrix membrane electrode is the calcium ion electrode, in which the cation exchange material is based upon a dialkyl phosphate such as didecyl hydrogenphosphate or better still dioctylphenyl hydrogenphosphate dissolved in dioctyl phenyl phosphonate. In contact with an aqueous solution containing calcium ions, reaction involving the loss of a proton from each of two molecules of ester occurs to form a calcium (dialkyl or dialkylphenyl) phosphate at the surface of the membrane, which thus acquires calcium ions that can equilibrate with any other solution containing calcium ions in which it may be placed. At the internal face of the membrane, a solution of some specified calcium ion concentration is present, so a definite potential is established. On the outer side of the membrane, the potential established will be determined by the calcium ion activity of the test solution, so the overall electrode potential can be related to the activity of calcium ions in the test solution. The electrode fails in acid solution because the reaction producing the calcium (dialkyl) phosphate is reversed. If the solvent used is changed to decanol, the electrode becomes responsive to other ions similar to calcium, including magnesium, and can be used as a 'water hardness' electrode.

An electrode sensitive to nitrate ions can be prepared by using the salt hexadecyl-(tridodecyl)-ammonium nitrate in the membrane, and a perchlorate (chlorate(VII) ion) electrode can be produced based upon a membrane containing tris (o-phenanthroline) iron(II) perchlorate.

The ionic organic ion exchangers used in the electrodes described above can in some cases be replaced by neutral organic ligands. A typical example is the potassium ion selective electrode based upon the antibiotic valinomycin; this substance contains a number of oxygen atoms that can form a ring and coordinate with a potassium ion by displacing its hydration shell. The electrode is set up with an internal potassium chloride solution of definite concentration from which some potassium ions are extracted by the valinomycin into the inner surface of the membrane. When the electrode is placed in an aqueous solution containing potassium ions, some of them are extracted into the external surface of the membrane, and the resultant electrode potential will depend on the potassium ion activity of the test solution. Many synthetic neutral organic ligands are now available which can be used as sensors in ion selective electrodes for a large range of cations.

A whole series of electrodes have also been developed for analysing solutions of gases such as ammonia, carbon dioxide, nitrogen dioxide, sulphur dioxide and hydrogen sulphide. For hydrogen sulphide, a sulphide ion responsive electrode is used, and for nitrogen dioxide a nitrate ion responsive electrode is used; the other gases are analysed using a glass pH electrode. To determine the proportion of any of these gases in a stream of gas, the gaseous mixture is passed through a scrubber, where the gas is dissolved in water and the resultant liquid is examined with the appropriate gas-sensing electrode.

The essential features of a gas-sensing electrode can be seen in Figure 13.8(a); only the central portion of this diagram, namely vessel B and its attachments, are now relevant. Membrane A is permeable to the dissolved gas in the test solution and may be a micro-porous membrane manufactured from either polytetrafluoroethylene or from polypropylene, both of which are water repellent and are not penetrated by aqueous solutions, but they allow gas molecules to pass through. This kind of membrane is used with ammonia, carbon dioxide and nitrogen dioxide. Alternatively, the membrane is a very thin homogeneous film, commonly of silicone rubber, through which the gas diffuses (sulphur dioxide, hydrogen sulphide). Electrode C is now a glass pH electrode or other suitable ion selective electrode and a silver–silver chloride reference electrode is also incorporated in B. The internal solution in B contains sodium chloride and an electrolyte appropriate to the gas which is being determined: for NH_3 use NH_4Cl, for CO_2 use $NaHCO_3$, for NO_2 use $NaNO_2$, for SO_2 use $K_2S_2O_5$, for H_2S use a citrate buffer. Membrane A is small in area, and the volume of liquid in B is also small, so it rapidly equilibrates with the test solution.

519

Table 13.7 *A selection of commercially available ion selective electrodes*

Ion	Lower limit of detection (M)	Ion	Lower limit of detection (M)
Na^+	1×10^{-6}	F^-	1×10^{-6}
K^+	1×10^{-6}	Cl^-	5×10^{-5}
NH_4^+	5×10^{-7}	Br^-	5×10^{-6}
Ag^+	1×10^{-7}	I^-	5×10^{-8}
Ag^+/S^{2-}	1×10^{-7}	CN^-	8×10^{-6}
Ca^{2+}	5×10^{-7}	ClO_4^-	8×10^{-6}
Ca^{2+}/Mg^{2+}	6×10^{-6}	NO_2^-	4×10^{-6}
Cd^{2+}	1×10^{-7}	NO_3^-	7×10^{-6}
Cu^{2+}	1×10^{-8}	SCN^-	5×10^{-6}
Pb^{2+}	1×10^{-6}		

One of the most well-known applications of a membrane-type gas-sensing electrode is the measurement of dissolved oxygen. The apparatus consists of a thin PVC membrane covering an electrolyte and two metallic electrodes. Oxygen diffuses through the membrane and is reduced at the electrode as a result of a fixed potential between the cathode and the anode. This means that the greater the partial pressure of oxygen, the greater the diffusion through the membrane in any fixed period of time, producing a current proportional to the oxygen concentration in the solution. The membrane thickness determines the response time of the electrode, and there can be depletion of the oxygen around the electrode during the measurement if the sample is not stirred. The basic design of an oxygen electrode is shown in Figure 13.8(b). Calibration of the electrode is necessary and is normally easy to carry out using oxygen-saturated water. Measurements of dissolved oxygen are of particular importance in environmental studies, wastewater treatment and brewing.

As a result of extensive development in recent years, a very wide range of ion selective electrodes is now available commercially. Table 13.7 gives just a small indication of what is available and the limits of detection that can be attained; they are being improved all the time. Two important considerations are the concentration range over which the electrode may be used, and the response time. If the e.m.f. of a given ion selective electrode is measured in a series of solutions containing the relevant ion at varying activities, then on plotting the e.m.f. against the logarithm of the ionic activity, a graph is obtained similar to Figure 13.9. The curve falls into three distinct parts: a straight-line portion AB, a curved portion BC and a nearly horizontal portion CD. The straight line AB has a slope equal to $2.303\ RT/nF = 59.1$ mV at 25 °C, when $n = 1$, and in this region the electrode is said to show a Nernstian response; point B may be regarded as the lower limit of measurement for practical purposes. Nevertheless, measurements can be made over the curved portion BC by making a series of readings with solutions of known activities falling within the required range and then plotting a calibration curve. IUPAC defines the lower detection limit as 'the concentration of the ion at which the extrapolated linear portion of the graph intersects the extrapolated Nernstian portion of the graph', i.e. point E in Figure 13.9. The detection limit will be influenced by the presence of interfering ions.

The response time of an electrode is defined as the time taken for the cell e.m.f. to a reach a value which is 1 mV from the final equilibrium value. The response time is

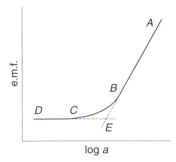

Figure 13.9 A plot of e.m.f. against log *a* falls into three distinct regions: straight line *AB*, curve *BC* and almost horizontal region *CD*

obviously affected by the type of electrode, particularly the nature of the membrane, as well as by the presence of interfering ions and any change in temperature. General details for the care and maintenance of ion selective electrodes are given in the literature.[17]

13.20 Solid-state ion selective detectors

A whole new world of ion selective electrodes has been developed with the use of semi-conductors to act as chemical-sensing field-effect transistors. These devices depend upon interfacing a thin ion selective membrane (C) with a modified metal–oxide–semiconductor (MOS) field-effect transistor (A) encased in a non-conducting shield (B) (Figure 13.10). When the membrane (C) is placed in contact with a test solution containing an appropriate ion, a potential is developed, and this potential affects the current flowing through the transistor between terminals T_1 and T_2.

By calibration against solutions containing known activities of the ion being determined, measurement of the current can be used to ascertain the activity of the ion in the test solution. These measurements can be carried out with very small volumes of liquid, and find application in biochemical analyses. However, the simpler ion selective electrodes discussed above can be readily adapted for dealing with small volumes, and even for intracellular measurements.

Ion sensitive field-effect transistor (ISFET) electrodes have been specially developed as non-glass, rugged, clog-resistant probes for working with substances ranging from liquids to semisolids, including meats and other foods. They are particularly useful for pH measurements under difficult conditions.

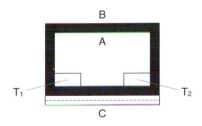

Figure 13.10 Solid-state ion selective detector: A = MOSFET, B = non-conducting shield, C = membrane, T_1, T_2 = MOSFET terminals

An alternative procedure designed to deal with minute volumes of liquid, employs a 'layer cell' based upon the technique employed in 'instant colour' photographic films. Designed to determine potassium ions, it used two layer assemblies terminating in valino-mycin electrodes. A standard potassium chloride solution was added to one assembly, and the solution under test was added to the other; the two assemblies were joined by a salt bridge to set up a concentration cell and the potassium ion concentration in the test solution could be calculated from the e.m.f. measured for the concentration cell.

13.21   Biochemical electrodes

An increasing number of electrodes are being developed which employ enzymes to convert substances in solution into ionic products which can be quantified using a known ion selective electrode. A typical example is the **urea electrode**, in which the enzyme urease is employed to hydrolyse urea:

$$CO(NH_2)_2 + H_2O + 2H^+ \xrightarrow{\text{urease}} 2NH_4^+ + CO_2$$

and the progress of the reaction can be followed by means of a glass electrode which is sensitive to ammonium ions. The final concentration of ammonium ions determined can be related to the urea present.

The urease is incorporated into a polyacrylamide gel which is allowed to set on the bulb of the glass electrode and may be held in position by nylon gauze. Preferably, the urease can be chemically immobilised on to bovine serum albumin or even on to nylon. When the electrode is inserted into a solution containing urea, ammonium ions are produced, diffuse through the gel and cause a response by the ammonium ion probe:

$$E_{cell} = k + 0.0295 \log a_{urea}$$

Penicillin can likewise be determined by using the enzyme penicillinase to destroy the penicillin with production of hydrogen ions which can be determined using a normal glass pH electrode. Many other organic materials can be determined by similar procedures,[20] including a procedure to monitor development of the digoxin antibody in rabbits.[21]

Instrumentation and measurement of cell e.m.f.

13.22   Using pH meters and selective ion meters

Direct-reading pH meters using solid-state circuitry have made it much easier and much more accurate to measure small d.c. potentials, such as those from ion selective electrodes. The modern pH meter is an electronic digital voltmeter, scaled to read pH directly, and may range from a comparatively simple hand-held instrument, suitable for use in the field, to more elaborate bench models, often provided with a scale expansion facility, with a resolution of 0.001 pH unit and an accuracy of ±0.001 unit.

A glass electrode has an asymmetry potential which makes it impossible to relate a measured electrode potential directly to the pH of the solution, and makes it necessary to calibrate the electrode. A pH meter therefore always includes a control (set buffer, stand-ardise or calibrate) so that with the electrode assembly (glass plus reference electrode or a combination electrode) placed in a buffer solution of known pH, the scale reading of the instrument can be adjusted to the correct value.

The Nernst equation shows that the glass electrode potential for a given pH value will depend on the temperature of the solution. A pH meter therefore includes a biasing control so that the scale of the meter can be adjusted to correspond to the temperature of the solution under test. This may take the form of a manual control, calibrated in degrees Celsius, which is set to the temperature of the solution as determined with an ordinary mercury thermometer. In some instruments, arrangements are made for automatic temperature compensation by inserting a temperature probe (a resistance thermometer) into the solution, and the output from this is fed into the pH meter circuit.

Some instruments also include a slope control. If a meter is calibrated at a certain pH (say pH 4.00) then when the electrode assembly is placed in a new buffer solution of different pH (say 9.20), the meter reading may not agree exactly with the known pH of the solution. In this event the slope control is adjusted so the meter reading in the second solution agrees with the known pH value. The meter is again checked in the first buffer solution, and provided the scale reading is correct (4.00), it is assumed the meter will give accurate readings for all pH values falling within the limits of the two buffer solutions.

Using a given glass electrode–reference electrode assembly, if the cell e.m.f. is measured over a range of pH, all measurements at the same temperature, and if the readings are then repeated for a series of different temperatures, then on plotting the results as a series of isothermal curves, we find that at some pH value (pH_i), the cell e.m.f. is independent of temperature; pH_i is called the isopotential pH. If the composition of the solution surrounding the inner silver–silver chloride electrode is altered, or if an entirely different external reference electrode is used, then the value of pH_i changes, and some pH meters include an isopotential control which can be used to take account of such changes in the electrode system.

Mode of operation

Before using a pH meter, become familiar with its instruction manual. But the general procedure for making a pH measurement is similar for all instruments;

1. Switch on and allow the instrument to warm up; the time for this will be quite short if the circuit is of the solid-state type. During the warm-up make certain the requisite buffer solutions for calibration of the meter are available, and if necessary prepare any required solutions; this is most conveniently done by dissolving an appropriate buffer tablet (obtainable from many suppliers of pH meters and from laboratory supply houses) in the specified volume of distilled water.
2. If the instrument is equipped with a manual temperature control, take the temperature of the solutions and set the control to this value; if automatic control is available, place the temperature probe into some of the first standard buffer solution contained in a small beaker which has been previously rinsed with a little of the solution.
3. Insert the electrode assembly into the same beaker, and if available, set the selector switch of the instrument to read pH.
4. Adjust the 'set buffer' control until the meter reading agrees with the known pH of the buffer solution.
5. Remove the electrode assembly (and the thermometer probe if used), rinse in distilled water, and place into a small beaker containing a little of the second buffer solution. If the meter reading does not agree exactly with the known pH, adjust the slope control until the required reading is obtained.
6. Remove the electrode assembly, rinse in distilled water, place in the first buffer solution and confirm that the correct pH reading is shown on the meter; if not, repeat the calibration procedure.

7. If the calibration is satisfactory, rinse the electrodes, etc., with distilled water, and introduce into the test solution contained in a smaller beaker. Read off the pH of the solution.
8. Remove the electrodes, etc., rinse in distilled water, and leave standing in distilled water.

Direct-reading meters suitable for use with ion selective electrodes are available from a number of manufacturers; they are sometimes known as ion activity meters. They are very similar in construction to pH meters, and most can in fact be used as a pH meter, but by virtue of the extended range of measurements for which they must be used (anions as well as cations, and doubly charged as well as singly charged ions), the circuitry is necessarily more complex and scale expansion facilities are included. They are commonly used in the millivolt mode.

As with a pH meter, the electrode appropriate to the measurement to be undertaken must be calibrated in solutions of known concentration of the chosen ion; at least two reference solutions should be used, differing in concentration by 2–5 units of pM according to the particular determination to be made. The general procedure for carrying out a determination with one of these instruments is outlined in Section 13.24.

Practical potentiometry

The use of a pH meter or an ion activity meter to measure the concentration of hydrogen ions or of some other ionic species in a solution is clearly an example of direct potentiometry. Practical uses of electrodes should follow fairly standard procedures, starting with the maintenance and standardisation of the systems according to the manufacturer's specifications. The following practical examples, with particular emphasis on the determination of pH, emphasise the essential features in obtaining reliable results.

13.23   Determination of pH

The original definition of pH $= -\log c_H$ (due to Sørensen, 1909; and which may be written as pcH) is not exact, and cannot be determined exactly by electrometric methods. The activity rather than the concentration of an ion determines the e.m.f. of a galvanic cell of the type commonly used to measure pH, hence pH may be defined as

$$pH = -\log a_{H^+}$$

where a_{H^+} is the activity of the hydrogen ion. But even this quantity, as defined, is not capable of precise measurement, since any cell of the type

$$H_2,Pt\,|\,H^+\,(\text{unknown})\,\|\,\text{salt bridge}\,\|\,\text{reference electrode}$$

used for the measurement inevitably involves a liquid junction potential of more or less uncertain magnitude. Nevertheless, the measurement of pH by the e.m.f. method gives values corresponding more closely to the activity than the concentration of hydrogen ion. It can be shown that the pcH value is nearly equal to $-\log 1.1 a_{H^+}$, hence

$$pH = pcH + 0.04$$

This equation is a useful practical formula for converting tables of pH based on the Sørensen scale to an approximate activity basis, in line with the practical definition of pH given below.

Table 13.8 *pH of IUPAC standards from 0 to 90 °C*

Temperature (°C)	RVS	Primary			Operational	
		P1	P2	P3	O1	O2
0	4.000	–	3.863	9.464	–	13.360
5	3.998	–	3.840	9.395	–	13.159
10	3.997	–	3.820	9.332	1.638	12.965
15	3.998	–	3.802	9.276	1.642	12.780
20	4.001	–	3.788	9.225	1.644	12.602
25	4.005	3.557	3.776	9.180	1.646	12.431
30	4.011	3.552	3.766	9.139	1.648	12.267
35	4.018	3.549	3.759	9.102	1.649	12.049
40	4.027	3.547	3.754	9.068	1.650	11.959
50	4.050	3.549	3.749	9.011	1.653	11.678
60	4.060	3.560	–	8.962	1.660	11.423
70	4.116	3.580	–	8.921	1.671	11.192
80	4.159	3.610	–	8.885	1.689	10.984
90	4.21	3.650	–	8.850	1.720	10.800

The modern definition of pH is an operational one and is based on the work of standardisation and the recommendations of the US National Bureau of Standards (NBS). In the 1987 IUPAC definition[22] the **difference** in pH between two solutions S (a standard) and X (an unknown) at the same temperature with the same reference electrode and with hydrogen electrodes at the same hydrogen pressure is given by

$$pH(X) - pH(S) = \frac{E_X - E_S}{2.3026RT/F}$$

where E_x is the e.m.f. of the cell

$H_2,Pt\,|\,$ solution $X\,\|\,3.5\,M$ KCl $|$ reference electrode

and E_s is the e.m.f. of the cell

$H_2,Pt\,|\,$ solution $S\,\|\,3.5\,M$ KCl $|$ reference electrode

The two hydrogen electrodes may be replaced by a **single** glass electrode which is transferred from one cell to the other. The pH difference thus determined is a pure number. The pH scale is defined by specifying the nature of the standard solution and assigning a pH value to it.

The **IUPAC definition of pH**[22] is based upon a 0.05 M solution of potassium hydrogenphthalate as the reference value pH standard (RVS). In addition, six further primary standard solutions are also defined which between them cover a range of pH values lying between 3.5 and 10.3 at room temperature. They are further supplemented by a number of operational standard solutions which extend the pH range covered to 1.5–12.6 at room temperature. The composition of the RVS solution, of three of the primary standard solutions and of two of the operational standard solutions is detailed below, and their pH values at various temperatures are given in Table 13.8. Note that the concentrations are expressed on a **molal** basis, i.e. moles of solute per kilogram of solution.

The British Standard (BS1647:1984 Parts 1 and 2) is also based upon potassium hydrogenphthalate and a number of reference solutions of a range of substances, and leads to results which are very similar to the figures given in Table 13.8. When applied to dilute solutions ($< 0.1\,M$) at pH between 2 and 12, it conforms approximately to the equation

$$\text{pH} = -\log\{c_{H^+} y_{1:1}\} \pm 0.02$$

where $y_{1:1}$ is the mean activity coefficient which a typical $1:1$ electrolyte would have in that solution.

Details for the preparation of the solutions referred to in the table are as follows (note that concentrations are expressed in molalities). All reagents must be of the highest purity. Freshly distilled water protected from carbon dioxide during cooling, having a pH of 6.7–7.3, should be used, and is essential for basic standards. Deionised water is also suitable. Standard buffer solutions may be stored in well-closed Pyrex or polythene bottles. If the formation of mould or sediment is visible, the solution must be discarded.

RVS: 0.05 molal potassium hydrogenphthalate Dissolve 10.21 g of the solid (dried below 130 °C) in water and dilute to 1 kg. The pH is not affected by atmospheric carbon dioxide; the buffer capacity is rather low. The solution should be replaced after 5–6 weeks, or earlier if mould growth is apparent.

P1: saturated potassium hydrogentartrate solution The pH is insensitive to changes of concentration and the temperature of saturation may vary from 22 to 28 °C; the excess of solid must be removed. The solution does not keep for more than a few days unless a preservative (crystal of thymol) is added.

P2: 0.025 molal phosphate buffer Dissolve 3.40 g of KH_2PO_4 and 3.55 g of Na_2HPO_4 (dried for 2 h at 110–113 °C) in CO_2-free water and dilute to 1 kg. The solution is stable when protected from undue exposure to the atmosphere.

P3: 0.01 molal Borax Dissolve 3.81 g of sodium tetraborate $Na_2B_4O_7 \cdot 10H_2O$ in CO_2-free water and dilute to 1 kg. The solution should be protected from exposure to atmospheric carbon dioxide, and replaced about a month after preparation.

O1: 0.05 molal potassium tetroxalate Dissolve 12.70 g of the dihydrate in water and dilute to 1 kg. The salt $KHC_2O_4 \cdot H_2C_2O_4 \cdot 2H_2O$ must not be dried above 50 °C. The solution is stable and the buffer capacity is relatively high.

O2: Saturated calcium hydroxide solution Shake a large excess of finely divided calcium hydroxide vigorously with water at 25 °C, filter through a sintered glass filter (porosity 3) and store in a polythene bottle. Entrance of carbon dioxide into the solution should be avoided. The solution should be replaced if turbidity develops. The solution is $0.0203\,M$ at 25 °C, $0.0211\,M$ at 20 °C, and $0.0195\,M$ at 30 °C.

Buffer tablets For most purposes it is not necessary to follow the procedures given above for the preparation of standard buffer solutions; the buffer tablets which are available from laboratory suppliers, when dissolved in the specified volume of distilled (deionised) water, produce buffer solutions suitable for the calibration of pH meters.

Measuring the pH of a given solution

The normal procedure is to use a glass electrode together with a saturated calomel reference electrode and to measure the e.m.f. of the cell with a pH meter. The procedure for using a pH meter is given in Section 13.22, but consult the instruction manual for details of minor variations in the controls supplied. The glass electrode supplied with the instrument should be standing in distilled water; if a new electrode is needed, leave it soaking in distilled water for at least 12 h before using it to make measurements. Never handle the bulb of the electrode – the glass is only ~0.1 mm thick. Remember that the assembly is necessarily somewhat fragile and treat it with great care; in particular, the electrode must always be supported within the measuring vessel (special electrode stands are usually supplied with pH meters) and not allowed to stand on the base of the vessel.

Prepare the buffer solutions for calibration of the pH meter if these are not already available; the potassium hydrogenphthalate buffer (pH 4) and the sodium tetraborate buffer (pH 9.2) are the most commonly used for calibration.

Check whether the instrument supplied is equipped for automatic temperature compensation, and if so, check that the temperature probe (resistance thermometer) is available. If not, the temperature of the solutions to be used must be measured, and the appropriate setting made on the manual temperature control of the instrument.

Proceed to measure the pH of the given solution, following the steps outlined in Section 13.22. Once the measurements are made, wash down the electrodes with distilled water and leave them to stand in distilled water.

13.24 Determining fluoride

This determination uses an ion selective electrode and an ion activity meter. As with the glass electrode used for pH measurements, the electrode must be calibrated using solutions containing the appropriate ion at known concentrations. For pH measurements it suffices to calibrate the glass electrode at two pH values, but for ion selective electrodes it is advisable to plot a calibration graph by making measurements with five to six standard solutions of varying concentration. This calibration graph can be used to ascertain the fluoride ion concentration in a test solution by measuring the e.m.f. of the calibrated electrode system when placed in the test solution.

As an alternative to plotting a calibration curve, the method of standard additions may be used. The appropriate ion selective electrode is first set up, together with a suitable reference electrode in a known volume (V_t) of the test solution, then the resultant e.m.f. (E_t) is measured. Applying the usual Nernst equation, we can say

$$E_t = k_e + k \log y_t C_t$$

where k_e is the electrode constant, k is theoretically 2.303 RT/nF but in practice is the experimentally determined slope of the E versus $\log C$ plot for the given electrode, y_t and C_t are the activity coefficient and the concentration respectively of the ion to be determined in the test solution. A known volume V_2 of a standard solution (concentration C_s) of the ion to be determined is added to the test solution, and the new e.m.f. E_2 is measured; C_s should be 50–100 times greater than the value of C_t. For the new e.m.f. E_2 we can write

$$E_2 = k_e + k \log y_t (V_t C_t + V_2 C_s)/(V_t + V_2)$$

where V_t is the original volume of the test solution.

Provided the first and second solutions are of similar ionic strength, the activity coefficients will be the same in each solution, and the difference between the two e.m.f. values can be expressed as

$$\Delta E = (E_2 - E_1) = k \log (V_t C_t + V_2 C_s)/C_t(V_t + V_2)$$

so that

$$C_t = \frac{C_s}{10^{\Delta E/k}(1 + V_t/V_2) - V_t/V_2}$$

Provided the value of the slope constant k is known, the unknown concentration C_t can be calculated.

Procedure

Set up the ion activity meter in accordance with the manual supplied with the instrument. The electrodes required are a fluoride ion selective electrode and a calomel reference electrode of the type supplied for use with pH meters. Prepare the following solutions.

Sodium fluoride standards Using analytical grade sodium fluoride and deionised water, prepare a standard solution which is approximately $0.05\,M$ ($2.1\,\mathrm{g\,L^{-1}}$), and of accurately known concentration (solution A). Take $10\,\mathrm{mL}$ of solution A and dilute to $1\,\mathrm{L}$ in a graduated flask to obtain solution B which contains approximately $10\,\mathrm{mg\,L^{-1}}$ fluoride ion. A $20\,\mathrm{mL}$ volume of solution B further diluted (graduated flask) to $100\,\mathrm{mL}$ gives a standard (solution C), containing approximately $2\,\mathrm{mg\,L^{-1}}$ fluoride ion, and by diluting $10\,\mathrm{mL}$ and $5\,\mathrm{mL}$ portions of solution B to $100\,\mathrm{mL}$, standards D and E are obtained, containing respectively 1 and $0.5\,\mathrm{mg\,L^{-1}}$.

Total Ionic Strength Adjustment Buffer (TISAB) Dissolve $57\,\mathrm{mL}$ ethanoic acid, $58\,\mathrm{g}$ sodium chloride and $4\,\mathrm{g}$ cyclohexane diaminotetra-acetic acid (CDTA) in $500\,\mathrm{mL}$ of deionised water contained in a large beaker. Stand the beaker inside a water bath fitted with a constant-level device, and place a rubber tube connected to the cold water tap **inside** the bath. Allow water to flow slowly into the bath and discharge through the constant level; this will ensure that in the subsequent treatment the solution in the beaker will remain at constant temperature.

Insert into the beaker a calibrated glass electrode–calomel electrode assembly which is joined to a pH meter, then with constant stirring and continuous monitoring of the pH, slowly add sodium hydroxide solution ($5\,M$) until the solution acquires a pH of 5.0–5.5. Pour into a $1\,\mathrm{L}$ graduated flask and make up to mark with deionised water. The buffering procedure is necessary because OH^- ions having the same charge and similar size to the F^- ion act as an interference with the LaF_3 electrode.

The resulting solution will exert a buffering action in the region pH 5–6, the CDTA will complex any polyvalent ions which may interact with fluoride, and by virtue of its relatively high concentration the solution will furnish a medium of high total ionic strength, thus obviating the possibility of e.m.f. variation owing to varying ionic strength of the test solutions.

Pipette $25\,\mathrm{mL}$ of solution B into a $100\,\mathrm{mL}$ beaker mounted on a magnetic stirrer and add an equal volume of TISAB from a pipette. Stir the solution to ensure thorough mixing, stop the stirrer, insert the fluoride ion–calomel electrode system and measure the e.m.f. The electrode rapidly comes to equilibrium, and a stable e.m.f. reading is obtained immediately. Wash down the electrodes and then insert into a second beaker containing a solution prepared from 25 mL each of standard solution C and TISAB; read the e.m.f. Carry out further determinations using the standards D and E.

Plot the observed e.m.f. values against the concentrations of the standard solutions, using a semilog graph paper which covers four cycles (i.e. spans four decades on the log scale): use the log axis for the concentrations, which should be in terms of fluoride ion concentration. A straight-line plot (calibration curve) will be obtained. With increasing dilution of the solutions there tends to be a departure from the straight line; with the electrode combination and measuring system in this experiment, it becomes apparent when the fluoride ion concentration is reduced to $\sim 0.2\,mg\,L^{-1}$.

Now take 25 mL of the test solution, add 25 mL TISAB and proceed to measure the e.m.f. as above. Using the calibration curve, the fluoride ion concentration of the test solution may be deduced. The procedure described is suitable for measuring the fluoride ion concentration of tap water in areas where fluoridation of the supply is undertaken. The result may be checked by adding four successive portions (2 mL) of standard solution C to the test solution of which the e.m.f. has already been determined, and measuring the e.m.f. after each addition; the calculation for this standard addition procedure is as described above.

An alternative to these calculations is the **Gran's plot** procedure to evaluate the initial concentration of the test solution. It was shown by Gran[23] that if antilog ($E_{cell} = nF/2.303\,RT$) is plotted against volume of reagent added, a straight line is obtained; when this line is extrapolated it cuts the horizontal axis at a point corresponding to the concentration of the test solution. Special graph paper is available (Gran's plot paper) which is a semi-antilog paper, in which the vertical axis is scaled to antilogarithm values, and the horizontal axis is a normal linear scale; using this paper the observed cell e.m.f. is plotted against the volume of reagent added. This plot is particularly useful for determining end points in potentiometric titrations.

13.25   Potentiometry in an oscillating reaction

During the course of some catalysed processes, the concentration of the catalysing species may fluctuate over a wide range of concentrations. This fluctuation can be measured potentiometrically and may sometimes be observed visually by changes in colour or luminescence.[24,25] Although the study of oscillating chemical dynamics is a very broad subject, the Belousov–Zhabotinskii reaction,[26,27] involving an organic compound such as malonic acid (propanedioic acid), a catalyst, bromate ions and an aqueous solution, has received extensive study, and the fluctuations in cerium catalyst concentration between oxidation states Ce^{3+} and Ce^{4+} can be followed visually by the yellow to colourless change as well as potentiometrically. The laboratory procedure for this has been well documented[28] and Rosenthal[29] has developed an improved visual form of the process in which he replaced cerium with ferroin. The reaction must be carried out free of chloride ions as they inhibit the oscillations. The process he has provided is as follows.

Apparatus
150 mL tall-form beaker
Platinum disc electrode
Ag–AgCl reference electrode
Magnetic stirrer
Photodiode detector linked to an amplifying circuit
Red filter to transmit light with a wavelength > 600 nm
Focused source of white light

Reagents *Solution A* 0.6 M sodium bromate in 0.6 M sulphuric acid.

Solution B 0.48 M aqueous solution of malonic acid (propanedioic acid).

Solution C Sodium bromide, 1 g in 10 mL deionised water.

Solution D A 25 mM aqueous solution of ferroin, freshly prepared from iron(II) ammonium sulphate and the stoichiometric amount of o-phenanthroline (i.e. 1 mol to 3 mol).

Chloride ions Use reagents free of chloride ions.

Procedure Set up the beaker with the stirrer bar on the magnetic stirring unit. Arrange the electrodes diametrically opposite each other in the beaker, so the light beam passes between them through the beaker, and the solution when it is later added, and through the filter arranged in a vertical plane to the photodiode. In a separate covered conical flask, mix 14 mL of solution A with 7 mL of solution B and 2 mL of solution C. Swirl the mixture until the orange colour (due to bromine) disappears from the solution and the vapour. Then add 1 mL of solution D and stir the mixture to give a homogeneous red colour. Pour this solution into the 150 mL beaker on the magnetic stirrer unit. Use the electrodes and the photodiode to record the changes in the potential of the iron(II) o-phenanthroline–iron(III) o-phenanthroline couple. It should be possible to follow the oscillations for several minutes.

Calculation The Fe^{2+}/Fe^{3+} concentration ratio can be calculated from the Nernst equation (Section 2.28), where

$$E_{cell} = E^{\ominus} - 0.0591 \log \frac{[Fe^{II}(o\text{-phen})_3^{2+}]}{[Fe^{III}(o\text{-phen})_3^{3+}]}$$

where E_{cell} is the measured potential and $E^{\ominus}$ is the standard reduction potential for the iron(II) o-phenanthroline–iron(III) o-phenanthroline complex redox couple (given by Rosenthal[29] as 0.950 V relative to the SCE).

The ratio can be calculated for both maximum and minimum potentials and the peaks and troughs of the plot compared with the corresponding results from the photodiode circuit. Other sources of data on oscillating reactions are given in the literature.

Voltammetry

13.26 Fundamentals of voltammetry

Voltammetry is concerned with the study of voltage–current–time relationships during electrolysis carried out in a cell. Normally it involves the determination of substances in solution which can be reproducibly reduced or oxidised at an electrode surface. This electrode, known as the working electrode, has a continuously variable potential applied to it and the current flow is monitored. The resulting current–voltage graph which may be drawn is known as a voltammogram. The components in solution are either oxidised or reduced depending upon the polarity of the impressed potential at characteristic potentials which may be used to identify the species. Although simple in concept, these experiments can only be performed reliably under certain conditions. The working electrode should be completely polarised so that the current which flows through the electrode is proportional to the concentration. This is normally obtained by using an electrode of small surface area (a microelectrode), and it should not be readily contaminated so that reproducible behaviour at its surface is preserved. These constraints were solved in an elegant manner by the Czechoslovakian electrochemist Jaroslav Heyrovsky[30] in the early 1920s, when he proposed the use of a continuously dropping mercury microelectrode. Voltammetry using this electrode system is called polarography and was until recently the most widely used form of voltammetry. Its use and application will be described in the next section.

13.27   Conventional or d.c. polarography

In order to obtain reproducible current voltage curves at an electrode immersed in a dilute solution of electrolytically active species, it is necessary to have a very small surface area at the electrode to induce polarisation but it is also essential that the electrode is not changed in any way, e.g. by surface contamination. With solid electrodes these two constraints are almost mutually exclusive and reliable voltammetric measurements were not possible until Heyrovsky and Shikata developed an apparatus which used a continuously replaceable mercury drop as the working electrode and a mercury pool as the counter-electrode. Since the curves obtained with this instrument are a graphical representation of the polarisation of the dropping electrode, the apparatus was called a polarograph. The current–voltage curves, which in Herovsky's original apparatus were recorded photographically, are polarograms. The basic apparatus for polarographic analysis is depicted in Figure 13.11.

The dropping mercury electrode is shown here as the cathode, which is the most common configuration, but it may be used with reversed polarity in some experiments when oxidisation rather than reduction occurs at the microelectrode. The counter-electrode,

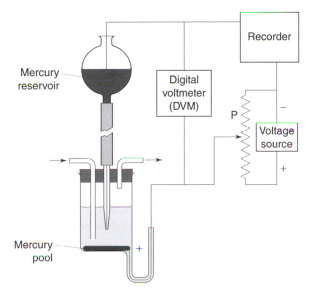

Figure 13.11   Polarography: basic apparatus

normally the anode, is a pool of mercury which, due to its relatively large surface area, has a low current density at its surface and is non-polarised, thus remaining at a constant potential. If the electrolyte solution contains anions capable of forming insoluble salts with mercury (Cl^-, SO_4^{2-}, etc.) then it acts as a reference electrode of constant potential which depends only on the nature and concentration of the anion in solution (these can be added to the solution in large excess over the analyte concentration so that the potential of the total cell only depends upon the reactions occurring at the working electrode).

The dropping mercury electrode in this system is constructed from a length of barometer capillary glass tubing about 10–15 cm long and connected by rubber tubing to a mercury reservoir whose height can be adjusted. By maintaining a suitable vertical distance between the capillary and the reservoir, a continuous stream of drops can be produced with a drop time of 1–5 s. Since polarography is characterised by a dropping mercury electrode, it is appropriate to consider the advantages and disadvantages that this electrode has compared to other microelectrodes:

(a)   A fresh, smooth, reproducible drop is produced at regular intervals. This limits the effects of contamination or poisoning of the electrode surface.

(b)   Many metals are reversibly reduced to amalgams at the surface.

(c)   Hydrogen has a high overvoltage at a mercury electrode. This means that a number of metals with high reduction potentials can still be reduced at the surface without reduction of water interfering with the experiment. In suitable electrolyte solutions the electrode may be used at negative potentials (relative to the SCE) of up to −2.6 V before hydrogen is produced. The positive potential that may be used is limited to about 0.4 V since at this value mercury is oxidised to Hg(I).

(d)   The surface area of the drops can be calculated from the weight of the drops.

These are all advantages of the dropping mercury electrode, but there are some disadvantages too, mainly related to the fact that, in the apparatus as shown, the size of the drop changes with time and this complicates the electrochemistry. Most modern apparatus reduces these limitations by using a series of hanging drops (or even a single hanging drop) rather than a continuous stream of drops.

However it is probably best to consider the processes that occur in the classical apparatus before describing more recent variants. Note that the cell in Figure 13.11 is fitted with a tube so that the solution can be purged with an inert gas prior to the experiment. This is essential in order to remove dissolved oxygen from solution. If this is not done then a large current will flow through the system as oxygen is reduced to water. This process actually occurs in two distinct steps:

$$O_2 + 2H^+ + 2e^- = H_2O_2$$

$$H_2O_2 + 2H^+ + 2e^- = 2H_2O$$

These two reduction processes would completely obscure most other reductions at the concentrations normally used in polarography. Fortunately, the oxygen can be effectively removed by bubbling either nitrogen or hydrogen through the solution for a short time before determining other species.

Suppose a dilute oxygen-free solution of cadmium chloride (10^{-3} to 10^{-5} M) is placed in the cell in Figure 13.11 and the potential of the dropping mercury electrode is gradually increased by moving the slider of the potentiometer. The current flowing through the system can then be monitored using the chart recorder and the current voltage curve or

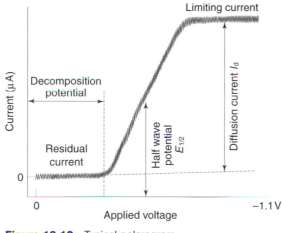

Figure 13.12   Typical polarogram

polarogram shown in Figure 13.12 will be observed. As the potential is slowly increased from zero to higher negative voltages, very little current flows through the system until the potential is high enough to initiate an electrochemical reaction. In this case, at about −0.6 V, a reduction of cadmium ions will occur to give cadmium metal:

$$Cd^{2+} + 2e^- = Cd$$

The potential at the electrode will be controlled by the Nernst equation:

$$E = E^\ominus + \frac{RT}{nF} \ln\left(\frac{a_{ox}}{a_{red}}\right) \tag{13.1}$$

Where $E^\ominus$ is the standard electrode potential. With the simple two-electrode system described here it is not possible to exactly determine the potential at the cathode, only the potential difference between the anode and cathode.

The current rises very steeply at this point but very quickly reaches a maximum value which remains constant for several hundred millivolts until a further sharp rise is seen at the end of the voltage scan. By convention if reduction is occurring, the measured current is plotted as a positive value. This polarographic wave can be predicted from a knowledge of standard electrode potentials. However, the significant measured voltage – which occurs at half the maximum current and is thus known as the half-wave potential $E_{1/2}$ – is normally not exactly the same value as the standard electrode potential of the reduced species. Reasons for the difference will be discussed later. It is more important at this point to explain why a limiting current flows through the system and hence produces a plateau.

13.28   Theoretical principles

A simple explanation suggests that the limiting current occurs because the reduction is limited by the rate at which ions reach the surface of the electrode; the reduction step itself is assumed to occur rapidly. There are three main mechanisms, hence we will consider three contributions to the limiting current $i_{limit} = i_d + i_c + i_m$.

The migration current i_m

Under the influence of an applied potential, ions in solution will tend to move in order to reduce the potential, i.e. cations will move towards the cathode and anions towards the anode. This process is used in some analytical systems such as electrophoresis, but in polarography it can lead to erratic results for low concentrations of analyte ions and it is normally reduced to a very low value by including in the same solution as the analyte ions, a high concentration of ions which can carry the migration current but which are not reduced at the electrode surface and therefore do not contribute to the limiting current. This indifferent electrolyte or supporting electrolyte is added in at least a hundredfold excess over the electroactive material, so that only a minute fraction of the migration current flow is due to the analyte ions. In the example shown this is $1\,M$ HCl. The role of the supporting electrolyte is more complex than just swamping the migration current and it can modify the chemistry of the overall reactions, but this will be discussed later.

The convection current i_c

If ions are transported toward the electrode surface by mechanical means such as stirring in the solution, then this will affect the limiting current. Traditional polarography simplifies the system by assuming that in an unstirred solution, as used with the dropping mercury electrode, convection processes are absent and thus i_c is zero. This places some constraints on the polarographer, who must ensure the polarography experiment is carried out in an unstirred solution. Since the liquid has probably been agitated by blowing inert gas through the cell just prior to the experiment, it is essential to wait several minutes before measuring the polarographic wave.

   An alternative and increasingly used system ensures that, rather than remaining stationary, the solution is in continual and constant motion. Hydrodynamic voltammetry, as this is known, is either performed in a stirred solution with a fixed microelectrode or uses a rotating microelectrode to stir the solution, when currents of one or more orders of magnitude higher are observed. A slight variant on this system maintains a constant flow of solution past the electrode. This is the basis of the slightly misnamed polarographic detector used in HPLC.

The diffusion current i_d

The rate of diffusion of an ion to an electrode surface is given by Fick's second law as

$$\frac{\partial c}{\partial t} = D\frac{\partial^2 c}{\partial x^2}$$

where D is the diffusion coefficient, c is the concentration, t is time and x is distance from the electrode surface.

   For classical polarography, using an unstirred solution which contains an excess of supporting electrolyte, both i_m and i_c tend to zero and the only mechanism for transporting ions to the electrode surface is the normal diffusion of ions in solution i_d. Thus, in polarography the limiting current is identical to the diffusion current and, in most polarographic traces, the top of the plateau is labelled i_d. It was quickly realised that the diffusion current is directly proportional to the concentration of ions in solution which are being reduced (or oxidised), but it was not until 1934 that Ilkovic[31] examined the various factors which govern the diffusion current and deduced the following equation:

$$i_d = 607nD^{1/2}Cm^{2/3}t^{1/6} \tag{13.2}$$

where

i_d = the average diffusion current (μA) during the life of the drop

n = the number of faradays of electricity required per mole of analyte*

D = a constant ($cm^2 s^{-1}$) known as the diffusion coefficient of the reducible or oxidisable species

C = the analyte concentration ($mmol\,L^{-1}$)

m = the mass of mercury (mg) dropping from the electrode per second

t = the drop time (s)

The constant 607 is a combination of several constants, including the Faraday constant; it is slightly temperature dependent and the value 607 is correct at 25 °C. If i is measured as the maximum current rather than the average current, the constant takes the value 706 rather than 607.

This rather complicated equation was important because it gave a theoretical basis for the observation that i_d is directly proportional to concentration. The original Ilkovic equation is in fact slightly inaccurate because it does not completely allow for the curvature of the mercury drop. This may be allowed for by multiplying the right-hand side of the equation by $(1 - AD^{1/2}t^{1/6}m^{-1/3})$ where A is a constant of value 39. The correction is not large since the term in parentheses usually has a value of between 1.05 and 1.15 and needs only to be taken into account for very accurate calculations.

In practice the diffusion current depends on a number of factors such as temperature (1.5–2.0% per degree), viscosity of the solution, the composition of the supporting electrolyte and the exact chemical form of the electroactive species. A metal ion complex usually yields a different diffusion current from the same concentration of the simple hydrated ion. Note also that the diffusion current depends on the drop time and the size (mass) of each drop. The product $m^{2/3}t^{1/6}$ is the capillary constant and enables the comparison of diffusion currents from different capillaries.

Since the diffusion current depends upon the size of the drop, classical polarography is characterised by the sawtooth shape of the trace, clearly seen in Figure 13.12 at the top of the plateau. The current increases as each drop grows, then it sharply decreases as each drop falls. At low concentrations which require a sensitive current range for the trace, these oscillations can tend to obscure the overall reduction wave as the analyte is reduced. A careful study of the polarographic trace shows there is another phenomena which can influence the measured currents.

The residual current

If a current–voltage curve is plotted for a properly degassed solution containing **only** a supporting electrolyte which is not reduced at the electrode surface, at potentials more negative than about −0.4 V a small current is still measured in the system. This is because the mercury drop and the solution act like a small condenser; the mercury drop stores a negative charge on its surface compared to the potential of the thin layer of solution surrounding it (the electrical double layer). As the mercury drops fall, they take this charge with them and a small positive current results. This condenser, or charging, current is unlike all the previous currents; it is a non-faradaic current, it is not due to electrochemical processes at the electrode surface. Below about −0.4 V the current flows in the other direction as the drop becomes positively charged with respect to the solution. The point at

---

* This is numerically the same as the number of electrons consumed or liberated in the reduction or oxidation of one molecule of the electroactive species.

which zero current flows through the system is known as the electrocapillary maximum and is where the mercury drop carries no charge with respect to the surrounding solution.

The condenser current increases almost linearly with applied potential and is seen even in very pure solutions of supporting electrolyte. It depends on the size of the mercury drop, increasing to a maximum as the drop grows and then decreasing as each drop falls off. In polarographic experiments the net current that is measured is the sum of the faradaic (electrochemical) and non-Faradaic charging components. At low concentrations the charging current is a significant proportion of the total and it limits the lowest concentration that can be determined by classical polarography to about $10^{-5} M$ solutions.

Polarographic maxima

Occasionally, instead of a flat-topped plateau in the polarogram, there is a maximum, perhaps a rounded hump or even a sharp peak; Figure 13.13 shows a typical example. Curve A is for copper ions in $0.1 M$ potassium hydrogen citrate solution and curve B is for $0.1 M$ potassium hydrogen citrate plus 0.005% acid fuschine solution. To measure the true diffusion current, the maxima must be eliminated or suppressed. Although the exact mechanism of formation for these maxima is not understood, they can normally be removed or reduced by adding a very small amount of **maximum suppressor**.

Maximum suppressors include colloids such as gelatine, dyestuffs or detergents such as Triton X-100 (a non-ionic detergent). Since all these materials are surface-active, they probably form some sort of adsorbed layer in the aqueous phase close to the mercury surface; this would prevent streaming of ions across the interface and thus prevent the formation of maxima. But maximum suppressors must be used with care since too high a concentration will suppress the normal diffusion current. Concentrations of gelatine between 0.002% and 0.1% are said to be effective, whereas Triton X-100 is used at 0.002–0.004%.

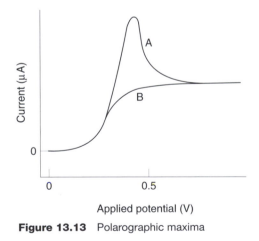

Figure 13.13   Polarographic maxima

The half-wave potential $E_{1/2}$

The potential at a polarised electrode obeys the Nernst equation and the concentration of electroactive species is directly related to the diffusion current. This allows one to relate the potential and the current to each other via the equation

$$E = E_{1/2} + \frac{RT}{nF} \ln\left(\frac{I_d - I}{I}\right)$$

which at 25 °C takes the form

$$E = E_{1/2} + \frac{0.0591}{n} \log\left(\frac{I_d - I}{I}\right) \tag{13.3}$$

where I is the measured current at any point (minus the residual current).

This equation, sometimes known as the equation of the polarographic wave, clearly shows that for a reversible reaction when $I = I_d/2$ the measured potential is $E_{1/2}$. A graph of E versus $\log[(I_d - I)/I]$ should produce a straight line with slope $= -0.00591/n$, thus enabling the number of electrons taking part in the reaction to be determined. It also shows why reactions involving more than one electron give sharper polarographic waves than single-electron processes. Notice that the equation predicts $E_{1/2}$ is concentration independent; however, several criteria must be satisfied if the equation is to be useful. Firstly, since it is derived from the Nernst equation, it will only apply to reversible processes. Also it is necessary to correct both I and I_d for residual currents and to correct the measured potential for any IR drop in the cell. The supporting electrolyte in polarographic experiments raises the overall conductance of the solution, thus reducing R.

Although the half-wave potential is independent of concentration in a given cell, it does depend on the exact nature of the reacting species and in most cases it will not have exactly the same value as the standard electrode potential of the species $E^{\ominus}$ (13.1). The most common situation is where a free ion in solution is reduced to a metal, which dissolves in mercury to form an amalgam. This is a reversible reaction, but the product is thermodynamically stabilised when it forms the amalgam so that $E_{1/2}$ is not equal to $E^{\ominus}$. The relationship is

$$E_{1/2} = E^{\ominus} - E_s + (RT/n)\log Cg \tag{13.4}$$

where E_s measures the tendency of electrons to flow between the solid analyte and amalgam, Cg is the activity of the metal in the amalgam. For cadmium $E^{\ominus} = -0.647$ (at 25 °C) and $E_{1/2} = -0.570$ V.

13.29   Complex ions

Where the reaction at the electrode involves complex ions rather than simple ions, if the complex dissociates rapidly at the electrode surface (compared to the diffusion rate) then satisfactory polarograms can still be obtained but the value of $E_{1/2}$ will be shifted. Generally the half-wave potential for the reduction of a complex is more negative than for the reduction of the free metal ion. This shift in half-wave potential of metal ions by complexation can be used to eliminate or reduce the interfering effect of one ion on another in a mixture of ions; separate reduction waves for each ion can then be observed, allowing the determination of each.

In the analysis of copper-based alloys for nickel, lead, etc., the reduction wave of copper(II) ions in most supporting electrolytes precedes those of the other metals and swamps those of the other metals present. By using a supporting electrolyte containing cyanide, the copper is converted into the cyanocuprate(I) complex, which is not reduced until after the waves for nickel, lead, etc., have been observed. Provided the electrode reaction is actually reversible, this shift in half-wave potentials on complexation may be used to determine both the composition of the complex and its formation constant.

Consider the general case of the dissociation of a complex ion:

$$MX_p^{(n-pb)+} = M^{n+} + pX^{b-}$$

The instability constant may be written as

$$K_{instab} = \frac{[M^{n+}][X^{b-}]^p}{[MX_p^{(n-pb)+}]}$$

strictly speaking, activities should be used rather than concentrations.

The electrode reaction, assuming amalgam formation, is

$$M^{n-} + ne + Hg = M(Hg)$$

Combining these equations gives

$$MX_p^{(n-pb)+} + ne + Hg = M(Hg) + pX^{b-}$$

It can be shown[32] that the expression for the electrode potential is

$$E_{1/2} = E^\ominus - \frac{0.0591}{n} \log K_{instab} - \frac{0.0591}{n} \log [X^{b-}]^p \qquad (13.5)$$

Here p is the coordination number of the complex ion formed, X^{b-} is the ligand and n is the number of electrons involved in the electrode reaction. Since the actual concentration of the complex does not appear in this equation, the observed half-wave potential will be constant and independent of complex concentration. (Note, however, that the smaller the value of K_{instab}, i.e. the more stable the complex, the more negative the value of the half-wave potential compared to the free ion.) But the potential does depend on the concentration of the ligand X^{b-}, and if this is determined at two different ligand concentrations, then it may be shown that

$$\Delta E_{1/2} = \left(\frac{-0.0591}{n}\right) p \, \Delta \log [X^{b-}] \qquad (13.6)$$

From this, the coordination number of the complex and hence its formula may be determined. It has also been shown that

$$(E_{1/2})_{complex} - E_{1/2} = \left(\frac{0.0591}{n}\right) \log K_{instab} - \left(\frac{0.0591}{n}\right) p \log [X^{b-}] \qquad \text{(at 25 °C)} \quad (13.7)$$

This equation is true, provided the concentration of the ligand is high enough to assume it is the same at the surface of the electrode and in the bulk of the solution. The formation constant may then be obtained by comparing the half-wave potential of the free ion with the half-wave potential at a given ligand concentration.

13.30   Quantitative techniques

One of the great advantages of classical or direct current polarography is that the processes discussed above can be observed in solution for a wide range of materials. Provided that the substance can be reduced or oxidised between the potentials +0.4 and −1.2 V, a polarographic wave should be observed that is proportional to the concentration of the electroactive species. In fact, if several different species are present in the same solution then a number of polarographic waves will be observed, each wave having a half-wave potential close to the standard electrode potential of the species, and the increase in diffusion current from one plateau to the next will be proportional to the concentration of that species. Generally a potential difference of about 100 mV between the electrode potentials is required to allow sufficient resolution of the two or more waves.

Substances which are oxidised at potentials higher than 0.4 V cannot be observed due to the oxidation of the mercury at this value, and at potentials higher than about -1.2 V water is reduced (Figure 13.14). This still means that the technique can be used for a large number of metals (cations) and most of the common anions. Organic substances with functional groups that can be oxidised or reduced can also be determined in aqueous solution by classical polarography, but many of these reactions are slow and thus kinetically irreversible, hence they lead to drawn-out and less well-defined waves.

In classical polarography the wave height is proportional to the concentration (13.2), so the technique is widely used for quantitative analysis. Two methods for determining concentrations of unknown solutions are frequently used: wave height–concentration plots and the method of standard additions.

Wave height–concentration plots

Solutions of several different concentrations of the ion under investigation are prepared; the composition of the supporting electrolyte and the amount of maximum suppressor added are the same for the comparison standards and for the unknown. The heights of the waves obtained are measured in any convenient manner and plotted as a function of the concentration. The polarogram of the unknown is produced exactly as the standards, and the concentration is read from the graph. The method is strictly empirical and no assumptions are made, except correspondence with the conditions of the calibration. The wave height need not be a linear function of the concentration, although this is frequently the case. For results of the highest precision, the unknown should be bracketed by standard solutions run consecutively.

Method of standard addition

The polarogram of the unknown solution is first recorded, after which a known volume of a standard solution of the same ion is added to the cell and a second polarogram is taken. The concentration of the unknown may be calculated from the heights of the two waves, the known concentration of ion added, and the volume of the solution after the addition. If I_1 is the observed diffusion current (equivalent to the wave height) of the unknown solution of volume V ml and of concentration C_u, and I_2 is the observed diffusion current after v ml of a standard solution of concentration C_s have been added, then according to the Ilkovic equation, we have

$$I_1 = kC_u$$

and

$$I_2 = k(VC_u + vC_s)/(V + v)$$

Thus

$$k = I_2(V + v)/(VC_u + vC_s)$$

thus

$$C_u = \frac{I_1 v C_s}{(I_2 - I_1)(V - v) + I_1 v} \tag{13.8}$$

The accuracy of the method depends upon the precision with which the two volumes of solution and the corresponding diffusion currents are measured. The material added should be contained in a medium of the same composition as the supporting electrolyte, so the supporting electrolyte is not altered by the addition. The assumption is made that the wave height is a linear function of the concentration in the range of concentration employed. The best results appear to be obtained when the wave height is about doubled by the addition of the known amount of standard solution. This procedure is sometimes called spiking.

13.31   The effect of oxygen

Using the basic apparatus in Figure 13.11, qualitative analysis may be conveniently carried out on solutions where the concentration of the electroactive species is 10^{-3} to $10^{-4} M$ and the total volume is 2–25 mL. However, concentrations 10 times larger or smaller than this can often be determined in volumes of less than 1 mL if special cells are used. But oxygen must be excluded from solution. This is because oxygen dissolved in electrolytic solutions is easily reduced at the dropping mercury electrode, producing a polarogram that shows two waves of approximately equal height and extending over a considerable voltage range. The position of the waves depends upon the pH of the solution; alkali displaces them to higher voltages. The concentration of oxygen in aqueous solutions that are saturated with air at room temperature is about $2.5 \times 10^{-4} M$, so its polarographic behaviour is of considerable practical importance. Figure 13.14 (curve A) shows a typical polarogram for air-saturated $1 M$ potassium chloride solution (in the presence of 0.01% methyl red).

The first wave (starting at about −0.1 V relative to SCE) is due to the reduction of oxygen to hydrogen peroxide:

$$O_2 + 2H_2O + 2e = H_2O_2 + 2OH^- \quad \text{(neutral or alkaline solution)}$$

$$O_2 + 2H^+ + 2e = H_2O_2 \quad \text{(acid solution)}$$

The second wave is ascribed to the reduction of the hydrogen peroxide either to hydroxyl ions or to water:

$$H_2O_2 + 2e = 2OH^- \quad \text{(alkaline solution)}$$

$$H_2O_2 + 2H^+ + 2e = 2H_2O \quad \text{(acid solution)}$$

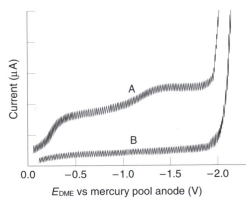

Figure 13.14   Polarography: effect of oxygen

Oxygen is easily removed by bubbling an inert gas (nitrogen or hydrogen) through the solution for about 10–15 min before determining the current–voltage curve. Figure 13.14 (curve B) was obtained after removing the oxygen by oxygen-free nitrogen from a cylinder of the compressed gas. The gas stream must be discontinued at least a minute before the actual measurements to prevent the stirring effect interfering with the normal formation of drops of mercury or with the diffusion process near the microelectrode. Measurement of the oxygen polarographic wave has been used as a method for determining dissolved oxygen: the given sample is first examined after adding some potassium chloride solution as supporting electrolyte, and the procedure is then repeated after subjecting the sample to deoxygenation by passing oxygen-free nitrogen through it for 5 min.

Ideally the electrolysis cell should be immersed in a thermostat bath maintained within $\pm0.2\,°C$, but for many purposes a temperature variation of $\pm0.5\,°C$ is permissible. A temperature of $25\,°C$ is usually employed. As a precautionary measure to prevent the appearance of maxima, sufficient gelatin to give a final concentration of 0.005% should be added. The gelatin should preferably be prepared fresh each day since bacterial action usually appears after a few days. Other maximum suppressors (e.g. Triton X-l00 and methyl cellulose) are sometimes used.

13.32 Simple polarography and classical d.c. polarography

Simple polarography

Two or more electroactive ions may be determined successively if their half-wave potentials differ by at least 0.4 V for singly charged ions and 0.2 V for doubly charged ions, provided the ions are present in approximately equal concentrations. If the concentrations differ considerably, the difference between the half-wave potentials must be correspondingly larger. If the waves of two ions overlap or interfere, various experimental devices may be employed. The half-wave potential of one of the ions may be displaced to more negative potentials by the use of suitable complexing agents which are incorporated in the supporting electrolyte; for example, Cu^{2+} ions may be complexed by the addition of potassium cyanide. Sometimes one ion may be removed by precipitation, e.g. with a mixture of lead and zinc, the lead will not interfere if it is precipitated as sulphate; the lead sulphate formed need not be removed by filtration. But the possibilities of adsorption or coprecipitation of part of the other ions must be borne in mind; electrolytic separations are also very useful.

Measurement of wave heights

With a well-defined polarographic wave where the limiting current plateau is parallel to the residual current curve, the measurement of the diffusion current is relatively simple. In the exact procedure (Figure 13.15) the actual residual current curve is determined separately with the supporting electrolyte alone; the diffusion current is obtained by subtracting the residual current from the value of the current at the diffusion current plateau (both measured at the same applied voltage). Notice that when employing polarograms produced with a chart recorder, a line is drawn through the midpoints of the recorder oscillations. For subsequent electroactive substances, the diffusion current would be evaluated by subtracting both the residual current and all preceding diffusion currents.

It is simpler, though less exact, to extrapolate the part of the residual current curve that precedes the initial rise of the wave; this is done by drawing a line parallel to it through the diffusion current plateau, as shown in Figure 13.15(b). For succeeding waves, the

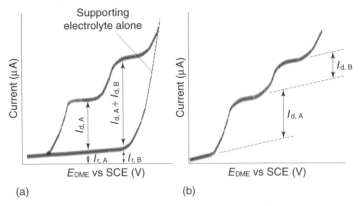

Current (μA) — $I_{d,A}$ — $I_{d,A} + I_{d,B}$ — Supporting electrolyte alone — $I_{r,A}$ — $I_{r,B}$ — E_{DME} vs SCE (V)

(a)

Current (μA) — $I_{d,A}$ — $I_{d,B}$ — E_{DME} vs SCE (V)

(b)

Figure 13.15 How to measure polarographic wave height

diffusion current plateau of the preceding wave is used as a pseudo-residual current curve. At low concentrations the saw-tooth fluctuations as each mercury drop is formed and then drops is larger, so it becomes very difficult to make accurate measurements of these curves.

Classical d.c. polarography

The essential requirements for producing polarographic current–voltage curves are a means of applying a variable known d.c. voltage of about 0–3 V to the electrolysis cell, a working DME and counter-electrode in contact with the analyte solution, and a method for recording the resultant current. The apparatus in Figure 13.11 is a manual polarograph; it can be used to study the basic techniques of polarography. Commercial polarographs normally perform the voltage scan automatically while a chart recorder plots the current–voltage curve. A countercurrent control applies a small opposing current to the cell which can be adjusted to compensate for the residual current; this leads to polarograms which are better defined. Most of these instruments also incorporate circuits which allow alternative, more sensitive types of polarography (Section 13.34).

A useful feature of many simple polarographs is the facility of plotting derivative polarograms, i.e. curves obtained by plotting dI/dE against E. These curves show a peak at the half-wave potential, and by measuring the height of the peak it is possible to obtain quantitative data on the reducible substance; the height of the peak is proportional to the concentration of the ion being discharged. Figure 13.16 shows a typical conventional polarogram for $0.003\,M$ cadmium sulphate in $1\,M$ potassium chloride, in the presence of 0.001% gelatin, and the corresponding derivative curve (I_{max} is the maximum current recorded in the derivative mode).

A derivative plot can be used to measure half-wave potentials which are closer than $150\,mV$; this is not possible with a normal polarogram owing to the disturbing effect of the diffusion current of the first ion to be discharged on the second step in the polarogram. When the element with the lower half-wave potential is present in much higher concentration, e.g. determining copper for cadmium content, it is almost impossible to interpret a conventional polarogram without previous chemical separation; but a derivative polarogram will have a series of peaks approximately in the positions of the half-wave potentials, and these peaks will allow the individual elements to be quantified. Figure 13.17(a) illustrates the polarogram obtained with copper and cadmium ions in the ratio 40 : 1, and Figure 13.17(b) shows the corresponding derivative polarogram where the two peaks are clearly visible.

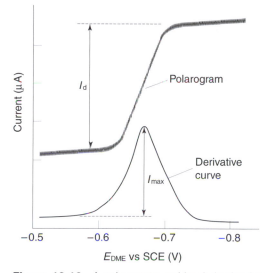

Figure 13.16   A polarogram and its derivative for $0.003\,M$ cadmium sulphate in $1\,M$ potassium chloride plus 0.001% gelatin

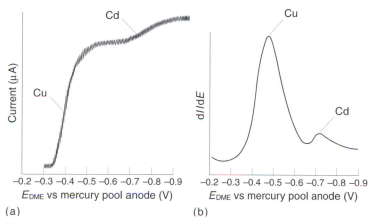

Figure 13.17   Determining copper for cadmium content: (a) polarogram, (b) derivative

13.33   The three-electrode polarograph: potentiostatic control

Many modern polarographs provide for potentiostatic control of the dropping electrode potential; this is particularly valuable when solutions of high resistance are involved, e.g. when using non-aqueous solvents (or mixtures of water and organic solvents). A high resistance in the polarographic cell leads to a large Ohm's law voltage drop (IR) across the cell, which not only influences the measured electrode potential but may also distort the polarogram; in extreme cases, with some non-aqueous solvents, a straightforward I–E plot may appear to be virtually a straight line, and only after correction of the ohmic voltage drop is a normal polarogram obtained. Potentiostatic control requires the introduction of a third electrode (a counter-electrode) into the polarographic cell.

The reference electrode must be sited as close as possible to the DME so that the resistance of the solution between the two electrodes is reduced to a minimum, and the potentiostat then maintains the e.m.f. of the DME/reference electrode combination at the correct value. This arrangement has the further advantage that virtually no current passes through the reference electrode, hence there is no possibility of polarising the reference electrode with consequent variation in potential.

The dropping mercury electrode

Although some polarographic measurements are still made using simple capillary electrodes, it is usually advantageous to replace the dropping mercury electrode by a hanging mercury drop electrode (HMDE) or a static mercury drop electrode (SMDE), in which a solenoid-controlled plunger produces drops of mercury at the tip of a rather wider capillary than is normally used in the DME. The solenoid is adjusted so that development of the mercury drop ceases before the point at which the drop would fall under its own weight (Figure 13.18). This establishes an electrode surface of fixed area, so that once the surface has been charged by the condenser current, it is only the diffusion current which continues to flow, and this can be measured with greater accuracy than is possible with the normal dropping electrode. At predetermined intervals the drop is dislodged from the capillary by a drop knocker, and a new mercury drop is then dispensed so a further reading can be made.

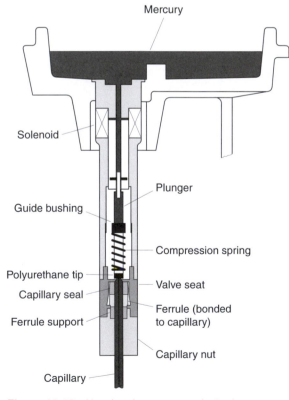

Figure 13.18 Hanging drop mercury electrode

A number of systems are commercially available.[33] They have made the mechanical aspects of polarography much more reliable and versatile than traditional gravity-operated capillary systems. Most allow the operator to determine the drop size and drop frequency – the drops are dispensed very rapidly compared to the slow growth and decay of the traditional DME – and most will also allow the production of a single stable and static drop for experiments such as stripping voltammetry (Section 13.43).

Polarographic cells

HMDE and SMDE equipment has the advantage that all the components required to perform a polarographic experiment (mercury electrode, reference electrode, counter-electrode and a degassing system) are mounted onto the main assembly. Only the analyte solution is required, held in a very simple vessel which can be attached or removed from the main assembly quickly and easily; this allows a number of different solutions to be measured. However, specially constructed cells may be used for some applications, such as the H-type cell devised by Lingane and Laitinen (Figure 13.19). A particular feature is the built-in reference electrode. A saturated calomel electrode is usually employed; but if the presence of chloride ion is harmful, a mercury(I) sulphate electrode may be used (Hg/Hg_2SO_4 in potassium sulphate solution, potential $\sim+0.40$ V relative to SCE). It is usually designed to contain 10–50 mL of the sample solution in the left-hand compartment, but it can be constructed to accommodate a smaller volume down to 1–2 mL. To avoid polarisation of the reference electrode, it should be made of tubing at least 20 mm in diameter, but the dimensions of the solution compartment can be varied over wide limits.

The compartments are separated by a cross member filled with a 4% agar-saturated potassium chloride gel, which is held in position by a medium-porosity sintered Pyrex glass disc (diameter at least 10 mm) placed as near the solution compartment as possible in order to facilitate deaeration of the test solution. By clamping the cell so that the cross member is vertical, the molten agar gel is pipetted into the cross member and the cell is allowed to stand undisturbed until the gel has solidified.

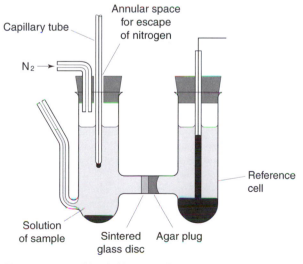

Figure 13.19   Simple H-type cell

In use, the solution compartment (either dried by aspiration of air through it or rinsed with several portions of the test solution) is charged with at least enough test solution to cover the entire sintered-glass disc. Dissolved air is removed by bubbling pure nitrogen through the solution via the side arm; by means of a two-way tap in the gas stream, the gas is then diverted over the surface of the solution. Measurements should not be attempted while gas is bubbling through the solution, for the stirring causes high and erratic currents. Finally, the dropping electrode is inserted through another hole in the stopper (which should be large enough for ease of insertion and removal of the capillary) and the measurements are made. When the H-cell is not in use, the left-hand compartment should be kept filled with water or with saturated potassium chloride solution (or other electrolyte appropriate to the reference electrode being used) to prevent the agar plug from drying out.

When potentiometric control is exercised, an additional electrode (counter-electrode) must be used, and under these conditions a suitably sized four-necked flask makes a convenient electrolysis vessel.

13.34 Modified voltammetry

Normal d.c. polarography gives perfectly adequate results at moderate concentrations, but it has limited application below about $10^{-5} M$ (~1 µg g^{-1}) due to the large rhythmic fluctuations seen in the trace as mercury drops continuously form and then fall. Also, the half-wave reduction potential difference between two ions must be at least 200 mV if the reduction waves are to be separated. These limitations are largely due to the condenser current associated with the charging of each mercury drop as it forms, and various procedures have been devised to overcome this problem. All can be performed using the same electrode stand described in the previous section, provided suitable electrical or electronic control is applied to the electrode systems.

Rapid scan polarography

In this technique the applied potential is swept rapidly through a range of up to 2 V during part of the lifetime of a single mercury drop. Typically the voltage sweep occurs during the last 2 s of the lifetime of a mercury drop which has a drop time of about 6 s. The resulting current–voltage curve has a peak in it reminiscent of a derivative polarogram. The peak potential is related to the half-wave potential of the ion being discharged by the expression

$$E_s = E_{1/2} = 1.1RT/nF$$

The peak current is greater than the diffusion current recorded with a conventional d.c. polarograph by a factor of 10 or even more. The method thus shows enhanced sensitivity and it can be used to make measurements with solutions having concentrations as low as 10^{-6} to $10^{-7} M$ and with a resolution of the order of 50 mV. An HMDE is normally used, although it is possible to use platinum, graphite or glassy carbon electrodes, in which case the procedure should be called voltammetry rather than polarography.

Sinusoidal a.c. polarography

In this procedure a constant sine wave a.c. potential of a few millivolts is superimposed upon a d.c. potential sweep. The applied d.c. potential is measured in the usual way and

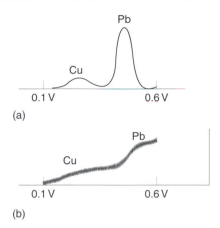

(a)

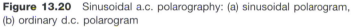

(b)

Figure 13.20   Sinusoidal a.c. polarography: (a) sinusoidal polarogram,
(b) ordinary d.c. polarogram

these results are coupled with measurements of the alternating current. If the values of the
a.c. current are plotted against the potential applied by the potentiometer, a series of peaks
are obtained as illustrated in Figure 13.20(a); the normal d.c. polarogram of the same
solution is shown in Figure 13.20(b).

The a.c. curve is similar in character to a derivative polarogram (Figure 13.17) but must
not be confused with this type of curve. Each peak in the a.c. curve corresponds to a step
in the normal polarographic record. The voltage of the peak is the same as the voltage of
the midpoint of the step, and the height of a peak above the baseline is proportional to the
concentration of the depolariser, hence it corresponds to the step height. With closely
separated waves, measurements are much more readily made from the a.c. polarogram than
from the d.c. polarogram; it is considered that peaks separated by 40 mV can be resolved
on a a.c. polarogram compared with the separation of 200 mV required on a d.c. polarogram. Because the residual current is rather large, the limit of sensitivity for a.c. polarography ($10^{-5}M$) is not very different from the limit for d.c. polarography. The large residual
current arises because the condenser current is relatively large compared with the diffusion
or faradaic current, so the curve is similar to a rapid scan polarogram.

Pulse polarography

Barker and Jenkins[34] attempted to solve the problem of a varying (sawtooth) baseline due
to the charging or capacitance current induced as the mercury drops grow and become
dislodged. They applied the polarising current in a series of pulses during the lifetime of
a single drop. Several different modifications are used.

Normal pulse polarography   In its simplest form, the potential of the DME is held
at a constant initial potential for most of the lifetime of the drop, but during the last
50–60 ms of its life a pulse of higher potential is applied to the drop. The current is then
measured during the last 20 ms of the drop life. The potential applied to each successive
drop is gradually increased to give the required voltage scan. Using this method, the onset
of the pulse is marked at first by a sudden rise in the total current passing; this is largely

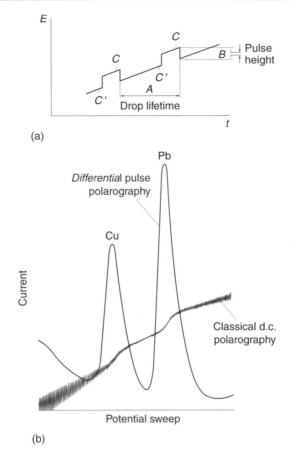

(a)

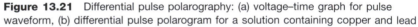

(b)

Figure 13.21   Differential pulse polarography: (a) voltage–time graph for pulse waveform, (b) differential pulse polarogram for a solution containing copper and lead

due to the condenser (charging) current which soon decays to zero, i.e. during the first 20 ms or so of the pulse. The current measured in the final stages of the drop lifetime is thus almost completely due to the faradaic process, which is proportional to analyte concentration. The resulting polarogram is similar to a conventional d.c. polarogram except that the characteristic sawtooth pattern of the conventional polarogram is replaced by a stepped curve.

Differential pulse polarography (DPP)   Widely used for many polarographic experiments, an even better approach is possible if modern electronic systems are available; the applied potential is varied with time as shown in Figure 13.21. The sloping line, periodically interrupted by the pulses of pulse height B (magnitude 5–100 mV), represents the normal steadily increasing d.c. voltage; this is normally scanned at between 1 and $10\,\mathrm{mV\,s}^{-1}$. The interval A between the termination of two succeeding mercury drops represents the drop time. The current is measured **twice** during the lifetime of each mercury drop: once just before the application of the pulse (points corresponding to C' in Figure 13.21) as well as at the usual point C near the end of the pulse. The current at C' is the current that would

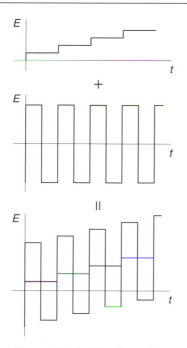

Figure 13.22   Waveforms for square wave pulse polarography

be observed in normal d.c. polarography; its value is stored in the polarograph. The onset of the pulse is then marked by a sudden rise in current; as in normal pulse polarography, the current soon settles down as the condenser current decays, and the current is read again near the end of the pulse (point C). This value is then compared with the value stored in the instrument (the value for point C'), and the difference between them is amplified and recorded.

If the measurements of current are made on the residual current curve, or on a plateau of the d.c. polarogram, the difference in current between points C and C' will be small; but if the measurements are made on a polarographic wave, an appreciable difference in current will be recorded, and it will reach a maximum value when the applied d.c. potential is equal to the half-wave potential. The difference in current plotted against the applied d.c. potential will therefore be a peaked curve with the height of the peak directly proportional to the concentration of the reducible substance in the solution.

Differential pulse polarography is a very satisfactory and widely used method for the determination of many substances with a detection limit of about $10^{-8}M$, and a resolution of about 50 mV. An important feature of pulse polarography is the sampling of the current at definite points in the lifetime of the mercury drop, and it is essential to establish an exact timing procedure. This is normally done using the HMDE system described earlier, where mechanical tapping of the capillary is used to dislodge the mercury drop at precise time intervals.

Square wave polarography   Figure 13.22 shows the waveforms used for square wave polarography, a modification of pulse polarography that is beginning to gain wide acceptance. A symmetrical square wave is superimposed on a staircase waveform where the forward pulse of the square wave is coincident with the staircase step in time and polarity, whereas the reverse cycle of the square wave occurs halfway through the staircase step.[35] The current is sampled twice during each square wave cycle, once at the end of the forward pulse and again at the end of the reverse pulse. Again the technique reduces the effect of capacitative currents by only measuring the current at the end of the pulses. Just as for DPP, the difference current is plotted against scan potential and gives a peak proportional to the concentration of electroactive species.

The scan rate (mV s^{-1}) in this technique can be varied over wide limits, where

$$\text{scan rate} = \frac{E_{step}}{\tau}$$

E_{step} (mV) is the staircase step size and τ (s) is the time for one cycle of the square wave ($1/\tau$ = frequency). Typical values are $E_{step} = 2$ mV and $1/\tau = 100$ Hz, giving a scan rate of 200 mV s^{-1}. These very fast scan rates, up to 100 times faster than conventional DPP, mean that complete analysis of a solution can be performed using a single drop, and this can considerably speed up sample throughput or allow signal averaging of multiple cycles to decrease signal-to-noise ratios with improved detection limits and precision.

Because of its speed, square wave voltammetry can be used as a rapid detector for HPLC and to study electrode kinetics by monitoring the variation of peak height for a given solution concentration as τ is varied. To maximise its effectiveness requires reliable and controllable mercury drop characteristics plus suitable fast and stable electronic control, both of which are now available in modern instrumentation.

Cyclic voltammetry

Rapid scan, sinusoidal and pulse polarography superimpose various waveforms on the main voltage, but the main voltage scan has been in one direction. In cyclic voltammetry a fast or very fast scan is carried out in two directions, i.e. from 0 to V and back down to 0 (Figure 13.23(a)). Normally this process is conducted using electrodes with a small surface area in unstirred solutions, producing a very small redox current. This means the ohmic drop IR is small, even for poorly conducting solutions, and the low capacitance allows fast scan rates of up to few $\times$ 10 kV s^{-1} in suitable cases, completely eliminating any contribution due to diffusion. Since the scan is in two directions, two curves are normally seen; a 'normal' cathodic reduction wave and an anodic or oxidation wave as the voltage reverses (Figure 13.23(b)). The curves shown in Figure 13.23(b) are equal in magnitude and approximately vertically aligned. This indicates that the electrode process is essentially a fast, reversible reaction. The peak-to-peak separation ΔE_p (mV) is in fact directly related to the number of electrons transferred in the reversible reaction:

$$\Delta E_p = \frac{2.22RT}{nF} = \frac{57.0}{n}$$

But if the kinetics of the reverse reaction are slow, the anodic wave moves further away from the cathodic wave, eventually disappearing for non-reversible reactions. This provides an elegant method for investigating the kinetics of quite fast reactions and often allows a determination of the number of species involved in a reaction sequence.

Cyclic voltammetry does have several limitations. The electronic equipment used for producing the rapidly changing waveform and for recording the resulting time–current

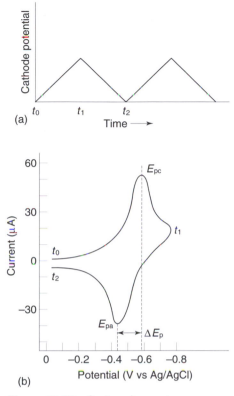

Figure 13.23   Cyclic voltammetry

waves must be capable of responding fast enough to display the correct waveform. Also, the magnitude of the waves produced depends only partially on the concentration of redox species involved and partly on other factors such as scan rate; the faster the scan, the larger the wave for a given solution. Cyclic voltammetry essentially provides qualitative information on reaction speeds and reaction mechanisms, and as such it is a valuable diagnostic tool which is often the first of the electrochemical techniques applied to a new system. However, for accurate quantitative measurements, one of the pulsed techniques will normally give more reliable data.

13.35   Quantitative applications of polarography

One great advantage of polarographic techniques over other forms of analysis is that they may be applied to a very wide range of sample types; the only requirement is that a measurable electrochemical reduction or oxidation takes place in the solution in a reasonable time. This means that the technique is likely to be suitable for most metals and anions and many organic species. Table 13.9 shows which analytes may be determined by voltammetry and also by stripping techniques. This versatility means that only some specific applications can be discussed here, but extensive listings of polarographic applications are available.

Table 13.9   *Periodic table showing elements which can be determined from voltammetry*

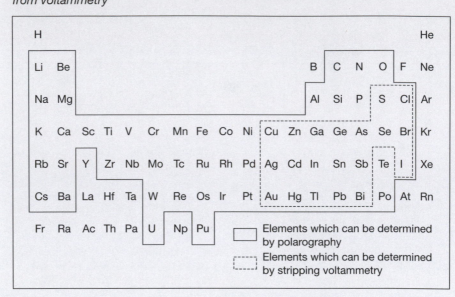

Inorganic applications

Many inorganic anions and cations, and some molecules can be determined polarographically. Cathodic reduction waves are particularly valuable for the determination of transition metal cations. The ease of determination frequently depends on selection of the appropriate supporting electrolyte. For example, with potassium chloride as supporting electrolyte the waves for copper(II) and iron(III) interfere with each other; but with potassium fluoride as supporting electrolyte, the iron(III) forms a complex with fluoride ions and the iron(III) wave becomes 50 mV more negative whereas the copper wave is practically unaffected, thus interference is no longer experienced. The choice of supporting electrolyte for a given determination is often the most critical step in the analysis, but again the literature will usually provide at least a starting point for this choice.

The polarographic determination of metal ions which are readily hydrolysed, such as Al^{3+}, can present problems in aqueous solution, but they can often be overcome by using non-aqueous solvents. Typical non-aqueous solvents, with appropriate supporting electrolytes shown in parentheses, include ethanoic acid (CH_3CO_2Na), methyl cyanide ($LiClO_4$), dimethylformamide (tetrabutylammonium perchlorate), methanol (KCN or KOH), and pyridine (tetraethylammonium perchlorate). In these media the dropping mercury electrode is normally replaced with a platinum microelectrode.

The polarographic method may also be used for the determination of inorganic anions such as cyanide, bromate, iodate, dichromate and vanadate. Hydrogen ions are involved in many of these reduction processes, and the supporting electrolyte must therefore be adequately buffered. Although not determined directly, nitrate and several sulphur species may also be measured at environmentally significant concentrations by appropriate choice of method (Section 13.42). Typical applications in the inorganic field are the analysis of minerals, metals (including alloys), plating solutions, foods and beverages, fertilisers, cosmetics and drugs, natural waters, industrial effluents and polluted atmospheres. The technique can also be used to establish the formulae of various complexes.

Organic applications

Provided the organic material can undergo a redox reaction, then in theory it may be determined using polarographic techniques. Many organic functional groups will undergo reduction or oxidation at a dropping electrode and may therefore be determined, although the more stable functional groups such as alkanes are not suitable for this technique. Remember that, in general, the reactions of organic compounds at the dropping electrode are slower and they are often more complex than for inorganic ions. Nevertheless, polarographic investigations can be useful for structure determination as well as for qualitative and quantitative analysis. Table 13.10 lists the functional groups that can often be determined using polarographic techniques. Reactions of organic substances at the dropping electrode usually involve hydrogen ions, a typical reaction can be represented by the equation

$$R + nH^+ + ne \rightleftharpoons RH_n$$

where RH_n is the reduced form of the reducible organic compound R. As hydrogen ions (supplied from the solution) are involved in the reaction, the supporting electrolyte must be well buffered. Change in the pH of the supporting electrolyte may even lead to the formation of different reaction products. Thus, in slightly alkaline solution, benzaldehyde is reduced at $-1.4\,V$ with formation of benzyl alcohol; but in acid solution (pH < 2), reduction takes place at $-1.0\,V$ with formation of hydrobenzoin:

$$2C_6H_5CHO + 2H^+ + 2e \rightleftharpoons C_6H_5CH(OH)CH(OH)C_6H_5$$

Many complex organic materials found in biological systems or foodstuffs can be easily and reliably determined in aqueous solution using an anodic scan rather than the more usual cathodic conditions appropriate for metals. This is easily performed provided the positive potential is not allowed to exceed 400 mV, where mercury metal is oxidised to Hg(I). Several vitamins, including vitamin C and the E group, can be determined in a wide range of matrices (Section 13.41).

Some organic compounds are best investigated in mixed solvent systems (aqueous and organic) which improve the solubility. Suitable water-miscible solvents include ethanol, methanol, ethane-1,2-diol, dioxan, methyl cyanide and ethanoic acid. Sometimes a purely organic solvent must be used, and anhydrous materials such as ethanoic acid, formamide and diethylamine have been employed; suitable supporting electrolytes in these solvents include lithium perchlorate and tetra-alkylammonium salts R_4NX (R = ethyl or butyl; X = iodide or perchlorate).

Polarographic methods can be used to examine pure organic chemicals, organics in food and food products, biological materials, herbicides, insecticides and pesticides, petroleum and petroleum products, polymers and pharmaceuticals. Blood and urine samples are frequently examined to establish the presence of drugs and to obtain quantitative results. Polarographic detectors are also finding increasing application in monitoring organic effluents from HPLC chromatography systems.

In all polarographic experiments, but especially organic analyses, when information regarding the polarographic behaviour of a substance is not available, remember that other factors may influence the observed current, factors besides the normal polarographic wave associated with reduction and diffusion of the species under investigation. These include the following:

13 Direct electroanalytical methods

Table 13.10 *Electroactive organic groups*

Acetylenes	$-C\equiv C-C=C\diagup$
Aldehydes	$R-\overset{\displaystyle O}{\overset{\|}{C}}-H$
Conjugated aromatics	$\text{Ph}-C=C\diagup$, $\text{Ph}-C\equiv C-$
Conjugated carboxylic acids	$HOOC-C=C-COOH$, $HOOC-COOH$
Diazo compounds	$-\overset{+}{N}\equiv N$
Dienes: conjugated double bonds	$\diagup C=C-C=C\diagup$
Disulphides	$-S-S-$
Halides	$-CO-\overset{\|}{C}-Cl$, $\diagup C-Cl$, $\diagup C=C-Br$
Heterocycles	(thiazole), (pyridine), (furanone)
Ketones	$R-\overset{\displaystyle O}{\overset{\|}{C}}-R$
Nitriles	$-CN-\overset{\|}{C}-\overset{+}{N}\diagup$, $CO-\overset{\|}{C}-\overset{+}{N}\diagup$
Nitro compounds	$-NO_2$, $\diagup C=C-NO_2$, $\text{Ph}-NO_2$
Nitroso compounds	$\diagdown N-N=O$
Peroxides	$-O-O-$
Quinones	$O=\bigcirc=O$
Sulphides	$-CO-CH_2-S-$

Kinetic currents   Kinetic currents occur when the rate of a chemical reaction exerts a controlling influence. Thus, a species S may not itself be electroactive, but it may be capable of conversion into a substance O which undergoes reduction at the dropping electrode:

$$S \rightleftharpoons O + ne \rightleftharpoons R$$

If the rate of formation of O is slower than the rate of diffusion, the diffusion current will be controlled by the rate at which O is formed.

Catalytic currents   Catalytic currents are observed when the product of reduction at the electrode is reconverted to the original species by interaction with another substance in solution which acts as an oxidising agent:

$$S + ne \rightleftharpoons R \qquad R + X \rightleftharpoons S$$

This is observed with a solution containing Fe(III) ions and hydrogen peroxide. Fe(II) ions formed at the electrode are converted back to Fe(III) ions by the hydrogen peroxide; it produces an enhanced diffusion wave which makes quantification difficult.

Adsorption currents   If either the oxidised or the reduced form of an electroactive substance is adsorbed on the surface of the electrode, the behaviour of the system is altered. If the oxidised form is adsorbed, reduction takes place at a more negative potential; if the reduced form is adsorbed, then the reduction is favoured and a 'prewave' is observed in the polarogram. When the electrode surface is completely covered by reductant, the current is controlled by a normal diffusion process and a normal polarographic wave is obtained.

Polarography experiments

! **Safety.** Before carrying out any experiments in this section, pay full attention to any safety warnings and make sure you adhere to national safety regulations.

13.36   Polarography experiments: introduction

Polarographic measurements may be made on a wide range of sample types to determine both the nature and amount of the electroactive material present. The following experiments have been chosen to illustrate some of the methods which can be used. Other applications may be found in the literature or in the detailed briefings which can usually be obtained free of charge from instrument manufacturers.[36] The first two experiments use a conventional dropping mercury electrode which can be assembled in the laboratory without special equipment, but all the other experiments require an HMDE with electronic control of the potentials.

13.37   Half-wave potential of the cadmium ion in 1 *M* KCl

Introduction   This experiment can be performed with a manual polarograph; it illustrates the general procedure for d.c. polarography. If a commercial polarograph which includes a potentiostat is employed, then the three-electrode procedure is conveniently used with the controlled potential applied between the dropping electrode and the calomel reference electrode, while the electrolysis current flows between the working (mercury) electrode and the auxiliary platinum electrode. An apparatus similar to Figure 13.11, using an H-type cell (Figure 13.19), will be satisfactory.

Procedure   Make sure the reservoir of the dropping electrode contains an adequate supply of mercury, and that mercury drops freely from the capillary when the tip is immersed in distilled water while the reservoir is raised to near the maximum height of the stand; allow the mercury to drop for 5–10 min. Replace the beaker of water by one containing $1 M$ potassium chloride solution and adjust the rate of dropping by varying the height of the mercury reservoir until the dropping rate is 20–24 per minute; then note the height of the mercury column. When the adjustment has been completed, rinse the capillary well with a stream of distilled water from a wash bottle then dry by blotting with filter paper. Insert the capillary through an inverted cone of quantitative filter paper and clamp vertically over a small beaker. Lower the levelling bulb until the mercury drops just cease to flow. Pipette 10 mL of a cadmium sulphate solution ($1.0 \, \mathrm{g \, L^{-1} \, Cd^{2+}}$) into a 100 mL graduated flask, add 2.5 mL of 0.2% gelatin solution, 50 mL of $2 M$ potassium chloride solution and dilute to the mark. The resulting solution (A) will contain $0.100 \, \mathrm{g \, L^{-1} \, Cd^{2+}}$ in a base solution (supporting electrolyte) of $1 M$ potassium chloride with 0.005% gelatin solution as suppressor.

Measurements   Place 5.0 mL of solution A in a polarographic cell equipped with an external reference electrode, a saturated calomel electrode (SCE). Pass pure nitrogen through the solution at a rate of about 2 bubbles per second for 10–15 min in order to remove dissolved oxygen. Raise the mercury reservoir to the previously determined height and insert the capillary into the cell so that the capillary tip is immersed in the solution. Connect the SCE to the positive terminal and the mercury in the reservoir to the negative terminal of the polarograph.

After about 15 min stop the passage of inert gas through the solution and wait 1 min for the solution to become stationary; the electrical measurements may now be commenced. Carry out a preliminary test, adjusting the sensitivity of the instrument so the trace on the recorder uses as much of the width of the chart paper as possible at the maximum applied potential; this must not exceed the decomposition potential of the supporting electrolyte.

Now commence the voltage sweep using a scan rate of $5 \, \mathrm{mV \, s^{-1}}$, or with a manual polarograph, increase the voltage in steps of 0.05 V; the recorder plot will take the form shown in Figure 13.12. Since the current oscillates as mercury drops grow and then fall away, the plot will have a sawtooth appearance, so a smooth curve must be drawn through the midpoint of the peaks of the plot. When the experiment has been completed, clean the capillary as described above then store it by inserting through a bored cork (or silicone rubber bung, normal rubber bungs which contain sulphur must be avoided) which is then placed in a test tube containing a little pure mercury. Lower the mercury reservoir until drops no longer issue from the capillary, then push the end of the capillary into the mercury pool. The mercury at the bottom of the electrolysis vessel must be carefully recovered, washed by shaking with distilled water, then stored under water in a recovered mercury bottle.

Determine the half-wave potential from the current–voltage curve as described in Section 13.27. The value in $1 M$ potassium chloride should be about -0.60 V relative to the

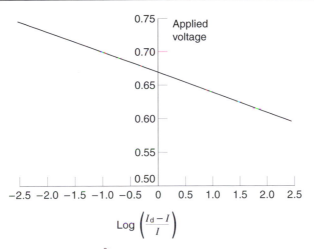

Figure 13.24 Cd^{2+} ions in $1\,M$ KCl: the graph of applied potential versus $\log[(I_d - I)/I]$ should have gradient $\approx -0.030\,V$ and its intercept on the vertical axis will give the half-wave potential of the Cd^{2+} ions relative to the SCE

SCE. Measure the maximum height of the diffusion wave after correction has been made for the residual current; this is the diffusion current I_d and is proportional to the total concentration of cadmium ions in the solution. Measure the height of the diffusion wave I, after correcting for the residual current at each increment of the applied voltage. Plot the applied voltage against $\log[(I_d - I)/I]$ as shown in Figure 13.24 (strictly speaking, the values of the applied voltage are negative). The graph should be a straight line with slope $-\sim 0.030\,V$; the intercept on the voltage axis is the desired half-wave potential of the cadmium ion relative to the SCE. As an additional exercise, the current–voltage curve of the supporting electrolyte ($1\,M$ potassium chloride) may be evaluated; this gives the residual current directly and no extrapolation is required for the determination of I and I_d.

13.38   Investigating the influence of dissolved oxygen

Introduction   The solubility of oxygen in water at the ordinary laboratory temperature is about $8\,mg\,L^{-1}$ (or $2.5 \times 10^{-4}\,mol\,L^{-1}$). Oxygen gives two polarographic waves ($O_2 \rightarrow H_2O_2 \rightarrow H_2O$) which occupy a considerable voltage range, and their positions depend upon the pH of the solution. Unless the test solution contains a substance which yields a large wave or waves compared with which those due to oxygen are negligible, dissolved oxygen will interfere. Particularly in dilute solution, dissolved oxygen must be removed by passing pure nitrogen or hydrogen through the solution.

Procedure   Place some $1\,M$ potassium chloride solution containing 0.005% gelatin in a polarographic cell immersed in a thermostat. Make the usual preliminary adjustments with regard to sensitivity of the recorder, then record the current–applied voltage curve. Now pass oxygen-free nitrogen through the solution for 10–15 min. Record the polarogram using the same sensitivity. Notice that the two oxygen waves are absent in the new polarogram (see Figure 13.14).

13.39 Copper and zinc in tap water using DPP

Introduction Most tap water mains contain sufficient copper and zinc for them to be determined directly using DPP, and since the sample contains very little organic or other interfering species, the experiment may be conducted without any sample preparation other than the addition of a supporting electrolyte to reduce the resistivity of the solution. The experiment may be conducted by comparing the response of the sample against a calibration curve prepared from standard solutions of the two metals or it may use the method of standard additions made directly to the sample. Here are some typical conditions (f.s.d. is full scale deflection):

Drop time	1 s
Starting potential	+0.1 V (vs SCE)
Scan range	1.4 V negative
Scan rate	$2\,mV\,s^{-1}$
Pulse modulation amplitude	50 mV
Current range	$2\,\mu A$ f.s.d.

Depending upon the exact equipment used, other parameters such as drop size and display direction, or various electronic filters may also need to be specified, but the exact parameters for this experiment are not normally critical, and default values may often be used.

Method 1 Prepare solutions which contain both Zn^{2+} and Cu^{2+} in the range 0.5 to $20\,\mu g\,g^{-1}$ in a supporting electrolyte of 0.01 M KCl. (A number of other supporting electrolyte solutions could also be used, such as an ethanoate or citrate buffer at 0.05 to 0.2 M, but in each case the material must be low in heavy metals.) Place a sample of the lowest concentration solution in the polarographic cell and degas for the recommended time, usually about 2–3 min; wait 1 min for the solution to become stationary then record the DPP trace using the conditions above. The copper peak may well be on the side of a larger peak at positive potentials but should be fairly well resolved, whereas the zinc peak should be completely symmetrical.

If the peak heights are satisfactory, record traces for each of the other standard solutions in the same way and construct a calibration curve of peak height against concentration for the two metals (Figure 13.25). If the peaks are too large or small, adjust the sensitivity of the apparatus so that all the traces can be conveniently observed. Obtain traces for the tap water sample using the same conditions and with the addition of sufficient 0.1 M KCl solution to the sample to give the same final concentration of supporting electrolyte as the standard solutions.

Measure the peak heights for both copper and zinc in this solution, and by comparison against the standard curve, estimate the concentration of the two metals in the tap water. When measuring these two metals in domestic water supplies, it is often good practice to compare water which has stood in the pipes overnight with water from the same source after it has been running for some time.

Method 2 This determination can also be conveniently carried out using the method of standard additions; here a known volume of the tap water sample, plus supporting electrolyte is analysed first and then known additions of both copper and zinc are added to this solution. Many commercial systems use cells of between 5 and 20 mL, and the required amount of addition is readily calculated. In the experiment shown in Figure 13.25 the cell had a volume of 5 mL and the expected concentration was $Cu = 5\,\mu g\,g^{-1}$ and $Zn = 1\,\mu g\,g^{-1}$.

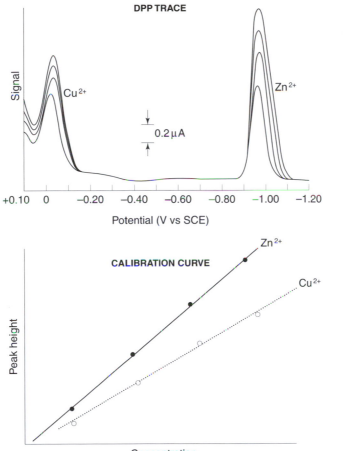

Figure 13.25   Copper and zinc in tap water: obtain the DPP trace for each of the standard solutions, measure the copper and zinc peak heights then transfer them to the calibration graph of peak height against concentration

An addition of 25 μL of $1000\,\mu g\,g^{-1}$ Cu solution adds 25 μg to the solution, which should double the peak height for copper. An addition of 25 μL of $200\,\mu g\,g^{-1}$ Zn solution will give the same result for the zinc peak. By adding small volumes of relatively concentrated solutions the dilution effect of the addition can be ignored. The levels of both of these metals vary considerably in water from different sources and it may be necessary to vary the sensitivity of the instrument and the level of the additions to reflect the actual levels in a given sample; if very low values are found, then stripping analysis (Section 13.43) may have to be used to give the required sensitivity.

Although designed to illustrate the simplicity of DPP measurements, this procedure will only give satisfactory results if all solutions are made up in the purest water available using reagents low in heavy metals and with apparatus which has been precleaned to remove contamination. The experiment can be used as a guide to a large number of similar metal ion determinations using DPP, where the sample pretreatment may be slightly more complex, such as wet or dry ashing of the sample prior to analysis to remove interferences. The supporting electrolyte may also be chosen to provide a suitable environment for the experiment, such as a stable pH, or to provide some complexing action on one or more ions in the sample. Literature available from several instrument manufacturers provides a good starting source for many determinations of this type.[37]

13.40   Copper and zinc in tap water using square wave polarography

With apparatus that can perform square wave polarography, the tap water experiment may be conducted very much more quickly, allowing increased sample throughput. Typical experimental conditions are:

Drop time	single hanging drop
Scan increment (E_{step})	2 mV
Square wave frequency	100 Hz
Starting potential	+0.1 V (vs SCE)
Scan range	1.4 V negative
Scan rate	200 mV s^{-1}
Pulse modulation amplitude	50 mV
Current range	2 µA f.s.d.

Clearly this experiment will take 7 s per scan compared with 700 s or nearly 12 min for the standard DPP scan.

13.41   Ascorbic acid (vitamin C) in fruit juice

Introduction   This experiment illustrates the use of DPP in the anodic mode, i.e. scanning in a positive direction to observe an oxidation rather than the more normal cathodic reduction process. The sample is a water-soluble organic compound. The supporting electrolyte used in this experiment can either be an ethanoate buffer at pH = 3 prepared by mixing equal volumes of 0.05 M ethanoic acid and 0.01 M $NaNO_3$ or preferably Britton–Robinson buffer at pH = 2.87. This is prepared by mixing equal volumes of 0.04 M ethanoic acid, 0.04 M phosphoric acid and 0.04 M boric acid. The pH is adjusted to 2.87 by adding approximately 17.5 mL of 0.2 M NaOH to each 100 mL of the solution. These solutions should be degassed to remove dissolved oxygen before addition of ascorbic acid, since the ascorbic acid is a relatively strong reducing agent and will tend to reduce dissolved oxygen. The low pH of these solutions tends to inhibit this process.

The juice of many fruits and particularly those of the citrus family contain appreciable quantities of ascorbic acid (vitamin C), and by simple dilution to reduce interferences from surface-active agents in the juice it is possible to determine the levels of vitamin C, even in highly coloured solutions which could not be examined spectroscopically or by conventional titrimetric methods. The experiment can either use a comparison against a calibration curve constructed from measurements on standard solutions or it can use the method of standard additions.

Procedure   For the construction of a calibration curve, a stock solution of ascorbic acid is prepared by dissolving 30 mg ascorbic acid in 100 mL of supporting electrolyte (300 µg g^{-1}). Successive 20 µL aliquots of this stock are then added to 10.0 mL of degassed supporting electrolyte in the polarographic cell and determined using the following conditions:

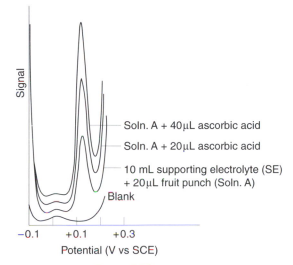

Soln. A + 40 μL ascorbic acid

Soln. A + 20 μL ascorbic acid

10 mL supporting electrolyte (SE)
+ 20 μL fruit punch (Soln. A)

Blank

Figure 13.26   Typical DPP traces for ascorbic acid in fruit juice: drop time = 1 s, starting potential = −0.1 V (vs SCE), scan range = 0.4 V (positive), scan rate = 2 mV s^{-1}, pulse modulation amplitude = 25 mV, current range = μA (f.s.d.)

Drop time	1 s
Starting potential	−0.1 V (vs SCE)
Scan range	0.4 V positive
Scan rate	2 mV s^{-1}
Pulse modulation amplitude	25 mV
Current range	1–2 μA f.s.d.

A 20–40 μL aliquot of fruit juice (Solution A) is then added to a fresh, degassed 10.0 mL portion of supporting electrolyte in the cell and the polarographic trace determined as before. Since most commercially produced fruit juices contain 200–700 mg per 100 mL, a calibration curve constructed from four standard solutions should provide a suitable graph. If the method of standard additions is used, then 20 μL additions of the ascorbic acid standard can be added to the diluted fruit juice in the cell. Typical traces are shown in Figure 13.26. The experiment is normally free of significant interferences from other components in the juice and determines **only** ascorbic acid, not the oxidised dehydro or dehydro ascorbic acid hydrate forms which may be present, especially at low pH.

13.42   Indirect determination of nitrate via *o*-nitrophenol

Introduction   Although nitrate ion cannot be determined directly using polarographic techniques; by reacting nitrate ion in solution with phenol in the presence of concentrated sulphuric acid, the resulting *o*-nitrophenol can be easily measured.[38] The method is suitable for a range of samples containing nitrate ion in the low μg g^{-1} range and has been applied to natural waters, foods, beverages and fertilisers. A standard solution of *o*-nitrophenol is prepared by weighing out 0.2266 g of pure reagent which is slurried in 50 mL of distilled water, then 40 mL of concentrated sulphuric acid is carefully added with stirring (heat is generated). When cool this solution is made up to 100 mL with distilled water; it is equivalent to a solution containing $1\,mg\,mL^{-1}$ of nitrate or $0.226\,mg\,mL^{-1}$ of nitrogen. The method of standard additions is used for this determination.

Procedure   To sufficient sample to contain $2\text{--}50\,\mu g$ nitrate (NO_3^-) add $1\,mL$ of liquefied phenol (80 wt% in water) plus $4\,mL$ of 98% sulphuric acid with mixing. Allow to cool and cautiously add $4\,mL$ of distilled water. This solution is added to the polarographic cell and degassed in the normal way. Typical conditions are:

Drop time	$1\,s$
Starting potential	$+0.1\,V$ (vs SCE)
Scan range	$1.5\,V$ negative
Scan rate	$5\,mV\,s^{-1}$
Pulse modulation amplitude	$50\,mV$
Current range	$1\text{--}2\,\mu A$ f.s.d.

Once a satisfactory trace has been obtained for this solution, one or two standard additions of the standard o-nitrophenol are made ($1\,\mu L$ of standard is equivalent to $1\,\mu g$ of nitrate). The method works well, provided the sample is contained in a relatively small volume (less than $2\,mL$) so that the acid reagent is not unduly diluted before it can cause reaction between the nitrate and phenol. Solutions containing dissolved solids should be filtered before reaction, and if significant quantities of insoluble proteins are present, they should be removed by precipitation with phosphotungstic acid. Nitrite ion may also be determined if the nitrite is first oxidised to nitrate using hydrogen peroxide prior to the reaction stage. The difference in nitrate concentration between two solutions, one treated with H_2O_2 and the other analysed directly, gives a measure of the nitrite concentration.

Stripping analysis

13.43   Stripping voltammetry: basic principles

When using normal d.c. polarographic techniques, the limit of detection is of the order of $10^{-5}\,M$, or approximately $1\,\mu g\,g^{-1}$, and this can be improved by a factor of 2–10 using differential pulse techniques to give 'best case' detection levels of approximately $100\,\mu g\,kg^{-1}$. However, the intrinsic limitation to these techniques is that the redox reactions which are measured occur in a small layer of liquid close to the working electrode, the diffusion region, which for low concentrations of solution will contain very few ions. The techniques of stripping analysis have been designed to allow more ions **from a given solution** to undergo reaction, allowing a dramatic improvement by 3 or even 4 orders of magnitude over classical techniques; this gives detection limits of less than $1\,\mu g\,kg^{-1}$. In fact, the limits imposed on this technique are usually due to impurities in the reagents or adsorption onto the apparatus rather than due to the analyte itself.

Stripping voltammetry is a two-step process in which the first step consists of concentrating the ions from the bulk of solution by electrodeposition onto an electrode surface of small area, and then a second step in which the species of interest are electrolytically stripped from this electrode back into solution. The deposition step can either be a cathodic process where metal ions are reduced at say a mercury electrode to give an amalgam, or an anodic process where mercury is oxidised to Hg^+ and this then reacts with certain anions to produce insoluble mercurous salts which form a film on the electrode surface. This is then stripped back into solution by a negative-going potential (cathodic) in the second, analytical, step. The first of these two processes is more commonly used and the steps involved are shown below:

Deposition Apply negative potential to solution, giving

$$M^{n+} + ne^- \rightarrow M_{(Hg\ amalgam)}$$

Stripping Scan from negative to positive potentials, when

$$M_{(Hg)} \rightarrow M^{n+} + ne^-$$

Since the analytical stage of this process is a positive-going potential, it is called anode stripping voltammetry (ASV). Similarly, cathode stripping voltammetry gives these steps:

Deposition $Hg \rightarrow Hg^+ + e^-$ then $2Hg^+ + 2X^- \rightarrow Hg_2X_2$
insoluble
salt

Stripping $Hg_2X_2 + 2e^- \rightarrow 2Hg + 2X^-$

The stripping step in each case is similar to the processes already discussed, and its potential–current trace contains a peak at the reduction (or oxidation) potential of each species as it returns to solution. These peaks are proportional to the **amount** of material which has been preconcentrated onto the electrode. In effect, the material present in a large volume of solution of low concentration is concentrated onto a small surface area, from which it is electrolytically removed later on. The amount of material deposited onto the electrode from a given solution must therefore be carefully controlled if it is to be related to the solution concentration. This is achieved by controlling the surface area of the electrode, the potential applied to it, the duration of the deposition step and the rate of stirring of the solution. This may seem a difficult task, but in fact only a small modification is required to the apparatus used for DPP, i.e. the HMDE and its associated electronics. The solution needs to be stirred at a constant rate.

Consider a conventional polarographic system fitted with an HMDE or an SMDE containing an oxygen-free solution of pure supporting electrolyte plus a low concentration of one or more ions reducible at a mercury cathode. If the potentiostat of the polarograph is then set to a **fixed value**, which is chosen to be 0.2–0.4 V more negative than the highest reduction potential encountered among the reducible ions, then electrolysis will occur, deposition of metals will take place on the HMDE cathode and amalgam formation will usually take place. The rate of amalgam formation will be governed by the magnitude of the current flowing, by the concentrations of the reducible ions, and by the rate at which the ions travel to the electrode; the rate of travel of the ions can be controlled by stirring the solution. Given sufficient time, virtually the whole of the reducible ion content of the solution may be transferred to the mercury cathode, but complete exhaustion of the solution is not really necessary for the present procedure, and in practice, electrolysis is carried out for a carefully controlled time interval so that a fraction (say 10%) of the reducible ions are discharged. This operation is often known as a concentration step; the metals become concentrated into the relatively small volume of the mercury drop.

The stirrer is then switched off, although the electrolysis current is continued, and the cell is allowed to stand for about 30 s; this allows the solution to become quiescent, so the stripping step can be performed on an unstirred solution. The potentiostat is then swept in reverse, starting from the potential used in the electrolysis. This means that a gradually increasing positive potential is applied to the HMDE, which is now the anode of the cell.

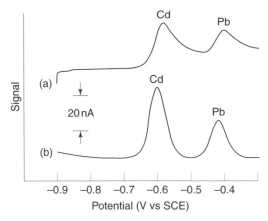

Figure 13.27 Stripping voltammetry: (a) d.c. stripping at $50\,mV\,s^{-1}$, (b) DPSAV at $2\,mV\,s^{-1}$

If the current is measured and plotted against the voltage, it initially produces a gradually increasing current; this corresponds to the residual current of conventional polarography and it is mainly due to the electrolyte solution. As the potential approaches the oxidation potential of one of the metals dissolved in the mercury, ions of that metal pass into solution from the amalgam, the current increases rapidly and it attains a maximum value when the potential has a value that is roughly equal to the appropriate oxidation potential.

The metal is said to be 'stripped' from the amalgam, and if the potential were held at the value corresponding to the maximum current, all of the metal would eventually be returned to the solution. In fact, the potential is not held stationary; as the potential sweep continues, the current declines from its maximum value and settles down to a new approximately steady value. In other words, the curve shows a peak. With continuing rise in the anodic potential, fresh peaks will be produced in the curve as the oxidation potentials of the different metals contained in the amalgam are reached (Figure 13.27(a)). Notice that although a steadily increasing potential is applied, the resulting curve shows peaks rather than plateaux. To some extent, the widths of these peaks depend on the size of the electrode and the rate of scan of the anodic potential, but under a given set of conditions the height is proportional to the original concentration of the metal ion in solution.

Instead of applying a relatively rapid d.c. scan ($10–100\,mV\,s^{-1}$), it is possible to obtain sharper and larger peaks (Figure 13.27(b)) by using the differential pulse mode (at $2–5\,mV\,s^{-1}$), and differential pulse anode stripping voltammetry (DPASV) is the method of choice for determining a number of metals at very low concentrations (Section 13.46). However, the technique cannot be used for all cations, since an amalgam must be produced between the metal and mercury during the deposition step, but the following metals have all been determined by DPASV:

antimony	gallium	**mercury**
arsenic	germanium	**silver**
bismuth	**gold**	thallium
cadium	indium	tin
copper	lead	zinc

With gold, silver and mercury a solid electrode **must be used** rather than an HMDE. Other metals such as nickel, cobalt, iron and aluminium can also be determined as complexes which can be potentiostatically adsorbed onto the mercury surface. DPASV is essentially a method for trace analysis, and solutions containing electroactive ions at concentrations higher than about $1\,\mu g\,g^{-1}$ should not be used. Fortunately, at $1\,\mu g\,g^{-1}$ and above, normal DPP techniques can be used in any case.

13.44   Electrodes used for stripping analysis

In the preceding section it was assumed that stripping analysis was conducted using a hanging mercury drop as the concentrating medium, and this is certainly the case in a large number of experiments, because the HMDE has several features which make it suitable for this purpose:

1. The mercury used can be highly purified to contain very low levels of impurities.
2. The drop size, hence its surface area, is controlled to close limits.
3. A fresh drop is used for each stripping step, thus avoiding any carry-over or memory effects.
4. The production and removal of each (identical) drop is readily automated.

However, a number of solid electrodes, such as gold, platinum and various types of carbon electrodes, may be used when the small surface area on which metal ions are deposited gives rise to high sensitivity and sharp, well-defined peaks. Since the same electrode is used for many determinations, it has to be conditioned before each new determination to avoid memory effects or contamination from the previous determination. This conditioning or cleaning can either take the form of mechanical or chemical cleaning by abrading the surface on a tissue or immersion in acidic solutions, or it can be achieved electrochemically by applying a positive potential to the electrode as the first stage of the cycle to remove contaminants from the electrode surface. This step is obviously performed in pure supporting electrolyte before immersing the electrode in the analyte solution. Although a number of solid metal electrodes have been used in the past, contamination and variation in overpotentials have caused some problems and it is now more common to use a thin film mercury electrode (TFME) where increased sensitivity is required. This has a very thin layer of mercury metal deposited on the surface of a glassy carbon electrode.

13.45   Apparatus for stripping analysis

In view of the limitations described in Section 13.44, particularly the influence of electrode characteristics on the peaks in the voltammogram, some care must be exercised in setting up an apparatus for stripping voltammetry. Two requirements should be satisfied to create optimum conditions:

(a) In the concentration step use a small mercury volume compared with the volume of solution to be electrolysed, and efficient stirring of the solution during the electrolysis, otherwise the deposition procedure may be unduly prolonged.
(b) In the stripping operation use a voltage scan that is as fast as possible but without causing peak tailing.

Electrodes

The hanging mercury drop electrode (HMDE) is traditionally associated with stripping voltammetry and its capabilities were investigated by Kemula and Kublik.[39] In view of the importance of drop size, it is essential to be able to set up **exactly** reproducible drops, and this can be done as explained in Section 13.33 for the SMDE. An alternative technique uses a platinum wire sealed into a glass tube. The wire is first thoroughly cleaned and is then used as anode during the electrolysis of pure perchloric acid; this treatment exerts a polishing effect. The current is then reversed, and the electrode used as cathode to ensure that no oxidised or adsorbed oxygen films remain on the surface of the electrode. Still using it as cathode, the electrode is now placed in a solution of mercury(II) nitrate solution and it thus becomes plated with mercury to give a mercury film electrode (MFE). This is claimed to possess advantages related to its rigidity and also to the greater surface area/volume ratio as compared with a mercury drop.

With the electrode based on platinum, however, should there be any bare platinum areas which have escaped plating in contact with the solution, then complications may arise owing to the smaller hydrogen overpotential on platinum compared with that on mercury, and metals having high positive electrode potentials may fail to deposit when the electrode is used for the concentration step. In recent years the use of mercury film electrodes based on substrates other than platinum has become more popular, and increased sensitivity and reliability are claimed for electrodes based on wax-impregnated graphite, on carbon paste and on vitreous carbon. A technique has also been developed for simultaneous deposition onto carbon electrodes of mercury and the metals to be determined.

Cells

The cell employed can be a suitable polarographic cell, or it can be specially constructed to fulfil the following requirements. Efficient and **reproducible** stirring of the solution is essential, and a magnetic stirrer is usually suitable. Exclusion of oxygen is important, so the cell must be provided with a cover, and provision should be made for passing pure, oxygen-free nitrogen through the solution before commencing the experiment, and over the surface of the liquid during the determination. The cover of the cell must provide a firm seating for the HMDE (or other type of electrode used), and must also have openings for the reference electrode (usually an SCE) and for a platinum counter-electrode if it is required to operate under conditions of controlled potential. If the solution under investigation is to be analysed for mercury, the reference electrode should be isolated from the solution using a salt bridge.

Reagents

In view of the sensitivity of the method, the reagents for preparing the primary solutions must be very pure, and the water used should be redistilled in an all-glass, or better, an all-silica apparatus; the traces of organic material sometimes encountered in normal deionised water make it unsuitable for this technique unless it is subsequently distilled. The common supporting electrolytes used in these experiments include potassium chloride, sodium ethanoate–ethanoic acid buffer solutions, ammonia–ammonium chloride buffer solutions, hydrochloric acid and potassium nitrate.

Normal analytical grades of these chemicals often contain trace impurities which are quite unimportant for most analytical purposes, but in terms of stripping voltammetry may represent serious contamination; this is especially true if heavy metals are involved. This makes it necessary to use reagents of very high purity (e.g. the BDH Aristar reagents or similar grade), or alternatively to subject the purest material available to an electrolytic purification process. The stirred solution is electrolysed with a small current (10 mA) for 24 h, using a pool of mercury at the bottom of a beaker as cathode and a platinum anode; pure nitrogen is passed through the solution before commencing the electrolysis so as to remove dissolved oxygen, and during the purification process, a current of pure nitrogen is maintained over the surface of the solution. It will usually be necessary to use some form of potentiostatic control during the electrolysis.

All glassware must be scrupulously cleaned, items used in preliminary treatment as well as the apparatus itself. It is usually recommended to soak glassware for some hours in **pure** nitric acid (6 M) or in a 10% solution of **pure** 70% perchloric acid, followed by washing with deionised water.

Concentration process

In the concentration process a fraction of the amount of a metal ion in solution is deposited at the mercury electrode, and the concentration of the metal in the mercury may become from 10 to 1000 times greater than the original ionic concentration in the solution; this preconcentration process is what gives the stripping method its great sensitivity.

Since the deposition is not exhaustive it is important that, in a given investigation, the same fraction of metal is incorporated in the mercury for each voltammogram recorded, and to achieve this, the electrode surface area, the deposition time and the rate of stirring must be carefully duplicated. Electrolysis times vary from 30 s to 30 min, depending on the concentration of the analyte in the solution under investigation, the electrode in use and the stripping procedure to be employed. The weaker (in analyte) the solution being analysed, the longer the electrolysis time. Electrolysis at a mercury film electrode is quicker than at an HMDE, and the preconcentration step may be shortened if differential pulse stripping is employed rather than direct current stripping.

During the electrolysis process, certain materials may give rise to species which form an insoluble salt with Hg(I) ions, thus producing an insoluble film on the mercury surface. Substances in this category include halides (other than fluorides), sulphides, thiocyanates, mercaptans and many organic thio compounds. In the subsequent stripping process, the voltage scan must be in the **negative** direction; the procedure is then known as cathodic stripping voltammetry. Many 'polarographic analysers' are available commercially which can be used for the preconcentration step and will then apply the requisite stripping procedure, displaying the results on a chart recorder. Many modern instruments carry out the preconcentration and stripping sequence automatically under microprocessor control.

13.46   Determination of lead in tap water

Introduction   This experiment describes SMDE and differential pulse stripping. All the glass apparatus must be rigorously cleaned; vessels should be filled with pure 6 M nitric acid and left standing overnight, then thoroughly cleaned with redistilled water.

Procedure   Prepare (1) a standard 0.01 M lead solution by dissolving 1.65 g (accurately weighed) analytical grade lead nitrate in redistilled water and adjusting to the mark in a 500 mL graduated flask; (2) a supporting electrolyte solution (0.02 M potassium nitrate) by dissolving 10 g of Aristar (or similar highly purified material) in 500 mL of redistilled water. Confirm this solution does not contain significant amounts of impurities (especially lead) by carrying out the preconcentration and stripping procedure described in detail below using 10 mL of the solution and 10 mL of redistilled water. If a significant lead peak is obtained, the solution must be discarded.

Place 10 mL of the tap water and 10 mL of the supporting electrolyte in the electrolysis vessel, add a magnetic stirring bar and mount the vessel on a magnetic stirrer. Insert an SCE, a platinum-plate auxiliary electrode, a nitrogen circulating tube and the capillary of the mercury electrode; operate the dispensing mechanism to form a mercury drop at the end of the capillary. Pass oxygen-free nitrogen through the solution for 5 min then adjust the nitrogen to pass over the surface of the liquid.

Make the connections to the polarographic analyser and adjust the applied voltage to −0.8 V, i.e. a value well in excess of the deposition potential of lead ions. Set the stirrer in motion, noting the setting of the speed controller, and after 15–20 s simultaneously switch on the electrolysis current and start a stopclock; allow electrolysis to proceed for 5 min. On completion of the electrolysis time, turn off the stirrer but leave the electrolysis potential applied to the cell. After 30 s to allow the liquid to become quiescent, replace the electrolysis current by the pulsed stripping potential and set the chart recorder in motion. When the lead peak at ~0.5 V has been passed, turn off the stripping current and the recorder. A suitable scan rate for stripping is $5\,mV\,s^{-1}$. Clean out the electrolysis cell and charge it with 10 mL of supporting electrolyte, 10 mL of redistilled water and 1 mL of the standard lead solution.

Set up a new mercury drop, pass nitrogen for 5 min and then record a new stripping voltammogram by repeating the above procedure with timings and stirring rate repeated exactly. Repeat with three further solutions containing 2, 3 and 4 mL of the standard lead solution. Measure the peak heights of the five voltammograms and thus deduce the lead content of the tap water. In some commercially available equipment, the concentration calculation can be performed automatically using built-in nomograms.

13.47   References

1. I M Kolthoff and P J Elving 1959 *Treatise on analytical chemistry*, Part I, Volume 4, John Wiley, New York
2. *Wilson and Wilson's comprehensive analytical chemistry*, Volume IIA, Elsevier, Amsterdam (1964)
3. A J Bard and L R. Faulkner 1980 *Electrochemical methods: fundamentals and applications*, John Wiley, New York
4. R Greef, R Peat, L M Peter, D Pletcher and J Robinson 1985 *Instrumental methods in electrochemistry*, Ellis Horwood, Chichester
5. J J Lingane and S Jones 1951 *Anal. Chem.*, **23**; 1804
6. R M Fuoss and F Accascina 1959 *Electrolytic conductance*, Van Nostrand, New York
7. J E Harrar 1975 In *Electroanalytical Chemistry*, Volume 8, A J Bard (ed), Marcel Dekker, New York
8. *Wilson and Wilson's comprehensive analytical chemistry*, Volume IID, Elsevier, Amsterdam (1975)
9. T Meites and L Meites 1955 *Anal. Chem.*, **27**; 1531. T Meites and L Meites 1956 *Anal. Chem.*, **28**; 103
10. A N Strohl and D J Curran 1979 *Anal. Chem.*, **53**; 353, 1050
11. J Ruzicka and E H Hansen 1981 *Flow injection analysis*, John Wiley, New York
12. C N Reilley 1968 *Rev. Pure Appl. Chem.*, **18**; 137
13. H Galster 1991 *pH measurement: fundamentals, methods, applications, instrumentation*, John Wiley, Chichester

14. I M Kolthoff and P J Elving 1978 *Treatise on analytical chemistry*, Part 1, Volume 1, John Wiley, New York

15. *Wilson and Wilson's comprehensive analytical chemistry*, Volume XXII, Elsevier, Amsterdam (1986)

16. S West and X Wen 1997 *Analysis Europa*, **4** (2); 14–19

17. J Marlow 1987 Care and maintenance of pH and redox electrodes, *Int. Lab.*, **XII** (2); J Marlow 1987 Care and maintenance of ion-selective electrodes, *Int. Lab.*, **XII** (2)

18. E Pungor, J Havas and K Toth 1965 *Zeit. Chem.*, **5**, 9

19. G J Moody, R B Oke and J D R Thomas 1970 *Analyst*, **75**; 910. A Craggs, G J Moody and J D R Thomas 1974 *J. Chem. Educ.*, **51**; 541

20. J Koryta (ed) 1980 *Use of enzyme electrodes in biomedical investigations*, John Wiley, Chichester

21. M Y Keating and G A Rechnitz 1984 *Anal. Chem.*, **56**; 801

22. *IUPAC manual of symbols and terminology for physicochemical quantities and units*, Butterworth, London (1969)

23. G Gran 1952 *Analyst*, **77**, 661

24. I R Epstein, K Kustin, P De Kepper and M Orban 1983 *Scientific American*, **248**; 96–108

25. R J Field and F W Schneider 1989 *J. Chem. Educ.*, **66** (3); 195–204

26. B P Belousov 1959 *Sb. ref. radiats. med. za. 1958*, Medgiz, Moscow

27. A M Zhabotinskii 1964 *Biofizika*, **9**; 306. A M Zhabotinskii 1964 *Dokl. Akad. Nauk. SSR*, **157**; 392

28. J A Pojman, R Craven abd D C Leard 1994 *J. Chem. Educ.*, **71** (1); 84–90

29. J Rosenthal 1991 *J. Chem. Educ.*, **68**; 794–95

30. J Heyrovsky 1922 *Chemicke Listy*, **16**; 256

31. D Ilkovic 1934 *Coll. Czech. Chem. Comm.*, **6**; 498

32. D R Crow 1969 *Polarography of metal complexes*, Academic Press, London

33. EG&G Princeton Applied Research, Princeton NJ

34. G C Barker and I L Jenkins 1952 *Analyst*, **77**; 685

35. J Krause, S Mathews and L Ramaley 1969 *Anal. Chem.*, **41**; 1365

36. Both Metrohm Ltd, CH-9101 Herisau, Switzerland and EG&G Princeton Applied Research provide free applications notes

37. Applications Notes P-2, S-6 and S-7 from EG&G Princeton Applied Research

38. Metrohm Ltd., CH-9101, Switzerland, Application Bulletin 70e (1979)

39. W Kemula and Z Kublik 1958 *Anal. Chim. Acta*, **18**; 104

13.48   Bibliography

D R Crow 1994 *Principles and applications of electrochemistry*, 4th edn, Blackie, London

R Hughes 1993 The future in direct ion measurement with ion selective electrodes, *Int. Lab. News*, **Dec**; 6

T Ishi 1994 The measurement of total organic halogen with coulometric titration, *Int. Lab. News*, **April**; 34

P T Kissinger and W R Heineman (eds) 1996 *Laboratory techniques in electroanalytical chemistry*, 2nd edn, Marcel Dekker, New York

T Riley, C Tomlinson and A M Jones 1987 *Principles of electroanalytical methods*, ACOL–Wiley, Chichester

D T Sawyer, A Sobowiak and J L Roberts 1995 *Electrochemistry for chemists*, 2nd edn, John Wiley, Chichester

B Z Shakkhashivi 1985 *Chemical demonstrations: a handbook for teachers*, University of Wisconsin, Madison WI

K R Trethewey and J Chamberlain 1995 *Corrosion for science and engineering*, 2nd edn, Addison Wesley Longman, Harlow

J Wang 1994 *Analytical electrochemistry*, Wiley–VCH, Chichester

K Yoshikawa, S Nakata, M Yamanaka and T Waki 1989 *J. Chem. Educ.*, **66** (3); 205–7

14

Nuclear magnetic resonance spectroscopy

14.1 Introduction

Nuclear magnetic resonance spectroscopy (NMR) measures the absorption of electromagnetic radiation in the radiofrequency region of roughly 4 MHz to 750 MHz, which corresponds to a wavelength of about 75 m to 0.4 m. Unlike ultraviolet, visible and infrared absorption, nuclei of atoms rather than outer electrons are involved in the absorption process. In order to cause nuclei to absorb radiation it is necessary to expose the sample containing the nuclei of interest to a magnetic field of several teslas (T). The magnetic field has the effect of developing the energy states required for absorption to occur. Sections 14.2 to 14.6 give a brief introduction to the theory and instrumentation of solution-state NMR spectroscopy. Solid-state NMR spectroscopy is a specialised area outside the scope of this book. For those applications which are described, enough detail is given to comprehend the processes behind them. They include polymer analysis, pharmaceutical purity determinations, in vivo studies of phosphorus metabolism and magnetic resonance imaging.

14.2 Theory

In order to account for some of the properties of nuclei, it is necessary to assume they rotate about an axis and therefore have the property of spin. The maximum spin component for a particular nucleus is its **nuclear spin quantum number** I; a nucleus will then have $(2I + 1)$ discrete states. In the absence of an external magnetic field, the various states have identical energies. The discussion will be restricted to those nuclei having a nuclear spin quantum number of $\frac{1}{2}$. Such nuclei include ^{1}H, ^{13}C and ^{19}F. For these nuclei two spin states exist corresponding to $I = \frac{1}{2}$ and $I = -\frac{1}{2}$. In the presence of a magnetic field these spin states are split as shown in Figure 14.1. The difference in energy ΔE between the two spin states can be shown to be $g_I\beta_N B_0$, where g_I is the nuclear g factor, which is characteristic of the nucleus, β_N is the nuclear magneton (= 5.050×10^{-27} J T^{-1}) and B_0 is the applied magnetic field in tesla. Thus

$$\Delta E = h\nu = g_I\beta_N B_0 \tag{14.1}$$

$$\nu = g_I\beta_N B_0/h$$

An alternative constant γ, called the magnetogyric ratio, can also be defined for each nucleus; it too combines g_I with β_N. Table 14.1 contains some data for a few of the important nuclei.

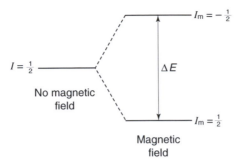

Figure 14.1 Splitting of nuclear energy levels in a magnetic field

---

Table 14.1 *Properties of some NMR nuclei having* $I = \frac{1}{2}$

Nucleus	Resonance frequency (MHz) in field of 2.35 T	g_I
^{1}H	100.0	5.585
^{13}C	25.14	1.404
^{15}N	10.13	-0.566
^{19}F	94.07	5.255
^{29}Si	19.87	-1.110
^{31}P	40.48	2.261
^{119}Sn	37.27	-2.082

Example 14.1

Calculate the frequency at which a ^{1}H nucleus would absorb in a magnetic field of 1.41 T, given that g_I for ^{1}H is 5.585.

Substituting into equation (14.1) we find

$$v = \frac{5.585 \times 5.05 \times 10^{-27} \times 1.41}{6.623 \times 10^{-34}} = 60.0 \times 10^6 \, \text{Hz} = 60.0 \, \text{MHz}$$

This example shows that a radiofrequency of 60 MHz will bring about a transition from $I_m = \frac{1}{2}$ to $I_m = -\frac{1}{2}$.

In a sample containing a very large number of nuclei at equilibrium, some of the nuclei will be in the lower ($I = +\frac{1}{2}$) state and some will be in the higher ($I = -\frac{1}{2}$) state. The Boltzmann distribution law can be used to calculate the relative numbers occupying each level.

Example 14.2

Use the information from Equation 14.1 to calculate the relative populations of the lower and upper energy levels at 300 K. (Boltzmann's constant = $1.38 \times 10^{-23} \, \text{J K}^{-1}$)

From Equation 14.1 the energy ΔE between the upper and lower energy levels is $g_I \beta_N B_0$, i.e.

$$5.585 \times 5.05 \times 10^{-27} \times 1.41 = 3.97 \times 10^{-26} \, \text{J}$$

From the Boltzmann equation

$$\frac{N_{\text{upper}}}{N_{\text{lower}}} = \exp\left(-\frac{\Delta E}{kT}\right)$$

where N_{upper} and N_{lower} are the numbers of nuclei in the upper and lower energy levels, respectively.

Substituting the appropriate values into this equation gives

$$\frac{N_{\text{upper}}}{N_{\text{lower}}} = \exp\left(-\frac{3.97 \times 10^{-26}}{1.38 \times 10^{-23} \times 300}\right) = \exp\left(-\frac{3.97 \times 10^{-26}}{4.2 \times 10^{-21}}\right) \approx \exp(-1 \times 10^{-5})$$

$$\approx 1 - (1 \times 10^{-5})$$

This means that the ratio is approximately unity; if the upper energy level has $1\,000\,000$ nuclei then the lower energy level will have about $1\,000\,010$ nuclei.

Net absorption of radiofrequency by the sample only arises because there is an excess of nuclei in the lower energy state at thermal equilibrium. When radiofrequency is absorbed the spin populations are no longer at thermal equilibrium, and if the input of radio-frequency energy is sufficient to cause the populations of the lower and upper energy states to be equal, there will be no further net absorption of energy and the system is said to be **saturated**. Consequently, the NMR absorption signal will no longer be observable. Net absorption can only be restored if some of the nuclei in the upper state **relax** back to the lower energy state.

There are two relaxation processes: spin–lattice and spin–spin. In spin–lattice relaxation, characterized by a time T_1, nuclei in the upper energy state lose energy to their surroundings. The second relaxation mechanism, spin–spin relaxation characterized by a time T_2, does not lead to a change in the relative populations of the upper and lower energy levels. A nucleus in the upper energy state transfers its energy to a nucleus in the lower energy level. The effect of spin–spin relaxation is to increase the lifetime of the upper energy state. The Heisenberg uncertainty principle shows that $\Delta E \, \Delta t \approx h/2\pi \approx 10^{-34} \, \text{J s}$, where ΔE is the uncertainty in energy of an energy level and Δt is the lifetime of the energy level. If the lifetime of an energy state is very short then the uncertainty in its energy is very large. Similarly, a very long lifetime for an energy level means that its energy can be defined very precisely. Uncertainties in the energy of the upper energy state in NMR spectroscopy contribute to the linewidths of the peaks observed in the spectrum. If the uncertainty is large then the peaks will be very broad.

14.3   The chemical shift

So far only an isolated nucleus in an applied field has been considered. In practice this situation does not occur since, in general, all nuclei are associated with electrons in atoms and molecules. When placed in a magnetic field, the surrounding electron cloud tends to circulate in such a direction as to produce a field which opposes that applied. The total field experienced by the nucleus is

$$B_{\text{effective}} = B_{\text{applied}} - B_{\text{induced}}$$

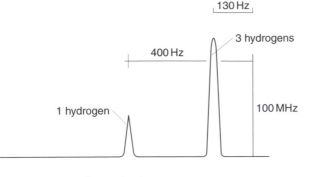

Figure 14.2   Methanol: proton NMR spectrum

Since the induced field is directly proportional to the applied field

$$B_{\text{induced}} = \sigma B_{\text{applied}}$$

where σ is a constant, called the shielding constant, and so

$$B_{\text{effective}} = B_0(1 - \sigma)$$

where B_0 is the applied magnetic field. Thus the nucleus is **shielded** from the applied field. The extent of the shielding will vary with the electron density about an atom or molecule, which will depend on the chemical environment of the nucleus. For a simple molecule like methanol (Figure 14.2) there are protons in two different chemical environments – three of them attached to a carbon in the methyl group and one attached to the oxygen of the hydroxyl group – and the methyl protons absorb at a lower frequency than the proton of the hydroxyl group. Note also that the integrated area of the larger peak is three times that of the smaller OH signal, showing that the signal intensity is proportional to the number of protons present in each of the chemical environments within the molecule.

Thus, if the applied magnetic field was 2.35 T, giving a reference radiofrequency of 100 MHz, using equation (14.1), the methyl protons would show a signal at a higher frequency than 100 MHz. This **chemical shift** can be quoted as 130 Hz or 1.3 ppm (that is 1.3 parts per million in 100 million hertz). If the NMR spectrum of this compound were recorded at 4.7 T (when the operating frequency would be 200 MHz) the chemical shift for the methyl protons would be 260 Hz but still 1.3 ppm. It is therefore more convenient to quote chemical shifts in parts per million from the reference frequency since values in hertz would be frequency dependent.

The preferred method for measuring chemical shifts is to reference them to an internal standard. For both proton and ^{13}C NMR spectroscopy the most convenient standard material is tetramethylsilane ($SiMe_4$). Tetramethylsilane (TMS) is chosen for several reasons: it can be used to reference ^{1}H and ^{13}C in the same sample; it contains 12 equivalent protons and 4 equivalent carbon atoms, hence only a very small amount of TMS is required to give a strong signal; it gives a sharp peak well away from most other proton and carbon resonances; it is chemically inert, soluble in most organic solvents and volatile. The chemical shift is given the symbol δ and is the difference in frequency in ppm from the TMS signal. It is defined by the equation

$$\delta = \frac{\nu_{\text{sample}} - \nu_{\text{reference}}}{\nu_{\text{reference}}} \times 10^6$$

When δ is positive the shift is to a higher frequency. TMS is not soluble in water and so for aqueous solutions the commonly used standards are sodium 4,4-dimethyl-4-silapentanesulphonate [14.A], usually called DSS, or the deuterium-substituted carboxylic acid salt [14.B].

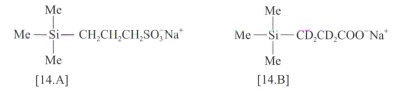

$$\text{Me} - \underset{\underset{\text{Me}}{|}}{\overset{\overset{\text{Me}}{|}}{\text{Si}}} - \text{CH}_2\text{CH}_2\text{CH}_2\text{SO}_3^-\text{Na}^+ \qquad \text{Me} - \underset{\underset{\text{Me}}{|}}{\overset{\overset{\text{Me}}{|}}{\text{Si}}} - \text{CD}_2\text{CD}_2\text{COO}^-\text{Na}^+$$

[14.A]   [14.B]

Figure 14.3 gives the approximate proton chemical shifts for some simple molecules and groups.[1]

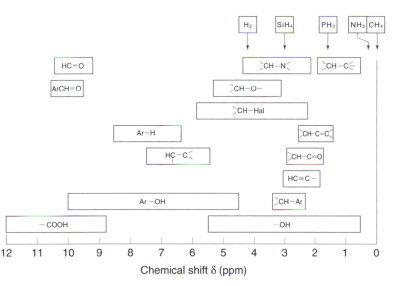

Figure 14.3 Simple molecules and functional groups: approximate proton NMR chemical shifts (Reprinted, with permission, from C. N. Banwell and E. M. McCash, 1994, *Fundamentals of molecular spectroscopy*, 4th edn, McGraw-Hill, Maidenhead)

14.4   Coupling of magnetic nuclei

Not all spectra consist of single-line absorption bands as in Figure 14.2, since splitting of the lines frequently occurs as a complicating factor. The splitting occurs because the spins of neighbouring nuclei **couple** (or interact) with one another. Only a very brief description will be given. Further details can be found in a textbook on NMR spectroscopy.[2] When the signal from any given nucleus is split, the degree of splitting is called the **multiplicity**. The multiplicity depends on the number of other nuclei to which the given nucleus is coupled. Note that the splitting can arise whether or not the coupling nuclei are of the same element. The splittings between peaks in related multiplets are equal. Figure 14.4 shows examples of splitting patterns. For nuclei having nuclear spin quantum numbers of $\frac{1}{2}$, the number of lines observed in a multiplet is $(n + 1)$, where n is the number of equivalent

575

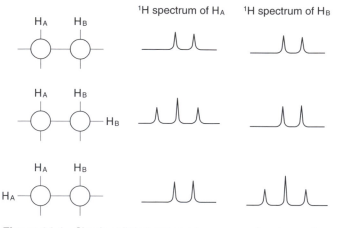

Figure 14.4   Simple splitting patterns for a range of proton environments

nuclei coupling to an adjacent nucleus. Spin coupling constitutes a major part of the study of NMR spectroscopy. Further details can be found in the literature.[2]

14.5   Instrumentation

14.5.1   Continuous wave instruments

The important components of a continuous wave spectrometer are shown in Figure 14.5. The sensitivity and resolution of a spectrometer are critically dependent on the strength and quality of the **magnet**. As the field strength of the magnet increases, both the sensitivity and resolution increase. The magnets used in NMR spectrometers are of three types: permanent magnets, electromagnets and superconducting magnets. Permanent magnets are highly sensitive to temperature and it is not possible to get large field strengths. Electromagnets are much less sensitive to temperature but require cooling systems and elaborate power supplies. Commercial electromagnets can produce fields up to about 2.3 T, corresponding to proton absorption frequencies of 100 MHz. Superconducting magnets are used in instruments having the highest resolution. Spectrometers having fields of 17.5 T,

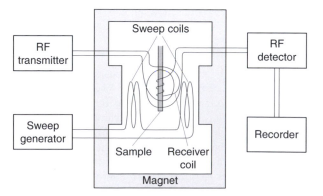

Figure 14.5   Principal components of a continuous wave NMR

corresponding to a proton frequency of 750 MHz, are now commercially available, and some go even higher.

The signal from a **radiofrequency transmitter** is fed into a pair of coils mounted at 90° to the magnetic field. A fixed frequency, dependent on the magnetic field strength, is normally used and a spectrum is obtained by the use of the **field sweep generator**, which permits alteration of the applied magnetic field over a very small range by varying the current through a pair of coils located parallel to the faces of the magnet. Usually the field strength is changed linearly with time, and this change is synchronised with the linear drive of a chart recorder. For a 60 MHz proton spectrometer the sweep range is 1000 Hz, corresponding to a field sweep of 2.3×10^{-5} T.

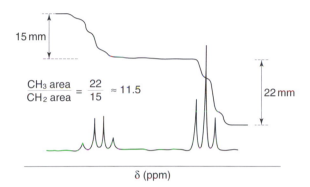

Figure 14.6 Bromoethane: absorption spectrum and integration

The amount of radiation passing through the sample is detected and the spectrum recorded. All modern spectrometers are equipped with electronic or digital integrators to provide information about areas under absorption peaks. Usually the integral data appears as step functions superimposed on the NMR spectrum (Figure 14.6). The **sample cell** usually consists of a 5 mm external diameter glass tube containing about 0.5 mL of liquid, although samples smaller than 0.5 mL can be analysed using a microcell. The **sample probe** holds the sample in a fixed place within the magnetic field. It also has an air-driven turbine so the sample can be rotated along its longitudinal length at about 4000 rpm. This reduces any effects due to inhomogeneities in the magnetic field, giving sharper lines and better resolution.

14.5.2 Pulsed Fourier transform instruments

An alternative to continuous wave methods in which individual molecular frequencies are excited in succession is to irradiate all the frequencies simultaneously by applying an intense pulse of radiofrequency radiation lasting only a few microseconds. This radiofrequency pulse causes saturation for all the absorbing nuclei. A single oscillator frequency is used, but because the pulse has a very short duration the Heisenberg uncertainty principle shows that a spread of frequencies will be obtained, covering the complete spectrum (Figure 14.7).

At the end of the excitation pulse, the radiofrequency receiver records the **free induction decay** signal as a function of time coming back from the sample. This signal decays exponentially and it contains information on all the frequencies included in the frequency sweep spectrum. But the problem is they arrive at the detector together, producing an

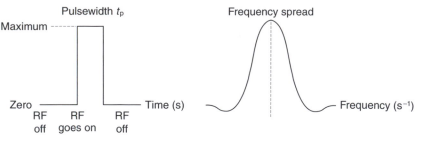

Pulsewidth t_p

Maximum ------

Frequency spread

Zero ——— Time (s)

RF off   RF goes on   RF off

Frequency (s^{-1})

TIME DOMAIN   **FREQUENCY DOMAIN**

Figure 14.7   The frequency spectrum produced by a short pulse of monochromatic radiation. (Reprinted, with permission, from W. Kemp, 1986, *NMR in chemistry: a multinuclear introduction*, Macmillan, Basingstoke)

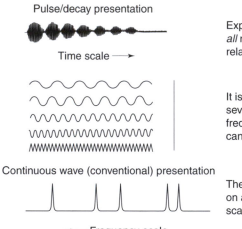

Pulse/decay presentation

Time scale ⟶

Exponential decay from *all* nuclei undergoing relaxation

It is a composite of several individual frequencies, which can be separated

Continuous wave (conventional) presentation

They can be presented on a linear frequency scale

◄—— Frequency scale

Figure 14.8   Free induction decay and its Fourier transform to the frequency domain (Reprinted, with permission, from W. Kemp, 1986, *NMR in chemistry: a multinuclear introduction*, Macmillan, Basingstoke)

interference pattern which has to be analysed and converted into the individual contributing frequencies using a procedure known as **Fourier transformation** (Figure 14.8). Fourier transforms are discussed in a little more detail in Section 18.5 but the details are outside the scope of this book.

The pulse and subsequent detection of the free induction decay (FID) can take as little as a second, and provided there is an interface and a computer to record and store the data, the process can be repeated again and again and the FIDs added together giving improvements in the signal-to-noise ratio. The delay between successive radiofrequency pulses depends on the longest relaxation times of the group of nuclei being irradiated and can vary from a few seconds to several minutes. Pulsed Fourier transform NMR spectroscopy has several advantages over the continuous wave method described in Section 14.5.1. The signal strength and hence the sensitivity depends on many variables, including the amount of magnetic nuclei present in the sample. In proton NMR spectroscopy the signal is inherently strong and the proton has a natural abundance of almost 100%, so spectra can be easily obtained unless the compound under investigation is available in only very small amounts or it is only sparingly soluble. The proton NMR spectrum of a typical organic compound can be obtained in less than one minute. For less abundant nuclei, such as ^{13}C, the sensitivity is such that spectra can only be achieved if many recordings of the spectrum are made and they are then added together. A typical ^{13}C spectrum may take several minutes or hours to obtain, depending on the number of scans collected and the relaxation times of the carbon-containing functional groups present.

14.6   Sample preparation

Only solution or liquid phase samples will be considered since solid-state and gas phase NMR studies, while important, are outside the scope of this book. A liquid can be examined as a pure liquid or more usually as a solution. Pure liquids may be viscous which often leads to line broadening. Similarly, solids must be dissolved in a suitable solvent. Solvents for proton NMR spectra, must be aprotic, so deuterated or naturally aprotic solvents, such as tetrachloromethane, are most commonly used. Most commercial deuterated solvents are at least 98% isotopically pure. Typical solvents with the positions of residual protons in NMR spectra are given in Table 14.2. Chloroform-d$_1$ is the most versatile and commonly used solvent. The solvent may also affect the spectrum of the solute.

Two important factors, which may be concentration dependent, are **hydrogen-bonding** and **proton exchange**. The chemical shifts of hydrogen-bonded protons, e.g. O—H or N—H in alcohols or amines can vary greatly (4–5 ppm) depending on concentration. Also chemically labile protons can exchange with each other. If the solvent contains a labile deuteron then, because the solvent is usually in excess, it can exchange with the labile protons of the sample, effectively removing them from the sample. This process of solvent exchange can be used to identify labile protons. A spectrum of the solute is firstly obtained using CDCl$_3$ as solvent and then a drop of D$_2$O is added and the solution shaken. Labile protons exchange for deuterons and their signal disappears. A small peak for HOD appears.

The concentration of the sample should be sufficient to obtain an adequate signal-to-noise ratio. Concentrations of 1–10% (w/v) are common and a volume of about 0.5 mL is required. Although Fourier transform spectrometers can cope with much smaller sample

Table 14.2   *Common solvents used for NMR spectroscopy*

Solvent	Approximate δ for ^{1}H equivalent as contaminant	Boiling point (°C)	Freezing point (°C)
Ethanoic (acetic) acid-d$_4$	2, 13	118	16.6
Acetone-d$_6$	2	56	−95
Acetonitrile-d$_3$	2	83	−44
Benzene-d$_6$	7.3	80	5.5
Tetrachloromethane	–	77	−23
Chloroform-d$_1$	7.3	61	−63
Deuterium oxide	4.7 to 5	101.5	3.8
Dimethylsulphoxide-d$_6$	2	189	18
Methanol-d$_4$	3.4	65	−98

sizes, there is a corresponding increase in the time required to obtain the spectrum, so high concentrations and large sample volumes are preferable. For ^{13}C NMR spectroscopy 10 mm diameter tubes are often used. If the sample does not dissolve completely, any solid residue should be removed by filtering the solution before the spectrum is obtained.

Experimental determinations

! **Safety.** Before carrying out any experiments in this section, pay full attention to any safety warnings and make sure you adhere to national laboratory and safety regulations.

14.7 Ethanol content of an alcoholic liquor[3]

Discussion Alcoholic liquors in the United Kingdom must contain at least 40% ethanol. The other major component of the spirit is water. The proton NMR spectrum of such a spirit (Figure 14.9) shows only the signals for the hydroxyl protons at about $\delta = 5.0$ ppm and the ethanol protons at 3.8 ppm (CH_2) and 1.2 ppm (CH_3). Note that the triplet at 1.2 ppm due to the CH_3 group is due to coupling with the two protons of the CH_2 group and the quartet at 3.8 ppm is due to coupling of the CH_2 group with the three protons of the CH_3 group. The other components of the liquor are in such low concentration that they do not contribute to the spectrum. It is possible to calculate ratios of the integrated signals CH_3/OH and CH_2/OH, but neither will give the ratio of ethanol to water since the signals come from groups having a different number of protons associated with them. In addition the peak at 5.0 ppm contains contributions from the two protons of the water as well as the proton from the OH group of the ethanol. However, from the integrations for the CH_2 and CH_3 groups (having two and three protons, respectively) the integration for a single proton (from the hydroxyl group of the ethanol) can be calculated and this value subtracted from the total integration at 5.0 ppm to give an integration for the H_2O.

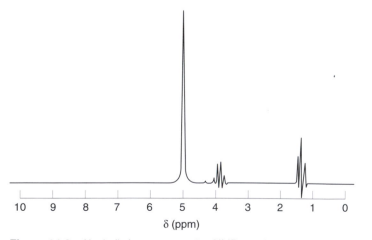

Figure 14.9 Alcoholic beverage: proton NMR spectrum

Method   Mix a sample of the liquor with $CDCl_3$ in order to give an approximately 10% solution and add one drop of tetramethylsilane standard. Run the spectrum and integrate the three peaks at 1.2, 3.8 and 5.0 ppm. Repeat the integration several times and take the average for each peak.

Calculation   Let the integration for the peaks at 1.2 ppm (due to CH_3) be x mm, the integration for the peaks at 3.8 ppm (due to CH_2) be y mm and the integration for the peak at 5.0 ppm (due to OH from ethanol and H_2O) be z mm.

For the CH_3 group the integration for 1 proton = $x/3$

For the CH_2 group the integration for 1 proton = $y/2 = x/3$

Therefore the integration corresponding to the hydroxyl group of the ethanol is $y/2$, and

Integration corresponding to $H_2O = (z - y/2)$

Ratio $H_2O : CH_3CH_2OH = (z - y/2) : y$, since both H_2O and CH_2 contain two protons. But this will not give the correct percentage composition for the liquor since the two components, water and ethanol, have different relative molecular masses (RMMs) and densities.

RMM (water)   = 18   density = $1.00\,g\,mL^{-1}$

RMM (ethanol) = 46   density = $0.96\,g\,mL^{-1}$

Hence $\dfrac{\text{water mass}}{\text{ethanol mass}} = \dfrac{[(z - y/2) \times 18]/1}{(y \times 46)/0.96}$

and   % ethanol = $\dfrac{46y/0.96}{(46y/0.96) \times 18(z - y/2)} \times 100$

14.8   Ethanol content of a beer by standard addition[3]

Discussion   In Section 14.7 the ethanol content was about 40% and it was possible to measure reasonably large integrations for both the ethanol and the water. For beer, having a low alcohol concentration (about 4%) measurements have to be made very carefully to avoid signal-to-noise problems which occur when comparing a very large signal (due to water) with a very small signal (from the ethanol). One possible way of overcoming this problem is to use the method of standard additions in which small, known amounts of ethanol are added to the sample.

Method   Pipette exactly 0.5 mL of the beer into five 1 mL graduated flasks and add 0.0, 10.0, 20.0, 50.0 and 100.0 mg of pure ethanol to the flasks. Make up to the mark with distilled water. Obtain the proton NMR spectrum of each sample and obtain the integration for the CH_3 peak at about 1.2 ppm. Plot a graph of integration against mass of ethanol added to the sample. A straight line should be obtained. Extrapolate the line until it reaches the x-axis. This is the amount of ethanol (mg) in the diluted beer sample. Since the beer has been diluted by a factor of 2, the original alcohol content will be twice the value obtained from the graph. This is the mass of ethanol in 0.5 mL of beer and so, assuming that the density of beer is $1\,g\,mL^{-1}$, the percentage ethanol in the beer can be calculated.

14.9 Aspirin, phenacetin and caffeine in an analgesic tablet[4]

Discussion   Some analgesic preparations contain a mixture of aspirin [14.C], phenacetin [14.D] and caffeine [14.E]. The three components can be conveniently analysed using ^{1}H NMR spectroscopy.

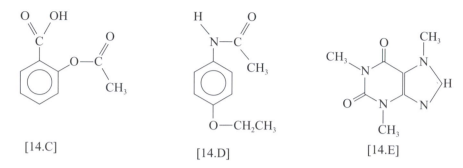

[14.C]   [14.D]   [14.E]

Method   Grind an analgesic tablet to a fine powder using a pestle and mortar. Weigh accurately into an NMR tube about 60 mg of the powder and use a micropipette to add exactly 0.500 mL of $CDCl_3$. Cap the sample tube, shake and gently warm the sample to dissolve the three components. Binder materials such as starch and lactose will not dissolve but they will not interfere with the analysis. Add one drop of TMS to the tube. Obtain the spectrum and the integration of the sample. The integration should be repeated and the average taken. Obtain the spectrum for a sample containing 50 mg mL^{-1} of pure caffeine and measure the integration of the peak at about 4.0 ppm. The sharp peak at 2.3 ppm can be used for the analysis of aspirin, the quartet at around 4.0 ppm is used for the phenacetin and the sharp peaks at 3.4 and 3.6 ppm for the caffeine. Note that the quartet at around 4 ppm due to phenacetin also has a contribution from the caffeine, for which a correction must be made.

Calculation   Let the integral of the quartet due to phenacetin at around 4.0 ppm be A; B is the integral of the two peaks at 3.4 and 3.6 ppm due to caffeine; and C is the integral of the peak at 2.3 due to aspirin. MW_{asp} is the relative molecular mass of aspirin; MW_{phen} is the molecular weight of phenacetin and MW_{caf} is the relative molecular mass of caffeine. I_{caf} is the integration of the peak at 4.0 ppm in the spectrum of pure caffeine. The number of milligrams of each component per milligram of sample can be calculated from the following equations:

Aspirin

$$\frac{\text{mg aspirin}}{\text{mg sample}} = \left(\frac{C}{I_{caf}}\right)\left(\frac{MW_{asp}}{MW_{caf}}\right)\left(\frac{\text{mg caffeine}}{\text{mg solvent}}\right)_{STD}\left(\frac{0.500\text{ mL}}{\text{mg sample}}\right)$$

Phenacetin

$$\frac{\text{mg phenacetin}}{\text{mg sample}} = \frac{A - \frac{1}{2}(B - 0.0055C)}{I_{caf}}\frac{3}{2}\left(\frac{MW_{phen}}{MW_{caf}}\right)\left(\frac{\text{mg caffeine}}{\text{mg solvent}}\right)_{STD}\left(\frac{0.500\text{ mL}}{\text{mg sample}}\right)$$

Caffeine

$$\frac{\text{mg caffeine}}{\text{mg sample}} = \frac{\frac{1}{2}(B - 0.0055C)}{I_{caf}}\left(\frac{\text{mg caffeine}}{\text{mg solvent}}\right)_{STD}\left(\frac{0.500\text{ mL}}{\text{mg sample}}\right)$$

Note   The term $0.0055C$ represents a correction for the ^{13}C sideband from the aspirin peak which falls beneath the caffeine peak at 3.4 ppm.

14.10   Keto–enol tautomerism in pentan-2,4-dione (acetylacetone)

Discussion   Keto–enol tautomerism is particularly marked for β-dicarbonyl compounds. When the interchange time in the keto–enol equilibrium is slow compared with the time-scale of the NMR experiment ($\sim 10^{-3}$ s) then signals will be observed from each component and integration of the signals, suitably weighted for the number of equivalent protons in each group may be used to determine the relative proportions.

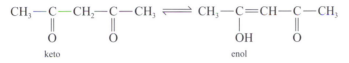

Keto

enol

Method   Mix a sample of pentan-2,4-dione with $CDCl_3$ to give a solution which contains a mole fraction of the analyte of approximately 0.2. Add one drop of TMS and run the proton NMR spectrum and integration. Measure the integrals at ~1.95 ppm, due to the methyl groups in the enol form (E), and 2.15 ppm, due to the methyl groups in the keto form (K), then

$$\text{percentage enol form} = \frac{E}{(E + K)} \times 100$$

14.11   References

1. C N Banwell and E M McCash 1994 *Fundamentals of molecular spectroscopy*, 4th edn, McGraw-Hill, Maidenhead
2. W Kemp 1996 *NMR in chemistry: a multinuclear introduction*, Macmillan, Basingstoke
3. D A R Williams 1986 *Nuclear magnetic resonance spectroscopy*, ACOL–Wiley, Chichester
4. D P Hollis 1963 *Anal. Chem.*, **35**; 1682

14.12   Bibliography

J W Akitt 1983 *NMR and chemistry: an introduction to NMR spectroscopy*, 2nd edn, Chapman and Hall

L D Field and S Sternhell (eds) 1989 *Analytical NMR*, John Wiley, Chichester

R K Harris 1986 *Nuclear magnetic resonance: a physiochemical review*, 2nd edn, Longman, Harlow

D E Leyden and R H Cox 1997 *Analytical applications of NMR*, John Wiley, Chichester

15

Atomic absorption
spectroscopy

15.1 Introduction

If a solution containing a metallic salt (or some other metallic compound) is aspirated into a flame (e.g. of acetylene burning in air), a vapour which contains atoms of the metal may be formed. Some of these gaseous metal atoms may be raised to an energy level which is sufficiently high to permit the emission of radiation characteristic of the metal, e.g. the characteristic yellow colour imparted to flames by compounds of sodium. This is the basis of **flame emission spectroscopy (FES)**, formerly known as **flame photometry** (Chapter 16).

However, a much larger number of the gaseous metal atoms will normally remain in an unexcited state or, in other words, the ground state. These ground state atoms are capable of absorbing radiant energy of their own specific resonance wavelength, which in general is the wavelength of the radiation that the atoms would emit if excited from the ground state. Hence if light of the resonance wavelength is passed through a flame containing the atoms in question, then part of the light will be absorbed, and the extent of absorption will be proportional to the number of ground state atoms present in the flame. This is the underlying principle of **atomic absorption spectroscopy (AAS)**. Another technique, **atomic fluorescence spectroscopy (AFS)**, is based on the re-emission of absorbed energy by free atoms.

The procedure by which gaseous metal atoms are produced in the flame may be summarised as follows. When a solution containing a suitable compound of the metal to be investigated is aspirated into a flame, the following events occur in rapid succession:

1. Evaporation of solvent, leaving a solid residue.
2. Vaporisation of the solid with dissociation into its constituent atoms, initially in the ground state.
3. Some atoms may be excited by the thermal energy of the flame to higher energy levels, and attain a condition in which they radiate energy.

The resulting emission spectrum thus consists of lines originating from excited atoms or ions. These processes are conveniently represented on a diagram (Figure 15.1).

15.2 Elementary theory

Consider the simplified energy level diagram shown in Figure 15.2, where E_0 represents the ground state in which the electrons of a given atom are at their lowest energy level and

584

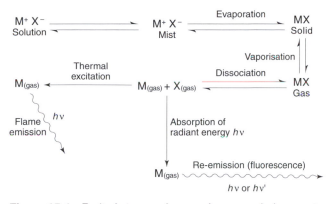

Figure 15.1   Excited atoms or ions produce an emission spectrum

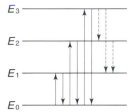

Figure 15.2   Simplified energy level diagram

E_1, E_2, E_3, etc., represent higher or excited energy levels. Transitions between two quantised energy levels, say from E_0 to E_1, correspond to the absorption of radiant energy, and the amount of energy absorbed ΔE is determined by Bohr's equation

$$\Delta E = E_1 - E_0 = h\nu = hc/\lambda$$

where c is the velocity of light, h is Planck's constant, and ν is the frequency and λ the wavelength of the radiation absorbed. The transition from E_1 to E_0 corresponds to the **emission** of radiation of frequency ν.

Since an atom of a given element gives rise to a definite, characteristic line spectrum, it follows there are different excitation states associated with different elements. The consequent emission spectra involve not only transitions from excited states to the ground state, e.g. E_3 to E_0, E_2 to E_0 (indicated by the full lines in Figure 15.2), but also transitions such as E_3 to E_2, E_3 to E_1 (indicated by the broken lines). Thus it follows that the emission spectrum of a given element may be quite complex. In theory it is also possible for absorption of radiation by already excited states to occur, e.g. E_1 to E_2, E_2 to E_3, but in practice the ratio of excited atoms to ground state atoms is extremely small, so the absorption spectrum of a given element is usually only associated with transitions from the ground state to higher energy states and is consequently much simpler in character than the emission spectrum.

The relationship between the ground state and excited state populations is given by the Boltzmann equation

$$N_1/N_0 = (g_1/g_0)e^{-\Delta E/kT}$$

where

N_1 = number of atoms in the excited state

N_0 = number of ground state atoms

g_1/g_0 = ratio of statistical weights for ground and excited states

ΔE = energy of excitation = $h\nu$

k = the Boltzmann constant

T = the temperature in kelvins

Notice that the ratio N_1/N_0 depends on the excitation energy ΔE and the temperature T. An increase in temperature and a decrease in ΔE (i.e. when dealing with transitions which occur at longer wavelengths) will each result in a higher value for the ratio N_1/N_0.

Calculation shows that only a small fraction of the atoms are excited, even under the most favourable conditions, i.e. when the temperature is high and the excitation energy low. This is illustrated by the data in Table 15.1 for some typical resonance lines.

Table 15.1 *Variation of atomic excitation with wavelength and with temperature*

Element	Wavelength (nm)	N_1/N_0	
		2000 K	4000 K
Na	589.0	9.86×10^{-6}	4.44×10^{-3}
Ca	422.7	1.21×10^{-7}	6.03×10^{-4}
Zn	213.9	7.31×10^{-15}	1.48×10^{-7}

Since the absorption spectra of most elements are simple in character, as compared with the emission spectra, it follows that atomic absorption spectroscopy is less prone to interelement interferences than flame emission spectroscopy. Furthermore, in view of the high proportion of ground state to excited atoms, it would appear that atomic absorption spectroscopy should also be more sensitive than flame emission spectroscopy. However, in this respect the wavelength of the resonance line is a critical factor; and in flame emission spectroscopy, elements whose resonance lines are associated with relatively low energy values are more sensitive than those whose resonance lines are associated with higher energy values. Thus sodium, emission line at 589.0 nm, shows great sensitivity in flame emission spectroscopy; whereas zinc, emission line at 213.9 nm, is relatively insensitive.

The integrated absorption is given by

$$K\mathrm{d}\nu = fN_0(\pi e^2/mc)$$

where

K = the absorption coefficient at frequency ν

e = the electronic charge

m = the mass of an electron

c = the velocity of light

f = the oscillator strength of the absorbing line

N_0 = the number of metal atoms per millilitre able to absorb the radiation

The oscillator strength f of the absorbing line is inversely proportional to the lifetime of the excited state. In this expression the only variable is N_0 and it is this which governs the extent of absorption. Thus it follows that the integrated absorption coefficient is directly proportional to the concentration of the absorbing species.

It would appear that measurement of the integrated absorption coefficient should furnish an ideal method of quantitative analysis. In practice, however, the absolute measurement of the absorption coefficients of atomic spectral lines is extremely difficult. The natural linewidth of an atomic spectral line is about 10^{-5} nm, but owing to the influence of Doppler and pressure effects, the line is broadened to about 0.002 nm at flame temperatures of 2000–3000 K. To measure the absorption coefficient of a line thus broadened would require a spectrometer with a resolving power of 500 000. This difficulty was overcome by Walsh,[1] who used a source of sharp emission lines with a much smaller half-width than the absorption line, and the radiation frequency of which is centred on the absorption frequency. In this way, the absorption coefficient at the centre of the line, K_{max}, may be measured. If the profile of the absorption line is assumed to be due only to Doppler broadening, then there is a relationship between K_{max} and N_0. Thus the only requirement of the spectrometer is that it shall be capable of isolating the required resonance line from all other lines emitted by the source.

Note that in atomic absorption spectroscopy, as with molecular absorption, the absorbance A is given by the logarithmic ratio of the intensity of the incident light signal I_0 to the intensity of the transmitted light I_t, i.e.

$$A = \log I_0/I_t = KLN_0$$

where

N_0 = the concentration of atoms in the flame (number of atoms per millilitre)
L = the path length through the flame (cm)
K = a constant related to the absorption coefficient

For small values of the absorbance, this is a linear function.

With flame emission spectroscopy, the detector response E is given by the expression

$$E = k\alpha c$$

where

k = a constant containing a variety of factors, including the efficiency of atomisation and the self-absorption
α = the efficiency of atomic excitation
c = the concentration of the test solution

It follows that any electrical method of increasing E, e.g. improved amplification, will make the technique more sensitive.

The basic equation for atomic fluorescence is given by

$$F = QI_0kc$$

where

Q = the quantum efficiency of the atomic fluorescence process
I_0 = the intensity of the incident radiation
k = a constant which is governed by the efficiency of the atomisation process
c = the concentration of the element concerned in the test solution

The more powerful the radiation source, the greater the sensitivity of the technique.

To summarise, in both atomic absorption spectroscopy and atomic fluorescence spectroscopy, the factors which favour the production of gaseous atoms in the ground state determine the success of the techniques. In flame emission spectroscopy there is an additional requirement, namely the production of excited atoms in the vapour state. Note that the conversion of the original solid MX into gaseous metal atoms (M_{gas}) will be governed by a variety of factors, including the rate of vaporisation, flame composition and flame temperature; furthermore, if MX is replaced by a new solid MY, then the formation of M_{gas} may proceed in a different manner, and with a different efficiency from that observed with MX.

15.3   Instrumentation

The three flame spectrophotometric procedures require the following essential apparatus.

Nebuliser–burner Flame emission spectroscopy requires a nebuliser–burner system which produces gaseous metal atoms by using a suitable combustion flame involving a fuel gas–oxidant gas mixture. But with so-called non-flame cells, the burner is not required.

Spectrophotometer The spectrophotometer system should include a suitable optical train, a photosensitive detector and appropriate display device for the output from the detector.

Resonance line source For both atomic absorption spectroscopy and atomic fluorescence spectroscopy, a resonance line source is required for each element to be determined; these line sources are usually modulated (Section 15.10).

A schematic diagram showing the disposition of these essential components for the different techniques is given in Figure 15.3. The boxed components on the diagram represent the apparatus required for flame emission spectroscopy. For atomic absorption spectroscopy and for atomic fluorescence spectroscopy there is the additional requirement of a resonance line source. In atomic absorption spectroscopy this source is placed in line with the detector, but in atomic fluorescence spectroscopy it is placed at right angles to the detector, as shown in the diagram. Chapter 16 gives more detail on the apparatus for flame spectrophotometry.

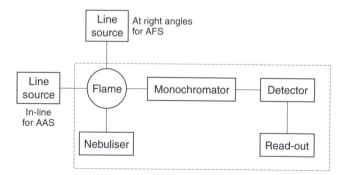

Figure 15.3   A flame spectrophotometer has three essential components: nebuliser–burner, spectrophotometer, resonance line source

15.4   Flames

An essential requirement of flame spectroscopy is a flame temperature greater than 2000 K. In most cases this can only be met by burning the fuel gas in an oxidant gas which is

Table 15.2 *Flame temperatures with various fuels*

Fuel gas	Temperature (K)	
	Air	Nitrous oxide
Acetylene	2400	3200
Hydrogen	2300	2900
Propane	2200	3000

usually air, nitrous oxide, or oxygen diluted with either nitrogen or argon. Table 15.2 gives the temperatures attained by the common fuel gases burning in air and burning in nitrous oxide. The flow rates of both the fuel gas and the oxidant gas should be measured, for some flames need to be rich in fuel gas, whereas other flames need to be lean in fuel gas; these requirements are discussed in Section 15.20. The concentration of gaseous atoms within the flame, both in the ground state and in the excited state, may be influenced by two factors:

Flame composition An acetylene–air mixture is suitable for the determination of some 30 metals, but a propane–air flame is preferred for metals which are easily converted into an atomic vapour state. For metals such as aluminium and titanium, which form refractory oxides, the higher temperature of the acetylene–nitrous oxide flame is essential, and the sensitivity is enhanced if the flame is fuel-rich.

Position in the flame In certain cases the concentration of atoms may vary widely if the flame is moved relative to the light path either vertically or laterally from the resonance line source. Rann and Hambly[2] have shown that with certain metals (e.g. calcium and molybdenum), the region of maximum absorption is restricted to specific areas of the flame, whereas the absorption of silver atoms does not alter appreciably within the flame, and is unaffected by the fuel gas/oxidant gas ratio. For the sake of brevity, the so-called cool flame techniques based upon the use of an oxidant-lean flame such as hydrogen/nitrogen–air, have not been included.

15.5 The nebuliser–burner system

The purpose of the nebuliser–burner system is to convert the test solution to gaseous atoms as indicated in Figure 15.2, and the success of flame photometric methods depends on its correct functioning. Section 16.3 describes the burner system for a flame photometer. The function of the nebuliser is to produce a mist or aerosol of the test solution. The solution to be nebulised is drawn up a capillary tube by the venturi action of a jet of air blowing across the top of the capillary; a gas flow at high pressure is necessary in order to produce a fine aerosol. The main type of burner system used in flame atomic absorption is the **premix or laminar flow burner**.

In the premix type of burner, the aerosol is produced in a vaporising chamber where the larger droplets of liquid fall out from the gas stream and are discharged to waste. The resulting fine mist is mixed with the fuel gas and the carrier (oxidant) gas, and the mixed gases then flow to the burner head. In atomic absorption spectroscopy the burner is a long horizontal tube with a narrow slit along its length. This produces a thin flame of long path

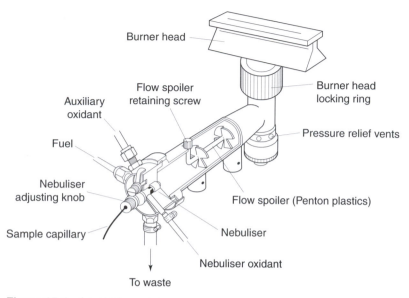

Figure 15.4   A typical premix burner

length which can be turned into or away from the beam of radiant energy. The flame path of a burner using air–acetylene, air–propane or air–hydrogen mixtures is about 10–12 cm in length, but with a nitrous oxide–acetylene burner it is usually reduced to about 5 cm because of the higher burning velocity of this gas mixture. In addition to a long light path, this type of burner has the advantages of being quiet in action and with little danger of incrustation around the burner head, since large droplets of solution have been eliminated from the stream of gas reaching the burner. Its disadvantages are that with solutions made up in mixed solvents, the more volatile solvents are evaporated preferentially; and a potential explosion hazard exists since the burner uses relatively large volumes of gas, but this hazard is minimised in modern versions.

A typical burner of this type is shown in Figure 15.4. In this particular burner (Perkin-Elmer Corp.) the mixing chamber is a steel casting lined with a plastic (Penton) which is extremely resistant to corrosion. The burner head is manufactured from titanium, thus avoiding the occasional high readings which are encountered when solutions containing iron and copper in presence of acid are examined with burners having a stainless steel head. The nebuliser is capable of adjustment so it can handle sample uptake rates from 1 to 5 mL min^{-1}. The burner can be adjusted in three directions, and horizontal and vertical scales are provided so that its position can be recorded. The head may be turned through an angle of 90° with respect to the light beam, so the path length of the flame traversed by the resonance line radiation may be varied considerably. By choosing a small path length it becomes possible to analyse solutions of relatively high concentration without the need for prior dilution.

In general terms, Thomerson and Thompson[3] have cited the following disadvantages of flame atomisation procedures:

1. Only 5–15% of the nebulised sample reaches the flame (in the case of the premix type of burner) and it is then further diluted by the fuel and oxidant gases so that the concentration of the test material in the flame may be extremely minute.

2.   A minimum sample volume of 0.5–1.0 mL is needed to give a reliable reading by aspiration into a flame system.

3.   Samples which are viscous (e.g. oils, blood, blood serum) require dilution with a solvent, or alternatively must be 'wet ashed' before the sample can be nebulised.

15.6   Graphite furnace technique

Instead of employing the high temperature of a flame to bring about the production of atoms from the sample, it is sometimes possible to use non-flame methods involving electrically heated graphite tubes or rods. This is graphite furnace atomic absorption spectroscopy (GFAAS).

The graphite tube furnace

A graphite tube furnace is illustrated in Figure 15.5. It consists of a hollow graphite cylinder about 50 mm in length and about 9 mm internal diameter, situated so the radiation beam passes along the axis of the tube. The graphite tube is surrounded by a metal jacket through which water is circulated and which is separated from the graphite tube by a gas space. An inert gas, usually argon, is circulated in the gas space, and enters the graphite tube through openings in the cylinder wall.

The solution of the sample to be analysed (1–100 μL) is introduced by inserting the tip of a micropipette through a port in the outer (water) jacket, and into the gas inlet orifice in the centre of the graphite tube. The graphite cylinder is then heated by the passage of an electric current to a temperature that is high enough to evaporate the solvent from the solution. The current is then increased so that firstly the sample is ashed, and then ultimately it is vaporised so that metal atoms are produced, typically at a temperature of about 3000 K. For reproducibility, the temperatures and the timing of the drying, ashing and atomisation processes must be carefully selected according to the metal which is to be determined. The absorption signals produced by this method may last for several seconds and can be recorded on a chart recorder. Each graphite tube can be used for 100–200 analyses, depending upon the nature of the material to be determined.

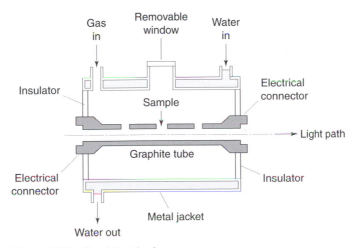

Figure 15.5   Graphite tube furnace

The main advantages are:

(a) Very small sample sizes can be used (as low as $0.5\,\mu L$).
(b) Often very little or no sample preparation is needed; some solid samples do not require prior dissolution.
(c) Sensitivity is enhanced; there may be hundred- or thousandfold improvements in the detection limits for furnace AAS compared with flame AAS.

The disadvantages are:

(a) Background absorption effects are usually more serious (Section 15.13).
(b) Analyte may be lost at the ashing stage, especially from volatile compounds, e.g. arsenic, selenium, tellurium and mercury compounds.
(c) The sample may not be completely atomised, which can produce 'memory effects' within the furnace.
(d) The precision was poorer than the flame methods, but the introduction of furnace autosamplers has enhanced the precision of furnace AAS.

Because of these disadvantages listed above, graphite furnace atomic absorption spectro-scopy (GFAAS) has gained the reputation of being a fairly difficult technique. Indeed, problems do arise from interferences and high background levels.

A modern instrument, e.g. the GF90 electrothermal analyser marketed by Thermo-Unicam, Cambridge, UK, incorporates features that allow the analyst to minimise the errors previously encountered using this technique. For example, a furnace autosampler is now considered mandatory in GFAAS. Features of a modern furnace autosampler include automatic matrix modification (see later), automatic standard preparation, reconcentration or dilution of samples and autorecalibration facilities.

The nature and design of graphite cuvettes is of great importance in GFAAS. A number of different graphite cuvettes are now available. The standard cuvette made from electro-graphite is suitable for the determination of volatile elements such as lead and cadmium. Those elements which form stable carbides, e.g. vanadium and tungsten, and elements of medium volatility, e.g. nickel and chromium, need a furnace with a pyrolytic graphite coating. Extended lifetime cuvettes can sustain faster heating rates and have longer useful lifetimes than either of the electrographite-based cuvettes. Extended lifetime cuvettes are used in the determination of refractory elements.

When a sample is volatilised from the cuvette wall, it is subjected to a relatively cooler and rapidly changing vapour temperature. This process, known as **non-isothermal atomisa-tion**, promotes the formation of analyte molecules rather than the required analyte atoms. This reduction in atom concentration is indicated by a signal depression. These interfer-ences may be eliminated by delaying the absorption signal until isothermal conditions have been reached by **platform atomisation**. The platform is a small piece of graphite located inside the graphite cuvette on to which the sample is located. The platform is heated mainly by radiation from the cuvette wall, hence the platform temperature lags behind the wall temperature. When the sample is eventually vaporised and atomised, it does so at a hotter temperature, thus reducing analyte molecule formation.

Several methods are available to reduce unacceptably high background levels. This may be achieved by simply diluting the sample or selecting another resonance wavelength line, ideally at a longer wavelength. However, the use of a **matrix modifier** is normally the major method used to reduce background effects. Matrix modification is effected by adding a reagent to the sample that may modify the behaviour of the matrix, or sometimes the analyte. There are three main reasons for adding a matrix modifier:

(a)   To stabilise the analyte during the ashing stage.

(b)   To convert the interfering matrix into a volatile compound that is removed in the ashing stage.

(c)   To obtain isothermal conditions in the graphite tube by delaying the analyte atomisation.

Recently, a technique termed **graphite probe atomisation** has been developed.[4] Initially the sample is automatically injected on to the probe inside the graphite cuvette, where it is dried and ashed. The spectrometer then automatically zeros, the probe is withdrawn from the cuvette and the furnace is heated to the preset atomisation temperature. When the cuvette temperature has been stabilised, the probe is inserted again into this constant-temperature environment, allowing isothermal atomisation and analytical measurement. Using this technique reduces the need for matrix modification, so it improves the control of contamination.

Great care must be taken in sample preparation with GFAAS. Contamination can arise from glassware and volumetric pipettes. All solutions for furnace work should be freshly prepared each day.

15.7   Cold vapour technique and hydride generation

The cold vapour technique is strictly confined to the determination of mercury[5] which, in its elemental state, has an appreciable vapour pressure at room temperature, so gaseous atoms exist without the need for any special treatment. As a method for determining mercury, the procedure consists in the reduction of a mercury(II) compound with either sodium borohydride or (more usually) tin(II) chloride to form elemental mercury:

$$Hg^{2+} + Sn^{2+} \rightleftharpoons Hg + Sn(IV)$$

A suitable apparatus is shown in Figure 15.6. The mercury vapour is flushed out of the reaction vessel by bubbling argon through the solution into the absorption tube.

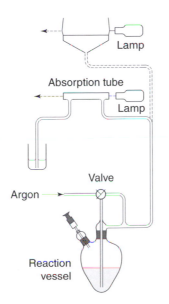

Figure 15.6   Apparatus for reducing a mercury(II) compound

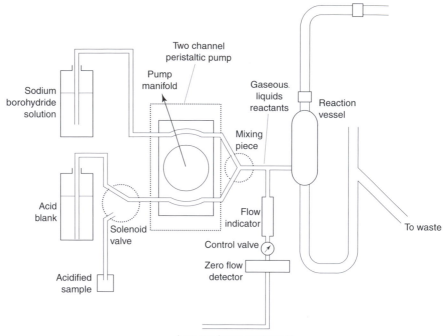

Figure 15.7 Continuous-flow vapour system (Courtesy of Thermo-Unicam, Cambridge, UK

This apparatus may also be adapted for what are called **hydride generation methods** (strictly speaking they are flame-assisted methods). Elements such as arsenic, antimony and selenium are difficult to analyse by flame AAS because it is difficult to reduce compounds of these elements (especially those in the higher oxidation states) to the gaseous atomic state.

Although electrothermal atomisation methods can be applied to the determination of arsenic, antimony and selenium, the alternative approach of hydride generation is often preferred. Compounds of these three elements may be converted to their volatile hydrides by the use of sodium borohydride as reducing agent. The hydride can then be dissociated into an atomic vapour by the relatively moderate temperatures of an air–acetylene flame.

The reaction sequence for arsenic may be represented as

$$\underset{(\text{sol})}{\text{As(V)}} \xrightarrow[\text{[H}^+]]{\text{NaBH}_4} \text{AsH}_3 \xrightarrow[\text{in flame}]{\text{heat}} \text{As}^0_{(\text{gas})} + \text{H}_2$$

The requisite additional apparatus is indicated by the broken lines in Figure 15.6.

Note that the hydride generation method may also be applied to the determination of other elements forming volatile covalent hydrides that are easily thermally dissociated. Thus, the hydride generation method has also been used for the determination of lead, bismuth, tin and germanium.

A continuous-flow vapour system, the Thermo-Unicam VP90, is shown in Figure 15.7. On the left-hand side of the system, the two reservoirs containing sodium borohydride and an acid blank are pumped to a reaction point in a continuous stream using a peristaltic pump. During an analysis the solenoid valve switches from the acid blank to the sample at a preset time and automatically switches back to the blank after the analytical measurement. The resulting gases and solutions are separated by a gas–liquid separator with a constant-head U-trap which allows the liquid to drain to waste automatically (right-hand side of diagram). The hydride is swept to the heated atom cell by a stream of nitrogen or argon. Atomisation is achieved usually by a flame (air–acetylene) heated silica cell.

The main advantage of the continuous flow principle is the steady signal that is produced. The use of normal integration is thus possible, instead of the peak height or area readings previously obtained. This procedure considerably improves the precision of measurement. A further advantage is that the reaction vessel is cleaned automatically between samples so they can be measured continuously without recourse to dismantling and washing out the system. A furnace accessory, the EC90, is available to achieve hydride atomisation through electrical heating of the silica cell. Electrical heating of the atomisation cell gives significantly higher sensitivity than flame heating.

For determination of mercury the same continuous-flow vapour system can be used, with tin(II) chloride as the reductant. One concentration system involves collecting the mercury as an amalgam for a defined period of time and then releasing it by heating the amalgam. This gives a stronger absorbance signal than obtained when mercury vapour is not subjected to a preconcentration stage.

15.8 Resonance line sources

As indicated in Figure 15.3, a resonance line source is required for both atomic absorption spectroscopy and atomic fluorescence spectroscopy, and the most important source is the **hollow cathode lamp**. The hollow cathode lamp has an emitting cathode made of the element being studied in the flame. The cathode is in the form of a cylinder, and the

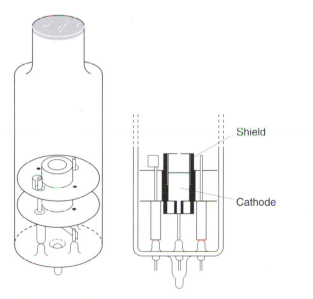

Shield

Cathode

Figure 15.8   Resonance line source

electrodes are enclosed in a borosilicate or quartz envelope which contains an inert gas (neon or argon) at a pressure of approximately 5 torr (670 Pa). The application of a high potential across the electrodes causes a discharge which creates ions of the noble gas. These ions are accelerated to the cathode and, on collision, excite the cathode element to emission. Multi-element lamps are available in which the cathodes are made from alloys, but in these lamps the resonance line intensities of individual elements are somewhat reduced.

Originally developed by Dagnall *et al.*[6] as radiation sources for AAS and AFS, **electrodeless discharge lamps** give radiation intensities which are much greater than those given by hollow cathode lamps. The electrodeless discharge lamp consists of a quartz tube 2–7 cm in length and 8 mm in internal diameter, containing up to 20 mg of the required element or a volatile salt of the element, commonly the iodide; the tube also contains argon at low pressure (about 270 Pa). Under operating conditions, the material placed in the tube must have a vapour pressure of about 1 mm at a temperature of 200–400 °C. A microwave frequency of 2000–3000 MHz applied through a waveguide cavity provides the energy of excitation.

15.9   Monochromators

A monochromator selects a given emission line and isolates it from other lines, and occasionally from molecular band emissions. In AAS the monochromator isolates the resonance line from all non-absorbed lines emitted by the radiation source. Most commercial instruments use diffraction gratings (Section 17.6) because the dispersion from a grating is more uniform than the dispersion from a prism, hence grating instruments can maintain a higher resolution over a longer range of wavelengths.

15.10   Detectors

A photomultiplier is now the universally used detector in atomic absorption spectrophotometers. The output from the detector is fed to a suitable read-out system, in order that the radiation received by the detector originates not only from the resonance line which has been selected, but also perhaps from emission within the flame. This emission can be due to atomic emission arising from atoms of the element under investigation and may also arise from molecular band emissions. Hence, instead of an absorption signal intensity I_A, the detector may receive a signal of intensity $(I_A + S)$ where S is the intensity of emitted radiation.

Since only the measurement arising from the resonance line is required, it is important to distinguish it from the effects of flame emission. This is achieved by **modulating** the emission from the resonance line source using a mechanical chopper device (Figure 15.9). In Figure 15.9(a) the beam is not blocked, so the detector receives the signal from both the lamp and the flame. When the beam is blocked (Figure 15.9(b)), the signal reaching the detector arises from flame emission. The difference between the two signals is the analyte signal (Figure 15.9(c)). An alternative form of modulation is to use an alternating current signal appropriate to the particular frequency of the resonance line. The detector amplifier is then tuned to this frequency so that signals arising from the flame, essentially d.c. signals, are effectively removed.

The read-out systems available include meters, chart recorders and digital display; meters have now been virtually superseded by the alternative methods of data presentation.

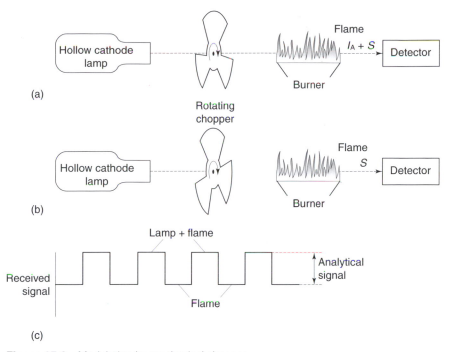

(a)

Rotating
chopper

(b)

(c)

Figure 15.9   Modulation by mechanical chopper

15.11   Interferences

Various factors may affect the flame emission of a given element, interfering with the determination of its concentration. These factors may be broadly classified as **spectral interferences** and **chemical interferences**.

Spectral interferences in AAS arise mainly from overlap between the frequencies of a selected resonance line with lines emitted by some other element; this occurs because in practice a chosen line has in fact a finite 'bandwidth'. Since the linewidth of an absorption line is about 0.005 nm, only a few cases of spectral overlap between the emitted lines of a hollow cathode lamp and the absorption lines of metal atoms in flames have been reported. Table 15.3 includes some typical examples of spectral interferences which have been observed.[7–10] However, most of these entries relate to relatively minor resonance lines and the only interferences which occur with preferred resonance lines are for copper and mercury. Copper would experience interference from europium at a concentration of about 150 mg L^{-1}; mercury would experience interference from cobalt at concentrations higher than 200 mg L^{-1}.

There is a greater likelihood with FES than with AAS of experiencing interferences when the element to be determined and the interfering substances have line emissions of similar wavelengths. Obviously some of these interferences may be eliminated by improved resolution of the instrument, e.g. by using a prism rather than a filter, but sometimes it may be necessary to select other, non-interfering, lines for the determination. It may even be necessary to separate the element to be determined from interfering elements by a separation process such as ion exchange or solvent extraction (Chapter 6).

Table 15.3 *Some typical spectral interferences*

Resonance source	Wavelength λ (nm)	Analyte	Wavelength λ (nm)
Aluminium	308.216	Vanadium	308.211
Antimony	231.147	Nickel	231.095
Copper	324.754	Europium	324.755
Gallium	403.307	Manganese	403.307
Iron	271.903	Platinum	271.904
Mercury	253.652	Cobalt	253.649

Apart from the interferences which may arise from other elements present in the substance to be analysed, some interference may arise from the emission band spectra produced by molecules or molecular fragments present in the flame gases; in particular, band spectra due to hydroxyl and cyanogen radicals arise in many flames. Although in AAS these signals are modulated (Section 15.10), in practice care should be taken to select an absorption line which does not correspond with the wavelengths due to any molecular bands because of the excessive 'noise' produced by the molecular bands; this leads to decreased sensitivity and poor analytical precision.

15.12 Chemical interferences

Stable compound formation

Stable compound formation leads to incomplete dissociation of the substance to be analysed when placed in the flame; another possibility is formation within the flame of refractory compounds which fail to dissociate into the constituent atoms. Examples of these types of behaviour are shown by the determination of calcium in the presence of sulphate or phosphate, and the formation of stable refractory oxides of titanium, vanadium and aluminium. Chemical interferences can usually be overcome in one of the following ways:

Increase in flame temperature This often leads to the formation of free gaseous atoms. For example, aluminium oxide is more readily dissociated in an acetylene–nitrous oxide flame than in an acetylene–air flame. And a calcium–aluminium interference arising from the formation of calcium aluminate can also be overcome by working at the higher temperature of an acetylene–nitrous oxide flame.

Use of releasing agents If we consider the reaction

$$M—X + R \rightleftharpoons R—X + M$$

then it is clear that an excess of the releasing agent R will lead to an enhanced concentration of the required gaseous metal atoms M; this will be especially true if the product R—X is a stable compound. Thus in the determination of calcium in the presence of phosphate, the addition of an excess of lanthanum chloride or of strontium chloride to the test solution will lead to formation of lanthanum (or strontium) phosphate, and the calcium can then be determined in an acetylene–air flame without any interference due to phosphate. The addition of EDTA to a calcium solution before analysis may increase the sensitivity of the subsequent flame spectrophotometric determination; this is possibly due to the formation of an EDTA complex of calcium which is readily dissociated in the flame.

Extraction of the analyte   Extraction of the analyte or extraction of the interfering element(s) is an obvious method of overcoming the effect of 'interferences'. It is frequently sufficient to perform a simple solvent extraction to remove the major portion of an interfering substance so, at its new concentration in the solution, the interference becomes negligible. If necessary, repeated solvent extraction will reduce the effect of the interference even further and, equally, a quantitative solvent extraction procedure may be carried out so as to isolate the substance to be determined from any interfering substances.

Ionisation

Ionisation of the ground state gaseous atoms within a flame

$$M = M^+ + e$$

will reduce the intensity of the emission of the atomic spectra lines in FES, or it will reduce the extent of absorption in AAS. It is therefore necessary to minimise the possibility of ionisation occurring, and an obvious precaution to take is to use a flame operating at the lowest possible temperature which is satisfactory for the element to be determined. Thus, the high temperature of an acetylene–air flame or an acetylene–nitrous oxide flame may result in the appreciable ionisation of elements such as the alkali metals and calcium, strontium and barium. The ionisation of the element to be determined may also be reduced by adding an excess of an **ionisation suppressant**; this is usually a solution containing a cation having a lower ionisation potential than the analyte. For example, a solution containing potassium ions at a concentration of $2000 \, mg \, L^{-1}$ added to a solution containing calcium, barium or strontium ions creates an excess of electrons when the resulting solution is nebulised into the flame, and this has the result that ionisation of the metal to be determined is virtually completely suppressed.

Other effects

Besides ionisation and compound formation, it is also necessary to take account of so-called **matrix effects**. These are predominantly physical factors which will influence the amount of sample reaching the flame, and are related in particular to factors such as the viscosity, density, surface tension and volatility of the solvent used to prepare the test solution. If we wish to compare a series of solutions, e.g. a series of standards to be compared with a test solution, it is essential to use the same solvent for each, and the solutions should not differ too widely in their bulk composition. This procedure is commonly known as **matrix matching**.

Interference may sometimes result from **molecular absorption**. Thus, in an acetylene–air flame, a high concentration of sodium chloride will absorb radiation at wavelengths in the neighbourhood of 213.9 nm, which is the wavelength of the major zinc resonance line; hence sodium chloride would represent an interference in the determination of zinc under these conditions. Such interferences can usually be avoided by choosing a different resonance line, or by using a hotter flame to increase the operating temperature, leading to dissociation of the interfering molecules.

The interference, known as **background absorption**, arises from the presence in the flame of gaseous molecules, molecular fragments and sometimes smoke; smoke may occur when organic solvents are used. It is dealt with instrumentally by the incorporation of a **background correction** facility. In addition, background effects can be caused by light scatter. The degree of scatter is inversely proportional to the fourth power of the radiation wavelength. Hence background effects due to scatter are a particularly important interference at high-energy wavelengths in the UV (between 185 and 230 nm). Background effects are a major problem in furnace AAS, especially smoke caused by particulate material. Note that background correction methods should always be used in furnace AAS. The background effect in this case may be as high as 85% of the total absorption signal.

Summary

Almost all interferences encountered in atomic absorption spectroscopy can be reduced, if not completely eliminated, by the following procedures:

1. Ensure if possible that standard and sample solutions are of similar bulk composition to eliminate matrix effects (matrix matching).
2. Alteration of flame composition or of flame temperature can be used to reduce the likelihood of stable compound formation within the flame.
3. Selection of an alternative resonance line will overcome spectral interferences from other atoms or molecules and from molecular fragments.
4. Separation, e.g. by solvent extraction or an ion exchange process, may occasionally be necessary to remove an interfering element; these separations are most frequently required when dealing with flame emission spectroscopy.
5. Use an appropriate background correction facility (Section 15.13).

15.13 Background correction methods

The measured sample absorbance may be higher than the true absorbance signal of the analyte to be determined. This elevated absorbance value can occur by molecular absorption or by light scattering. Three techniques can be used for background correction: the deuterium arc, the Zeeman effect and the Smith–Hieftje system.

Deuterium arc background correction

Deuterium arc background correction uses two lamps, a high-intensity deuterium arc lamp producing an emission continuum over a wide wavelength range and the hollow cathode lamp of the element to be determined. The deuterium arc continuum travels the same double-beam path as the light from the resonance source (Figure 15.10). The background absorption affects both the sample beam and the reference beam, so when the ratio of the intensities of the two beams is taken, the background effects are eliminated.

The deuterium arc method is widely used in many instruments and is often satisfactory for background correction. It suffers, however, from the difficulty of achieving perfect alignment of the two lamps along an identical light path through the sample cell. In addition due to the low output of a deuterium lamp in the visible region of the spectrum, its use is limited to wavelengths of less than 340 nm. The background absorption should ideally be measured as near as possible to that of the analyte line. This approach has been achieved in the Zeeman and Smith–Hieftje methods.

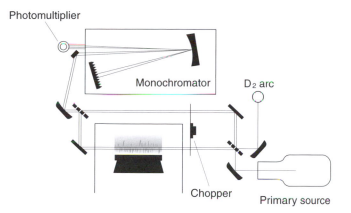

Figure 15.10   Deuterium arc background correction

Zeeman background correction

In a strong magnetic field the electronic energy levels of atoms may be split, producing several absorption lines for each electronic transition (the Zeeman effect). The simplest form of splitting pattern is shown in Figure 15.11. In the presence of the magnetic field three components are observed, the π-component having the same energy as the transition in the absence of the magnetic field, and the σ-components observed at lower and higher energies, typically 0.01 nm distant from the π-component. The π-peak is plane-polarised parallel to the direction of the magnetic field, whereas the σ-peaks are polarised perpendicular to the direction of the magnetic field (Figure 15.12).

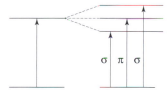

Figure 15.11   Zeeman splitting: the simplest pattern

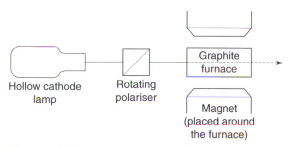

Figure 15.12   Zeeman background correction

In practice the emission line is split into three peaks by the magnetic field. The polariser is used to isolate the central line, and the absorption A_π, which includes absorption of

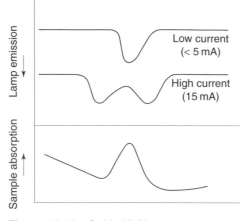

Figure 15.13   Smith–Hieftje system: the sample absorption (bottom) is the difference between the low current trace (top) and the high current trace (centre) followed by an inversion

radiation by the analyte is measured. The polariser is then rotated and the absorption of the background A_σ is measured. The analyte absorption is given by $A_\pi - A_\sigma$. A detailed discussion of the application of the Zeeman effect in atomic absorption is given in the literature.[11]

The Smith–Hieftje system

The Smith–Hieftje system[12] is based on the principle of self-absorption. If a hollow cathode lamp is operated at low current, the normal emission line is obtained. At high lamp currents the emission band is broadened with a minimum appearing in the emission profile that corresponds exactly to the wavelength of the absorption peak. Hence at low current the total absorbance due to the analyte and the background is measured, whereas at high current essentially only the background absorbance is obtained. The hollow cathode lamp is run alternatively at low and high current (Figure 15.13). Thus, the analyte absorption is given by the difference between the absorbance measured at low lamp current and the absorbance measured at high lamp current.

Both the Zeeman and the Smith–Hieftje systems have the advantages that only one light source is used and the background is measured very close to the sample absorption. Unlike the deuterium arc, neither technique is restricted to operating solely in the ultraviolet region of the spectrum. But the Zeeman background correction is normally limited to use in furnace AAS and can suffer from a lack of sensitivity. The Smith–Hieftje system, although less expensive than the Zeeman method, suffers the disadvantage that the lifetime of hollow cathode lamps may be shortened, particularly those incorporating the more volatile elements.

15.14   **Atomic absorption spectrophotometers**

Many commercial instruments are now available and are based either on a single- or double-beam design. Here are some instrumental features of a modern atomic absorption instrument:

1. It should have a lamp turret capable of holding at least four hollow cathode lamps with an independent current-stabilised supply to each lamp.

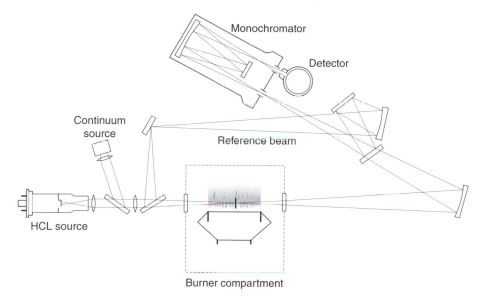

Monochromator

Detector

Continuum source

Reference beam

HCL source

Burner compartment

Figure 15.14   The Stockdale optical system used in the Thermo-Unicam Solaar AA

2. The sample area should be able to incorporate an autosampler which can work with both flame and furnace atomisers. Improved analytical precision is obtained when an autosampler is used in conjunction with a furnace atomiser.
3. The monochromator should be capable of high resolution, typically 0.04 nm. This feature is most desirable if the AAS is adapted for flame emission work; good resolution is also desirable for many elements in atomic absorption.
4. The photomultiplier should be able to function over the wavelength range 188–800 mm.
5. All instruments should be equipped with a background correction facility.
6. An integral video screen makes the instrument much easier to operate; analytical methods become much easier to develop and understand. A modern software package, e.g. the SOLAAR software used in conjunction with the Thermo-Unicam instruments, includes help facilities, full graphical data presentation, complete data storage and flexible report generation. Self-optimisation of flame and spectrometer parameters, even furnace temperature programmes, may be optimised using this software package.

The optical design in Thermo-Unicam SOLAAR AA spectrometers is shown in Figure 15.14. This double-beam system has the unique facility that the reference optics are completely removed throughout the measurement period. Thus it manages to combine enhanced signal-to-noise performance, an advantage of the single-beam instrument, with the drift stability of a double-beam system. In addition, the SOLAAR flame systems provide a very high-energy deuterium source to ensure good background correction.

Experimental preliminaries

15.15   Calibration curve procedure

A calibration curve for use in atomic absorption measurements is plotted by aspirating into the flame samples of solutions containing known concentrations of the element to be determined, measuring the absorption of each solution, and then constructing a graph in which the measured absorption is plotted against the concentration of the solutions. If we are dealing with a test solution which contains a single component, then the standard solutions are prepared by dissolving a weighed quantity of a salt of the element to be determined in a known volume of distilled (deionised) water in a graduated flask. But if other substances are present in the test solution, they should also be incorporated in the standard solutions and at a similar concentration to whatever exists in the test solution.

At least four standard solutions should be used covering the optimum absorbance range 0.1–0.4; and if the calibration curve proves to be non-linear (this often happens at high absorbance values), then measurements with additional standard solutions should be carried out. In common with all absorbance measurements, the readings must be taken after the instrument zero has been adjusted against a blank, which may be distilled water or a solution of similar composition to the test solution but minus the component to be determined. It is usual to examine the standard solutions in order of increasing concentration, and after making the measurements with one solution, distilled water is aspirated into the flame to remove all traces of solution before proceeding to the next solution. At least two, and preferably three, separate absorption readings should be made with each solution, and an average value taken. If necessary, the test solution must be suitably diluted using a pipette and a graduated flask, so it too gives absorbance readings in the range 0.1–0.4.

Using the calibration curve it is a simple matter to interpolate from the measured absorbance of the test solution the concentration of the relevant element in the solution. The working graph should be checked occasionally by making measurements with the standard solutions, and if necessary a new calibration curve must be drawn. All modern instruments include a microcomputer which stores the calibration curve and allows a direct read-out of concentration. For details on the errors in the slope and intercept of a regression line, see Sections 4.17 and 4.18; for errors in the estimate of concentration, see Section 4.19. Be aware of the likely sources of error using the calibration curve procedure.

The standard addition technique

When dealing with a test solution that is complex in character, or whose exact composition is unknown, it may be very difficult and even impossible to prepare standard solutions having a similar composition to the sample. Then the method of standard addition can be employed. It is described in detail in Section 4.20, along with how to estimate the error in concentration. Use the following experimental procedure.

Add **known** amounts of the analyte solution to a number of aliquots of the sample solution; the resulting solutions should all be diluted to the same final volume. If the absorbance of the test solution is too high, a quantitative dilution must be carried out and the measurements made with this diluted solution. Measure the absorbance of the test solution then examine each of the prepared solutions in turn, leading up to the solution of highest concentration; remember to aspirate distilled water into the flame between each solution. Plot the absorbance values against the added concentration values; a straight-line plot should result, and the straight line can be extrapolated to the concentration axis – the point where the axis is cut gives the concentration of the test solution. If the graph is **non-linear** then extrapolation is not possible. Remember that an extrapolation procedure is never as reliable as interpolation, so always choose interpolation if possible.

15.16   Preparation of sample solutions

For application of flame spectroscopic methods the sample must be prepared in the form of a suitable solution unless it is already presented in this form; exceptionally, solid samples can be handled directly in some of the non-flame techniques (Section 15.6).

Aqueous solutions may sometimes be analysed directly without any pretreatment, but it is a matter of chance that the given solution should contain the correct amount of material to give a satisfactory absorbance reading. If the existing concentration of the element to be determined is too high, then the solution must be diluted quantitatively before commencing the absorption measurements. Conversely, if the concentration of the metal in the test solution is too low, then a concentration procedure must be carried out by one of the methods outlined at the end of this section.

With certain reservations, solutions in organic solvents may be used directly, provided the viscosity of the solution is not very different from the viscosity of an aqueous solution. The important consideration is that the solvent should not lead to any disturbance of the flame; an extreme example is carbon tetrachloride, which may extinguish an air–acetylene flame. In many cases, suitable organic solvents, e.g. 4-methylpentan-2-one (methyl isobutyl ketone) and the hydrocarbon mixture sold as white spirit, give enhanced production of ground state gaseous atoms and lead to about three times the sensitivity achieved with aqueous solutions. Observe all relevant safety procedures (Section 15.18).

Solid samples will need some form of dissolution procedure prior to measurement; it is more acceptable for both flame and graphite furnace to use a liquid sample. Many dissolution procedures are available; here are some of them.

Wet ashing

The usual method is to treat the solid sample by acid digestion, producing a clear solution with no loss of the element to be determined. Hydrochloric acid, nitric acid or aqua regia (3 : 1 hydrochloric acid : nitric acid) will dissolve many inorganic substances. Hydrofluoric acid must be used to decompose silicates, and perchloric acid is often used to break up organic complexes. The instruction manual normally supplied with the instrument will give guidance on acceptable acid concentrations. Biological samples usually only require simple dilution prior to measurement, or they can be measured directly using furnace atomic absorption.

Safety. Hydrofluoric acid and perchloric acid must be handled with care. Strict adherence to safety precautions when using these acids must be observed. Generally speaking, the final solution should not contain acid at a concentration more than about 1 M, since the aspiration of corrosive solutions into the burner should be avoided as far as possible.

Fusions

A weighed sample is mixed with a flux in a metal or graphite crucible. The sample and flux mixture is heated over a flame, or in a furnace, and the resulting fused material is leached with either water or an appropriate acid or an alkali. The most widely used flux is sodium peroxide (**care**). Fusions with this substance are normally carried out in a zirconium crucible and the cooled melt is then leached with dilute mineral acid. Lithium metaborate (Section 3.31) is a good flux for silicate rocks. The extra salt content produced by the fusion process may create problems in flame atomic absorption where the salt concentration should ideally be less than 3% (w/v). This high salt concentration can be even more critical in furnace atomic absorption.

Dry ashing

The sample is weighed into a crucible, heated in a muffle furnace and then the residue is dissolved in a suitable acid. This technique is often used to remove organic substances from the analyte material. Care must be taken to ensure that volatile elements such as mercury, arsenic and even lead are not removed in the ashing process.

Microwave dissolution

Microwave ovens have been used for sample dissolution. The sample is sealed in a specially designed microwave digestion vessel with a mixture of the appropriate acids. The high-frequency microwave temperature, typically $100-250\,°C$, and the increased pressure assist in the considerable reduction in the time taken for sample dissolution. The method has been used for the dissolution of samples of coal, fly ash, biological and geological materials.[13]

Concentration procedures

Separation techniques may have to be applied if the analyte sample contains substances which act as interferences (Section 15.10) or if the concentration of the test element to be determined is too low to give satisfactory absorbance readings. The separation methods most commonly used with flame spectrophometric methods are solvent extraction and ion exchange (Chapter 6), where ion exchange chromatography has been used in the separation of gallium from aluminium and indium.[14]

15.17 Preparation of standard solutions

In flame spectrophotometric measurements we are concerned with solutions having very small concentrations of the element to be determined. It follows that the standard solutions which will be required for the analyses must also contain very small concentrations of the relevant elements, and it is rarely practicable to prepare the standard solutions by directly weighing out the required reference substance. The usual practice, therefore, is to prepare stock solutions which contain about $1000\,\mu g\,mL^{-1}$ of the required element, and then the working standard solutions are prepared by suitable dilution of the stock solutions. Solutions which contain less than $10\,\mu g\,mL^{-1}$ are often found to deteriorate on standing, owing to adsorption of the solute on to the walls of glass vessels. Consequently, standard solutions in which the solute concentration is of this order should not be stored for more than 1–2 days. The stock solutions are ideally prepared from the pure metal or from the pure metal oxide by dissolution in a suitable acid solution; the solids used must be of the highest purity.

15.18 Safety practices

Before commencing any experimental work with an atomic absorption spectrophotometer, the following guidelines on safety practices should be studied. These recommendations are a summary of the Code of Practice recommended by the Scientific Apparatus Makers' Association (SAMA) of the USA; for full details see Ref. 15.

1. Ensure that the laboratory in which the apparatus is housed is well ventilated and is provided with an adequate exhaust system having air-tight joints on the discharge side; some organic solvents, especially those containing chlorine, give toxic products in a flame.

2. Gas cylinders must be fastened securely in an adequately ventilated room well away from any heat or ignition sources. The cylinders must be clearly marked so that the contents can be immediately identified.

3. When the equipment is turned off, close the fuel gas cylinder valve tightly and bleed the gas line to the atmosphere via the exhaust system.

4. The piping which carries the gases from the cylinders must be securely fixed in such a position that it is unlikely to suffer damage.

5. Make periodic checks for leaks by applying soap solution to joints and seals.

6. The following special precautions should be observed with acetylene.

    (a) Never run acetylene gas at a pressure higher than 15 psi ($103 \, kN \, m^{-2}$); at higher pressures acetylene can explode spontaneously.

    (b) Avoid the use of copper tubing. Use tubing made from brass containing less than 65% copper, from galvanised iron or from any other material that does not react with acetylene.

    (c) Avoid contact between gaseous acetylene and silver, mercury or chlorine.

    (d) Never run an acetylene cylinder after the pressure has dropped to 50 psi ($3430 \, kN \, m^{-2}$); at lower pressures the gas will be contaminated with acetone.

7. A nitrous oxide cylinder should not be used after the regulator gauge has dropped to a reading of 100 psi ($6860 \, kN \, m^{-2}$).

8. A burner which utilises a mixture of fuel and oxidant gases and which is attached to a waste vessel (liquid trap) should be provided with a U-shaped connection between the trap and the burner chamber. The head of liquid in the connecting tube should be greater than the operating pressure of the burner: if this is not achieved, mixtures of fuel and oxidant gas may be vented to the atmosphere and form an explosive mixture. The trap should be made of a material that will not shatter in the event of an explosive flash-back in the burner chamber.

9. Care must be exercised when using volatile flammable organic solvents for aspiration into the flame. A container fitted with a cover which is provided with a small hole for the sample capillary is recommended.

10. Never view the flame or hollow cathode lamps directly; protective eye wear should always be worn. Safety spectacles will usually provide adequate protection from ultraviolet light, and will also provide protection for the eyes in the event of the apparatus being shattered by an explosion.

15.19 Detection limits

Sections 15.20 to 15.22 describe a few applications of atomic absorption; they have been chosen to illustrate the general procedures involved, including the manner in which certain interferences may be overcome. Table 15.4 lists the wavelength of the most widely used resonance line for each of the common elements, together with the normal composition of the flame gases. The optimum working range of concentrations is quoted, and although it can vary with the instrument used, the cited values may be regarded as typical.

The term 'sensitivity' used in atomic absorption spectroscopy is defined as the concentration of an aqueous solution of the elements which absorbs 1% of the incident resonance

15 Atomic absorption spectroscopy

Table 15.4 *FAAS data for the common elements*

Element	Wavelength of main resonance line λ (nm)	Flame[a]	Working range (μg mL^{-1})
Ag	328.1	AA(L)	1–5
Al	309.3	NA(R)	40–200
As[b]	193.7	AH(R)	50–200
B	249.8	NA(R)	400–600
Ba	553.6	NA(R)	10–40
Be	234.9	NA(R)	1–5
Bi	223.1	AA(L)	10–40
Ca	422.7	NA(R)	1–4
Cd	228.8	AA(L)	0.5–2
Co	240.7	AA(L)	3–12
Cr	357.9	AA(R)	2–8
Cs	852.1	AP(L)	5–20
Cu	324.7	AA(L)	2–8
Fe	248.3	AA(L)	2.5–10
Ga	294.4	AA(L)	50–200
Ge[b]	265.2	NA(R)	70–280
Hg[c]	253.7	AA(L)	100–400
In	303.9	AA(L)	15–60
Ir	208.9	AA(R)	40–160
K	766.5	AP(L)	0.5–2
Li	670.8	AP(L)	1–4
Mg	285.2	AA(L)	0.1–0.4
Mn	279.5	AA(L)	1–4
Mo	313.3	NA(R)	15–60
Na	589.0	AP(L)	0.15–0.60
Ni	232.0	AA(L)	3–12
Os	290.9	NA(R)	50–200
Pb	217.0	AA(L)	5–20
Pd	244.8	AA(L)	4–16
Pt	265.9	AA(L)	50–200
Rb	780.0	AP(L)	2–10
Rh	343.5	AA(L)	5–25
Ru	349.9	AA(L)	30–120
Sb[b]	217.6	AA(L)	10–40
Sc	391.2	NA(R)	15–60
Se[b]	196.0	AH(R)	20–90
Si	251.6	NA(R)	70–280
Sn	224.6	AH(R)[d]	15–60
Sr	460.7	NA(L)	2–10
Te	214.3	AA(L)	10–40
Ti	364.3	NA(R)	60–240
Tl	276.8	AA(L)	10–50
V	318.5	NA(R)	40–120
W	255.1	NA(R)	250–1000
Y	410.2	NA(R)	200–800
Zn	213.9	AA(L)	0.4–1.6

[a] L = fuel-lean; R = fuel-rich; AA = air–acetylene; AP = air–propane; NA = nitrous oxide–acetylene; AH = air–hydrogen
[b] The hydride generation method (Section 15.7) is far more sensitive for determining the listed elements.
[c] The non-flame mercury cell (Section 15.7) is far more sensitive for determining mercury.
[d] If there are many interferences then NA is to be preferred.

radiation; in other words, it is the concentration which gives an absorbance of 0.0044. Note that sensitivity depends on the reaction occurring in a flame; it is not strictly a characteristic of a given instrument. Remember that the sensitivity of a technique is a different concept.[16] It is defined as the slope of the calibration graph. The use of this definition has been applied to estimating a detection limit (see below). Since both these definitions are encountered in the literature, be sure to make a clear distinction between them. Another widely quoted value is the **detection limit**.

Clarification of detection limit and sensitivity

There has been considerable confusion in the definition and the use of the term 'detection limit'. The limit of detection may be defined as the lowest concentration of an analyte that can be distinguished with **reasonable confidence** from a **field blank** (a sample containing zero concentration of the analyte). What exactly was understood by 'reasonable confidence' and what was taken to be a 'blank' were the main causes of the disparate definitions of detection limit. The International Union of Pure and Applied Chemists (IUPAC)[17] definition is expressed in terms of concentration C_L or amount q_L; it is related to the smallest measure of the response Y_L (e.g. absorbance) that can be detected with reasonable certainty in an analytical procedure, where

$$Y_L = \bar{Y}_B + kS_B \tag{15.1}$$

and
$\bar{Y}_B$ = mean of blank measures
S_B = standard deviation of the blank measures
k = numerical constant

A value of $k = 3$ was strongly recommended by IUPAC. In a detailed paper the Analytical Methods Committee of the Royal Society of Chemistry sought to clarify the IUPAC definition.[19] It gives a full explanation of all aspects of the detection limit.

The estimation of the detection limit is best understood by considering a calibration graph. Using the linear regression method, it is possible to obtain the intercept on the y-axis and the slope of the best-fit line (Section 4.17). The calculated intercept can be used as an estimate of $\bar{Y}_B$, and the statistic $S_{y/x}$ is acceptable as a measure of S_B (Section 4.18). Thus, from equation (15.1), given $k = 3$, the value of Y_L can be calculated, hence the detection limit C_L may be evaluated from

$$C_L = 3\left(\frac{S_B}{S}\right) \tag{15.2}$$

where S, the **sensitivity** of the technique, is defined as the **slope** of the calibration line. The following example shows how the detection limit may be estimated.

Example 15.1

Standard aqueous solutions of Ca^{2+}, each containing $1000\,mg\,L^{-1}$ of lanthanum chloride as releasing agent, were measured using flame atomic absorption spectroscopy. The following absorbance values were obtained. Estimate the detection limit for Ca using this analytical procedure.

Absorbance	0.015	0.081	0.152	0.230	0.306
Ca^{2+} concentration (mg L^{-1})	0.0	1.0	2.0	3.0	4.0

From the data, using the method of Section 4.17, the equation of the line is found to be

$$y = 0.073x + 0.11$$

The intercept, 0.11, is an estimate of $\bar{Y}_B$ the mean of the field blank measures. And the statistic $S_{y/x} = 0.0077$, obtained using the method of Section 4.18, is a measure of S_B. The slope of the line is 0.073. From equation (15.2) the detection limit is

$$C_L = 3\left(\frac{S_B}{S}\right) = 3\left(\frac{0.0077}{0.073}\right) = 0.32 \, \text{mg L}^{-1}$$

Strictly, the detection limit of an **analytical system** should be estimated, taking into account a wide number of factors that can influence the method response. Many textbooks define detection limits in terms of the standard deviation of the blank with the value $k = 2$. However, the blank is simply the solvent in which the sample is conveyed to the instrument; it is not a field blank which contains the sample matrix. This instrumental detection limit can describe the instrument performance; nevertheless, a value of $k = 3$ should now be adopted.

A more detailed table of the resonance lines is given in Appendix 9. The data presented in Table 15.4 in conjunction with the experimental details given in Sections 15.20 to 15.22 will enable the determination of most elements to be carried out successfully. For detailed accounts of the determination of individual elements by atomic absorption spectroscopy, consult the works in Section 15.25. And most instrument manufacturers supply applications handbooks for their apparatus, giving full experimental details.

Selected determinations by atomic absorption spectroscopy

!   **Safety.** Before carrying out any experiments in this section, pay full attention to any safety warnings and make sure your adhere to national laboratory and safety regulations.

15.20   Magnesium and calcium in tap water

The determination of magnesium in potable water is very straightforward; very few interferences are encountered when using an acetylene–air flame. However, the determination of calcium is more complicated; many chemical interferences are encountered in the acetylene–air flame and it requires the use of releasing agents such as strontium chloride, lanthanum chloride or EDTA. Using the hotter acetylene–nitrous oxide flame, the only significant interference arises from the ionisation of calcium, and under these conditions an ionisation buffer such as potassium chloride is added to the test solutions.

Determination of magnesium

Preparation of the standard solutions   A magnesium stock solution $(1000 \, \text{mg L}^{-1})$ is prepared by dissolving 1.000 g magnesium metal in 50 mL of 5 M hydrochloric acid.

After dissolution of the metal, the solution is transferred to a 1 L graduated flask and made up to the mark with distilled water. An intermediate stock solution containing $50\,mg\,L^{-1}$ is prepared by pipetting 50 mL of the stock solution into a 1 L graduated flask and diluting to the mark. Dilute accurately four portions of this solution to give four standard solutions of magnesium with known magnesium concentrations lying within the optimum working range of the instrument to be used (typically $0.1–0.4\,\mu g\,mL^{-1}\ Mg^{2+}$).

Procedure   Although the precise mode of operation may vary according to the particular instrument used, the following procedure may be regarded as typical. Place a magnesium hollow cathode lamp in the operating position, adjust the current to the recommended value (usually 2–3 mA), and select the magnesium line at 285.2 nm using the appropriate mono-chromator slit width. Connect the appropriate gas supplies to the burner following the instructions detailed for the instrument, and adjust the operating conditions to give a fuel-lean acetylene–air flame.

Starting with the least concentrated solution, aspirate in turn the standard magnesium solutions into the flame, and for each take three readings of the absorbance; between each solution, remember to aspirate deionised water into the burner. Finally, read the absorbance of the sample of tap water; this will usually require considerable dilution in order to give an absorbance reading lying within the range of values recorded for the standard solutions. Plot the calibration curve and use this to determine the magnesium concentration of the tap water. If the magnesium content of the water is greater than $5\,\mu g\,mL^{-1}$ it might be considered preferable to work with the less sensitive magnesium line at wavelength 202.5 nm.

Determination of calcium

Preparation of the standard solutions   For procedure 1 it is necessary to incorporate a releasing agent in the standard solutions. Three different releasing agents may be used for calcium, (a) lanthanum chloride, (b) strontium chloride and (c) EDTA; of these (a) is the preferred regent, but (b) or (c) make satisfactory alternatives.

(a)   Prepare a lanthanum stock solution ($50\,000\,mg\,L^{-1}$) by dissolving 67 g of lanthanum chloride ($LaCl_3 \cdot 7H_2O$) in 100 mL of 1 M nitric acid. Warm gently to dissolve the salt, then cool the solution and make up to 500 mL in a graduated flask.

(b)   A strontium stock solution is prepared by dissolving 76 g of strontium chloride ($SrCl_2 \cdot 6H_2O$) in 250 mL of deionised water and then making up to 500 mL in a graduated flask.

(c)   An EDTA stock solution is prepared by dissolving 75 g of EDTA disodium salt (analytical grade) in 800 mL of deionised water. Warm gently until the salt is dissolved, then cool and make up to 1 L in a graduated flask.

For procedure 2 an ionisation buffer is required and this involves preparing a potassium stock solution ($10\,000\,mg\,L^{-1}$). Dissolve 9.6 g of potassium chloride in deionised water and make up to 500 mL in a graduated flask.

Prepare a calcium stock solution ($1000\,mg\,L^{-1}$) by dissolving 2.497 g of dried calcium carbonate in a minimum volume of 1 M hydrochloric acid; about 50 mL will be required. When dissolution is complete, transfer the solution to a 1 L graduated flask and make up to the mark with deionised water. An intermediate calcium stock solution is prepared by pipetting 50 cm of the stock solution into a 1 L flask and making up to the mark with deionised water.

The working standard solutions for procedure 1 contain 1–$5\,\mu g\,mL^{-1}$ Ca^{2+} and are prepared by mixing appropriate volumes of the intermediate stock solution (measured using a grade A pipette), with suitable volumes of the chosen releasing agent solution, and then making up to 50 mL in a graduated flask; the releasing agent solution is measured in a 25 mL measuring cylinder. Five standard solutions are prepared containing respectively 1.0, 2.0, 3.0, 4.0 and 5.0 mL of the intermediate stock solution and 10 mL of releasing agent (a) or 5 mL of either reagent (b) or (c). A blank solution is similarly prepared but without the addition of any of the intermediate calcium stock solution. For procedure 2 the working standard solutions are prepared as for procedure except the releasing agent solution is replaced by 10 mL of the stock potassium solution.

The unknown calcium solution (the tap water) will normally need diluting so its absorbance reading will lie on the calibration curve, and the same amount of releasing agent (procedure 1) or ionisation buffer (procedure 2) must be added as in the standard solutions. So, if the tap water contains about $100\,\mu g\,mL^{-1}$ of calcium, 25 mL of it are pipetted into a 100 mL graduated flask and made up to the mark with deionised water. Then 5 mL of this solution is pipetted into a 50 mL graduated flask, and if procedure is being followed, 10 mL of reagent (a) is added, or 5 mL of either reagent (b) or (c) and then the solution is made up to the mark. If procedure 2 is being followed, then 10 mL of the stock potassium solution are used in place of the releasing agent. If any cloudiness should develop during the preparation of the final solution, add 1 mL of $1\,M$ hydrochloric acid before making up to the mark.

Procedure 1 Set up a calcium hollow cathode lamp selecting the resonance line of wavelength 422.7 nm, and a fuel-lean acetylene–air flame following the details given in the instrument manual. The calibration procedure is similar to that described above for magnesium, but the aspiration of deionised water into the burner after taking the readings for each solution is even more important in this case owing to the relatively high concentrations of salts present as releasing agent; remember that deionised water should be aspirated into the burner for a few minutes after each series of readings.

Procedure 2 Make certain the instrument is fitted with the correct burner for an acetylene–nitrous oxide flame, then set the instrument up with the calcium hollow cathode lamp, select the resonance line of wavelength 422.7 nm, and adjust the gas controls as specified in the instrument manual to give a fuel-rich flame. Take measurements with the blank, and the standard solutions, and with the test solution, all of which contain the ionisation buffer; adequate treatment with deionised water is required after each measurement (see procedure 1). Plot the calibration graph and ascertain the concentration of the unknown solution.

15.21 Vanadium in lubricating oil

Discussion The oil is dissolved in white spirit and the absorption of this solution is compared with the absorption of standards made up from vanadium naphthenate dissolved in white spirit.

Preparation of the standard solutions The standard solutions are prepared from a solution of vanadium naphthenate in white spirit which contains about 3% of vanadium. Weigh out accurately about 0.6 g of the vanadium naphthenate into a 100 mL graduated flask

612

and make up to the mark with white spirit; this stock solution contains about $180 \, \mu g \, mL^{-1}$ of vanadium. Dilute portions of this stock solution measured with the aid of a grade A 50 mL burette to obtain a series of working standards containing $10-40 \, \mu g \, mL^{-1}$ of vanadium.

Procedure   Weigh out accurately about 5 g of the oil sample, dissolve in a small volume of white spirit and transfer to a 50 mL graduated flask; using the same solvent, wash out the weighing bottle and make up the solution to the mark. Set up a vanadium hollow cathode lamp selecting the resonance line of wavelength 318.5 nm, and adjust the gas controls to give a fuel-rich acetylene–nitrous oxide flame in accordance with the instruction manual. Aspirate successively into the flame the solvent blank, the standard solutions, and finally the test solution, in each case recording the absorbance reading. Plot the calibration curve and ascertain the vanadium content of the oil.

15.22   Trace elements in contaminated soil

Discussion   The procedure followed describes methods for determining **total** levels, and in certain cases, **available** amounts of trace elements in soils.

Sampling   Incremental samples (Section 5.1.2) of approximately 50 g should be taken from specified sampling points on the site. The sampling points should include surface soil and two further samples taken at depth, typically at 0.5 and 1.0 m. The exact location of these points should be noted, for it may be necessary to take further samples subsequently. The individual samples, carefully labelled, must be stored in separate containers to avoid cross-contamination. When received by the laboratory the samples are air-dried for a period to remove excess moisture. The individual dried samples are passed through a 0.5 mm sieve. The soil passing through the sieve is mixed and used to obtain the analytical sample.

Sample treatment for total element determination   Weigh out accurately about 1 g of the sieved soil and transfer to a 100 mL tall-form beaker. Add, from a measuring cylinder, about 20 mL of 1 : 1 nitric acid (Spectrosol grade) and boil **gently** on a hotplate until the volume of nitric acid is reduced to about 5 mL. Then add about 20 mL of deionised water and boil gently again until the volume is approximately 10 mL. Cool the suspension, and filter through a Whatman no. 540 filter paper, washing the beaker and the filter paper with small portions of deionised water until a volume of about 25 mL is obtained. Transfer the filtrate to a 50 mL graduated flask and make up to the mark with deionised water.

Sample treatment for 'available' metals   The metals zinc, copper and nickel are phytotoxic and it is necessary to ascertain, in addition to total levels, the available amounts of these metals that can be taken up by plants. The procedure adopted is as follows. Add from a measuring cylinder about 25 mL of approximately $0.05 \, M$ EDTA solution to about 1 g of an accurately weighed soil sample. Shake the suspension mechanically for a period of about 4 h. Continue with the filtering procedure described above under element determination.

Analysis of total metals by flame AAS   *Lead*   Use a fuel-lean acetylene–air flame with either the 217.0 nm resonance line (for samples containing a low lead concentration) or the 283.3 nm resonance line. Standard lead solutions containing $1-10 \, mg \, Pb \, mL^{-1}$ are suitable for the measurement at 217.0 nm, and $10-30 \, \mu g \, Pb \, mL^{-1}$ solutions at 283.3 nm. If

613

the lead concentration is too high to be measured directly using the 283.3 nm resonance line then further dilution of the sample solution is necessary. It is advisable to employ background correction (Section 15.12), especially when the 217.0 nm line is used.

Cadmium, copper, zinc and nickel   These can all be determined using an acetylene–air flame with the appropriate resonance lines and working range for the standard solutions given in Table 15.4. Again, further dilution of the sample solutions and background correction may be necessary.

Zinc, copper and nickel   These can be determined as available metals using the conditions given in Table 15.4. The standard solutions in this case should contain 0.05 M EDTA.

Arsenic in soil by hydride generation

Discussion   This procedure uses an atomic absorption spectrometer fitted with a hydride generation accessory. It is important that the atomic absorption spectrometer has a background correction system.

Reagents   *Reagent quality*   All reagents should be of analytical and spectroscopic quality.

Sodium tetrahydroborate(III), 1% (w/v)   Dissolve sodium hydroxide pellets (5.0 g) in 300 mL deionised water and cool. Add 5.0 g of sodium tetrahydroborate(III) directly to the sodium hydroxide solution and make up the total volume to 500 mL with deionised water. Shake the solution thoroughly and filter through a Whatman no. 541 filter paper. (The resulting solution is stable for at least one week.)

Hydrochloric acid, 4 M   Dilute Spectrosol hydrochloric acid (365 mL) to 1 L using deionised water.

Arsenic standards   Prepare a 1 mg L^{-1} working standard solution from a 1000 mg L^{-1} Spectrosol solution of arsenic trichloride in a 4 M hydrochloric acid.

Sample digestion procedure (aqua regia method)   Place an accurately weighed sample (about 1 g) of the soil into a Pyrex tube (50 mL) and a small quantity (2–3 mL) of deionised water to obtain a slurry. Then add Spectrosol hydrochloric acid (7.5 mL) followed by Spectrosol nitric acid (2.5 mL). Cover the tube with cling film overnight and then digest the sample in a Techne block digester for 2 h under reflux conditions using a cold-finger condenser. Filter the cooled solution through a Whatman no. 540 filter paper into a 50 mL graduated flask, and wash the residue with warm nitric acid (2 M). Make the filtrate up to the mark with deionised water.

Procedure   Follow the conditions recommended by the instrument manufacturers for the determination of arsenic by hydride generation. Typical instrumental parameters are resonance line at 193.7 nm with deuterium arc background correction. For exact operating conditions consult the instrument manufacturer's instructions. To obtain the calibration standards, take aliquots of 50–300 μL from the standard arsenic working solution, using an Eppindorf micropipette.

15.23   Tin in canned fruit juice

Discussion   The traditional flame atomic spectrometric determination of tin is relatively insensitive, and accurate quantification at low concentrations is difficult. But tin can be determined successfully by graphite furnace atomic absorption (GFAAS). The following procedure is an outline of the method described by Thermo-Unicam using the 939 QZ atomic absorption spectrometer fitted with the GF90 graphite furnace and a furnace autosampler.

Reagents
Spectrosol hydrochloric acid
Spectrosol tin standard ($1000\,mg\,L^{-1}$) from BDH Laboratory Supplies, Merck Ltd, Poole, Dorset, UK
Specpure ammonium nitrate from Johnson Matthey Chemicals, Royston, Herts, UK

Sample preparation   Pipette 20.0 mL of the fruit juice into a 100 mL beaker and add 10 mL of Spectrosol hydrochloric acid. Heat the mixture to boiling, cool and transfer to a 100 mL volumetric flask, then make up to the mark with deionised water. An aliquot of this solution is centrifuged before analysis, and the clear supernatant liquid is transferred to the autosampler cup.

Procedure   Prepare standard solutions containing 25, 50 and $100\,\mu g\,L^{-1}$ of tin in 10 vol% hydrochloric acid, together with an acid blank. Since tin is lost at quite low temperatures, $10\,\mu L$ of a 2% solution of ammonium nitrate is used as a matrix modifier. Set up the spectrometer with deuterium background correction and use the tin resonance line at 224.6 nm. To ensure good precision, use a relatively slow two-step drying stage (100 °C at 10 s, then 450 °C at 15 s). Set the ashing temperature at 800 °C (15 s) and the atomisation temperature at 2500 °C (3 s). Determine the tin present in the fruit juice by comparing the peak height measurements.

15.24   References

1. A Walsh 1955 *Spectrochim Acta*, **7**; 108
2. C S Rann and A N Hambly 1965 *Anal Chim.*, **37**; 879
3. D R Thomerson and K C Thompson 1975 *Chemistry in Britain*, **11**; 316
4. *Design considerations for a graphite probe in graphite furnace atomic absorption spectrometry*, Unicam Cambridge
5. W R Hatch and W L Ott 1968 *Anal Chem.*, **40**; 2085
6. R M Dagnall, K C Thompson and T S West 1967 *Talanta*, **14**; 551
7. C W Frank, W G Schrenk and C E McLean 1966 *Anal Chem.*, **38**; 1005
8. V A Fassel, J A Rasmuson and T G Cowley 1968 *Spectrochim Acta*, **23B**; 579
9. J E Allen 1969 *Spectrochim Acta*, **24B**; 13
10. D C Manning and F Fernandez 1968 *Atom. Absorption Newsletter*, **7**; 24
11. F J Fernandez, S A Myers and W Slavin 1980 *Anal Chem.*, **52**; 741
12. S Smith, R G Schleicher and G M Hieftje 1982 New atomic absorption background correction technique. Paper 442, 33rd Pittsburgh Conference on Analytical Chemistry and Applied Spectroscopy, Atlantic City NJ
13. R A Nadkarni 1984 *Anal Chem.*, **56**; 2233
14. J Anderson *et al.* 1985 *Geostandards Newsletter*, **9**; 17

15. Anon 1974 Safety practices for atomic absorption spectrophotometers, *International Laboratory*, **May/June**; 63
16. J C Miller and J N Miller 1993 *Statistics for analytical chemistry*, 3rd edn, John Wiley, Chichester
17. Anon 1978 Nomenclature, symbols, units and their usage in spectrochemical analysis II, *Spectrochim Acta*, **33B**; 242
18. Analytical Methods Committee 1987 Recommendations for the definition, estimation and use of the detection limit, *Analyst*, **112**; 199

15.25   Bibliography

M S Cresser 1995 *Flame spectrometry in environmental chemical analysis: a practical guide*, Royal Society of Chemistry, Cambridge

J R Dean 1997 *Atomic absorption and plasma spectroscopy*, 2nd edn, ACOL–Wiley, Chichester

L Ebdon, E H Evans, A Fisher and S J Hill 1998 *An introduction to analytical atomic spectroscopy*, John Wiley, Chichester

S Haswell (ed) 1991 *Atomic absorption spectrometry*, Elsevier, New York

G F Kirkbright and M Sargent 1997 *Atomic absorption and fluorescence spectroscopy*, 2nd edn, Academic Press, London

J Sneddon (ed) 1975 *Advances in atomic spectrometry*, Volumes 1–3, JAI Press, Greenwich CT

16

Atomic emission spectroscopy

16.1 Introduction

This chapter describes the basic principles and practice of atomic emission spectroscopy. Following a general discussion of the technique, the first part of this chapter is devoted to flame emission spectroscopy. Sections 16.6 to 16.11 deal predominantly with emission spectroscopy based on plasma sources, now the most important mode of excitation.

16.2 Emission spectra

When certain metals are introduced as salts into the Bunsen flame, characteristic colours are produced; this procedure has long been used for detecting elements qualitatively. If the light from such a flame is passed through a spectroscope, several lines may be seen, each having a characteristic colour; calcium produces red, green and blue radiation, of which the red is largely responsible for the typical calcium flame colour. A definite wavelength can be assigned to each radiation, corresponding with its fixed position in the spectrum. Although the flame colours of calcium, strontium and lithium are very similar, it is possible to differentiate them with certainty by observing their spectra, and it is possible to detect each in the presence of the others. By extending and amplifying the principles inherent in the qualitative flame test, analytical applications of emission spectrography have been developed. With electric spark or electric arc excitation, the spectra are recorded photographically by means of a spectrograph, and since the characteristic spectra of many elements occur in the ultraviolet, the optical system used to disperse the radiation is generally made of quartz. But these techniques have been virtually replaced by plasma emission (Section 16.6).

A detailed discussion on the origin of emission spectra is beyond the scope of this book, but a simplified treatment is given in Sections 15.1 and 15.2. There are three kinds of emission spectra: continuous spectra, band spectra and line spectra. The continuous spectra are emitted by incandescent solids, and sharply defined lines are absent. The band spectra consist of groups of lines that come closer and closer together as they approach a limit, the head of the band; they are caused by excited molecules. Line spectra consist of definite, usually widely and seemingly irregularly spaced, lines; they are characteristic of atoms or atomic ions which have been excited and emit their energy as light of definite wavelengths.

The quantum theory predicts that each atom or ion possesses definite energy states in which the various electrons can exist; in the normal or ground state the electrons have the lowest energy. Upon the application of sufficient energy by electrical, thermal or other

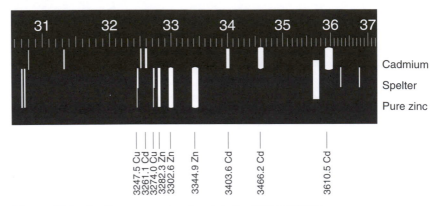

Figure 16.1   Contiguous spectra of cadmium, spelter and zinc

means, one or more electrons may be removed to a higher energy state further from the nucleus; these excited electrons tend to return to the ground state, and in so doing emit the extra energy as a photon of radiant energy. Since there are definite energy states and since only certain changes are possible according to the quantum theory, there are a limited number of wavelengths possible in the emission spectrum. The greater the energy of the exciting source, the higher the energy of the excited electrons, hence the greater the number of lines that may appear.

The intensity of a spectral line depends largely upon the probability of the required energy transition or 'jump' taking place. The intensity of some of the stronger lines may occasionally be decreased by self-absorption caused by reabsorption of energy by the cool gaseous atoms in the outer regions of the source. With high-energy sources the atoms may be ionised by the loss of one or more electrons; the spectrum of an ionised atom is different from the spectrum of a neutral atom and, indeed, the spectrum of a singly ionised atom resembles that of the neutral atom with an atomic number one less than its own.

The lines in the spectrum from any element always occur in the same positions relative to each other. When sufficient amounts of several elements are present in the source of radiation, each emits its characteristic spectrum; this is the basis for qualitative analysis by the spectrochemical method. Elements in an unknown spectrum may be identified by comparing the unknown spectrum with the spectrum of a known element. Figure 16.1 shows part of the contiguous spectra of cadmium, spelter and zinc. Examination of the three spectra will reveal whether cadmium is present or absent in the sample of spelter. In this example, cadmium is not detected in spelter, as no lines in the cadmium spectrum correspond with lines in the spelter spectrum.

Quantitative analysis was originally performed using an electric arc or spark as the excitation source. The light emitted by the sample was then passed through the slits of a spectrograph and dispersed by a prism before using a camera of long focal length to record the spectrum on a photographic plate. By keeping the excitation conditions constant and varying the sample composition over a narrow range, the energy emitted for a given spectral line of an element could be made proportional to the number of atoms excited, hence to the concentration of the element in the sample. The energy emitted (i.e. the intensity of the light) was usually determined photographically; the concentration of the unknown was determined from the blackening of the photographic plate for certain lines in the spectrum.

The quantitative determination of the blackening of the individual lines was made with a microphotometer. Measurements were made of i and i_0, respectively the light transmitted by the line in question and the light transmitted by the clear portion of the plate. The density D, strictly the density of blackening (also represented by B), may then be defined as $D = \log_{10}(i_0/i)$. This assumes that the galvanometer deflection obtained on the micro-photometer is directly proportional to the light falling on the photocell.

16.3   Flame emission spectroscopy (FES)

Nowadays two main methods are used for flame emission spectroscopy. The original method, known as flame photometry, is now used mainly for the analysis of alkali metals, particularly in biological fluids and tissues. At present, however, the usual flame emission method is obtained by simply operating a flame atomic absorption spectrometer in the emission mode (Figure 15.3). In this case the flame now acts as the source of radiation, hence the hollow cathode lamp and signal modulation are no longer necessary. Flame emission spectroscopy can be more sensitive than flame atomic absorption spectroscopy. This is true for elements whose resonance lines are associated with relatively low energy values (typically with wavelengths > 400 nm). Thus, for example, sodium (emission line at 589.0 nm) and lithium (at 670.8 nm) show great sensitivity with flame emission spectroscopy (Section 15.1).

Flame photometers

The flame is a much lower-energy source than the electrical means of excitation outlined in Section 16.2. Thus, the flame produces a simpler emission spectrum with fewer lines. Furthermore, relatively cool flames, e.g. air–propane, are normally used in flame photometers. In a flame photometer the emitted radiation is isolated by an optical filter (usually an interference filter) and then converted to an electrical signal by the photodetector, a photomultiplier. Figure 16.2 shows the layout of a simple flame photometer, the Chiron Diagnostics Model 410.

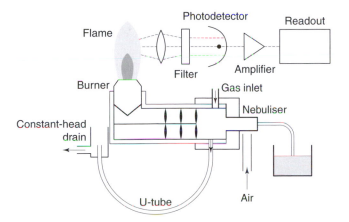

Figure 16.2   Simple flame photometer (Courtesy Chiron Diagnostics, Sudbury)

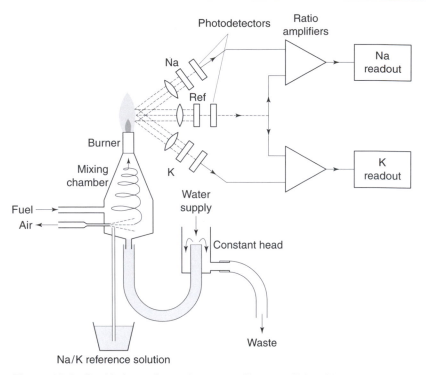

Figure 16.3   Double-beam flame photometer (Courtesy Chiron Diagnostics, Sudbury)

Basic components

Air at a given pressure is passed into an atomiser, and the suction this produces draws a solution of the sample into the atomiser, where it joins the airstream as a fine mist and passes into the burner. Here, in a small mixing chamber, the air meets the fuel gas (usually propane) supplied to the burner at a given pressure and the mixture is burnt. Radiation from the resulting flame passes through a lens and finally through an optical filter (usually an interference filter), which permits only the radiation characteristic of the element under investigation to pass through to the photodetector (a photomultiplier cell). The output is measured on a digital read-out system.

The precision of the technique can be improved by using a double-beam (internal standard) flame photometer. Figure 16.3 shows a line flow diagram of the Chiron Diagnostics 480 double-beam clinical flame photometer. The internal standard solution (Section 16.4), based on a lithium salt, is continuously monitored to ensure within-run precision. The internal reference optics (Ref) use a lithium interference filter. The ratio of the line intensities of either Na and Li or K and Li can be obtained from the appropriate photodetectors; the electronic circuit is designed to give a direct reading of the sodium and potassium concentrations. Besides these facilities, it incorporates an integral dilutor, which automatically dilutes all sample types (serum/plasma and urine), thus eliminating time-consuming manual pre-dilution processes.

16.4 Evaluation methods

In order to convert the measured emission values into the concentration of the analyte, the following methods may be used:

(a) calibration curve
(b) the standard addition procedure (Section 15.15)
(c) the internal standard method

The internal standard method involves the addition of a fixed amount of a reference material (the internal standard) to the sample solution and the standard solution. Upon excitation, the emitted energy of the analyte element and the internal standard are measured simultaneously by two photodetectors. With double-beam instruments, the ratio is given directly, and can then be plotted against the concentration of the analyte element (Section 16.3). The internal standard method compensates for any variations in the nebuliser uptake and changes in the flow rates of both the fuel gas and the oxidant.

16.5 Evaluating flame emission spectroscopy

The demand for clinical flame photometers has diminished in the last decade. They are now being largely replaced by electrochemical techniques, particularly ion selective electrodes (Section 13.17). Nevertheless, flame photometry has the following advantages:

(a) It is a well-understood technique.
(b) There are low running costs and maintenance costs.
(c) Measurement is possible in a wide range of fluid systems.

The determination of over 60 elements is possible when an atomic absorption spectrometer is used in the emission mode. As explained in Section 15.2, flame emission spectroscopy is often as sensitive or sometimes more sensitive than flame atomic absorption spectroscopy.

A major problem in flame atomic spectroscopy is due to **self-absorption**. This arises because there is a lower population of excited atoms in the outer region of the flame than in the hotter central area of the flame. Consequently, emission from the hot central part is absorbed by the relatively cooler outer region. At high analyte concentrations, self-absorption leads to a negative non-linear calibration curve. Thus, the linear operating range is limited when using flame methods.

16.6 Plasma emission spectroscopy

The use of a plasma as an atomisation source for emission spectroscopy has been developed largely in the last 25 years.[1,2] As a result, the scope of atomic emission spectroscopy has been considerably enhanced by the application of plasma techniques. A plasma may be defined as a cloud of highly ionised gas, composed of ions, electrons and neutral particles. Typically, in a plasma, over 1% of the total atoms in a gas are ionised.

In plasma emission spectroscopy the gas, usually argon, is ionised by the influence of a strong electrical field either by a direct current or by radio frequency. Both types of discharge produce a plasma, the **direct current plasma (DCP)** or the **inductively coupled plasma (ICP)**. Plasma sources operate at high temperatures somewhere between 7000 and

15 000 K. Especially in the ultraviolet region, the plasma source produces a greater number of excited emitted atoms than obtained by the comparatively low temperatures used in flame emission spectroscopy (Section 16.3).

Furthermore, the plasma source is able to reproduce atomisation conditions with a far greater degree of precision than obtained by classical arc and spark spectroscopy. As a result, spectra are produced for a large number of elements, which makes the plasma source suitable for simultaneous elemental determinations. This feature is especially important for the multi-element determinations over a wide concentration range.

16.7 Direct current plasma (DCP)

The DCP plasma source is shown in Figure 16.4. It consists of a high-voltage discharge between two graphite electrodes. The recent design employs a third electrode arranged in an inverted Y-shape which improves the stability of the discharge. The sample is nebulised (Section 16.9) at a flow rate of 1 mL min^{-1} using argon as the carrier gas. The argon ionised by the high-voltage discharge is able to sustain a current of ~20 A indefinitely.[3,4] The DCP generally has inferior detection limits to the ICP. Although the DCP is comparatively less expensive than the ICP, the graphite electrodes need replacing after a few hours of use.

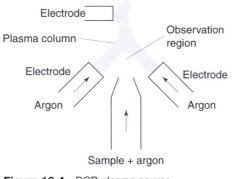

Figure 16.4 DCP plasma source

16.8 Inductively coupled plasma (ICP)

The inductively coupled plasma source (Figure 16.5) comprises three concentric silica quartz tubes, each of which is open at the top. The argon stream that carries the sample, in the form of an aerosol, passes through the central tube. The excitation is provided by two or three turns of a metal induction tube through which flows a radiofrequency current (frequency ~27 MHz). The second flow of argon, rate 10–15 L min^{-1}, maintains the plasma. It is this gas stream that is excited by the radiofrequency power. The plasma gas flows in a helical pattern which provides stability and helps to thermally isolate the outermost quartz tube. The plasma is initiated by a spark from a Tesla coil probe and is thereafter self-sustaining. The plasma itself has a characteristic torroidal (doughnut) shape in which the sample is introduced into the relatively cool central hole of the torus.

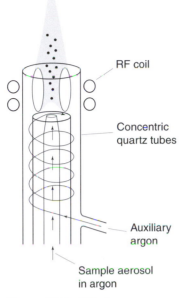

Figure 16.5   ICP plasma source

16.9   Sample introduction

The sample, usually in the form of a solution, is carried into the hot plasma by a nebuliser similar to the system for flame spectroscopy (Section 15.5) although a much slower flow rate of $1 \, mL \, min^{-1}$ is used for ICP. The most widely used nebuliser system in ICP is the crossed-flow nebuliser (Figure 16.6). The sample is forced into the mixing chamber at a flow rate of $1 \, mL \, min^{-1}$ by the peristaltic pump and nebulised by the stream of argon flowing at about $1 \, L \, min^{-1}$.

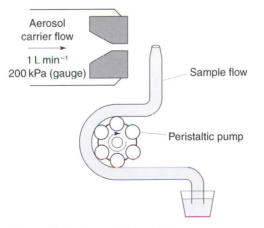

Figure 16.6   Crossed-flow nebuliser

623

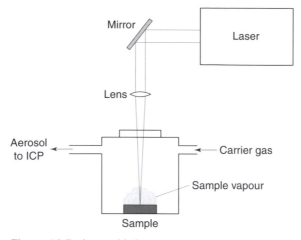

Figure 16.7   Laser ablation

Another kind of nebuliser is designed for handling slurries and can generate aerosols from aqueous samples containing a high percentage of solids (up to 20%). In a typical design, the sample solution is pumped along a V-shaped channel. A small orifice in the middle of this channel allows the argon carrier gas to escape. When the sample passes over the orifice, the evolved carrier gas produces a coarse aerosol.

Aerosols generated from crossed flow or V-channel nebulisers cannot be introduced directly into the plasma; this is because they could cool the plasma or even extinguish it. In order to avoid matrix interferences that would otherwise occur, a spray chamber is added prior to the plasma. The function of the spray chamber is to reduce the particle size of the aerosol to an ideal 10 µm.

A particulate aerosol may be produced from a solid sample by an ablation method. In **laser ablation** the sample is vaporised by a laser directed on to the solid surface. The mobilised sample is then transferred directly into the plasma (Figure 16.7). Another technique is **spark ablation**, which can be used for conductive solids such as steels and a variety of metallurgical samples (Section 16.10).

16.10   ICP instrumentation

Simultaneous multi-element spectrometer

Figure 16.8 shows the light path of an early simultaneous multi-element system. The radiation from the plasma enters through a single slit and is then dispersed by a concave reflection grating; the component wavelengths reach a series of exit slits which isolate the selected emission lines for specific elements. The entrance and exit slits and the diffraction grating surface are positioned along the circumference of the Rowland circle, whose curvature equals the radius of curvature of the concave grating. The light from each exit slit is directed to fall on the cathode of a photomultiplier tube, one for each spectral line isolated. The light falling on the photomultiplier gives an output which is integrated on a capacitor; the resulting voltages are proportional to the concentrations of the elements in the sample.

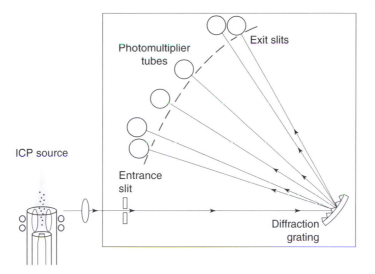

Figure 16.8   Simultaneous ICP: an early multi-element system

Multichannel instruments are capable of measuring the intensities of the emission lines of up to 60 elements simultaneously. To overcome the effects of possible non-specific background radiation, one or more additional wavelengths may be measured and background correction can be achieved (Section 15.13).

It is also possible to subtract background at each specific wavelength by using vibrating quartz plates which produce a small shift of the wavelength. The emission of the shifted wavelength is then subtracted from the original value. The main advantage of photoelectric multichannel instruments is that, with the modern computational facilities, rapid simultaneous analysis may be achieved with a better precision than obtained from the spectrograph using a photographic plate. Thus, results for 25 elements may be obtained within about 1–2 min.

But there are two major disadvantages:

1. The exit slits are preset; this prevents the use of other wavelengths, inhibiting the analysis of further elements.
2. An instrument with, say, 25 exit slits requires 25 photomultiplier tubes; this makes it complex and costly.

If an alternative detector system were designed that avoided the use of a battery of photomultiplier tubes, then a more versatile instrument would result. The Thermo Jarrell Ash IRIS simultaneous ICP has overcome these disadvantages by using an optical system that, in combination with a sensitive multiwavelength detector, enables full spectral information of the sample to be obtained. A diagram of this simultaneous ICP is shown in Figure 16.9.

High resolution is achieved by an echelle-based optical system. The resolution R of a diffraction grating is directly related to the groove density N (the number of lines per millimetre) and the spectral order n. In a conventional grating (Section 17.6) the spectral order n is usually 1 or 2, whereas in an echelle grating n may be as high as 100. Although the groove density N may be only 300 lines per millimetre in an echelle grating compared with a typical value of 1200 lines per millimetre, the considerably higher value of n

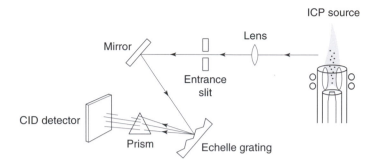

Figure 16.9 Simultaneous ICP: a contemporary multi-element system (Courtesy of Thermo Jarrell Ash, Franklin MA)

achieved in an echelle grating gives rise to the higher resolution. In order to prevent overlapping of high-order spectra (Section 17.6), further dispersion is required, usually by a prism. The prism is placed so that the spectral data is sorted into a two-dimensional presentation, the spectral order n on the vertical axis and the wavelength on the horizontal axis. The detector is a **charge injection device (CID)**, which offers high sensitivity and continuous wavelength coverage. The CID consists essentially of an array of closely spaced metal–insulator–semiconductor diodes in which the incident light is converted into a signal. A major advantage of the CID detector is that the optimum wavelength can be selected for each element in every type of sample.

The IRIS spectrometer is also capable of routinely analysing electrically conductive samples. An electric spark ablates material from the solid sample, generating a particulate aerosol which is automatically transferred into the ICP, where it is excited to analytical emission. This solid sampling accessory (SSA) can determine elemental concentrations in solids such as stainless steels, aluminium, brass, gold, nickel and other metallurgical samples. Direct solid sampling eliminates the need for acid digestion, reducing costly and time-consuming sample preparation.

Sequential instruments

The sequential instrument is a low-cost alternative to simultaneous ICP. A diagram of the light path of the Thermo Jarrell Ash Atom Scan 16 sequential ICP spectrometer is shown in Figure 16.10. The galvanometer grating drive scans the entire wavelength range from 165 to 800 nm in 20 ms. All monochromator-related electronics are sealed in a purged optical system to enhance vacuum ultraviolet performance. This enables elements to be determined which give emission lines in the vacuum ultraviolet region (< 195 nm), e.g. sulphur.

In order to achieve high detector sensitivity over its whole wavelength range, the diffracted light is monitored by two photomultiplier tubes each with different spectral ranges. Before each wavelength scan, the grating automatically locates itself by finding the 365 nm triplet of a built-in mercury source and then moves sequentially to the wavelengths for the analyte elements. The system provides a resolution of 0.018 nm comparable with emission linewidth. Microprocessor control of instrumental parameters, such as sample introduction, argon gas flow and RF power, guarantees uniform results even with the most difficult samples.

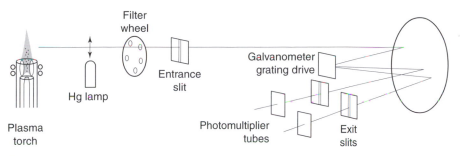

Figure 16.10   Light path of a sequential spectrometer (Courtesy Thermo Jarrell Ash, Franklin MA)

16.11   Evaluating ICP AES

There are four major advantages of ICP AES over AAS methods:

1. There is a wide linear working range for ICP AES, typically from 0.1 to $1000\,\mu g\,mL^{-1}$ (five decades of concentration). AAS usually ranges over one decade, typically 1 to $10\,\mu g\,mL^{-1}$.
2. It can perform simultaneous multi-element analyses or rapid sequential analyses. Flame AAS is normally sequential and furnace AAS always sequential.
3. With a simultaneous instrument, precision can be improved by using an internal standard, typically 0.1–1.0% relative standard deviation (RSD). With flame AAS the precision is usually 1–2% RSD and with furnace AAS it is 1–3% RSD;
4. Ablation and other vaporisation methods enable it to make rapid measurements of a variety of solid samples.

Interferences that arise in both atomic emission and atomic absorption are now well documented and suitable background correction methods are used. However, ICP AES is more expensive than AAS and generally has higher operating costs than AAS methods. A significant advance over recent years is the efficient coupling of an ICP with a mass spectrometer (MS), leading to ICP-MS (Chapter 19).

16.12   Determining alkali metals by flame photometry

Although flame emission measurements can be made by using an atomic absorption spectrometer in the emission mode, this section is based on a simple flame photometer, the Chiron Diagnostics Model 410. Before attempting to use the instrument, read the instruction manual supplied by the manufacturers. Prepare standard solutions of sodium, potassium, calcium and lithium by following the directions given. Produce the corresponding calibration curves then carry out the four determinations described.

Standard solutions

Sodium   Dissolve 2.542 g sodium chloride in 1 L deionised water in a graduated flask. This solution contains the equivalent of 1.000 mg Na per millilitre. Dilute this solution to give four solutions containing 10, 5, 2.5 and $1\,\mu g\,mL^{-1}$ of sodium ions.

Potassium   Dissolve 1.909 g potassium chloride in 1 L deionised water. This solution contains the equivalent of 1.000 mg K per millilitre. Dilute this stock solution to give four solutions containing 20, 10, 5 and $2\,\mu g\,mL^{-1}$ of potassium ions.

Calcium   Dissolve 2.497 g calcium carbonate in a little dilute hydrochloric acid, and dilute to 1 L with deionised water. This stock solution contains the equivalent of 1.000 mg Ca per millilitre. Dilute this solution to give solutions containing 100, 50, 25 and 10 μg mL^{-1} of calcium ions.

Lithium   Dissolve 5.324 g pure lithium carbonate in a little dilute hydrochloric acid and dilute to 1 L with deionised water. This solution contains 1.000 mg Li per millilitre. Dilute the stock solution to give solutions containing 20, 10, 5 and 2 μg mL^{-1} of lithium ions.

Determinations

Potassium in potassium sulphate   Weigh out accurately about 0.20 g potassium sulphate and dissolve it in 1 L deionised water. Dilute 10.0 mL of this solution to 100 mL, and determine the potassium with the flame photometer using the potassium filter.

Potassium and sodium in a mixture   Mix suitable volumes of the above stock solutions so that the resulting solution contains, say, 4–10 μg mL^{-1} Na and 10–15 μg mL^{-1} K. Determine the sodium and potassium with the aid of the appropriate filters. Compare the results obtained with the true values.

Sodium, potassium and calcium in a mixture   Mix appropriate volumes of the above stock solutions so that the test solution contains, say, 5 μg mL^{-1} Na, 10 μg mL^{-1} K and 40 μg mL^{-1} Ca. Determine the sodium, potassium and calcium with the aid of appropriate filters. Compare the results obtained with the true values.

Potassium and sodium in a mixture   If a double-beam flame photometer is available, e.g. the Chiron Diagnostics Model 482, then lithium can be used as an internal standard. Prepare sodium and potassium standards as described above, but each standard and each unknown must contain 100 μg mL^{-1} of lithium. Compare the results obtained using this method to the results obtained in the second determination.

16.13   References

1. V A Fassel 1978 *Science*, **208**; 183
2. M Thompson and J N Walsh 1989 *A handbook of inductively coupled plasma spectrometry*, 2nd edn, Blackie, Glasgow
3. G W Johnson, H E Taylor and R K Skogerboe 1979 *Anal. Chem.*, **51**; 2403
4. J Reednick 1979 *Am. Lab.*, **11** (3); 53

16.14   Bibliography

P W J M Bouman's 1987 *Inductively coupled plasma emission spectrometry*, Parts 1 and 2, John Wiley, New York

M S Cresser 1995 *Flame spectrometry in environmental chemical analysis: a practical approach*, Royal Society of Chemistry, Cambridge

J R Dean 1997 *Atomic absorption and plasma spectroscopy*, 2nd edn, ACOL–Wiley, Chichester

R K Fassel *et al.* 1985 *Inductively coupled plasma emission spectroscopy: an atlas of spectral information*, Elsevier, New York

A Montaser and D W Golightly (eds) 1992 *Inductively coupled plasmas in analytical atomic spectrometry*, 2nd edn, VCH, New York

G L Moore 1989 *Introduction to inductively coupled plasma atomic emission spectro-scopy*, Elsevier, Amsterdam

M Thompson and J N Walsh 1989 *A handbook of inductively coupled plasma spectro-metry*, 2nd edn, Blackie, Glasgow

17

Molecular electronic spectroscopy

17.1 General discussion

The variation of the colour of a system with change in concentration of some component forms the basis of what the chemist commonly terms **colorimetric analysis**. The colour is usually due to the formation of a coloured compound by the addition of an appropriate reagent, or it may be inherent in the desired constituent itself. The intensity of the colour may then be compared with the intensity obtained by treating a known amount of the substance in the same manner. **Fluorimetric analysis** describes a method of analysis in which the amount of radiation emitted by an analyte is used to measure the concentration of that analyte. In **spectrophotometric analysis** a source of radiation is used that extends into the ultraviolet region of the spectrum. From this, definite wavelengths of radiation are chosen possessing a bandwidth of less than 1 nm. This requires a more complicated and consequently more expensive instrument – a spectrophotometer.

An optical spectrometer is an instrument possessing an optical system which can produce dispersion of incident electromagnetic radiation, and with which measurements can be made of the quantity of transmitted radiation at selected wavelengths of the spectral range. A photometer is a device for measuring the intensity of transmitted radiation or a function of this quantity. When combined in the spectrophotometer, the spectrometer and photometer produce a signal that corresponds to the difference between the transmitted radiation of a reference material and the transmitted radiation of a sample, at selected wavelengths. The chief advantage of colorimetric and spectrophotometric methods is that they provide a simple means for determining minute quantities of substances. In general, the upper limit of colorimetric methods is the determination of constituents which are present in quantities of less than 1% or 2%. Fluorimetry, besides being two to three orders of magnitude more sensitive than colorimetric and spectrophotometric methods, has the further advantage of greater selectivity.

The selectivity of both spectrophotometric and fluorimetric techniques can be further improved by using **derivative spectrophotometry** (Section 17.13). In this chapter we are concerned with analytical methods that are based upon the absorption of electromagnetic radiation and, in the case of fluorimetry, its subsequent emission. Light consists of radiation to which the human eye is sensitive; waves of different wavelengths give rise to light of different colours, and a mixture of these wavelengths constitutes white light. White light covers the entire visible spectrum, 400–760 nm. The approximate wavelength ranges of colours are given in Table 17.1. The visual perception of colour arises from the selective

Table 17.1 *Approximate wavelength of colours (nm)*

Ultraviolet	Violet	Blue	Green	Yellow	Orange	Red	Infrared
< 400	400–450	450–500	500–570	570–590	590–630	620–760	> 760

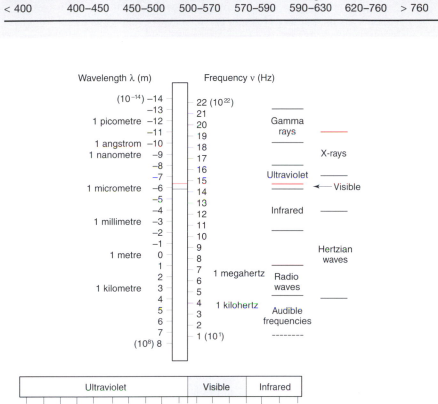

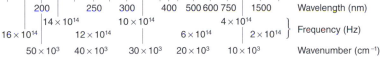

Figure 17.1   The electromagnetic spectrum

absorption of certain wavelengths of incident light by the coloured object. The other wavelengths are either reflected or transmitted, according to the nature of the object, and they are perceived by the eye as the colour of the object. If a solid opaque object appears white, all wavelengths are reflected equally; if the object appears black, very little light of any wavelength is reflected; if it appears blue, the wavelengths that give the blue stimulus are reflected, etc.

Note that the range of electromagnetic radiation extends considerably beyond the visible region. The approximate limits of wavelength and frequency for the various types of radiation, including the frequency range of sound waves, are shown in Figure 17.1 (not drawn to scale); this may be regarded as an electromagnetic spectrum. Notice that γ-rays and X-rays have very short wavelengths, whereas ultraviolet, visible, infrared and radio waves

631

have progressively longer wavelengths. For fluorimetry, colorimetry and spectrophotometry, the visible region and the adjacent ultraviolet region are of major importance. Electromagnetic waves are usually described in terms of wavelength λ, wavenumber $\bar{\nu}$ and frequency ν. Wavelength is the distance between two adjacent wave crests; unless otherwise stated, its units are centimetres (cm). Wavenumber is the number of waves per centimetre. Frequency is the number of waves per second. The three quantities are related as follows:

$$\frac{1}{\text{wavelength}} = \text{wavenumber} = \frac{\text{frequency}}{\text{velocity of light}}$$

$$\frac{1}{\lambda} = \bar{\nu} = \frac{\nu}{c} \qquad c = 2.997\,93 \times 10^8\,\text{m s}^{-1}$$

The following units are in common use:

$$1 \text{ angstrom} = 1\,\text{Å} = 10^{-10}\,\text{m} = 10^{-8}\,\text{cm}$$

$$1 \text{ nanometre} = 1\,\text{nm} = 10\,\text{Å} = 10^{-7}\,\text{cm}$$

$$1 \text{ micrometre} = 1\,\mu\text{m} = 10^4\,\text{Å} = 10^{-4}\,\text{cm}$$

Two useful relations:

Wavenumber $\bar{\nu} = 1/\lambda$ waves per centimetre

Frequency $\nu = c/\lambda \approx 3 \times 10^{10}/\lambda$ waves per second

To fully comply with SI units, these functions should be calculated using the metre as the basic unit. But it is still common practice to use centimetres for this purpose.

17.2 Theory of spectrophotometry* and colorimetry

When light, either monochromatic or heterogeneous, falls upon a homogeneous medium, a portion of the incident light is reflected, a portion is absorbed within the medium, and the rest is transmitted. The light intensities are expressed as follows: I_0 for the incident light, I_a for the absorbed light, I_t for the transmitted light, and I_r for the reflected light. Then

$$I_0 = I_a + I_t + I_r$$

For air – glass interfaces, which occur when using glass cells, about 4% of the incident light is reflected. I_r is usually eliminated by the use of a control, such as a comparison cell, hence

$$I_0 = I_a + I_t \tag{17.1}$$

Credit for investigating the change of absorption of light with the thickness of the medium is frequently given to Lambert,[1] although he really extended concepts originally developed by Bouguer.[2] Beer[3] later applied similar experiments to solutions of different concentrations and published his results just before those of Bernard.[4] This very confusing story has been explained by Malinin and Yoe.[5] The two separate laws governing absorption are usually known as Lambert's law and Beer's law. In the combined form[6] they are known as the Beer–Lambert law.

---

* Spectrophotometry proper is mainly concerned with the following regions of the spectrum: ultraviolet 185 to 400 nm; visible 400 to 760 nm; infrared 0.76 to 15 μm. Colorimetry is concerned with the visible region of the spectrum.

Lambert's law

Lambert's law states that when monochromatic light passes through a transparent medium, the rate of decrease in intensity with the thickness of the medium is proportional to the intensity of the light. This is equivalent to stating that the intensity of the emitted light decreases exponentially as the thickness of the absorbing medium increases arithmetically, or that any layer of given thickness of the medium absorbs the same fraction of the light incident upon it. The law may be expressed by the differential equation

$$-\frac{\mathrm{d}I}{\mathrm{d}l} = kI \tag{17.2}$$

where I is the intensity of the incident light of wavelength λ, l is the thickness of the medium, and k is a proportionality factor. Integrating equation (17.2) and putting $I = I_0$ when $l = 0$, we obtain

$$\ln\frac{I_0}{I_t} = kl$$

or, put another way,

$$I_t = I_0 e^{-kl} \tag{17.3}$$

where I_0 is the intensity of the incident light falling upon an absorbing medium of thickness l, I_t is the intensity of the transmitted light, and k is a constant for the wavelength and the absorbing medium used. By changing from natural to common logarithms we obtain

$$I_t = I_0 \times 10^{-0.4343kl} = I_0 \times 10^{-Kl} \tag{17.4}$$

where $K = k/2.3026$ and is usually called the **absorption coefficient**. The absorption coefficient is generally defined as the reciprocal of the thickness (1 cm) required to reduce the light to 1/10 of its intensity. This follows from equation (17.4) since

$$I_t/I_0 = 0.1 = 10^{-Kl} \quad \text{or} \quad Kl = 1 \quad \text{and} \quad K = 1/l$$

The ratio I_t/I_0 is the fraction of the incident light transmitted by a thickness l of the medium and is called the **transmittance** T. Its reciprocal I_0/I_t is the **opacity**, and the **absorbance** A of the medium (formerly called the optical density D or extinction E) is given by

$$A = \log(I_0/I_t) \tag{17.5}$$

Thus a medium with absorbance 1 for a given wavelength transmits 10% of the incident light at that wavelength.

Beer's law

We have so far considered the light absorption and the light transmission for monochromatic light as a function of the thickness of the absorbing layer only. In quantitative analysis, however, we are mainly concerned with solutions. Beer studied the effect of concentration of the coloured constituent in solution upon the light transmission or absorption. He found the same relation (17.3) between transmission and concentration as Lambert had discovered between transmission and thickness of the layer, i.e. the intensity of a beam of monochromatic light decreases exponentially as the concentration of the absorbing substance increases arithmetically. This may be written in the from

$$I_t = I_0 e^{-k'c}$$
$$= I_0 \times 10^{-0.4343k'c} = I_0 \times 10^{-K'c} \tag{17.6}$$

where c is the concentration, and k' and K' are constants. Combining equations (17.4) and (17.5), we have[6]

$$I_t = I_0 \times 10^{-acl} \tag{17.7}$$

or

$$\log(I_0/I) = acl \tag{17.8}$$

This is the fundamental equation of colorimetry and spectrophotometry, and is often known as the **Beer–Lambert law** or, more recently, as **Beer's law**. The value of a will depend on how the concentration is expressed. If c is expressed in $\mathrm{mol\,L^{-1}}$ and l in cm then a is given the symbol ε and is called the **molar absorption coefficient** or molar absorptivity (formerly the molar extinction coefficient).

The specific absorption (or extinction) coefficient E_s (sometimes termed absorbancy index) may be defined as the absorption per unit thickness (path length) and unit concentration. Where the molecular weight of a substance is not definitely known, it is not possible to write down the molecular absorption coefficient, then it is usual to write the unit of concentration as a superscript and the unit of length as a subscript, e.g.

$$E_{1\,cm}^{1\%}\,325\,nm = 30$$

This means that for the substance in question, at a wavelength of 325 nm, a solution of length 1 cm, and concentration 1% (1 g of solute per 100 mL of solution) $\log(I_0/I_t)$ has a value of 30.

Notice there is a relationship between the absorbance A, the transmittance T and the molar absorption coefficient, since

$$A = \varepsilon cl = \log\frac{I_0}{I_t} = \log\frac{1}{T} = -\log T \tag{17.9}$$

The scales of spectrophotometers are often calibrated to read directly in absorbances, and frequently also in percentage transmittance. For colorimetric measurements I_0 is usually understood as the intensity of the light transmitted by the pure solvent, or the intensity of the light entering the solution; I_t is the intensity of the light emerging from the solution, or transmitted by the solution. The following terms are used:

Absorption coefficient (or extinction coefficient) is the absorbance for unit path length:

$$K = A/t \quad \text{or} \quad I_t = I_0 \times 10^{-Kt}$$

Specific absorption coefficient (or absorbancy index) is the absorbance per unit path length and unit concentration:

$$E_s = A/cl \quad \text{or} \quad I_t = I_0 \times 10^{-E_s cl}$$

Molar absorption coefficient is the specific absorption coefficient for a concentration of $1\,\mathrm{mol\,L^{-1}}$ and a path length of 1 cm:

$$\varepsilon = A/cl$$

Application of Beer's law

Consider two solutions of a coloured substance with concentrations c_1 and c_2. They are placed in an instrument that allows the thickness of the layers to be altered and measured easily; it also allows a comparison of the transmitted light. When the two layers have the same colour intensity, then

$$I_{t_1} = I_0 \times 10^{-\varepsilon l_1 c_1} = I_{t_2} = I_0 \times 10^{-\varepsilon l_2 c_2} \qquad (17.10)$$

Here l_1 and l_2 are the lengths of the solution columns with concentrations c_1 and c_2 respectively when the system is optically balanced. Hence, under these conditions and when Beer's law holds, we have

$$l_1 c_1 = l_2 c_2 \qquad (17.11)$$

A colorimeter can therefore be employed in a dual capacity: (a) to investigate the validity of Beer's law by varying c_1 and c_2 and noting whether equation (17.11) applies, and (b) for the determination of an unknown concentration c_2 of a coloured solution by comparison with a solution of known concentration c_1. Note that equation (17.11) is valid only if Beer's law is obeyed over the concentration range employed and the instrument has no optical defects.

When a spectrophotometer is used it is unnecessary to make comparison with solutions of known concentration. The intensity of the transmitted light or, better, the ratio I_t/I_0 (the transmittance) is found directly at a known thickness l. By varying l and c the validity of the Beer–Lambert law (17.9) can be tested and the value of ε may be evaluated. When the value of ε is known, the concentration c_X of an unknown solution can be calculated from the formula:

$$c_X = \frac{\log I_0/I_t}{\varepsilon l} \qquad (17.12)$$

The molar absorption coefficient ε depends upon the wavelength of the incident light, the temperature and the solvent employed. In general, it is best to choose the wavelength of the incident light so it is approximately the same as the wavelength where the solution exhibits a maximum selective absorption (or minimum selective transmittance); this will give the maximum sensitivity.

For matched cells (i.e. l constant) the Beer–Lambert law may be written

$$c \propto \log \frac{I_0}{I_t} \quad \text{or} \quad c \propto \log \frac{1}{T}$$

or

$$c \propto A \qquad (17.13)$$

Hence plotting A, or $\log(1/T)$, against concentration will produce a straight line, and this will pass through the point $c = 0$, $A = 0$ ($T = 100\%$). This calibration line may then be used to determine unknown concentrations of solutions of the same material after measurement of absorbances.

Deviation from Beer's law

Beer's law will generally hold over a wide range of concentration if the structure of the coloured ion or of the coloured non-electrolyte in the dissolved state does not change with concentration. Small amounts of electrolytes, which do not react chemically with the coloured components, do not usually affect the light absorption; large amounts of electrolytes may produce a shift of the maximum absorption, and may also change the value of the molar absorptivity. Discrepancies are usually found when the coloured solute ionises, dissociates, or associates in solution, since the nature of the species in solution will vary with the concentration. The law does not hold when the coloured solute forms complexes, the composition of which depends upon the concentration. Discrepancies may occur when monochromatic light is not used. The behaviour of a substance can always be tested by plotting $\log(I_0/I_t)$ or $\log(1/T)$ against the concentration; a straight line passing through the origin indicates conformity to the law.

For solutions which do not follow Beer's law, it is best to prepare a calibration curve using a series of standards of known concentration. Instrumental readings are plotted against concentrations in, say, mg per 100 mL or 1000 mL. For the most precise work, each calibration curve should cover the dilution range likely to be encountered in the actual comparison. Instrumental effects may also cause deviations from Beer's law; for example, if the photomultiplier tube is faulty, a straight line will be obtained for absorbance versus concentration but the line will cut the concentration axis at a value other than zero. Similarly, if the cuvettes (cells) are dirty the line will intersect the absorbance axis at a value greater than zero.

17.3 Fluorimetry (theory)

Fluorescence is caused by the absorption of radiant energy and the emission of some of this energy in the form of light. The emitted light almost always has a higher wavelength than the absorbed light (Stokes' law). In true fluorescence the absorption and emission take place in a short but measurable time, of the order of 10^{-12} to 10^{-9} s. If the light is emitted with a time delay ($> 10^{-8}$ s), because the transition is forbidden, the phenomenon is known as **phosphorescence**; this time delay may range from a fraction of a second to several weeks. Fluorescence and phosphorescence are types of **photoluminescence**; 'photoluminescence' is the general term applied to absorption and re-emission of light energy.

At present the most widely used type of photoluminescence in analytical chemistry is fluorescence, which is distinguished from other forms of photoluminescence by the fact that the excited molecule returns to the ground state immediately after excitation. When a molecule absorbs a photon of ultraviolet radiation, it undergoes a transition to an excited electronic state and one of its electrons is promoted to an orbital of higher energy. There are two important types of transition for organic molecules:

(a) n → π* in which an electron in a non-bonding orbital is promoted to a π-antibonding orbital.

(b) π → π* in which an electron in a π-bonding orbital is raised to a π-antibonding orbital.

It is the π → π* type of excitation which leads to significant fluorescence; the n → π* type produces only a weak fluorescence. The electronic transitions corresponding to charge transfer bands also lead to strong fluorescence. However, electronic energy is not the only type of energy affected when a molecule absorbs a photon of UV radiation. Organic molecules have a large number of vibrations and each of them contributes a series of nearly equally spaced vibrational levels to each of the electronic states. The various energy states available to a molecule may be represented by means of an energy level diagram; more details can be found in the items listed in Section 17.42.

Before a molecule can emit radiation by fluorescence it must first be able to absorb radiation. Not all molecules which absorb UV or visible radiation are fluorescent, and it is useful to quantify the extent to which a particular molecule fluoresces. The **quantum efficiency** is defined as the fraction of the incident radiation which is re-emitted as fluorescence at a specific wavelength:

$$\phi_f(\leq 1) = \frac{\text{number of photons emitted}}{\text{number of photons absorbed}} = \frac{\text{quantity of light emitted}}{\text{quantity of light absorbed}}$$

A proportion of the excited molecules may lose their excess energy by undergoing bond dissociation, leading to a photochemical reaction, or may return to the ground state by other mechanisms. The quantum efficiency will then be less than unity and may be extremely small. The value of ϕ_f is an inherent property of a molecule and is determined to a large extent by its structure. In general a high value of ϕ_f is associated with molecules possessing an extensive system of conjugated double bonds with a relatively rigid structure due to ring formation. This is illustrated by the intense fluorescence exhibited by organic molecules such as anthracene, fluorescein and other condensed-ring aromatic structures. The number of simple inorganic species which are fluorescent is more limited; the chief examples are lanthanide and actinide compounds, some organometallic compounds and compounds such as the tris(bipyridyl)ruthenium(II) ion, $Ru(bpy)_3^{2+}$, which has a quantum yield of 0.1. In metal ions this limitation may be overcome by the formation of a complex with an appropriate organic ligand, e.g. many of the metal ion complexes formed with the complexing agent 8-hydroxyquinoline are fluorescent.

Quantitative aspects

The total fluorescence intensity F is given by the equation $F = I_a \phi_f$ where I_a is the intensity of light absorption and ϕ_f is the quantum efficiency of fluorescence. Since $I_0 = I_a + I_t$, where I_0 is the intensity of incident light and I_t is the intensity of transmitted light, then

$$F = (I_0 - I_t)\phi_f$$

and since $I_t = I_0 e^{-\varepsilon cl}$ (Beer's law) we have

$$F = I_0(1 - e^{-\varepsilon cl})\phi_f \tag{17.14}$$

For weakly absorbing solutions, when εcl is small, the equation becomes

$$F = 2.3 I_0 \varepsilon cl \phi_f \tag{17.15}$$

so that for very dilute solutions ($\leq$ few $\mu g\,g^{-1}$) the total fluorescence intensity F is proportional to the concentration of the sample and the intensity of the excitation energy. It is instructive to compare the sensitivity which may be achieved by absorption and fluorescence methods. The overall precision with which absorbance can be measured is certainly not better than 0.001 units using a 1 cm cell. Since for most molecules the value of ε_{max} is rarely greater than 10^6, then on the basis of Beer's law, the minimum detectable concentration is given by

$$c_{min} > 10^{-3}/10^6\,M = 10^{-9}\,M$$

With fluorescence the sensitivity is limited in principle only by the maximum intensity of the exciting light source, so that under ideal conditions $c_{min} = 10^{-12}\,M$. In general the limit of detection of the fluorescence technique is of the order of 10^3 times lower than the limit for UV absorption spectrometry.

Selectivity may also be superior using fluorescence methods since (a) not all absorbing species fluoresce, and (b) the analyst can select two wavelengths (excitation and emission) as compared with one for absorption methods. This inherent selectivity may be inadequate, however, and must often be enhanced by chemical separation, e.g. solvent extraction (Chapter 6). Improved selectivity can also be achieved using derivative techniques, i.e. by basing the measurement of a sample component on the derivative spectrum rather than the original fluorescence emission spectrum; thus, weak shoulders on the original spectrum convert into easily quantified peaks in the derivative spectrum.

It is important to distinguish between **emission** and **excitation** fluorescence spectra. Emission spectra are produced using an excitation of fixed wavelength and recording the emission intensity as a function of emitted wavelength. Whereas excitation spectra are obtained by measuring the fluorescence intensity at a fixed wavelength while the wavelength of the exciting radiation is varied. Excitation is not the same as absorption. Factors such as dissociation, association or solvation, which produce deviations from Beer's law, can be expected to have a similar effect in fluorescence. Any material that causes the intensity of fluorescence to be less than the expected value given by equation (17.15) is known as a quencher, and the effect is called quenching; it is normally caused by the presence of foreign ions or molecules. Fluorescence is affected by the pH of the solution, by the nature of the solvent, the concentration of the reagent which is added in the determination of inorganic ions, and sometimes by temperature. The time taken to reach the maximum intensity of fluorescence varies considerably with the reaction.

An important aspect of quenching in analysis is that the fluorescence exhibited by the analyte may be quenched by the molecules of some compound present in the sample – an example of a matrix effect. If the concentration of the quenching species is constant, this may be allowed for by using suitable standards (i.e. containing the same concentration of quenching species), but difficulties occur when there is an unpredictable variation in the concentration of quenching species.

17.4 Methods of 'colour' measurement

The basic principle of most colorimetric measurements consists in comparing, under well-defined conditions, the colour produced by the substance in unknown amount with the same colour produced by a known amount of the material being determined. It is not essential to prepare a series of standards with a spectrophotometer; the molar absorption coefficient can be calculated from one measurement of the absorbance or transmittance of a standard solution, and the unknown concentration can then be calculated with the aid of the molar absorption coefficient and the observed value of the absorbance or transmittance; see equations (17.12) and (17.13). Two important methods, absorptiometer and spectrophotometer, are outlined here then described in more detail in Sections 17.5 and 17.6. For a complete treatment of visual comparison, see previous editions of this book.

Photoelectric photometer method

In this method the human eye, used for older, visual comparison techniques, is replaced by a suitable photoelectric cell; the photoelectric cell gives a direct measure of the light intensity, and hence of the absorption. Instruments incorporating photoelectric cells measure the light absorption not the colour of the substance; that is why the term 'photoelectric colorimeters' is a misnomer; better names are photoelectric comparators, photometers, or best of all, **absorptiometers**. Most absorptiometers consist of a light source, a suitable light filter to secure an approximation to monochromatic light (hence the name photoelectric filter photometer), a glass cell for the solution, a photoelectric cell to receive the radiation transmitted by the solution, and a measuring device to determine the response of the photoelectric cell. The comparator is first calibrated in terms of a series of solutions of known concentration, and the results plotted in the form of a curve connecting concentrations and readings of the measuring device employed. The concentration of the unknown solution is then determined by noting the response of the cell and referring to the calibration curve.

Absorptiometers are available in a number of different forms incorporating one or two photocells. With the one-cell type, the absorption of light by the solution is usually measured directly by determining the current output of the photoelectric cell relative to the value obtained with the pure solvent. It is of the utmost importance to use a light source of constant intensity, and if the photocells exhibit a 'fatigue effect' it is necessary to allow them to attain their equilibrium current after each change of light intensity. The two-cell type of filter photometer is usually regarded as the more trustworthy (provided the electrical circuit is appropriately designed) in that any fluctuation of the intensity of the light source will affect both cells alike if they are matched for their spectral response. Here the two photocells, illuminated by the same source of light, are balanced against each other through a galvanometer; the test solution is placed before one cell and the pure solvent before the other. The galvanometer indicates the difference between the current outputs of the two photocells.

Spectrophotometric method

This is undoubtedly the most accurate method for determining, among other things, the concentration of substances in solution, but the instruments are, of necessity, more expensive. A spectrophotometer may be regarded as a refined filter photoelectric photometer which permits the use of continuously variable and more nearly monochromatic bands of light. The essential parts of a spectrophotometer are (1) a source of radiant energy, (2) a monochromator, i.e. a device for isolating monochromatic light or, more accurately, narrow bands of radiant energy from the light source, (3) glass or silica cells for the solvent and for the solution under test, and (4) a device to receive or measure the beam or beams of radiant energy passing through the solvent or solution.

17.5   Photoelectric photometer method

Photoelectric colorimeters (absorptiometers)

One of the greatest advances in the design of colorimeters has been the use of photoelectric cells to measure the intensity of the light, thus eliminating the errors due to the personal characteristics of each observer. The photovoltaic or barrier-layer cell, in which light striking the surface of a semiconductor such as selenium mounted upon a baseplate (usually iron) leads to the generation of an electric current, the magnitude of which is governed by the intensity of light beam, has been widely used in many absorptiometers. But it does suffer from two defects: (1) amplification of the current produced by the cell is difficult to achieve, which means the cell does not have a very high sensitivity for low levels of light; and (2) the cell tends to become fatigued. For these reasons it has now been largely superseded by the photomultiplier and by silicon diode detectors.

Photoemissive cells

In the simplest form of photoemissive cell (also called a phototube) a glass bulb is coated internally with a thin, sensitive layer, such as caesium or potassium oxide and silver oxide (i.e. one which emits electrons when illuminated); a free space is left to permit the entry

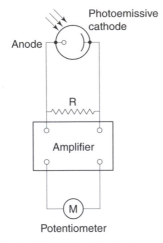

Figure 17.2 Photoemissive cell

of light. This layer is the cathode. A metal ring inserted near the centre of the bulb forms the anode, and is maintained at a high voltage by means of a battery. The interior of the bulb may be evacuated or, less desirably, filled with an inert gas at low pressure (e.g. argon at about 0.2 mm). When light, penetrating the bulb, falls on the sensitive layer, electrons are emitted, causing a current to flow through an outside circuit; this current may be amplified by electronic means, and is taken as a measure of the light striking the photo-sensitive surface. Otherwise expressed, the emission of electrons leads to a potential drop across a high resistance (R) in series with the cell and the battery; the fall in potential may be measured by a suitable potentiometer (M), and is related to the amount of light falling on the cathode. The action of the photoemissive cell is shown in Figure 17.2.

The sensitivity of a photoemissive cell (phototube) may be considerably increased by means of the so-called photomultiplier tube. This consists of an electrode covered with a photoemissive material and a series of positively charged plates, called photodynodes, each charged at a successively higher potential. The plates are covered with a material which emits between two and five electrons for each electron collected on its surface. When the electrons hit the first plate, secondary electrons are emitted in greater number than initially struck the plate, with the net result of a large amplification (up to 10^6) in the current output of the cell. The output of a photomultiplier tube is limited to several milliamps, and for this reason only low intensites of incident radiant energy can be employed. Compared with an ordinary photoelectric cell and amplifier, it can measure intensities about 200 times weaker. Phototube detectors are normally sensitive to radiation of wavelength 200–650 nm or wavelength 600–1000 nm. To scan a complete spectral range, an instrument must there-fore contain two photocells: a 'red' cell (600–800 nm) and a 'blue' cell (200–600 nm).

The **silicon diode** (photodiode) detector is a strip of p-type silicon on a chip of *n*-type silicon. By applying a biasing potential with the silicon chip connected to the positive pole of the biasing source, electrons and holes are caused to move away from the p–n junction. This creates a depletion region in the neighbourhood of the junction, which effectively becomes a capacitor. Light striking the chip surface creates free electrons and holes which migrate to discharge the capacitor; the magnitude of the resultant current is a measure of the light intensity. This detector has a greater sensitivity than a single phototube, but less than a photomultiplier. With modern technology it is possible to form a large number of photodiodes on the surface of a single silicon chip. This chip also contains an integrating circuit which can scan each photodiode in turn to give a signal that is transmitted to a microprocessor. Each photodiode can be programmed to respond to a certain small band of wavelengths, so the complete spectrum can be scanned virtually instantaneously.[7] This detector is known as a **diode array**.

When using a spectrophotometer equipped with a diode array, an absorption spectrum is obtained by electronic scanning rather than the mechanical scanning used in a conventional spectrophotometer. This produces a virtually instantaneous recording of the absorption curve; it may be accomplished in 1–5 s. Samples are therefore exposed to radiation for such short periods of time that there is little possibility of photochemical reactions taking place, and the effects of fluorescence in the samples are minimised. The speed of operation makes such instruments useful for the investigation of fast chemical reactions, and for monitoring the eluate from liquid chromatographs. But diode detectors do suffer from somewhat limited resolution: about 1 nm in the ultraviolet region and about 2 nm in the visible region. Figure 17.3 shows a typical diode array detector.

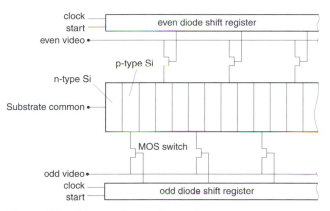

Figure 17.3   Linear self-scanning diode array

17.6   Wavelength selection

The colour of a substance is related to its ability to absorb selectively in the visible region of the electromagnetic spectrum. Now we can measure the intensity of light with a high degree of accuracy, if we wish to analyse a solution by measuring the extent to which some coloured component absorbs light, the accuracy will be improved if measurements are made at the wavelength which is being absorbed. Remember that the observed colour is due to the radiation which is **not** absorbed, or in other words, the radiation which is transmitted by the coloured solution. The colour of this radiation is **complementary** to the colour of the radiation which is being absorbed. Complementary colours are listed in Table 17.2. Several procedures can be used to select specific regions of the visible spectrum.

Light filters

Optical filters are used in colorimeters (absorptiometers) for isolating any desired spectral region. They consist of coloured glass or thin films of gelatin containing different dyes.

Table 17.2 *Complementary colours*

Wavelength (nm)	Hue (transmitted)	Complementary hue
400–435	Violet	Yellowish green
435–480	Blue	Yellow
480–490	Greenish blue	Orange
490–500	Bluish green	Red
500–560	Green	Purple
560–580	Yellowish green	Violet
580–595	Yellow	Blue
595–610	Orange	Greenish blue
610–750	Red	Bluish green

Interference filters (transmission type)

Interference filters have somewhat narrower transmitted bands than coloured filters and are essentially composed of two highly reflecting but partially transmitting films of metal (usually silver separated by a spacer film of transparent material). The amount of separation of the metal films governs the wavelength position of the passband, hence the colour of the light the filter will transmit. This is the result of an optical interference effect which produces a high transmission of light when the optical separation of the metal films is effectively one half-wavelength or a multiple of half-wavelengths. Light which is not transmitted is for the most part reflected. The wavelength region covered is either 253–390 nm or 380–1100 nm, peak transmission is 25–50% and the bandwidth is less than 18 nm for the narrowband filters suitable for colorimetry. Absorptiometers equipped with filters are now rarely used, but they are inexpensive and can be very satisfactory for certain measurements.

Prisms

To obtain improved resolution of spectra in both the visible and ultraviolet regions of the spectrum, it is necessary to employ a better optical system than is possible with filters. In many instruments, both manual and automatic, this is achieved by using prisms to disperse the radiation obtained from incandescent tungsten or deuterium sources. Dispersion occurs because the refractive index n of the prism material varies with wavelength λ; the dispersive power is given by $dn/d\lambda$. The separation achieved between different wavelengths depends on the dispersive power and the apical angle of the prism.

In instruments where the radiation is only passed through the prism in a single direction, it is common to use a 60° prism. In some cases double dispersion is achieved by reflecting the radiation back through the prism using a mirrored surface placed behind the prism, as in the Littrow mounting (Figure 17.4). Monochromatic radiation of different wavelengths is focused on the instrument slit by rotating the prism.

Unfortunately, no single material is entirely suitable for use over the full range of 200–1000 nm, although fused silica is the favourite compromise. Glass prisms can be employed between 400 and 1000 nm for the visible region, but they are not transparent to ultraviolet radiation. Quartz or fused silica prisms are required for the region below 400 nm. If quartz is used for a 60° single-pass prism, it is necessary to make the prism in

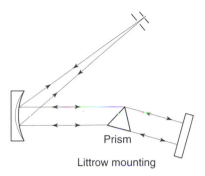

Figure 17.4   Littrow mounting

two halves, one half from right-handed quartz and the other from left-handed quartz, then polarisation effects introduced by one half will be cancelled out by the other half. Prisms have the advantage that, unlike diffraction gratings, they only produce a single-order spectrum.

Diffraction gratings

Diffraction gratings have almost completely superseded the dispersion methods; the incident radiation is diffracted by a series of closely spaced lines marked on a surface. Early diffraction gratings arranged a beam of light to pass through a piece of glass marked with the slits; they are known as transmission gratings. But to achieve the diffraction of ultra-violet radiation, modern grating spectrophotometers use metal reflection gratings where the radiation is reflected from the surfaces of a series of parallel grooves. They are often known as echelette gratings.

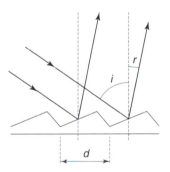

Figure 17.5   Diffraction grating

The principle of diffraction depends on the differences in path length experienced by a wavefront incident at an angle to the individual surfaces of the grooves of the grating. If i is the angle of incidence and r the angle of reflection, the path difference between rays from adjacent grooves is given by

$$d \sin i - d \sin r$$

where d is the distance between the grooves (Figure 17.5). This path difference means the new wavefronts interfere with each other, unless the path difference is an integral number of wavelengths, i.e. when

$$n\lambda = d(\sin i \pm \sin r) \tag{17.16}$$

When polychromatic radiation is incident upon the diffraction grating, equation (17.16) can usually only be satisfied for a single wavelength at a time. Rotation of the grating to change the angle of incidence i will bring each wavelength in turn to a position to satisfy the equation, thus serving as a method of monochromation.

Diffraction gratings suffer from the disadvantage that they produce second-order and higher-order spectra which can overlap the desired first-order spectrum. This overlap is most commonly seen between the long-wavelength region of the first-order spectrum and the shorter-wavelength region of the second-order spectrum. The difficulty is overcome by using carefully positioned filters in the instrument to block the undesired wavelengths. For UV/visible spectrophotometers the gratings employed have between 10 000 and 30 000 lines per centimetre. This very fine ruling means that the value of d in equation (17.16) is small and produces high dispersion between wavelengths in the first-order spectrum. Only a single grating is required to cover the region between 200 and 900 nm. The groove density is much lower for the **echelle grating**, only about 800 cm^{-1}. Coupled with its geometry, this makes it particularly useful for multi-element emission analysis. (Chapter 16).

A more recent development, found in most modern instruments, is the holographic grating. An interference pattern is produced by two monochromatic laser beams then allowed to impinge on a layer of photoresist material; photographic development turns the pattern into a series of parallel grooves which constitute the grating. A reflective coating is applied, and if the grating is mounted on a flexible base, it can be produced in curved forms which collimate light beams. This means the spectrophotometer can be made with fewer lenses. Holographic gratings also have smoother lines than machined gratings, so less light is scattered and the beam has a higher intensity.

Monochromation and bandwidth

For simple absorptiometers (colorimeters) where only the visual spectrum is involved, filters or prisms will suffice for selecting the appropriate spectral region. But with the diffraction gratings used in spectrophotometers, a much wider range of wavelengths extending into the ultraviolet may be examined, and the instrument is of much greater sensitivity. The appropriate spectral region is selected using a **monochromator**; besides the prism or diffraction grating, it contains an entrance slit which reduces the incident beam of radiation to a suitable area, and an exit slit which selects the wavelength of the radiation presented to the sample.

An important feature is the range of wavelengths present in the beam that goes to the sample; it is measured at the point where the intensity of the beam is one-half of its maximum value and it is called the **spectral half-bandwidth** (Figure 17.6). The width of the slits in the monochromator can be adjusted, and the narrower the slit the better the resolution of an absorption band. But a decrease in slit width necessarily reduces the intensity of the beam reaching the detector. In practice a compromise may need to be made between resolution and adequate intensity of radiation to permit accurate absorption readings.

Focusing of radiation within the instrument used to be performed by lenses, but lenses suffer from chromatic aberration, particularly where the ultraviolet meets the visible part of the spectrum. Nowadays focusing is usually performed by suitably curved mirrors, their reflecting surfaces coated with aluminium and protected by a silica film.

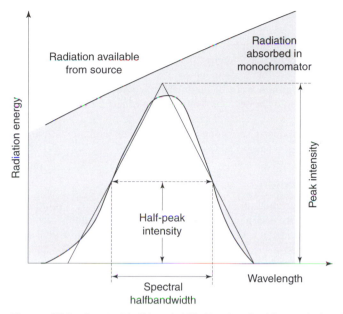

Figure 17.6   Spectral half-bandwidth (Reprinted, with permission, from J. E. Seward (ed), 1985, *Introduction to ultraviolet and visible spectrophotometry*, 2nd edn, Philips/Pye Unicam, Cambridge)

17.7   Radiation sources

For simple absorptiometers a **tungsten lamp** is the usual source of illumination. But to cover their full wavelength range, spectrophotometers have two lamps. The first is usually a tungsten–halogen (or quartz–iodine) lamp which covers wavelengths from the red end of the visible spectrum (750–800 nm) to the near-ultraviolet (300–320 nm); the lamp has a quartz outer sheath which transmits the ultraviolet wavelengths. For measurements in the ultraviolet (down to 200 nm), a **hydrogen or deuterium lamp** is used; deuterium is preferred because it produces more intense radiation. This lamp must also have a quartz envelope. Xenon arc lamps with a spectral range of 250–600 nm may also be used.

17.8   Standard cells

To investigate the absorption of radiation by a given solution, it must be placed in a suitable container called a cell (or cuvette) which can be accurately located in the beam of radiation. The instrument is provided with a cell carrier which sites the cells correctly. Standard cells are rectangular with a 1 cm light path; larger cells are available for solutions of low absorbance, and smaller cells for solutions of high absorbance (semimicro- or microcells). For aqueous solutions it is possible to obtain comparatively cheap cells made of polystyrene for visible work and polymethylmethacrylate(PMMA) for ultraviolet work. Standard cells are made of glass to cover the wavelength range 340–1000 nm, but for lower wavelengths (down to 220 nm) they must be made of silica, and for the lowest wavelength (down to 185 nm) a special grade of silica must be used. All standard cells are supplied with a lid to prevent spillage, but if volatile solvents are to be used, special cells with a well-fitting stopper should be employed. Besides the usual rectangular cell, it is possible to obtain continuous-flow cells (useful as chromatographic detectors) and sampling cells (fitted with tubes so they can be emptied and filled without removing them from the instrument).

Standard cells are produced in three grades: grade A cells have a path length tolerance of 0.1%; grade B have a tolerance of 0.5% and are regarded as suitable for routine use; grade C can have a tolerance of up to 3%. Even the highest-quality cells differ slightly from each other, and in work of the highest accuracy it is usual to select a matching pair of cells, one to hold the test solution and the other, a blank or reference solution. If unmatched cells are used, then a correction must be applied for the differing transmissions, but with many modern spectrophotometers which embody microprocessor control, the necessary correction can be done automatically.

17.9   Data presentation

The most common method of recording the absorbance of a solution was to use a micro-ammeter to measure the output of the photoelectric cell, and this method is still applied to the simplest absorptiometers. The meter is usually provided with a dual scale calibrated to read the percentage absorbance and the percentage transmittance. For quantitative measurements it is more convenient to work in terms of absorbance rather than transmittance; this is emphasised by the two graphs in Figure 17.7. In more sophisticated instruments, under microprocessor control, the microammeter tends to be replaced by a digital read-out, and there is an increasing trend towards visual displays for showing the results. These instruments may also indicate the sequence of operations required to take a measurement, perhaps the absorbance of a solution at a fixed wavelength or the absorption spectrum of a sample. Alternatively the whole procedure may be automated; the adsorption spectrum will come up on the display screen and a permanent record may be made by downloading it to a printer.

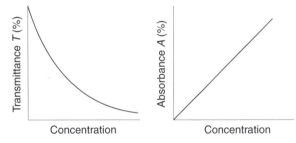

Figure 17.7   How transmission and absorbance vary with concentration. (Reprinted, with permission, from J. E. Seward (ed), 1985, *Introduction to ultraviolet and visible spectrophotometry*, 2nd edn, Philips/Pye Unicam, Cambridge)

17.10   Instrumental layout

Single-beam

The make-up of a single-beam spectrophotometer is shown in Figure 17.8. An image of the light source (A) is focused by the condensing mirror (B) and the diagonal mirror (C) on the entrance slit (D); the entrance slit is the lower of two slits placed vertically, one above the other. Light falling on the collimating mirror (E) is rendered parallel and reflected to the quartz prism (F). The back surface of the prism is aluminised, so the light refracted at the first surface is reflected back through the prism, undergoing further refraction as it

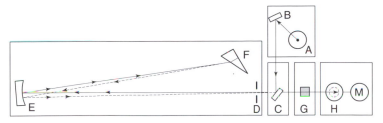

Figure 17.8   Single-beam spectrophotometer: A = light source, B = mirror, C = diagonal mirror, D = entrance slit, E = collimating mirror, F = prism, G = absorption cell, H = photocell, M = meter

emerges from the prism. The collimating mirror focuses the spectrum in the plane of the slits (D), and light of the selected wavelength passes out of the monochromator through the exit (upper) slit, through the absorption cell (G) to the photocell (R). The photocell response is amplified and is registered on the meter (M).

The light source (A) in fact consists of two lamps, as explained in Section 17.7; a tungsten–halogen lamp for the visible and near-ultraviolet part of the spectrum and a deuterium lamp for the far-ultraviolet. They are mounted on a movable arm which allows each lamp to be brought to the correct operating position as required. Likewise there are two phototubes (Section 17.5) and the appropriate one is brought to the focal point as the lamps are changed over. In modern versions of the instrument, the prism (F) will be replaced by a diffraction grating.

Double-beam

Most modern general-purpose UV/visible spectrophotometers are double-beam instruments which cover the range between about 200 and 800 nm by a continuous automatic scanning process, producing the spectrum on a display screen. The monochromated beam of radiation, from tungsten or deuterium lamps, is divided into two identical beams, one of which passes through the reference cell and the other through the sample cell. The absorption signal produced by the reference cell is automatically subtracted from the absorption signal produced by the sample cell, giving a net signal that corresponds to the absorption of the sample solution. The splitting and recombination of the beam is accomplished by two rotating sector mirrors, geared to the same electric motor so they work in unison (Figure 17.9). The instrument's microprocessor will automatically correct for the dark current of the photocell, i.e. the small current which passes even when the cell is not exposed to radiation.

Stray light needs to be minimised. As indicated in Section 17.6, diffraction gratings can give rise to spectra of different orders, some of which may overlap the main or first-order beam; their effect can be overcome by suitable filters correctly sited within the instrument. Although the interior of the monochromator is blackened, some stray light arising from reflections within the monochromator may pass through the exit slit, and when the photodetector reading is small, the stray light, perhaps containing wavelengths that are not absorbed by the sample solution, may well account for a significant proportion of the reading. When this happens, a calibration plot (Section 17.2) departs from the expected straight line, and with increasing absorbance of the solution, it becomes curved towards the concentration axis. This complicates quantitative determinations and every effort is therefore made to reduce the amount of stray light. One method is to include an extra monochromator before the main monochromator.

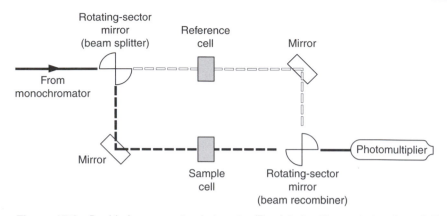

Figure 17.9 Double-beam spectrophotometer (Reprinted, with permission, from J. E. Seward (ed), 1985, *Introduction to ultraviolet and visible spectrophotometry*, 2nd edn, Philips/Pye Unicam, Cambridge)

17.11 Instruments for fluorimetry

Instruments for measuring fluorescence are known as fluorimeters or spectrofluorimeters. The essential parts of a simple fluorimeter are shown in Figure 17.10. The light from a mercury vapour lamp (or other source of ultraviolet light)* is passed through a condensing lens, a primary filter (to permit the light band required for excitation to pass), a sample container,[†] a secondary filter (selected to absorb the primary radiant energy but transmit the fluorescent radiation), a receiving photocell placed in a position at right angles to the incident beam (so it is not affected by the primary radiation), and a sensitive galvanometer or other device for measuring the output of the photocell.

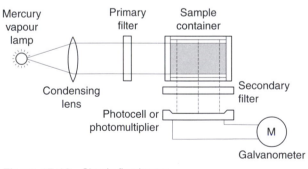

Figure 17.10 Simple fluorimeter

---

* **Mercury and xenon vapour lamps emit intense UV radiation which is damaging to the eyes. Never look at the unshielded lamp when it is on. Take care when handling high-pressure xenon lamps; if dropped they may shatter and explode.**

[†] Fluorescence cells are usually made of glass or silica with all four faces polished. For precise quantitative work, always insert them into the cell holder the same way round. Wash them after use and store them carefully.

Since fluorescence intensity is proportional to the intensity of irradiation, the light source must be very stable if fluctuations in its intensity are not compensated. It is usual, therefore, to employ a two-cell instrument; the galvanometer is used as a null instrument, and readings are taken on a potentiometer which balances the photocells against each other. Since the two photocells are selected to be similar in spectral response, it is assumed that fluctuations in the intensity of the light source are minimised.

The simpler fluorimeters are manual instruments operating only at a single wavelength at any one time. Despite this they are perfectly suitable for quantitative measurements, as these are almost always carried out at a fixed wavelength. The experiments listed at the end of this chapter have all been carried out at single fixed wavelengths. The more advanced spectrofluorimeters are capable of automatically scanning fluorescent spectra between about 200 and 900 nm, and producing a chart record of the spectrum obtained. They can also operate at a fixed wavelength and are equally suitable for carrying out quantitative work; their main application tends to be the detection and determination of organic substances in small concentrations.

Some commercial spectrophotometers have fluorescence attachments which allow the sample to be irradiated from an ancillary source and the resulting fluorescence to pass through the monochromator for spectral analysis.

17.12   The origins of absorption spectra

Molecules absorb radiation because they contain electrons that can be raised to higher energy levels by the absorption of energy. The energy can sometimes be supplied at visible wavelengths, producing an absorption spectrum in the visible region, but it may require the higher energy associated with ultraviolet radiation. Besides the change in electronic energy which follows the absorption of radiation, there are also changes associated with variation in the vibrational energy of the atoms in the molecule and changes in rotational energy. This means that many different amounts of energy will be absorbed, depending upon the varying vibrational levels to which the electrons may be raised, and the result is that we do not observe a sharp absorption line, but a comparatively broad absorption band.

Electrons in a molecule can be classified into three different types:

1. Electrons in a single covalent bond (σ-bond) are tightly bound and radiation of high energy (short wavelength) is required to excite them.
2. Electrons may be attached to atoms as lone pairs, e.g. in chlorine, oxygen and nitrogen. These non-bonding electrons can be excited at a lower energy (longer wavelength) than tightly bound bonding electrons.
3. Electrons in double or triple bonds (π-orbitals) can be excited relatively easily. In molecules containing a series of alternating double bonds (conjugated systems), the π-electrons are delocalised and require less energy for excitation, so the absorption rises to higher wavelengths.

The absorption of a given substance is greatly affected if it contains a **chromophore**. A chromophore is a functional group which has a characteristic absorption spectrum in the visible or ultraviolet region. These groups invariably contain double or triple bonds and include the C=C linkage (and therefore the benzene ring), the C≡C bond, the nitro and nitroso groups, the azo group, the carbonyl and thiocarbonyl groups. If the chromophore is conjugated with another group of the same kind or a different kind, then the absorption is enhanced and a new absorption band appears at a higher wavelength. Many of these

Table 17.3 *Some electronic transitions in selected organic molecules*

Compound	Transition	λ_{max} (nm)	ε (m^2 mol^{-1})
CH_4	$\sigma \rightarrow \sigma^*$	122	–
CH_3Cl	$n \rightarrow \sigma^*$	173	200
$CH_2{=}CH_2$	$\pi \rightarrow \pi^*$	162	1500
$Me_2C{=}O$	$\pi \rightarrow \pi^*$	185	95
	$n \rightarrow \pi^*$	277	2
$H_2C{=}CH{-}CH{=}CH_2$	$\pi \rightarrow \pi^*$	180	2100
	$\pi \rightarrow \pi^*$	200	800
	$\pi \rightarrow \pi^*$	255	22

features can be identified in Table 17.3, which also includes the molar absorption coefficients for the λ_{max} absorption.

The absorption of a given molecule may also be enhanced by the presence of groups called **auxochromes**. Auxochromes do not absorb significantly in the ultraviolet region but may have a profound effect on the absorption of the molecule to which they are attached. Important examples are OH, NH$_2$ CH$_3$ and NO$_2$ groups, and their effect, which is to displace the absorption maximum to a longer wavelength, is called a low-energy or **red** shift; it is related to the electron-donating properties of the auxochromes.

Many complexes of metals with organic ligands absorb in the visible part of the spectrum and are important in quantitative analysis. The colours arise from d–d transitions within the metal ion, which usually produce absorptions of low intensity; and n $\rightarrow \pi^*$ and $\pi \rightarrow \pi^*$ transitions within the ligand. Another type of transition, known as charge transfer, may also operate, where an electron is transferred between an orbital in the ligand and an unfilled orbital of the metal, or vice versa. Charge transfers give rise to more intense absorption bands which are of analytical importance.

For detailed consideration of the relationships between chemical constitution and the absorption of UV/visible radiation, consult textbooks of physical chemistry or spectroscopy.[8-13] A table of λ_{max} and ε_{max} values is given in Appendix 9.

17.13 Derivative spectrophotometry

As shown in Section 15.17, the end point of a potentiometric titration can often be located more exactly from the first or second derivative of the titration curve than from the titration curve itself. Similarly, absorption and emission observations will often yield more information from derivative plots than from the original absorption or emission curve. This technique was used as long ago as 1955,[14] but with the development of microcomputers which permit rapid generation of derivative curves, the method has acquired great impetus.[15,16]

If we consider an absorption band showing a normal (Gaussian) distribution (Figure 17.11(a)), we find that the first and third derivative plots (Figure 17.11(b) and (d)) are disperse functions that are unlike the original curve, but they can be used to accurately fix the wavelength of maximum absorption λ_{max} (point *M*). The second and fourth derivatives (Figure 17.11(c) and (e)) have a central peak which is sharper than the original band but

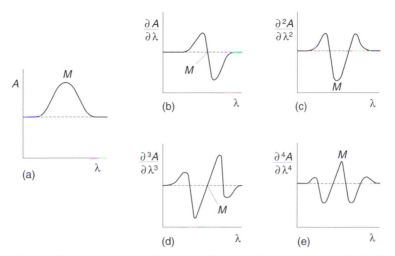

Figure 17.11 Derivatives of an absorption band having a normal distribution

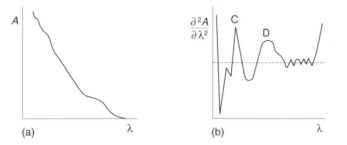

Figure 17.12 Mixture of two components C and D: (a) zero-order spectrum, (b) second-order derivative spectrum

of the same height; its sign alternates with increasing order. It is clear that resolution is improved in the even-order derivative spectra, and this offers the possibility of separating two absorption bands which may in fact merge in the zero-order spectrum. Thus, a mixture of two substances C and D gave a zero-order spectrum (Figure 17.12(a)) showing no well-defined absorption bands, but the second-order derivative spectrum deduced from this curve showed peaks at 280 nm and 330 nm (Figure 17.12(b)).

The influence of an impurity Y on the absorption spectrum of a substance X can often be eliminated by considering derivative curves (Figure 17.13); the second-order derivative plot of the mixture is identical with that of pure X. When the interference spectrum can be described by an nth order polynomial, the interference is eliminated in the $(n + 1)$th derivative. For quantitative measurements, peak heights (expressed in millimetres) are usually measured for the long-wave peak satellite of either the second- or fourth-order derivative curves, or for the short-wave peak satellite of the same curves. This is illustrated in Figure 17.14(a) for a second-order derivative: D_L is the long-wave peak height and D_S the short-wave peak height. Some workers[17] have preferred to use the peak tangent baseline (D_B) or the derivative peak zero (D_Z) measurements (Figure 17.14(b)).

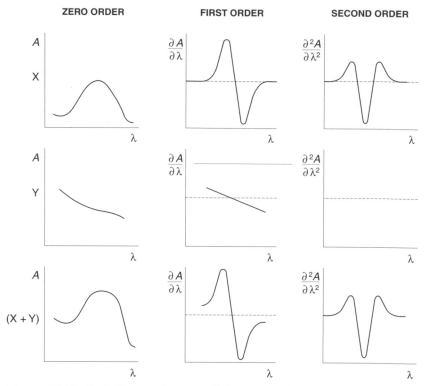

Figure 17.13 Derivative spectroscopy eliminates the effects of impurity Y on analyte X

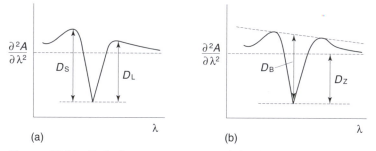

Figure 17.14 Derivative spectroscopy: peak height measurements

Derivative spectra can be recorded by means of a wavelength modulation device in which beams of radiation differing in wavelength by a small amount (1–2 nm) fall alternately on the sample cell and the difference between the two readings is recorded. In an alternative procedure, a derivative unit involving resistance/capacitance circuits, filters and operational amplifier are attached to the spectrophoptometer, but derivative curves are most readily obtained by computer-based calculations. The derivative techniques described can also be applied to fluorescence spectroscopy.

Colorimetry

17.14   General remarks

For most determinations, visual methods have been virtually superseded by methods that depend on photoelectric cells (filter photometers or absorptiometers, and spectrophotometers), and this has reduced the experimental errors. The so-called photoelectric colorimeter is a comparatively inexpensive instrument, and should be available in every laboratory. The use of spectrophotometers has enabled determinations to be extended into the ultraviolet region of the spectrum, and the use of chart recorders means the analyst is not limited to working at a single fixed wavelength.

The choice of a colorimetric procedure for the determination of a substance will depend on considerations like these:

(a)   A colorimetric method will often give more accurate results at low concentrations than the corresponding titrimetric or gravimetric procedure. It may also be simpler to carry out.

(b)   A colorimetric method may frequently be applied under conditions where no satisfactory gravimetric or titrimetric procedure exists, e.g. for certain biological substances.

(c)   Colorimetric procedures possess advantages for routine determinations on several similar samples, simply because the samples can be made so quickly. There is often no serious sacrifice of accuracy over the corresponding gravimetric or titrimetric procedures, provided the experimental conditions are rigidly controlled.

A satisfactory colorimetric analysis should satisfy six criteria:

Specificity of the colour reaction Very few reactions are specific for a particular substance, but many give colours for a small group of related substances only, i.e. they are selective. By introducing other complex-forming compounds, by altering the oxidation states, and by controlling the pH, close approximation to specificity may often be obtained. This is discussed in detail below.

Proportionality between colour and concentration For visual colorimeters it is important that the colour intensity should increase linearly with the concentration of the substance to be determined. This is not essential for photoelectric instruments, since a calibration curve may be constructed to relate the instrumental reading of the colour with the concentration of the solution. Put another way, it is desirable that the system should follow Beer's law even when photoelectric colorimeters are used.

Stability of the colour The colour produced should be sufficiently stable to permit an accurate reading to be taken. This also applies to those reactions in which colours tend to reach a maximum after a time; the period of maximum colour must be long enough for precise measurements to be made. The influence of other substances must be known, and the influence of experimental conditions (temperature, pH, stability in air, etc.).

Reproducibility The colorimetric procedure must give reproducible results under specific experimental conditions. The reaction need not necessarily represent a stoichiometrically quantitative chemical change.

Clarity of the solution The solution must be free from precipitate if comparison is to be made with a clear standard. Turbidity scatters light as well as absorbing it.

High sensitivity The colour reaction should be highly sensitive, particularly when minute amounts of substances are to be determined. It is also desirable that the reaction product absorbs strongly in the visible rather than in the ultraviolet; the interfering effect of other substances in the ultraviolet is usually more pronounced.

In view of the selective character of many colorimetric reactions, it is important to control the operational procedure so that the colour is specific for the component being determined. This may be achieved by isolating the substance by the ordinary methods of inorganic analysis; double precipitation is frequently necessary to avoid errors due to occlusion and coprecipitation. These methods of chemical separation may be tedious and lengthy, and if minute quantities are under consideration, appreciable loss may occur owing to solubility, supersaturation and peptisation effects. Any of the following processes may be used to render colour reactions specific and/or to separate the individual substances:

(a)   The action of interfering substances may be suppressed by the formation of complex ions or non-reactive complexes.

(b)   Many reactions take place within well-defined limits of pH, so the specificity may be improved by adjusting the pH.

(c)   Interfering substances may be removed by extraction with an organic solvent, sometimes after suitable chemical treatment.

(d)   The substance to be determined may be isolated by forming an organic complex, which is then removed by extraction with an organic solvent. This method may be combined with (a), in which an interfering ion is prevented from forming a soluble organic complex by converting it into a complex ion that remains in the aqueous layer (Chapter 6).

(e)   Separation by volatilisation is of limited application, but it does give good results, e.g. distillation of arsenic as the trichloride in the presence of hydrochloric acid.

(f)   Electrolysis with a mercury cathode or a controlled cathode potential.

(g)   Use physical methods such as selective absorption, chromatographic separations and ion exchange separations.

Standard curves

A filter photometer or spectrophotometer usually requires a standard curve (reference curve or calibration curve) to be constructed for the constituent being determined. Suitable quantities of the constituent are taken and treated in the same way as the sample solution for the development of colour and the measurement of the transmission (or absorbance) at the optimum wavelength. The absorbance, $\log(I_0/I_t)$, is plotted against the concentration; a straight-line plot is obtained if Beer's law is obeyed. The curve may then be used for future determinations of the constituent under the same experimental conditions. When the absorbance is directly proportional to the concentration, only a few points are required to establish the line; when the relation is not linear, a greater number of points will be necessary.

The standard curve should be checked at intervals. When a filter photometer is used, the characteristics of the filter and the light source may change with time. When plotting the standard curve it is customary to assign a transmission of 100% to the blank solution (reagent solution plus water); this represents zero concentration of the constituent. Some coloured solutions have an appreciable temperature coefficient of transmission, and the temperature of the determination should not differ appreciably from the temperature at which the calibration curve was prepared.

17.15   Choice of solvent

A solvent for use in colorimetry or spectrophotometry must be a good solvent for the substance under determination, must not interact with the solute, and must not show significant absorption at the wavelength employed in the determination. For inorganic

Table 17.4 *Cut-off wavelengths for some common solvents*

Solvent	Cut-off wavelength (nm)
Water	190
Hexane	199
Heptane	200
Diethylether	205
Ethanol	207
Methanol	210
Cyclohexane	212
Dichloromethane	233
Trichloromethane (chloroform)	247
Tetrachloromethane (carbon tetrachloride)	257
Benzene	280
Pyridine	306
Propanone (acetone)	331

compounds, water usually meets these requirements, but for the majority of organic compounds it is necessary to use an organic solvent. An immediate complication which arises is that polar solvents such as alcohols, esters and ketones (water also falls into this category) tend to obliterate the fine structure of absorption spectra, structure related to vibrational effects. To preserve these details, the absorption measurements must be carried out in a hydrocarbon (non-polar) solvent. For example, a solution of phenol in cyclohexane gives an absorption curve showing three sharply defined peaks in the ultraviolet, but an aqueous solution of the same concentration gives a single broad absorption band covering the same wavelength range as observed in the hydrocarbon solvent. There is the further complication with spectrophotometry that all solvents show absorption at some point in the ultraviolet, and care must be taken to choose a solvent that is transparent in the region required for the determination.

Organic solvents are listed in order of 'cut-off wavelength', which is the wavelength at which the transmittance is reduced to 25% when measured in a 10 cm cell against water as reference material. Values for some typical solvents are given in Table 17.4. Any impurities in the solvents may affect the cut-off value, so it is essential to use materials of the highest purity. Most major suppliers of laboratory chemicals offer products which have been specially purified and carefully tested to ensure they are suitable for spectrophotometric determinations. They are usually identified by a special name, e.g. the Spectrosol materials supplied by Merck BDH. But it often suffices to take the purest available solvent and run it through the spectrophotometer. If there is no appreciable absorption over the spectral range required for the determination, then the solvent may be used; otherwise careful purification will be needed.[18]

17.16   Colorimetric determinations: general procedure

In any colorimetric determination the exact procedure will be governed partly by the specification of the instrument and partly by the nature of the sample. Nevertheless, there are certain general principles which are universally applicable, and they are also relevant to spectrophotometric determinations in general. It is important to ensure the colorimeter (or spectrophotometer) is functioning correctly and that no adjustments or replacements are necessary. The exact procedures to be followed will be detailed in the operating manual supplied by the manufacturer, and many modern instruments can be programmed to check the wavelength setting and make appropriate corrections. In the absence of such automated facilities, manual checks should be made on the illumination, the wavelength scale, the absorbance scale and the cells. For instruments in regular use, only the illumination and the cells are likely to require attention, but the other two items should be checked periodically and after any apparent misuse of the instrument, or if it has been left unused for a long period.

Illumination   The emission of any lamp tends to decrease with age, and the lamp must be replaced when the maximum usage time is reached; a log should be kept of the burning time for each lamp.

Wavelength scale   The wavelength scale must be correctly adjusted. Many instruments are supplied with test filters; check the scale by measuring the λ_{max} values for the absorption peaks of the test filters. This can be done more precisely for an instrument containing a hydrogen or deuterium lamp by checking the red (656.1 nm) and blue-green (486.0 nm) lines in the hydrogen spectrum. Alternatively, a neon or mercury lamp can be substituted for the normal illumination of the instrument; the resultant spectra contain a number of lines of accurately known wavelength with which to test the scale.

Absorbance scale   The absorbance scale can be checked by using one or more standard solutions which have been carefully prepared; examples include potassium dichromate in either acid or alkaline solution, and postassium nitrate solution. Full details of recommended standard solutions and their standard absorption values are given in the literature.[19]

Cells   Unless the two cells are matched cells (Section 17.8), they must be tested to ensure that, under identical conditions, they show the same transmission within very narrow limits. This is done by setting up the instrument at a selected wavelength, preferably a wavelength to be used in the determination, and with no cells in position, adjusting the controls so that the scale reads 0% absorbance (or 100% transmittance). The chosen cells, which must be clean and dry, are filled with the solvent to be used and the outsides dried with paper tissue; care must be taken to handle them by the ground surfaces only and to avoid touching the polished faces. Check the transmission of one cell using the second cell as a blank and, after noting the reading, empty the cell, refill with solvent and repeat the reading, which should be consistent with the first. Now replace the solvent in the second cell and again repeat the reading. If the difference in absorbance between the two cells is less than 0.02, the cells can be accepted for use; many modern instruments are capable of correcting automatically for a small difference between the cells. If a double-beam instrument is in use, then the difference in the absorbance of the two cells is obtained immediately, and the checking procedure is more straightforward.

After use the cells must be carefully washed. For determinations that use water as the solvent, wash the cells in distilled water. For determinations that use an organic solvent which is not miscible with water, rinse the cells with a cleaning solvent that is miscible both with water and with the determination solvent, then wash the cells thoroughly with distilled water. Finally, rinse the cells with ethanol followed by drying, which may be conveniently done in a vacuum desiccator. Cells which have become contaminated can usually be cleaned by soaking in a detergent solution such as Teepol.

Preparing the solutions

The determination itself will normally require (i) a weighed quantity of the material under investigation in an appropriate solvent; (ii) a standard solution of the compound being determined in the same solvent; (iii) the requisite reagent; and (iv) any ancillary reagents such as buffers, acids or alkalis necessary to establish the correct conditions for formation of the required coloured product. If ultraviolet measurements are to be carried out, it is often possible to make the determination on the solution of the substance without the need for adding other reagents. In view of the sensitivity of colorimetric and spectrophotometric methods, the absorbance measurements are usually made on very dilute solutions. In order to take sufficient material for an accurate weight to be achieved when preparing the original solution of analyte and the corresponding standard solution, it is commonly necessary to prepare solutions which are too concentrated for the absorbance measurements, and these must then be diluted accurately to the appropriate strength.

Remember that solutions of colour-producing reagents are frequently unstable and normally should not be stored for more than a day or so. Even in the solid state, many of these materials tend to deteriorate slowly and it is generally advisable to store only small quantities so that fresh supplies are obtained at frequent intervals. A little-used reagent which may have been in stock for some months should be subjected to a trial run with the appropriate substance before commencing the actual determination.

The absorbance of the test and standard solutions will be measured in the manner described above for comparing cells, but the wavelength is chosen to suit the substance being determined. The blank solution will have a similar composition to the test solution, but without any of the determinand. If necessary, the prepared solution of the test material must be diluted so the absorbance lies in the 0.2–1.5 region. If the determination is a routine procedure, then a calibration plot will be available and it will be a simple matter to ascertain the concentration of the test solution; in this case it is unnecessary to prepare a standard solution of the compound being determined. If a calibration plot is not available, then a series of solutions containing say 5.0, 7.5, 10.0 and 15.0 mL of the standard solution are diluted to, say, 100 mL in graduated flasks. The absorbance of each of these solutions is measured and the results plotted against concentration.

For measurements in the ultraviolet, concentrations are frequently calculated by using the following relationship (Section 17.2):

$$\text{molar absorption coefficient } \varepsilon = A/cl$$

where A is the measured absorbance of a solution of concentration c in a cell of path length l. It follows that if ε for the compound being determined is known, then the concentration of the solution can be calculated. In some cases, especially with natural products, it is not possible to give ε values because relative molecular masses are not known. Recourse may then be made to $E_{1\,cm}^{1\%}$ values, which represent the absorbance of a 1% solution in a cell of path length 1 cm (Section 17.2); these values are recorded for many materials, including numerous pharmaceutical preparations.

A few typical examples of spectrophotometric determinations of selected analytes are given in Sections 17.20 to 17.30. Further examples of analytes which can be determined colorimetrically or spectrophotometrically are given in Table 17.5. The detailed experimental procedures for these analytes can be found in earlier editions of this book.

Table 17.5 *Conditions for determining a range of analytes by spectrophotometry*

Analyte	Reagent	pH	λ_{max} (nm)
Al	Eriochrome cyanine R	5.9–6.1	535
Sb	Potassium iodide		425 or 330
Be	4-nitrobenzenazo-orcinol		520
Bi	Potassium iodide–hypophosphorous acid		460
Co	Nitroso-R-salt	5.5	425
Cu	Bicyclohexanone oxalylhydrozone		570–600
Fe	Potassium thiocyanate		480
	1,10-phenanthroline	2.5–4.5	515
Pb	Dithizone	9.5	510
Mg	Solochrome black	10.1	520
Mo	Toluene-3,4-dithiol		670
Ni	Dimethylglyoxime	>7.5	445
Sn	Toluene-3,4-dithiol		630
V	Phosphoric acid–sodium tungstate		400
F^-	Thorium chloranilate	4.5	540 or 330
NO_2^-	Sulphanilamide–N-(1-naphthyl)ethylenediamine dihydrochloride	7.0	550
Silicate	Ammonium molybdate	4.5–5.0	815

17.17 Enzymatic analysis[20]

Routine enzyme analysis originated in biochemical and clinical analysis. These methods have been adapted for food analysis and bioanalysis. Compounds that occur in nature, e.g. sugars, acids or their salts and alcohols, can be analysed enzymatically since living cells contain enzymes capable of synthesising or decomposing these substances. Thus, if such an enzyme and a suitable measuring system can be made available, the compound can be determined enzymatically. Most enzymatic methods use test kits supplied by manufacturers, e.g. Boehringer Mannheim, which are sufficient for a given number of determinations of the analyte(s).

Many of the ultraviolet methods of analysis are based on the measurement of the increase or decrease in absorbance of the coenzymes NADH (nicotinamide adenine dinucleotide, reduced form) or NADPH (nicotinamide adenine dinucleotide phosphate, reduced form), which absorb light with λ_{max} at 340 nm. The measurement of analytes with dehydrogenases or with reductases via the formation or consumption of NAD(P)H is advantageous because the molar absorptivities used for the calculation of results are well known and the methods are largely free from interferences.

Colorimetric methods are based on the formation of a light-absorbing dye in the visible spectrum. Oxidases react with a substrate (analyte) to form hydrogen peroxide as an intermediate. This can react with a leuco-dye in the presence of a peroxidase enzyme and a dye is formed which can be measured in the visible region of the spectrum. Two examples of the use of test kits for the analysis of foodstuffs are given in Sections 17.37 and 17.38. The Boehringer publication *Methods of enzymic bioanalysis and food analysis* lists over 40 analyses that can be carried out using enzyme-based test kits.[20]

17.18   Some applications of fluorimetry

Fluorimetry is generally used if there is no colorimetric method sufficiently sensitive or selective for the substance to be determined. In inorganic analysis the most frequent applications are for the determination of metal ions as fluorescent organic complexes. Many of the complexes of 8-hydroxyquinoline (oxine) fluoresce strongly; aluminium, zinc, magnesium and gallium are sometimes determined at low concentrations by this method. Aluminium forms fluorescent complexes with the dyestuff eriochrome blue black RC (pontachrome blue black R), whereas beryllium forms a fluorescent complex with quinizarin. The analysis of non-metallic elements and anionic species may present a problem since many do not readily form suitable derivatives for fluorimetric analysis. The best-known fluorimetric methods for non-metals are those for boron and selenium; both involve derivatisation reactions leading to ring closure, e.g. the derivative [17.A] formed by the condensation reaction between boric acid and benzoin.

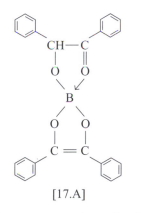

[17.A]

Important organic applications are for the determination of quinine and the vitamins riboflavin (vitamin B_2) and thiamine (vitamin B_1). Riboflavin fluoresces in aqueous solution; thiamine must first be oxidised with alkaline hexacyanoferrate(III) solution to thiochrome, which gives a blue fluorescence in butanol solution. Under standard conditions, the net fluorescence of the thiochrome produced by oxidation of the vitamin B_1 is directly proportional to its concentration over a given range. The fluorescence can be measured either by reference to a standard quinine solution in a null-point instrument or directly in a spectrofluorimeter.[21]

Quenching is described in Section 17.3, and the application of quenching methods will be briefly considered here. The principle is that the emission of a fluorescent species is quenched by the analyte, so the intensity of fluorescence decreases as the analyte concentration increases. However, a major limitation of quenching is that, since it is completely non-specific, applications are restricted to analyses in which only the analyte is able to quench the fluorescence. Probably the most important application of quenching is in the determination of molecular oxygen, a paramagnetic species particularly effective as a quenching agent for molecules with relatively long fluorescence lifetimes; with eosin ($\tau \sim 10^{-3}$ s) about $10\,mg\,L^{-1}$ oxygen produces 50% quenching. Fluorescence quenching thus provides a useful method for monitoring low levels of oxygen, e.g. in a supply of 'oxygen-free' nitrogen.

The intensity and colour of the fluorescence of many substances depend upon the pH of the solution; indeed, some substances are so sensitive to pH that they can be used as pH indicators; they are termed **fluorescent or luminescent indicators**. Substances which

Table 17.6 *Some fluorescent indicators*

Indicator	Approximate pH range	Colour change
Acridine	5.2–6.6	Green to violet blue
Chromotropic acid	3.0–4.5	Colourless to blue
2-Hydroxycinnamic acid	7.2–9.0	Colourless to green
3,6-Dihydroxyphthalimide	0.0–2.5	Colourless to yellowish green
	6.0–8.0	Yellowish green to green
Eosin	3.0–4.0	Colourless to green
Erythrosin-B	2.5–4.0	Colourless to green
Fluorescein	4.0–6.0	Colourless to green
4-Methylaesculetin	4.0–6.2	Colourless to blue
	9.0–10.0	Blue to light green
2-Naphthoquinoline	4.4–6.3	Blue to colourless
Quinine sulphate	3.0–5.0	Blue to violet
	9.5–10.0	Violet to colourless
Quininic acid	4.0–5.0	Yellow to blue
Umbelliferone	6.5–8.0	Faint blue to bright blue

Table 17.7 *Fluorimetric methods of analysis*

Analyte	Reagent	λ_{ex} (nm)	λ_{em} (nm)
Al	Eriochrome blue black RC	480	590
Ca	Calcein	330 or 480	540
Vitamin A	None	330	500

fluoresce in ultraviolet light and change in colour, or which have their fluorescence quenched by a change in pH can be used as fluorescent indicators in acid–base titrations. The merit of these indicators is that they can be employed in the titration of coloured (and sometimes intensely coloured) solutions in which the colour changes of the usual indicators would be masked. Titrations are best performed in a silica flask. Examples of fluorescent indicators are given in Table 17.6. Methods illustrating the application of fluorimetry follow in Sections 17.39 and 17.40. Table 17.7 gives further examples with their excitation and emission wavelengths.

17.19 Flow injection analysis

The original concept of flow injection analysis was introduced by Ruzicka and Hansen.[22] The method was developed from **continuous segmented flow sequential analysis** used widely in clinical laboratories. In segmented flow systems, samples and reagents were transported to the detector by a flowing stream segmented by closely spaced air bubbles. The air bubbles ensured that successive samples were separated, thus avoiding sample dispersion

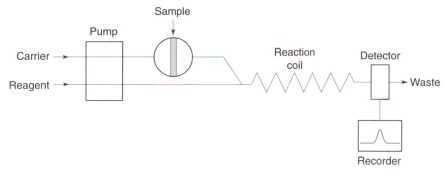

Figure 17.15   Flow injection analysis: simple system

and the likelihood of cross-contamination. Ruzicka and Hansen showed, however, that air bubble segmentation was not necessary, providing conditions were carefully controlled. Flow injection analysis (FIA) is able to achieve a faster sample throughput using simpler and more flexible equipment, and has now virtually replaced the air segmented systems.

A simple FIA system is shown in Figure 17.15. The carrier and the reagent are transported by a peristaltic pump at a constant flow rate. The liquid sample, typically a few microlitres, is injected then mixed with the reagent. The mixture passes through a reaction coil and then on to a flow cell detector. The analytical signal thus obtained is normally displayed on a recorder. The injected sample forms a so-called sample zone, which is dispersed when mixed with the carrier and reagent solutions. The extent of dispersion and/or any necessary dilution of the sample zone, prior to measurement, may be controlled by a number of factors, including volume of injected sample, flow rate, tube length and tube diameter of the reaction coil. The flow cell detector continuously records changes in the analyte concentration using absorbance, fluorescence intensity, potentiometric measurements, etc. The analyte concentration may be determined by interpolation with a series of standards which are injected under the same conditions and in an identical manner as the samples.

FIA has five main advantages:

1.   Fast sample throughput, typically 200 samples per hour.
2.   Very short response times, between 20 and 60 s between sample injection and detector response.
3.   Small sample volume and use of reagent.
4.   The system can be set up ready for analysis in a few minutes.
5.   Enhanced flexibility allows changes in the system to be performed simply and rapidly.

Applications of FIA

The scope and performance of FIA have been extended by using a separation step which enhances the selectivity of the determination. A full review of sample preparations and separatory methods in FIA is given by Clark and co-workers.[23] Separations by reaction columns, by dialysis, by gaseous diffusion or by solvent extraction may be incorporated directly in line with FIA systems. Reaction columns may be used prior to the sample injection stage, e.g. pretreatment of the sample using an ion exchange column can remove ions that interfere with the analyte.[24] Dialysis is widely used in FIA where there is a need to

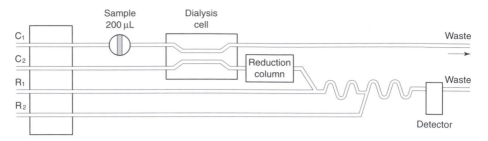

Figure 17.16   Determining nitrite and nitrate by FIA

separate ions and small organic molecules from particulate matter or macromolecules such as proteins. Dialysis separations are used in the determination of ions in blood or serum samples. The determination of nitrite and nitrate in dairy products[25] is shown in Figure 17.16.

The carrier solutions C_1 and C_2 are ammonium chloride solutions at pH 9.6 and pH 6.1 respectively. The reduction column contains cadmium granules to reduce nitrate to nitrite; the reduction column is bypassed for nitrite determinations. The colorimetric determination is based on the reaction of the nitrite ions under acidic conditions to diazotise 4-aminobenzenesulphonamide (sulphanilamide) R_1 which is then coupled with N-(1-naphthyl)-ethylenediamine dihydrochloride solution R_2. The absorbance of the product is measured at 550 nm.

Separation can also be achieved by production of a gas within the FIA system. The gaseous analyte then diffuses through a gas-permeable membrane (such as Teflon) into a stream containing a reagent enabling the determination of the gas. For example, a solution containing NH_4^+ ions is injected into a stream of sodium hydroxide solution which is directed into a gas diffusion system where the ammonia gas diffuses into an acceptor stream containing Nessler's reagent. The yellow-coloured solution produced can be used for the colorimetric determination of ammonia at 530 nm.

Solvent extraction is a further separatory method that can be used in FIA. The organic phase and the aqueous phase are separated in a separator containing a Teflon membrane. The analyte in the organic phase is passed through a detector in the normal manner. An example is the colorimetric determination of iron by extraction from the aqueous solution of the sample with a 1% solution of 8-hydroxyquinoline (oxine) in chloroform.

The determination of calcium in serum and in water has been described by Hanson and co-workers.[26] Calcium ions form a bright red chelated complex when reacted with phthalein complexone (o-cresolphthalein complexone) buffered to a pH between 10 and 11 using sodium tetraborate. For FIA the phthalein complexone and sodium tetraborate solutions are pumped to a mixing coil before injection of the sample containing calcium. The calcium chelate, formed in a second mixing coil, is then determined colorimetrically. The precision of the method was enhanced by duplicate injections of each of four calcium standards and triplicate injections of each analyte sample.

Many manual procedures may be readily adapted into a flow injection system, so FIA has a considerable range of applications. It extends from simple neutralisation titrations to the determination of caffeine in drugs.[27] Samples and standards must be subject to the same analytical operating conditions during FIA, since none of the separatory reactions necessarily reaches completion.

Experimental section
Cations

> **!** **Safety.** Before carrying out any experiments in this section, pay full
> attention to any safety warnings and make sure you adhere to national
> laboratory and safety regulations.

17.20   Ammonia

Discussion   J. Nessler in 1856 first proposed an alkaline solution of mercury(II) iodide
in potassium iodide as a reagent for the colorimetric determination of ammonia. Various
modifications of the reagent have since been made. When Nessler's reagent is added to a
dilute ammonium salt solution, the liberated ammonia reacts with the reagent fairly rapidly
but not instantaneously to form an orange-brown product, which remains in colloidal
solution, but flocculates on long standing. The colorimetric comparison must be made
before flocculation occurs. The reaction with Nessler's reagent, an alkaline solution of
potassium tetraiodomercurate(II), may be represented as

$$2K_2[HgI_4] + 2NH_3 = NH_2Hg_2I_3 + 4KI + NH_4I$$

The reagent is employed for the determination of ammonia in very dilute ammonia solutions and in water. In the presence of interfering substances, it is best to separate the
ammonia first by distillation under suitable conditions. The method is also applicable to the
determination of nitrates and nitrites; these are reduced in alkaline solution by Devarda's
alloy to ammonia, which is removed by distillation. The procedure is applicable to concentrations of ammonia as low as $0.1 \, mg \, L^{-1}$.

Reagents   *Nessler's reagent*   Dissolve 35 g potassium iodide in 100 mL water, and
add 4% cent mercury(II) chloride solution, with stirring or shaking, until a slight red
precipitate remains (about 325 mL are required). Then introduce with stirring, a solution
of 120 g sodium hydroxide in 250 mL water and make up to 1 L with distilled water. Add
a little more mercury(II) chloride solution until there is a permanent turbidity. Allow the
mixture to stand for one day and decant from the sediment. Keep the solution stoppered in
a dark-coloured bottle.

  Here is an alternative method of preparation. Dissolve 100 g mercury(II) iodide and 70 g
potassium iodide in 100 mL ammonia-free water. Add slowly, and with stirring, to a cooled
solution of 160 g sodium hydroxide pellets (or 224 g potassium hydroxide) in 700 mL
ammonia-free water, and dilute to 1 L with ammonia-free distilled water. Allow the precipitate to settle, preferably for a few days, before using the pale yellow supernatant liquid.

Ammonia-free water   Ammonia-free water may be prepared in a conductivity water
still, or by means of a column charged with a mixed cation and anion exchange resin (e.g.
Permutit Bio-Deminrolit or Amberlite MB-1), or as follows. Redistil 500 mL of distilled
water in a Pyrex apparatus from a solution containing 1 g potassium permanganate and 1 g
anhydrous sodium carbonate; reject the first 100 mL portion of the distillate and then
collect about 300 mL.

Procedure   For practice in this determination, employ either a very dilute ammonium
chloride solution or ordinary distilled water which usually contains sufficient ammonia for

the exercise. Prepare a standard ammonium chloride solution as follows. Dissolve 3.141 g ammonium chloride, dried at 100 °C, in ammonia-free water and dilute to 1 L with the same water. This stock solution is too concentrated for most purposes. A standard solution is made by diluting 10 mL of this solution to 1 L with ammonia-free water: 1 mL contains 0.01 mg of NH_3. If necessary, dilute the sample to give an ammonia concentration of $1 \, mg \, L^{-1}$.

When 1 mL of the Nessler reagent is added to 50 mL of the sample, measurements in the wavelength region 400–425 nm with a 1 cm path allows concentrations of 20–250 µg of ammonia to be determined. Nitrogen concentrations approaching up to 1 mg can be determined in the wavelength range near 525 nm. The calibration curve should be prepared under exactly the same conditions of temperature and reaction time adopted for the sample. Ammonia can also be determined colorimetrically by flow injection analysis.

17.21   Arsenic

Discussion   Of the numerous procedures available for the determination of minute amounts of arsenic, only one will be described, the molybdenum blue method. It possesses great sensitivity and precision, and is readily applied colorimetrically or spectrophotometrically.

Molybdenum blue method   When arsenic, as arsenate, is treated with ammonium molybdate solution and the resulting heteropolymolybdoarsenate (arsenomolybdate) is reduced with hydrazinium sulphate or with tin(II) chloride, a blue soluble complex 'molybdenum blue' is formed. The constitution is uncertain, but it is evident that the molybdenum is present in a lower oxidation state. The stable blue colour has a maximum absorption at about 840 nm and shows no appreciable change in 24 h. Various techniques for carrying out the determination are available, but only one can be given here. Phosphate reacts in the same manner as arsenate (and with about the same sensitivity) and must be absent.

Both macro and micro quantities of arsenic may be isolated by distillation of arsenic(III) chloride from hydrochloric acid solution in an all-glass apparatus in a stream of carbon dioxide or nitrogen; a reducing agent, such as hydrazinium sulphate, is used to reduce arsenic(V) to arsenic(III). The distillate may be collected in cold water. Germanium accompanies arsenic in the distillation; if phosphate is present in large amounts, the distillate should be redistilled under the same conditions. Another method of isolation involves volatilisation of arsenic as arsine by the action of zinc in hydrochloric or sulphuric acid solution. Appreciable amounts of certain reducible heavy metals, such as copper, nickel and cobalt, slow down the evolution of arsine, as do also large amounts of metals that are precipitated by zinc. Copper in more than small quantities prevents complete evolution of arsine; the error amounts to 20% (for 5–10 µg As) with 50 mg of copper. The arsine which is evolved may be absorbed in a sodium hydrogencarbonate solution of iodine. The absorption apparatus should be designed so the arsine is completely absorbed.

Reagents[*]   *Potassium iodide solution*   Dissolve 15 g of solid in 100 mL water.

Tin(II) chloride solution   Dissolve 40 g hydrated tin(II) chloride in 100 mL concentrated hydrochloric acid.

Zinc   Use 20–30 mesh or granulated; arsenic-free.

---

[*] Special pure, arsenic-free reagents are available from chemical supply houses (e.g. Merck BDH) and are indicated by AST after the name of the compound; as far as possible, they should be used in the determination and for the preparation of the reagents listed here.

Iodine–potassium iodide solution   Dissolve 0.25 g iodine in a small volume of water containing 0.4 g potassium iodide, and dilute to 100 mL.

Sodium disulphite solution (sodium metabisulphite)   Dissolve 0.5 g of the solid reagent ($Na_2S_2O_5$) in 10 mL water. Prepare fresh daily.

Sodium hydrogencarbonate solution   Dissolve 4.2 g of the solid in 100 mL water.

Ammonium molybdate–hydrazinium sulphate reagent   Solution A: dissolve 10 g ammonium molybdate in 10 mL water and add 90 mL of 3 M sulphuric acid. Solution B: dissolve 0.15 g pure hydrazinium sulphate in 100 mL water. Mix 10.0 mL each of solutions A and B just before use.

Hydrochloric acid   This must be arsenic-free.

Standard arsenic solution   Dissolve 1.320 g arsenic(III) oxide in the minimum volume of 1 M sodium hydroxide solution, acidify with dilute hydrochloric acid, and make up to 1 L in a graduated flask: 1 mL contains 1 mg of As. A solution containing 0.001 mg mL^{-1} As is prepared by dilution.

Procedure   The arsenic must be in the arsenic(III) state; this may be secured by first distilling in an all-glass apparatus with concentrated hydrochloric acid and hydrazinium sulphate, preferably in a stream of carbon dioxide or nitrogen. Another method consists in reducing the arsenate (obtained by the wet oxidation of a sample) with potassium iodide and tin(II) chloride; the acid concentration of the solution after dilution to 100 mL must not exceed 0.2–0.5 M. Then 1 mL of 50% potassium iodide solution and 1 mL of a 40% solution of tin(II) chloride in concentrated hydrochloric acid are added, and the mixture heated to boiling.

Transfer an aliquot of the arsenate solution, having a volume of 25 mL and containing not more than 20 µg of arsenic, to the 50 mL Pyrex evolution vessel (A) shown in Figure 17.17, and add sufficient concentrated hydrochloric acid to make the total volume present

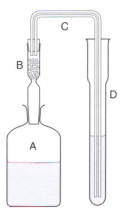

Figure 17.17   Distillation of $AsCl_3$ from HCl solution: A = Pyrex solution vessel, B = tube loosely packed with cotton wool soaked in lead ethanoate, C = capillary tube with 0.5 mm internal diameter and 4 mm external diameter, D = absorption tube

in the solution 5–6 mL, followed by 2 mL of the potassium iodide solution and 0.5 mL of the tin(II) chloride solution. Allow to stand at room temperature for 20–30 min to permit the complete reduction of the arsenate.

The tube (B) is loosely packed with cotton wool soaked in lead ethanoate solution (to remove hydrogen sulphide and trap acid spray); if connects to capillary tube (C) with 4 mm external diameter and 0.5 mm internal diameter. Place 1.0 mL iodine–potassium iodide solution and 0.2 mL of the sodium hydrogencarbonate solution in the narrow absorption tube (D). Mix with the end of the delivery tube.

Rapidly add 2.0 g of zinc to vessel A, immediately insert the stopper, and allow the gases to bubble through the solution for 30 min. At the end of this time the solution in D should still contain some iodine. Disconnect the delivery tube C and leave it in the absorption tube. Add 5.0 mL of the ammonium molybdate–hydrazine reagent and a drop or two of sodium disulphite solution. Heat the resulting colourless solution in a water bath at 95–100 °C, cool, transfer to a 10 mL graduated flask, and make up to volume with water.

Measure the absorbance of the solution at 840 nm. Charge the reference cell with a solution obtained by taking the iodine–iodide–hydrogencarbonate mixture and treating it with molybdate–hydrazinium sulphate–disulphite as in the actual procedure. Construct the calibration curve by taking say 0, 2.5, 5.0, 7.5 and 10.0 μg As (for a final volume of 10 mL), mixing with iodine–iodide–hydrogencarbonate solution, adding molybdate–hydrazinium sulphate–disulphite, and heating to 95–100 °C.

The following procedure has been recommended by the Analytical Methods Committee of the Society for Analytical Chemistry; it is suitable for determining small amounts of arsenic in organic matter.[28] Organic matter is destroyed by wet oxidation, and the arsenic, after extraction with diethylammonium diethyldithiocarbamate in chloroform, is converted into the arsenomolybdate complex; this complex is reduced by means of hydrazinium sulphate to a molybdenum blue complex, determined spectrophotometrically at 840 nm and referred to a calibration graph in the usual manner.

17.22   Boron

Discussion   Minute amounts of boron are usually separated by distillation from an acid solution as methyl borate. Borosilicate glass should be avoided, even for the storage of chemicals. The apparatus should be made of fused silica,* but a platinum dish receiver may also be used. Distillation may be made from a strong acid solution, e.g. sulphuric or phosphoric(V) acid. In the simplest apparatus, methanol vapour is passed through a flask containing the solution of the sample and is condensed and collected in an excess of either calcium hydroxide or sodium hydroxide solution in a silica or platinum dish. In a more efficient apparatus, the methanol is made to cycle between the sample dissolved in the acid medium and a flask containing calcium or sodium hydroxide solution: distillation can thus be continued for several hours with only a small amount of methanol. At the end of the distillation, the contents of the receiver in which the methyl borate was collected (which must be strongly alkaline, a minimum of four times the theoretical amount of base) are evaporated to dryness. The residue is used for the colorimetric determination.

Most of the reagents, e.g. quinalizarin (1,2,5,8-tetrahydroxyanthraquinone) or 1,1′-dianthrimide (1,1′-iminodianthraquinone) react only in concentrated sulphuric acid solution.

* Corning Vycor glass containing 96% silica is usually suitable.

With quinalizarin the absorption maxima for the reagent and its boron complex lie close together; with dianthrimide the maximum absorption for the reagent is below 400 nm and for the boron complex is at 620 nm. The use of dianthrimide will accordingly be described. The colour change of 1,1'-dianthrimide from greenish yellow to blue in the presence of borates in concentrated sulphuric acid is the basis of a trustworthy method for the determination of micro amounts of boron; the effective range of the reagent is 0.5 to 6 μg and the colour is stable for several hours.

Interferences in the distillation method are fluoride and large amounts of gelatinous silica. Fluoride interference may be overcome by the addition of calcium chloride. Strong oxidising agents, such as chromate and nitrate, interfere because they destroy the reagent. Boron in natural waters can be determined without separation; the residue obtained after evaporation to dryness with a little calcium hydroxide solution may be used directly in the colour formation. In the analysis of steel by dissolution in sulphuric acid, no oxidising compounds are formed which can interfere with the reaction.

Reagents   *Dianthrimide reagent solution*   Dissolve 150 mg of 1,1'-dianthrimide in 1 L concentrated sulphuric acid (~96% w/v). Keep in the dark and protected from moisture.

Standard boron solution   Dissolve 0.7621 g boric acid in water and dilute to 1 L. Take 50 mL of this solution and dilute to 1 L; the resulting solution contains 6.667 μg mL^{-1} boron.

Dilute sulphuric acid   Prepare a 25% v/v solution.

Procedure (boron in steel)   Dissolve about 3 g of the steel (B content ≤ 0.02%), accurately weighed, in 40 mL dilute sulphuric acid in a 150 mL Vycor or silica flask fitted with a reflux condenser. Heat until dissolved. Filter through a quantitative filter paper into a 100 mL graduated flask. Wash with hot water, cool to room temperature, and dilute to the mark with water. This flask (A) contains the acid-soluble boron.

Ignite the filter in a platinum crucible, fuse with 2.0 g of anhydrous sodium carbonate, dissolve the melt in 40 mL of dilute sulphuric acid, and add 1 mL of sulphurous acid solution (about 6%) to reduce any iron(III) salt, etc., formed in the fusion, and filter if necessary. Transfer the solution to a 100 mL graduated flask, dilute to the mark, and mix. This flask (B) contains the acid-insoluble boron.

Transfer 3.0 mL of solutions A and B to two dry, glass-stoppered conical flasks (Vycor or silica). Add 25 mL of dianthrimide reagent solution to each with shaking, and insert the glass stoppers loosely. For the blank, use 3.0 mL of solutions A and B in two similar 50 mL conical flasks and add 25 mL concentrated sulphuric acid (98% w/v). Heat all four flasks in a boiling water bath for 60 min. Cool to room temperature and measure the absorbance of each of the solutions at 620 nm against pure concentrated sulphuric acid in 1 cm or 2 cm cells. Correct for the blanks.

To construct the calibration curve, run 5–50 mL of the standard boron solution by means of a burette into 100 mL graduated flasks, add 30 mL of dilute sulphuric acid, and make up to volume. These solutions contain 1–10 μg of boron per 3 mL. Use 3 mL of each solution and 3 mL of a boron-free comparison solution and proceed as above. Plot a calibration curve relating absorbance and boron content. Calculate the total boron content of the steel (i.e. acid-soluble plus acid-insoluble boron). An alternative method for determination of boron as borate is listed in Table 11.2.

17.23 Chromium

Discussion   Small amounts of chromium (up to 0.5%) may be determined colorimetrically in alkaline solution as chromate; uranium and cerium interfere, but vanadium has little influence. The absorbance of the solution is measured at 365–370 nm. The standard solution used for the preparation of the reference curve should have the same alkalinity as the sample solution, and should preferably have the same concentration of foreign salts. Standards may be prepared from analytical grade potassium chromate. A more sensitive method is to employ 1,5-diphenylcarbazide $CO(NH.NHC_6H_5)_2$; in acid solution ($\sim 0.2\,M$), chromates give a soluble violet compound with this reagent.

Molybdenum(VI), vanadium(V), mercury, and iron interfere; permanganates, if present, may be removed by boiling with a little ethanol. If the ratio of vanadium to chromium does not exceed 10 : 1, nearly correct results may be obtained by allowing the solution to stand for 10–15 min after the addition of the reagent, since the vanadium diphenylcarbazide colour fades fairly rapidly. Vanadate can be separated from chromate by adding 8-hydroxyquinoline (oxine) to the solution and extracting at about pH 4 with chloroform; chromate remains in the aqueous solution. Vanadium as well as iron can be precipitated in acid solution with cupferron and thus separated from chromium(III).

Procedure   Prepare a 0.25% solution of diphenylcarbazide in 50% acetone (propanone) as required. The test solution may contain 0.2–0.5 ppm chromate. To about 15 mL of this solution add sufficient $3\,M$ sulphuric acid to make the concentration about $0.1\,M$ when subsequently diluted to 25 mL; add 1 mL of the diphenylcarbazide reagent and make up to 25 mL with water. Match the colour produced against standards prepared from $0.0002\,M$ potassium dichromate solution. A green filter having transmission maximum at about 540 nm may be used.

Chromium in steel

Discussion   The chromium in the steel is oxidised by perchloric acid to the dichromate ion; the colour of the dichromate ion is intensified by iron(III) perchlorate, itself colourless. The coloured solution is compared with a blank in which the dichromate is reduced with ammonium iron(II) sulphate. The method is not subject to interference by iron or by moderate amounts of alloying elements usually present in steel.

Procedure   Place a 1.000 g sample of the steel (Cr content < 0.1%) in a 100 mL beaker and dissolve it in 10 mL of dilute nitric acid (1 : 1) and 20 mL of perchloric acid (specific gravity 1.70; 70–72%). If the Cr content is 0.1–1.0% then dissolve a 0.5000 g sample in 10 mL of dilute nitric acid (1 : 1) and 15 mL of perchloric acid (specific gravity 1.70). Evaporate to dense fumes of perchloric acid and boil gently for 5 min to oxidise the chromium. **Care!** Cool the beaker and contents rapidly, dissolve soluble salts by adding 20 mL of water, transfer the solution quantitatively to a glass-stoppered 50 mL graduated flask, and dilute to the mark. Remove an aliquot portion to a cuvette, reduce it with a little (~ 20 mg) ammonium iron(II) sulphate, and adjust the colorimeter or spectrophotometer so the reading is zero at about 450 nm with this solution. Discard the solution in the cuvette, and refill it with an equal volume of the oxidised solution; the reading is a measure of the colour due to the dichromate.

Standardisation may be carried out by using solutions prepared from a chromium-free standard steel and standard potassium dichromate solution. After dissolution of the standard steel, the solution is boiled with perchloric acid (**Great care and full protective measures must be taken when using the perchloric acid**), potassium dichromate is added and the resulting solution is diluted to volume; the measurements are carried out as above. The chromium content of any unknown steel may then be deduced from the colorimeter reading.

17.24   Titanium

Discussion   Hydrogen peroxide produces a yellow colour with an acidic titanium(IV) solution.* With small amounts of titanium ($\leq 0.5\,\mathrm{mg\,mL^{-1}}$ TiO_2) the intensity of the colour is proportional to the amount of the element present. Comparison is usually made with standard titanium(IV) sulphate solutions, and a method for their preparation from potassium titanyl oxalate is described below. The hydrogen peroxide solution should be about 3% strength (ten volume) and the final solution should contain sulphuric acid having a concentration of 0.75–$1.75\,M$ in order to prevent hydrolysis to a basic sulphate and to prevent condensation to metatitanic acid. The colour intensity increases slightly with rise of temperature, hence the solutions to be compared should have the same temperature, preferably 20–$25\,^\circ C$. Interfering species fall into three categories:

(a)   Iron, nickel, chromium, etc., because of the colour of their solutions.
(b)   Vanadium, molybdenum and sometimes chromium, because they form coloured compounds with hydrogen peroxide.
(c)   Fluorine (even in minute amount) and large quantities of phosphates, sulphates and alkali salts.

The influence of sulphates and alkali salts is progressively reduced the greater the concentration of sulphuric acid present, up to 10%. The influence of class (a) is overcome, if present in small amount, through matching the colour by adding like quantities of the coloured elements to the standard before hydrogen peroxide is added. When large amounts of iron are present, as in the analysis of cast irons and steels, two methods may be adopted: (1) phosphoric(V) acid can be added in like amount to both unknown and standard, after the addition of hydrogen peroxide; (2) the iron content of the unknown solution is determined, and a quantity of standard ammonium iron(III) sulphate solution, containing the same amount of iron, is added to the standard solution. Large quantities of nickel, chromium, etc., must be removed.

Elements of class (b) must also be removed; vanadium and molybdenum are most easily separated by precipitation of the titanium with sodium hydroxide solution in the presence of a little iron. Fluoride has the most powerful effect in bleaching the colour; it must be removed by repeated evaporation with concentrated sulphuric acid. The bleaching effect of phosphoric acid is overcome by adding a like amount to the standard, or by adding $1\,\mathrm{mL}$ of 0.1% uranyl acetate solution for each $0.1\,\mathrm{mg}$ of Ti present.

Preparation of standard titanium solution   Weigh out $3.68\,\mathrm{g}$ potassium titanyl oxalate $K_2TiO(C_2O_4)_2\cdot 2H_2O$ into a Kjeldahl flask; add $8\,\mathrm{g}$ ammonium sulphate and $100\,\mathrm{mL}$ concentrated sulphuric acid. Gradually heat the mixture to boiling and boil for $10\,\mathrm{min}$. **Care!** Cool, pour the solution into $750\,\mathrm{mL}$ of water, and dilute to $1\,\mathrm{L}$ in a graduated flask; $1\,\mathrm{mL} \equiv 0.50\,\mathrm{mg}$ of Ti. If there is any doubt concerning the purity of the potassium titanyl oxalate, standardise the solution by precipitating the titanium with ammonia solution or with cupferron solution, and igniting the precipitate to TiO_2.

* One proposed formula for the coloured species is $[TiO(SO_4)_2]^{2-}$; similar ions have also been suggested. Another proposal is $[Ti(H_2O_2)]^{4+}$; analogous complexes have also been suggested.

Procedure The sample solution should preferably contain titanium as sulphate in sulphuric acid solution, and be free from the interfering constituents mentioned above. The final acidity may vary from 0.75 to 1.75 M. If iron is present in appreciable amounts, add dilute phosphoric(V) acid from a burette until the yellow colour of the iron(III) is eliminated; the same amount of phosphoric(V) acid must be added to the standards. If alkali sulphates are present in the test solution in appreciable quantity, add a like amount to the standards. Add 10 mL of 3% hydrogen peroxide solution and dilute the solution to 100 mL in a graduated flask; the final concentration of Ti may conveniently be 2–25 ppm. Following any of the usual methods, compare the colour produced by the unknown solution with the colour of standards having similar composition. A wavelength of 410 nm is employed when using a spectrophotometer. In this case the effect of iron, nickel, chromium(III) and other coloured ions not reacting with hydrogen peroxide may be compensated by using a solution of the sample, not treated with hydrogen peroxide, in the reference cell.

17.25 Tungsten

Discussion Toluene-3,4-dithiol (dithiol) may be used for the colorimetric determination of tungsten; it forms a slightly soluble coloured complex with tungsten(VI) which can be extracted with butyl or pentyl ethanoate and other organic solvents. Molybdenum reacts similarly and must be removed before tungsten can be determined. The molybdenum complex can be preferentially developed in cold weak acid solution and selectively extracted with pentyl ethanoate before developing the tungsten colour in a hot solution of increased acidity. The procedure will be illustrated by describing the determination of tungsten in steel.

Reagents *Dithiol reagent solution* Dissolve 1 g toluene-3,4-dithiol in 100 mL pentyl ethanoate. This should be prepared immediately before use.

Standard tungsten solution Dissolve 0.1794 g sodium tungstate $Na_2WO_4 \cdot 2H_2O$ in water and dilute to 1 L : 1 mL $\equiv$ 0.1 g W. For use, dilute 100 mL of this solution to 1 L : 1 mL $\equiv$ 0.01 mg W.

Mixed acid Mix 15.0 mL concentrated sulphuric acid and 15.0 mL orthophosphoric acid (specific gravity 1.75) and dilute to 100 mL with distilled water by adding the acid slowly with stirring to the water.

Procedure (tungsten in steel) Dissolve 0.5 g of the steel, accurately weighed, in 30 mL of the 'mixed acid' by heating, oxidise with concentrated nitric acid, and evaporate to fuming. Extract with 100 mL water, boil, transfer to a 500 mL graduated flask, cool, dilute to the mark with water, and mix. Pipette a 15 mL aliquot into a 50 mL flask, evaporate to fuming, cool, add 5 mL dilute hydrochloric acid (specific gravity 1.06), warm until the salts dissolve, and cool to room temperature. Add 5 drops of 10% aqueous hydroxylammonium sulphate solution, 10 mL of the dithiol reagent, and allow to stand in a bath at 20–25 °C for 15 min with periodic shaking. Transfer the contents quantitatively to a 25 mL separatory funnel, using 3–4 mL portions of pentyl ethanoate for washing. Shake and allow the layers to separate.

Run off the lower acid layer containing the tungsten and reserve it in the original 50 mL flask. Wash the pentyl ethanoate layer twice consecutively with 5 mL portions of hydrochloric acid (specific gravity 1.06) and combine the acid washings with the original acid layer. Discard the molybdenum-containing pentyl ethanoate layer. Evaporate the acid tungsten solution carefully to fuming (to expel dissolved pentyl ethanoate), then add a few drops of concentrated nitric acid during fuming to clear up any charred organic matter. Add 5 mL of 10% tin(II) chloride solution (in concentrated hydrochloric acid) and heat to 100 °C for 4 min: add 10 mL of the dithiol reagent and heat at 100 °C for 10 min longer with periodic shaking. Transfer to a 25 mL stoppered separatory funnel, and rinse three times with 2 mL portions of pentyl ethanoate. Shake, separate, then draw off the lower acid layer and discard it. Add 5 mL concentrated hydrochloric acid to the organic layer, repeat the extraction and again discard the lower layer. Draw off the pentyl ethanoate layer containing the tungsten complex into a 50 mL graduated flask and dilute to volume with pentyl ethanoate. Measure the absorbance with a spectrophotometer at 630 nm in 4 cm cells. Refer the readings to a calibration curve prepared from a solution containing spectroscopically pure iron to which suitable amounts of standard sodium tungstate solution have been added.

Anions

▮
■ **Safety.** Before carrying out any experiments in this section, pay full attention to any safety warnings and make sure you adhere to national laboratory and safety regulations.

17.26   Chloride

Mercury(II) chloranilate method

Discussion   The mercury(II) salt of chloranilic acid (2,5-dichloro-3,6-dihydroxy-*p*-benzoquinone) may be used for the determination of small amounts of chloride ion. The reaction is

$$HgC_6Cl_2O_4 + 2Cl^- + H^+ = HgCl_2 + HC_6Cl_2O_4^-$$

The amount of reddish purple acid-chloranilate ion liberated is proportional to the chloride ion concentration. Methyl cellosolve (2-methoxyethanol) is added to lower the solubility of mercury(II) chloranilate and to suppress the dissociation of the mercury(II) chloride; nitric acid is added (concentration $0.05\,M$) to give the maximum absorption. Measurements are made at 530 nm in the visible region or 305 nm in the ultraviolet region. Bromide, iodide, iodate, thiocyanate, fluoride and phosphate interfere, but sulphate, acetate, oxalate and citrate have little effect at the $25\,mg\,L^{-1}$ level. The limit of detection is $0.2\,mg\,L^{-1}$ of chloride ion; the upper limit is about $120\,mg\,L^{-1}$. Most cations, but not ammonium ion, interfere and must be removed. Silver chloranilate cannot be used in the determination because it produces colloidal silver chloride.

Procedure   Remove interfering cations by passing the aqueous solution containing the chloride ion through a strongly acidic ion exchange resin in the hydrogen form (e.g. Zerolit 225 or Amberlite 120) contained in a tube 15 cm long and 1.5 cm in diameter. Adjust the pH of the effluent to 7 with dilute nitric acid or aqueous ammonia and pH paper. To an aliquot containing not more than 1 mg of chloride ion in less than 45 mL of water in a 100 mL graduated flask, add 5 mL $1\,M$ nitric acid and 50 mL methylcellosolve. Dilute the mixture to volume with distilled water, add 0.2 g mercury(II) chloranilate, and shake the flask intermittently for 15 min. Separate the excess of mercury(II) chloranilate by filtration through a fine ashless filter paper or by centrifugation. Measure the absorbance of the clear solution with a spectrophotometer at 530 nm against a blank prepared in the same manner. Construct a calibration curve using standard ammonium chloride solution (1–100 mg L^{-1} Cl$^-$) and deduce the chloride ion concentration of the test solution with its aid.

Mercury(II) chloranilate may be prepared by adding dropwise a 5% solution of mercury(II) nitrate in 2% nitric acid to a stirred solution of chloranilic acid at 50 °C until no further precipitate forms. Decant the supernatant liquid, wash the precipitate three times by decantation with ethanol and once with diethyl ether, and dry in a vacuum oven at 60 °C.

Mercury(II) thiocyanate method

Discussion   This second procedure for the determination of trace amounts of chloride ion depends upon the displacement of thiocyanate ion from mercury(II) thiocyanate by chloride ion; in the presence of iron(III) ion a highly coloured iron(III) thiocyanate complex is formed, and the intensity of its colour is proportional to the original chloride ion concentration:

$$2Cl^- + Hg(SCN)_2 + 2Fe^{3+} = HgCl_2 + 2[Fe(SCN)]^{2+}$$

The method is applicable to the range 0.5 to 100 µg of chloride ion.

Procedure   Place a 20 mL aliquot of the chloride solution in a 25 mL graduated flask, add 2.0 mL of 0.25 M ammonium iron(III) sulphate [Fe(NH$_4$)(SO$_4$)$_2$·12H$_2$O] in 9 M nitric acid, followed by 2.0 mL of a saturated solution of mercury(II) thiocyanate in ethanol. After 10 min measure the absorbance of the sample solution and the absorbance of the blank; use 5 cm cells in a spectrophotometer at 460 nm and put water in the reference cell. The amount of chloride ion in the sample corresponds to the difference between the two absorbances and is obtained from a calibration curve. Construct a calibration curve using a standard sodium chloride solution containing 10 µg mL^{-1} Cl$^-$; cover the range 0–50 µg as above. Plot absorbance against micrograms of chloride ion.

17.27 Phosphate

Molybdenum blue method

Discussion   Orthophosphate and molybdate ions condense in acidic solution to give molybdophosphoric acid (phosphomolybdic acid), which upon selective reduction (perhaps with hydrazinium sulphate) produces a blue colour, due to molybdenum blue of uncertain composition. The intensity of the blue colour is proportional to the amount of phosphate initially incorporated in the heteropolyacid. If the acidity at the time of reduction is 0.5 M in sulphuric acid and hydrazinium sulphate is the reductant, the resulting blue complex exhibits maximum absorption at 820–830 nm.

Phosphovanadomolybdate method

Discussion   This method is considered slightly less sensitive than the previous one, but it has been particularly useful for phosphorus determinations carried out using the Schöniger oxygen flask method (Section 3.32). The phosphovanadomolybdate complex formed between the phosphate, ammonium vanadate and ammonium molybdate is bright yellow in colour and its absorbance can be measured between 460 and 480 nm.

Reagents   *Ammonium vanadate solution*   Dissolve 2.5 g ammonium vanadate (NH_4VO_3) in 500 mL hot water, add 20 mL concentrated nitric acid and dilute with water to 1 L in a graduated flask.

Ammonium molybdate solution   Dissolve 50 g ammonium molybdate ($(NH_4)_6Mo_7O_{24}\cdot 4H_2O$) in warm water and dilute to 1 L in a graduated flask. Filter the solution before use.

Procedure   Dissolve 0.4 g of the phosphate sample in 2.5 M nitric acid to give 1 L in a graduated flask. Place a 10 mL aliquot of this solution in a 100 mL graduated flask, add 50 mL water, 10 mL of the ammonium vanadate solution, 10 mL of the ammonium molybdate solution and dilute to the mark. Determine the absorbance of this solution at 465 nm against a blank prepared in the same manner, using 1 cm cells. Prepare a series of standards from potassium dihydrogenphosphate covering the range 0–2 mg phosphorus per 100 mL and containing the same concentration of acid, ammonium vanadate and ammonium molybdate as the previous solution. Construct a calibration curve and use it to calculate the concentration of phosphorus in the sample.

17.28   Sulphate

Discussion   The barium salt of chloranilic acid (2,5-dichloro-3,6-dihydroxy-p-benzoquinone) illustrates the principle of a method which may find wide application in the colorimetric determination of various anions. In the reaction

$$Y^- + MA(solid) = A^- + MY(solid)$$

where Y^- is the anion to be determined and A^- is the coloured anion of an organic acid, MY must be so much less soluble than MA that the reaction is quantitative. MA must be only sparingly soluble so that the blanks will not be too high. Sulphate ion in the range 2–400 mg L^{-1} may be readily determined by using the reaction between barium chloranilate with sulphate ion in acid solution to give barium sulphate and the acid-chloranilate ion:

$$SO_4^{2-} + BaC_6Cl_2O_4 + H^+ = BaSO_4 + HC_6Cl_2O_4^-$$

The amount of acid chloranilate ion liberated is proportional to the sulphate ion concentration. The reaction is carried out in 50% aqueous ethanol buffered at an approximate pH of 4. Most cations must be removed because they form insoluble chloranilates; this is simply effected by passing the solution through a strongly acidic ion exchange resin in the hydrogen form (Section 6.5). Chloride, nitrate, hydrogencarbonate, phosphate and oxalate do not interfere at the 100 mg L^{-1} level. The pH of the solution governs the absorbance of chloranilic acid solutions at a particular wavelength; chloranilic acid is yellow, the acid-chloranilate ion is dark purple and the chloranilate ion is light purple. At pH 4 the acid-chloranilate ion gives a broad peak at 530 nm, and this wavelength is employed for measurements in the visible region. A much more intense absorption occurs in the ultraviolet; a sharp band at 332 nm enables the limit of detection of sulphate ion to be extended to 0.06 mg L^{-1}.

Procedure   Pass the aqueous solution containing sulphate ion (2–400 mg L^{-1}) through an ion exchange column 1.5 cm in diameter and 15 cm long; use Zerolit 225 or an equivalent cation exchange resin in the hydrogen form. Adjust the effluent to pH 4 with dilute hydrochloric acid or ammonia solution. Make up to volume in a graduated flask. To an aliquot containing up to 40 mg of sulphate ion in less than 40 mL in a 100 mL graduated flask, add 10 mL of a buffer (pH 4; a 0.05 M solution of potassium hydrogenphthalate) and 50 mL of 95% ethanol. Dilute to the mark with distilled water, add 0.3 g of barium chloranilate and shake the flask for 10 min. Remove the precipitated barium sulphate and the excess of barium chloranilate by filtering or centrifuging. Measure the absorbance of the filtrate with a colorimeter or a spectophotometer at 530 nm against a blank prepared in the same manner. Construct a calibration curve using standard potassium sulphate solutions prepared from the analytical grade salt.

Organic compounds

! **Safety.** Before carrying out any experiments in this section, pay full
■ attention to any safety warnings and make sure you adhere to national laboratory and safety regulations.

17.29 Primary amines

The determination of primary amines on the macro scale is most conveniently carried out by titration in non-aqueous solution (Section 10.41), but for small quantities of amines spectroscopic methods of determination are very valuable. In some cases the procedure is applicable to aromatic amines only, and the diazotisation method for the determination of nitrite can be adapted as a method for the determination of aromatic primary amines. On the other hand, the naphthaquinone method can be applied to both aliphatic and aromatic primary amines.

Diazotisation method

Discussion   In this procedure the amine is diazotised and then coupled with N-(1-naphthyl)ethylenediamine. This leads to formation of a coloured product whose concentration can be determined with an absorptiometer or a spectrophotometer.

Reagents   N-*(1-naphthyl) ethylenediamine dihydrochloride*   Dissolve 0.3 g of the solid in 100 mL of 1 % v/v hydrochloric acid (solution A).

Sodium nitrite   Dissolve 0.7 g sodium nitrite in 100 mL distilled water (solution B).

Other reagents   1 M hydrochloric acid and 90% ethanol (rectified spirit).

Procedure   Weigh out 10–15 mg of the amine sample and dissolve in 1 M hydrochloric acid in a 50 mL graduated flask. Place 2.0 mL of this solution in a small conical flask clamped in a 400 mL beaker containing tap water, and then add 1 mL of solution B; allow to stand for 5 min. Now add 5 mL of 90% ethanol and after waiting a further 3 min, add 2 mL of solution A. A red coloration develops rapidly, the absorbance of which can be measured against a blank solution containing all the reagents except the amine. The measurement should be made at a wavelength of about 550 nm; the exact value for λ_{max} varies slightly with the nature of the amine. A calibration curve can be prepared using a series of solutions of the pure amine of appropriate concentrations, which are treated in the manner described above.

Naphthaquinone method

Discussion   Many primary amines develop a blue colour when treated with *ortho*-quinones; the preferred reagent is the sodium salt of 1,2-naphthaquinone-4-sulphonic acid.

Reagents   *1,2-Naphthaquinone-4-sulphonic acid sodium salt*   Dissolve 0.4 g of the solid in 100 mL distilled water (solution C).

Buffer solution   Dissolve 4.5 g of disodium hydrogenphosphate in 1 L of distilled water and carefully add 0.1 M sodium hydroxide solution to give a pH of 10.2–10.4 (pH meter).

Procedure   Place 25 mL of an aqueous solution of the amine, 10 mL of solution C and 1 mL of buffer solution in a 100 mL stoppered conical flask; the amount of amine should not exceed 10 μg. After 1 min add 10 mL chloroform and stir on a magnetic stirrer for 15–20 min. Transfer to a separatory funnel, and after the phases have separated, run off the chloroform layer. Measure the absorbance at a wavelength of 450 nm against a chloroform blank.

17.30   Anionic detergents

For detergents based on sodium salts of the higher homologues of alkanesulphonic acids, an early assay was to take an aqueous solution and treat it with methylene blue in the presence of chloroform.[29] Reaction takes place between the ionic dyestuff (which is a chloride) and the detergent:

$$(MB^+)Cl^- + RSO_3Na \rightarrow (MB^+)(RSO_3^-) + NaCl$$

where MB^+ indicates the cation of methylene blue. The reaction product can be extracted by chloroform, whereas the original dyestuff is insoluble in this medium, and the intensity of the colour in the chloroform layer is proportional to the concentration of the detergent. The method is especially valuable for determination of small concentrations of detergents and is therefore useful in pollution studies.

Reagents   *Methylene blue solution*   Dissolve 0.1 g of the solid (use redox indicator quality) in 100 mL distilled water.

Other reagents   6 M hydrochloric acid and chloroform.

Procedure   Weigh sufficient solid material to contain 0.001–0.004 mmol detergent and dissolve in 100 mL distilled water. Place 20 mL of this solution in a 150 mL separatory funnel and neutralise by adding 6 M hydrochloric acid dropwise; when neutrality is achieved (use a test paper), add 3–4 further drops of acid. Add 20 mL chloroform (trichloromethane) and 1 mL of the methylene blue solution. Shake for 1 min, allow to stand for 5 min, then run the chloroform layer into a 100 mL separatory funnel; retain the aqueous layer (solution A). Add 20 mL distilled water to the chloroform solution, shake for 1 min, allow to stand for 5 min, then transfer the chloroform layer through a small filter funnel with a cotton wool plug at the apex, into a 100 mL graduated flask. Repeat the extraction of solution A three more times, following the above procedure exactly. After the final extraction rinse the filter funnel and plug with a little chloroform, and finally make the solution in the graduated flask up to the mark with chloroform.

Measure the absorbance of this solution at 650 nm in a 1 cm cell against a blank of distilled water, and carry out the above procedure on the reagents, using distilled water in place of the sample solution; if necessary, use the resultant absorbance reading to correct the sample reading. If not already available, a calibration curve must be constructed for the detergent being measured, using four appropriate dilutions of a standard solution and carrying out the procedure on each one.

An obvious modification of the procedure will permit the determination of long-chain amines or quaternary ammonium salts (cationic surfactants):

$$R_3NH^+X^- + R'SO_3^- \rightarrow (R_3NH^+)(R'SO_3^-) + X^-$$

In this case the sulphonic acid group is present in a sulphon–phthalein dye, the indicator bromophenol blue. As in the previous example, the species $(R_3NH^+)(R'SO_3^-)$ can be extracted into chloroform but the indicator itself is not extracted, and the colour of the extract is proportional to the quantity of surfactant in the material under test.

UV/visible spectrophotometry

! **Safety.** Before carrying out any experiments in this section, pay full attention to any safety warnings and make sure you adhere to national laboratory and safety regulations.

17.31 Absorption curve and concentration of potassium nitrate

Discussion Potassium nitrate is an example of an inorganic compound which absorbs mainly in the ultraviolet, and can be employed to obtain experience in the use of a manually operated UV/visible spectrophotometer. Some of the exercise can also be carried out using an automatic recording spectrophotometer (Section 17.16).

The absorbance and the percentage transmittance of an approximately $0.1\,M$ potassium nitrate solution are measured over the wavelength range 240–360 nm at 5 nm intervals and at smaller intervals in the vicinity of the maxima or minima. Manual spectrophotometers are calibrated to read both absorbance and percentage transmittance on the dial settings, whereas the automatic recording double-beam spectrophotometers usually use chart paper printed with both scales. The linear conversion chart (Figure 17.18) is useful for visualising the relationship between these two quantities.

The three normal means of presenting the spectrophotometric data are described below; by far the most common procedure is to plot absorbance against wavelength (measured in nanometres). The wavelength corresponding to the absorbance maximum (or transmittance minimum) is read from the plot and used to prepare the calibration curve. This point is chosen for two reasons: (1) it is the region in which the greatest difference in absorbance between any two different concentrations will be obtained, thus giving the maximum sensitivity for concentration studies, and (2) as it is a turning point on the curve it gives the least alteration in absorbance value for any slight variation in wavelength. No general rule can be given concerning the strength of the solution to be prepared, as this will depend upon the spectrophotometer used for the study. Usually a 0.01–$0.001\,M$ solution is sufficiently concentrated for the highest absorbances, and other concentrations are prepared by dilution. The concentrations should be selected such that the absorbance lies between about 0.3 and 1.5.

For the determination of the concentration of a substance, select the wavelength of maximum absorption for the compound (e.g. 302.5 to 305 nm for potassium nitrate) and construct a calibration curve by measuring the absorbances of four or five concentrations

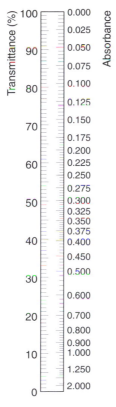

Figure 17.18 Linear correlation chart showing relationship between absorbance and transmittance

of the substance (e.g. 2, 4, 6, 8 and $10 \, g \, L^{-1}$ KNO$_3$) at the selected wavelength. Plot absorbance against concentration. If the compound obeys Beer's law, the result will be a linear calibration curve passing through the origin. If the absorbance of the unknown solution is measured, the concentration can be obtained from the calibration curve. If it is known that the compound obeys Beer's law, the molar absorption coefficient ε can be determined from one measurement of the absorbance of a standard solution. The unknown concentration is then calculated using the value of the constant ε and the measured value of the absorbance under the same conditions.

Procedure Dry some pure potassium nitrate at 110 °C for 2–3 h and cool in a desiccator. Prepare an aqueous solution containing $10.000 \, g \, L^{-1}$. With the aid of a spectrophotometer* and matched 1 cm rectangular cells, measure the absorbance and the percentage transmittance over a series of wavelengths covering the range 240–350 nm. Plot the data in three different ways: (a) absorbance against wavelength, (b) percentage transmittance against wavelength, and (c) log ε (molecular decadic absorption coefficient) against wavelength. The curves obtained for potassium nitrate are shown in Figure 17.19.

---

* When reporting spectrophotometric measurements, details should be given of the concentration used, the solvent employed, the make and model of the instrument, as well as the slit widths employed, together with any other pertinent information.

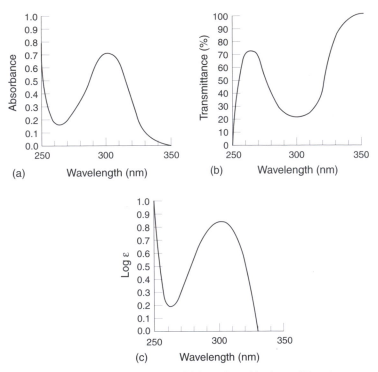

Figure 17.19 Potassium nitrate: UV data plotted in three different ways

From the curves, evaluate the wavelength of maximum absorption (or minimum transmission). Use this value of the wavelength to determine the absorbance of solutions of potassium nitrate containing 2.000, 4.000, 6.000 and 8.000 g L^{-1} KNO$_3$. Run a blank on the two cells, filling them both with distilled water; if the cells are correctly matched, no difference in absorbance should be discernible. Plot absorbance against concentration for each cell. Determine the absorbance of an unknown solution of potassium nitrate and read the concentration from the calibration curve.

17.32 How substituents affect the absorption spectrum of benzoic acid

Section 17.12 considered how various substituents in the benzene ring influence the absorption of ultraviolet radiation. This experiment looks at the effect in benzoic acid by comparing the absorption spectra of benzoic acid, 4-hydroxybenzoic acid and 4-aminobenzoic acid.

Reagents
Hydrochloric acid ($0.1 M$)
Sodium hydroxide solution ($0.1 M$)
Benzoic acid (A$_1$)
4-Hydroxybenzoic acid (A$_2$)
4-Aminobenzoic acid (A$_3$)

Procedure   Prepare a solution of benzoic acid in distilled water by dissolving 0.100 g in a 100 mL graduated flask and making up to the mark (solution A$_1$). Prepare similar solutions of the other two acids, giving solutions A_2 and A_3. Now take 10.0 mL of solution A_1 and dilute to the mark in a 100 mL graduated flask with distilled water, giving a solution B_1 containing $0.01\,mg\,mL^{-1}$ benzoic acid. Prepare similar solutions B_2 and B_3 from solutions A_2 and A_3. Use a recording spectrophotometer to plot the absorption curves of the three separate solutions, in each case using distilled water as the blank. Use silica cells and record the spectra over the range 210–310 nm.

Now prepare a new solution of 4-hydroxybenzoic acid (solution C_2) by placing 10.0 mL of solution A_2 in a 100 mL graduated flask and making up to the mark with $0.1\,M$ hydrochloric acid solution. Prepare a similar solution C_3 for the 4-aminobenzoic acid and plot the absorption curves of these two solutions. Prepare another 4-hydroxybenzoic acid solution (D_2) by placing 10.0 mL of solution A_2 in a 100 mL graduated flask and making up to the mark with $0.1\,M$ sodium hydroxide solution. Prepare a similar solution D_3 from solution A_3 and record the absorption spectra of these two new solutions.

Examine the seven absorption spectra, record the λ_{max} values of absorption peaks; comment on how —OH and —NH$_2$ groups affect the absorption spectrum of benzoic acid, and how hydrochloric acid and sodium hydroxide affect the spectra of the two substituted benzoic acids.

17.33   Simultaneous determinations (chromium and manganese)

Discussion   This section is concerned with the simultaneous spectrophotometric determination of two solutes in a solution. The absorbances are additive, provided there is no reaction between the two solutes. We may write

$$A_{\lambda_1} = {}_{\lambda_1}A_1 + {}_{\lambda_1}A_2 \tag{17.17}$$

$$A_{\lambda_2} = {}_{\lambda_2}A_1 + {}_{\lambda_2}A_2 \tag{17.18}$$

where A_1 and A_2 are the **measured** absorbances at the two wavelengths λ_1 and λ_2; the subscripts 1 and 2 refer to the two different substances, and the subscripts λ_1 and λ_2 refer to the different wavelengths. The wavelengths are selected to coincide with the absorption maxima of the two solutes; the absorption spectra of the two solutes should not overlap appreciably (Figure 17.20), so that substance 1 absorbs strongly at wavelength λ_1 and weakly at wavelength λ_2, and substance 2 absorbs strongly at λ_2 and weakly at λ_1. Now $A = \varepsilon cl$, where ε is the molar absorption coefficient (molar absorptivity) at any particular

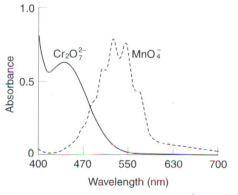

Figure 17.20   Visible spectra of $Cr_2O_7^{2-}$ and MnO_4^-

679

wavelength, c is the concentration (mol L^{-1}) and l is the thickness, or length, of the absorbing solution (cm). If we set $l = 1$ cm then

$$A_{\lambda_1} = \,_{\lambda_1}\varepsilon_1 c_1 + \,_{\lambda_1}\varepsilon_2 c_2 \tag{17.19}$$

$$A_{\lambda_2} = \,_{\lambda_2}\varepsilon_1 c_1 + \,_{\lambda_2}\varepsilon_2 c_2 \tag{17.20}$$

Solution of these simultaneous equations gives

$$c_1 = \frac{\,_{\lambda_2}\varepsilon_2 A_{\lambda_1} - \,_{\lambda_1}\varepsilon_2 A_{\lambda_2}}{\,_{\lambda_1}\varepsilon_1 \,_{\lambda_2}\varepsilon_2 - \,_{\lambda_1}\varepsilon_2 \,_{\lambda_2}\varepsilon_1} \tag{17.21}$$

$$c_2 = \frac{\,_{\lambda_1}\varepsilon_1 A_{\lambda_2} - \,_{\lambda_2}\varepsilon_1 A_{\lambda_1}}{\,_{\lambda_1}\varepsilon_1 \,_{\lambda_2}\varepsilon_2 - \,_{\lambda_1}\varepsilon_2 \,_{\lambda_2}\varepsilon_1} \tag{17.22}$$

The values of the molar absorption coefficients ε_1 and ε_2 can be deduced from measurements of the absorbances of pure solutions of substances 1 and 2. By measuring the absorbance of the mixture at wavelengths λ_1 and λ_2, the concentrations of the two components can be calculated.

These considerations will be illustrated by the simultaneous determination of manganese and chromium in steel and other ferroalloys. Figure 17.18 shows the absorption spectra of $0.001\,M$ permanganate and dichromate ions in $1\,M$ sulphuric acid, determined with a spectrophotometer and against $1\,M$ sulphuric acid in the reference cell. For permanganate, the absorption maximum is at 545 nm, and a small correction must be applied for dichromate absorption. Similarly the peak dichromate absorption is at 440 nm, at which permanganate only absorbs weakly. Absorbances for these two ions, individually and in mixtures, obey Beer's law provided the concentration of sulphuric acid is at least $0.5\,M$. Iron(III), nickel, cobalt and vanadium absorb at 425 nm and 545 nm; they should be absent, else corrections must be made.

Reagents   *Potassium dichromate* $0.002\,M$, $0.001\,M$ and $0.0005\,M$ in $1\,M$ sulphuric acid and $0.7\,M$ phosphoric(V) acid, prepared from the analytical grade reagents.

Potassium permanganate $0.002\,M$, $0.001\,M$ and $0.0005\,M$ in $1\,M$ sulphuric acid and $0.7\,M$ phosphoric(V) acid, prepared from the analytical grade reagents. All flasks must be scrupulously clean.

Procedure   *Molar absorption coefficients*   The molar absorption coefficients must be determined for the particular set of cells and the spectrophotometer employed. We may write

$$A = \varepsilon c l$$

where ε is the molar absorption coefficient, c is the concentration (mol L^{-1}), and l is the cell thickness or length (cm). Measure the absorbance A for the three solutions of potassium dichromate and the three solutions of potassium permanganate; determine each solution separately at both 440 nm and 545 nm in 1 cm cells. Calculate ε in each case and record the mean values for $Cr_2O_7^{2-}$ and MnO_4^- at the two wavelengths.

Mix $0.001\,M$ potassium dichromate and $0.0005\,M$ potassium permanganate in the following amounts (plus 1.0 mL of concentrated sulphuric acid), and prepare a set of results similar to those in Table 17.8, which is a set of typical results included for guidance only. Measure the absorbance of each of the mixtures at 440 nm. Calculate the absorbance of the mixtures from

$$A_{440} = \,_{440}\varepsilon_{Cr} c_{Cr} + \,_{440}\varepsilon_{Mn} c_{Mn}$$

Table 17.8 *Testing the additivity principle with $Cr_2O_7^{2-}$ and MnO_4^- mixtures at 440 nm*

$K_2Cr_2O_7$ solution (mL)	$KMnO_4$ solution (mL)	A (observed)	A (calculated)
50	0	0.371	–
45	5	0.338	0.340
40	10	0.307	0.308
35	15	0.277	0.277
25	25	0.211	0.214
15	35	0.147	0.151
5	45	0.086	0.088
0	50	0.057	–

Chromium and manganese in an alloy steel*   Accurately weigh out about 1.0 g of the alloy steel into a 300 mL Kjeldahl flask, add 30 mL of water and 10 mL of concentrated sulphuric acid (also 10 mL of 85% phosphoric(V) acid if tungsten is present). Boil gently until decomposition is complete or the reaction subsides. Then add 5 mL of concentrated nitric acid in several small portions. If much carbonaceous residue persists, add a further 5 mL of concentrated nitric acid, and boil down to copious fumes of sulphuric acid. Cool and dilute to about 100 mL and boil until all salts have dissolved. Cool, transfer to a 250 mL graduated flask, and dilute to the mark.

Pipette a 25 mL or 50 mL aliquot of the clear sample solution into a 250 mL conical flask, add 5 mL concentrated sulphuric acid, 5 mL of 85% phosphoric(V) acid, and 1–2 mL of 0.1 M silver nitrate solution, and dilute to about 80 mL. Add 5 g potassium persulphate, swirl the contents of the flask until most of the salt has dissolved, and heat to boiling. Keep at the boiling point for 5–7 min. Cool slightly and add 0.5 g pure potassium periodate. Again heat to boiling and maintain at the boiling point for about 5 min. Cool, transfer to a 100 mL graduated flask, and measure the absorbances at 440 nm and 545 nm in 1 cm cells.

Table 17.9 *Corrections for interfering substances*

Substance	Cr correction at 440 nm (%)	Mn correction at 545 nm (%)
$Cr_2O_7^{2-}$	–	0.0025
MnO_4^-	0.490	–
VO_2^+	0.0266	–
Co^{2+}	0.0072	0.0011
Ni^{2+}	0.0039	0.0001
Fe^{3+}	0.0005	–

* British Chemical Standard BCS-CRM 225/2 *Ni–Cr–Mo steel* is suitable for practice in this determination.

Calculate the percentage of chromium and manganese in the sample. Use equations (17.21) and (17.22) and values of the molar absorption coefficients ε determined above; these will give concentrations expressed in $mol\,L^{-1}$, from which the percentages can readily be calculated. Using Table 17.9 correct each value for the amounts of vanadium, cobalt, nickel and iron which may be present. The values in the table are the equivalent percentages of the respective constituent to be subtracted from the apparent Cr and Mn percentages for each 1% of the element in question. From the known (or determined) molar absorption coefficients

$$_{545}\varepsilon_{Cr} = 0.011 \qquad _{545}\varepsilon_{Mn} = 2.35 \qquad _{440}\varepsilon_{Cr} = 0.369 \qquad _{440}\varepsilon_{Mn} = 0.095$$

the following relationships can be obtained:

$$\%\ Mn = \frac{0.005\,49\,V}{W}(0.426A_{545} - 0.013A_{440})$$

$$\%\ Cr = \frac{0.010\,40\,V}{W}(2.71A_{440} - 0.110A_{545})$$

for a sample of weight W (g) in a volume V (mL).

Note   At high concentrations of dichromate and permanganate, the additivity of the absorbances may not obey Beer's law. Derivative spectroscopy may be used in this case.

17.34   Aromatic hydrocarbons and binary mixtures

Discussion   This experiment provides the opportunity to examine the absorption spectra of typical aromatic hydrocarbons and to investigate the possibility of analysing mixtures of hydrocarbons by ultraviolet spectrophotometry.

Reagents   Methanol and benzene (Spectrosol or equivalent purity); toluene (analytical grade). **Avoid inhalation or skin contact with benzene; it is carcinogenic.**

Procedure   Using a 0.1 mL capillary micropipette with 0.005 mL graduations, place 0.05 mL of the benzene in a 25 mL graduated flask and prepare a stock solution by diluting to the mark with methanol. **Work in a fume cupboard**. Prepare a series of five dilutions of the stock solution; use a 2 mL graduated pipette to transfer 0.25, 0.50, 0.75, 1.00 and 1.50 mL of the solution into a series of 10 mL graduated flasks then make up to the mark with methanol.

Using stoppered quartz cells, use solution 5 (i.e. the most concentrated of the test solutions) to plot an absorption curve using pure methanol as the blank. Take absorbance readings over the wavelength range 200–300 nm, but preferably use a spectrophotometer equipped with a chart recorder. Make a note of the λ_{max} values for the peaks observed in the curve. There is a well-developed peak at approximately 250 nm, and using each of the test solutions in turn, measure the absorption at the observed peak wavelength and test the validity of Beer's law.

Now, starting with 0.05 mL toluene, repeat the procedure to obtain five working solutions, 1' to 5'. Use solution 5' to plot the absorption curve of toluene; again record the λ_{max} values for the peaks of the curve. There is a well-developed peak at approximately 270 nm, and using the five new test solutions, measure the absorbance of each one at the observed peak wavelength and test the application of Beer's law. Next measure solution 5' at the wavelength used for benzene, and solution 5 at the wavelength used for toluene.

Prepare a benzene–toluene mixture; place 0.05 mL of each liquid in a 25 mL graduated flask then make up to the mark with methanol. Take 1.5 mL of this solution, place in a 10 mL graduated flask and dilute to the mark with methanol; this solution contains benzene at the same concentration as solution 5, and toluene at the same concentration as solution 5'. Measure the absorbances of this solution at the two wavelengths selected for the Beer law plots of both benzene and toluene. Then use the procedure detailed in Section 17.33 to evaluate the composition of the solution and compare it with the composition calculated from the amounts of benzene and toluene taken. A similar procedure can be applied to mixtures of 1,2-dimethylbenzene (*o*-xylene) and 1,4-dimethylbenzene (*p*-xylene).

17.35   Phenols in water

Phenols show an ultraviolet absorption spectrum with a band between 270 and 280 nm, and its intensity is greatly increased by working in alkaline solution so the phenol is predominantly in the form of the phenoxide ion. At the same time, there is a shift in the absorption band and many phenols under these conditions show a well-developed peak at a wavelength of 287–296 nm. Using an average value of λ_{max} of 293 nm, the molar absorption coefficients of a number of common phenols in alkaline solution have been determined at this wavelength,[30] and they can be used for quantitative measurements. Alternatively, calibration curves may be prepared using a pure sample of the phenol to be determined. Water samples showing contamination by phenols are best examined by extracting the phenol into an organic solvent; tri-*n*-butyl phosphate is very suitable for this purpose. Photometric measurements can be carried out on the extract, and the requisite alkaline conditions are achieved by the addition of tetra-*n*-butylammonium hydroxide.

Reagents   *Stock phenol solution*   Weigh out 0.5 g phenol, dissolve in distilled water and make up to the mark in a 500 mL graduated flask; try to use freshly boiled and cooled distilled water.

Standard phenol solution (0.025 mg L^{-1})   Dilute 25.0 mL of the stock solution to 1.0 L using freshly boiled and cooled distilled water. This solution must be freshly prepared.

*Tetra-*n*-butylammonium hydroxide (0.1 M solution in methanol)*   Prepare an anion exchange column using a resin such as Duolite A113 or Amberlite IRA-400, convert to the hydroxide form, wash the column with water then pass through 300–400 mL of methanol to remove the water (Section 6.5). Dissolve 20 g of tetra-*n*-butylammonium iodide in 100 mL of dry methanol and pass this solution through the column at a rate of about 5 mL min^{-1}; the effluent must be collected in a vessel fitted with a Carbosorb guard tube to protect it from atmospheric carbon dioxide. Then pass 200 mL of dry methanol through the column. Standardise the methanolic solution by carrying out a potentiometric titration of an accurately weighed portion of benzoic acid (about 0.3 g). Calculate the molarity of the solution and add sufficient dry methanol to make it approximately 0.1 *M*.

Other reagents   5 *M* hydrochloric acid and tri-*n*-butyl phosphate.

Procedure   Prepare four test solutions of phenol by placing 200 mL of boiled and cooled distilled water in each of four stoppered 500 mL bottles, and adding to each 5 g of sodium chloride; this assists the extraction procedure by 'salting out' the phenol. Add 5.0, 10.0, 15.0 and 20.0 mL of the standard phenol solution to the four bottles, then adjust the pH of each solution to about 5 by the careful addition of 5 *M* hydrochloric acid (use a test paper).

Add distilled water to each bottle to make a total volume of 250 mL and then add 20.0 mL tri-*n*-butyl phosphate. Stopper each bottle securely (the stoppers should be wired on), and then shake in a mechanical shaker for 30 min. Transfer to separatory funnels, and when the phases have settled, run off and discard the aqueous layers.

Prepare an alkaline solution of the phenol concentrate by placing 4.0 mL of a tri-*n*-butyl phosphate layer in a 5 mL graduated flask and adding 1.0 mL of the tetra-*n*-butylammonium hydroxide; do this for each of the four solutions. The reference solution consists of 4 mL of the organic layer (in which the phenol is undissociated) plus 1 mL of methanol. Measure the absorbance of each of the extracts from the four test solutions and plot a calibration curve. The unknown solution (which should contain 0.5–$2.0\,\mathrm{mg\,L^{-1}}$ phenol) is treated in the manner described above, and by reference to the calibration curve the absorbance reading will determine the phenol content of the unknown sample.

If the sample supplied contains organic substances which can be extracted into tri-*n*-butyl phosphate, a preliminary extraction with carbon tetrachloride must be carried out. Solid potassium hydroxide is added to a 600–700 mL portion of the sample to raise the pH to about 12; use a test paper. A 20 mL portion of carbon tetrachloride is added followed by shaking for 30 min using a mechanical shaker. Using a separatory funnel, the organic layer is discarded and a 200 mL portion of the aqueous layer is transferred to a 500 mL bottle with a well-fitting stopper. Add 5 *M* hydrochloric acid to bring the pH of the solution to about 5, then follow the above procedure.

17.36 Active constituents in a medicine by derivative spectroscopy

Discussion Use a spectrophotometer that can produce derivative curves. Actifed is a medicinal preparation in which the active constituents are the two drugs pseudoephedrine hydrochloride and triprolidine hydrochloride. The absorption spectrum of Actifed tablets dissolved in 0.1 *M* hydrochloric acid is similar to Figure 17.12(a), of no value for quantitative determinations. But a second-order derivative spectrum is similar to Figure 17.12(b) – peak C corresponds to the pseudoephedrine hydrochloride and peak D corresponds to the triprolidine hydrochloride – and this can be used for quantitative measurements. Experience shows it is advisable to use different response times for the two peaks. With the instrument used, a response setting of 3 gave the best results for pseudoephedrine hydrochloride, whereas a setting of 4 was best for the triprolidine hydrochloride.

Reagents
Pseudoephedrine hydrochloride
Triprolidine hydrochloride
Hydrochloric acid (0.1 *M*)

Procedure Prepare standard solutions of pseudoephedrine hydrochloride by weighing out accurately about 60 mg into a 500 mL graduated flask. Add about 50 mL of 0.1 *M* hydrochloric acid to dissolve the solid and then make up to the mark with 0.1 *M* hydrochloric acid. Place 25, 30 and 40 mL of this solution in a series of 50 mL graduated flasks and dilute to the mark with the hydrochloric acid. This gives three solutions plus the original (undiluted) solution, a total of four standard solutions.

Prepare standard solutions of triprolidine hydrochloride by weighing accurately about 0.1 g of the solid into a 100 mL graduated flask, adding about 50 mL of 0.1 *M* hydrochloric acid and swirling the flask until the solid has dissolved. Make up to the mark with the hydrochloric acid. Dilute 10 mL of this solution in a 100 mL graduated flask using the

0.1 M hydrochloric acid to make up the volume. Pipette 25, 30 and 40 mL of this diluted solution into a series of 50 mL graduated flasks, and make up to the mark with the hydrochloric acid, thus giving four standard solutions.

Weigh 8–10 Actifed tablets, grind to a fine powder using a pestle and mortar, and then accurately weigh an amount of the powder equivalent to the weight of one tablet into a 500 mL graduated flask. Add about 200 mL of the 0.1 M hydrochloric acid, stopper the flask, and shake for 5 min to dissolve the tablets; dilute to the mark with the hydrochloric acid. Filter the turbid liquid through a dry filter paper and reject the first 20 mL of filtrate. Collect the succeeding filtrate in a dry vessel as the test solution.

Set up the spectrophotometer to plot the second-order derivative spectrum and record the results for the four standard triprolidine hydrochloride solutions. Use quartz cells, with 0.1 M hydrochloric acid in the reference cell and scan between 210 and 350 nm. For each spectrum measure the long-wave peak heights D_L (Figure 17.14(a)) covering the wavelength range 290–310 nm; plot the results against the concentrations of the solutions and confirm that a straight line results.

Likewise, record the second-order derivative spectra of the four standard pseudoephedrine hydrochloride solutions and measure the peak heights D_L at 258–259 nm; plot the results against concentration and confirm that a straight line is obtained. Now record the second-order derivative spectrum of the Actifed solution, determine the long-wave peak heights for both components, and by comparison with the calibration plots of the individual components, deduce their proportions in the tablets.

17.37   Glycerol in fruit juice

Discussion   Glycerol is phosphorylated by adenosine-5′-triphosphate (ATP) to L-glycerol-3-phosphate in the reaction catalysed by glycerokinase (GK):

$$\text{glycerol} + \text{ATP} \xrightarrow{\text{GK}} \text{L-glycerol-3-phosphate} + \text{ADP}$$

The adenosine-5′-diphosphate (ADP) formed in this reaction is reconverted by phosphoenolpyruvate (PEP) with the aid of pyruvate kinase (PK) into ATP with the formation of pyruvate:

$$\text{ADP} + \text{PEP} \xrightarrow{\text{PK}} \text{ATP} + \text{pyruvate}$$

In the presence of the enzyme lactate dehydrogenase (L-LDH) pyruvate is reduced to L-lactate by reduced nicotinamide adenine dinucleotide (NADH) with the oxidation of NADH to NAD.

$$\text{pyruvate} + \text{NADH} + \text{H}^+ \xrightarrow{\text{L-LDH}} \text{L-lactate} + \text{NAD}^+$$

The amount of NADH oxidised in this reaction is stoichiometric to the amount of glycerol. NADH is determined by means of its light absorption at 340 nm.

Reagents   The test combination obtained from Boehringer Mannheim contains enough reagents for 30 determinations. It comes in five bottles, three of type 1 and one each of types 2 and 3:

Bottle 1   contains approximately 2 g coenzyme/buffer mixture, consisting of glycylglycine buffer (pH 7.4), NADH ~7 mg, ATP ~22 mg, PEP-CHA ~11 mg, magnesium sulphate, stabilisers.

Bottle 2   contains approximately 0.4 mL suspension, consisting of pyruvate kinase and lactate dehydrogenase.

Bottle 3   contains approximately 0.4 mL glycerokinase suspension.

Solutions for ten determinations   Dissolve the contents of one of the coenzyme/ buffer bottles in 11 mL of redistilled water. Allow the solution to stand for about 10 min before use. The contents of bottle 2 and bottle 3 are used undiluted.

Procedure   Dilute the fruit juice sample to yield a glycerol concentration of less than $0.4\,g\,L^{-1}$. If the juice is turbid, filter and use the clear solution for the assay. When analysing **strongly coloured** juices, decolorise the sample as follows. Take 10 mL of the juice and add about 0.1 g of polyamide powder or polyvinylpolypyrrolidine. Stir for 1 min and filter. Use the clear solution, which may be slightly coloured for the determination. Pipette into a cuvette 1.000 mL of diluted bottle 1 solution, 2.000 mL of distilled water, 0.100 mL of the sample solution and 0.010 mL of bottle 2 suspension. In a separate cuvette prepare a blank using the same reagents as for the sample, but do not add the sample solution. Mix each solution well and record its absorbance at 340 nm against either an air or distilled water reference when the reaction is complete (5–7 min).

Start the reaction by adding 0.010 mL of bottle 3 suspension to both cuvettes. Mix well and wait for completion of the reaction (about 5–10 min). Measure the absorbances of sample and blank immediately, one after the other, at 340 nm. If the reaction has not stopped after 15 min, continue to record the absorbances at 2 min intervals until the absorbance decreases constantly over 2 min. Extrapolate the absorbances to the time the bottle 3 suspension was added.

Calculation   Let the absorbance before adding bottle 3 be A_1 and the absorbance after adding bottle 3 be A_2. Determine the absorbance differences $(A_1 - A_2)$ for both blank and sample:

$$\Delta A = (A_1 - A_2)_{sample} - (A_1 - A_2)_{blank}$$

The measured absorbance differences should be at least 0.1 to achieve sufficiently accurate results. If ΔA_{sample} is greater than 1.000, the concentration of glycerol in the sample solution is too high and should be diluted.

The concentration c $(g\,L^{-1})$ is given by

$$c = \frac{V \times MW}{1000\varepsilon dv} \times \Delta A$$

where
V = final volume (3.020 mL)
v = sample volume (0.100 mL)
MW = relative molecular mass of glycerol (92.1)
d = path length (1 cm)
ε = absorption coefficient ($6.3\,L\,mmol^{-1}\,cm^{-1}$)

i.e $c = 0.4414\,\Delta A$

17.38   Cholesterol in mayonnaise

Discussion   Cholesterol is oxidised by cholesterol oxidase:

$$cholesterol + O_2 \xrightarrow{\text{cholesterol oxidase}} \Delta^4\text{-cholesterone} + H_2O_2$$

In the presence of catalase, the hydrogen peroxide produced in this reaction oxidises methanol to methanal (formaldehyde):

$$\text{methanol} + H_2O_2 \xrightarrow{\text{catalase}} \text{methanal} + 2H_2O$$

The methanal reacts with acetylacetone (pentan-2,4-dione) to form a yellow lutidine dye in the presence of ammonium ions:

$$\text{methanal} + NH_4^+ + 2 \text{ acetylacetone} \rightarrow \text{lutidine dye} + 3H_2O$$

The concentration of the lutidine dye formed is stoichiometric to the amount of cholesterol and is measured by the increase of light absorbance in the visible range at 405 nm.

Reagents *Test kit* The test combination supplied by Boehringer Mannheim is sufficient for about 25 determinations; it comes in three bottles:

Bottle 1 contains approximately 95 mL solution consisting of ammonium phosphate buffer (pH 7.0), methanol 2.6 mol L^{-1}, catalase, stabilisers.
Bottle 2 contains approximately 60 mL solution consisting of acetylacetone 0.05 mol L^{-1}, methanol 0.3 mol L^{-1}, stabilisers.
Bottle 3 contains approximately 0.8 mL cholesterol oxidase suspension.

Cholesterol reagent Mix three parts of the solution from bottle 1 with two parts of the solution from bottle 2 in a brown bottle adjusted to room temperature. Allow the mixture to stand at room temperature for 1 h before use.

Solution 3 Use the contents of bottle 3 undiluted.

Other reagents Freshly prepared $1.0 M$ methanolic potassium hydroxide.

Procedure Weigh accurately approximately 1 g of mayonnaise and 1 g of sea sand into a 50 mL round-bottomed flask. Add 10 mL of methanolic potassium hydroxide and heat under reflux for 25 min, stirring continuously. Transfer the supernatant solution into a 25 mL graduated flask with a pipette. Boil the residue twice with portions of 6 mL propan-2-ol each under reflux for 5 min. Collect the solutions in the volumetric flask and allow to cool. Dilute the contents of the flask to the mark with propan-2-ol and mix. If the solutions are turbid, filter through a fluted filter paper. Use the clear solution for the assay. The sample solution should contain between 0.07 and 0.4 g L^{-1} cholesterol.

Prepare a sample blank by pipetting into a **glass** test tube 5.000 mL of the cholesterol reagent and 0.400 mL of the sample solution; mix thoroughly. Into a separate glass test tube pipette 2.500 mL of the sample blank and add 0.020 mL of solution 3. This is the sample. Mix thoroughly. Cover both tubes and incubate in a water bath at 37–40 °C for 60 min. Allow to cool to room temperature. Read the absorbances of the sample blank and the sample one after the other in the same cuvette against an air reference at 405 nm. Subtract the absorbance of the blank from the absorbance of the sample ($= \Delta A$). In order to achieve sufficient accuracy, ΔA should be at least 0.100.

Calculation The concentration c (g L^{-1}) is given by

$$c = \frac{V_f \times MW}{1000 \varepsilon d v} \times \Delta A$$

where
- V = final volume (5.400 mL)
- v = sample volume (0.400 mL)
- MW = relative molecular mass of cholesterol (386.64)
- d = path length (1 cm)
- ε = absorption coefficient (7.4 L mmol^{-1} cm^{-1})
- f = dilution factor = 2.52/2.5 = 1.008

i.e. $c = 0.711\,\Delta A$

So the cholesterol content of mayonnaise (in mg/100 g) is

$$c\left(\frac{100 \times 25}{w}\right)$$

where w is the weight of the mayonnaise sample in grams.

Fluorimetry

! **Safety.** Before carrying out any experiments in this section, pay full attention to any safety warnings and make sure you adhere to national laboratory and safety regulations.

17.39   Quinine in tonic water

Discussion   This determination is ideal for gaining experience in quantitative fluorimetry. It is particularly used to determine the amount of quinine in samples of tonic water.

Reagents   *Dilute sulphuric acid, (~0.05 M)*   Add 3.0 mL concentrated sulphuric acid to 100 mL water, and dilute to 1 L with distilled water.

Standard solution of quinine   Weigh out accurately 0.100 g quinine and dissolve it in 1 L 0.05 *M* sulphuric acid in a graduated flask. Dilute 10.0 mL of this solution to 1 L with 0.05 *M* sulphuric acid. The resulting solution contains 0.001 00 mg mL^{-1} quinine. With the aid of a calibrated burette, run 10.0, 17.0, 24.0, 31.0, 38.0, 45.0, 52.0 and 62.0 mL of the above dilute standard solution into separate 100 mL graduated flasks and dilute each to the mark with 0.05 *M* sulphuric acid.

Procedure   Measure the fluorescence of each of the above solutions at 445 nm, using the solution containing 62.0 mL of the dilute quinine solution as a standard for the fluorimeter. Use LF2 or an equivalent primary filter (λ_{ex} = 350 nm) and gelatin as the secondary filter if using a simple fluorimeter. Now prepare test solutions containing say 0.000 25 and 0.000 45 mg mL^{-1} quinine. Determine their concentrations by measuring the fluorescence on the instrument and using the calibration curve (see note).

To determine the quinine content of tonic water, it is first necessary to de-gas the sample either by leaving the bottle open to the atmosphere for a prolonged period or by stirring it vigorously in a beaker for several minutes. Take 12.5 mL of the degassed tonic water and make up to 25 mL in a graduated flask with 0.1 *M* sulphuric acid. From this solution prepare other dilutions with 0.05 *M* sulphuric acid until a fluorimeter reading is obtained that falls on the calibration line previously prepared. From the value obtained, calculate the concentration of quinine in the original tonic water.

Note   It is good practice to make the fluorescence measurements for samples and standards as close together as possible to minimise any drift in instrument response.

17.40   Codeine and morphine in a mixture

Discussion   This experiment[31] illustrates how adjustment of pH may be used to control fluorescence and so make the determination more specific. The alkaloids codeine and morphine can be determined independently because, although both fluoresce strongly at the same wavelength in dilute sulphuric acid solution, morphine gives a generally negligible fluorescence in dilute sodium hydroxide. The fluorescence intensities of the two compounds are assumed to be additive.

Solutions   Prepare the following series of standard solutions of codeine and morphine, each of which should cover the range 5–20 mg L^{-1}:

(a)   Codeine in H_2SO_4 (0.05 M)
(b)   Codeine in NaOH (0.1 M)
(c)   Morphine in H_2SO_4 (0.05 M)
(d)   Morphine in NaOH (0.1 M)

Prepare solutions of the accurately weighed sample (codeine–morphine mixture) in H_2SO_4 (0.05 M) and in NaOH (0.1 M).

Procedure   Measure the fluorescence intensities of each of the series of standard solutions at 345 nm, with excitation at 285 nm. Construct a calibration graph for each of the four series (a) to (d). Measure the fluorescence intensity of the sample in NaOH solution using the above emission and excitation wavelengths. Read off the codeine concentration from the appropriate calibration graph (b). Calculate the fluorescence intensity which corresponds to this concentration of codeine in H_2SO_4 using calibration graph (a). Now measure the fluorescence intensity of the sample in H_2SO_4 solution and subtract the fluorescence intensity due to codeine. The value obtained gives the fluorescence intensity due to morphine in H_2SO_4 and its concentration can be deduced from calibration graph (c). Calibration graph (d) may be used to correct for the small fluorescence intensity due to morphine in NaOH; this is not negligible when the morphine concentration is high and the codeine concentration low.

17.41   References

1.   H Lambert 1760 *Photometria de Mensura et Gradibus Luminus, Colorum et Umbrae*, Augsberg. Reprinted in W Ostwald 1892 *Klassiker der Exakten Wissenschaften*, No. 32; 64
2.   M Bouguer 1729 *Essai d'Optique sur la Graduation de la Lumière*, Paris. See also W Ostwald 1891 *Klassiker der Exakten Wissenschaften*, No. 33, 38; M Bouguer 1760 *Traite d'Optique sur la Graduation de la Lumière*, Lacaille (published posthumously)
3.   A Beer 1852 *Ann. Physik. Chem. (J C Poggendorff)*, **86**; 78. See also H G Pfeiffer and H A Liebhafsky 1951 *J. Chem. Educ.*, **23**; 123

4.　F Bernard 1852 *Ann. Chim. Phys.*, **35**; 385
5.　D R Malinin and J H Yoe 1961 *J. Chem. Educ.*, **38**; 129
6.　F H Lohman 1955 *J. Chem. Educ.*, **32**; 155
7.　J Talmi 1982 *Appl. Spectrosc.*, **36**; 1
8.　A T Giese and C S French 1955 *Appl. Spectrosc.*, **9**; 78
9.　T C O'Haver 1979 *Anal. Chem.*, **51**; 91A
10.　J E Cahill and F C Padera 1980 *Am. Lab.*, **12** (4); 101
11.　T C O'Haver and G L Green 1976 *Anal. Chem.*, **48**; 312
12.　R A Albery 1987 *Physical chemistry*, 7th edn, John Wiley, New York
13.　C N Banwell and E A McCall 1995 *Fundamentals of molecular spectroscopy*, 4th edn, McGraw-Hill, London
14.　J M Hollas 1987 *Modern spectroscopy*, John Wiley, Chichester
15.　D L Pavia, G M Lampman and G S Kriz 1979 *Introduction to spectroscopy*, Holt, Rhinehart and Winston, New York
16.　D H Williams and I Fleming 1987 *Spectroscopic methods in organic chemistry*, 4th edn, McGraw-Hill, London
17.　D A Skoog 1984 *Principles of instrumental analysis*, 3rd edn, CBS College Publishing, Philadelphia PA
18.　J Coetzee (ed) 1982 *Recommended methods for purification of solvents and tests for impurities*, Pergamon, Oxford
19.　C Burgess and A Knowles 1981 *Standards in absorption spectroscopy*, Chapman Hall, London
20.　Boehringer 1995 *Methods of enzymic bioanalysis and food analysis*, Boehringer Mannheim Biochemical, Mannheim
21.　H Egan, R Sawyer and R S Kirk 1981 *Pearson's chemical analysis of foods*, 8th edn, Longman, Harlow, p. 240
22.　J Ruzicka and E H Hansen 1975 *Anal. Chim. Acta*, **78**; 145
23.　G D Clark, D A Whitman, G D Christian and J Ruzicka 1990 *Crit. Rev. Anal. Chem.*, **21**; 357
24.　B C Madsen and J R Murphy 1981 *Anal. Chem.*, **53**; 1924
25.　K H Croner and M R Kula 1984 *Anal. Chim. Acta*, **163**; 3
26.　E H Hansen, J Ruzicka and A K Ghose 1978 *Anal. Chim. Acta*, **100**; 151
27.　B Karlberg and S Thelander 1978 *Anal. Chim. Acta*, **98**; 2
28.　Analytical Methods Committee 1960 *Determination of arsenic in organic materials*, Society for Analytical Chemistry, London
29.　J H Jones 1945 *J. Assoc. Offic. Anal. Chemists*, **28**; 398
30.　J M Martin Jr, C R Orr, C B Kincannon and J L Bishop 1967 *J. Water Pollution Control*, **39**; 21
31.　R A Chelmers and G A Wadds 1970 *Analyst*, **95**; 234

17.42　Bibliography

C Burgess and A Knowles 1981 *Techniques in visible and ultraviolet absorption spectroscopy*, Chapman and Hall, London

C T Cottrell, D Irish, V M Masters and J E Steward (eds) 1985 *Introduction to ultraviolet and visible spectrophotometry*, 2nd edn, Pye Unicam, Cambridge

A F Fell and G Smith 1982 *Anal. Proc.*, **19**; 28

G G Guilbault 1967 *Fluorescence – theory, instrumentation and practice*, Edward Arnold, London, and Marcel Dekker, New York

Z Marczenko 1986 *Separation and spectrophotometric determination of elements*, 2nd edn, John Wiley, Chichester

E B Sandell and H Onishi 1978 *Colorimetric determination of traces of metals*, 4th edn, Interscience, New York

S G Schulman 1985 *Molecular luminescence spectroscopy*, John Wiley, New York

F D Snell 1978–81 *Photometric and fluorometric methods of analysis*, Parts 1/2, *Metals*; Part 3, *Non-metals*, John Wiley, New York

L C Thomas and G J Chamberlin (revised by G Shute) 1970 *Colorimetric chemical analytical methods*, 9th edn, Tintometer Ltd, Salisbury

M J K Thomas 1996 *Ultraviolet and visible spectroscopy*, 2nd edn, ACOL–Wiley, Chichester

Analytical Chemistry

The journal *Analytical Chemistry* publishes biennial reviews of fluorimetric analysis.

18

Vibrational spectroscopy

The term 'vibrational spectroscopy' is used to describe the techniques of infrared spectroscopy and Raman spectroscopy. Raman and infrared spectroscopy give the same kind of molecular information, and each method can be used to supplement or complement the other.

18.1 Infrared spectroscopy

The infrared region of the electromagnetic spectrum may be divided into three main sections:[1]

Near-infrared (overtone region) 0.8–2.5 µm (12 500–4000 cm^{-1})
Middle-infrared (vibration–rotation region) 2.5–50 µm (4000–200 cm^{-1})
Far-infrared (rotation region) 50–1000 µm (200–10 cm^{-1})

The main region of interest for analytical purposes is from 2.5 to 25 µm (micrometres), i.e. wavenumber 4000 to 400 cm^{-1}; the wavenumber is the number of waves per centimetre. Normal optical materials such as glass or quartz absorb strongly in the infrared, so instruments for carrying our measurements in this region differ from those used for the electronic (UV/visible) region. Infrared spectra originate from the different modes of vibration and rotation of a molecule. At wavelengths below 25 µm the radiation has sufficient energy to cause changes in the vibrational energy levels of the molecule, and these are accompanied by changes in the rotational energy levels. The pure rotational spectra of molecules occur in the far-infrared region and are used for determining molecular dimensions.

For simple diatomic molecules it is possible to calculate the vibrational frequencies by treating the molecule as a harmonic oscillator. The frequency of vibration is given by

$$v = \frac{1}{2\pi}\left(\frac{f}{\mu}\right)^{1/2} \text{s}^{-1}$$

where v is the frequency (vibrations per second), f is the force constant (N m^{-1}), i.e. the stretching or restoring force between two atoms in newtons per metre, and μ is the reduced mass per molecule (kg); μ is defined by the relationship

$$\mu = \frac{m_1 m_2}{m_1 + m_2} = \frac{A_{r1} A_{r2}}{1000 L (A_{r1} + A_{r2})} \text{ kg}$$

where m_1 and m_2 are the masses of the individual atoms, and A_{r1} and A_{r2} are the relative atomic masses; L is Avogadro's constant.

But it is customary to quote absorption bands in units of wavenumber ($\bar{\nu}$) which are expressed in reciprocal centimetres (cm^{-1}); wavelengths (λ) measured in micrometres (μm) are sometimes used. The relationship between these quantities is given by

$$\bar{\nu} = \frac{1}{\lambda} = \frac{\nu}{c}$$

so

$$\bar{\nu} = \frac{1}{2\pi c}\left(\frac{f}{\mu}\right)^{1/2} cm^{-1} \tag{18.1}$$

There is usually good agreement between calculated and experimental values for wavenumbers. As an example, we may take the C—O bond in methanol (CH_3OH). For this, $f = 5 \times 10^2\,N\,m^{-1}$, $\mu = 6.85\,m_u$ kg (m_u is the unified atomic mass constant $= 1.660 \times 10^{-27}$ kg), and the velocity of light $c = 2.998 \times 10^{10}\,cm\,s^{-1}$. So

$$\bar{\nu} = \frac{1}{2\pi \times 2.998 \times 10^{10}}\left(\frac{5 \times 10^2}{6.85 \times 1.66 \times 10^{-27}}\right)^{1/2}$$

$$= \frac{20.97 \times 10^{13}}{18.84 \times 10^{10}} = 1113\,cm^{-1}$$

The observed C—O band for methanol is at $1034\,cm^{-1}$.

This simple calculation has not taken into consideration any possible effects arising from other atoms in the molecule. More sophisticated methods of calculation which take account of these interactions have been developed but are outside the scope of this book; consult appropriate texts[2] to study this subject further. For a vibrational mode* to appear in the infrared spectrum, and therefore absorb energy from the incident radiation, it is essential that a change in dipole moment occurs during the vibration. Vibration of two similar atoms against each other, e.g. oxygen or nitrogen molecules, will not produce a change of electrical symmetry or dipole moment of the molecule, and these molecules will not absorb in the infrared region.

In many of the normal modes of vibration for a molecule, the main participants in the vibration will be two atoms held together by a chemical bond. These vibrations have frequencies which depend primarily on the masses of the two vibrating atoms and on the force constant of the bond between them. The frequencies are also slightly affected by other atoms attached to the two atoms concerned. These vibrational modes are characteristic of the groups in the molecule and are useful in identifying a compound, particularly in establishing the structure of an unknown substance. Some of these group frequencies are listed in Table 18.1, and a more complete correlation table is provided in Appendix 10. This, of course, is a very simplified picture, as many bands of much weaker intensities occur at shorter wavelengths (they are known as overtone bands and combination bands), but these are unlikely to be confused with the much more intense fundamental bands originating from normal modes of vibration.

---

* Bond vibrations are divisible into two distinct modes, stretching and bending (deformation). Stretching modes constitute the periodic stretchings of the bond along the bond axis. Bending modes are displacements occurring at right angles to the bond axis. For further information, consult a textbook on infrared spectroscopy.

Table 18.1 *Approximate positions of some infrared absorption bands*

Group	Wavenumber (cm^{-1})	Wavelength (µm)
C—H (aliphatic)	2700–3000	3.33–3.70
C—H (aromatic)	3000–3100	3.23–3.33
O—H (phenolic)	3700	2.70
O—H (phenolic, hydrogen bonding)	3300–3700	2.70–3.03
S—H	2570–2600	3.85–3.89
N—H	3300–3370	2.97–3.03
C—O	1000–1050	9.52–10.00
C=O (aldehyde)	1720–1740	5.75–5.8
C=O (ketone)	1705–1725	5.80–5.86
C=O (acid)	1650	6.06
C=O (ester)	1700–1750	5.71–5.88
C—N	1590–1660	6.02–6.23
C—C	750–1100	9.09–13.33
C=C	1620–1670	5.99–6.17
C≡C	2100–2250	4.44–4.76
C≡N	2100–2250	4.44–4.76
CH$_3$—, —CH$_2$—	1350–1480	6.76–7.41
C—F	1000–1400	7.14–10.00
C—Cl	600–800	12.50–16.67
C—Br	500–600	16.67–20.00
C—I	500	20.00

Infrared absorption spectra can be used to identify pure compounds or for the detection and identification of impurities. Most of the applications are concerned with organic compounds, primarily because water, the chief solvent for inorganic compounds, absorbs strongly beyond 1.5 µm. Moreover, inorganic compounds often have broad absorption bands, whereas organic substances may give rise to numerous narrower bands. The infrared absorption spectrum of a compound may be regarded as a sort of fingerprint of that compound (Figure 18.1). Thus, for the identification of a pure compound, the spectrum of the unknown substance is compared with the spectra of a limited number of possible substances suggested by other properties. When a match between spectra is obtained, identification is complete. This procedure is especially valuable for distinguishing between structural isomers[3] (but not optical isomers).

The spectrum of a mixture of compounds is essentially the sum of the spectra of the individual components, provided association, dissociation, polymerisation or compound formation does not take place. In order to detect an impurity in a substance, the spectrum of the substance can be compared with the spectrum of the pure compound; impurities will cause extra absorption bands to appear in the spectrum. The most favourable case will occur when the impurities present possess characteristic groupings not present in the main constituent.

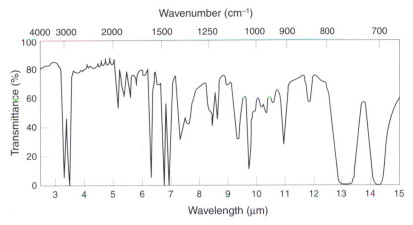

Figure 18.1   Polystyrene: infrared spectrum

18.2   Raman spectroscopy

The Raman effect was discovered in 1928 by the Indian physicist C. V. Raman, and during the 1930s Raman measurements were more widely used than infrared measurements because Raman measurements could be directly recorded using a photographic plate whereas infrared measurements had to be recorded manually. Raman spectroscopy has the following advantages over infrared spectroscopy.

1.   Water is an excellent solvent for Raman spectroscopy whereas it cannot be used in infrared studies.
2.   Glass cells can be used in Raman spectroscopy.
3.   Raman spectra are usually simpler than the corresponding infrared spectra, so overlapping bands are much less common in Raman spectroscopy.
4.   Totally symmetric modes of vibration can be studied by the Raman effect, whereas they are not observed in infrared spectroscopy.
5.   The polarization of Raman spectra gives extra information.
6.   Because of the nature of the Raman effect, one instrument and a single continuous scan can be used to cover the entire range of molecular vibration frequencies.
7.   The intensity of a Raman line is directly proportional to concentration, whereas Beer's law has to be applied in infrared spectroscopy. Thus quantitative analysis is often more convenient in Raman spectroscopy and often more accurate.

18.3   The Raman effect

Raman found that when molecules are irradiated with monochromatic light, a portion of the light is scattered; most of this scattered radiation (about 99%) has the original frequency (Rayleigh scattering), but a small portion (less than 1%) is found at other frequencies. The **difference** in frequency between these new frequencies (Raman bands) and the original frequency is characteristic of the molecule irradiated and numerically identical with certain of the vibrational and rotational frequencies of the molecule. Figure 18.2 depicts a portion of the Raman spectrum of $CHCl_3$ which was obtained by irradiating the

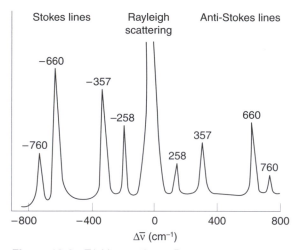

Figure 18.2 Trichloromethane: Raman spectrum

sample with the intense beam of a helium-neon laser having a wavelength of 632.8 nm. The scattered radiation, observed at 90° to the incident beam is of three types: Rayleigh, Stokes and anti-Stokes.

The **Rayleigh-scattered** radiation is considerably more intense than either of the other two types. As is usually the case for Raman spectra, the horizontal axis in Figure 18.2 is the wavenumber shift between the observed radiation and the source radiation. Note that three Raman peaks are found on either side of the Rayleigh peak, and the pattern of the peaks is the same on both sides. **Stokes** lines are found at wavenumbers **smaller** than the Rayleigh peak; **anti-Stokes** lines are found at wavenumbers **greater** than the Rayleigh peak. Because the pattern is the same on both sides of the Rayleigh peak, the lines occur in Stokes–anti-Stokes pairs. For the two lines in each pair, the size of the wavenumber shift is the same but the shifts occur in opposite directions. The magnitudes of the Raman shifts are **independent of the excitation wavelength**. Anti-Stokes lines are generally much less intense than the corresponding Stokes lines, so usually it is only the Stokes lines which are used. The horizontal axis is often labelled in wavenumber rather than wavenumber shift; negative signs for Stokes shifts are sometimes omitted.

The appearance of Stokes and anti-Stokes lines can be explained by Figure 18.3. In the quantum mechanical treatment of the Raman effect, radiation is pictured as a stream of photons scattered by collisions between them and the molecules of the sample. Most of these collisions are elastic in the sense that no net transfer of energy occurs; a few collisions are inelastic. Here the vibrational energy of a bond is added to or subtracted from the energy of the incident photon, changing its frequency. Note that the molecule is not generally excited into its first excited electronic energy level, but instead the energy of the molecule can assume any of an infinite number of **virtual states** between the ground state and excited electronic energy levels. Notice that for anti-Stokes behaviour to occur, the molecule must already be in an excited vibrational energy level. At a given temperature the probability of this can be calculated by assuming a Boltzmann distribution. Thus, as the temperature increases, the population of the first excited vibrational energy level increases and the relative intensity of the anti-Stokes lines increases, compared with the Stokes lines.

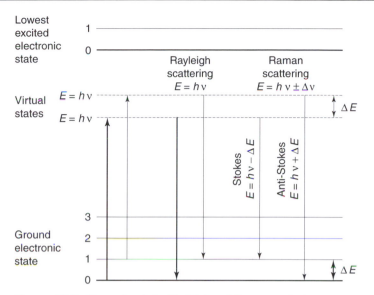

Figure 18.3   Energy levels for Rayleigh and Raman scattering

For a particular vibrational mode to appear in the Raman spectrum (i.e. for it to be Raman-active), the molecule's **polarisability** must change during the course of the vibration. The polarisability of a molecule is the ability of the molecule to be polarised under the action of an electric field such as the electric field of a light wave. It can be defined in terms of the dipole moment μ induced by an electric field E:

$$\mu = \alpha E$$

where α is the polarisability.

18.4   Correlation between IR and Raman

Infrared and Raman spectra tend to be complementary because of their differing **selection rules** for activity. Whereas a change in polarisability must occur for a vibration to be Raman-active, the requirement for a vibration to be infrared-active is that there must be a change in dipole moment (Section 18.1). The vibrations of many molecules will generally be both infrared-active and Raman-active, but for molecules with a centre of symmetry the mutual exclusion principle applies. This states:

> For all molecules with a centre of symmetry, transitions that are allowed in the infrared are forbidden in the Raman spectrum and conversely transitions that are allowed in the Raman spectrum are forbidden in the infrared.

Thus, if the Raman and infrared spectra of a molecule have peaks at the same frequencies then the molecule cannot be centrosymmetric (i.e. have a centre of symmetry).

The Raman shift $\bar{\nu}$ is related to the force constant f and the reduced mass μ by the same expression as for infrared spectroscopy:

$$\bar{\nu} = \frac{1}{2\pi c}\left(\frac{f}{\mu}\right)^{1/2} \text{cm}^{-1} \tag{18.2}$$

Table 18.2   *Transmission ranges of materials for cells and windows*

Material	Transmission range	
	μm	cm⁻¹
Lithium fluoride	2.5–5.9	4000–1695
Calcium fluoride	2.4–7.7	4167–1299
Sodium chloride	2.0–15.4	5000–649
Potassium bromide	9.0–26.0	1111–385
Caesium bromide	9.0–26.0	1111–385
KRS-5 (TlBr + TlI)	25.0–40.0	400–250

18.5   Infrared instruments

For measurements in the middle-infrared region, 2.5 to 50 μm, there are several differences between the instruments used for UV/visible spectrophotometry and those designed for infrared determinations. These changes are mainly dictated by the fact that glass and quartz absorb strongly in the infrared region and photomultipliers are insensitive to the radiation. Front-surfaced mirrors are largely employed to avoid the necessity of radiation passing through glass or quartz layers as reflection from metallic surfaces is generally very efficient in the infrared region. But absorption cells and windows must be fabricated from infrared-transparent materials. The substances most commonly used with infrared radiation and their useful transmission ranges are given in Table 18.2.

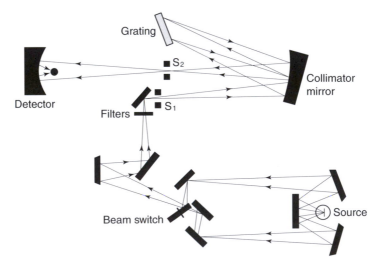

Figure 18.4   Infrared spectrophotometer with diffraction grating monochromator (Reprinted, with permission, from R. C. J. Osland, 1985, *Principles and practices of infrared spectroscopy*, 2nd edn, Philips Ltd)

The main sources of infrared radiation used in spectrophotometers are (1) a Nichrome wire wound on a ceramic support, (2) the Nernst glower, which is a filament containing zirconium, thorium and cerium oxides held together by a binder, (3) the Globar, a bonded silicon carbide rod. These are heated electrically to temperatures within the range 1200–2000 °C when they will glow and produce the infrared radiation approximating to that of a black body. Traditional infrared spectrophotometers were constructed with monochromation being carried out using sodium chloride or potassium bromide prisms, but they had the disadvantage that the prisms are hygroscopic and the middle-infrared region normally necessitated the use of two different prisms in order to obtain adequate dispersion over the whole range. That is why diffraction gratings have displaced prisms as the main means of monochromation in the infrared region. Gratings provide higher resolving powers than prisms and can be designed to operate effectively over a wider spectral range. Even so, most grating instruments operate with two gratings and there is an automatic change of grating at around 2000 cm^{-1}. The layout of a typical grating infrared spectrophotometer is shown in Figure 18.4.

More advanced infrared spectrophotometers produce the infrared spectra by a procedure based upon interferometry. This is known as Fourier transform infrared spectroscopy (FT-IR).[4] The instruments are normally based upon the Michelson interferometer, which takes the radiation from an infrared source and splits it into two beams using a half-silvered 45° mirror so that the resulting beams are at right angles to each other. If an absorbing material is placed in one of the beams, the resulting interferogram will carry the spectral characteristics of the sample in the beam. The resulting interference pattern is shown in Figure 18.5(a) for a source of monochromatic radiation and in Figure 18.5(b) for a source of polychromatic radiation. The monochromatic radiation produces a simple cosine curve,

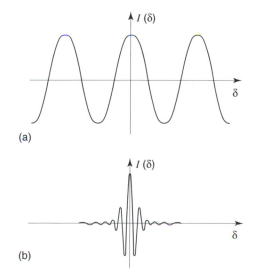

(a)

(b)

Figure 18.5   Interferograms for (a) monochromatic radiation and (b) polychromatic radiation (Reprinted, with permission, from B. Stuart, 1996, *Modern infrared spectroscopy*, ACOL–Wiley, Chichester)

but the polychromatic radiation produces a more complicated pattern because it contains all the spectral information falling onto the detector. Two essential equations, a cosine transform pair, relate the intensity $I(\delta)$ of radiation falling onto the detector to the spectral power density $B(v)$ at a particular wavenumber $\bar{v}$:

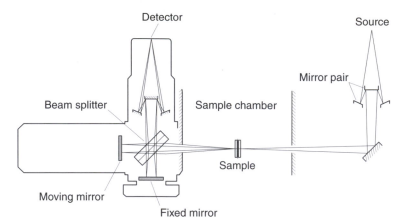

Figure 18.6   FT-IR spectrometer (Courtesy Lloyd Instruments plc, Southampton)

$$I(\delta) = \int_0^\infty B(v) \cos 2\pi v \delta \, dv$$

$$B(v) = \int_{-\infty}^\infty I(\delta) \cos 2\pi v \delta \, d\delta$$

The first equation shows the variation in power density as a function of difference in path length δ, which is an interference pattern. The second shows the variation in intensity as a function of wavenumber. Each can be converted into the other by a Fourier transformation. The actual conversion of the information from the interferogram into an infrared spectrum is very complex and has only been possible by the development of computers, but there are great advantages in using FT-IR. All the frequencies are recorded simultaneously, there is an improvement in signal-to-noise (S/N) ratios, and it is easier to study small samples or materials with weak absorptions. Besides this, the time taken for a full spectral scan is less than one second, which makes it possible to obtain improved spectra by carrying out repetitive scans and averaging the collected signals. This is because the signal-to-noise ratio is directly related to $\sqrt{n}$, where n is the number of scans. Thus 16 repeat scans give a fourfold enhancement of the S/N ratio. Quantitative infrared analysis has benefited greatly from the development of FT-IR. The layout of a typical FT-IR spectrophotometer is shown in Figure 18.6.

The development of FT-IR has also led to its use in combination with other analytical techniques. Gas chromatography–infrared spectroscopy (GC-IR) allows the identification of the components eluting from a gas chromatograph and thermogravimetry combined with FT-IR can give qualitative as well as quantitative information about thermal decomposition products (Section 18.14).

Detection of the infrared signal is of prime importance. A range of detectors is available; the choice depends on the type and quality of the spectrophotometer. A **thermocouple** is made by welding together two wires of metals 1 and 2 in such a manner that a segment of metal 1 is connected to two terminal wires of metal 2. One junction between metals 1 and 2 is heated by the infrared beam, and the other junction is kept at constant temperature; small changes in ambient temperature are thus minimised. To avoid losses of energy by convection, the couples are enclosed in an evacuated vessel with a window transparent to infrared radiation. The metallic junctions are also covered with a black deposit to decrease reflection of the incident beam.

A **bolometer** is essentially a thin strip of blackened platinum in an evacuated glass vessel with a window transparent to the infrared rays; it is connected as one arm of a Wheatstone bridge. Any radiation absorbed raises the temperature of the strip and changes its resistance. Two identical elements are usually placed in the opposite arms of a bridge; one of the elements is in the path of the infrared beam and the other compensates for variations in ambient temperature. Thermocouples and bolometers give a very small direct current, which may be amplified by special methods to drive a recorder.

The **Golay pneumatic detector** is sometimes used; it consists of a gas-filled chamber which undergoes a pressure rise when heated by radiant energy. Small pressure changes cause deflections of one wall of the chamber. This movable wall also functions as a mirror and reflects an incident light beam towards a photocell; the amount of light reflected is directly related to the expansion of the gas chamber, hence to the radiant energy of the light from the monochromator. This detector responds to the total light energy received as distinct from energy received per unit area (thermocouples and bolometers).

The **pyroelectric detectors** fitted in many modern instruments use ferroelectric materials operating below their Curie temperatures. When infrared radiation is incident on the detector there is a change in polarisation which can be employed to produce an electrical signal. The detector will only produce a signal when the intensity of the incident radiation changes. These detectors are of especial value in FT-IR, where rapid response times are needed; they use deuterium triglycine sulphate as the detecting medium in an evacuated chamber. For high sensitivity a mercury cadmium telluride (MCT) detector is used, cooled by liquid nitrogen.

All infrared spectrophotometers are provided with a display which will present the complete infrared spectrum on a single continuous sheet, usually with wavelength or wavenumber scales for the horizontal axis and absorbance or percentage transmittance as the vertical axis. More advanced instruments also possess visual display units to output the spectra, and new spectra can be compared with earlier spectra saved in memory or with spectra drawn from an extensive database. Compared with just a few years ago, quantitative IR spectrophotometry is now much more useful as an analytical procedure, largely due to modern developments in computers.

18.6   Dedicated process analysers

One very important group of infrared instruments consists of spectrometers used for quantitative measurements either as part of a continuous industrial monitoring process or for environmental studies. These instruments are normally purpose-made, dedicated machines designed to run virtually automatically, and they are normally intended to measure only a single compound or family of compounds.

This type of instrument is typified by the non-dispersive continuous stream analyser used for the detection of carbon monoxide (Figure 18.7). Identical infrared beams are passed through the reference cell and the sample cell; a diaphragm detector balances the two signals against each other after detection. The diaphragm detector consists of two small compartments of equal volume filled with the pure gas which has to be determined. When an increase in the carbon monoxide level in the sample flow chamber occurs, the infrared radiation at 4.2 μm is absorbed and the strength of the infrared beam on the detector is diminished. As a result, the diaphragm becomes distended due to the unbalanced heating effect on the cell, and a signal is recorded whose magnitude is proportional to the quantity of carbon monoxide in the sample; the recording is usually made on a continuous sheet or tape.

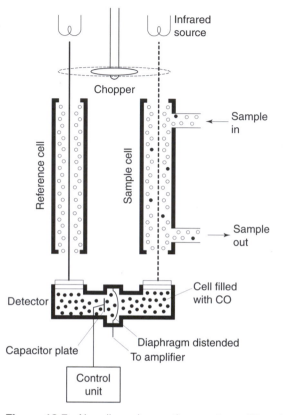

Figure 18.7 Non-dispersive continuous stream IR analyser (Courtesy Beckman Instrument Co.)

Another dedicated application of quantitative infrared spectrometry which has aroused much interest in recent years is the measurement of ethanol in the breath of motorists who are suspected of having had alcoholic drinks prior to driving motor vehicles. The arguments surrounding this application have been well documented elsewhere,[5] but these analysers are used in many countries throughout the world in order to curb drink-driving. Typical of the infrared analysers used for this purpose is the Lion Intoximeter 3000 (Figure 18.8). Unlike the carbon monoxide instrument discussed above, the Lion Intoximeter 3000 uses an interference filter to produce monochromatic radiation of 3.39 μm, which corresponds to the C—H stretching frequency for ethanol. The infrared radiation source is a Nichrome filament helix-wound around a ceramic rod and heated to 800 °C. The beam from the Nichrome source is divided into two before passing through the fixed path length, double-chambered gas sample cell.

Under normal circumstances the composition of the atmosphere in the two cells will be identical; the resulting signals will be balanced, and the ratios of the two energy levels can be used to set and establish the baseline conditions. When ethanol is passed into the machine, either from a simulator or from someone with ethanol in their breath, the infrared beam in the sample cell is partially absorbed by the alcohol. The amount of radiation

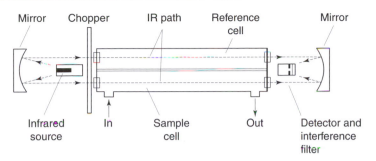

Figure 18.8 Lion Intoximeter 3000 breath alcohol analyser (Courtesy Lion Laboratories Ltd, Barry, Wales)

reaching the detector from the sample cell will depend upon the concentration of ethanol in the cell. The infrared detector again measures the ratio of the reference and sample chamber beams, and the resulting signal is converted into a corresponding breath alcohol value. In this instrument the monochromation takes place after the beams have passed through the cells and before they are measured on the solid-state photoconductive detector.[6] The Intoximeter 3000 is designed to take a 70 mL sample of deep lung air only after at least an initial 1.5 L volume has been expelled through the sample tube by the subject. Condensation of alcohol and water from the breath is prevented by maintaining the sample flow unit at 45 °C.

So long as a compound has a fairly intense absorption that is unlikely to overlap with the absorptions of other substances it may well be mixed with, it is possible to monitor that compound on a continuous basis with a dedicated infrared detector. Gases such as carbon monoxide, nitrogen oxides, ethylene oxide and ammonia can now be measured and regulated using these devices.

18.7 Infrared cells for liquid samples

As many solutions for infrared spectrophotometry involve organic solvents, it is necessary to use sample cells which can not only be stoppered to prevent solvent evaporation, but also be readily dismantled for cleaning and polishing. Cells used for accurate quantitative work must have a fixed path length and window surfaces that are smooth, polished and parallel. Cells are available commercially with fixed path lengths from 0.025 to 1.0 mm, and with variable path lengths to 6.0 mm. The cells have transparent windows made from plates of potassium bromide, or less commonly sodium chloride. Cut from large crystals, the plates are held in place by a stainless steel former and a lead or polytetrafluoroethylene spacer provides the fixed separation between them. An obvious limitation to the use of potassium bromide or sodium chloride plates is that they cannot be used to study aqueous solutions. If water is to be used as the solvent, the plates should be made of calcium fluoride or barium fluoride. An accurate measurement of the path length can be made by the procedure given in Section 18.8. The construction of two typical infrared cells is shown in Figure 18.9. They have to be carefully filled using a syringe or Pasteur pipette to ensure no air is trapped inside. To prevent evaporation the ports should be plugged with small plastic stoppers once the cell has been filled with the solution.

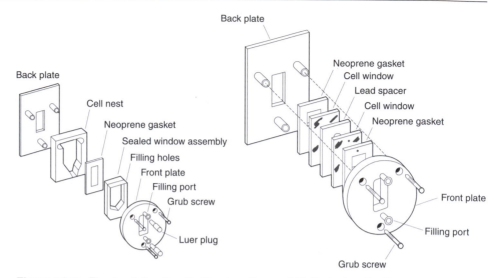

Figure 18.9   Fixed path length cells (Courtesy Specac Ltd, Orpington, Kent)

18.8   Measuring cell path length

When a beam of monochromatic radiation is passed through the windows of an infrared cell, some reflection occurs on the window surfaces and interference takes place between radiation passing from the internal surface of the first window and radiation reflected back from the internal surface of the second window. This interference is at a maximum when $2d = (n + \frac{1}{2})\lambda$, where d is the distance (μm) between the inner surfaces of the two cell windows, λ is the wavelength (μm), and n is any integral number. If the wavelength λ of the monochromatic radiation is varied continuously, the result is an interference pattern consisting of a series of waves (Figure 18.10). A value for the cell path length d can be calculated from the formula

$$d = \frac{\Delta n}{2(\bar{\nu}_1 - \bar{\nu}_2)} \text{ cm}$$

where n is the number of complete interference fringes between wavenumbers $\bar{\nu}_1$ and $\bar{\nu}_2$.

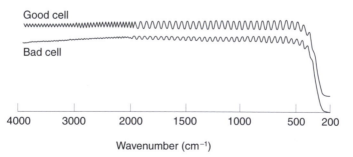

Figure 18.10   Interference patterns from an empty fixed path length cell (Reprinted with permission, from R. C. J. Osland, 1985, *Principles and practices of infrared spectroscopy*, 2nd edn, Philips Ltd)

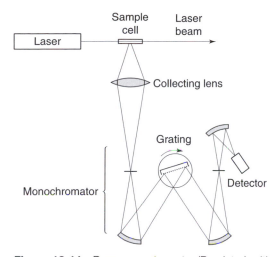

Figure 18.11   Raman spectrometer (Reprinted, with permission, from C. N. Banwell and E. M. McCash, 1994, *Fundamentals of molecular spectroscopy*, 4th edn, McGraw-Hill, Maidenhead)

18.9   Raman instruments

Instrumentation for modern Raman spectrometers consists of three components: an intense source, a sample illumination system and a suitable spectrometer. A schematic diagram of a Raman spectrometer is shown in Figure 18.11.

Sources

Before the advent of the laser, the most common source for Raman spectroscopy was the mercury arc. Today, however, continuous gas lasers have replaced the mercury lamp. Lasers have the following advantages: laser radiation is highly monochromatic and intense; laser beams can be focused so that very small sample sizes can be examined; more precise corrections for reflection losses in the optics of the spectrometer can be made since the laser beam is very well collimated. The most common laser sources are the helium-neon and argon ion lasers.

Sample illumination systems

Sample handling for Raman spectroscopy tends to be simpler than for infrared studies because the measured wavenumber differences are between two visible frequencies. This means that glass can be used for windows, lenses and other optical components. A common sample holder for liquid samples is an ordinary glass melting-point capillary. In modern Raman spectrometers a microscope is often used for sample illumination and collection of the scattered radiation so that very small samples can be studied. The first Raman micro-probe was called the MOLE (molecular optical laser examiner).[7] It consisted of four parts (Figure 18.12): the optical microscope, the coupling optics, the optical filter and the detector.

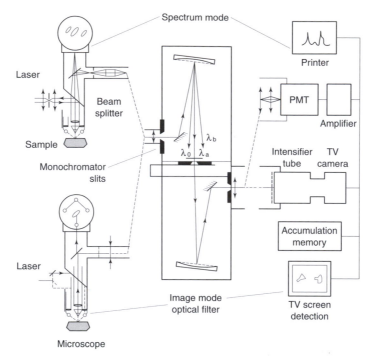

Figure 18.12   Raman microscope (Reprinted, with permission, from J. Corset, P. Dhamelincourt and J. Barbillat, 1989, *Chemistry in Britain*, **25**; 612)

The optical microscope comprises a partly transparent beam splitter which reflects a portion of the incident laser beam towards the sample. The microscope objective then focuses the beam on the sample. The Rayleigh and Raman scattered radiation is collected by the same microscope objective then transmitted through the beam splitter towards the transfer optics and the monochromator.

The instrument can operate in two separate modes: the spectrum mode and the image mode. In the spectrum mode the laser beam is focused onto a fine point on the sample and the spectrum of the scattered radiation can be scanned. In the image mode the laser beam is swept over the sample with the monochromator set at the wavelength of a strong Raman line for one of the components of the sample. This process can then be repeated with the monochromator set at a strong Raman line of a second component. In this way the variation in composition across a sample can be determined.

Spectrometer

In Raman spectroscopy it is necessary to separate the Raman-scattered radiation from the Rayleigh-scattered incident wavelength. This is usually achieved using holographic gratings or double and triple monochromators. The scattered radiation can then be detected using, in early instruments, a photomultiplier tube. Multichannel detectors, in which up to 1000 spectral elements are recorded simultaneously, have made it possible to obtain spectra very quickly.

Fourier transform Raman spectroscopy

Fourier transform Raman spectroscopy uses excitation wavelengths in the near-infrared, making it possible to avoid problems of sample fluorescence which can occur using lasers operating in the visible region.

18.10 Measuring IR absorption bands

As with electronic spectra, the use of infrared spectra for quantitative determinations depends upon measuring the intensity of either the transmission or absorption of the infrared radiation at a specific wavelength, usually the maximum of a strong, sharp, narrow, well-resolved absorption band. Most organic compounds will possess several peaks in their spectra which satisfy these criteria and which can be used so long as there is no substantial overlap with the absorption peaks from other substances in the sample matrix.

The background to any spectrum does not normally correspond to a 100% transmittance at all wavelengths, so measurements are best made by what is known as the baseline method.[8] This involves selecting an absorption peak to which a tangential line can be drawn, as shown in Figure 18.13. This is then used to establish a value for I_0 by measuring vertically from the tangent through the peak to the wavenumber scale. Similarly, a value for I is obtained by measuring the corresponding distance from the absorption peak maximum. So, for any peak, the absorbance will not be the value corresponding to the height of the absorption, measured from the horizontal axis of the chart paper; instead it will be the value of A_{calc} obtained from the equation

$$A_{calc} = \log \frac{1}{T} = \log \frac{I_0}{I}$$

where I_0 and I are values measured using the tangential baseline.

This procedure has the great advantage that some potential sources of error are eliminated. The measurements do not depend upon accurate wavelength positions as they are made with respect to the spectrum itself, and any cell errors are avoided by using the same cell of fixed path length. Measuring A_{calc} eliminates any variations in the source intensity, the instrument optics or the sensitivity.

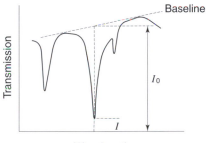

Figure 18.13 Tangent baseline measurement

18.11 Beer's law: quantitative IR spectra

Infrared spectra are recorded using either or both absorbance (A) and percentage transmittance (T) just as they are in visible ultraviolet electronic spectra, and Beer's law,

$$A = \varepsilon cl = \log \frac{1}{T} = \log \frac{I_0}{I}$$

as given in Section 17.2, applies equally to infrared spectra as it does to electronic spectra. Similarly, for a mixture of compounds the observed absorbance at a particular wavelength (or frequency) will be the sum of the absorbances for the individual constituents of the mixture at that wavelength:

$$A_{observed} = A_1 + A_2 + A_3 = \varepsilon_1 c_1 l + \varepsilon_2 c_2 l + \varepsilon_3 c_3 l$$

as the path length l is constant for the mixture.

It has taken a long time for quantitative infrared spectrophotometry to become a commonly used procedure; there are several reasons for this:

(a)   The molar absorption coefficients (molar absorptivities) are usually 10 times smaller than coefficients in the electronic region. So the infrared procedure is usually less sensitive.
(b)   The most accurate range for quantitative measurements is $T = 55\%$ down to $T = 20\%$ ($A = 0.26$ to $A = 0.70$) with the accuracy diminishing rapidly outside these values.
(c)   Older null-balance spectrophotometers possessed an instrumental error in T of $\pm 1\%$.

Although nothing can improve on the disadvantage of low molar absorptivities, instrumental designs and improvements with ratio recording and FT-IR instruments have virtually overcome the accuracy and instrumental limitations. As a result, quantitative infrared procedures are now much more widely used and are frequently applied in quality control and materials investigations. Applications fall into three distinct groups.

Measurements using Beer's law

Where a compound gives a strong, narrow, well-defined absorption band, this can be used as the basis for quantitative measurements simply by comparing the magnitude of the absorbance A_u of the unknown concentration c_u with the corresponding absorbance A_s of a standard solution of known concentration c_s using a cell of measured path length. The absorbance of the standard solution at the chosen wavelength is given by

$$A_s = \varepsilon c_s l$$

and the absorbance for the unknown concentration at the same wavelength is

$$A_u = \varepsilon c_u l$$

Therefore

$$\frac{A_s}{c_s} = \frac{A_u}{c_u} = \varepsilon l$$

Hence the concentration of the unknown is given by

$$c_u = \frac{c_s A_u}{A_s}$$

Note that the calculation assumes a linear absorbance/concentration relationship, and this may only apply over short concentration ranges.

Use of a calibration graph

A calibration curve overcomes any problems created due to non-linear absorbance or concentration features, and it means that any unknown concentration run under the same

Table 18.3 *Determining antioxidant concentration using IR spectroscopy*

Antioxidant concentration (% w/v)	*A*
0.129	0.024
0.251	0.044
0.514	0.094
0.755	0.138
1.016	0.186
1.273	0.233
unknown	0.104

conditions as the series of standards can be determined directly from the graph. The procedure requires that all standards and samples are measured in the same cell of fixed path length, although the dimensions of the cell and the molar absorptivity for the chosen absorption band are not needed; they are constant for all the measurements. The results in Table 18.3 are typical for the concentration of an antioxidant, measured at $3655\,\text{cm}^{-1}$, used as an additive in oil. The results are presented in Figure 18.14.

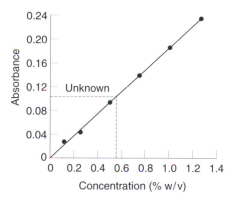

Figure 18.14   Calibration graph for antioxidant determination

Standard addition methods

Standard addition methods are not widely applied in quantitative infrared spectrophotometry; they are limited to determinations of low-concentration components in multicomponent mixtures. The procedure involves preparing a series of solutions in a solvent which does not absorb at the wavelength of the chosen absorption band. The solutions are made from a series of increasing concentrations of the pure analyte (similar to a normal set of calibration graph concentrations) but to each is added a constant, known amount of the sample containing the unknown concentration. All the solutions are diluted to a fixed volume and their absorbances measured in a cell of fixed path length by scanning over the chosen absorption band.

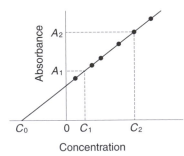

Figure 18.15   Calibration graph for standard additions

A plot of the absorbance against the concentration of the pure analyte does not pass through zero as all the absorbance values are enhanced by an equal amount due to the presence of the unknown concentration in the added sample. Extrapolation of the graph back to the the horizontal axis gives the concentration of the unknown as a negative value. Alternatively it can be determined from the slope of the line by taking any two points on the line, as shown in Figure 18.15. From this it can be seen that

$$\frac{A_2}{A_1} = \frac{c_u + c_2}{c_u + c_1}$$

Hence

$$A_2 c_u - A_1 c_u = A_1 c_2 - A_2 c_1$$

$$c_u = \frac{A_1 c_2 - A_2 c_1}{A_2 - A_1}$$

18.12   Measurements using compressed discs

Infrared spectra for solid organic compounds are frequently obtained by mixing and grinding a small sample of the material with specially dry and pure potassium bromide (the carrier), then compressing the powder in a special metal die under a pressure of 15–30 tonnes to produce a transparent potassium bromide disc. As the potassium bromide has virtually no absorption in the middle-infrared region, a very well-resolved spectrum of the organic compound is obtained when the disc is placed in the path of the infrared beam.

In theory, increased quantities of the organic compound finely ground with constant quantities of potassium bromide should give infrared spectra of increasing intensity. However, good quantitative results by this direct procedure are difficult to obtain due to problems associated with the non-quantitative transfer of powder from the small ball-mill grinder (or pestle and mortar) into the compression die. These problems are only partially overcome by using a micrometer to measure the final disc thickness.

To use potassium bromide discs for quantitative measurements it is best to employ an internal standard procedure in which a substance possessing a prominent isolated infrared absorption band is mixed with the potassium bromide (KBr). The substance most commonly used is potassium thiocyanate (KSCN), which is intimately mixed and ground to give a uniform concentration of usually 0.1–0.2% in the KBr. A KBr/KSCN disc will give a characteristic absorption band at $2125\,\mathrm{cm}^{-1}$. Before quantitative measurements can be carried out, it is necessary to prepare a calibration curve from a series of standards made using different amounts of the pure organic compound with the KBr/KSCN. A practical application is given in Section 18.16.

18.13   Reflectance methods[8]

Attenuated total reflectance

Reflectance methods can be used to analyse samples which are difficult to analyse by standard transmission techniques. Attenuated total reflectance (ATR) spectroscopy uses the phenomenon of internal reflection. A typical ATR cell is illustrated in Figure 18.16. A beam of radiation entering a crystal will undergo total internal reflection when the angle of incidence at the interface between the sample and the crystal is greater than the critical angle, which is a function of the refractive indices of the two surfaces. The beam penetrates a short distance (a fraction of a wavelength) beyond the reflecting surface, and if a sample which selectively absorbs radiation is in close contact with the reflecting surface, the beam loses energy at the characteristic wavelengths where the material absorbs. The resultant attenuated reflection is measured as a function of wavelength by the spectrometer, and the spectrum obtained is equivalent to the absorption spectrum of the sample.

The crystals used in ATR cells are made from materials which have low solubility in water and which have very high refractive indices. Such materials include zinc selenide (ZnSe), germanium (Ge) and thallium iodide (KRS-5). Different designs of ATR cell allow both liquid and solid samples to be examined. A less intense solvent contribution to the overall infrared spectrum is usually observed, so solvent spectra can easily be subtracted from the sample spectrum of interest. Multiple internal reflectance (MIR) is a similar technique to ATR but it produces more intense spectra due to multiple reflections.

Specular reflectance

Specular reflectance spectroscopy measures the radiation reflected from a surface. The material must therefore be reflective or attached to a reflective backing. A particularly useful application for this technique is the study of surface coatings such as paints and polymers.

Diffuse reflectance

Diffuse reflectance spectroscopy measures the radiation which penetrates one or more particles in a sample and which is reflected in all directions. In the diffuse reflectance technique, commonly called DRIFT, a powdered sample is mixed with KBr powder. The

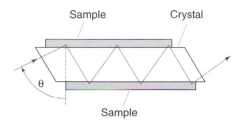

Figure 18.16   Attenuated total reflectance: an ATR cell (Reprinted, with permission, from B. Stuart, 1996, *Modern infrared spectroscopy: analytical chemistry by open learning*, ACOL–Wiley, Chichester)

DRIFT cell reflects radiation to the powder and collects the energy reflected back over a wide angle. Diffusely scattered light can be collected directly from a sample or by using an abrasive sampling pad. The technique is particularly suitable for sampling powders or fibres. The spectrum obtained cannot be used directly for quantitiative analysis. Kubelka and Munk developed an expression which relates the sample concentration to the scattered radiation intensity:

$$\frac{(1 - R_\infty)^2}{2R_\infty} = \frac{c}{k}$$

where R_∞ is the absolute reflectance of the layer, c is the concentration and k is the molar absorptivity.

18.14   GC-FTIR systems

The combination of gas chromatography with Fourier transform infrared spectroscopy (GC-FTIR) has the potential to provide information on molecular structure for individual components in a complex mixture, even giving the ability to distinguish between positional isomers – this cannot be done with gas chromatography–mass spectroscopy (GC-MS) – or indicating functional group classes within the mixture. But although well established in many laboratories, GC-FTIR often requires compromises in either the chromatography or the spectroscopy, in order to interface the instruments. If the combination is designed from the onset as a dedicated GC-FTIR spectrometer and the computer software is written to control this combination, then optimum performance and cost can be achieved. In one commercially available system the infrared detector plus GC occupy only slightly more bench space than a standard GC of similar performance. A typical layout for a GC-FTIR system is shown in Figure 18.17.

The eluent from a conventional capillary column is passed directly into a heated flow-through infrared cell with a pathlength of 12 cm and 1 mm internal diameter (<0.1 mL volume), where it is irradiated with a conventional black-body source and the resulting absorption dispersed onto a cooled mercury cadmium telluride (MCT) detector. Direct coupling and a small volume cell reduce the loss of chromatographic resolution, sometimes troublesome in earlier hyphenated systems. The detector is also claimed to be much more sensitive than conventional systems with a limit of detection under favourable circumstances of less than 5 ng injected into the system.

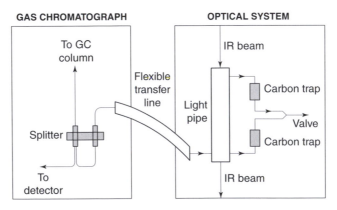

Figure 18.17   Typical GC-FTIR system

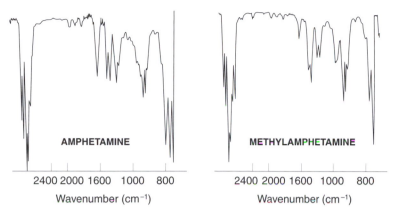

Figure 18.18   Vapour phase IR spectra of amphetamine and methylamphetamine

The instrument is under full computer control, and a number of software packages can be used to achieve a variety of data acquisition and report modes. Just as for GC-MS, library search and match programs are available for the identification of unknowns. The combined system is a powerful analytical tool, capable of identifying a large number of compounds in a wide range of sample types, even distinguishing between compounds which, although different from each other, would give identical mass spectra. This can be used to advantage in the analysis of pharmaceutical compounds, where drugs which have different functional groups and thus different pharmacological activity will give almost identical fragmentation patterns in mass spectrometry. For example, the commonly abused 1-amphetamine (speed) cannot be easily distinguished from methylamphetamine by mass spectroscopic methods, but the IR spectra are completely different (Figure 18.18).

There are, however, problems with the use of GC-FTIR due to the fact that the spectra are vapour phase spectra, and are often significantly different from those obtained in the condensed phase. Vapour phase spectra often show aspects of rotational fine structure which are normally broadened out in liquids or solids. Equally, bands which are seen in the condensed phase due to intermolecular interactions, particularly hydrogen bonding, and which can be highly diagnostic, are absent in the vapour phase. This caused some difficulties when searching large infrared databases that contained mainly condensed phase spectra, but a number of computer databases are becoming available for vapour phase spectra.[9]

18.15   Near-infrared spectroscopy

Near-infrared (NIR) spectra result from overtones or combination bands of fundamental vibrations. Overtone transitions are 'forbidden' but they can be observed because of the anharmonicity of real oscillators. So-called forbidden bands are between 10 and 1000 times weaker than fundamental bands. A broadband quartz halogen lamp is used to give the radiation in the near-infrared region, $0.8-2.5\,\mu m$ ($12\,500-4000\,cm^{-1}$) and wavelength dispersion is achieved using laser-etched holographic gratings moved by stepper motors. Two types of detector are used to cover the complete wavelength range: silicon for $0.8-1.1\,\mu m$ and lead sulphide for $1.1-2.5\,\mu m$. Fourier transform NIR spectroscopy can also be used, giving a highly reproducible wavenumber scale, better resolution and higher sensitivity when compared with a dispersive instrument.

NIR spectroscopy is widely used in the pharmaceutical, agricultural and food industries for both qualitative and quantitative applications. In the pharmaceutical industry it is typically used for quality control, e.g. to assay an active component in tablets, usually by presentation of a single whole tablet, or to monitor blending processes. In the agricultural industry it has been used to determine the amount of crude protein in dried grass silage. In the food industry it is typically used to determine fat, protein and lactose in milk; to determine nicotinamide in flour premixes; and for the non-invasive monitoring of ethanol in fermentation processes. NIR spectroscopy is most commonly carried out in the reflectance mode. The higher scattering/absorption ratio in the NIR region of the spectrum makes diffuse reflectance more quantitative, and little sample preparation is required since samples can be analysed directly in glass vials or in blister packs by spectral subtraction of the container.

The use of NIR spectroscopy as a quantitative tool requires an appropriate set of calibration samples, ideally based on real samples that contain enough concentration variation in the component(s) to be analysed. One of the problems with this method, however, is the choice of wavelength or wavelengths. Bands in the NIR region are broad and overlap. It is therefore necessary to use chemometric techniques (Chapter 4) when carrying out multivariate calibrations. A method such as **partial least squares (PLS)** uses all the information in the spectrum to determine an analyte concentration; it combines principal component analysis and linear regression in one algorithm. The details are outside the scope of this book.

Experimental determinations

!
■   **Safety.** Before carrying out any experiments in this section, pay full attention to any safety warnings and make sure you adhere to national laboratory and safety regulations.

18.16   Purity of commercial benzoic acid by compressed discs

To obtain a calibration curve for benzoic acid, six discs should be prepared using potassium bromide containing 0.1% potassium thiocyanate, as described in Section 18.12, and increasing quantities of pure benzoic acid using the following quantities.

KBr/KSCN (g)	1.000	1.000	1.000	1.000	1.000	1.000
Benzoic acid (g)	0.000	0.050	0.075	0.100	0.150	0.200

Note that the weighed amount of KBr/KSCN is constant and although the problem of non-quantitative transfer of powder from the ball-mill grinder still exists, it affects both the carrier and the organic compound equally. When the infrared spectra for the six discs have been obtained, the calibration curve is prepared by plotting the ratio of the intensity of the selected benzoic acid band (carbonyl $1695\,\mathrm{cm}^{-1}$) and the KSCN $2125\,\mathrm{cm}^{-1}$ peak against the benzoic acid concentration in the discs.

The result should be a calibration plot of the type shown in Figure 18.19. This can then be used to assay an impure benzoic acid sample by weighing say 0.125 g of the sample, grinding it with 1.000 g of the KBr/KSCN carrier and preparing a compressed disc as before. The peak ratio of the measured absorbances can then be referred to the calibration curve to give a value for the true amount of benzoic acid in the sample.

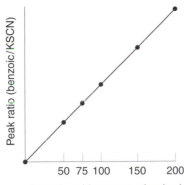

Figure 18.19   Benzoic acid: calibration graph for internal standard procedure

18.17   A calibration curve for cyclohexane

Run infrared spectra for pure cyclohexane and pure nitromethane. From the spectra select a cyclohexane absorption which is not affected by, or overlapping with, those of the nitromethane. Prepare a series of solutions of known concentrations of cyclohexane in nitromethane covering the range from 0% to 20% (w/v). Using a cell of fixed path length 0.1 mm, measure the absorbances for the solutions at the chosen peak absorption using the baseline method (Section 18.10) and plot the calibration graph. Use this graph to determine the unknown concentration of cyclohexane in the sample.

18.18   2-, 3-, 4-Methylphenols (cresols) in a mixture

Prepare solutions containing weighed amounts of the individual pure cresols (0.5 g) in cyclohexane (20 mL) and use them to obtain infrared spectra for the three cresols in the cyclohexane. Also, prepare a single solution containing the three cresols by mixing together 5 mL of each of the individual solutions. Record the infrared spectra for each of the four solutions using a cell of fixed path length 0.1 mm or 0.25 mm. From the spectra select absorption bands suitable for the separate measurement of each isomer. You should find the most appropriate to be 2-methylphenol 750 cm^{-1}, 3-methylphenol 773 cm^{-1}, 4-methylphenol 815 cm^{-1}.

Use the solutions of the individual cresols to prepare a series of calibration standards at appropriate dilutions with cyclohexane for each of the individual cresols and construct the three calibration curves. Take some crude cresol mixture (1 g) and dissolve it in cyclohexane (20 mL). Obtain the infrared spectrum for the mixture; if necessary, dilute the solution further with cyclohexane to obtain absorbances which will lie on the calibration graphs. From the selected absorption peaks, calculate the absorbances for the three individual isomers and use the calibration graphs to calculate the percentage composition of the cresol mixture.

18.19   Propanone (acetone) in propan-2-ol

Due to atmospheric oxidation it is common for commercial propan-2-ol to contain a small amount of propanone:

$$CH_3-\underset{\underset{OH}{|}}{C}-CH_3 \xrightarrow{O_2} CH_3-\underset{\overset{O}{\|}}{C}-CH_3$$

As the greatest accuracy is achieved in any determination by measuring the smaller component, in this case it is the propanone which is determined quantitatively rather than the propan-2-ol. Run infrared spectra of pure propanone and of pure propan-2-ol. From them select an absorption band for propanone which does not overlap significantly with any of those for the propan-2-ol. The best band is most probably at $1718\,cm^{-1}$, the carbonyl stretching frequency.

Prepare a $10\,\%$ v/v bulk solution by dissolving pure propanone ($25\,mL$) in tetrachloromethane and diluting to $250\,mL$ in a graduated flask. From this prepare a series of dilutions of propanone in tetrachloromethane covering the concentration range 0.1–2.5 vol%. **Carry out this work in a fume cupboard**. Measure the percentage transmittance for each solution at $1718\,cm^{-1}$ using a cell of fixed path length $0.1\,mm$. Use the baseline method (Section 18.10) on the spectra to calculate the absorbance for each concentration and plot a calibration curve of absorbance against concentration.

Take $10\,mL$ of commercial propan-2-ol and dilute to $100\,mL$ with tetrachloromethane in a graduated flask. Record the infrared spectrum and calculate the absorbance for the peak at $1718\,cm^{-1}$. Obtain a value for the propanone concentration from the calibration graph. The true value for the propanone in the propan-2-ol will be 10 times the figure obtained from the graph (this allows for the dilution) and the % by volume value can be converted to a molar concentration ($mol\,L^{-1}$) through dividing by 7.326; e.g. $1.25\,\%$ v/v becomes $1.25/7.326 = 0.171\,mol\,L^{-1}$.

18.20   References

1. R C Denney 1982 *A dictionary of spectroscopy*, 2nd edn, Macmillan, London, p. 89
2. J M Hollas 1982 *High resolution spectroscopy*, Butterworth, London
3. R C J Osland 1985 *Principles and practices of infrared spectroscopy*, 2nd edn, Philips, Eindhoven
4. P R Griffiths and J A de Haseth 1986 *Fourier transform infrared spectroscopy*, John Wiley, Chichester
5. R C Denney 1986 *Alcohol and accidents*, Sigma Press, Wilmslow and John Wiley, Chichester
6. Anon 1982 *Lion Intoximeter 3000 – operators' handbook*, Lion Laboratories, Barry
7. J Corset, P Dhamelincourt and J Barbillat 1989 *Chemistry in Britain*, **25**; 612
8. B Stuart 1996 *Modern infrared spectroscopy: analytical chemistry by open learning*, ACOL–Wiley, Chichester
9. Sadtler 1986 *The Sadtler standard gas chromatography retention index library*, Volume 4, Sadtler Research Laboratories, Philadelphia PA

18.21   Bibliography

C N Banwell and E M McCash 1994 *Fundamentals of molecular spectroscopy*, 4th edn, McGraw-Hill, Maidenhead

L J Bellamy 1980 *The infrared spectra of complex molecules*, Volumes I and II, Chapman and Hall, London

P R Griffiths 1975 *Fourier transform infrared spectrometry*, 2nd edn, John Wiley, New York

P J Hendra, C Jones and G Warnes 1991 *Fourier transform Raman spectroscopy: instrumentation and chemical applications*, Ellis Horwood, Chichester

D A Long 1977 *Raman spectroscopy*, McGraw-Hill, Maidenhead

E D Olsen 1975 *Modern optical methods of analysis*, McGraw-Hill, New York

19

Mass spectrometry

19.1 Introduction

This chapter is not an exhaustive explanation of the theory and practice of mass spectrometry, but some details have been included as it is the first time mass spectrometry has been included in Vogel. Direct input of samples to any form of mass spectrometer will seldom give data that can be considered quantitative, even if that sample is 'pure' and consists only of one component. This is due to the high sensitivity of the technique (normally only micrograms are required) and the sometimes variable ionisation efficiency of certain ion sources. The great power of mass spectroscopy is as a tool for identifying substances. However, a mass spectrometer often receives input from another technique, usually GC or HPLC, so it acts as a detector for the chromatographic front end. Small, reproducible amounts of sample are introduced into the mass spectrometer as they elute from the column and, under these conditions, quantitative analysis is possible.

In these hyphenated techniques the analyst has the facility of separating complex mixtures, identifying them and also quantifying the amounts all in one system. Hyphenated systems are now increasingly being used, even in routine analytical laboratories, where they provide data on number of components, type and identity of each compound, and amount of each, in a way that was just not possible a few decades ago. But their power as analytical tools is partly offset by their complexity and cost, and occasionally when the analyst misunderstands their limitations. Before considering the applications of mass spectrometry coupled to GC or HPLC the reader should become familiar with the sections explaining the basic principles of mass spectrometry. These should be studied along with the theory and practice of chromatography in Chapters 6 to 9. Once these concept are understood, then it will be possible to make much more valid evaluations on the hyphenated techniques in Section 19.10.

Mass spectrometry is essentially the process of taking individual atoms or molecules, causing them to acquire a charge and possibly fragment, and then sorting them, in the gas phase, into a spectrum according to their mass/charge ratio. Since most ions acquire a single charge, this represents a selection according to mass, and in theory allows identification of the original particle. Besides giving atomic and molecular weights, the technique can now provide information on structure, mechanism, kinetics of reaction and mixture analysis.

Mass spectrometry is performed on both organic, inorganic and biological samples, which may initially be present in the gas, liquid or solid phase, either as bulk solids or as surfaces. The first mass spectrometry experiments were performed in 1910 by J. J. Thompson, who showed that neon consists of two different types of atom (isotopes). This first mass spectrometer could resolve ions differing in mass by 1 part in 15; today resolution of 125 000 is possible. By 1919–20 F. W. Aston had introduced electrostatic and magnetic

Table 19.1 *Atomic masses and abundances of some elements*

Element	Isotopes	Mass	Abundance (%)	Abundance (100 : minor)[a]	Chemical or atomic weight
C	^{12}C	12.000 00	98.903		12.011
	^{13}C	13.003 35	1.103	1.15	
H	^{1}H	1.007 825	99.985		1.007 94
	^{2}H	2.0140	0.015	0.015	
O	^{16}O	15.994 915	99.76		15.9994
	^{17}O	16.999 131	0.04	0.04	
	^{18}O	17.999 160	0.20	0.20	
N	^{14}N	14.003 074	99.63		14.006 74
	^{15}N	15.000 108	0.37	0.37	
F	^{19}F	18.998 403	100		18.998 403
Cl	^{35}Cl	34.968 852	75.77		35.4527
	^{37}Cl	36.965 903	24.23	31.98	
Br	^{79}Br	78.918 336	50.69		79.904
	^{81}Br	80.916 289	49.31	97.28	
Si	^{28}Si	27.976 927	92.23		
	^{29}Si	28.976 495	4.67	5.06	28.0855
	^{30}Si	29.973 770	3.10	3.36	
S	^{32}S	31.972 070	95.02		32.066
	^{33}S	32.971 456	0.75	0.79	
	^{34}S	33.967 866	4.21	4.43	
I	^{127}I	126.904 47	100		126.904 47
P	^{31}P	30.973 762	100		

[a] The major isotope is assigned abundance 100 and the other abundances are calculated by proportion.

focusing, which are still in use, and which increased the resolution to 1 part in 100, enabling the isotopic composition of many of the elements to be determined.

Aston introduced the term 'mass spectrum' to describe the output of mass/charge versus intensity, normally displayed as the output. But the term 'mass spectrum' is in many ways a misnomer. The words 'spectrum' and 'spectroscopy' are reserved for those processes involving the interaction of electromagnetic radiation with matter. In all forms of spectroscopy considered in this text, UV, IR, NMR, etc., this is true, but it is not true for mass spectroscopy and it would be more accurate to describe the process as 'mass analysis' or as a 'mass filter'; unfortunately, 'mass spectrum' has been adopted and it would be difficult to change. What about 'mass'? To most chemists the statement 'the relative molecular mass of chloroethane is 64.5' is both straightforward and accurate. This value is based upon the masses of carbon = 12.0111 u (unified mass unit), hydrogen = 1.0079 u and chlorine = 35.4527 u, themselves defined[1] by the IUPAC convention of unified atomic mass based upon ^{12}C having atomic mass = 12.00 u (Table 19.1). A mass of 1 u = 1/12 the mass of a single atom of carbon $12 = 1.660\,540 \times 10^{-24}$ g. Occasionally the dalton is used to express mass, especially in the life sciences; 1 u or 1 amu = 1 dalton.

But the mass spectrum of chloroethane (Figure 19.1b) clearly shows a number of different mass spectral lines, in particular two lines at 64 and 66 with intensities of approximately 3 : 1. This observation is due to the fact that, in mass spectrometry, individual ions are examined and the isotopic composition of each atomic or molecular ion is important. For naturally occurring chlorine, the two isotopes ^{35}Cl and ^{37}Cl are present in the ratio 100 : 32.4, giving an average molecular mass or chemical atomic mass of 35.4527, but for every hundred molecules, 75.77 will contain a chlorine-35 atom and 24.23 will contain a chlorine-37 atom, **and each type will give a unique mass spectrum**, in this case differing by two mass units. This is shown more clearly by the chlorine isotope combinations in dichloromethane (Figure 19.1(b)).

The study of isotopic differences was the object of the first experiments in the early 1900s and today, at the end of the century, it is still an important and almost unique property of the technique even though we are generally observing molecular species. Because very few elements are monoisotopic, it means the mass obtained from observation of the 'molecular ion' is normally different from the chemical mass, derived from averaged atomic masses, of the same substance. This property is examined later in the chapter when the use of isotopes is described in more detail, but at this stage it is important to distinguish between the different masses that are used (Table 19.2).

Although the principles of mass spectrometry are simple and easily understood, the very wide diversity, complexity and high cost of much of the apparatus can appear somewhat daunting. The block diagram in Figure 19.2 relates the main components and also illustrates how computers may be interfaced for both data acquisition, processing and computer control of the actual instrument. This topic is expanded later in the chapter, but even the most complex mass spectrometers need to satisfy the original requirement, the production of gas phase ions, analysing them by mass and detecting the resulting pattern. The way this is done, the resolution that may be obtained, and the limits of detection (10 fg or less) are very far removed from the original experiments on atoms with masses of less than 100. (The highest molecular masses studied today are > 200 000 u.)

19.2   Vacuum systems

Nearly all mass spectrometry experiments are conducted in a high-vacuum environment; in order to convert most molecules into gas phase ions with a lifetime long enough to be measured. The ideal system should be capable of providing and maintaining a vacuum of 10^{-4} to 10^{-8} torr (10 mPa to 1 μPa) even when volatile organic molecules are introduced into the system and heated up to 300 °C. It should also be able to be pumped down from atmospheric pressure to an acceptable vacuum in a reasonable time after cleaning or maintenance. And it should be possible to achieve a working vacuum in the system in as little as 5 min after a 'cold' start. A number of different approaches are used to provide this vacuum in commercial spectrometers but they all have some common characteristics. The first stage or 'rough' vacuum is produced using direct-drive two-stage rotary pumps which are reliable and capable of high pumping speeds; they are coupled in series to the high-vacuum stage, where two different systems are in use.

Conventional oil diffusion pumps may be used; they are extremely reliable and can give high vacuums in relatively large enclosures (they have pumping speeds of 200–750 L s^{-1}) but the running and maintenance costs are high. The special oil used in these pumps is expensive and needs to be replaced at regular intervals to maintain performance. Also there

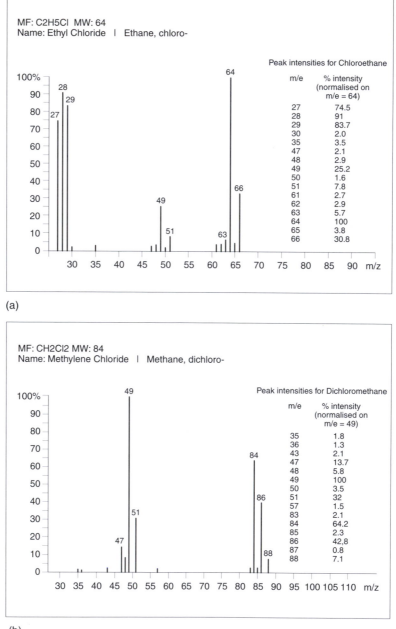

MF: C2H5Cl MW: 64
Name: Ethyl Chloride | Ethane, chloro-

Peak intensities for Chloroethane

m/e	% intensity (normalised on m/e = 64)
27	74.5
28	91
29	83.7
30	2.0
35	3.5
47	2.1
48	2.9
49	25.2
50	1.6
51	7.8
61	2.7
62	2.9
63	5.7
64	100
65	3.8
66	30.8

(a)

MF: CH2Cl2 MW: 84
Name: Methylene Chloride | Methane, dichloro-

Peak intensities for Dichloromethane

m/e	% intensity (normalised on m/e = 49)
35	1.8
36	1.3
43	2.1
47	13.7
48	5.8
49	100
50	3.5
51	32
57	1.5
83	2.1
84	64.2
85	2.3
86	42,8
87	0.8
88	7.1

(b)

Figure 19.1 Electron impact spectra of (a) chloroethane and (b) dichloromethane

Table 19.2   *Types of mass used in mass spectrometry*

	Carbon	Hydrogen	Nitrogen	Oxygen	Chlorine
Nominal mass	12	1	14	16	35.5
Atomic mass	12.011	1.008	14.007	15.994	35.453
Exact mass	^{12}C 12.000 ^{13}C 13.003	^{1}H 1.008 ^{2}H 2.014	^{14}N 14.003 ^{15}N 15.000	^{16}O 15.995 ^{18}O 17.999	^{35}Cl 34.969 ^{37}Cl 36.959

is always a low risk of oil contamination within the main spectrometer. The second, and increasingly popular, approach is to use turbomolecular pumps which are purely mechanical devices that do not rely on heated sources or contain oil, and thus are intrinsically safe in the sense that they cannot contaminate the main mass spectrometer. Recent engineering developments have led to pumps which have a significant advantage over oil diffusion pumps except where large pumping speeds are required.

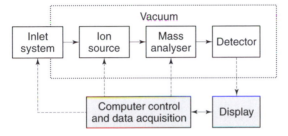

Figure 19.2   Block diagram of a mass spectrometer

Many mass spectrometers provide differential pumping so that parts of the system can be evacuated quickly or can be maintained at a different pressure to the main system, perhaps where chemical ionisation (CI) sources are deliberately operated at pressures of up to 1 atm (100 kPa) while the mass analyser of the same system will be operated four or five orders of magnitude lower. Most commercial mass spectrometers have automated pumping chains which are controlled by the data system. These will maintain the appropriate vacuum in the system by a series of interlocks and valves, thus preventing the operator from destroying the vacuum by mistake and causing extensive damage to the system; they also enable parts of the spectrometer to be vented to air in order to introduce a sample or change the type of ion source without losing the main vacuum in the system.

19.3   Sample inlet systems

Early mass spectrometers often analysed either gases and vapours or relatively volatile liquids. They were introduced via an inlet system containing a number of vacuum taps and reservoirs, which may have been as large as 5 L capacity, and connected to the mass spectrometer by a 'molecular leak' which allowed only small amounts of material in to the source in a given time, thus maintaining the required vacuum. The molecular leak was either a very small pinhole in a piece of gold foil, or a fritted glass disc. The inlet system could often be heated and was made almost entirely from glass to reduce chemical reaction

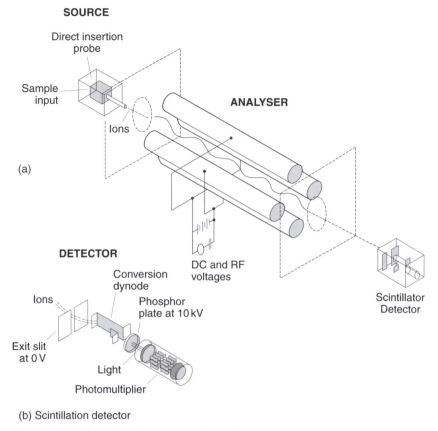

SOURCE

Direct insertion probe

Sample input

Ions

(a)

ANALYSER

DETECTOR

Conversion dynode

DC and RF voltages

Ions

Phosphor plate at 10 kV

Exit slit at 0 V

Light

Photomultiplier

Scintillator Detector

(b) Scintillation detector

Figure 19.3 A quadrupole mass spectrometer

and adsorption effects which could occur on metal surfaces. Such all-glass heated-inlet systems (AGHIS) are not suitable for introducing solid material, unless it has a high vapour pressure, or for introducing mixtures. Currently they are not widely used, except for special applications such as isotope ratio experiments with $^{12}C/C^{13}$ (as CO_2). Where liquids need to be introduced, they may be admitted using a hypodermic syringe via a simpler device called a septum inlet. This inlet is much smaller and easier to pump free of sample than the AGHIS. Some calibrating materials may also be admitted this way.

Solid or liquid material can be admitted to the ion source using a direct insertion probe. This probe has at its tip a small sample container, often in the form of a removable silica capillary tube, in which a few micrograms of sample can be placed or deposited. The probe is inserted through a vacuum lock to within a few millimetres of the ion beam, where precise and controllable internal heating can be used to gently evaporate the sample with minimum decomposition. Some probes can be heated to temperatures in excess of 800 °C and can be programmed to give some separating capability for simple mixtures.

A modification of the direct insertion probe is the desorption chemical ionisation (DCI) probe, which has a probe tip in the form of a platinum filament that may be heated very rapidly to high temperatures (incandescence). As the substance is heated it evaporates and partially ionises; it has been demonstrated that acceptable mass spectra may be obtained from relatively high-mass, polar or thermally labile materials which are difficult to ionise by other means.[2] Since the filament is heated to red heat, it is self-cleaning and at least one commercial autosampling system is available for unattended operation of up to 100 samples. This type of probe has useful application in the polymer and pharmaceutical industries.

The range and complexity of sample types examined by mass spectrometry is continually increasing and this places severe constraints on the inlet system. For large involatile and thermally labile materials, a number of novel ionisation sources have been developed such as MALDI-TOF, APCI, Frit-FAB and LSIMS, which require more specialised sample preparation and handling; they are described later on. Complex mixtures often require some sort of preseparation, and it is becoming increasingly common to find the mass spectrometer coupled to other techniques such as GC or HPLC; these hyphenated systems have special requirements for input of sample, just as GC and HPLC on their own, but because small controlled amounts of material can be introduced to the system, it is possible to quantify the separated components.

19.4   The ion source

At present there exist more than a dozen different commercial ionisation methods for mass spectrometry (Table 19.3); but before examining some in detail, consider the nature of an

Table 19.3 *Ion sources used in mass spectrometry*

Name	Abbreviation	Phase	Fragmentation[a]
Electron impact	EI	gas	M^+ and fragments
Chemical ionisation	CI	gas	$(M + 1)^+$ and some fragments
Field ionisation	FI	gas	M^+ or $(M + 1)^+$
Field desorption	FD	solid/liquid	M^+ or $(M + 1)^+$
Desorption chemical ionisation	DCI	solid/liquid	$(M + 1)^+$
Fast atom bombardment	FAB	liquid	$(M - 1)^+$, $(M + 1)^+$, $(M + A)^+$
Plasma desorption	PD	liquid	$(M + 1)^+$, $(M + 2)^{2+}$, $(M + 3)^{3+}$
Laser desorption	MALDI	liquid	$(M + 1)^+$, $(M + 2)^{2+}$, $(M + 3)^{3+}$
Secondary ion mass Spectrometry	SIMS	solid	element ion or M^+
Ion microprobe	IMP	solid	element ion
Spark source	SS	solid	element ion
Inductively coupled Plasma	ICP	liquid	element ion plus multiple charged ions
Atmospheric pressure Ionisation	API	gas/liquid	$(M + 1)^+$ and $(M + A)^+$
Thermospray	TSP	liquid	$(M + 1)^+$, $(M + A)^+$
Plasmaspray	PS	liquid	$(M + 1)^+$ plus fragments
Electrospray	ES	liquid	$(M + 1)^+$ to $(M + NH)^{n+}$ where $n = 1$–60
Particle beam interface	PBI	liquid	EI or CI spectra

[a] $(M + 1)^+$ is a singly protonated, singly charged ion, i.e. MH^+; A = adduct.

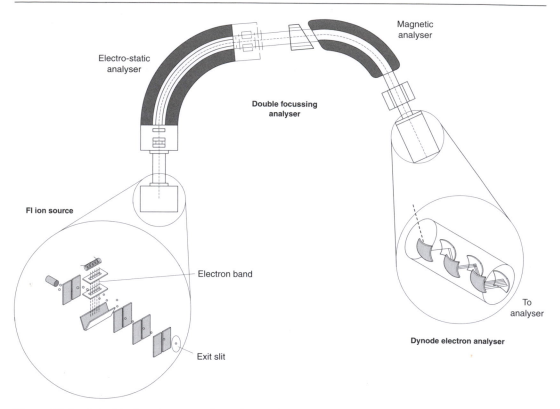

Figure 19.4   A double-focusing magnetic sector spectrometer

ion source. In all mass spectrometers the ion source is required only to provide gas phase ions derived from the starting material. The problem with this simple definition is that the starting material may be a gas, liquid or solid, it may be highly polar or have a high molar mass, making it hard to exist in the gas phase, or it may be a very labile material which tends to fragment into smaller components rather than simply ionise. Each of these problems can be solved to some extent by appropriate choice of ion source, although this choice may be limited by what is available on a particular instrument.

19.4.1   Electron impact ionisation sources

Electron impact ionisation (EI) remains probably the most widely used source, even though it was the earliest to come into common use; it was developed in the 1920s. The source is essentially an ion gun, similar to an ordinary television gun. A heated tungsten or rhenium filament emits electrons which are accelerated across a small chamber by a voltage imposed between the filament and an anode (Figure 19.4). The sample, which must be in the gas phase, is admitted to this chamber at right angles to the electron beam, and a number of interactions are possible:

$$M + e \rightarrow M^{\cdot+} + 2e \qquad M + e \rightarrow M^{\cdot-} \qquad M + e \rightarrow M^*$$

These processes have been written in this order because this is the observed probability of interaction; loss of an electron to form a positive ion is about 100 times more likely than electron capture forming a negative ion, and simple excitation of the particle is even less favoured. Even for the first process only a small proportion of the sample is ionised in the ion beam when the electron potential, hence the energy, is optimised; this gives about 1% ionisation for a beam energy of 70 eV (produced by applying 70 V between the filament and the anode). The ionisation efficiency drops markedly below this value, although there may be very good reasons for operating the system at potentials as low as 10 eV (see below). Typical ion beam currents at 70 eV are of the order of 10^{-10} to 10^{-15} A.

Assuming for the moment that only positive ions are formed and assuming they are stable, they may be directed out of the cavity or ion chamber by applying a potential between the repeller and a slit, thus producing a beam of positive ions which enter the mass analyser. Although this sounds a very simple process, which it is, the construction of a working ion source is very much more complex than this, often containing more than a hundred small components providing magnetic or electrostatic fields in various regions of the source. This complexity is required to focus the beam of ions into the analyser. In magnetic sector instruments it is also needed to give them sufficient kinetic energy to pass through the analyser; potentials in excess of 10 kV may be used to provide the ions with sufficient velocity.

Complexity also means that when the source becomes contaminated, as sample material is passed into it, clean-up is a delicate and time-consuming process. This is gradually becoming less of a problem in more modern instruments where the manufacturer mounts the whole ion source on a removable flange which can be quickly swapped for a clean source; in fact, on one instrument the source is described as disposable! Other reasons for maintaining a high vacuum in this ion source should now be obvious; at atmospheric pressure the filament would burn out very quickly and there is an increased possibility of arcing between closely spaced components with high applied potentials.

Unfortunately, it is not the mechanical or electrical problems within the source which cause the greatest difficulty, but the chemical reactions which may occur. In an EI ion source operated at 70 eV, approximately 1% of the sample – generally a molecular species – is ionised by the beam. However, before it can pass from the source, through the analyser and on to the detector, a number of reactions usually occur. The primary product or molecular ion normally has an excess of energy, and this must be lost via a number of fragmentation processes:

$$\begin{array}{ccccc}
& F_2^{\cdot+} & + & N_2 & \longrightarrow \text{Further fragmentation} \\
& \text{radical cation} & & \text{neutral molecule} & \\
M^{\cdot+} & & & & \\
\text{molecular ion} & & & & \\
\text{(radical cation)} & & & & \\
& F_1^{+} & + & N_1^{\cdot} & \\
& \text{cation} & & \text{neutral radical} & \\
& & & \text{Further fragmentation} &
\end{array}$$

Note that both the charge and unpaired electron are maintained in one or both fragments. This fragmentation enables mass spectrometry to provide a wealth of data on individual species, since the fragmentation pattern is usually diagnostic of the sample structure. However, apart from the resulting complexity of the mass spectrum, this fragmentation also leads to other problems. Many molecules fragment so readily that few if any molecular ions, which are very important diagnostically, are detected. Also, the pattern of fragmented ions depends on the ionising energy of the electron beam, the temperature in the ion source, the pressure and even the geometry of the rest of the system. The first of these problems can sometimes be reduced by operating the electron beam at less than 70 eV down to as low as 10 eV; fragmentation is greatly reduced (the energy of most organic bonds is in the range 14–20 eV), giving a simplified mass spectrum, but the sensitivity of determination also drops drastically at low voltages.

The second problem is also an advantage as valuable kinetic information can be deduced from the fragmentation pattern obtained under different experimental conditions. However, for routine analytical determinations, the fragmentation pattern obtained at $100 \, \mu Pa$ (low enough so that molecular collisions are rare) and 70 eV is generally sufficiently similar on any spectrometer to enable comparisons to be made between a sample spectrum and a reference spectrum, often using computerised databases. This assumes the molecule has not decomposed before ionisation as it is made to volatilise in the inlet system.

Most EI spectra are recorded from positive ions; but by reversing the voltages in the ion source, it is possible to record negative ion spectra. However, for most compounds, they will be much less intense, thereby reducing one of the primary advantages of EI – its high sensitivity compared to other ionisation techniques. Even so, for electrophilic species such as halogenated compounds, negative ion spectra can be used to great effect.

19.4.2 Chemical ionisation sources

It is often desirable to reduce the amount of fragmentation that occurs in the ion source, simplifying the interpretation of data for large molecules. This can be accomplished in many cases by using chemical ionisation (CI), in which molecular collisions in the ion cell are actually encouraged rather than minimised. The most common construction for a CI source is to modify an EI chamber so it can maintain a pressure of 0.1–1.0 torr (10–100 Pa) within the source while still operating at low pressures in the rest of the system. Often this is done in such a way that by actuating a valve the source can be switched from EI to CI very easily. Alternating CI/EI (ACE) is then possible. In the CI mode a continuous supply of a reagent gas is fed into the source, where the following reactions can occur using methane as reagent gas:

$$CH_4 + e \rightarrow CH_4^{\cdot +} + 2e$$

$$CH_4^{\cdot +} \rightarrow CH_3^+ + H^{\cdot} \quad \text{or} \quad CH_4^{\cdot +} \rightarrow CH_2^{\cdot +} + H_2$$

This is the normal ionisation process described above, except that an electron energy of 200–500 eV is used. This primary ionisation is immediately followed by a second process in which the ion collides with another molecule of reagent gas:

$$CH_4^{\cdot +} + CH_4 \rightarrow CH_5^+ + CH_3^{\cdot} \quad \text{or} \quad CH_3^+ + CH_4 \rightarrow C_2H_5^+ + H_2$$

producing a stable 'ion plasma' in the source, which can then react with sample molecules (MH) by either proton transfer or occasionally hydride transfer. Here are two reactions for proton transfer:

$$CH_5^+ + M \rightarrow MH^+ + CH_4 \quad \text{or} \quad C_2H_5^+ + M \rightarrow MH^+ + C_2H_4$$

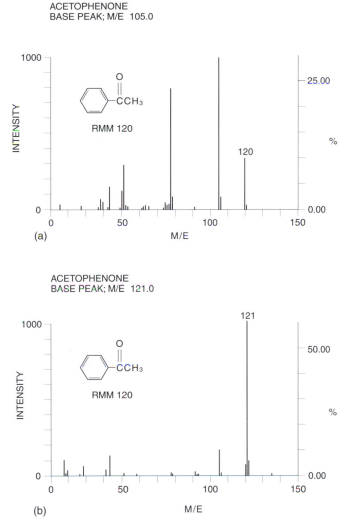

ACETOPHENONE
BASE PEAK; M/E 105.0

(a)

ACETOPHENONE
BASE PEAK; M/E 121.0

(b)

Figure 19.5   Acetophenone: (a) EI spectrum; (b) CI spectrum using methane

Aliphatic hydrocarbons can also react via hydride transfer:

$$C_2H_5^+ + M \rightarrow (M - H)^+ + C_2H_6$$

Note that these processes give rise to either an $(M + 1)^+$ ion or an $(M - 1)^+$ ion and it is these pseudomolecular ions which are observed rather than the true molecular ion. Addition can also occur to give adduct ions with values $(M + 29)$ and $(M + 41)$ for addition of C_2H_5 and C_3H_5. CI is a softer ionisation process than direct EI and gives a simpler and often more readily interpreted spectrum, as can be seen in Figure 19.5 for the spectra of acetophenone.

Although methane was used as the reagent gas in the above examples, other gases such as ammonia or isobutane are equally effective and may be preferred for some samples. The efficiency of energy transfer from the conjugate base of the reagent gas to the sample depends on the relative proton affinity (acidity) of the sample to the conjugate base. Ammonia is best for basic molecules such as amines and amides but does not give CI spectra for hydrocarbons and other non-polar molecules; it tends to give high intensities of adduct ions $(M + NH_4)^+$. At the high pressures used in the CI cell, electron capture becomes as likely as electron abstraction; negative ions can therefore be produced in high yields, making negative ion chemical ionisation (NCI) an attractive technique for highly electronegative molecules such as pesticides.

Typical reagent gases are hexafluorobenzene, dichloromethane, nitrous oxide and nitrous oxide/methane, giving the negative ions F^-, Cl^-, O^- and OH^-. These gases are introduced into the ion source at pressures of at least 1 torr (100 Pa) where reactions such as proton abstraction readily occur

$$MH + OH^- \rightarrow M^- + H_2O$$

$\quad$ sample $\qquad\qquad$ negative

$\quad$ molecule $\qquad\qquad$ ion

Some commercial spectrometers can perform both positive and negative CI very rapidly, and both techniques can be applied to the same sample even when it is eluted from a chromatographic column. For favourable samples, positive ion CI can give sensitivities comparable to EI, whereas negative ion CI can be an order of magnitude more sensitive than EI. Each of the ionisation methods described so far requires the sample to be in the gas or vapour phase **before** ionisation. This is a severe limitation if large, polar, labile molecules are to be studied, as found in the pharmaceutical and life sciences. The following methods have the common feature that they are all capable of producing ions from samples in the condensed phase (either liquid or solid); we begin with fast atom bombardment.

19.4.3 Fast atom bombardment

Fast atom bombardment (FAB) was developed independently from mass spectrometry, but has become a very reliable technique for species with high molecular weight. A beam of fast atoms, usually xenon or argon, is produced in a saddle gun or atom gun then directed onto a solid target. A beam of argon ions is first produced by a hot-wire ion source in an atmosphere of argon gas; the ions are then accelerated by a field into a second chamber containing neutral gas atoms at about 10^{-5} torr (1 mPa); collision occurs between the ions and (thermal) atoms and some atoms acquire extra energies of the order of 20–30 keV, becoming fast atoms. The ions are removed by a charged deflector plate, leaving a beam of high-energy atoms which can be directed onto a target; the target holds the sample dissolved in a non-volatile liquid matrix such as glycerol or carbowax.

The beam of atoms produces intense localised heating and some energy is transferred from the atom beam to the sample molecules, producing ionisation and thus ions in the gas phase. The matrix liquid plays an important role in this process, probably both reducing the lattice energy of the sample and allowing transfer of energy from the beam to the sample, as well as providing a mobile surface so that fresh sample is continuously exposed to the beam. Both positive and negative ions are sputtered from the surface in this desorption process, and all can be directed into the analyser of the spectrometer. Mass spectra of molecules in excess of 10 000 u have been obtained by this method.[3] Generally the spectra are relatively simple, containing high intensities of molecular ions or pseudomolecular ions resulting from proton transfer $(M + 1)^+$, hydride transfer $(M - 1)^+$ or adduct species $(M + G)^+$ where G is glycerol, but also giving some fragmentation (Figure 19.6). Some

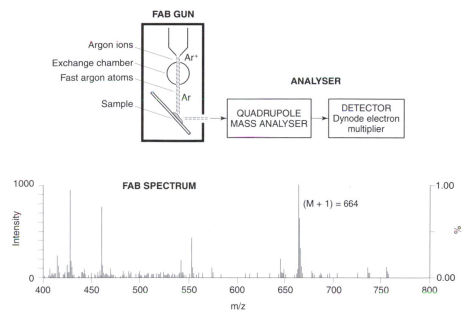

Figure 19.6 A FAB source and the FAB spectrum for NAD ($C_{21}H_{27}N_7O_{14}P_2$)

spectra also show a cationised species where a metal atom, often sodium, has added to the ion, giving $(M + 23)^+$.

A disadvantage is that spectra from the matrix are also seen, but they are usually at much lower mass than the sample. Fast atom guns are now readily available on most spectrometers and can even be fitted to some older spectrometers, enabling determinations to be made on high-mass species. Dynamic versions of this ion source are also available which can be used as continuous or on-line sources coupled to HPLC (Section 19.11.2).

19.4.4 Ion guns

Similar to FAB and also 'borrowed' from the physics laboratory, ion guns bombard the sample with a beam of ions rather than a beam of atoms. When used on solid inorganic targets, this is generally known as secondary ion mass spectrometry (SIMS); it has been a powerful tool in surface studies for a number of years. Recently, however, the ion gun has been directed at high-mass organic species dissolved in a liquid matrix, generally glycerol; a number of titles have been suggested for this technique, although the acronym LSIMS (liquid secondary ion mass spectrometry) appears the most logical. Some manufactures do not distinguish between a beam of atoms or a beam of ions, calling both FAB. Although a number of ions have been used, including the inert gases such as argon, highly electropositive caesium ions with energies of 30–40 keV seem to be the most promising for organic mass spectrometry. SIMS spectra are similar to FAB spectra but they are claimed to have higher sensitivity.

19.4.5 Laser desorption mass spectrometry

Similar results may be obtained by directing a series of pulses from a high-power laser onto a sample which has been mixed with a suitable matrix and allowed to dry. Since the laser pulse only lasts a few milliseconds, a time-of-flight (TOF) system is normally used; a series of results are obtained and stored, then averaged to obtain the required spectrum. The system is rapid (5 min is a typical time from sample loading to final printout), sensitive (down to 1 pmol) and may be fully automated with robotic sample loading and removal. Since the analyser is a TOF analyser, there is theoretically no limit to the upper mass that can be determined, and claims have been made for determinations up to 200 000 u,[4] but more typical measurements tend to be in the low tens of thousands u. First reported in 1988, laser desorption systems seem to have great promise and they are beginning to revolutionise the determination of proteins, peptides and other species with a high molecular weight.

Sample preparation is not critical; normally a simple organic matrix is prepared by mixing 0.1 to 1 μL of sample with 0.5 μL of solution and then allowing it to dry. For peptide analysis, 2,5-dihydroxybenzoic acid (DHB) and α-cyano-4-hydroxy cinnamic acid (4-HCCA) are often used since they provide good ionisation of the large sample molecule but do not produce any ions above about 400 or form adducts with the biomolecular ion. The water is then evaporated and the mixture irradiated under vacuum with light from a pulsed laser. For most biopolymers this technique will give intense spectra from as little as a few picomoles of sample. The mass spectra tend to be simple, with either the singly charged molecular ion or a simple alkali metal adduct as the predominant ion, plus some lines from the matrix, but matrix lines can easily be filtered out as they are of much lower mass. However, it appears that the matrix in which the sample is mixed plays a critical role in the ionisation step, allowing the transfer of energy from the laser to the sample, so the acronym MALDI or MALDI-TOF is widely used, where the initials stand for matrix-assisted laser desorption ionisation.

Because of the relative simplicity and ease of obtaining a spectrum from say a protein solution, MALDI-TOF is already beginning to rival more conventional methods such as electrophoresis for molecular weight determinations of these species. Several commercial systems using MALDI-TOF are already available, and at least one semi-automatic system removes any ions produced by the matrix through including a pulsed electrostatic filter, synchronised with the laser, before the TOF. It is claimed that, even when using only a few picomoles of sample, a mass accuracy of 0.01% may be obtained using an internal calibration. It is also possible to interface MALDI-TOF to HPLC for the direct determination of mixtures. A recent interesting application of this type of system to non-biological samples was the first observation of C_{60} or buckminsterfullerene, discovered by irradiating graphite in a helium atmosphere using a pulsed Nd: YAG laser.[5]

19.4.6   Atmospheric pressure ionisation (API)

The sources described so far have all been contained within a reduced pressure zone (high vacuum) of the mass spectrometer, so any ions can be directed immediately to the analyser section, also at reduced pressure. But it has long been known that if ionisation can be induced at atmospheric pressure, then ionisation efficiency can be increased by a factor of up to 10^4. Two design problems have to be solved and they are by no means straightforward: how to produce the ions and how to introduce them into the analyser, which typically operates at 10^{-4} to 10^{-8} torr (10 mPa to 1 μPa). But the potential advantages in sensitivity and the ability to interface directly with an HPLC column have spurred the development of several systems.

Since the majority of the applications involve hyphenated techniques, such as ICP-MS and HPLC-MS, detailed explanations will be given in the relevant sections, but in most cases the sample solution is converted into an aerosol (mist) via pneumatic nebulisation at a capillary and simultaneously molecules are ionised by application of either heat or a high potential. The resulting ion cloud is then introduced into the reduced pressure zone of the mass spectrometer via a very small opening or series of openings (sampling or skimmer cones) with high-capacity pumps removing the excess material.

19.5 Mass analysers

A mass analyser separates the ions formed in the source into their different mass/charge ratios; since most ions carry a single charge, this will represent the 'mass spectrum'. Different mass spectrometers place different constraints upon the analyser, with some requiring separation only of unit masses up to about 500 u and others requiring separation of ions differing by only a few parts per million. So different analysers can be compared by defining a quantity called the resolving power R. The most common definition is

$$R = m/\Delta m$$

where Δm is the mass difference between two adjacent peaks that are just resolved and m is the nominal mean mass of the two peaks being considered. The phrase 'just resolved' also needs to be defined, and although there is no strict convention, most spectroscopists use the 10% valley criterion, in which peaks are considered to be resolved if the height of the valley between them is less than 10% (Figure 19.7).

A spectrometer with resolving power of 1000 (10% valley) could make the following distinctions:

At masses of ~1000	peaks separated by 1 unit
	e.g. 1000 and 1001
At masses of ~100	peaks separated by 0.1 unit
	e.g. 99.9 and 100

The value 1000 was chosen as this is typical of the resolving power for many mass spectrometers, but the value can vary widely with high-resolution systems capable of resolutions up to 125 000. The resolving power is independent of mass for these systems.

Note that high resolution and high mass capability are not directly related; for example, to distinguish between CO^+, N_2^+ and $C_2H_4^+$ which all have nominal mass of 28 u but actual masses of respectively 27.9949 u, 28.0062 u and 28.0313 u, this requires a resolution of about 3000. Although resolution is an important parameter, it is not the only one, and

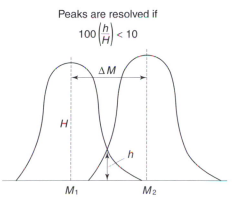

Peaks are resolved if

$$100\left(\frac{h}{H}\right) < 10$$

Figure 19.7   Resolution: the 10% valley definition

sensitivity, highest measurable mass and speed of scan should all be considered when assessing an analyser for its task.

19.5.1 Magnetic sector analysers (single-focusing)

The earliest experiments in mass spectrometry used a combination of magnetic and electric fields to separate one ion from another, and for many purposes these analysers are still the most suitable. If the ions produced in the ion source are made to pass through a series of slits which have increasing potentials applied to them, the ion accelerates and acquires an energy related to the potential gradient, the mass and charge of the ion:

$$\text{kinetic energy} = \tfrac{1}{2}mv^2 = zeV \tag{19.1}$$

where V is the potential gradient, v is the ion velocity, e is the charge on the ion ($e = 1.60 \times 10^{-19}$ C) and z is the number of charges.

If these ions then pass between the poles of an electromagnet with the field at right angles to the direction of motion, they are deflected in to a circular path which has a radius r proportional to the momentum of the ion and inversely proportional to the field strength B:

$$r = \frac{mv}{zeB} \tag{19.2}$$

Eliminating v from equations (19.1) and (19.2) leads to

$$\frac{m}{z} = \frac{B^2 r^2 e}{2V} \tag{19.3}$$

All parameters should be expressed in SI units: m (kg), r (m), B (T), ze (C) and V (V). But normally m is expressed in unified mass units (u) or daltons (1 u = 1.66×10^{-27} kg), r in centimetres and ze is the number of elementary electronic charges; B remains in teslas (1 tesla = 10 000 gauss).

Equation (19.3) shows that ions can be separated according to mass (m/z) by varying one of the three variables B, V or r, while holding the other two constant. Most modern magnetic sector analysers hold r and V constant and vary B, the magnetic field, by altering the current in the electromagnet. This is because the sensitivity and resolution decrease with a decrease in accelerating voltage. This form of scanning gives rise to a non-linear mass scale since m/z is proportional to B^2. Although, in theory, both small magnetic fields and acceleration voltages could be used, in practice it is necessary to use high voltages (8–15 kV) hence high magnetic fields, generated by large and expensive laminated-core electromagnets.

A simple calculation based on equation (19.3) shows the typical field strength in a modern magnetic sector instrument. For $m/z = 500$ u, magnetic radius $r = 30$ cm and accelerating voltage $V = 5$ kV

$$\frac{B^2 (30 \times 0.01)^2 (1.6 \times 10^{-19})}{2 \times 5000} = 500 \times 1.66 \times 10^{-27}$$

$$B^2 = \frac{500 \times 5000}{900} \times \frac{2(1.66 \times 10^{-27})}{(0.01)^2 (1.6 \times 10^{-19})}$$

$$= \frac{500 \times 5000}{900} \times 2.075 \times 10^{-4}$$

$$= 0.576$$

$$B = 0.76 \text{ T}$$

The resolution of the best of these single-focusing systems is limited to about 2000. This is because, when they are first produced in the source, the ions do not all have the same velocity. They are produced with a range of energies and this gives them a range of velocities. So the beam of ions leaving the analyser becomes spread out, limiting the resolution. This broadening is smallest when the kinetic component of velocity is small compared to the velocity imparted by the accelerating voltage; that is why large acceleration voltages are used, but there is still a limit on resolution.

19.5.2 Double-focusing analysers

The variable initial energy of ions can mainly be overcome by using two focusing elements in series; the so-called double-focusing mass spectrometer. In this configuration the ions are accelerated from the source into an electrostatic analyser (ESA) which consists of two curved metal plates with a d.c. potential applied across them, effectively a capacitor. This causes the ions to experience a force F acting at right angles to their direction of flight.

The magnitude of the force is given by

$$F = zeE$$

This is balanced by an opposite centrifugal force generated as the ions acquire a circular path, so

$$zeE = mv^2/R$$

i.e. the centrifugal force acting on a mass m moving in a circle of radius R.

Since $\frac{1}{2}mv^2 = zeV$, these equation can be rearranged to give

$$R = 2V/E$$

This shows that all ions of a given energy accelerated through a potential V and passing through an electric field E follow paths with the same radius of curvature R, irrespective of their m/z values; this implies that ions passing through a field E will follow paths of different radii of curvature depending upon their kinetic energies, i.e. velocity focusing. An energy filter is produced by placing a slit at the exit of the electrostatic analyser that allows only one ion beam to leave. The ions then entering the second, magnetic analyser, can be separated according to their momentum only. There are several different ways of coupling the ESA and magnetic sectors. A common configuration is known as the Nier–Johnson geometry (Figure 19.4); it usually has a 90° ESA followed by a 90° magnetic sector, but sectors of 30°, 60° and 90° are common. One problem with this system is that ions are only brought to focus through the system at one point and the detector must be positioned accurately at this focus.

Reverse-geometry instruments, where the magnetic sector precedes the ESA, do have certain advantages, especially when studying metastable ions; several systems are available. The great advantage of double-focusing analysers is the very high resolution that may be obtained (125 000) in the most sophisticated spectrometers. But this resolution is only obtained by using large and expensive electrostatic analysers followed by magnetic sectors; and they often require additional focusing devices or ion lenses based on the quadrupole filter.

19.5.3 Quadrupole mass filters

The quadrupole filter (Figure 19.3) is much smaller and more rugged than the magnetic sector instrument; it is also easier and cheaper to build. The instrument contains four parallel metal rods, arranged in two pairs, across which are applied a combination of d.c. and a.c. voltages. Ions produced in the source are passed down the gap between the four rods to a detector at the far end. In this type of analyser the ions do not need to pass through a high field to gain acceleration and normally a potential of 5–15 V is sufficient to propel the ions into the analyser; it is capable of accepting a fairly wide range of initial ion energies (velocities) so a much wider, usually circular, slit may be placed between the source and the analyser, which gives the quadrupole higher intrinsic sensitivity.

Opposite rods are connected electrically, one pair to the positive side of a variable d.c. source and the other pair to the negative side; an r.f. generator supplies an a.c. potential $(+V \cos t)$ to one pair and the same frequency but 180° out of phase (i.e. $-V \cos t$) to the other pair. This combination of potentials causes the ions to take an erratic path down the centre of the quadrupole. In the plane defined by the positive pair of rods, light ions will be deflected and collide with the rods, but heavier ions will be deflected less and, with the help of the a.c. component, they will tend to pass straight through, giving a high-mass filter. In the plane defined by the negative pair of rods, all positive ions will tend to be attracted to the rods and neutralised, but the a.c. component will tend to prevent this preferentially for the lighter ions than for the heavier ions, thus giving a low-mass filter. The combined effect is a narrowband mass filter and by controlling the ratio of V_{dc}/V_{rf}, a stable trajectory is established for ions of one m/z value only; the others collide with the rods and become neutralised. (The maximum resolution is obtained when $V_{dc}/V_{rf} = 0.168$.)

By ramping the combined d.c. and r.f. voltages while keeping the ratio constant, a complete mass spectrum may be scanned. A great deal of research has gone into the design of modern quadrupoles. The resolution depends on the length of the rod, the degree of parallelism and the frequency of the r.f. component, whereas the sensitivity is increased by increasing the rod diameters. However, a number of factors tend to work in opposition, and although decreasing the rod diameter does decrease the sensitivity, it also increases the mass range. Typical high-quality commercial systems have rods of the order of 20 cm long and 5 mm diameter with r.f. frequencies of about 10^8 Hz, giving a maximum mass of the order of 1500 u and with unit resolution, i.e. resolution of 1500 at mass 1500 u.

Newer designs have a number of refinements. The rods are deposited onto a rigid ceramic matrix, thereby avoiding any dimensional changes even at elevated temperatures. There is also a short quadrupole prefilter supplied only with d.c. voltages prior to the main quadrupole; this protects the analyser from contamination and maintains optimum performance.

Although both resolution and maximum mass range appear rather restricted compared to a double-focusing magnetic sector instrument, the quadrupole does have one great advantage apart from its much smaller size and cost – its ability to scan spectra very rapidly. Since the scan is controlled only by altering applied potentials to the rods and the scan is a linear function of potential, an entire mass spectrum can be scanned (very accurately and reproducibly) in a few milliseconds. A magnetic sector instrument is normally limited to about 0.1 s decade by the eddy currents produced in the electromagnet as the current is increased. (A decade is 1–10 or 10–100 mass units, and this way of expressing scanning speeds in terms of time per decade stems from the non-linear scan obtained by magnetic scanning.) As shown in Section 19.10, this rapid scanning ability makes the quadrupole mass filter an attractive choice for coupling to chromatographic front ends. The quadrupole is also ideally suited for negative ion mass spectrometry since the filter does not depend on the polarity of the ions passing through, and it is even possible to perform alternate positive/negative ion scans in a very short time interval, even on a single chromatographic peak from a combined GC-MS.

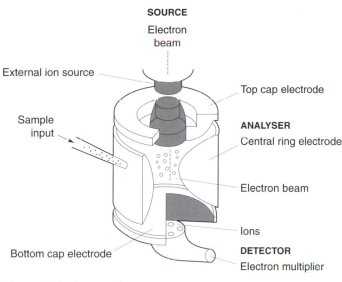

Figure 19.8 Ion trap detector

19.5.4 Ion trap detectors

A closely related analyser is the ion trap detector (ITD),[6] in which a variable radiofrequency voltage is used to trap ions in the cavity formed between a central toroidal electrode
and two grounded endcap electrodes (Figure 19.8). This mechanically simple system combines the functions of the ion source and the analyser, since in operation electrons from a
conventional heated filament are pulsed into the ion trap by means of a gate electrode, the
electrons interact with sample molecules to produce fragments identical to normal EI but
they are trapped in the cavity region by the r.f. field until the voltage of this field is scanned
upwards. Ions are then ejected from the cavity according to their m/z values, when they are
detected as a mass spectrum. The ion trap has a number of features that make it ideal to
couple to a gas chromatograph, and currently it is mainly used as a detector for GC-MS,
although HPLC-MS or LCQ systems are becoming more widely available, especially when
coupled to an API source.

The optimum performance of the device is obtained when the pressure in the cavity is
of the order of 10^{-2} to 10^{-3} torr (1 Pa to 100 mPa) for a low-mass gas, either hydrogen or
helium, because operation at this pressure damps the motion of ions in the cavity and prevents
them from colliding with the walls. An operating frequency of 1 MHz at about 20 V squeezes
the ions in the xy plane so they can only move in the z-direction; as the voltage is increased
up to about 15 kV, ions of increasing mass are ejected through a hole in one endcap. This
voltage control of scan can be achieved very quickly (even faster than in a quadrupole) and
normally computer control of the system provides a very large number of microscans,
which are averaged to produce the final spectrum. These microscans are synchronised with
the gate voltage in such a way that the optimum number of ions are trapped to produce a
spectrum, but even for very small sample sizes this process takes less than 30 ms per scan
and for large amounts of material in the cavity this can be as little as 75 ms per scan. This
results in very high sensitivity, and systems are designed to give unit resolution in the range
50–650 u for samples as small as 1 pg and at lower cost than other types of analyser.

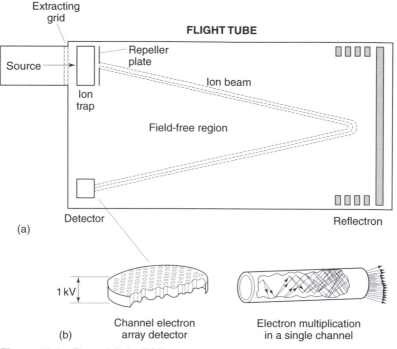

Figure 19.9 Time-of-flight (TOF) mass spectrometer

Although normally used in the EI mode, it is also possible to perform chemical ionisation experiments at high sensitivity, and the change from one mode to the other can be effected very rapidly, allowing alternating EI/CI measurements even on rapidly eluting GC peaks.

19.5.5 Time-of-flight analysers

A long-established mass analyser is the time-of-flight (TOF) analyser, another very simple mechanical device (Figure 19.9) controlled by sophisticated electronics. A pulse of ions is produced, originally by a californium-252 plasma desorption, but now more generally either by one of the beam techniques, particularly a laser, or by an API source and then accelerated by a potential of 1–10 kV between repeller and extractor plates, into an evacuated drift tube 15–100 cm long. This tube is not in either a magnetic or electric field, and is sometimes known as a field-free zone. Since all the ions initially pass through the same potential gradient, they acquire the same kinetic energy zV, but they travel with velocities inversely proportional to the square root of their masses, and thus reach the far end of the drift tube, housing the detector, at different times. Typical flight times are of the order of 1–30 ms depending upon mass, and data is acquired and processed for a large number of pulses to produce the final mass spectrum.

The detected signal is of mass versus time, and provided the time taken for two or more masses of known value can be computed, the entire mass spectrum may be easily calibrated. The ability to calibrate the spectrum easily up to high masses is a great advantage because TOF is the only type of analyser which is not limited in its mass range and the major application of the technique is for study of large biomolecules where m/z values in excess of 300 000 u have been measured using MALDI-TOF (Section 19.4.5). However, the simple system described here has no energy focusing of the primary ion beam, so the resolution is restricted to less than 1000. This has the unexpected effect of not resolving isotopic clusters for large molecules, so the observed ions are not the nominal mass ions (isotopic mass) normally seen in mass spectrometry but the isotopically averaged or chemical masses. The two masses differ by about 1 u for every 1500 u.

A recent development known as time-lag focusing allows an initial separation of ions according to velocity. The ions produced are first allowed to expand into a field-free region between the repeller and the extracting grid. After a short delay (hundreds of nanoseconds to several microseconds) an extraction pulse is then applied to the repeller to extract the ions into the flight tube. Appropriate adjustment of the delay time and the pulse voltage produces energy focusing of the ions for increased resolution, similar to the double-focusing magnetic sector system (Section 19.5.2). If this is combined with an ion mirror or reflectron in the flight tube, which effectively doubles the length of the flight tube and also gives some velocity focusing, then unit resolution for masses in excess of 10 000 u may be observed with a mass accuracy of better than 500 ppm.

19.6   Detectors

Five main types of detector have been used in mass spectrometry: photographic plates, Faraday cup, electron multipliers, channel electron multipliers and scintillation detectors. The photographic plate, although important historically, is now seldom used even though it is capable of higher resolution and speed than electronic devices (it can detect ions of all masses simultaneously, provided a reverse-geometry analyser is used). Equally the Faraday cup detector, which is simply a metal cup into which all the ions are directed, is only used for specialist applications. The disadvantage of this system is that the currents induced by the average ion beam in the circuit connected to the cup are very small, leading to low sensitivity, but the signal produced is very stable and reproducible. The Faraday cup is almost invariably used on spectrometers where quantitative data is important (Section 19.9). The majority of mass spectrometers in general use are fitted with an electron multiplier or scintillation detector; both devices are very versatile and sensitive. Two types of electron multiplier are in common use. The discrete-dynode device is very similar to the photomultipliers used for detection of electromagnetic radiation in conventional spectrometers. The ion beam impinges onto a Cu/Be surface and ejects electrons which are accelerated by a high potential onto the second plate or dynode, where they cause a larger number of secondary electrons to be emitted. By having a series of dynodes (up to 20) arranged in a venetian blind pattern (Figure 19.4), each with a higher potential applied to it, a cascade process occurs giving up to 10^7 electrons per ion at the final electrode.

The continuous dynode electron multiplier works in an entirely analogous way and has similar gain characteristics, but this time the individual electrodes are replaced by a curved horn-shaped glass tube made from a high lead glass doped with metal oxides (Figure 19.8). A high potential is again applied, which causes the cascade process to occur each time the electron hits the inner surface of the tube. Both of these systems require the ions to reach them with high velocity and thus high energy. This is the case if the ions have been separated by a magnetic analyser, but if a quadrupole filter or other low-energy analyser is used, then the ion beam is often directed onto a conversion dynode at a potential of up to 20 kV prior to entry into the detector. This provides the necessary energy to displace sufficient electrons from the first stage of the detector. These detectors are reliable, rugged and sensitive devices, but their performance and lifetime are affected by the vacuum maintained in the main spectrometer and they may become contaminated.

An attractive alternative is to use a scintillation or Daly detector (Figure 19.3) very similar to the detectors used in radiochemical applications. This consists of a conventional end-window photomultiplier on which is mounted a thin disc of crystalline phosphor maintained at a potential of +10 kV. When ions strike the phosphor, photons of light are emitted which are detected by the photomultiplier tube. Since the main electron amplification occurs in an enclosed glass envelope containing its own high vacuum, the detector has an extended life and is more stable than the conventional electron multiplier; it also opens the possibility of using photon counting techniques which are much more sensitive than normal amplification.

The most modern instruments are fitted with a channel electron multiplier array detector (Figure 19.9); this consists of a plate whose ~1 mm thickness is perforated by a large number of pores or channels each with a diameter of 10–25 mm and coated with an emissive material. Each channel acts as an electron multiplier, and very high gains are possible by connecting two or more arrays in tandem, one in front of the other. Their advantage is they allow the output from all the channels to be monitored independently so that ions of different mass, which strike different parts of the detector, are observed without the need for an m/z scan. These devices are the electron equivalent to the diode array detectors which are making such a large impact in optical spectroscopy.

19.7   Data handling

Almost all mass spectrometers now employ computer control of some functions, and also use a computerised display and output. The range and complexity of data handling systems can be a daunting challenge to anybody contemplating the purchase of a mass spectrometer, since 30–80% of the total cost will be spent in this area. The actual hardware or computer engine can be based on personal computers, more expensive workstations, or it can be a custom-built computer containing a number of individual processors or transputers, each performing a specific task. The software is generally customised for a particular machine and may control all the normal operating parameters of the system as well as providing the output of spectra, or it may act as an interface to a laboratory information management system (LIMS). In each case it is probably true to say that whichever system is finally chosen, it will be outdated almost as soon as it is purchased. The amount of data generated even by a fairly modest mass spectrometer is very large indeed; a single run may have to digitise and store data for up to 100 fragments from each type of molecule, and if GC-MS or HPLC-MS analysis is being performed, a complete mass spectrum is generated and stored every second for up to 90 min. Obviously some data reduction or processing must be undertaken before the analyst can cope with this amount of material, and most data systems have a choice of output formats.

The most widely used output is a mass spectrum in normalised form (Figure 19.1); that is to say, the most intense peak, the base peak, is assigned an intensity of 100 or 1000 then the intensities of all other peaks are related to this. Although popular and very useful as a display, it does have some problems. Firstly, the mass axis must be correctly assigned by some calibration procedure – often using a known standard material such as perfluorokerosene (PFK) with a number of known and well-defined ions in its mass spectrum – otherwise the displayed masses may be one or more units high or low, which can cause confusion in the interpretation process. Secondly, if the intensity of the base peak is not correctly measured, then the intensity of all subsequent peaks will be incorrect. It is sometimes preferable to use an alternative display, where the intensity of each peak is recorded relative to the total ion current over a defined mass range. Tabular listings in either format are often available.

Even though the dynamic range of the total system is as high as six orders of magnitude, since the normal output described above is essentially displayed over a fairly restricted range – the intensity axis is 0–1000, and the mass axis is often 50–500 in unit steps – sometimes quite important information can be lost or at least not observed. Peaks of low intensity, which may be chemically important in terms of establishing a fragmentation pathway, may not be displayed if they are below a certain threshold value. Multiply charged ions which may appear at non-unit masses may also not be displayed and therefore missed. Normalised spectra are extremely useful for comparison purposes, either with spectra produced by analysts working in the same field but with different spectrometers and in different continents, or for library searching of spectral matches against a number of commercially available libraries.[7] Several of these libraries have more than 80 000 mass spectra stored in computer readable format, and most software packages are able to compare the spectrum from a substance with the spectra in the library to give a series of possible named compounds.

Most packages also allow the spectroscopist to perform a presearch, by compound name, formula or Chemical Abstracts (CAS) number, to reduce the number of inappropriate matches. An alternative is to use reverse searching, where a predetermined number of known spectra are compared to the data being generated. This is obviously a much faster process than normal searching. At present most of these programs are based simply on comparison of mass–intensity patterns and do not use any 'chemical' constraints on the listed possibilities.

Besides its role of gathering and manipulating data, the computer can normally control the spectrometer. This enables very rapid, repetitive scanning for GC-MS, or more sophisticated techniques such as single or multiple ion monitoring (SIM). Here the spectrometer is made to 'jump' to a number of predetermined masses rather than scanning over a range. Since the detector spends more time observing the ions of interest than it would in a normal scan, the sensitivity can often be increased by a factor of up to 1000. This is widely used in GC-MS and HPLC-MS trace analysis, where the limit of detection is often considerably lower than when using a conventional GC or HPLC detector.

As the processing power of computers increases, the manipulation of data becomes easier; nowadays many programs allow spectral averaging of a variable number of scans, to reduce or eliminate electrical noise. Background subtraction can also be used to reduce the effects of small amounts of contamination on the spectrum. Here the spectrum obtained without sample in the ion source is subtracted from the sample spectrum, removing lines due to the background. If the computer is fast enough and has sufficient data storage, it may be used in a multi-tasking mode, often known as foreground/background. Stored data can be accessed and manipulated at the same time as new data is being acquired, which increases the overall speed of the process.

19.8   Inorganic mass spectrometry

Apart from determinations involving large organic molecular species, there are many applications of mass spectrometry which are concerned only with the elemental composition of the sample. For convenience these methods have been grouped into two areas of study: surface methods and elemental analysis in solution.

19.8.1 Surface methods using mass spectrometry

It was shown in Section 19.4 that if a beam of sufficiently high energy is directed onto a surface, then ions are sputtered from the surface and could be separated and detected by a variety of analysers and detectors. The techniques FAB and LSIMS are often presented as new and powerful ways of looking at large species, particularly biomolecules. They are actually modifications of techniques which have been used to investigate inorganic surfaces for a number of years. Often these techniques are discussed under a blanket term of secondary ion mass spectrometry (SIMS) because the primary beam is usually an ion beam, though not always.

If a high-energy (5–20 keV) beam of ions such as Ar^+, Cs^+ or occasionally O_2^+ or N_2^+ is directed on a surface such as a metal film or semiconductor, even a mineral surface, then a number of secondary particles are ejected, including both positive and negative ions of the elements making up the surface. These ions can be passed into a magnetic quadrupole or other type of analyser for separation into individual m/z values and then detected. The resulting mass spectrum is normally very simple, containing one peak for each isotope of the elements present on the surface, and can give measurable signals from as little as 10^{-15} g of material. All elements from hydrogen upwards can be determined in this way, but truly quantitative data is difficult to obtain. Since the primary ion beam produces a small crater at the point of impact, it is also possible to perform depth profile studies in the surface layers of the sample.

A number of systems are available which can provide positive SIMS data, useful for elements on the left-hand side of the periodic table, and negative SIMS data, more useful for elements on the right-hand side of the periodic table.

Ion microprobe mass analysers

A modification of normal SIMS, ion microprobe mass analysers (IMPMAs) use much more sophisticated and expensive ion beams. Two or more electrostatic lenses focus the beam down from the 0.2 to 5 mm diameter typical of SIMS so it becomes only 1 to 2 mm, and with the aid of an ordinary optical microscope it can be directed onto specific areas of a surface. Again secondary ions are produced and normally directed through a double-focusing E-B analyser to give high resolution and high sensitivity (10^{-15} g). Microprobe analysers can map changes in the surface composition as the beam is traversed across the surface. They are powerful research tools for surface studies of semiconductors or deposited metal films, but their high sensitivity requires great care in sample preparation and presentation, and they normally use very high vacuums to ensure the surface is not contaminated by oxidation or a similar surface film prior to analysis. Together with their high capital cost, these demanding technical requirements have restricted their use.

Laser microprobe analysers

Rather than use a beam of primary ions, it is now possible to use a focused beam of light from a pulsed Nd: YAG laser; this can provide extremely high power at the surface, giving up to 10^{11} W cm^{-2} on a spot of diameter 0.5 μm. Detection limits may go down to 10^{-20} g and minute samples such as individual dust particles can be studied, or specific areas on a surface can be spatially differentiated with a resolution of about 1 μm.[8] Laser microprobe analysers are beginning to find application in the life sciences as well as more traditional areas. Element variation across small parts of a tissue can be mapped, providing insights into the distribution and function of the common electrolyte ions (Na, K, Ca, Mg, etc.) in neurones.

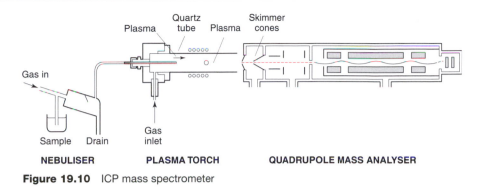

Figure 19.10   ICP mass spectrometer

19.8.2   Inductively coupled plasma mass spectrometry

Inductively coupled plasma mass spectrometry (ICP-MS) is more fully discussed in Section 19.11.5. It is a powerful and rapidly developing technique for multi-element determination at very low levels in liquid or solution samples. A conventional ICP plasma source is used to produce the primary ions, which are then analysed in a quadrupole mass spectrometer with mass range up to 300 and unit resolution across the range (Figure 19.10). The plasma source works at atmospheric pressure and is therefore an API system. Extensive use of computer control and optimised geometry means that a full elemental analysis can be made in one scan with measurement times of about 10 s per element even for concentrations as low as $1\,\mu g\,kg^{-1}$. Effectively it uses a quadrupole as detector for the ICP, so earlier sections on quantification of metals in solution (Chapters 15 and 16) also apply here.

19.9   Isotope ratio measurements

Most common elements naturally exist in more than one isotopic form (non-radioactive) and since most forms of mass spectrometer can distinguish between isotopes, this leads to a complication of the observed mass spectrum, whether for atomic species or molecular species. This isotopic pattern can often be of great help in interpreting the spectrum and obtaining molecular or structural information. Measurement of isotope ratios, either naturally occurring or in an enriched sample, can be used for a number of specific and important applications.

19.9.1   Natural isotope ratios

Due to the decay products of a number of common radionuclides, it is possible to determine the age of certain rocks and minerals by measuring the isotopic ratio of the stable elements at the end of the decay path. Geochemical dating has been performed using $^{40}K/^{40}Ar$, $^{87}Rb/^{87}Sr$ and $^{238}U/^{206}Pb$; and measurements on $^{235}U/^{238}U$ are important in the nuclear fuel industries. The ions are often generated from solid samples by using a thermal evaporation source where the powdered sample is very rapidly heated on a wire filament or bead.

Gas samples

Several other stable isotope ratios can be more easily measured if the sample is introduced to the spectrometer as a gas. Possibly the most widely known example of this type is the determination of ^{14}C (as CO_2) for radiocarbon dating of organic historical artefacts, but $^{13}C/^{12}C$, H/D, $^{15}N/^{14}N$, $^{18}O/^{16}O$ and $^{34}S/^{32}S$ are also determined. As for all isotope ratio experiments, the detection system must be capable of giving reliable results for these gaseous experiments, but there is an additional problem of providing a suitable inlet system for these materials.

Most measurements are made on purpose-built machines which normally have two gas inlet systems, one for sample and one for standard, constructed as two reservoirs of 2–50 mL volume. These bleed into a conventional EI ion source via a viscous flow capillary. The standard reservoir is often of variable volume so the ion currents at the detector can be optimised to give equal signals for sample and standard. But the most sophisticated systems of this type can now be coupled to a gas chromatograph via a combustion furnace so that volatile organic species can be separated, burned into CO_2 ($^{12}C/^{13}C$) and fed directly into the ion source without loss or contamination.[9] The analyser employed is a 90° magnetic sector with the ions projected at an angle of 26.5° to provide stigmatic focusing. Although earlier instruments used permanent magnets and accelerator voltage scanning, more modern machines employ electromagnets generating a field of up to 1 T, with a mass range of 280 u and a resolution of 500.

The detector must be capable of giving a very stable ion current for a given number of ions. Currently the only device capable of this is the Faraday cup, often used as a multiple detector with one cup at the focus of each ion beam and computer control to provide ion counting rather than total current measurement. To improve the accuracy of ratio measurement still further, the system can measure ratios for a given pair of ions in a sample and then in a known reference alternately and repeatedly. Isotope ratios measured in this way are normally expressed by a δ value where $\delta = 1000(R/R_s - 1)$, R is the ratio measured for the sample, and R_s is the ratio of a known standard.

A widely used standard for $^{12}C/^{13}C$ ratios is $CaCO_3$ derived from Southern Californian marine fossils, *Belemnitella americana*, which has very high values of ^{13}C and thus gives negative values of δ for most other materials.[10] (Unfortunately, this standard is not currently commercially obtainable.) The value of δ can range from 0 in marine carbon through −8 to −35 for land plants to −50 in some natural gases. Although very small, the variation in abundance of ^{13}C versus ^{12}C can be accurately measured even for modern species, and some important studies of adulteration in foodstuffs may be performed using a knowledge of the ratios in different foods, in particular vegetable material such as sugars and fruit juices. Isotope measurements of this type are rapidly becoming the standard technique for determining some types of food adulteration. For example, the substitution of low-cost syrups derived from a C_4 photosynthetic cycle in place of more expensive fruit sugars which utilise a C_3 cycle and discriminate against ^{13}C during photosynthesis will alter the $^{12}C/^{13}C$ ratio of sweetened foods to which they are added.[11]

^{14}C studies

Modern methods for ^{14}C determination in old materials do not in fact rely on gaseous forms of the sample because, if present, ^{14}N would interfere with measurements; nowadays the accepted technique is to extract and purify some organic material from the sample then convert this via CO_2 and acetylene into pure graphitic carbon. This carbon sample in the solid state is then ionised using a Cs^+ ion source, first to give $^{14}C^-$ ions which are converted to C^{3+} ions that are ultimately analysed on a modified TOF analyser. Since $^{14}N^-$ ions are unstable, they will not interfere with the determination. Carbon-14 gives reliable dates for artefacts older than 40 000 years, the age limit using conventional techniques.[12]

19.9.2   Isotope enrichment (isotope labelling)

A rapidly growing area for isotope studies is the use of stable isotopes to tag chemical or biochemical systems, either to gain information on structure, or to investigate the reaction pathways. This is sometimes known as stable isotope ratio mass spectrometry (SIRMS). Nuclear magnetic resonance (NMR) provides a simple example, where chemicals with active hydrogen can have it replaced by deuterium (D); but mass spectrometry can give information on smaller samples with greater clarity. Labelling of this type is also useful for unravelling fragmentation pathways where complex mass spectra are obtained, although ^{13}C labels often have to be used here and this can be expensive. Labelled molecules can be used to investigate metabolic pathways in medicine, biochemistry and toxicology.[13] These experiments can use labelled equivalents of normally occurring body substances or they can be used to follow the fate of exogenous substances such as pollutants, pesticides and drugs. The material is labelled with either D or ^{13}C at a site which will neither affect nor be affected by the subsequent reactions and then its path can be followed by mass spectrometry. Compared with earlier techniques which relied on radioactive trace materials, isotopic labelling is much safer for both analyst and subject, and it allows the use of ^{18}O and ^{15}N, which is impossible using radioactive tracers. However, incorporating the label into a specific group of the molecule can be a skilled and slow process.

19.9.3   Isotope dilution analysis

An extension of isotope enrichment often enables direct quantitative measurements to be made. Suppose a sample contains compound M contained within a matrix. If a known amount A_i of an isotopically modified analogue of the compound (denoted by I) is added to an amount A_m of the sample and a proportion introduced into a mass spectrometer, then the mass spectrum will contain lines due to both substance M and substance I. In the simple case, where an intense line derived from M and the analogous line derived from I (separated by the mass number of the difference between the original and modified compounds) can be seen, **and** there is no contribution to either line from the other compound, then

$$\frac{\text{intensity of line due to I}}{\text{intensity of line due to M}} = \frac{\text{quantity of I in the matrix}}{\text{quantity of M in the matrix}}$$

This is just a special application of the general internal standard method, discussed in the chapters on chromatography. Wherever this method is used, it overcomes the problems of measuring how much sample is introduced into the system, since both sample and reference are introduced together. Relatively large amounts of matrix can have precise amounts of internal standard (I) added before analysis of a small proportion of the mixture, so high precision can be obtained even when the sample is subjected to a number of manipulations before measurement. In the special case of the mass spectrometry experiment, it also removes many of the errors associated with variable ion production in the source (which often cause problems of quantification), since only one mass spectrum is produced and both M and I experience the same ionising conditions.

The sensitivity can be extended considerably if single ion monitoring (SIM) is used to continually monitor only the two lines of interest, and it is often possible to combine SIM with one of the chromatographic techniques (GC-MS or HPLC-MS), when both the compound of interest and its isotopically modified marker can be separated from other possible interferences before SIM is performed. In this case, since they are chemically identical, both substances will elute as a single peak (separated from other compounds in the matrix);

but if SIM is conducted repeatedly as the peak elutes, the abundances of M and I can be determined with high precision even at low concentrations.

19.10   Interpretation of spectra

Once the analyst has decided upon the most suitable configuration of components for a mass spectrometer, with luck a mass spectrum of the sample can be obtained, then all that has to be done is to interpret it! This can be a daunting task at first sight but a systematic approach to the problem will always help. Assuming the sample is organic, the analyst would ideally like information on the molecular formula, which may in turn give molecular structure and then identity. Obviously there are short cuts which can be used, and a perfect spectrum match between an authentic sample run under the same conditions as an unknown is often convincing proof of identity. This is not a guarantee of correct identification, because it is always possible for different positional isomers to give identical fragmentation patterns.

For a systematic approach, the first and most important rule is that the mass spectrum of a single substance is always **much** easier to interpret than the composite spectrum of a mixture, So try to obtain the purest sample possible or to separate the sample mixture before mass spectrometry. Once the mass spectrum is obtained, the analyst can use one of three approaches:

1. A mathematical or 'numbers' approach to identification
2. Chemical knowledge and fragmentation patterns to build up a possible structure
3. Spectral matching or correlation with compiled spectral libraries

Usually no single technique will give complete assurance of identity; the most sensible approach is usually an intelligent combination of all the methods. But to begin with, let us consider them separately.

19.10.1   The numbers approach

Mass spectrometry is simply a method of sorting ions of different mass from each other, and this can be done very accurately even for ions containing different isotopes, thus the data in Table 19.1 can be a very useful starting point. Potentially the ion of most interest is the molecular ion, because from this the molecular weight of the species can be obtained. But identification of the molecular ion can sometimes be difficult. For EI spectra the fragmentation may be so diverse that only fragment ions are seen with the molecular ion at very low intensity or absent altogether, and for other ionisation modes the ion may be as $(M + 1)^+$ or $(M - 1)^+$. The solution to the first problem is to use a lower ionisation energy than the standard $70\,eV$; this will lower the sensitivity of the experiment considerably, but often enables the molecular ion to be identified since less fragmentation will occur. The second problem of seeing ions other than the molecular ion with some ionisation techniques has to be solved by experience.

At this stage it is as well to define the molecular ion as 'the ion of highest mass for a given species, which contains all the elements present in their most abundant isotopic form'. Thus the molecular ion of dichloromethane (Figure 19.1) shows a peak at $m/z = 84\,u$ which is due to $^{12}C^1H_2^{35}Cl_2$, but it is obvious there are several other peaks around this value also due to the ion $CH_2Cl_2^+$ but with different isotopic composition, and the largest peak (the base peak) is at $m/z = 49\,u$. Only a few elements of interest to the organic mass spectroscopist are monoisotopic (Table 19.1); although this complicates the appearance of the mass spectrum, the isotopic peaks can also be of great help in elucidating structures. The isotope effects of C, H, N and O are considered further when the molecular ion is studied in more detail, but since the effects of some isotopes are large and immediately obvious, we will discuss them first.

The 'heavy' isotope effects that are most often seen occur in compounds containing chlorine or bromine, sulphur or silicon. Chlorine and bromine each have two naturally occurring isotopes, sulphur and silicon each have three. But when studying organometallic compounds, remember that many of them are polyatomic (Hg has seven natural isotopes with significant abundances).

A molecule containing one chlorine atom such as chloroethane (Figure 19.1) will have two peaks at the molecular weight region separated by two mass units due to some molecules containing chlorine-35 isotope and others containing chlorine-37. The intensities of these two peaks can be predicted from the natural abundance of each isotope (Table 19.1), but it is more normal when considering this problem to express the abundances as a percentage where the most abundant isotope takes the value 100. This would predict that the two peaks at 64 u and 66 u have intensities of 100 : 32 or roughly 3 : 1. As can be seen from the actual mass spectrum (Figure 19.1) this is exactly what is observed. This 3 : 1 pattern separated by two mass units is a very good indication of the presence of one chlorine atom in a molecule, and masses 49 u and 50 u in Figure 19.1 show how this does not just apply to the molecular ion but to any ion fragment containing one chlorine atom. For bromine the pattern is of two peaks of almost equal intensity (100 : 98) two mass units apart. Since both fluorine and iodine are monoisotopic, they will not give a characteristic pattern.

It is a fairly easy matter to calculate the expected intensities of ions having more than one chlorine atom; in general the isotopic pattern is found by expanding the binomial expression $(a + b)^n$ where a is the abundance of the lighter isotope, b is the abundance of the heavier isotope and n is the number of atoms of that element in the ion. This predicts that for two chlorine atoms one would observe peaks at M, (M + 2) and (M + 4) with intensities 9 : 6 : 1, and this is again observed in the cluster at 84, 86 and 88 in dichloromethane (Figure 19.1). This general expression can in fact be extended to the more general case for two different polyisotopic elements in a fragment by using $(a + b)^n (c + d)^m$ where c, d and m refer to the second element. Computer programs are now becoming available which can calculate these intensities accurately, and in many cases they form part of the standard operating software supplied with the instrument.

This mathematical approach can also be useful for substances containing only the more common organic elements C, H, N and O; and many texts explain that because of the presence of ^{13}C in all organic molecules, one would expect a peak one mass unit higher than the molecular ion and with normalised intensity 1.15 × (the number of carbon atoms) compared to the ^{12}C molecular ion expressed as 100; this gives an approximate way of calculating the number of carbon atoms in a molecule from measurements of the molecular ion and (M + 1) ion intensities. (This is only applicable for ionisation modes where (M + 1) ions are not formed in the ionisation process.) But with molecules containing only the four elements above, a more accurate calculation of relative intensities is obtained from the following two formulae:

$$\frac{M + 1}{M} = (1.15w + 0.015x + 0.37y + 0.04z)\%$$

$$\frac{M + 2}{M} = [(1.15w)^2/200 + 0.2z]\%$$

for all molecules of formula $C_wH_xN_yO_z$.

745

Table 19.4   *Possible formulae for mass 100: generated by computer*[a]

Possible compounds[b] *with mass = 100 (no valence rules)*

$C_2H_2N_3O_2$	$C_4H_4O_3$	$C_5H_{12}N_2$
$C_2H_4N_4O$	$C_4H_6NO_2$	$C_6H_{12}O$
$C_3H_2NO_3$	$C_4H_8N_2O$	$C_6H_{14}N$
$C_3H_4N_2O_2$	$C_4H_{10}N_3$	C_7H_{16}
$C_3H_6N_3O$	$C_5H_8O_2$	C_7H_2N
$C_3H_8N_4$	$C_5H_{10}NO$	C_8H_4

Probable compounds[c] *(using valence rules)*

Compound	Accurate mass	Intensities[d]		
		M	M + 1	M + 2
$C_2H_4N_4O$	100.038 511	100	3.7038	0.0329
$C_3H_4N_2O_2$	100.027 278	100	4.0702	0.4236
$C_3H_8N_4$	100.074 896	100	4.9330	0.0552
$C_4H_4O_3$	100.016 04	100	4.5561	0.6037
$C_4H_8N_2O$	100.063 683	100	5.2994	0.0391
$C_5H_8O_2$	100.052 43	100	5.6658	0.5286
$C_5H_{12}N_2$	100.100 048	100	6.4686	0.1755
$C_6H_{12}O$	100.088 815	100	6.8350	0.1965
C_7H_{16}	100.1252	100	8.0042	0.2720
C_8H_4	100.0313	100	8.8733	0.3445

[a] The computer programs were developed by R. J. Barnes.
[b] Possible compounds are all those combinations of $C_wH_xN_yO_z$ which give nominal mass 100, provided the following conditions are obeyed: $x > 1$, $y < 4$, $z < 4$, $x < 2w + y + 2$.
[c] Probable compounds are found using a recursive algorithm that finds all legitimate compounds of a given nominal mass which also obey valence rules.
[d] Relative peak intensities are found using an algorithm that expands the full polynomial probability terms. All values for mass and abundance are from Table 19.1.

If the molecular ion can be identified, it is possible to measure the molecular weight and also measure the intensities of the (M + 1) and (M + 2) ions. Tables of possible structures for all combinations of these four elements up to $m/z = 500$ u are available together with relative intensities of the two masses above the molecular ion, and it is often possible to reduce the number of possible alternatives for the molecule down to a relatively small number.[14]

This technique is ideally suited to computer-based calculations, and programs are available which can perform calculations relatively easily for a given molecular ion. A typical calculation is illustrated in Table 19.4, where all possible combinations of C, H, N and O were calculated for a mass of 100 assuming only the values 12, 1, 14 and 16, and a few simple restrictions to remove obvious anomalies such as H_{100}. Valency rules for each atom were then applied to reduce this to a list of possible molecules. Even this simple approach gives only 10 molecular formulae for mass 100. The nitrogen rule (Section 19.10.2) is a consequence of these valency rules. Chemical common sense will show that some combinations are not feasible.

The intensities for (M + 1) and (M + 2) for these molecules are also listed. Notice that for a molecular ion of $m/z = 100$ u with intensities of 6.9 and 0.2 relative to 100 for the molecular ion, the most likely formula is $C_6H_{12}O$. Unfortunately, at least 19 different structures are known for this formula, but at least the possibilities have been drastically reduced. (The Wiley eight-peak index lists 122 entries for mass = 100 u.) It is often possible to make measurements at high resolution, so the mass of the molecular ion can be measured to a few parts per million, i.e. to 0.0001 u, and then this identification becomes even more positive. Using the tabulated accurate masses of each isotope, one can calculate the exact mass to four decimal places for each of the combinations with nominal mass 100 u; this reveals that each has a different mass, and in the example chosen (mass 100.0888 u) it shows that the probable formula deduced from low-resolution measurements is almost certainly correct.

A formula has been derived from the accurate masses of the four most common elements in organic molecules:

$$\text{exact mass difference from nearest integral mass} + \frac{0.0051z - 0.0031y}{0.0078} = \text{number of hydrogens}$$

which again may be used as a check of molecular formula if high-resolution measurements are available. The software of some systems contains the necessary algorithms for several of these calculations. But it is not always possible to make measurements of accurate mass even if a high-resolution spectrometer is available, since it may take too long, especially with coupled techniques such as GC-MS. For many instruments with only moderate resolution, this technique is not possible under any conditions.

Another useful guide to identification of molecular formula is also possible at high resolution, again based on the simple numerical properties of the atoms within the ion. Most elements found in organic molecules have accurate atomic masses which are slightly higher than a whole number, they are mass proficient, but fluorine and iodine which are both monoisotopic have masses less than their nominal mass, they are mass deficient. This mass deficiency allows fluorine and iodine to be readily picked out when making accurate mass measurements. It also explains the widespread use of polyfluorokerosene as a calibrating material, especially at high resolution, where its fragments, which are numerous and spread across a wide mass range, are separated from those of most organic compounds.

19.10.2   Chemical interpretation

No matter how accurately one can measure a single ion or small group of isotope peaks, the data will not provide information on possible structures within a given molecular formula; for this the analyst has to interpret the major peaks in the mass spectrum. Accurate interpretation takes practice, but using a knowledge of organic chemistry, a few 'rules' can be formulated to help simplify the task. The starting point is again, where possible, the molecular ion which is always an odd-electron species and will have an even molecular mass, for C, H, N and O species, unless it contains an odd number of nitrogens. In its simplest form, this **nitrogen rule** says that molecules with odd numbers of nitrogens have odd molar mass and those with even numbers of nitrogens (including 0), will have even mass. Never violated, it proves a useful guide to identifying nitrogen in molecules, as well as a useful check that the molecular ion has been correctly identified.

For electron ionisation at 70 eV the molecular ion initially formed contains a large excess of energy, and since most organic bonds have energies in the region of 10–12 eV, it is very likely that the molecular ion will fragment into smaller charged species or fragment ions which will be observed in the mass spectrum. The exact mechanism will depend on a number of factors and a detailed explanation is not given here, but fragmentation normally occurs by one of two routes:

1. The charge and unpaired electron are retained on one fragment, and a neutral molecule is ejected. This is only possible if some rearrangement takes place in the ion just before fragmentation.
2. Simple bond cleavage takes place, giving a cation as one fragment and a neutral radical as the other. This process, which gives an even-electron ion, can then continue, giving smaller and smaller fragments until the excess energy is dissipated.

The two routes may be shown in a diagram:

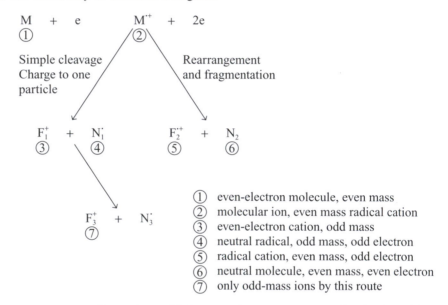

① even-electron molecule, even mass
② molecular ion, even mass radical cation
③ even-electron cation, odd mass
④ neutral radical, odd mass, odd electron
⑤ radical cation, even mass, odd electron
⑥ neutral molecule, even mass, even electron
⑦ only odd-mass ions by this route

The mass spectrum of most ions will thus contain an even-mass molecular ion with fragments predominantly of odd mass unless a rearrangement has occurred. The exact fragments that break off the parent molecule will depend on the relative strengths of bonding in the molecule, and any stabilising factors such as conjugation in the fragments so formed. A knowledge of the organic chemistry of the species is useful in predicting likely fragmentation patterns for the molecule. If it is known that heteroatoms are present, then cleavage at the adjacent carbon is most probable; this is because the electron is easily abstracted from the lone pair. If the molecule contains a branched chain, then the predicted outcome is cleavage at the branch with the charge residing on the fragment that is most branched. Carbon–carbon bonds are generally weaker than carbon–hydrogen bonds, so for saturated hydrocarbons the fragmentation occurs by loss of C_nH_{2n+1}.

These simple rules should, however, be used with caution because fragmentation of even relatively simple molecules gives a complex pattern due to primary fragmentation, secondary fragmentation (daughter ions) and rearrangement of fragments. This last phenomenon, although not always understood, often gives quite characteristic ions in many cases. The most common example is the very strong $m/z = 91$ u ion seen in aromatic molecules with a carbon side chain; the initial cleavage to the ring gives $C_6H_5CH_2^+$, which is resonance-stabilised as the $C_7H_7^+$ tropilium ion:

Another very important and often diagnostic rearrangement is known as the McLafferty rearrangement after the spectroscopist who first recognised it;[15] it was first observed in methyl esters and these will be used as the example, but the same process can occur in any molecule with a γ-hydrogen to a doubly bonded heteroatom such as $-C{=}O$, $-S{=}O$ or $-P{=}O$.

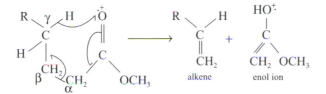

Note that six centres are involved and the fragment ion, the enol ion with $m/z = 74$ u, will be formed from any methyl ester larger than three carbons. The so-called McL peak, which is always at mass 74 u is a large peak because this is an energetically favoured process; it is therefore diagnostic of **all** methyl esters. This process will occur with a very wide range of compounds which have a six-centre relationship, each class of compounds giving rise to a McL peak of characteristic mass (Table 19.5). The fragment ions are always of opposite mass to fragments produced directly, so for molecules that contain C, H and O where the fragments predominantly have odd mass, the McL fragments have even mass and can therefore be seen more easily; whereas for molecules containing odd numbers of nitrogens, the fragments will have odd mass.

Although the particular fragmentation pathways of different molecules can only be understood by experience, two tabulations are often helpful in trying to interpret an unknown mass spectrum:

A table of common fragment ions Table 19.5 can be diagnostic for particular groups, but it should be used with care since it is selective rather than comprehensive, and many different compounds can give similar fragments.

A list of common losses Table 19.6 lists common losses from the molecular ion. It is very useful if used with care, since it approaches the problem from the other end of the mass scale. Instead of just identifying an ion of given (often quite small) mass, Table 19.6 allows the spectroscopist to count down from the molecular ion to see if peaks at a specified lower mass seem probable in a given situation. Again, this is a selective table, but notice that there are **no** common losses between 2 and 14 mass units, and if any appear to be present then it is almost certain the molecular ion has not been correctly identified.

This approach of counting down from the molecular ion is a mathematical process, yet it is often invaluable for providing an insight into an unknown structure. But it must be conducted properly, starting from the molecular ion each time rather than counting down from one fragment to the next. Chemical interpretation still relies on a mathematical approach, but reinforced by mechanistic concepts in organic chemistry.

Table 19.5   *Common fragment ions*

m/z value	Ion commonly associated with the mass	Possible inference
15	CH_3^+	–
18	$H_2O^{+\cdot}$	–
26	$C_2H_2^{+\cdot}$	–
27	$C_2H_3^+$	–
28	$CO^{+\cdot}$, $C_2H_4^{+\cdot}$, $N_2^{+\cdot}$	–
29	CHO^+, $C_2H_5^+$	–
30	$CH_2NH_2^+$	Amine
	NO^+	Nitro compound
31	CH_2OH^+	Primary alcohol
36/38 (3 : 1)	$HCl^{+\cdot}$	Chloro compound
39	$C_3H_3^+$	–
40	$Ar^{+\cdot}$, $C_3H_4^{+\cdot}$	–
41	$C_3H_5^{+\cdot}$	–
	$C_2H_3N^{+\cdot}$	Aliphatic nitrile
43	$C_3H_7^+$	Propyl compound
	CH_3CO^+	Acetyl compound (e.g. methyl ketone)
	$C_2H_6N^+$	Aliphatic amine
	$^+NH_2{=}C{=}O$	Primary amide
45	$CH_3OCH_2^+$	Methyl ether
	CH_3CHOH^+	Alcohol
47	$CH_2{=}SH^+$, CH_3S^+	Thio compound
50	$C_4H_2^{+\cdot}$	Aromatic compound
51	$C_4H_3^+$	Aromatic compound
55	$C_4H_7^+$	
56	$C_4H_8^+$	
57	$C_4H_9^+$	Butyl compound
	$C_2H_5CO^+$	Ethyl ketone, propionate ester
58	$CH_2{=}C(OH)CH_3^{+\cdot}$	Ketone[a]
	$C_3H_8N^+$	Amine
59	$COOCH_3^+$	Methyl ester
	$CH_2{=}C(OH)NH_2^{+\cdot}$	Primary amide[a]
	$C_2H_5CH{=}OH^+$	$C_2H_5CH(OH){-}R$
	$CH_2{=}O{-}C_2H_5^+$ + isomers	Ether
60	$CH_2{=}C(OH)_2^{+\cdot}$	Carboxylic acid,[a] sugar
61	$CH_3CO(OH_2)^+$	Alkyl acetate
	$C_2H_5S^+$	Thio compound
65	$C_5H_5^+$	Aromatic compound
66	$H_2S_2^{+\cdot}$	
69	CF_3^+	Fluorocarbon
70	$C_5H_{10}^+$	
71	$C_5H_{11}^+$	$C_5H_{11}{-}R$
	$C_3H_7CO^+$	Propyl ketone, butyl ester
72	$CH_2{=}C(OH)C_2H_5^{+\cdot}$	Ketone[a]
	$C_4H_{10}N^+$	Amine

Table 19.5   *(cont'd)*

m/z value	Ion commonly associated with the mass	Possible inference
73	$C_4H_9O^+$	Alcohol, ether
	$(CH_3)_3Si^+$	Trimethylsilyl derivative
	$COOC_2H_5^+$	Ethyl ester
74	$C_3H_6O_2^{+\cdot}$	Methyl ester[a]
		Carboxylic acid
75	$C_6H_3^+$	Disubstituted benzene
	$(CH_3)_2Si{=}OH^+$	Trimethylsilyl ether
76	$C_6H_4^+$	Benzene derivative
77	$C_6H_5^+$	Monosubstituted benzene
78	$C_6H_6^{+\cdot}$	Monosubstituted benzene
79/81 (1 : 1)	Br^+	Bromo compound
80/82 (1 : 1)	$HBr^{+\cdot}$	Bromo compound
83/85/87 (9 : 6 : 1)	$HCCl_2^+$	Chloroform
85	$C_4H_9CO^+$	$C_4H_9CO{-}R$
91	$C_7H_7^+$	Aromatic (possibly benzylic) compound
92	$C_6H_6N^+$	Substituted pyridine
91/93 (3 : 1)	$C_4H_8Cl^+$	Alkyl chloride
93/95 (1 : 1)	CH_2Br^+	Bromo compound
94	$C_6H_6O^{+\cdot}$	$C_6H_5O{-}R$
99	$C_5H_7O_2^{+\cdot}$	Ethylene ketal
105	$C_6H_5CO^+$	Benzoyl compound
122	$C_6H_5COOH^{+\cdot}$	Alkyl benzoate
123	$C_6H_5COOH_2^+$	Alkyl benzoate
127	I^+	Iodo compound
128	$HI^{+\cdot}$	Iodo compound
135/137 (1 : 1)	$C_4H_8Br^+$	Bromo compound
141	CH_2I^+	Iodo compound
147	$(CH_3)_2Si{=}O{-}Si(CH_3)_3$	Trimethysilyl ether
149	$C_8H_5O_3^+$	Dialkyl phthalate

[a] McL fragment.

19.10.3   Spectral correlation and computer-matched spectra

Although it can be enormously rewarding to make interpretations from first principles using the earlier guidelines (Sections 19.10.1 and 19.10.2), constraints of time and money often mean that alternative procedures are required. Matching the mass spectrum of an unknown sample to a library spectrum is an attractive option, especially if both the unknown and the library are in computer-readable form so that a large number of possibilities can be compared before deciding the best match. This is becoming increasingly true, as several libraries are available with large numbers of entries. The two most commonly used databases are the Wiley mass spectral library with 130 000 entries and the NIST base with 62 000 entries.[7] Even quite modest mass spectrometers are now able to search these

Table 19.6 *Common losses from the molecular ion*

Ion	Species often associated with mass lost	Possible inference
M-1	H$\cdot$	Aldehydes, acetals, alkynes
M-15	CH$_3^\cdot$	Acetals, methyl derivatives
M-16	O$\cdot$	Aromatic nitro compounds, *N*-oxides, sulphoxides
	NH$_2^\cdot$	Aromatic amides
M-17	OH$\cdot$	Carboxylic acids
	NH$_3$	Amines (rare)
M-18	H$_2$O	Alcohols, acids, aliphatic aldehydes, steroid alcohols and ketones
M-19	F$\cdot$	Fluoro compounds
M-20	HF	Fluoro compounds
M-26	C$_2$H$_2$	Aromatic compounds
M-27	HCN	Aromatic nitriles, nitrogen heterocycles
M-28	C$_2$H$_4$	Ethyl ester or ether, *n*-propyl ketone
	CO	Quinone, phenol, oxygen heterocycle
M-29	C$_2$H$_5^\cdot$	Ethyl group
	CHO$\cdot$	Aromatic aldehydes, phenols
M-30	CH$_2$O	Aromatic methyl esters
	NO	Aromatic nitro compounds
M-31	CH$_3$O$\cdot$	Methyl ester, dimethyl acetal or ketal
M-32	CH$_3$OH	Methyl ester
M-33	H$_2$O + CH$_3$	Short-chain unbranched primary alcohols, steroid alcohols
	HS$\cdot$	Thiols
M-34	H$_2$S	Thiols
M-35/37	Cl$\cdot$	Chloro compounds
M-36/38	HCl	Chloro compounds
M-40	$\cdot$CH$_2$CN	Aliphatic nitriles and dinitriles
M-41	C$_3$H$_5^\cdot$	Propyl ester
M-42	C$_3$H$_6$	Butyl ketone, propyl ether
	CH$_2$CO	Methyl ketone, acetates, *N*-acetyl compounds
M-43	C$_3$H$_7^\cdot$	Propyl ketone
	CH$_3$CO$\cdot$	Methyl ketone
	HNCO	Purines, dioxopiperazine
M-44	CO$_2$	Anhydrides, cyclic imide, unsaturated esters, carbonates
M-45	$\cdot$COOH	Carboxylic acids
	C$_2$H$_5$O$\cdot$	Ethyl ester, acetal or ketal
M-46	C$_2$H$_5$OH	Ethyl ester
	NO$_2$	Aromatic nitro compounds
	CO + H$_2$O	Carboxylic acids
M-47	CH$_2$SH, CH$_3$S	Aliphatic thiols
M-48	CH$_3$SH	Methyl thioester
	SO	Aromatic sulphoxides
M-55	C$_4$H$_7^\cdot$	Butyl ester
M-56	C$_4$H$_8$	Butyl compounds, pentyl ketones
	CO + CO	Quinone

two databases in a few seconds. However, there are limits to even the most advanced programs. Nearly 200 000 items available for comparison may sound impressive, but it represents at best only 10% of all the known organic compounds. Obviously databases cannot contain the spectra of every novel compounds produced in the analyst's laboratory, although most computer-based libraries do have provision for creating a user subset of chemicals in addition to the main library.

Most libraries are compiled from EI spectra; only a few contain entries produced by alternative ionisation methods. The exact fragmentation pattern resulting from a given compound partly depends on the temperature and pressure in the ion source and may even depend on the geometry of the mass spectrometer. So a complete match between library and sample is seldom possible, even if the sample spectrum is not complicated by peaks from impurities and background substances.

The data for these large databases is normally stored in some compressed form, not the m/z and the intensity value for every line in every spectrum. One popular method is to use the eight-peak approach which, as its name suggests, simply stores intensity data for the eight most intense peaks within a specified spectral range (say from 50 up to the value of the molecular ion). The normal output from such a spectrum search routine is a list of three to ten probable compounds, often with some sort of probability factor against each compound. But since this is actually a simple, albeit large, mathematical comparison, some of the probable substances can very quickly be discounted from a knowledge of the likely chemical or chemicals involved in the experiment. This process of matching spectra can be done without the aid of a computer, and the Wiley eight-peak index is available in book form as tabulations of ions in increasing molecular mass, or as increasing base ion mass. These tables may be accessed almost as quickly as computer-based searches.

19.11   Hyphenated systems

Although an excellent technique for the identification of unknowns, mass spectrometry requires pure substances for optimum results, even simple mixtures tend to produce complicated spectra that may sometimes be difficult or impossible to interpret. Quantification of directly introduced samples is often difficult because it may be hard to introduce known quantities into the ion source and there may be variations in ion abundances produced by some sources. For this reason a number of coupled or hyphenated techniques have been developed which combine a separating method as the front end of a mass spectrometer and allow controlled small amounts of sample to be introduced. The most obvious are the chromatographic methods, where interfaces are available for GC-MS, HPLC-MS and SFC-MS, but they are not the only ways of separating one species from another and capillary electrophoresis–mass spectrometry (CE-MS) has some promise. The techniques of MS-MS, or tandem mass spectrometry, have also been included in this section because in many ways they are better than conventional chromatography for the analysis of mixtures. The use of mass spectrometry as a detector for chromatographic methods not only allows identification of each compound as it elutes from the chromatograph, a huge advantage in itself, but through techniques specific to MS, e.g. single ion monitoring (SIM), it can also improve on the sensitivity of conventional detectors.

19.11.1   Gas chromatography–mass spectrometry (GC-MS)

Gas chromatography–mass spectrometry (GC-MS) was one of the earliest examples of hyphenated techniques and is still one of the most widely used and powerful applications of 'combined' mass spectrometry. The earlier systems had considerable difficulty reconciling the different requirements of the two techniques; MS requires high vacuum and high voltage with small sample batch input, whereas GC operates at atmospheric pressure in a continuous mode. Complicated interfaces between the two systems were required, but on modern instruments these problems have largely been solved, and very efficient hyphenated systems are available for routine laboratory use.

The first breakthrough came when capillary GC columns became widely available; these operate at flow rates of about $1 \, mL \, min^{-1}$ rather than the $30–80 \, mL \, min^{-1}$ of traditional columns. This meant that instead of a separator between the GC and the mass spectrometer to remove the carrier gas, the entire eluent from the column was led straight into the ion source of the mass spectrometer, giving direct coupling of the two systems. This is almost universally used for most GC-MS systems, and the direct inlet will be the only interface described here. If information is required on the older Rhyage (jet) or Watson–Biemann (effusion) separators, this may be obtained from the general references listed at the end of the chapter. Now a large percentage of all gas chromatography is performed on capillary columns, predominantly quartz capillaries, with the stationary phase chemically bonded to the walls of the column to prevent bleeding. These may be easily and directly coupled to most ion sources by simply passing the end of the column through a compression coupling straight in to the ion source.

Normally the column is guided to the correct position by passing it along a stainless steel guide tube, independently heated to prevent condensation in the transfer line. Once installed the pumping system of the mass spectrometer is able to handle a flow of about $1 \, mL \, min^{-1}$ eluting from the column and thus maintain the desired ion source pressure. Many dedicated GC-MS systems can even handle the flow rates of $3–10 \, mL \, min^{-1}$ used in wide-bore capillaries. So the physical interfacing of GC to MS is generally straightforward and many different types of mass spectrometer may be considered for hyphenation.

The main difficulty with this combination is to decide which will be most appropriate for the demands of a given laboratory. Since gas chromatography will only be able to separate volatile (predominantly organic) molecules, the mass spectrometer needs to have only a fairly modest mass range of up to say $750 \, u$, and resolution of less than 1000 is adequate for many purposes. However, it must have a fast response and be sensitive so it can detect and obtain a spectrum of even the smallest GC peaks. Of all the mass spectrometers described earlier in this chapter, the quadrupole mass filter (Q) and the ion trap (IT) are ideally suited to meet these requirements and a very large choice of GC-Q and GC-IT mass spectrometry systems are available commercially, although almost any mass spectrometer can be coupled to GC.

The systems are so closely optimised for each other that it is probably true to describe them as mass selective detectors for gas chromatography rather than hyphenated systems. This does not imply any criticism of these instruments, since they perform extremely well for their purpose and tend to be less than a quarter the cost of standard mass spectrometers (which can also have a GC system as one form of sample inlet). But remember that they do one job only and cannot be expected to provide the range of capabilities of more conventional mass spectrometers. A GC-Q and a GC-IT can be easily operated in EI or CI mode (positive or negative ion), but other external ionisation sources including the API sources can be used if desired.

All dedicated mass selective detectors have computerised data acquisition and control, since the sheer volume of data which is generated in a single GC-MS experiment would overwhelm any other system, and it is in this part of the system that differences between commercial instruments become most apparent. Would-be purchasers of GC-MS systems will find a very wide range of options, and the final choice of instrument depends upon the exact application and budget available. Most larger and more expensive double-focusing magnetic sector instruments also have the capability of GC input, and may have an advantage over dedicated GC-Q or GC-IT systems in that they have much higher resolution and can often make accurate mass measurements of each component eluted from the column. But their scan speed of ~1 full scan per second is a limitation for modern high-speed chromatography columns, so the quadrupole, which can scan at up to 8 full scans per second, is often first choice for GC-MS.

19.11.2   High-performance liquid chromatography–mass spectrometry (HPLC-MS)

In contrast to GC-MS, the interface between HPLC and MS was until recently a problem to be solved. The large quantities of liquid eluent often containing involatile components must be separated from the sample if satisfactory mass spectra are to be obtained. Earlier attempts to solve this problem used moving belts or wires on which the eluent was coated continuously; the solvent was removed by passing the wire or belt through an oven and then via a vacuum lock into the mass spectrometer.[16]

These systems have not been totally satisfactory and a number of other approaches are currently being used. The simplest is to use a direct inlet system similar to GC-MS, where microbore HPLC columns with relatively low flow rates are either fed directly into the ion source, or a portion of the stream is split from the main flow to enter the source where the solvent acts as a CI reagent. For some samples and some applications this is an acceptable approach, but problems of contamination and build-up of involatiles in the source are more severe than for GC-MS, where the material entering the ion source has already passed as a vapour through the column. Another general approach which is currently favoured is to adopt a technique widely used in atomic spectroscopy – converting the liquid eluent into a fine mist or spray. A number of these spray interfaces are available, and several have considerable potential, especially for studies of biological molecules which tend to be chromatographed in buffered liquid phases. These interfaces are derived from earlier atmospheric pressure ionisation (API) sources.

Thermospray interfaces

The earliest of the spray interfaces used for HPLC is the thermospray (TS) system, in which the column eluent (up to $2\,\mathrm{mL\,min^{-1}}$) containing a volatile electrolyte such as ammonium ethanoate is sprayed through a very fine vaporiser with a heated ruby tip into a heated chamber pumped by an auxiliary vacuum pump. The solvent rapidly evaporates from the supersonic jet of fine droplets, leaving behind the sample as a sheath of solid electrolyte ions. Because the particles are so small, the electrostatic field induced by the electrolyte causes a soft ionisation of the sample and the ions may then be directed straight into an analyser by a repeller. This system acts both as an interface and as an ion source; it has been used for large polar or ionic samples in ionic solvents, giving very simple mass spectra that often contain only the molecular ion or pseudo-molecular ion. This simplicity of spectra is to some extent a limitation of the method, since lack of fragmentation means interpretation or library searching is rather limited. The greatest problem is that it only works for polar molecules in polar solvents (largely aqueous solutions).

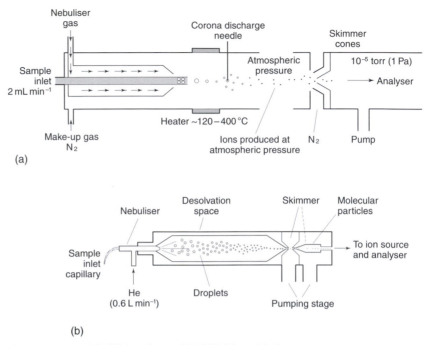

Figure 19.11   LC-MS interfaces: (a) APCI, (b) particle beam

Electrospray ionisation

In an electrospray ionisation (ESI) source the eluent from an HPLC is passed through a metallic capillary needle into a warm zone at atmospheric pressure. A high potential of up to 8 kV relative to a counter-electrode close to the capillary produces a strong electric field that induces charges on the droplets as they emerge from the capillary. The charged droplets are then caused to evaporate by a stream of heated (80–150 °C) nitrogen gas. As the liquid evaporates from the charged droplets, they shrink until the surface charge acquired on passing through the capillary is sufficient to break down the cohesion forces in the molecule and ionisation occurs (still at atmospheric pressure). The ions are then transmitted via a small orifice or skimmer into the high-vacuum mass spectrometer. ESI is best suited to polar molecules, although species with non-ionisable sites can still be ionised via adduct formation. Highest efficiencies are observed for relatively small flow rates (as low as a few $\mu L\,min^{-1}$), which can cause problems with interfacing to a standard HPLC system, but assisted nebulisation using a high flow of nitrogen gas can extend this up to $1\,mL\,min^{-1}$. The high ionisation efficiency of the process can give extremely low detection limits for suitable samples. Using flow rates of a few $nL\,min^{-1}$ μESI is capable of the highest sensitivity currently available in HPLC-MS, with detection limits of $pg\,\mu L^{-1}$.

Due to the nature of the processes which occur during atmospheric ionisation, ESI normally gives high yields of multiply charged ions with up to 100 charges on a single ion. Since mass spectrometry measures m/z, this means that a particle of say 25 000 u carrying 50 charges will appear at the same place as a singly charged ion with mass of 500 u. Thus, by coupling an electrospray source to a conventional (low-mass) mass spectrometer, it is quite possible to make measurements on high-mass molecules indirectly. The technique is already widely used for determinations on a number of high-mass biopolymers such as proteins and protein digests. The website given in Section 19.14 explains how to calculate molecular masses of these large molecules.

Although perhaps not having quite the mass capabilities of MALDI-TOF, ESI does bring the capability of high-mass measurements to laboratories only equipped with low-mass analysers. Each of the systems has advantages and disadvantages, and the exact requirements for a particular determination will determine which is most suitable for that application.

Atmospheric pressure chemical ionisation

Atmospheric pressure chemical ionisation (APCI) uses a corona discharge around an electrode in the heated zone beyond the capillary (Figure 19.11a). The eluent from the HPLC (up to $2\,mL\,min^{-1}$) passes through a heated capillary, where pneumatic nebulisation occurs via a flow of warm gas (normally nitrogen) to produce a fine mist of droplets. These droplets pass into a heated desolvation/vaporisation zone which can be heated up to 450 °C. While in this zone a primary ionisation of the mobile phase molecules is induced through proximity with an electrode at high d.c. potential, producing a corona discharge. Since these ions remain in the desolvation zone for several seconds, they are able to interact with the sample molecules, creating a secondary ionisation analogous to the low-pressure CI source described in Section 19.4.2. The source can be operated in the positive or negative CI mode and tends to give pseudomolecular ions: $(M + H)^+$ in the positive mode and $(M - H)^-$ in the negative mode. All these processes occur at atmospheric pressure, so the source is interfaced to the mass spectrometer analyser, which as for ESI, is at high vacuum via a small orifice or skimmer.

Unlike ESI, APCI is not well suited to the analysis of polar compounds, nor does it give multiply charged ions, so it cannot be used for high-mass substances. However, it does ionise non-polar compounds much more effectively than ESI and can be used with normal-phase HPLC systems as well as reversed-phase (aqueous base) systems. The device is extremely tolerant of 'dirty' solutions, and this relative freedom from contamination coupled with acceptance of high flow rates means that APCI is rapidly becoming a standard source for HPLC-MS applications, especially environmental monitoring of non-polar substances at low concentrations.

Particle beam interface

Another spray interface which can be used to couple HPLC to MS is the particle beam interface (PBI). Eluent is mixed with helium before being sprayed through a nebuliser into a heated chamber; the solvent, which rapidly evaporates, is again pumped away by an auxiliary pumping system, and the desolvated particles impinge on a skimmer and then into a conventional EI or CI source (Figure 19.11b). This is not APCI, it has to be attached to a conventional ionisation source. Compared with other systems, PBI is claimed to have much greater tolerance to viscosity or surface tension changes in the HPLC eluent and can handle the full range of HPLC solvents, although its performance can vary considerably with solvent type. The other advantage is that conventional spectra are obtained which may be compared against standard computer libraries.

A very large number of sample types can be separated by HPLC compared to GC, especially in the important research areas of pharmaceuticals and life sciences; that is why instrument manufacturers have made and continue to make considerable efforts in improving the specification, application and ease of use of their interfaces and sources. It seems likely that what today appears the ideal interface in terms of sensitivity, reliability and ease of quantification will soon be replaced by even more versatile, sensitive and stable systems in the near future.

19.11.3   Supercritical fluid chromatography–mass spectrometry (SFC-MS)

Supercritical fluid chromatography offers the advantages of both GC and HPLC but with few of the disadvantages. This is particularly true when coupled to mass spectrometry, since SFC columns may be directly coupled to the ion source without the need for complicated interfaces such as those described for HPLC. The most commonly used fluid, carbon dioxide, presents few difficulties for the mass spectrometer or its pumping system, although the spectra obtained are often a type of CI where the large excess of CO_2 acts as a reagent gas. Although a large amount of interest has been shown in this technique and interesting applications have been reported,[17] it has been a little slow to develop commercially, and SFC-MS has yet to justify its potential.

19.11.4   Tandem mass spectrometry (MS-MS)

Before discussing this technique, it is necessary to consider not only what can happen to produce ions in a mass spectrometer but how fast these processes occur. Until now it has been assumed that molecules in the ion source are converted into a molecular ion, and then ions with high enough energy break up to give fragment ions. This mixture of ions is then separated in the analyser and detected as the mass spectrum. The timescale for this whole process in a typical double-focusing spectrometer is about 5–30 µs. But if the ion fragments are produced slowly (decomposition rates of 10^4 to $10^6 \, s^{-1}$) then some of these daughter ions will be produced after leaving the ion source but before reaching the detector. These are known as metastable ions and have been observed in conventional mass spectra for many years, usually as broad non-integral-mass peaks. (Most computerised systems will not display metastable peaks in their normal modes of operation.)

The exact form and intensity of these metastable ions depends on a number of instrumental factors such as the temperature and pressure in various parts of the system. Their observation also depends on the type of analyser employed; for example, they are not distinguished from normal product ions in a quadrupole filter, and only those ions formed after the electric sector in conventional double-focusing spectrometers will reach the detector, those formed earlier will not pass through the electrostatic field. It appears that metastable ions are only formed under defined conditions, are often not transmitted through the system, and even if they are, they may not be detected! However, they provide the basis for one of the most interesting and important developments, not just in mass spectrometry, but in analysis generally. This is tandem mass spectrometry (MS-MS), the separation and identification of samples in a mixture by using mass spectrometry.[18]

The simplest introduction to this technique is to consider the experiment known as MIKES or mass-analysed ion kinetic energy spectroscopy, which may be performed on a reverse-geometry (B-E) double-focusing instrument. The field of the electromagnet is fixed at a value which allows only ions M of one particular mass m to pass from the source to the region between the magnet and the electric sector; this is known as the second field-free region, the first is in the source. Some of these ions may decompose here to give metastable daughter ions of masses m_1, m_2, m_3, etc., which have the same velocity as the parent ion but different kinetic energy. By scanning the voltage of the electrostatic sector, each of these ions may be focused at the detector in turn. The electrostatic sector is then acting as a mass filter rather than just an energy filter. The net result is that all the daughters of a given parent may be determined, and by altering the magnetic and electrostatic fields appropriately, a series of mass spectra are obtained for each parent ion.

This gives the same type of analytical capability as combined chromatography and mass spectrometry, i.e. separation then analysis, but very much faster than conventional chromatographic techniques using physical separation; the whole process only takes fractions of a second rather than minutes or hours. Often a collision chamber containing an inert gas such as helium or argon is introduced into the second field-free region to enhance the production of daughter ions by collision processes. This is known as collision-induced dissociation (CID), and in these circumstances the sensitivity is better than conventional GC-MS. This experiment may even be performed on instruments with conventional E-B geometry by linking the operations of each sector in some predefined manner. The most widely used method is linked B/E. The magnetic field B and the electrostatic field E are set to pass a particular ion M and then are both scanned simultaneously keeping the ratio B/E constant. This again results in all of the daughter ions for a given parent being detected. A complementary experiment, linked B^2/E, enables the analyst to identify all the parent ions that can give rise to a given daughter.

Although demonstrably a very powerful technique, it has a number of limitations. The ion composition produced by the ion source may not accurately represent the composition of the original mixture, due to ion suppression effects of one component upon another. Even with conventional ion sources, simple mixtures will give rise to a large number of parent ions, hence to a complex set of data (each parent ion M giving a number of daughter ions, some of which may be produced by other parents). One solution is to provide even more computing power to handle and simplify the data, but a more elegant method is to use a softer ionisation source such as CI, which gives a simpler set of parent ions, coupled to two or more spectrometers joined in series with a collision cell or cells between each component.

A number of possible multiple combinations can be used for MS-MS experiments, including BEEB, BEQQ, QQQ and QQTOF, where B, E and Q refer to the type of analyser. Although a **single** quadrupole filter cannot detect metastable ions, the combination QQQ is currently favoured for these tandem systems since the first Q can act as the parent ion separator; the second Q, which has an r.f. voltage but no d.c. voltage applied to the poles, acts as the collision cell and the third Q as the daughter ion filter (Figure 19.12). Apart

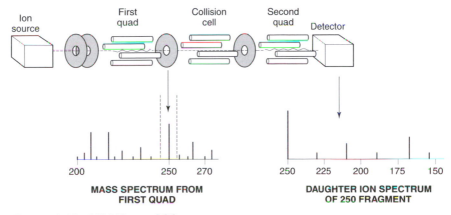

Figure 19.12   MS-MS on a QQQ

from the obvious advantage of using the simple, small, easily controlled Q in place of E or B sectors, this also means that low-energy ions (low acceleration voltages) can be passed through the system, and these ions are more readily collisionally activated giving extremely high sensitivity to the method.

This sensitivity and indeed selectivity can be further enhanced by using selected reaction monitoring (SRM). Here the system is set to determine only one specific ion produced in the reaction region. This ion is the result of a decomposition reaction undergone by an ion originally derived from the compound to be determined. Since only two ions are monitored, the original fragment and an ion produced on further reaction (in the reaction region) of this fragment, high sensitivity is possible for the same reasons as with single ion monitoring (Section 19.7), i.e. the spectrometer spends all its time observing ions of interest rather than scanning over a range of masses. Very high selectivity is also introduced into the measurement because two separate characteristics of the compound have to be satisfied; it must initially fragment to give a specific ion, and that ion must further react to give a second ion of defined mass.

In order to perform this type of experiment, the operating conditions of the tandem system have to be set for a particular compound, based on its known fragmentation and decomposition behaviour, with both the original fragment and its reaction daughter being carefully chosen. However, for selected systems, this process allows confirmation and quantification of a given compound in a complex matrix without prior chromatography and at high sensitivity. The quoted sensitivity of these QQQs is of the order of 100 fg, much lower than obtainable with conventional chromatographic front ends. The generation of large amounts of data, and a non-standard display of the components in the mixture still remain, but many analysts believe that MS-MS can supplant traditional GC-MS, HPLC-MS or SFC-MS for a number of applications.

19.11.5 Inductively coupled plasma–mass spectrometry (ICP-MS)

At first sight a rather unlikely combination, inductivey coupled plasma–mass spectrometry has proved to be so powerful that many laboratories now use it as the standard method for elemental analysis, and sales of commercial instruments are growing quickly despite high cost. The normal requirements for the mass spectrometry experiment, i.e. a single substance present in small discrete quantities in a high-vacuum zone, are rather different from those of ICP, i.e. continuous input of sample (usually aqueous) into a stream of argon gas at slightly above atmospheric pressure and subsequent heating to 8000 °C (Chapter 16). However, the plasma produced satisfies the basic requirements for mass spectrometry, in that it contains a high concentration of gas phase ions. A great deal of research has gone into the design of a suitable interface for these two techniques, and although far from perfect, the commercial instruments are efficient at transferring ions from the plasma to the analyser of the spectrometer (Figure 19.10).

The conventional ICP torch is normally turned on its side and directed onto a water-cooled nickel alloy sampling cone with a small orifice at its apex. Ions extracted through this orifice pass into a pumped chamber to reduce the pressure to about 1 torr (100 Pa) then they are pumped through another small aperture in a sampling cone into the analyser operating at 5×10^{-4} torr (70 mPa). The quadrupole filter separates the ions, which are then detected at a channel electron array detector. Using this type of system, a complete element analysis at concentrations of less than $1 \, \mu g \, kg^{-1}$ can be performed on a wide range of aqueous samples at about 10 s per element over a wide dynamic range. The technique does have some limitations, espcially interference from multiply charged ions of one element overlapping the line of another; the combination is very expensive and still under active development, but it has enormous potential in the fields of geological and environmental analysis.[19]

19.12   Future developments

Mass spectrometry has progressed from being a tool in the physics laboratory to become the most diversely used analytical technique. It can make measurements on biological molecules with masses of $100\,000\,u$ or it can determine dilute aqueous solutions of metal ions at parts per trillion concentrations. It may even replace gas chromatography as a separation technique. How the technique will evolve is difficult to predict, but several possibilities may become established techniques.

Additional methods for ionising samples are being investigated, in particular photo-ionisation using helium discharge lamps or lasers operating in the UV region, and these may well be used to give selective ionisation of some components in a mixture. New and better interfaces for coupling chromatography to MS, such as the spray interfaces, will become more reliable, more sensitive and more widely applicable. Dedicated hyphenated systems such as GC-MS and HPLC-MS will become cheaper and thus more widely used even in non-instrument laboratories, and special-purpose portable instruments will be developed. Already MS-MS is used to detect drugs and explosives in goods and luggage at customs points, by sampling headspace atmospheres, or to detect low levels of nerve gases in real time.[20]

Computer control and data handling will continue to enhance the capabilities of even the most modest systems, and computerised spectral interpretation rather than simple library search and compare, is under active development. Increases in both mass range and resolution will continue with dedicated high-mass instruments, based on TOF/laser ionisation, for biological studies becoming more widely available. Inorganic mass spectrometers designed to detect and quantify smaller samples at lower concentration will continue to extend the range of mass spectral analysis.

19.13   References

1. *Handbook of Chemistry and Physics*, 75th edn, CRC Press, Boca Raton FL (1994)
2. A Guarini, G Guglielmetti, I M Vincent, P Guarda and G Marchionni 1993 *Anal. Chem.*, **65** (8); 970
3. E L Esmans, D Broes, I Hoes, F Lemiere and K Vanhoutte 1998 *J. Chrom. A*, **794**; 109
4. Jie J Wei 1996 The application of MALDI and tandem reflectron TOF mass spectrometry to the analysis of biomolecular ions, University of Glasgow
5. H W Kroto, J R Heath, S C O'Brien, R F Curl and R E Smalley 1985 *Nature*, **318**; 162
6. R E March and J F J Todd (eds) 1995 *Practical applications of ion trap mass spectrometry*, CRC Press, London
7. F W McLafferty and A Staufer 1989 *The Wiley/NBS registry of mass spectral data*, 7 volumes, John Wiley, New York; *Eight peak index of mass spectra*, 4th edn, Royal Society of Chemistry, Cambridge (1992); *NIST/EPA/NIH mass spectral database*, v. 4.0, Royal Society of Chemistry, Cambridge (1992)
8. W Jambers and R Vangrieken 1996 *Trends in Analytical Chemistry*, **15** (3); 114
9. W Kulik, J Meesterburrie, C Jakobs and K deMeer 1998 *J. Chrom. B*, **710**; 37

10. H Craig 1957 *Geochim. Cosmochim. Acta*, **12**; 133
11. A Rossman, J Koziet, G J Martin and M J Dennis 1997 *Anal. Chim. Acta*, **340**; 21
12. M Warner 1989 *Anal. Chem.*, **101A**; 61
13. J R Turnlund 1991 *Critical Reviews in Food Science and Nutrition*, **30** (4); 387
14. J H Benyon 1960 *Mass spectrometry and its applications to organic chemistry*, Elsevier, Amsterdam
15. R W McLafferty 1973 *Interpretation of Mass Spectra*, 2nd edn, Benjamin, Reading MA
16. P Arpino 1989 *Mass Spectrom. Rev.*, **8**; 35
17. S V Olesik 1991 *J. High Resolut. Chromatogr.* **14**; 5
18. K L Busch, G L Glish and S A McLuckey 1988 *Mass spectrometry/mass spectrometry: techniques and applications of tandem mass spectrometry*, VCH, New York
19. A Montaser (ed) 1998 *Inductively coupled mass spectrometry*, Wiley–VCH, Chichester
20. S N Ketkar, S Penn and W L Fite 1991 *Anal Chem.*, **63**; 457

19.14 Bibliography

J R Chapman 1993 *Practical organic mass spectrometry: a guide for chemical and biochemical analysis*, 2nd edn, John Wiley, Chichester

E Constantin and A Schnell 1990 *Mass spectrometry*, Ellis Horwood, London

R J Cotter 1997 *Time of flight mass spectrometry: instrumentation and applications in biological research*, American Chemical Society, Washington DC

K E Jarvis and A L Gray 1992 *Handbook of inductively coupled plasma mass spectrometry*, Blackie, London

R Johnstone and M E Rose 1996 *Mass spectrometry for chemists and biochemists*, 2nd edn, Cambridge University Press, Cambridge

F G Kitson, B S Larsen and C N McEwen 1996 *Gas chromatography and mass spectrometry: a practical guide*, Academic Press, London

I T Platzner *et al.* 1997 *Modern isotope ratio mass spectrometry*, John Wiley, Chichester

Journals and periodicals

Advances in Mass Spectrometry
Analytical Chemistry (biennial reviews from 1992 onwards)
Journal of Chromatography
Journal of Mass Spectrometry, (formerly *Organic Mass Spectrometry*)
Journal of the American Chemical Society
Journal of the American Society of Mass Spectrometry
Mass Spectrometry Reviews
Rapid Communications in Mass Spectrometry

Web software for MS calculations

http://userwww.service.emory.edu/~kmurray/mslist.html

Appendices

Appendix 1 Relative atomic masses 1994

Element	Symbol	Atomic no.	Atomic weight	Element	Symbol	Atomic no.	Atomic weight
Actinium	Ac	89	(227)	Mercury	Hg	80	200.59
Aluminium	Al	13	26.981 539	Molybdenum	Mo	42	95.94
Americium	Am	95	(243)	Neodymium	Nd	60	144.24
Antimony	Sb	51	121.760	Neon	Ne	10	20.179 7
Argon	Ar	18	39.948	Neptunium	Np	93	(237)
Arsenic	As	33	74.921 59	Nickel	Ni	28	58.693 4
Astatine	At	85	(210)	Niobium	Nb	41	92.906 38
Barium	Ba	56	137.327	Nitrogen	N	7	14.006 74
Berkelium	Bk	97	(247)	Nobelium	No	102	(259)
Beryllium	Be	4	9.012 182	Osmium	Os	76	190.23
Bismuth	Bi	83	208.980 37	Oxygen	O	8	15.999 4
Boron	B	5	10.811	Palladium	Pd	46	106.42
Bromine	Br	35	79.904	Phosphorus	P	15	30.973 762
Cadmium	Cd	48	112.411	Platinum	Pt	78	195.08
Caesium	Cs	55	132.905 43	Plutonium	Pu	94	(244)
Calcium	Ca	20	40.078	Polonium	Po	84	(209)
Californium	Cf	98	(251)	Potassium	K	19	39.098 3
Carbon	C	6	12.011	Praseodymium	Pr	59	140.907 65
Cerium	Ce	58	140.115	Promethium	Pm	61	(145)
Chlorine	Cl	17	35.452 7	Protactinium	Pa	91	231.035 88
Chromium	Cr	24	51.996 1	Radium	Ra	88	(226)
Cobalt	Co	27	58.933 20	Radon	Rn	86	(222)
Copper	Cu	29	63.546	Rhenium	Re	75	186.207
Curium	Cm	96	(247)	Rhodium	Rh	45	102.905 50
Dysprosium	Dy	66	162.50	Rubidium	Rb	37	85.467 8
Einsteinium	Es	99	(252)	Ruthenium	Ru	44	101.07
Erbium	Er	68	167.26	Rutherfordium	Rf	104	(261)
Europium	Eu	63	151.965	Samarium	Sm	62	150.36
Fermium	Fm	100	(257)	Scandium	Sc	21	44.955 910
Fluorine	F	9	18.998 403 2	Selenium	Se	34	78.96
Francium	Fr	87	(223)	Silicon	Si	14	28.085 5
Gadolinium	Gd	64	157.25	Silver	Ag	47	107.868 2
Gallium	Ga	31	69.723	Sodium	Na	11	22.989 768
Germanium	Ge	32	72.61	Strontium	Sr	38	87.62
Gold	Au	79	196.966 54	Sulphur	S	16	32.066
Hafnium	Hf	72	178.49	Tantalum	Ta	73	180.947 9
Hahnium	Ha	105	(262)	Technetium	Tc	43	(98)
Helium	He	2	4.002 602	Tellurium	Te	52	127.60
Holmium	Ho	67	164.930 32	Terbium	Tb	65	158.925 34
Hydrogen	H	1	1.007 94	Thallium	Tl	81	204.383 3
Iodine	I	53	126.904 47	Thulium	Tm	69	168.934 21
Indium	In	49	114.818	Thorium	Th	90	232.038 1
Iridium	Ir	77	192.217	Tin	Sn	50	118.710
Iron	Fe	26	55.845	Titanium	Ti	22	47.867
Krypton	Kr	36	83.80	Tungsten	W	74	183.84
Lanthanum	La	57	138.905 5	Uranium	U	92	238.028 9
Lawrencium	Lr	103	(262)	Vanadium	V	23	50.941 5
Lead	Pb	82	207.2	Xenon	Xe	54	131.29
Lithium	Li	3	6.941	Ytterbium	Yb	70	173.04
Lutetium	Lu	71	174.967	Yttrium	Y	39	88.905 85
Magnesium	Mg	12	24.305 0	Zinc	Zn	30	65.39
Manganese	Mn	25	54.938 05	Zirconium	Zr	40	91.224
Mendelevium	Md	101	(258)				

Notes: This table is scaled to the relative atomic mass $A_r(^{12}C) = 12$. Values in parentheses refer to the isotope of longest known half-life for radioactive elements.
Source: Based mainly on the Report of the Commission on Relative Atomic Masses 1994 *Pure and Applied Chemistry*, **66** (12); 2423–44.

Appendix 2   Aqueous concentrations: common acids and ammonia

Reagent	Approximate			Volume (mL) to make 1 L ~1 M solution
	Wt%	Specific gravity	Molarity	
Ethanoic (acetic) acid	99.5	1.05	17.4	58
Hydrochloric acid	35	1.18	11.3	89
Hydrofluoric acid	46	1.15	26.5	38
Nitric acid	70	1.42	16.0	63
Perchloric acid	70	1.66	11.6	86
Phosphoric(V) acid	85	1.69	13.7	69
Sulphuric acid	96	1.84	18.0	56
Aqueous ammonia	27(NH_3)	0.90	14.3	71

Appendix 3   Saturated solutions of some reagents at 20 °C

Reagent	Formula	Specific gravity	Molarity	Quantities for 1 L saturated solution	
				Grams of reagent	mL of water
Ammonium chloride	NH_4Cl	1.075	5.44	291	784
Ammonium nitrate	NH_4NO_3	1.312	10.80	863	449
Ammonium oxalate	$(NH_4)_2C_2O_4,H_2O$	1.030	0.295	48	982
Ammonium sulphate	$(NH_4)_2SO_4$	1.243	4.06	535	708
Barium chloride	$BaCl_2,2H_2O$	1.290	1.63	398	892
Barium hydroxide	$Ba(OH)_2$	1.037	0.228	39	998
Barium hydroxide	$Ba(OH)_2,8H_2O$	1.037	0.228	72	965
Calcium hydroxide	$Ca(OH)_2$	1.000	0.022	1.6	1000
Mercury (II) chloride	$HgCl_2$	1.050	0.236	64	986
Potassium chloride	KCl	1.174	4.00	298	876
Potassium chromate	K_2CrO_4	1.396	3.00	583	858
Potassium dichromate	$K_2Cr_2O_7$	1.077	0.39	115	962
Potassium hydroxide	KOH	1.540	14.50	813	727
Sodium carbonate	Na_2CO_3	1.178	1.97	209	869
Sodium carbonate	$Na_2CO_3,10H_2O$	1.178	1.97	563	515
Sodium chloride	$NaCl$	1.197	5.40	316	881
Sodium ethanoate	CH_3COONa	1.205	5.67	465	740
Sodium hydroxide	$NaOH$	1.539	20.07	803	736

Appendix 4   Sources of analysed samples

Throughout this book the use of a number of standard analytical samples is recommended in order that practical experience may be gained on substances of known composition. In addition, standard reference materials of environmental samples for trace analysis are used for calibration standards, and pure organic compounds are employed as standard materials for elemental analysis.

BAS   The Bureau of Analytical Samples Ltd (BAS), Newham Hall, Newby, Middlesborough, Cleveland, UK, supplies samples suitable for metallurgical, chemical and spectroscopic analysis. A detailed list of British Chemical Standard (BCS) and EURONORM Certified Reference Materials (ECRM) is available. The Bureau of Analytical Samples distributes in the UK from the following overseas sources:

Alcan International, Arvida Laboratories, Canada
Bundesanstalt für Materialforschung und prüfung (BAM), Germany
Canada Centre for Mineral and Energy Technology (CANMET), Canada
Centre Technique des Industries de la Fonderie (CTIF), France
Institut de Recherches de la Sidérurgie Francaise (IRSID), France
National Bureau of Standards (NBS), United States
Research Institute CKD, Czechoslovakia
South African Bureau of Standards (SABS), South Africa
Swedish Institute for Metal Research (Jernkontoret), Sweden
SKF Steel (SKF), Sweden
Vasipari Kutato es Fejleszto Vallalat (VASKUT), Hungary

MBH   Elements, alloys, ceramics and oils of high purity and known compositions are marketed by MBH Analytical Ltd, Holland House, Queen's Road, Barnet EN5 4DJ, UK.

LGC   A reference materials advisory service is provided by the Laboratory of the Government Chemist (LGC), Queen's Road, Teddington TW11 0LY, UK.

BCR   The Community Bureau of Reference (BCR), 200 rue de la Loi, B-1049 Brussels, Belgium, supples geological, environmental and organic compounds for elemental analysis and artificated samples for trace metal analysis.

United States   In the United States a wide range of standards may be obtained from the US Department of Commerce, National Bureau of Standards, Washington DC 20234.

Japan   Environmental reference materials are available from the National Institute for Environmental Studies, Yatabe-machi Tsukuba, Ibarasi 305, Japan.

Appendix 5   Buffer solutions and secondary pH standards

The British standard for the pH scale is a $0.05\,M$ solution of potassium hydrogenphthalate (British Standard 1647: 1984, Parts 1, 2) which has a pH of 4.001 at $20\,°C$. The IUPAC standard is $0.05\,M$ potassium hydrogenphthalate. Subsidiary pH standards at $25\,°C$ include the following.

Appendix 5   Buffer solutions and secondary pH standards

	pH
0.05 M HCl + 0.09 M KCl	2.07
0.1 M Potassium tetroxalate	1.48
0.1 M Potassium dihydrogen citrate	3.72
0.1 M Ethanoic acid + 0.1 M sodium ethanoate	4.64
0.01 M Ethanoic acid + 0.01 M sodium ethanoate	4.70
0.01 M KH_2PO_4 + 0.01 M Na_2HPO_4	6.85
0.05 M Borax	9.18
0.025 M $NaHCO_3$ + 0.025 M Na_2CO_3	10.00
0.01 M Na_3PO_4	11.72

The following table covering the pH range 2.6–12.0 (18 °C) is included as an example of a universal buffer mixture. A mixture of 6.008 g of citric acid, 3.893 g of potassium dihydrogenphosphate, 1.769 g of boric acid and 5.226 g of pure diethylbarbituric acid is dissolved in water and made up to 1 L. The pH values at 18 °C of mixtures of 100 mL of this solution with various volumes X of 0.2 M sodium hydroxide solution (free from carbonate) are tabulated below.

pH	X (mL)	pH	X (mL)	pH	X (mL)
2.6	2.0	5.9	36.5	9.0	72.7
2.8	4.3	6.0	38.9	9.2	74.0
3.0	6.4	6.2	41.2	9.4	75.9
3.2	8.3	6.4	43.5	9.6	77.6
3.4	10.1	6.6	46.0	9.8	79.3
3.6	11.8	6.8	48.3	10.0	80.8
3.8	13.7	7.0	50.6	10.2	82.0
4.0	15.5	7.2	52.9	10.4	82.9
4.2	17.6	7.4	55.8	10.6	83.9
4.4	19.9	7.6	58.6	10.8	84.9
4.6	22.4	7.8	61.7	11.0	86.0
4.8	24.8	8.0	63.7	11.2	87.7
5.0	27.1	8.2	65.6	11.4	89.7
5.2	29.5	8.4	67.5	11.6	92.0
5.4	31.8	8.6	69.3	11.8	95.0
5.6	34.2	8.8	71.0	12.0	99.6

For many purposes an exact pH value is not required; it suffices if the pH of the solution lies within an appropriate pH range, and the text gives details of buffer solutions required for particular procedures. Buffer solutions covering a range of pH values are given below:

	pH range
Hydrochloric acid–sodium citrate	1.0– 5.0
Citric acid–sodium citrate	2.5– 5.6
Ethanoic acid–sodium ethanoate	3.7– 5.6
Disodium hydrogenorthophosphate–sodium dihydrogenorthophosphate	6.0– 9.0
Aqueous ammonia–hydrochloric acid	8.2–10.2
Sodium tetraborate–sodium hydroxide	9.2–11.0

The simplest method of preparing these solutions is to take a $0.1\,M$ solution of the acid component, and then to add $0.1\,M$ sodium hydroxide solution until the appropriate pH value is attained (use a pH meter). For the hydrochloric acid–sodium citrate buffer and the aqueous ammonia–hydrochloric acid buffer, $0.1\,M$ hydrochloric acid is added to the second component.

Appendix 6a   Dissociation constants of some acids in water at 25 °C

Acid		$pK_a = -\log K_a$	Acid		$pK_a = -\log K_a$
Aliphatic acids					
methanoic (Formic)		3.75	Succinic	K_1	4.21
ethanoic (Acetic)		4.76		K_2	5.64
Propanoic		4.88	Glutaric	K_1	4.34
Butanoic		4.82		K_2	5.27
3-Methylpropanoic		4.85	Adipic	K_1	4.43
Pentanoic		4.84		K_2	5.28
Fluoroethanoic		2.58	Methylmalonic	K_1	3.07
Chloroethanoic		2.86		K_2	5.87
Bromoethanoic		2.90	Ethylmalonic	K_1	2.96
Iodoethanoic		3.17		K_2	5.90
Cyanoethanoic		2.47	Dimethylmalonic	K_1	3.15
Diethylethanoic		4.73		K_2	6.20
Lactic		3.86	Diethylmalonic	K_1	2.15
Pyruvic		2.49		K_2	7.47
Acrylic		4.26	Fumaric	K_1	3.02
Vinylacetic		4.34		K_2	4.38
Tetrolic		2.65	Maleic	K_1	1.92
trans-Crotonic		4.69		K_2	6.23
Furoic		3.17	Tartaric	K_1	3.03
Oxalic	K_1	1.27		K_2	4.37
	K_2	4.27	Citric	K_1	3.13
Malonic	K_1	2.85		K_2	4.76
	K_2	5.70		K_3	6.40
Aromatic acids					
Benzoic		4.20			
Phenylethanoic		4.31	2-Benzoylbenzoic		3.54
Sulphanilic		3.23	Phthalic K_1		2.95
Phenoxyethanoic		3.17	K_2		5.41
Mandelic		3.41	*cis*-Cinnamic		3.88
1-Naphthoic		3.70	*trans*-Cinnamic		4.44
2-Naphthoic		4.16	Phenol		10.00
1-Naphthylethanoic		4.24	1-Nitroso-2-naphthol		7.77
2-Naphthylethanoic		4.26	2-Nitroso-1-naphthol		7.38

Appendix 6a   (cont'd)

	pK_a = −log K_a		
	ortho (2-)	meta (3-)	para (4-)
Fluorobenzoic	3.27	3.86	4.14
Chlorobenzoic	2.94	3.83	3.98
Bromobenzoic	2.85	3.81	3.97
Iodobenzoic	2.86	3.85	3.93
Hydroxybenzoic	3.00	4.08	4.53
Methoxybenzoic	4.09	4.09	4.47
Nitrobenzoic	2.17	3.49	3.42
Aminobenzoic	4.98	4.79	4.92
Toluic	3.91	4.24	4.34
Chlorophenol	8.48	9.02	9.38
Nitrophenol	7.23	8.40	7.15
Methylphenol (cresol)	10.29	10.09	10.26
Methoxyphenol	9.98	9.65	10.21

Acid		pK_a = −log K_a	Acid		pK_a = −log K_a
Arsenious		9.22	Nitrous		3.35
Arsenic	K_1	2.30	Phosphoric(V)	K_1	2.12
	K_2	7.08		K_2	7.21
	K_3	9.22		K_3	12.30
Boric		9.24	Phosphorous	K_1	1.8
Carbonic	K_1	6.37	(phosphonic)	K_2	6.15
	K_2	10.33	Sulphuric	K_1	1.92
Hydrocyanic		9.14	Sulphurous	K_1	1.92
Hydrofluoric		4.77		K_2	7.20
Hydrogen sulphide	K_1	7.00	Thiosulphuric	K_1	1.7
	K_2	14.00		K_2	2.5
Chloric(I) (hypochlorous)		7.25			

Appendix 6b   Acidic dissociation constants of bases in water at 25 °C

The constants for bases are expressed as acidic dissociation constants, e.g. for ammonia the value pK_a = 9.24 is given for the ammonium ion:

$$NH_4^+ + H_2O \rightleftharpoons NH_3 + H_3O^+$$

This means that bases are considered in terms of ionisation of the conjugated acids. The basic dissociation constant for the reaction

$$NH_3 + H_2O \rightleftharpoons NH_4^+ + OH^-$$

may then be obtained from the relation

$$pK_a \text{ (acidic)} + pK_b \text{ (basic)} = pK_w \text{ (water)}$$

where pK_w is 14.00 at 25 °C.* For simplicity, the name of the base will be expressed in the 'basic' form, e.g. ammonia for ammonium ion, propylamine for propylammonium ion, piperidine for piperidinium ion, aniline for anilinium ion, etc., although this is not strictly correct. No difficulty should be experienced in writing down the correct name, if required.

Base		pK_a	Base		pK_a
Ammonia		9.24	Hydrazine		7.93
Methylamine		10.64	Hydroxylamine		5.82
Ethylamine		10.63	Benzylamine		9.35
Propylamine		10.57	Aniline		4.58
Butylamine		10.62	o-Toluidine		4.39
Cyclohexylamine		10.64	m-Toluidine		4.68
Dimethylamine		10.77	p-Toluidine		5.09
Diethylamine		10.93	2-Chloroaniline		2.62
Monoethanolamine		9.50	3-Chloroaniline		3.32
Triethanolamine		7.77	4-Chloroaniline		3.81
Trimethylamine		9.80	N-Methylaniline		4.85
Triethylamine		10.72	N,N-Dimethylaniline		5.15
Tris(hydroxymethyl)aminomethane		8.08	Pyridine		5.17
Piperidine		11.12	2-Methylpyridine		5.97
Ethylenediamine	K_1	7.50	3-Methylpyridine		5.68
	K_2	10.09	4-Methylpyridine		6.02
1,3-Propylenediamine	K_1	8.64	Benzidine	K_1	4.97
	K_2	10.62		K_2	3.75
1,4-Butylenediamine	K_1	9.35	1,10-Phenanthroline		4.86
	K_2	10.80			

* The values at 20 °C and 30 °C are 14.17 and 13.83, respectively.

Appendix 7   Polarographic half-wave potentials

Ion	Supporting electrolyte	$E_{1/2}$ (volts vs SCE)
Ba^{2+}	0.1 M N(CH$_3$)$_4$Cl	−1.94
Bi^{3+}	1 M HCl	−0.09
	0.5 M H$_2$SO$_4$	−0.04
	0.5 M Tartrate + 0.1 M NaOH	−1.0
	0.5 M Sodium hydrogentartrate, pH 4.5	−0.23
Cd^{2+}	0.1 M KCl	−0.64
	1 M NH$_3$ + 1 M NH$_4^+$	−0.81
	1 M HNO$_3$	−0.59
	1 M KI	−0.74
	1 M KCN	−1.18
Co^{2+}	0.1 M KCl	−1.20
	0.1 M Pyridine + 0.1 M pyridinium ion	−1.07

Appendix 7 (cont'd)

Ion	Supporting electrolyte	$E_{1/2}$ (V vs SCE)
Cu^{2+}	0.1 M KCl	+0.04
	1 M NH_3 + 1 M NH_4Cl	−0.24 (1st wave)
		−0.50 (2nd wave)
	0.5 M Sodium hydrogentartrate, pH 4.5	−0.09
Fe^{2+}	0.1 M KCl	−1.3
Fe^{3+}	0.5 M Tartrate, pH 9.4	−1.20 (1st wave)
		−1.73 (2nd wave)
	0.1 M EDTA + 2 M CH_3COONa	−0.13 (1st wave)
		−1.3 (2nd wave)
K^+	0.1 M $N(CH_3)_4OH$ in 50% ethanol	−2.10
Li^+	0.1 M $N(CH_3)_4OH$ in 50% ethanol	−2.31
Mn^{2+}	1 M KCl	−1.51
	0.2 M $H_2P_2O_7^{2-}$, pH 2.2	+0.1
Na^+	0.1 M $N(CH_3)_4Cl$	−2.07
Ni^{2+}	1 M KCl	−1.1
	1 M KSCN	−0.70
	1 M KCN	−1.36
	1 M Pyridine + HCl, pH 7.0	−0.78
	1 M NH_3 + 0.2 M NH_4^+	−1.06
O_2	Most buffers, pH 1–10	−0.05 (1st wave)
		−0.9 (2nd wave)
Pb^{2+}	0.1 M KCl	−0.40
	1 M HNO_3	−0.40
	1 M NaOH	−0.75
	0.5 M Sodium hydrogentartrate, pH 4.5	−0.48
	0.5 M Tartrate + 0.1 M NaOH	−0.75
Sn^{2+}	1 M HCl	−0.47
Sn^{4+}	1 M HCl + 4 M NH_4^+	−0.25 (1st wave)
		−0.52 (2nd wave)
Zn^{2+}	0.1 M KCl	−1.00
	1 M NaOH	−1.53
	1 M NH_3 + 1 M NH_4^+	−1.33
	0.5 M Tartrate, pH 9	−1.15

In stripping voltammetry the stripping potential of a given ion is generally close to the polarographic half-wave potential of that ion in solutions with similar supporting electrolytes. Thus, typical stripping potentials in a 0.05 M potassium chloride base solution are as follows: Zn −1.00 V, Cd −0.07 V, Pb −0.45 V, Bi −0.10 V, Cu(II) −0.05 V. The half-wave potentials observed with some organic compounds are shown below. Where a percentage is quoted against a single organic solvent, the other solvent is water.

Compound	Supporting electrolyte	$E_{1/2}$ (V vs SCE)
Acids		
Ethanoic	Et_4NClO_4; CH_3CN	−2.3
Ascorbic	Ethanoic acid–ethanoate buffer (pH 3.4)	+0.17
Fumaric	0.1 M NH_4OH; 0.1 M NH_4Cl (pH 8.2)	−1.57
Maleic	0.1 M NH_4OH; 0.1 M NH_4Cl (pH 8.2)	−1.36
Thioglycollic	Phosphate buffer (pH 6.8)	−0.38
Carbonyl		
Ethanal	0.6 M LiOH; 0.07 M LiCl	−1.89
Acetone	0.1 M Bu_4NCl; 0.1 M Bu_4NOH; 80% EtOH	−2.53
Benzaldehyde	0.1 M Bu_4NCl; 0.1 M Bu_4NOH; 80% EtOH	−1.57
Fructose	0.1 M LiCl	−1.76
Glucose	0.1 M KCl	−1.55
Anthraquinone	NH_4OH/NH_4Cl buffer (pH 7.4); 40% dioxan	−0.54
Benzoquinone	Ethanoic acid–ethanoate buffer (pH 5.4); 50% MeOH	+0.15
Benzoyl peroxide	0.3 M LiCl; 50% MeOH; 50% C_6H_6	0.0
Nitro		
Nitrobenzene	Ethanoic acid–ethanoate buffer (pH 3); 50% EtOH	−0.43
2-Nitrophenol	Ethanoic acid–ethanoate buffer (pH 3); 50% EtOH	−0.23
3-Nitrophenol	Ethanoic acid–ethanoate buffer (pH 3); 50% EtOH	−0.37
4-Nitrophenol	Ethanoic acid–ethanoate buffer (pH 3); 50% EtOH	−0.35
Unsaturated hydrocarbons		
Naphthalene	0.18 M Bu_4NI; 75% dioxan	−2.50
Anthracene	0.18 M Bu_4NI; 75% dioxan	−1.94
Phenylacetylene	0.18 M Bu_4NI; 75% dioxan	−2.37
Stilbene	0.18 M Bu_4NI; 75% dioxan	−2.26
Styrene	0.18 M Bu_4NI; 75% dioxan	−2.35
Heterocyclic		
Pyridine	0.1 M HCl; 50% EtOH	−1.49
Quinoline	0.2 M Me_4NOH; 50% MeOH	−1.50
Miscellaneous		
Chloromethane	0.05 M Et_4NBr; dimethylformamide	−2.23
Diethyl sulphide	0.025 M Bu_4NOH; MeOH, PrOH, H_2O (2:2:1)	−1.78

Appendix 8   Resonance lines for atomic absorption

Element	Symbol	Absorbing lines (nm)		Element	Symbol	Absorbing lines (nm)	
		Most sensitive	Alternatives			Most sensitive	Alternatives
Aluminium	Al	396.2	308.2	Iridium	Ir	208.9	264.0
			309.3				266.5
			394.4	Iron	Fe	248.3	248.4
Antimony	Sb	217.6	206.8				372.0
			217.9				386.0
			231.2				392.0
Arsenic	As	193.7	189.0	Lanthanum	La	550.1	403.7
			197.2	Lead	Pb	217.0	261.4
Barium	Ba	553.5	455.4				283.3
			493.4	Lithium	Li	670.8	323.3
Beryllium	Be	234.9	–	Lutetium	Lu	335.9	356.7
Bismuth	Bi	223.1	222.8				337.6
			227.7	Magnesium	Mg	285.2	202.5
			306.8	Manganese	Mn	279.5	279.8
Boron	B	249.8	208.9				280.1
Cadmium	Cd	228.8	326.1				403.1
Caesium	Cs	852.1	455.6	Mercury	Hg	253.7	–
Calcium	Ca	422.7	–	Molybdenum	Mo	313.3	320.9
Cerium	Ce	520.0	569.7	Neodymium	Nd	492.5	463.4
Chromium	Cr	357.9	425.4	Nickel	Ni	232.0	231.1
			427.5				341.5
			429.0				351.5
			520.4				352.4
			520.8				362.5
Cobalt	Co	240.7	304.4	Niobium	Nb	334.9	405.9
			346.6				408.0
			347.4				412.4
			391.0	Osmium	Os	290.9	305.9
Copper	Cu	324.8	217.9				426.0
			218.2	Palladium	Pd	247.6	244.8
			222.6				340.5
			244.2	Phosphorus	P	213.6	214.9
			249.2	Platinum	Pt	265.9	264.7
			327.4				299.8
Dysprosium	Dy	419.5	404.6				306.5
Erbium	Er	400.8	389.3	Potassium	K	766.5	404.4
Europium	Eu	459.4	462.7				769.9
Gadolinium	Gd	368.4	405.8	Praseodymium	Pr	495.1	513.3
			407.9	Rhenium	Re	346.0	346.5
Gallium	Ga	287.4	403.3	Rhodium	Rh	343.5	328.1
			417.2				369.2
Germanium	Ge	265.1	271.0	Rubidium	Rb	780.0	794.8
Gold	Au	242.8	267.6	Ruthenium	Ru	349.9	392.6
Hafnium	Hf	307.8	268.2	Samarium	Sm	429.7	476.0
Holmium	Ho	410.4	425.4	Scandium	Sc	391.2	390.8
			405.4	Selenium	Se	196.0	204.0
Hydrogen	H	Continuum		Silicon	Si	251.6	250.7
Indium	In	303.9	325.6				251.4
			410.2				252.4
			451.1				288.1

Appendix 8   (cont'd)

Element	Symbol	Absorbing lines (nm)		Element	Symbol	Absorbing lines (nm)	
		Most sensitive	Alternatives			Most sensitive	Alternatives
Silver	Ag	328.1	338.3	Titanium	Ti		399.0
Sodium	Na	589.0	330.2				399.8
			589.6	Tungsten	W	255.1	294.7
Strontium	Sr	460.7	407.8				400.9
Tantalum	Ta	271.5	275.8				407.4
Tellurium	Te	214.3	225.9	Uranium	U	358.5	356.6
Terbium	Tb	432.7	431.9				351.4
			433.8	Vanadium	V	318.5	306.6
Thallium	Tl	276.7	258.0				318.4
Thorium	Th	371.9	–	Ytterbium	Yb	398.8	346.4
Thulium	Tm	371.8	436.0				246.4
			410.6	Yttrium	Y	410.2	414.2
Tin	Sn	233.5	224.6	Zinc	Zn	213.9	307.6
			266.1	Zirconium	Zr	360.1	468.7
Titanium	Ti	364.3	365.4				354.8

Appendix 9   Common chromophores: electronic absorption characteristics

Chromophore	λ_{max} (nm)	ε_{max}
Acetylide (—C≡C—)	175–180	6000
Aldehyde (—CHO)	210	strong
Amine (—NH$_2$)	195	2800
Bromide (—Br)	208	300
Carboxyl (—COOH)	200–210	50–70
Ethylene (—C=C—)	190	8000
Esters (—COOR)	205	50
Ether (—O—)	185	1000
Ketone (C=O)	195	1000
Iodide (—I)	260	400
Nitrate (—ONO$_2$)	270	12
Nitrite (—ONO)	220–230	1000–2000
Nitro (—NO$_2$)	210	strong
Nitroso (—N=O)	302	100
Oxime (—NOH)	190	5000
Aromatic and heterocyclic species[a]		
Benzene	184, 202	46 700, 6900
Naphthalene	220, 275	112 000, 5600
Pyridine	174, 195	80 000, 6000
Quinoline	227, 270	37 000, 3600

[a] The first λ_{max} in each pair corresponds with the first ε_{max}, and the second λ_{max} with the second ε_{max}.

Appendix 10   Characteristic infrared absorption bands[a]

		4000	3000	2000	1600	1200	800	400 cm⁻¹
O—H	Phenols, alcohols							
	Free		□m					
	H-bonded		m□					
	Carboxylic acids		m▭					
N—H	Amides, primary		m□		□m-s			
	and secondary amines							
C—H	Aromatic							
	(stretch)		s□					
	(out-of-plane bend)						▭s	
	Alkanes (stretch)		□s					
	CH₃— (bend)				m□ □m			
	—CH₂— (bend)				m□			
	Alkenes							
	(stretch)		m▯					
	(out-of-plane bend)						▭s	
C≡C	Alkyne			□m-w				
C≡N	Nitriles			m□				
C=O[b]	Aldehyde				□s			
	Ketone				□s			
	Acid				□s			
	Ester				□s			
	Amide				□s			
	Anhydride				s□□s			
C=C	Alkene				□m-w			
	Aromatic				m-w▯ □m-w			
C—O	Esters, ethers,					▭s		
	Anhydrides, alcohols					▭s		
	Carboxylic acids					▭s		
N=C	Nitro (RNO₂)				□s	□s		
C—Hal.	Fluoride					▭s		
	Chloride						▭s	
	Bromide, iodide							▭s

| | | 4000 3000 2000 | | | 1600 | 1200 | 800 | 400 cm⁻¹ |

[a] Key: w = weak, m = medium, s = strong.

[b] C=O stretching frequencies are typically lowered by about 20–30 cm⁻¹ from the values given when the carbonyl group is conjugated with an aromatic ring or an alkene group.

Source: R. Davis and C. H. J. Wells 1984 *Spectral problems in organic chemistry*, Chapman & Hall, New York.

Appendix 11   Percentage points of the *t*-distribution

The table gives the value of $t_{\alpha,v}$ – the 100α percentage point of the *t*-distribution for v degrees of freedom.

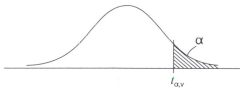

$t_{\alpha,v}$

The tabulation is for one tail only, i.e. for positive values of *t*. For $|t|$ the column headings for α must be doubled.

v	α						
	0.10	0.05	0.025	0.01	0.005	0.001	0.0005
1	3.078	6.314	12.706	31.821	63.657	318.31	636.62
2	1.886	2.920	4.303	6.965	9.925	22.326	31.598
3	1.638	2.353	3.182	4.541	5.841	10.213	12.924
4	1.533	2.132	2.776	3.747	4.604	7.173	8.610
5	1.476	2.015	2.571	3.365	4.032	5.893	6.869
6	1.440	1.943	2.447	3.143	3.707	5.208	5.959
7	1.415	1.895	2.365	2.998	3.499	4.785	5.408
8	1.397	1.860	2.306	2.896	3.355	4.501	5.041
9	1.383	1.833	2.262	2.821	3.250	4.297	4.781
10	1.372	1.812	2.228	2.764	3.169	4.144	4.587
11	1.363	1.796	2.201	2.718	3.106	4.025	4.437
12	1.356	1.782	2.179	2.681	3.055	3.930	4.318
13	1.350	1.771	2.160	2.650	3.012	3.852	4.221
14	1.345	1.761	2.145	2.624	2.977	3.787	4.140
15	1.341	1.753	2.131	2.602	2.947	3.733	4.073
16	1.337	1.746	2.120	2.583	2.921	3.686	4.015
17	1.333	1.740	2.110	2.567	2.898	3.646	3.965
18	1.330	1.734	2.101	2.552	2.878	3.610	3.922
19	1.328	1.729	2.093	2.539	2.861	3.579	3.883
20	1.325	1.725	2.086	2.528	2.845	3.552	3.850
21	1.323	1.721	2.080	2.518	2.831	3.527	3.819
22	1.321	1.717	2.074	2.508	2.819	3.505	3.792
23	1.319	1.714	2.069	2.500	2.807	3.485	3.767
24	1.318	1.711	2.064	2.492	2.797	3.467	3.745
25	1.316	1.708	2.060	2.485	2.787	3.450	3.725
26	1.315	1.706	2.056	2.479	2.779	3.435	3.707
27	1.314	1.703	2.052	2.473	2.771	3.421	3.690
28	1.313	1.701	2.048	2.467	2.763	3.408	3.674
29	1.311	1.699	2.045	2.462	2.756	3.396	3.659
30	1.310	1.697	2.042	2.457	2.750	3.385	3.646
40	1.303	1.684	2.021	2.423	2.704	3.307	3.551
60	1.296	1.671	2.000	2.390	2.660	3.232	3.460
120	1.289	1.658	1.980	2.358	2.617	3.160	3.373
∞	1.282	1.645	1.960	2.326	2.576	3.090	3.291

Appendix 12   *F*-distribution

Probability level	ϕ_2	ϕ_1 (corresponding to greater mean square)											
		1	2	3	4	5	6	7	8	9	10	15	∞
0.10	1	39.9	49.5	53.6	55.8	57.2	58.2	58.9	59.4	59.9	60.2	61.2	63.3
0.05		161.4	199.5	215.7	224.6	230.2	234.0	236.8	238.9	240.5	241.9	246.0	254.3
0.01		4052	4999	5403	5625	5764	5859	5928	5981	6023	6056	6157	6366
0.10	2	8.53	9.00	9.16	9.24	9.29	9.33	9.35	9.37	9.38	9.39	9.42	9.49
0.05		18.5	19.0	19.2	19.2	19.3	19.3	19.4	19.4	19.4	19.4	19.4	19.5
0.01		98.5	99.0	99.2	99.2	99.3	99.3	99.4	99.4	99.4	99.4	99.4	99.5
0.10	3	5.54	5.46	5.39	5.34	5.31	5.28	5.27	5.25	5.24	5.23	5.20	5.13
0.05		10.1	9.55	9.28	9.12	9.01	8.94	8.89	8.85	8.81	8.79	8.70	8.53
0.01		34.1	30.8	29.5	28.7	28.2	27.9	27.7	27.5	27.3	27.2	26.9	26.1
0.10	4	4.54	4.32	4.19	4.11	4.05	4.01	3.98	3.95	3.94	3.92	3.87	3.76
0.05		7.71	6.94	6.59	6.39	6.26	6.16	6.09	6.04	6.00	5.96	5.86	5.62
0.01		21.2	18.0	16.7	16.0	15.5	15.2	15.0	14.8	14.7	14.5	14.2	13.5
0.10	5	4.06	3.78	3.62	3.52	3.45	3.40	3.37	3.34	3.32	3.30	3.24	3.10
0.05		6.61	5.79	5.41	5.19	5.05	4.95	4.88	4.82	4.77	4.74	4.62	4.36
0.01		16.3	13.3	12.1	11.4	11.0	10.7	10.5	10.3	10.2	10.1	9.72	9.02
0.10	6	3.78	3.46	3.29	3.18	3.11	3.05	3.01	2.98	2.96	2.94	2.87	2.72
0.05		5.99	5.14	4.76	4.53	4.39	4.28	4.21	4.15	4.10	4.06	3.94	3.67
0.01		13.7	10.9	9.78	9.15	8.75	8.47	8.26	8.10	7.98	7.87	7.56	6.88
0.10	7	3.59	3.26	3.07	2.96	2.88	2.83	2.78	2.75	2.72	2.70	2.63	2.47
0.05		5.59	4.74	4.35	4.12	3.97	3.87	3.79	3.73	3.68	3.64	3.51	3.23
0.01		12.2	9.55	8.45	7.85	7.46	7.19	6.99	6.84	6.72	6.62	6.31	5.65
0.10	8	3.46	3.11	2.92	2.81	2.73	2.67	2.62	2.59	2.56	2.54	2.46	2.29
0.05		5.32	4.46	4.07	3.84	3.69	3.58	3.50	3.44	3.39	3.35	3.22	2.93
0.01		11.3	8.65	7.59	7.01	6.63	6.37	6.18	6.03	5.91	5.81	5.52	4.86
0.10	9	3.36	3.01	2.81	2.69	2.61	2.55	2.51	2.47	2.44	2.42	2.34	2.16
0.05		5.12	4.26	3.86	3.63	3.48	3.37	3.29	3.23	3.18	3.14	3.01	2.71
0.01		10.6	8.02	6.99	6.42	6.06	5.80	5.61	5.47	5.35	5.26	4.96	4.31
0.10	10	3.29	2.92	2.73	2.61	2.52	2.46	2.41	2.38	2.35	2.32	2.24	2.06
0.05		4.96	4.10	3.71	3.48	3.33	3.22	3.14	3.07	3.02	2.98	2.85	2.54
0.01		10.0	7.56	6.55	5.99	5.64	5.39	5.20	5.06	4.94	4.85	4.56	3.91
0.10	12	3.18	2.81	2.61	2.48	2.39	2.33	2.28	2.24	2.21	2.19	2.10	1.90
0.05		4.75	3.89	3.49	3.26	3.11	3.00	2.91	2.85	2.80	2.75	2.62	2.30
0.01		9.33	6.93	5.95	5.41	5.06	4.82	4.64	4.50	4.39	4.30	4.01	3.36
0.10	15	3.07	2.70	2.49	2.36	2.27	2.21	2.16	2.12	2.09	2.06	1.97	1.76
0.05		4.54	3.68	3.29	3.06	2.90	2.79	2.71	2.64	2.59	2.54	2.40	2.07
0.01		8.68	6.36	5.42	4.89	4.56	4.32	4.14	4.00	3.89	3.80	3.52	2.87
0.10	16	3.05	2.67	2.46	2.33	2.24	2.18	2.13	2.09	2.06	2.03	1.94	1.72
0.05		4.49	3.63	3.24	3.01	2.85	2.74	2.66	2.59	2.54	2.49	2.35	2.01
0.01		8.53	6.23	5.29	4.77	4.44	4.20	4.03	3.89	3.78	3.69	3.41	2.75
0.10	24	2.93	2.54	2.33	2.19	2.10	2.04	1.98	1.94	1.91	1.88	1.78	1.53
0.05		4.26	3.40	3.01	2.78	2.62	2.51	2.42	2.36	2.30	2.25	2.11	1.73
0.01		7.82	5.61	4.72	4.22	3.90	3.67	3.50	3.36	3.26	3.17	2.89	2.21
0.10	60	2.79	2.39	2.18	2.04	1.95	1.87	1.82	1.77	1.74	1.71	1.60	1.29
0.05		4.00	3.15	2.76	2.53	2.37	2.25	2.17	2.10	2.04	1.99	1.84	1.39
0.01		7.08	4.98	4.13	3.65	3.34	3.12	2.95	2.82	2.72	2.63	2.35	1.60
0.10	∞	2.71	2.30	2.08	1.94	1.85	1.77	1.72	1.67	1.63	1.60	1.49	1.00
0.05		3.84	3.00	2.60	2.37	2.21	2.10	2.01	1.94	1.88	1.83	1.67	1.00
0.01		6.63	4.61	3.78	3.32	3.02	2.80	2.64	2.51	2.41	2.32	2.04	1.00

Appendix 13   Critical values of *Q* (*P* = 0.05)

Sample size	Critical value
4	0.831
5	0.717
6	0.621
7	0.570
8	0.524
9	0.492
10	0.464

Source: E. P. King 1958 *J. Am. Statist. Assoc.*, **48**; 531.
Reprinted by permission of the American Statistical Association.

Appendix 14   Critical values of the correlation coefficient ρ (*P* = 0.05)

No. of data pairs (*x*, *y*)	Critical value
5	0.88
6	0.82
7	0.76
8	0.71
9	0.67
10	0.64
11	0.61
12	0.58

Appendix 15   Wilcoxon signed rank test: critical values (*P* = 0.05)

n	One-tailed test	Two-tailed test
5	0	NA
6	2	0
7	3	2
8	5	3
9	8	5
10	10	8
11	13	10
12	17	13
13	21	17
14	25	21
15	30	25

Notes: The null hypothesis can be rejected when the test statistic is less than or equal to the tabulated value. NA indicates that the test cannot be applied.
Source: E. Lord 1947 *Biometrika*, **34**; 66.
Reprinted by permission of the *Biometrika* trustees.

Appendix 16   Critical values for T_d ($P = 0.05$)

$n_1 = n_2$	T_d
2	3.43
3	1.27
4	0.81
5	0.61
6	0.50
7	0.43
8	0.37
9	0.33
10	0.30

Source: E. Lord 1947 *Biometrika*, **34**; 66.
Reprinted by permission of the *Biometrika* trustees.

Appendix 17   Critical F_R values for one- and two-tailed tests ($P = 0.05$)

Number of measurements in numerator and denominator	One-tailed test	Two-tailed test
2	12.7	25.5
3	4.4	6.3
4	3.1	4.0
5	2.6	3.2
6	2.3	2.8
7	2.1	2.5
8	2.0	2.3
9	1.9	2.2
10	1.9	2.1

Source: F. R. Link 1949 *Ann. Math. Statist.*, **20**, 257.
Reprinted by permission of the Institute of Mathematical Statistics.

Appendix 18   Equivalents and normalities

Throughout the main text of this book, standard solutions and quantities have all been expressed in terms of molarities, moles and relative molecular masses. However, there are still many chemists who have traditionally used what are known as normal solutions and equivalents as the basis for calculations, especially in titrimetry. This appendix defines the terms and illustrates how they are employed in the various types of determination. The International Union of Pure and Applied Chemistry (IUPAC) has given this definition:

The **equivalent** of a substance is that amount of it which, in a specified reaction, combines with, releases or replaces that amount of hydrogen which is combined with 3 grams of carbon-12 in methane $^{12}CH_4$. (*Information Bulletin*, No. 36, August 1974)

From this it follows that a **normal solution** is a solution containing one equivalent of a defined species per litre according to the specified reaction. In this definition, the amount of hydrogen may be replaced by the equivalent amount of electricity or by one equivalent of any other substance, but the reaction to which the definition is applied must be clearly specified.

The most important advantage of the equivalent system is that the calculations of titrimetric analysis are rendered very simple, since at the end point the number of equivalents of the substance titrated is equal to the number of equivalents of the standard solution employed. We may write

$$\text{normality} = \frac{\text{number of equivalents}}{\text{number of litres}}$$

$$= \frac{\text{number of milliequivalents}}{\text{number of millilitres}}$$

Hence the number of milliequivalents = number of millilitres × normality. If the solution volumes of two different substances A and B which exactly react with one another are V_A mL and V_B mL respectively, then these volumes contain the same number of equivalents or milliequivalents of A and B. Thus

$$V_A \times \text{normality}_A = V_B \times \text{normality}_B \tag{A18.1}$$

In practice V_A, V_B and normality$_A$ (the standard solution) are known, hence normality$_B$ (the unknown solution) can be readily calculated.

Example A18.1

How many mL of $0.2\,N$ hydrochloric acid are required to neutralise 25.0 mL of $0.1\,N$ sodium hydroxide?

Substituting in equation (A18.1) we obtain

$0.2x = 25.0 \times 0.1 \quad \text{so} \quad x = 12.5\,\text{mL}$

Example A18.2

How many mL of N hydrochloric acid are required to precipitate completely 1 g of silver nitrate?

The equivalent of $AgNO_3$ in a precipitation reaction is 1 mol or 169.89 g

Hence 1 g of $AgNO_3 = 1 \times 1000/169.89 = 5.886$ milliequivalents

Since 1 milliequivalent of HCl = 1 milliequivalent of $AgNO_3$

Hence $\quad 1x = 5.886 \quad \text{so} \quad x = 5.90\,\text{mL}$

Example A18.3

25 mL of an iron(II) sulphate solution react completely with 30.0 mL of 0.125 N potassium permanganate. Calculate the strength of the iron solution in g L^{-1} of $FeSO_4$.

A normal solution of $FeSO_4$ as a reductant contains 1 mol L^{-1} or 151.90 g. Let the normality of the iron solution be n_A. Then

$$25n_A = 30 \times 0.125$$

$$n_A = 30 \times 0.125/55 = 0.150\,N$$

Hence the solution will contain $0.150 \times 151.90 = 22.78$ g L^{-1} $FeSO_4$.

Example A18.4

What volume of 0.127 N reagent is required for the preparation of 1000 mL of 0.1 N solution?

$$V_A \times \text{normality}_A = V_B \times \text{normality}_B$$

$$V_A \times 0.127 = 1000 \times 0.1$$

$$V_A = 1000 \times 0.1/0.127 = 787.4\,\text{mL}$$

In other words, the required solution can be obtained by diluting 787.4 mL of 0.127 N reagent to 1 L.

The IUPAC definition of normal solution uses the term 'equivalent'. This quantity varies with the type of reaction, and since it is difficult to give a clear definition of 'equivalent' which will cover all reactions, it is proposed to discuss this subject in some detail below. It often happens that the same compound possesses different equivalents in different chemical reactions. The situation may therefore arise in which a solution has normal concentration when employed for one purpose, and a different normality when used in another chemical reaction.

Neutralisation reactions

The equivalent of an acid is that mass of it which contains 1.008 g (more accurately 1.0078 g) of replaceable hydrogen. The equivalent of a monoprotic acid, such as hydrochloric, hydrobromic, hydriodic, nitric, perchloric or ethanoic acid, is identical with the mole. A normal solution of a monoprotic acid will therefore contain one mole per litre of solution. The equivalent of a diprotic acid (e.g. sulphuric or oxalic acid) is one-half of a mole; and the equivalent of a triprotic acid (e.g. phosphoric(V) acid), is one-third of a mole.

The equivalent of a base is that mass of it which contains one replaceable hydroxyl group, i.e. 17.008 g of ionisable hydroxyl; 17.008 g of hydroxyl are equivalent to 1.008 g of hydrogen. The equivalent of sodium hydroxide and potassium hydroxide is one mole; the equivalent of calcium hydroxide, strontium hydroxide and barium hydroxide is one-half of a mole.

Salts of strong bases and weak acids possess alkaline reactions in aqueous solution because of hydrolysis. A solution containing 1 mol of sodium carbonate, with methyl orange as indicator, reacts with 2 mol of hydrochloric acid to form 2 mol of sodium chloride; hence its equivalent is 0.5 mol. Sodium tetraborate, under similar conditions, also reacts with 2 mol of hydrochloric acid, and its equivalent is also 0.5 mol.

Complex formation and precipitation reactions

Here the equivalent is the mass of the substance which contains or reacts with 1 mol of a singly charged cation M^+ (equivalent to 1.008 g of hydrogen), 0.5 mol of a doubly charged cation M^{2+}, 0.33 mol of a triply charged cation M^{3+}, etc. For the cation, the equivalent is the mole divided by the charge number. For a reagent which reacts with this cation, the equivalent is the mass of reagent which reacts with one equivalent of the cation. The equivalent of a salt in a precipitation reaction is the mole divided by the total charge number of the **reacting** ion. Thus, the equivalent of silver nitrate in the titration of chloride ion is the mole.

In a complex formation reaction the equivalent is most simply deduced by writing down the ionic equation of the reaction. For example, the equivalent of potassium cyanide in the titration with silver ions is 2 mol, since the reaction is

$$2CN^- + Ag^+ \rightleftharpoons [Ag(CN)_2]^-$$

In the titration of zinc ion with potassium hexacyanoferrate(II) solution:

$$3Zn^{2+} + 2K_4Fe(CN)_6 = 6K^+ + K_2Zn_3[Fe(CN)_6]_2$$

the equivalent of the hexacyanoferrate(II) is 0.33 mol. For other examples of complex formation reactions, see Chapter 10. In many complexation reactions it is preferable to work in moles rather than equivalents.

Oxidation–reduction reactions

The equivalent of an oxidising or reducing agent is most simply defined as that mass of the reagent which reacts with or contains 1.008 g of available hydrogen or 8.000 g of available oxygen. The word 'available' means capable of being used in oxidation or reduction. The amount of available oxygen may be indicated by the equation

$$MnO_4^- + 8H^+ + 5e \rightarrow Mn^{2+} + 4H_2O$$

Hence the equivalent is $KMnO_4/5$. For potassium dichromate in acid solution, the equation is

$$Cr_2O_7^{2-} + 14H^+ + 6e \rightarrow 2Cr^{3+} + 7H_2O$$

The equivalent is $K_2Cr_2O_7/6$.

A more general and fundamental view is obtained by considering (a) the number of electrons involved in the partial ionic equation representing the reaction, and (b) the change in the 'oxidation number' of a significant element in the oxidant or reductant. Both methods will be considered in some detail.

In quantitative analysis we are chiefly concerned with reactions which take place in solution, i.e. ionic reactions, so we shall limit our discussion to these. The oxidation of iron(II) chloride by chlorine in aqueous solution may be written

$$2FeCl_2 + Cl_2 = 2FeCl_3$$

or it may be expressed ionically

$$2Fe^{2+} + Cl_2 = 2Fe^{3+} + 2Cl^-$$

The ion Fe^{2+} is converted into ion Fe^{3+} (oxidation), and the neutral chlorine molecule into negatively charged chloride ions Cl^- (reduction). The conversion of Fe^{2+} into Fe^{3+} requires the loss of one electron; the transformation of the neutral chlorine molecule into chloride ions requires the gain of two electrons. So for reactions in solution, oxidation involves a loss of electrons, as in

$$Fe^{2+} - e = Fe^{3+}$$

and reduction involves a gain of electrons, as in

$$Cl_2 + 2e = 2Cl^-$$

In the actual oxidation–reduction process electrons are transferred from the reducing agent to the oxidising agent:

Oxidation is the process which results in the loss of one or more electrons by atoms or ions.
Reduction is the process which results in the gain of one or more electrons by atoms or ions.

An oxidising agent is one that gains electrons and is reduced; a reducing agent is one that loses electrons and is oxidised.

In all oxidation–reduction processes (or redox processes) there will be a reactant undergoing oxidation and one undergoing reduction, since the two reactions are complementary to one another and occur simultaneously – one cannot take place without the other. The reagent suffering oxidation is called the reducing agent or reductant, and the reagent undergoing reduction is called the oxidising agent or oxidant. The study of the electron changes in the oxidant and reductant forms the basis of the ion–electron method for balancing ionic equations. The equation is first divided into two balanced partial equations, representing the oxidation and the reduction. Remember that the reactions take place in aqueous solution, so besides the ions from the oxidant and reductant, there are molecules of water H_2O, hydrogen ions H^+ and hydroxide ions OH^-; all may be used in balancing the partial equations. The unit change in oxidation or reduction is a charge of one electron, denoted by e. This may be illustrated by the reaction between iron(III) chloride and tin(II) chloride in aqueous solution. The partial equation for the reduction is

$$Fe^{3+} \rightarrow Fe^{2+} \tag{A18.1}$$

and the partial equation for the oxidation is

$$Sn^{2+} \rightarrow Sn^{4+} \tag{A18.2}$$

The equations must be balanced not only with regard to the number and kind of atoms, but also electrically, i.e. the net electric charge on each side must be the same. Equation [A18.1] can be balanced by adding one electron to the left-hand side:

$$Fe^{3+} + e \rightleftharpoons Fe^{2+} \tag{A18.3}$$

and equation [A18.2] by adding two electrons to the right-hand side:

$$Sn^{2+} \rightleftharpoons Sn^{4+} + 2e \tag{A18.4}$$

These partial equations must then be multiplied by coefficients which result in the number of electrons used in one reaction being equal to those liberated in the other. Thus equation [A18.3] must be multiplied by two, and we have:

$$2Fe^{3+} + 2e \rightleftharpoons 2Fe^{2+} \tag{A18.5}$$

Adding equations [A18.4] and [A18.5]

$$2Fe^{3+} + Sn^{2+} + 2e \rightleftharpoons 2Fe^{2+} + Sn^{4+} + 2e$$

and by cancelling the electrons common to both sides, the simple ionic equation is obtained:

$$2Fe^{3+} + Sn^{2+} = 2Fe^{2+} + Sn^{4+} \qquad\qquad [A18.6]$$

The partial equations for a number of oxidising and reducing agents are collected in the table below. For oxidation–reduction reactions there are five steps to follow:

1. Ascertain the products of the reaction.
2. Set up a partial equation for the oxidising agent.
3. Set up a partial equation for the reducing agent.
4. Multiply each partial equation by a factor so that when the two are added the electrons just compensate each other.
5. Add the partial equations and cancel out substances which appear on both sides of the equation.

By following these procedures the concept of equivalents and normalities can be readily applied to most reactions. There are other approaches to these concepts, and more complete treatments can be found in traditional chemistry textbooks.

Substance	Partial ionic equation
Oxidants	
Potassium permanganate (acid)	$MnO_4^- + 8H^+ + 5e \rightleftharpoons Mn^{2+} + 4H_2O$
Potassium permanganate (neutral)	$MnO_4^- + 2H_2O + 3e \rightleftharpoons MnO_2 + 4OH^-$
Potassium permanganate (strongly alkaline)	$MnO_4^- + e \rightleftharpoons MnO_4^{2-}$
Cerium(IV) sulphate	$Ce^{4+} + e \rightleftharpoons Ce^{3+}$
Potassium dichromate	$Cr_2O_7^{2-} + 14H^+ + 6e \rightleftharpoons 2Cr^{3+} + 7H_2O$
Chlorine	$Cl_2 + 2e \rightleftharpoons 2Cl^-$
Bromine	$Br_2 + 2e \rightleftharpoons 2Br^-$
Iodine	$I_2 + 2e \rightleftharpoons 2I^-$
Iron(III) chloride	$Fe^{3+} + e \rightleftharpoons Fe^{2+}$
Potassium bromate	$BrO_3^- + 6H^+ + 6e \rightleftharpoons Br^- + 3H_2O$
Potassium iodate (dilute acid solution)	$IO_3^- + 6H^+ + 6e \rightleftharpoons I^- + 3H_2O$
Sodium hypochlorite	$ClO^- + H_2O + 2e \rightleftharpoons Cl^- + 2OH^-$
Hydrogen peroxide	$H_2O_2 + 2H^+ + 2e \rightleftharpoons 2H_2O$
Manganese dioxide	$MnO_2 + 4H^+ + 2e \rightleftharpoons Mn^{2+} + 2H_2O$
Sodium bismuthate	$BiO_3^- + 6H^+ + 2e \rightleftharpoons Bi^{3+} + 3H_2O$
Nitric acid (conc.)	$NO_3^- + 2H^+ + e \rightleftharpoons NO_2 + H_2O$
Nitric acid (dilute)	$NO_3^- + 4H^+ + 3e \rightleftharpoons NO + 2H_2O$
Reductants	
Hydrogen	$H_2 \rightleftharpoons 2H^+ + 2e$
Zinc	$Zn \rightleftharpoons Zn^{2+} + 2e$
Hydrogen sulphide	$H_2S \rightleftharpoons 2H^+ + S + 2e$
Hydrogen iodide	$2HI \rightleftharpoons I_2 + 2H^+ + 2e$
Oxalic acid	$C_2O_4^{2-} \rightleftharpoons 2CO_2 + 2e$
Iron(II) sulphate	$Fe^{2+} \rightleftharpoons Fe^{3+} + e$
Sulphurous acid	$H_2SO_3 + H_2O \rightleftharpoons SO_4^{2-} + 4H^+ + 2e$
Sodium thiosulphate	$2S_2O_3^{2-} \rightleftharpoons S_4O_6^{2-} + 2e$
Titanium(III) sulphate	$Ti^{3+} \rightleftharpoons Ti^{4+} + e$
Tin(II) chloride	$Sn^{2+} \rightleftharpoons Sn^{4+} + 2e$
Tin(II) chloride (in presence of hydrochloric acid)	$Sn^{2+} + 6Cl^- \rightleftharpoons SnCl_6^{2-} + 2e$
Hydrogen peroxide	$H_2O_2 \rightleftharpoons 2H^+ + O_2 + 2e$

Index

The following abbreviations are used:
aa = atomic absorption, fu = fluorimetric, sepn. = separation, am = amperometry, g = gravimetric, soln. = solution, ch = chromatographic, hf = high frequency, stdn. = standardisation, cm = coulometric, ir = infrared, temp. = temperature, cm = conductimetric, p = potentiometric, T = table, D = determination, prep. = preparation, th = thermal, eg = electrogravimetric, s = spectrophotometric, ti = titrimetric, em = emission spectrographic, se = solvent extraction, V = voltammetry, fi = flame emission, NMR = nuclear magnetic resonance.

Index

790

Index

Index